UNIVARIATE & MULTIVARIATE

General Linear Models:

THEORY AND APPLICATIONS USING SAS® SOFTWARE

Neil H. Timm

Tammy A. Mieczkowski

Comments or Questions?

The authors assume complete responsibility for the technical accuracy of the content of this book. If you have any questions about the material in this book, please write to the authors at this address:

SAS Institute Inc.
Books by Users
Attn: Neil H. Timm
SAS Campus Drive
Cary, NC 27513

If you prefer, you can send e-mail to sasbbu@unx.sas.com with "comments for Neil H. Timm" as the subject line, or you can fax the Books by Users program at (919) 677-4444.

The correct bibliographic citation for this manual is as follows: Timm, Neil H. and Mieczkowski, Tammy A., *Univariate and Multivariate General Linear Models: Theory and Applications Using SAS® Software*, Cary, NC: SAS Institute Inc., 1997. 619 pp.

Univariate and Multivariate General Linear Models: Theory and Applications Using SAS® Software

ISBN 1-55544-987-5

SAS Institute Inc., SAS Campus Drive, Cary, North Carolina 27513.

1st printing, August 1997

Neil dedicates this book
to his loving wife, Verena

Tammy dedicates this book
to Jesus

Contents

List of Tables

Tables (*continued*)

List of Figures

List of Graphics

Preface

The general linear model is often first introduced to graduate students during a course on multiple linear regression, analysis of variance, or experimental design; however, most students do not fully understand the generality of the model until they have taken several courses in applied statistics. Even students in graduate statistics programs do not fully appreciate the generality of the model until well into their program of study. This is due in part to the fact that theory and applications of the general linear model are discussed in discrete segments throughout the course of study rather than within a more general framework. In this book, we have tried to solve this problem by reviewing the theory of the general linear model using a general framework. Additionally, we use this general framework to present analyses of simple and complex models, both univariate and multivariate, using data sets from the social and behavioral sciences and other disciplines.

Audience

The book is written for advanced graduate students in the social and behavioral sciences and in applied statistics who are interested in statistical analysis using the general linear model. The book may be used to introduce students to the general linear model; at the University of Pittsburgh it is used in a one-semester course on linear models. The book may also be used as a supplement to courses taught in regression analysis, analysis of variance, and applied multivariate statistics. Additionally, it may serve as a reference for practitioners who are interested in applications of the general linear model using SAS software.

Students studying this text should have previously taken elementary courses in applied statistical methods covering the essentials of estimation theory and hypothesis testing, simple linear regression, and analysis of variance. They should also have some familiarity with matrix algebra and with running SAS procedures.

Overview

Each chapter of this book is divided into two sections: theory and applications. Standard SAS procedures are used to perform most of the analyses. When standard SAS procedures are not available, PROC IML code to perform the analysis is discussed. Because SAS is not widely used in the social and behavioral sciences, SAS code for analyzing general linear model applications is discussed in detail. The code can be used as a template.

Chapter 1 provides an overview of the general linear model using matrix algebra and an introduction to the multivariate normal distribution as well as to the general theory of hypothesis testing. Applications include the use of graphical methods to evaluate univariate and multivariate normality and the use of transformations to normality. In Chapter 2 the general linear model without restrictions is introduced and used to analyze multiple regression and ANOVA designs. In Chapter 3 the general linear model with restrictions is discussed and used to analyze ANCOVA designs and repeated measurement designs.

Chapter 4 extends the concepts of the first three chapters to general linear models with heteroscedastic errors and illustrates how the model may be used to perform weighted least squares regression and to analyze categorical data. Chapter 5 extends the theory of Chapter 2 to the multivariate case; applications include multivariate regression analysis, MANOVA, MANCOVA, and analyses of repeated measurement data. This chapter also extends "standard" hypothesis testing to extended linear hypotheses. In Chapter 6, the double multivariate linear model is discussed.

Chapter 7 extends the multivariate linear model to include restrictions and considers the growth curve model. In Chapter 8, the seemingly unrelated regression (SUR) and the restricted GMANOVA models are analyzed. Many of the applications in this chapter involve PROC IML code. Finally, Chapter 9 includes analyses of hierarchical linear models, and Chapter 10 treats the analysis of incomplete repeated measurement data.

While the coverage given the general linear model is extensive, it is not exhaustive. Excluded from the book are Bayesian methods, nonparametric procedures, nonlinear models, and generalized linear models, among others.

Acknowledgments

We would like to thank the reviewers at SAS Institute Inc. and the technical reviewer Vernon M. Chinchilli for their helpful comments and suggestions on the book. We thank Hanna Hicks Schoenrock and Caroline Brickley for making the process of producing this book run so smoothly. We also appreciate the helpful suggestions made on an early draft by doctoral students in the linear models course.

We would especially like to extend our gratitude to Roberta S. Allen. Ms. Allen expertly typed every draft of the book from inception through every revision, including all equations. Thank you for your excellent work and patience with us, Roberta. The authors also want to thank the authors and publishers of copyrighted material for permission to reproduce tables and data sets used in the book.

This book was completed while Neil H. Timm was on sabbatical leave during the 1996-1997 academic year. He would like to thank his colleagues for their support and the School of Education for this opportunity. He would also like to thank his wife, Verena, for her support and encouragement.

Tammy A. Mieczkowski was a full-time doctoral student in the Research Methodology program in the School of Education when this book was completed. She also works as a Graduate Student Researcher at the University of Pittsburgh School of Medicine, Department of Family Medicine and Clinical Epidemiology. She would like to thank her family, especially her sister and friend, Karen, for their continued support, encouragement, understanding, and love.

Neil H. Timm

Tammy A. Mieczkowski

1. Overview of the General Linear Model

Theory

Applications

1.1 Introduction

In this chapter we introduce the structure of the general linear model and use the structure to classify the linear models discussed in this book. The multivariate normal distribution which forms the basis for most of the hypothesis testing theory for the linear model is reviewed, along with a general approach to hypothesis testing. Graphical methods for assessing univariate and multivariate normality are also reviewed. The chapter illustrates multivariate normal data generation, Q-Q plots, chi-square plots, scatter plots, and data transformation procedures to evaluate normality.

1.2 The General Linear Model

Data analysis in the social and behavioral sciences and numerous other disciplines is associated with the model statisticians call the general linear model (GLM). If we employ matrix notation, univariate and multivariate linear models may be represented using the general structure

$$\mathbf{y} = \mathbf{X}\beta + \mathbf{e} \tag{1.1}$$

where $\mathbf{y}_{n\times 1}$ is a vector of n observations, $\mathbf{X}_{n\times k}$ is a known design matrix of full column rank k, $\beta_{k\times 1}$ is a vector of k fixed parameters, and $\mathbf{e}_{n\times 1}$ is a random vector of errors with mean zero, $E(\mathbf{e}) = \mathbf{0}$, and covariance matrix $\mathbf{\Omega} = \text{cov}(\mathbf{e})$. If the design matrix is not of full rank, one may reparameterize the model to create an equivalent model of full rank. In this book, we systematically discuss the GLM specified by (1.1) with various structures for $\mathbf{X}$, the design matrix, and $\mathbf{\Omega}$, the covariance matrix of errors.

Depending on the structure of $\mathbf{X}$ and $\mathbf{\Omega}$, the model in (1.1) has many names in the literature. To illustrate, if $\mathbf{\Omega} = \sigma^2 \mathbf{I}_n$ in (1.1), the model is called the classical linear regression model or the standard linear regression model. If we partition $\mathbf{X}$ to have the form $\mathbf{X} = (\mathbf{X}_1, \mathbf{X}_2)$ where $\mathbf{X}_1$ is associated with fixed effects and $\mathbf{X}_2$ is associated with random effects, and if the covariance matrix $\mathbf{\Omega}$ has the form

$$\mathbf{\Omega} = \mathbf{X}_2 \mathbf{V}\, \mathbf{X}_2' + \mathbf{\Psi}$$

where $\mathbf{V}$ and $\mathbf{\Psi}$ are covariance matrices, then (1.1) becomes the general linear mixed model (GLMM). If we let $\mathbf{X}$ and $\mathbf{\Omega}$ take the general form

$$\mathbf{X} = \begin{pmatrix} \mathbf{X}_1 & \cdots & \cdots & \mathbf{0} \\ \vdots & \mathbf{X}_2 & \cdots & \vdots \\ \vdots & \cdots & \ddots & \vdots \\ \mathbf{0} & \cdots & \cdots & \mathbf{X}_p \end{pmatrix}$$

$$\mathbf{\Omega} = \mathbf{\Sigma} \otimes \mathbf{I}_n$$

where $\mathbf{\Sigma}_{p\times p}$ is a covariance matrix, and $\mathbf{A} \otimes \mathbf{B}$ denotes the Kronecker product of two matrices $\mathbf{A}$ and $\mathbf{B}$ ($\mathbf{A} \otimes \mathbf{B} = a_{ij}\mathbf{B}$), then (1.1) is called Zellner's seemingly unrelated regression (SUR) model or the multiple design multivariate (MDM) model. The SUR model may also be formulated as p separate linear regression models that are not independent:

$$\begin{aligned} \mathbf{y}_i &= \mathbf{X}_i \beta_{ii} + \mathbf{e}_i \\ \text{cov}(\mathbf{y}_i, \mathbf{y}_j) &= \sigma_{ij} \mathbf{I}_n \end{aligned}$$

for $i, j = 1, 2, \ldots, p$ where $\mathbf{y}$, β and $\mathbf{e}$ in (1.1) are partitioned:

$$\begin{aligned} \mathbf{y}' &= (\mathbf{y}_1', \mathbf{y}_2', \ldots, \mathbf{y}_p') \quad \text{where} \quad \mathbf{y}_i : n^* \times 1 \\ \beta' &= (\beta_1', \beta_2', \ldots, \beta_p') \quad \text{where} \quad \beta_{ii} : k_i \times 1 \\ \mathbf{e}' &= (\mathbf{e}_1', \mathbf{e}_2', \ldots, \mathbf{e}_p') \quad \text{where} \quad \mathbf{e}_i : n^* \times 1 \end{aligned}$$

and $\mathbf{\Sigma}_{p\times p} = (\sigma_{ij})$. Alternatively, we may express the SUR model as a restricted multivariate regression model. To do this, we write

$$\mathbf{Y}_{n^*\times p} = \mathbf{X}_{n^*\times k} \tilde{\mathbf{B}}_{k\times p} + \mathbf{E}_{n^*\times p}$$

where $\mathbf{Y} = (\mathbf{y}_1, \mathbf{y}_2, \ldots, \mathbf{y}_p)$, $\mathbf{X} = (\mathbf{X}_1, \mathbf{X}_2, \ldots, \mathbf{X}_p)$, $\mathbf{E} = (\mathbf{e}_1, \mathbf{e}_2, \ldots, \mathbf{e}_p)$ and

$$\tilde{\mathbf{B}} = \begin{pmatrix} \beta_{11} & \cdots & \cdots & \mathbf{0} \\ \vdots & \beta_{22} & \cdots & \vdots \\ \vdots & \cdots & \ddots & \vdots \\ \mathbf{0} & \cdots & \cdots & \beta_{pp} \end{pmatrix}$$

Letting $\mathbf{X}_1 = \mathbf{X}_2 = \ldots = \mathbf{X}_p = \tilde{\mathbf{X}}_{n^* \times k}$ and $\mathbf{B} = (\beta_{11}, \beta_{22}, \ldots, \beta_{pp})$ in the SUR model, (1.1) becomes the classical multivariate regression model or the MANOVA model. Finally, letting

$$\mathbf{X} = \mathbf{X}_1 \otimes \mathbf{X}_2'$$
$$\mathbf{\Omega} = \mathbf{I}_n \otimes \mathbf{\Sigma}$$

model (1.1) becomes the generalized multivariate analysis of variance (GMANOVA) or the generalized growth curve model. All these models with some further extensions are special forms of the general linear model discussed in this book.

The model in (1.1) is often termed the "classical" model since its orientation is subjects or observations by variables where the number of variables is one. An alternative orientation for the model is to assume that $\mathbf{y}' = (y_1, y_2, \ldots, y_n)$ is a $(1 \times n)$ vector of observations where the number of variables is one. For each observation y_i, we may assume that there are $\mathbf{x}_i' = (\mathrm{x}_1, \mathrm{x}_2, \ldots, \mathrm{x}_k)$ or k independent (possibly dummy) variables. With this orientation, (1.1) becomes

$$\mathbf{y} = \mathbf{X}'\beta + \mathbf{e} \tag{1.2}$$

where $\mathbf{X} = (\mathbf{x}_1, \mathbf{x}_2, \ldots, \mathbf{x}_n)$, $\mathbf{e}' = (e_1, e_2, \ldots, e_n)$ and each $\mathbf{x}_i$ contains k independent variables for the ith observation. Model (1.2) is often called the "responsewise" form. Model (1.1) is clearly equivalent to (1.2) since the design matrix has the same order for either representation; however, in (1.2) $\mathbf{X}$ is of order $k \times n$. Thus, $\mathbf{X}'\mathbf{X}$ using (1.2) becomes $\mathbf{X}\mathbf{X}'$ for the responsewise form of the general linear model.

The simplest example of the GLM is the simple linear regression model:

$$y = \beta_o + \beta_1 \mathrm{x} + e \tag{1.3}$$

where x represents the independent variable, y the dependent variable and e a random error. Model (1.3) states that the observed dependent variable for each subject is hypothesized to be a function of a common parameter β_o for all subjects and an independent variable x for each subject that is related to the dependent variable by a weighting (i.e., regression) coefficient β_1 plus a random error e. For $k = m + 1$ with m variables, (1.3) becomes (1.4)

$$y = \beta_o + \beta_1 \mathrm{x}_1 + \beta_2 \mathrm{x}_2 + \ldots + \beta_m \mathrm{x}_m + e \tag{1.4}$$

or using matrix notation, (1.4) is written as

$$y = \mathbf{x}'\beta + e \tag{1.5}$$

where $\mathbf{x}' = (\mathrm{x}_1, \mathrm{x}_2, \ldots, \mathrm{x}_k)$ denotes k independent variables, and x_1 is a dummy variable in the vector $\mathbf{x}'$. Then for a sample of n observations, (1.5) has the general form (1.2) where $\mathbf{y}' = (y_1, y_2, \ldots, y_n)$, $\mathbf{e}' = (e_1, e_2, \ldots, e_n)$, and $\mathbf{X} = (\mathbf{x}_1, \mathbf{x}_2, \ldots, \mathbf{x}_n)$ of order $k \times n$ since each column vector $\mathbf{x}_i$ in $\mathbf{X}$ contains k variables. When using the classical form (1.1), $\mathbf{X} \equiv \mathbf{X}'$, a matrix of order $n \times k$. In discussions of the GLM, many authors use either the classical or the responsewise version of the GLM, while we in general prefer (1.1). In some applications (e.g., repeated measurement designs) the form (1.2) is preferred.

1.3 The Restricted General Linear Model

In specifying the GLM using (1.1) or (1.2), we have not restricted the k-variate parameter vector β. A linear restriction on the parameter vector β will affect the characterization of the model. Sometimes it is necessary to add restrictions to the GLM of the general form

$$\mathbf{R}\beta = \theta \tag{1.6}$$

where $\mathbf{R}_{s \times k}$ is a known matrix with full row rank, rank($\mathbf{R}$) = s, and θ is a known parameter vector, often assumed to be zero. With (1.6) associated with the GLM, the model is commonly called the restricted general linear model (RGLM). Returning to (1.3), we offer an example of this in the simple linear regression model with a restriction

$$\begin{aligned} y &= \beta_o + \beta_1 x + e \\ \beta_o &= 0 \end{aligned} \tag{1.7}$$

so that the regression of y on x is through the origin. For this situation, $\mathbf{R} \equiv (1,0)$ and $\theta = (0, 0)$. Clearly, the estimate of β_1 using (1.7) will differ from that obtained using the general linear model (1.3) without the restriction.

Assuming (1.1) or (1.4), one first wants to estimate β with $\hat{\beta}$ where the estimator $\hat{\beta}$ has some optimal properties like unbiasedness and minimal variance. Adding (1.6) to the GLM, one obtains a restricted estimator of β, $\hat{\beta}_r$, which in general is not equal to the unrestricted estimator. Having estimated β, one may next want to test hypotheses regarding the parameter vector β and the structure of $\mathbf{\Omega}$. The general form of the null hypothesis regarding β is

$$H\colon\ \mathbf{C}\beta = \xi \tag{1.8}$$

where $\mathbf{C}_{g \times k}$ is a matrix of full row rank g, rank($\mathbf{C}$) = g and $\xi_{g \times 1}$ is a vector of known parameters, usually equal to zero. The hypothesis in (1.8) may be tested using the GLM with or without the restriction given in (1.6). Hypotheses in the form (1.8) are in general testable, provided that β is estimable; however, testing (1.8) by assuming (1.6) is more complicated since the matrix $\mathbf{C}$ may not contain a row identical, inconsistent or dependent on the rows of $\mathbf{R}$ and since the rows of $\mathbf{C}$ must remain independent. Thus, the rank of the augmented matrix must be greater than s, $\binom{\mathbf{R}}{\mathbf{C}} = s + g > s$.

Returning to (1.3), we may test the null hypotheses

$$H\colon\ \beta = \begin{pmatrix} \beta_o \\ \beta_1 \end{pmatrix} = \xi = \begin{pmatrix} \xi_o \\ \xi_1 \end{pmatrix} \tag{1.9}$$

where ξ is a known parameter vector. The hypothesis in (1.9) may not be inconsistent with the restriction $\beta_o = 0$. Thus, given the restriction, we may test

$$H\colon\ \beta_1 = \xi_1 \tag{1.10}$$

so that (1.10) is not inconsistent or dependent on the restriction.

1.4 The Multivariate Normal Distribution

To test hypotheses of the form given in (1.8), one must usually make distributional assumptions regarding the observation vector **y** or **e**, namely the assumption of multivariate normality. To define the multivariate normal distribution, recall that the definition of a standard normal random variable y is defined by the density

$$f(y) = (2\pi)^{-1/2} \exp(-y^2/2) \tag{1.11}$$

denoted by $y \sim N(0,1)$. A random variable y has a normal distribution with mean μ and variance $\sigma^2 > 0$ if y has the same distribution as the random variable

$$\mu + \sigma e \tag{1.12}$$

where $e \sim N(0,1)$. The density for y is given by

$$\begin{aligned} f(y) &= \frac{1}{\sigma\sqrt{2\pi}} \exp[-(y-\mu)^2/2\sigma^2] \\ &= (2\pi\sigma^2)^{-1/2} \exp[-(y-\mu)^2/2\sigma^2] \end{aligned} \tag{1.13}$$

With this as motivation, the definition for a multivariate normal distribution is as follows.

Definition 1.1 A p-dimensional random vector **y** is said to have a multivariate normal distribution with mean μ and covariance matrix $\mathbf{\Sigma}$ [$\mathbf{y} \sim N_p(\mu, \mathbf{\Sigma})$] if **y** has the same distribution as $\mu + \mathbf{F}\mathbf{e}$ where $\mathbf{F}_{pxp}$ is a matrix of rank r, $\mathbf{\Sigma} = \mathbf{FF}'$ and each element of **e** is distributed: $e_i \sim N(0,1)$. The density of **y** is given by

$$f(\mathbf{y}) = (2\pi)^{-p/2} |\mathbf{\Sigma}|^{-1/2} \exp\left[-\frac{1}{2}(\mathbf{y}-\mu)'\mathbf{\Sigma}^{-1}(\mathbf{y}-\mu)\right] \tag{1.14}$$

For a random sample of n independent p-vectors $\mathbf{y}_1, \mathbf{y}_2, \ldots, \mathbf{y}_n$ from a multivariate normal distribution, $\mathbf{y}_i \sim \text{IN}_p(\mu, \mathbf{\Sigma})$, we shall in general write the data matrix $\mathbf{Y}_{nxp}$ in the classical form

$$\mathbf{Y}_{nxp} \equiv \begin{pmatrix} \mathbf{y}_1' \\ \mathbf{y}_2' \\ \vdots \\ \mathbf{y}_n' \end{pmatrix} = \begin{pmatrix} y_{11} & y_{12} & \cdots & y_{1p} \\ y_{21} & y_{22} & \cdots & y_{2p} \\ \vdots & \vdots & \vdots & \vdots \\ y_{n1} & y_{n2} & \cdots & y_{np} \end{pmatrix} \tag{1.15}$$

The corresponding responsewise representation for **Y** is

$$\mathbf{Y}_{pxn} \equiv (\mathbf{y}_1, \mathbf{y}_2, \ldots, \mathbf{y}_n) = \begin{pmatrix} y_{11} & y_{12} & \cdots & y_{1n} \\ y_{21} & y_{22} & \cdots & y_{2n} \\ \vdots & \vdots & \vdots & \vdots \\ y_{p1} & y_{p2} & \cdots & y_{pn} \end{pmatrix} \tag{1.16}$$

The joint probability density function (pdf) for $\mathbf{y}_1, \mathbf{y}_2, \ldots, \mathbf{y}_n$ or the likelihood function is

$$L = L(\mu, \mathbf{\Sigma} | \mathbf{y}) = \prod_{i=1}^{n} f(\mathbf{y}_i) \tag{1.17}$$

Substituting $f(\mathbf{y})$ in (1.14) for each $f(\mathbf{y}_i)$, the *pdf* is

$$[(2\pi)^p|\boldsymbol{\Sigma}|]^{-n/2} \exp\left[-\frac{1}{2}\sum_{i=1}^{n}(\mathbf{y}_i-\mu)'\boldsymbol{\Sigma}^{-1}(\mathbf{y}_i-\mu)\right] \tag{1.18}$$

Using the property of the trace of a matrix, $Tr(\mathbf{x}'\mathbf{A}\mathbf{x}) = Tr(\mathbf{A}\mathbf{x}\mathbf{x}')$, (1.18) becomes

$$[(2\pi)^p|\boldsymbol{\Sigma}|]^{-n/2} \operatorname{etr}\left\{-\frac{1}{2}\boldsymbol{\Sigma}^{-1}\left[\sum_{i=1}^{n}(\mathbf{y}_i-\mu)(\mathbf{y}_i-\mu)'\right]\right\} \tag{1.19}$$

where etr stands for the exponential of the trace of a matrix.

If we let the sample mean be represented by

$$\overline{\mathbf{y}} = n^{-1}\sum_{i=1}^{n}\mathbf{y}_i \tag{1.20}$$

then the sum of squares and products (*SSP*) matrix, using the classical form (1.15), is

$$\mathbf{E} = \sum_{i=1}^{n}(\mathbf{y}_i-\overline{\mathbf{y}})(\mathbf{y}_i-\overline{\mathbf{y}})' = \mathbf{Y}'\mathbf{Y} - n\overline{\mathbf{y}}\overline{\mathbf{y}}' \tag{1.21}$$

or if we use the responsewise form (1.16), then the SSP matrix is

$$\mathbf{E} = \mathbf{Y}\mathbf{Y}' - n\overline{\mathbf{y}}\,\overline{\mathbf{y}}' \tag{1.22}$$

In either case, we may write (1.19)

$$\left[(2\pi)^p|\boldsymbol{\Sigma}|\right]^{-n/2} \operatorname{etr}\left\{-\frac{1}{2}\boldsymbol{\Sigma}^{-1}\left[\mathbf{E} + n(\overline{\mathbf{y}}-\mu)(\overline{\mathbf{y}}-\mu)'\right]\right\} \tag{1.23}$$

so that by Neyman's factorization criterion $(\mathbf{E}, \overline{\mathbf{y}})$ are sufficient statistics for estimating $(\boldsymbol{\Sigma}, \mu)$, (see, e.g., Lehmann, 1994, p. 16).

Theorem 1.1 Let $\mathbf{y}_i \sim IN_p(\mu, \boldsymbol{\Sigma})$ of size n. Then $\overline{\mathbf{y}}$ and $\mathbf{E}$ are sufficient statistics for μ and $\boldsymbol{\Sigma}$.

It can also be shown that $\overline{\mathbf{y}}$ and $\mathbf{E}$ are independently distributed. The distribution of $\mathbf{E}$ is known as the Wishart distribution, a multivariate generalization of the chi-square distribution, with $v = n-1$ degrees of freedom. The density of the Wishart distribution is

$$c|\boldsymbol{\Sigma}|^{-v/2}|\mathbf{E}|^{(v-p-1)/2} \operatorname{etr}\left(-\frac{1}{2}\boldsymbol{\Sigma}^{-1}\mathbf{E}\right) \tag{1.24}$$

where c is an appropriately chosen constant so that the probability over the entire sample space is equal to one. We write this as $\mathbf{E} \sim W_p(v, \boldsymbol{\Sigma})$. The expectation of $\mathbf{E}$ is $v\boldsymbol{\Sigma}$.

Given a random sample of observations from a multivariate normal distribution, we usually estimate the parameters μ and $\boldsymbol{\Sigma}$.

Theorem 1.2 Let $\mathbf{y}_i \sim IN_p(\mu, \boldsymbol{\Sigma})$, then the maximum likelihood estimators of μ and $\boldsymbol{\Sigma}$ are $\overline{\mathbf{y}}$ and $\mathbf{E}/n = \hat{\boldsymbol{\Sigma}}$.

Furthermore, $\overline{\mathbf{y}}$ and

$$\mathbf{S} = \left(\frac{n}{n-1}\right)\hat{\mathbf{\Sigma}} = \mathbf{E}/(n-1) \tag{1.25}$$

are unbiased estimators of μ and $\mathbf{\Sigma}$, respectively, so that $E(\overline{\mathbf{y}}) = \mu$ and $E(\mathbf{S}) = \mathbf{\Sigma}$. Hence, the sample distribution of $\mathbf{S}$ is Wishart, $(n-1)\mathbf{S} \sim W_p(v = n-1, \mathbf{\Sigma})$ or $\mathbf{S} \sim W_p[n-1, \mathbf{\Sigma}/(n-1)]$.

Linear combinations of multivariate normal random variables are again normally distributed. If $\hat{\beta}$ is an unbiased estimate of β in (1.5) such that $\hat{\beta} = (\mathbf{X}'\mathbf{X})^{-1}\mathbf{X}'\mathbf{y}$ then

$$\hat{\beta} \sim N_k[\beta, \operatorname{cov}\hat{\beta}] \tag{1.26}$$

so that $\hat{\beta}$ has a multivariate normal distribution with covariance structure $\mathbf{\Psi} = \operatorname{cov}\hat{\beta}$ when one assumes that $\mathbf{y}$ is multivariate normal, $\mathbf{y} \sim N_n(\mathbf{X}\beta = \mu, \mathbf{\Omega} = \mathbf{\Sigma})$.

1.5 Hypothesis Testing

Having assumed a linear model for a random sample of observations, used the observations to obtain estimates of the population parameters, and decided upon the structure of the restriction $\mathbf{R}$ (if any) and the hypothesis test matrix $\mathbf{C}$, one next tests hypotheses. Two commonly used procedures for testing hypotheses are the likelihood ratio (LR) and union-intersection (UI) tests. To construct an LR test, two likelihood functions are compared for a random sample of observations, $L(\hat{\omega})$, the likelihood function maximized under the hypothesis H in (1.8), and the likelihood $L(\hat{\mathbf{\Omega}}_o)$, the likelihood function maximized over the entire parameter space $\mathbf{\Omega}_o$ unconstrained by the hypothesis. Defining λ as the ratio

$$\lambda = \frac{L(\hat{\omega})}{L(\hat{\mathbf{\Omega}}_o)} \tag{1.27}$$

the hypothesis is rejected for small values of λ since $L(\hat{\omega}) < L(\hat{\mathbf{\Omega}}_o)$ does not favor the hypothesis. The test is said to be of size α if, for a constant λ_o, the

$$P(\lambda < \lambda_o \mid H) = \alpha \tag{1.28}$$

where α is the size of the type I error rate, the probability of rejecting H given H is true. For large sample sizes and under very general conditions, Wald (1943) showed that $-2\ln\lambda$ converges in distribution to a chi-square distribution as $n \to \infty$, where the degrees of freedom v is equal to the number of independent parameters estimated under $\mathbf{\Omega}_o$ minus the number of independent parameters estimated under ω.

To construct a UI test according to Roy (1953), we write the null hypothesis H as an intersection of an infinite number of elementary tests

$$H: \bigcap_i H_i \tag{1.29}$$

and each H_i is associated with an alternative A_i such that

$$A: \bigcup_i A_i \tag{1.30}$$

The null hypothesis H is rejected if any elementary test of size α is rejected. The overall rejection region is defined as the union of all the rejection regions of the elementary tests of H_i vs. A_i. Similarly, the region of acceptance for H is the intersection of the acceptance regions. If T_i is a test statistic for testing H_i vs. A_i, the null hypothesis H is accepted or rejected if the $T_i \lesseqgtr T_\alpha$ where the

$$P(\sup_i T_i \le T_\alpha | H) = 1 - \alpha \tag{1.31}$$

and T_α is chosen such that the type I error is α.

1.6 Generating Multivariate Normal Data

In the hypothesis testing of both univariate and multivariate linear models, the assumption of normally distributed errors is made. The multivariate normal distribution of **y** has density given by (1.14), written $\mathbf{y} \sim N_p(\mu, \mathbf{\Sigma})$. When p = 1, (1.14) reduces to a univariate normal distribution (1.13).

Some important properties of the multivariate normal distribution include

(1) If **y** is distributed multivariate normal, then each subset of the components of **y** has a normal distribution. Thus, each y_i has a univariate normal distribution. However, the converse is not true. If each y_i has a normal distribution, this does not imply that **y** has a multivariate normal distribution.

(2) All linear combinations of the y_i are normally distributed.

(3) The conditional distributions of the components of **y** are normally distributed, Timm (1975, p. 114-123).

To generate data having a multivariate distribution with mean and covariance matrix

$$\mu = \begin{bmatrix} \mu_1 \\ \mu_2 \\ \vdots \\ \mu_p \end{bmatrix} \quad \text{and} \quad \mathbf{\Sigma} = \begin{bmatrix} \sigma_{11} & \sigma_{12} & \cdots & \sigma_{1p} \\ \sigma_{21} & \ddots & & \vdots \\ \vdots & & \ddots & \vdots \\ \sigma_{p1} & \cdots & & \sigma_{pp} \end{bmatrix} \tag{1.32}$$

we use Definition 1.1. Program 1_6.sas uses the IML procedure to generate 50 observations from a multivariate normal distribution with

$$\mu = \begin{bmatrix} 10 \\ 20 \\ 30 \\ 40 \end{bmatrix} \quad \text{and} \quad \mathbf{\Sigma} = \begin{bmatrix} 3 & 1 & 0 & 0 \\ 1 & 4 & 0 & 0 \\ 0 & 0 & 1 & 4 \\ 0 & 0 & 4 & 20 \end{bmatrix}$$

Program 1_6.sas

```
/* Program 1_6.sas */
/* Program to create a multivariate normal data set */

options ls=80 ps=60 nodate nonumber;
filename appl 'c:\1_6.dat';
title1 'Output 1.6: Generating a Multivariate Normal Data Set';
```

```
proc iml;
   seed=30195;
   z=normal(repeat(seed,50,4));
   u={10,20,30,40};
   s={3 1 0  0,
      1 4 0  0,
      0 0 1  4,
      0 0 4 20};
   a=root(s);
   uu=repeat(u`,50,1);
   y=(z*a) + uu;
   print y;

file app1;
   do i=1 to nrow(y);
      do j=1 to ncol(y);
        put (y[i,j]) 10.2 +2 @;
      end;
     put;
   end;
closefile app1;
quit;
```

In Program 1_6.sas, PROC IML is used to produce a matrix $\mathbf{Z}$ that contains $n = 50$ row vectors each with $p = 4$ observations. Each is generated from a standard normal distribution $N(0,1)$. Next, new variables $\mathbf{y}_i = \mathbf{z}_i\mathbf{A} + \mu$, must be created where $\mathbf{A}$ is such that $\text{cov}(\mathbf{y}) = \mathbf{A}' \text{ cov}(\mathbf{z})\mathbf{A} = \mathbf{A}'\mathbf{IA} = \mathbf{A}'\mathbf{A} = \mathbf{\Sigma}$ and $E(\mathbf{y}) = \mu$. The Cholesky factorization procedure can be used to obtain $\mathbf{A}$ from $\mathbf{\Sigma}$; the ROOT function of PROC IML is used to perform the Cholesky decomposition and produce the resultant matrix we named a. Next, the matrix uu is created by repeating the row vector u 50 times to produce a 50×4 matrix. Finally, the multivariate normal random variables are created in the statement $y = (z * a) + uu$. The observations are printed and then output to the file named 1_6.dat.

Result and Interpretation 1_6.sas

This is the output of Program 1_6.sas. The data are saved in the file 1_6.dat.

Output 1.6: *Generating a Multivariate Normal Data Set*

```
Output 1.6: Generating a Multivariate Normal Data Set

                 Y
      10.277241 18.515521 29.670682 35.682188
      11.929652 20.275932 28.911167 36.047376
      8.2156513 17.403286 28.721552 35.264579
      8.4058699 21.693912 32.140953 49.366754
      10.034067 19.206552  31.13305 41.664014
      12.852388 20.338035 29.292826 41.662329
      7.7777858 19.422314 30.000313 34.771545
      10.551284 17.678773 31.285126 44.347948
      10.105934 21.405632 29.357324 35.774756
      12.249358 21.003614 30.638821 44.324676
      8.7131087 23.276061 30.084822 40.906227
      12.086015 18.060828 30.285533 41.081205
      9.7895428 19.771547 32.692761  49.95753
      11.620772 19.016928 28.224879  32.90378
      8.9938978 20.103495 29.681585 39.858339
      10.863083 20.553824 31.148947  45.86955
```

Output 1.6 *(continued)*

```
12.484503 19.623805  29.90136 38.509333
11.402297 22.778862 28.018182 31.069522
11.821743 22.568622 29.524165 36.428279
11.318856 20.383911 29.374577 35.644656
11.381783 21.328723 31.799191 44.314454
13.771969 20.979002 30.335263 38.938178
9.5884659 22.248491 29.881163 41.837656
8.6288368 16.496882 29.921442 40.953565
11.304779  20.34674 29.286882  34.35696
12.304391 18.544842 29.169465 36.569666
13.019728 22.288911 28.364377 35.729171
 8.877829 19.951443 30.726531 44.400961
12.226032 20.373853 30.003996 43.396151
 8.532775 18.091436 32.765839 51.091007
8.8049882 20.945813 29.453855 34.251643
8.3840127 19.969099  29.94458 41.713592
9.5339415 20.027807 29.632425 39.880545
11.233662 19.164775 29.175422 38.209971
10.407781 21.653868 31.753313 49.197686
12.102968  21.75833 29.168331  38.05888
12.228035  19.13339 29.878038 36.567964
9.6612809  21.85244  29.61101 37.617794
9.8002753 23.859038  29.01271 38.490044
 9.138417 18.084905 28.812975 36.371415
11.259012 22.395983 29.811016 39.231856
7.8569112 21.512452 30.526549 40.452245
8.8971645 18.458986 30.576723 43.875282
7.8857517 21.186746 28.030963 34.526299
11.684868 23.639538 30.187498 40.505556
9.7260651 20.968905 28.914758 38.067146
 11.66679 23.005702 28.672645 30.684908
7.9976197   20.6001 29.810047  39.05618
10.134748 19.093801 29.053383 34.423978
 11.74687   20.7355 30.442122 43.669143
```

1.7 Assessing Univariate Normality with Q-Q Plots

Before we perform hypothesis tests on data, we should examine the distributional assumptions. Hypothesis testing in the presence of assumption violations may result in errors in statistical inference.

By the properties of the multivariate normal distribution of **y**, the components y_i are univariate normally distributed. Thus, one step in evaluating the multivariate normality assumption of **y** is to evaluate the univariate normality of each y_i. One can construct and examine histograms, stem-and-leaf plots, box plots, and Quantile-Quantile (Q-Q) probability plots.

Q-Q plots are plots of the observed, ordered quantile versus the quantile values expected if the observed data are normally distributed. Departures from a straight line are evidence against the assumption that the population from which the observations are drawn is normally distributed. Outliers may be detected from these plots as points well separated from the other observations. The behavior at the ends of the plots can provide information about the length of the tails of the distribution, and the symmetry or asymmetry of the distribution (Singh, 1993).

1.7.1 Normally Distributed Data

Program 1_7_1.sas produces a Q-Q plot for the univariate normally distributed random variable *y*1 from the data set generated by Program 1_6.sas.

Program 1_7_1.sas

```
/* Program 1_7_1.sas */
/* Program to create Q-Q plot of data */

options ls=80 ps=60 nodate nonumber;
filename appl 'c:\ 1_6.dat';
title1 'Output 1.7.1: Q-Q plot of 1.6 Data (y1)';

data ex171;
   infile appl;
   input y1-y4;
proc sort;
   by y1;
proc univariate noprint;
   var y1;
   output out=stats n=nobs mean=mean std=std;
data quantile;
   set ex171;
   if _n_=1 then set stats;
   i+1;
   p=(i - .5) / nobs;
   z=probit(p);
   normal = mean + z*std;
proc print;
proc corr;
   var y1 z;
run;

filename out 'c:\ 1_7_1.cgm';
goptions device=cgmmwwc gsfname=out gsfmode=replace
   colors=(black) hsize=5in vsize=4in;
proc gplot data=quantile;
   plot y1*z normal*z /overlay frame;
   symbol1 v=;
   symbol2 i=join v=none l=1;
run;
```

First the DATA step reads in the data and then PROC SORT sorts the values of *y*1 in ascending order. PROC UNIVARIATE outputs the number of observations (nobs), the mean of *y*1 (MEAN), and the standard deviation of *y*1 (STD). In the next DATA step the observed ordered quantiles are defined (P), and then the PROBIT function defines the quantiles expected if *z* is a normally distributed random variable with mean 0 and standard deviation of 1. Next the variable NORMAL is created so it can be compared to *y*1 in the plots. A filename for the plot is specified as well as options for the graphics plot. The option DEVICE = CGMMWWC specifies a cgm plot that can be accessed with Microsoft Word software. PROC GPLOT overlays the plots of *y*1 versus *z* and normal versus *z* where the plot of normal versus *z* is represented as the straight line (since we specified *I* = JOIN). The plotting symbol is specified by the v = option; when no value is specified, the default symbol is used.

Result and Interpretation 1_7_2.sas

Output 1.7.1 contains results of Program 1_7_1.sas and the resulting Q-Q plot.

Output 1.7.1: *Q-Q plot of 1.6 Data (y1)*

Output 1.7.1: Q-Q plot of 1.6 Data (y1)

OBS	Y1	Y2	Y3	Y4	NOBS	MEAN	STD	I	P	Z	NORMAL
1	7.78	19.42	30.00	34.77	50	10.4254	1.60353	1	0.01	-2.32635	6.6950
2	7.86	21.51	30.53	40.45	50	10.4254	1.60353	2	0.03	-1.88079	7.4095
3	7.89	21.19	28.03	34.53	50	10.4254	1.60353	3	0.05	-1.64485	7.7878
4	8.00	20.60	29.81	39.06	50	10.4254	1.60353	4	0.07	-1.47579	8.0589
5	8.22	17.40	28.72	35.26	50	10.4254	1.60353	5	0.09	-1.34076	8.2755
6	8.38	19.97	29.94	41.71	50	10.4254	1.60353	6	0.11	-1.22653	8.4586
7	8.41	21.69	32.14	49.37	50	10.4254	1.60353	7	0.13	-1.12639	8.6192
8	8.53	18.09	32.77	51.09	50	10.4254	1.60353	8	0.15	-1.03643	8.7634
9	8.63	16.50	29.92	40.95	50	10.4254	1.60353	9	0.17	-0.95417	8.8954
10	8.71	23.28	30.08	40.91	50	10.4254	1.60353	10	0.19	-0.87790	9.0177
11	8.80	20.95	29.45	34.25	50	10.4254	1.60353	11	0.21	-0.80642	9.1323
12	8.88	19.95	30.73	44.40	50	10.4254	1.60353	12	0.23	-0.73885	9.2406
13	8.90	18.46	30.58	43.88	50	10.4254	1.60353	13	0.25	-0.67449	9.3438
14	8.99	20.10	29.68	39.86	50	10.4254	1.60353	14	0.27	-0.61281	9.4427
15	9.14	18.08	28.81	36.37	50	10.4254	1.60353	15	0.29	-0.55338	9.5380
16	9.53	20.03	29.63	39.88	50	10.4254	1.60353	16	0.31	-0.49585	9.6303
17	9.59	22.25	29.88	41.84	50	10.4254	1.60353	17	0.33	-0.43991	9.7200
18	9.66	21.85	29.61	37.62	50	10.4254	1.60353	18	0.35	-0.38532	9.8075
19	9.73	20.97	28.91	38.07	50	10.4254	1.60353	19	0.37	-0.33185	9.8933
20	9.79	19.77	32.69	49.96	50	10.4254	1.60353	20	0.39	-0.27932	9.9775
21	9.80	23.86	29.01	38.49	50	10.4254	1.60353	21	0.41	-0.22754	10.0605
22	10.03	19.21	31.13	41.66	50	10.4254	1.60353	22	0.43	-0.17637	10.1426
23	10.11	21.41	29.36	35.77	50	10.4254	1.60353	23	0.45	-0.12566	10.2239
24	10.13	19.09	29.05	34.42	50	10.4254	1.60353	24	0.47	-0.07527	10.3047
25	10.28	18.52	29.67	35.68	50	10.4254	1.60353	25	0.49	-0.02507	10.3852
26	10.41	21.65	31.75	49.20	50	10.4254	1.60353	26	0.51	0.02507	10.4656
27	10.55	17.68	31.29	44.35	50	10.4254	1.60353	27	0.53	0.07527	10.5461
28	10.86	20.55	31.15	45.87	50	10.4254	1.60353	28	0.55	0.12566	10.6269
29	11.23	19.16	29.18	38.21	50	10.4254	1.60353	29	0.57	0.17637	10.7082
30	11.26	22.40	29.81	39.23	50	10.4254	1.60353	30	0.59	0.22754	10.7903
31	11.30	20.35	29.29	34.36	50	10.4254	1.60353	31	0.61	0.27932	10.8733
32	11.32	20.38	29.37	35.64	50	10.4254	1.60353	32	0.63	0.33185	10.9575
33	11.38	21.33	31.80	44.31	50	10.4254	1.60353	33	0.65	0.38532	11.0433
34	11.40	22.78	28.02	31.07	50	10.4254	1.60353	34	0.67	0.43991	11.1308
35	11.62	19.02	28.22	32.90	50	10.4254	1.60353	35	0.69	0.49585	11.2205
36	11.67	23.01	28.67	30.68	50	10.4254	1.60353	36	0.71	0.55338	11.3128
37	11.68	23.64	30.19	40.51	50	10.4254	1.60353	37	0.73	0.61281	11.4081
38	11.75	20.74	30.44	43.67	50	10.4254	1.60353	38	0.75	0.67449	11.5070
39	11.82	22.57	29.52	36.43	50	10.4254	1.60353	39	0.77	0.73885	11.6102
40	11.93	20.28	28.91	36.05	50	10.4254	1.60353	40	0.79	0.80642	11.7185
41	12.09	18.06	30.29	41.08	50	10.4254	1.60353	41	0.81	0.87790	11.8331
42	12.10	21.76	29.17	38.06	50	10.4254	1.60353	42	0.83	0.95417	11.9554
43	12.23	20.37	30.00	43.40	50	10.4254	1.60353	43	0.85	1.03643	12.0874
44	12.23	19.13	29.88	36.57	50	10.4254	1.60353	44	0.87	1.12639	12.2316
45	12.25	21.00	30.64	44.32	50	10.4254	1.60353	45	0.89	1.22653	12.3922
46	12.30	18.54	29.17	36.57	50	10.4254	1.60353	46	0.91	1.34076	12.5753
47	12.48	19.62	29.90	38.51	50	10.4254	1.60353	47	0.93	1.47579	12.7919
48	12.85	20.34	29.29	41.66	50	10.4254	1.60353	48	0.95	1.64485	13.0630
49	13.02	22.29	28.36	35.73	50	10.4254	1.60353	49	0.97	1.88079	13.4413
50	13.77	20.98	30.34	38.94	50	10.4254	1.60353	50	0.99	2.32635	14.1558

Output 1.7.1: Q-Q plot of 1.6 Data (y1)

Correlation Analysis

2 'VAR' Variables: Y1 Z

Output 1.7.1 (*continued*)

Simple Statistics

Variable	N	Mean	Std Dev	Sum	Minimum	Maximum
Y1	50	10.42540	1.60353	521.27000	7.78000	13.77000
Z	50	0	0.99740	0	-2.32635	2.32635

Pearson Correlation Coefficients / Prob > |R| under Ho: Rho=0 / N = 50

	Y1	Z
Y1	1.00000 0.0	0.98033 0.0001
Z	0.98033 0.0001	1.00000 0.0

Output 1.7.1: Q–Q plot of 1.6 Data (y1)

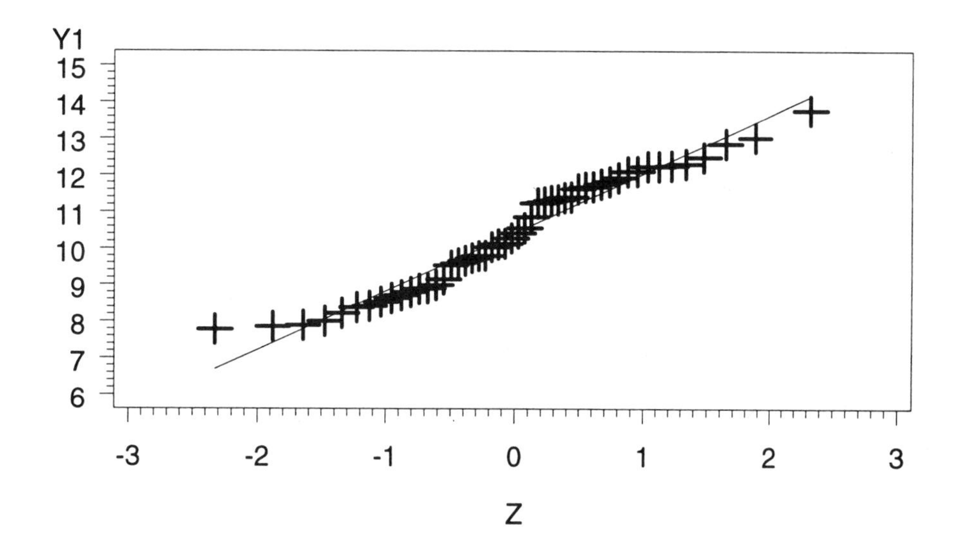

The plot shows that the observations lie close to the line, but not exactly; the tails, especially, fall off from the line. Recall that we know that these data are a random sample from a normal distribution. Thus, when using these plots for diagnostic purposes, we cannot expect that even normally distributed data will lie exactly on a straight line.

1.7.2 Nonnormally Distributed Data

The following example presents a Q-Q plot in which the data are not normally distributed: $y1$ is transformed by $1/y^2$ and plotted. Program 1_7_2.sas contains the SAS code.

Program 1_7_2.sas

```
/* Program 1_7_2.sas */
/* Program to create Q-Q plot of 1/(y1**2) Data */
```

```
options ls=80 ps=60 nodate nonumber;
filename app1 'c:\1_6.dat';
title1 'Output 1.7.2: Q-Q Plot of 1/(y1**2)';

data ex172;
   infile app1;
   input y1-y4;
   ty1=1/(y1**2);
proc sort;
   by ty1;
proc univariate noprint;
   var ty1;
   output out=stats n=nobs mean=mean std=std;
data quantile;
   set ex172;
   if _n_=1 then set stats;
   i+1;
   p=(i - .5) / nobs;
   z=probit(p);
   normal = mean + z*std;
proc print;
proc corr;
   var ty1 z;
run;

filename out 'c:\1_7_2.cgm';
goptions device=cgmmwwc gsfname=out gsfmode=replace
     colors=(black) hsize=5in vsize=4in;

proc gplot data=quantile;
   plot ty1*z normal*z /overlay frame;
   symbol1 v=;
   symbol2 i=join v=none l=1;
run;
```

The program is the same as Program 1_7_1.sas with the exception that the variable *ty*1 is created. This transformed variable is then used for plotting.

Result and Interpretation 1_7_2.sas

Output 1.7.2: Q−Q Plot of 1/(y1**2)

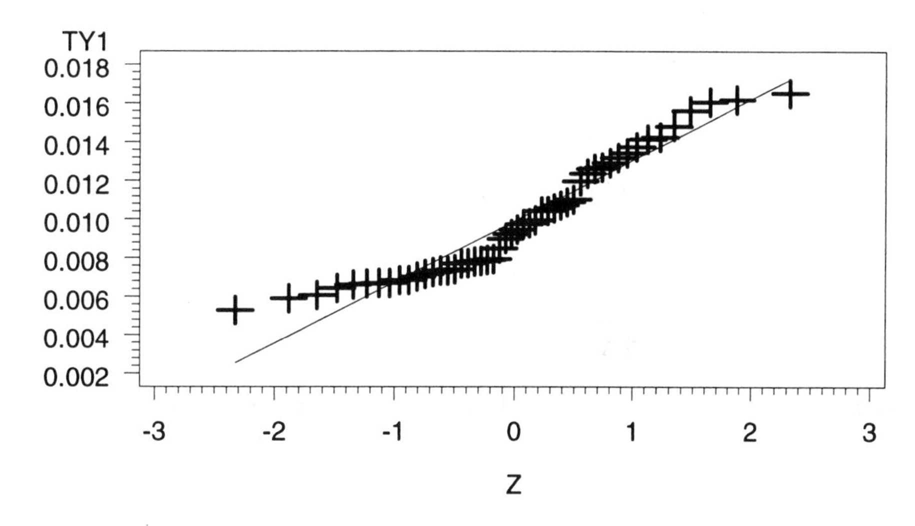

Observe that the observations show a marked curvilinear pattern and are not close to the line.

1.7.3 Outliers

The following example presents a Q-Q plot with an outlier. Program 1_7_3.sas is used to create an outlier in y_i and to plot the data.

Program 1_7_3.sas

```
/* Program 1_7_3.sas */
/* Program to create Q-Q plot of y1 data with Outlier */

options ls=80 ps=60 nodate nonumber;
filename appl 'c:\ 1_6.dat';
title1 'Output 1.7.3: Q-Q plot of y1 data with an Outlier';

data ex173;
   infile appl;
   input y1-y4;
   if y1 ge 13.7 then y1 = 17;
proc sort;
   by y1;
proc univariate noprint;
   var y1;
   output out=stats n=nobs mean=mean std=std;
data quantile;
   set ex173;
   if _n_=1 then set stats;
   i+1;
   p=(i - .5) / nobs;
   z=probit(p);
   normal = mean + z*std;
proc print;
proc corr;
   var y1 z;
run;
```

```
filename out 'c:\ 1_7_3.cgm';
goptions device=cgmmwwc gsfname=out gsfmode=replace
   colors=(black) hsize=5in vsize=4in;

proc gplot data=quantile;
   plot y1*z normal*z /overlay frame;
   symbol1 v=;
   symbol2 i=join v=none l=1;
run;
```

Again, the program is the same as Program 1_7_1.sas with exception that the *y*1 value of 13.77 is changed to the value of 17 by using an IF-THEN statement in the DATA step.

Result and Interpretation 1_7_3.sas

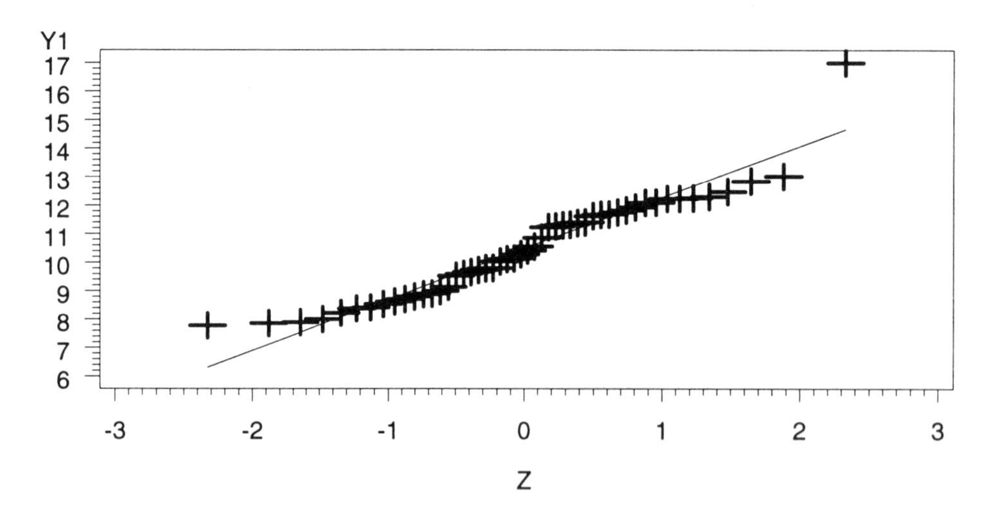

The extreme observation lying far from the line at the top of the plot indicates an outlier.

1.7.4 Real Data Example

To illustrate the application of Q-Q plots utilizing real data, we use the data from Rhower given in Timm (1975, p. 281 and p. 345). Rhower was interested in predicting the performance on three standardized tests (Peabody Picture Vocabulary (Y1), Student Achievement Test (Y2), and the Raven Progressive Matrices Test (Y3)) given five paired-associate, learning-proficiency tasks (named (X1), still (X2), named still (X3), named action (X4) and sentence still (X5)) for 32 randomly selected school children in an upper-class, white residential school.

A data set of the residuals resulting after the three dependent variables were regressed on the five independent variables is contained in the file named ycondx.dat. Program 1_7_4.sas produces three Q-Q plots for the residuals of the three dependent variables and then to produce five Q-Q plots of the independent variables.

Program 1_7_4.sas

```
/* Program 1_7_4.sas                                    */
/* Program to create Q-Q plot of a dataset              */
/* To run this program on your own dataset change       */
/* the name of the file in the file=_____statement      */
/* and the number of columns in the p=___ statement     */

options ls=80 ps=60 nodate nonumber;

%let file = ycondx.dat;
%let p = 3;

  /* macro to expand the string of variables that are processed */
%macro expand(cols);
   %do j=1 %to &cols;
      v&j
   %end;
%mend expand;

  /* macro to perform Q-Q plotting of the variables    */
%macro qq(cols);
   %do i=1 %to &cols;
proc sort data=ex174;
   by v&i;
proc univariate noprint data=ex174;
   var v&i;
   output out=stats n=nobs mean=mean std=std;
data quantile;
   set ex174;
   if _n_=1 then set stats;
   k+1;
   pr=(k - .5) / nobs;
   z=probit(pr);
   normal = mean + z*std;
proc print data=quantile;
   title "Output 1.7.4: Q-Q plot, variable V&i, &file";
proc corr data=quantile;
   var v&i z;

filename out "1_7_4_&i'.cgm";
goptions device=cgmmwwc gsfname=out gsfmode=replace
   colors=(black) hsize=5in vsize=4in;

proc gplot data=quantile;
   title "Output 1.7.4: Q-Q plot, variable V&i, &file";
   plot v&i*z normal*z /overlay frame;
   symbol1 v=;
   symbol2 i=join v=none l=1;
run;
%end;
%mend qq;

data ex174;
   infile "&file";
   input %expand(&p);
proc print data=ex174;
   title "Output 1.7.4: &file";
%qq(&p);
```

This program produces one Q-Q plot for each column of data in the data file. The name of the data file is contained in the statement `%let file` = and the number of columns must be specified in the statement `%let p=`. Here, to produce the output that follows we first ran the program with `file = ycondx.dat` and `p=3` and then we ran it with `file=rhowerx.dat` and `p=5`. To run this program on your own data set, simply change the name of the data file and the number of columns to correspond to your data set. Q-Q plots are output to files named 1_7_4&*i*.cgm where the &*i* is equal to 1 for the first column of data, 2 for the second column, and so on.

Result and Interpretation 1_7_4.sas

Output 1.7.4: Q – Q plot, variable V1, ycondx.dat

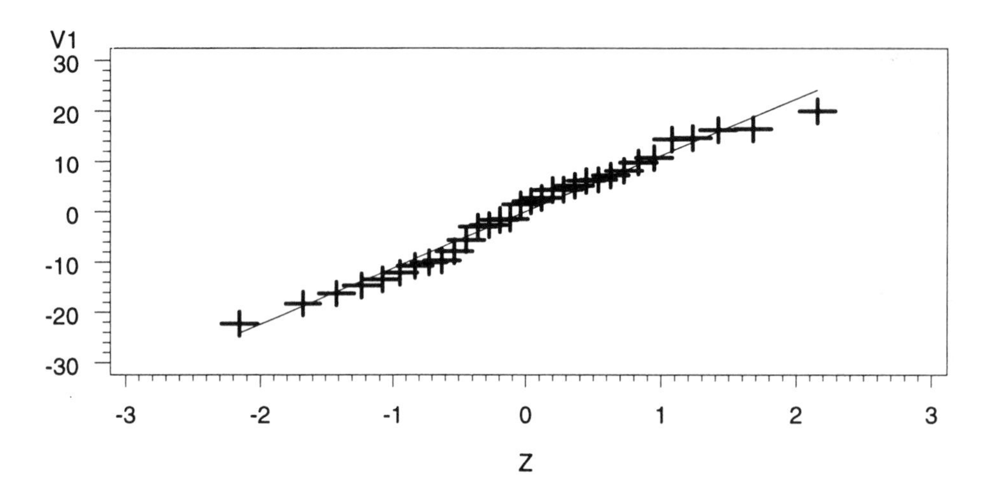

Output 1.7.4: Q – Q plot, variable V2, ycondx.dat

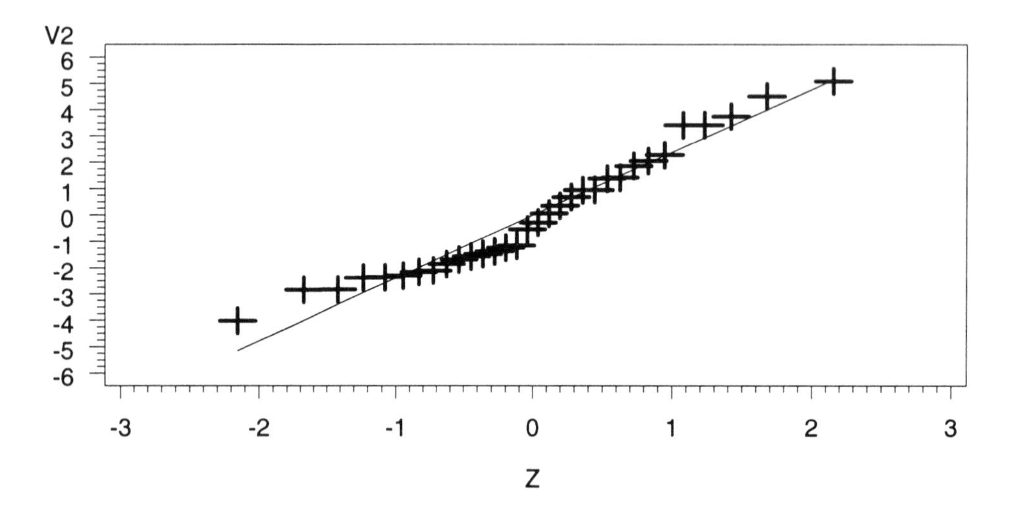

Output 1.7.4: Q – Q plot, variable V3, ycondx.dat

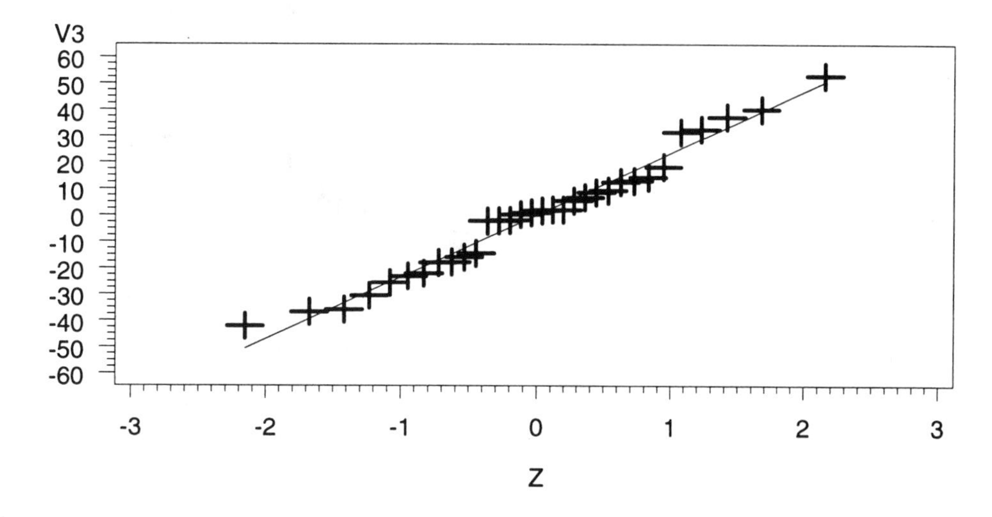

Output 1.7.4: Q – Q plot, variable V1, rhowerx.dat

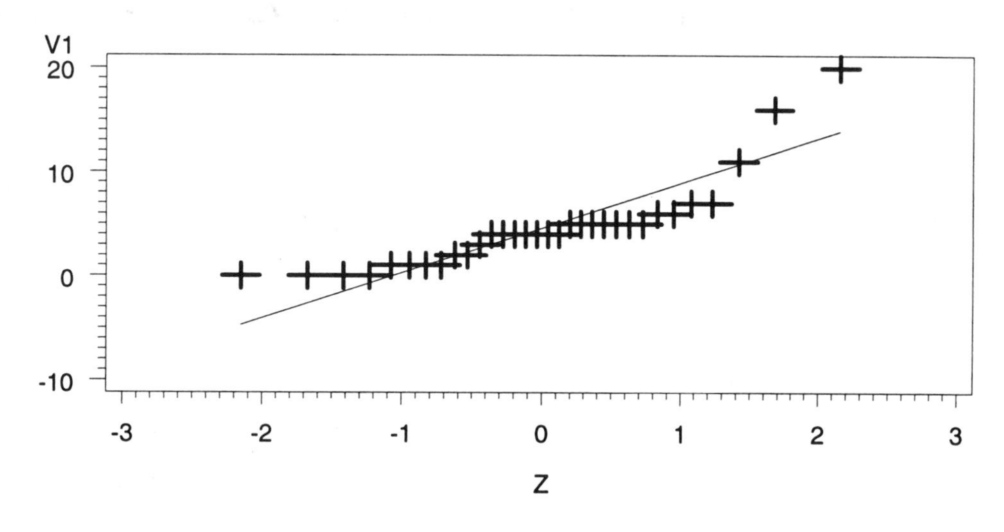

Output 1.7.4: Q − Q plot, variable V2, rhowerx.dat

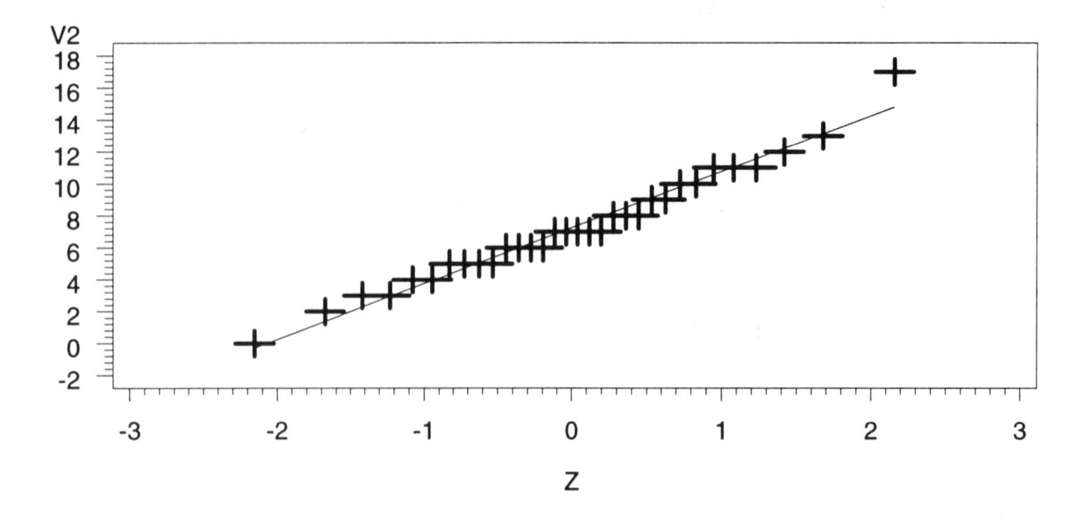

Output 1.7.4: Q − Q plot, variable V3, rhowerx.dat

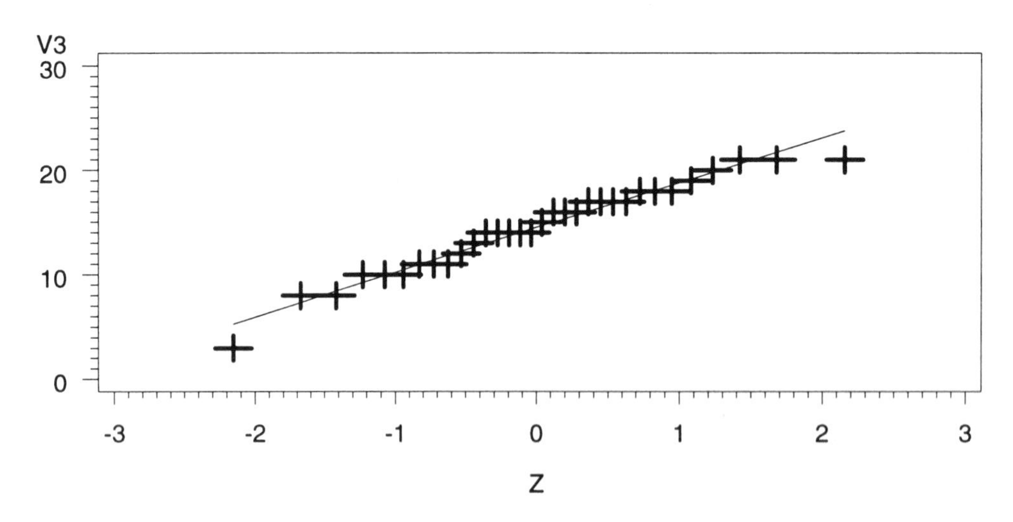

Output 1.7.4: Q − Q plot, variable V4, rhowerx.dat

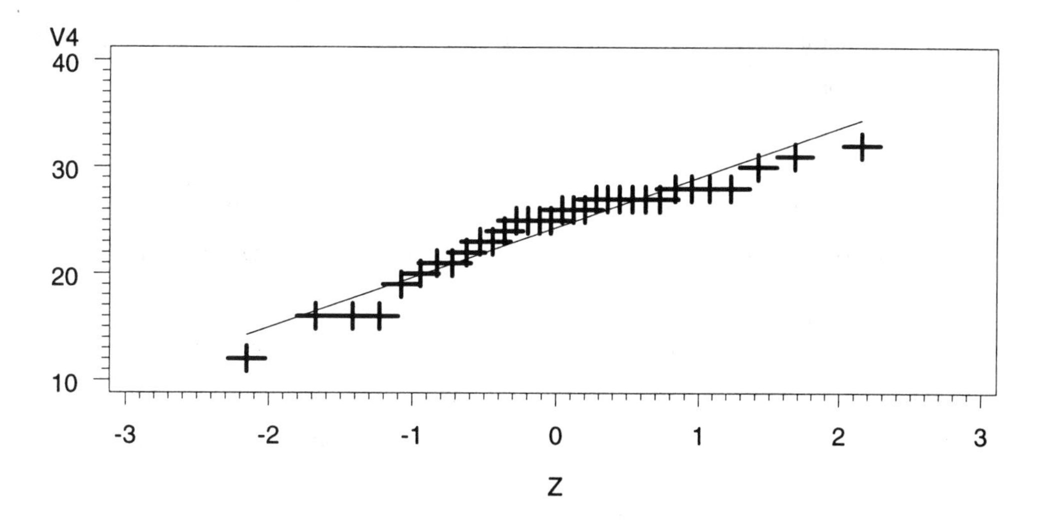

Output 1.7.4: Q − Q plot, variable V5, rhowerx.dat

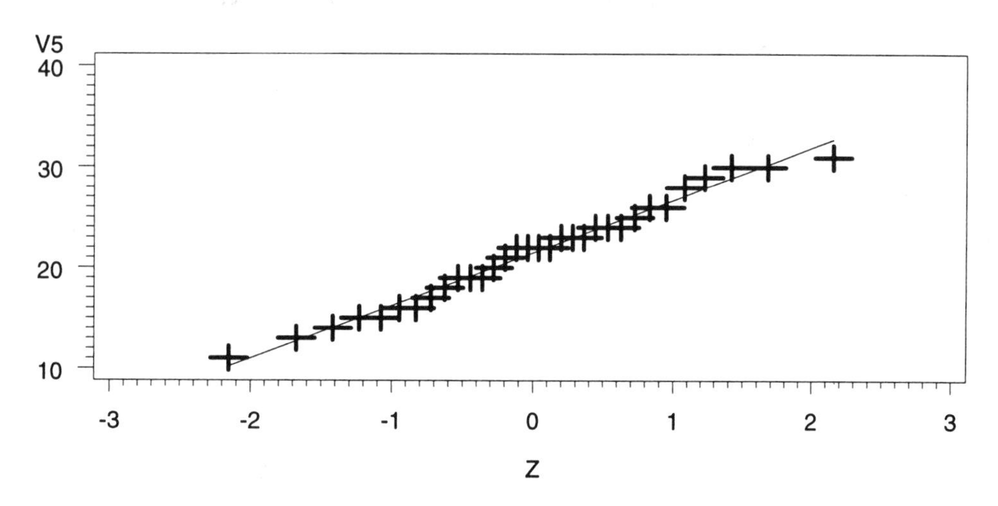

From the plots generated using the ycondx.dat file we see that the residuals for variables Y1 and Y3 appear to be univariate normally distributed, but that Y2 does not appear to satisfy the assumption of normality. With a sample size as small as 32, however, it is difficult to be certain. In general, for Q-Q plots a sample size of at least 50 observations is desirable.

The plots of the independent variables using the file rhowerx.dat appear to reveal a possibility of several outliers for variable 1, a possible outlier for variables 2 and 3, and a curved pattern for variable 4. Variable 5 appears to lie close to the line. With a sample size of only 32 it is difficult to be certain if the plots show departures from normality.

1.8 Assessing Multivariate Normality with Chi-Square Plots

Recall that marginal normality does not ensure multivariate normality. To evaluate multivariate normality, one may compute the Mahalanobis distance for the ith observation

$$D_i^2 = (\mathbf{x}_i - \overline{\mathbf{x}})' \mathbf{S}^{-1} (\mathbf{x}_i - \overline{\mathbf{x}}) \tag{1.33}$$

and plot these distances against the ordered chi-square percentile value $\chi_p^2[(i-(1/2)/n)]$. If the data are multivariate normal, the plotted pairs should be close to a straight line. As with Q-Q plots, points far from the line may be multivariate outliers.

Singh (1993) constructed probability plots resembling Shewart-type control charts, where warning points were placed at the α 100% critical value of the distribution of Mahalanobis distances, and a maximum point limit was also defined. Thus, any observation falling beyond the maximum limit was considered an outlier, and any point between the warning limit and the maximum limit required further investigation.

Singh (1993) constructed multivariate probability plots with the ordered Mahalanobis distances versus quantiles from a beta distribution, rather than the chi-square distribution. The exact distribution of $nD_i^2/(n-1)^2$ is in fact a beta distribution $B(p/2,(n-p-1)/2)$, Gnandesikan and Kettenring (1972). The chi-square distribution is only an approximation, which may not be good enough as p gets large, Small (1978).

1.8.1 Normally Distributed Data

For an example of a chi-square plot of multivariate normally distributed data, see the chi-square plot of the multivariate normally distributed data, **y**, produced by Program 1_8_1.sas.

Program 1_8_1.sas

```
/* Program 1_8_1.sas                          */
/* Program to create Chi-Square Plot of Y Data  */
/* Data set from 1_6.sas is used, with a column */
/* of observation numbers added to the file     */

options ls=80 ps=60 nodate nonumber;
filename appl 'c:\ 1_6.da2';
title1 'Output 1.8.1: Chi-Square Plot of the 1.6 Dataset';

data ex181;
   infile appl;
   input tag $ y1 - y4;
   label tag = 'id'
      y1 = 'var1'
      y2 = 'var2'
      y3 = 'var3'
      y4 = 'var4';
   %let id=tag;
   %let var=y1 y2 y3 y4;
```

```
proc iml;
   reset;
   start dsquare;
      use _last_;
      read all var {&var} into y [colname=vars rowname=&id];
      n=nrow(y);
      p=ncol(y);
      r1=&id;
      print y;
      m=y[ :,];
      d=y - j(n,1) * m;
      s=d` * d / (n-1);
      dsq=vecdiag(d * inv(s) * d`);
      r=rank(dsq);
      val=dsq; dsq [r, ] = val;
      val=r1; &id [r] = val;
      z=((1:n)` - .5) / n;
      chisq = 2 * gaminv(z,p/2);
      result = dsq||chisq;
      cl={'dsq' 'chisq'};
      create dsquare from result [colname=cl rowname=&id];
      append from result [rowname=&id];
   finish;
run dsquare;
quit;
proc print data=dsquare;
   var tag dsq chisq;
run;

filename out 'c:\ 1_8_1.cgm';
goptions device=cgmmwwc gsfname=out gsfmode=replace
   colors=(black) hsize=5in vsize=4in;

proc gplot data=dsquare;
   plot dsq*chisq /frame;
   symbol1 v=;
run;
```

The data set used is the file output by Program 1_6.sas but with a column of observation numbers added as the first column; the new file is named 1_6.da2. Program 1_8_1.sas first reads the data with the DATA step and labels the variables. PROC IML then computes a vector of Mahalanobis distances, named dsq. Next, the ordered chi-square percentile values are computed, named chisq. The values are printed and then plotted with PROC GPLOT. The graphics file is output to the file 1_8_1.cgm.

Result and Interpretation 1_8_1.sas

Output 1.8.1: Chi – Square Plot of the 1.6 Dataset

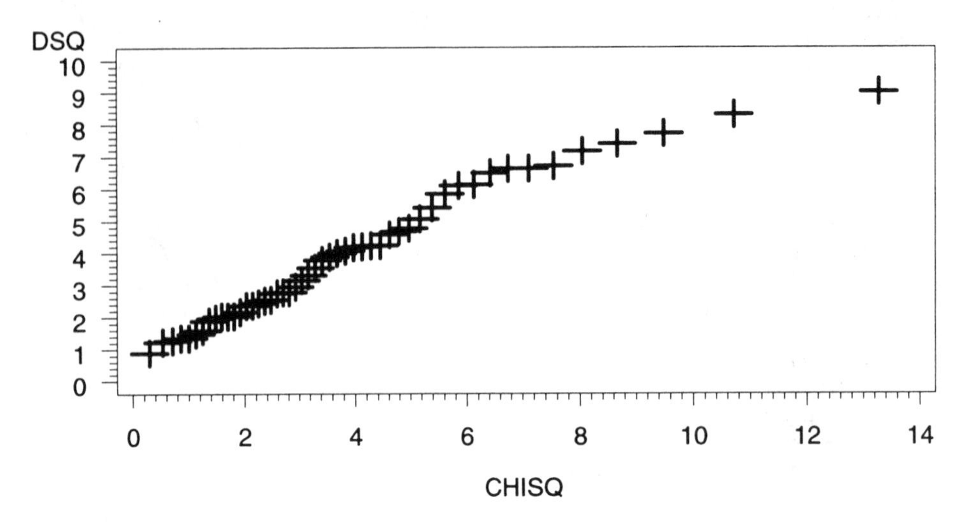

Upon examining the plot one can see that the observations lie somewhat close to a straight line, but not exactly, even for data known to be from a multivariate normal distribution. For multivariate chi-square plots, one would like to have approximately 100 observations.

1.8.2 Real Data Example

For an example of a chi-square plot using real data, see Program 1_8_2.sas, which produces a multivariate chi-square plot of the Rhower data (from Section 1.7.4). Namely, this is a plot of the residuals resulting after the three dependent variables were regressed on the five independent variables.

Program 1_8_2.sas

```
/* Program 1_8_2.sas                  */
/* Program to create Chi-Square Plot */
/* of Residuals from Rhower data      */

options ls=80 ps=60 nodate nonumber;
filename rhower 'c:\ycondx.da2';
title1 'Output 1.8.2: Chi-Square Plot of Residuals';

data ex182;
   infile rhower;
   input tag $ yc1-yc3;
   label tag = 'id'
      yc1 = 'var1'
      yc2 = 'var2'
      yc3 = 'var3';
   %let id=tag;
   %let var=yc1 yc2 yc3;
proc iml;
   reset;
   start dsquare;
      use _last_;
      read all var {&var} into y [colname=vars rowname=&id];
      n=nrow(y);
```

```
        p=ncol(y);
        r1=&id;
        print y;
        m=y[ :,];
        d=y - j(n,1) * m;
        s=d` * d / (n-1);
        dsq=vecdiag(d * inv(s) * d`);
        r=rank(dsq);
        val=dsq; dsq [r, ] = val;
        val=r1; &id [r] = val;
        z=((1:n)` - .5) / n;
        chisq = 2 * gaminv(z,p/2);
        result = dsq||chisq;
        cl={'dsq' 'chisq'};
        create dsquare from result [colname=cl rowname=&id];
        append from result [rowname=&id];
    finish;
    run dsquare;
    quit;
proc print data=dsquare;
    var tag dsq chisq;
run;

filename out 'c:\ 1_8_2.cgm';
goptions device=cgmmwwc gsfname=out gsfmode=replace
    colors=(black) hsize=5in vsize=4in;

proc gplot data=dsquare;
    plot dsq*chisq /frame;
    symbol1 v=;
run;
```

Result and Interpretation 1_8_2.sas

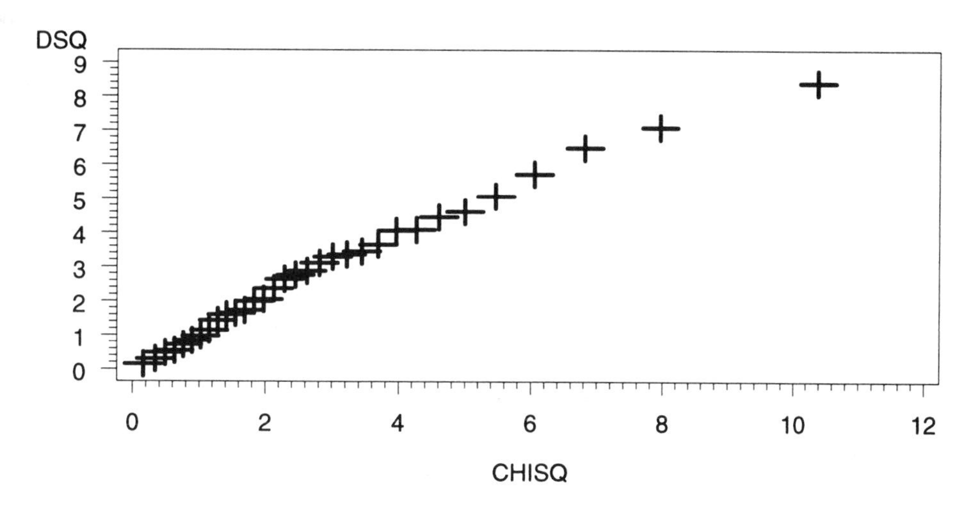

Recall that variable 2 appeared nonnormal from the univariate Q-Q plot. According to the multivariate chi-square plot shown in Output 1.8.2, the points appear to lie close to a straight line, thus indicating that the data are normally distributed.

1.9 Scatter Plots

Two-dimensional and three-dimensional scatter plots of variables may be used to detect possible skewness, kurtosis, and outlying observations.

1.9.1 Two-Dimensional Plots

PROC PLOT may be used to produce simple bivariate scatter plots. Program 1_9_1.sas produces scatter plots of the Peabody Picture Vocabulary variable ($y1$) with each of the independent variables in the Rhower data set.

Program 1_9_1.sas

```
/* Program 1_9_1.sas */
/* Program to produce bivariate scatter plot of Rhower Data */

options ls=80 ps=60 nodate nonumber;
filename rhower 'c:\5_6.dat';
title 'Output 1.9.1: Bivariate Scatter Plots of Rhower Data';

data ex191;
   infile rhower;
   input y1-y3 x0-x5;
proc plot;
   plot y1*x1;
   plot y1*x2;
   plot y1*x3;
   plot y1*x4;
   plot y1*x5;
run;
```

Result and Interpretation 1_9_1.sas

The output is given in Output 1.9.1.

Output 1.9.1 *Bivariate Scatter Plots of Rhower Data*

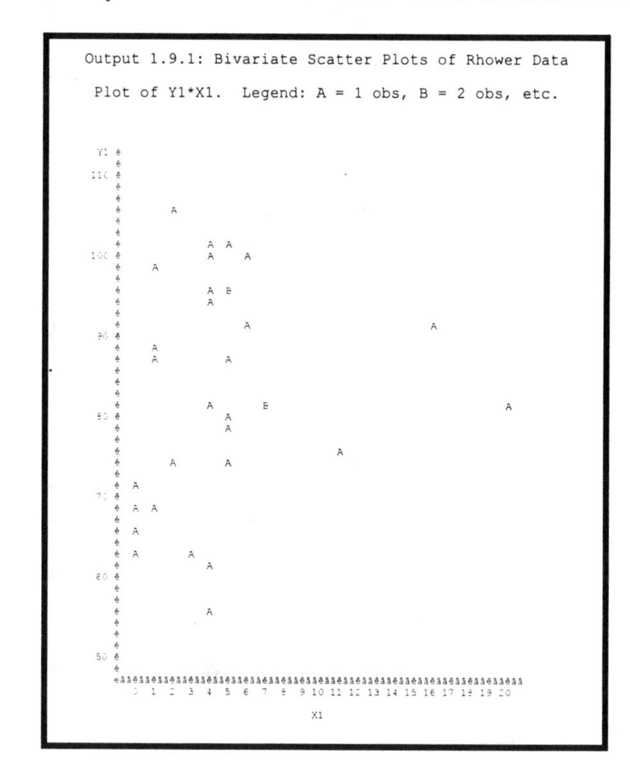

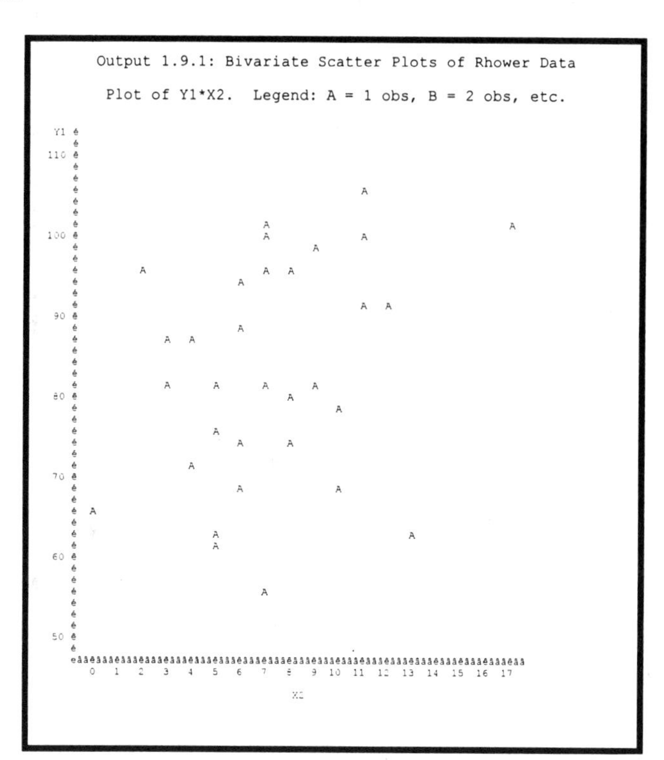

Output 1.9.1 *(continued)*

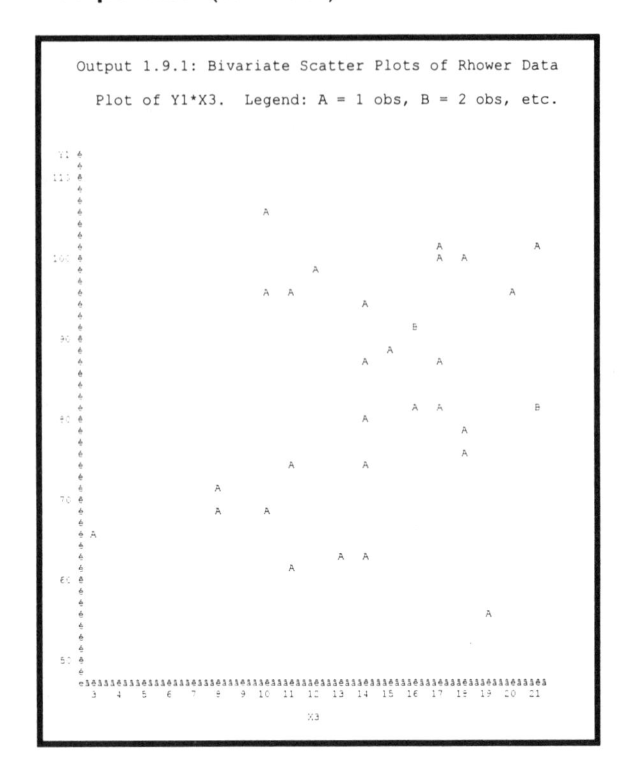

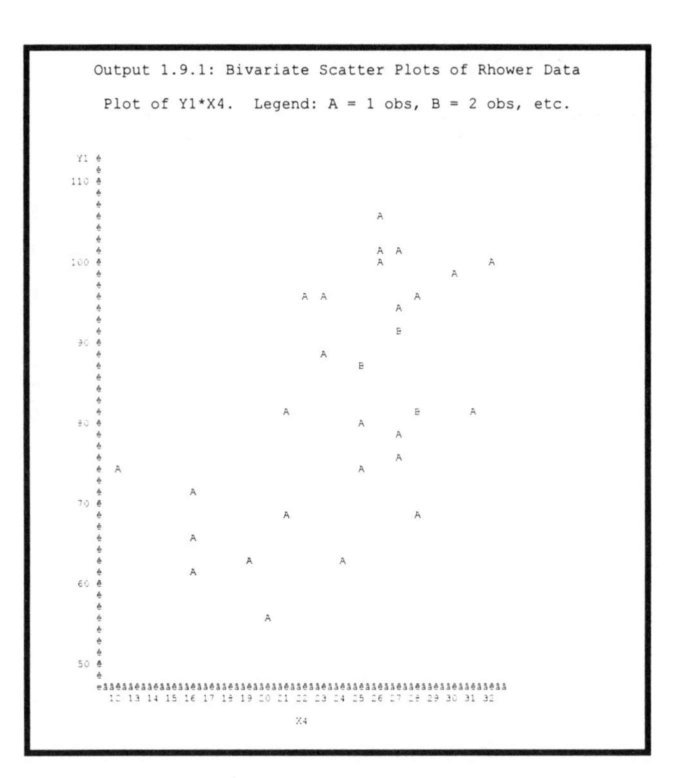

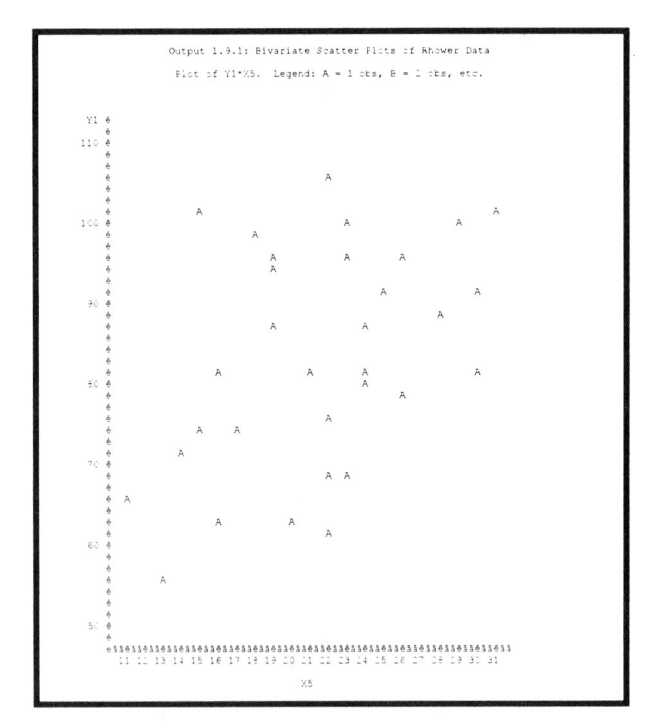

From the plot of Y1 versus X1, three of the observations appeared to be outliers. The plot of Y1 versus X3 reveals a possible outlier as well. The other plots show no obvious outliers.

1.9.2 Three-Dimensional Plots

Three-dimensional scatter plots may be generated using the G3D procedure. The first part of Program 1_9_2.sas (adapted from Khatree and Naik, 1995, p. 65) produces a plot of a bivariate normal distribution with a mean and covariance matrix

$$\mu = \begin{pmatrix} 0 \\ 0 \end{pmatrix}, \quad \boldsymbol{\Sigma} = \begin{pmatrix} 3 & 1 \\ 1 & 4 \end{pmatrix}$$

This is the covariance matrix of variables $y1$ and $y2$ from the simulated multivariate normal data generated by Program 1_6.sas.

Program 1_9_2.sas

```
/* Program 1_9_2.sas */
/* Program to create 3-D Plots of bivariate normal distributions*/

options ls=80 ps=60 nodate nonumber;
title 'Output 1.9.2: Bivariate Normal Distribution';

title2 'with u=(0, 0), var(y1)=3, var(y2)=4, cov(y1,y2)=1, r=.289';
data bivar;
   vy1=3;
   vy2=4;
   r=.289;
   keep y1 y2 z;
   cons=1/(2*3.14159265*sqrt(vy1*vy2*(1-r*r)));
   do y1=-10 to 10 by .2;
      do y2=-10 to 10 by .2;
         zy1=y1/sqrt(vy1);
         zy2=y2/sqrt(vy2);
         d=((zy1**2)+(zy2**2)-2*r*zy1*zy2)/(1-r**2);
         z=cons*exp(-d/2);
         if z > .001 then output;
     end;
   end;
run;

filename out1 'c:\ 1_9_2_1.cgm';
goptions device=cgmmwwc gsfname=out1 gsfmode=replace
   colors=(black) hsize=6in vsize=5in;

proc g3d data=bivar;
   plot y1*y2=z;
run;

/* A Second Plot*/

title2 'with u=(0, 0), var(y1)=3, var(y2)=20, cov=0, r=0';
data bivar2;
   vy1=3;
   vy2=20;
   r=0;
   keep y1 y2 z;
   cons=1/(2*3.14159265*sqrt(vy1*vy2*(1-r*r)));
```

```
      do y1=-10 to 10 by .2;
         do y2=-10 to 10 by .2;
            zy1=y1/sqrt(vy1);
            zy2=y2/sqrt(vy2);
            d=((zy1**2)+(zy2**2)-2*r*zy1*zy2)/(1-r**2);
            z=cons*exp(-d/2);
            if z > .001 then output;
         end;
      end;
   run;

   filename out2 'c:\ 1_9_2_2.cgm';
   goptions device=cgmmwwc gsfname=out2 gsfmode=replace
       colors=(black) hsize=6in vsize=5in;

   proc g3d data=bivar2;
      plot y1*y2=z;
   run;
```

Result and Interpretation 1_9_2.sas

The three-dimensional plot is given in Output 1.9.2.

Output 1.9.2: Bivariate Normal Distribution

with u=(0, 0), var(y1)=3, var(y2)=4, cov(y1,y2)=1, r=.289

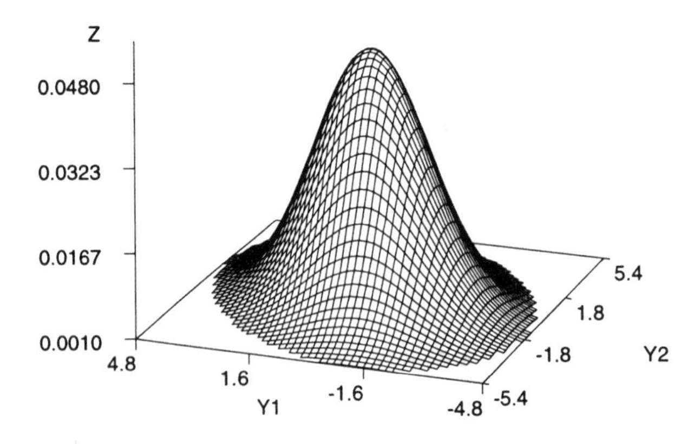

To see how plots vary, see the three-dimensional plot of the bivariate normal distribution having the same covariance matrix as variables $y1$ and $y4$ of 1_6.dat that is also plotted using Program 1_9_2.sas, here

$$\Sigma = \begin{pmatrix} 3 & 0 \\ 0 & 20 \end{pmatrix}$$

The output is given in the continuation of Output 1.9.2 that follows.

Output 1.9.2: Bivariate Normal Distribution

with u=(0, 0), var(y1)=3, var(y2)=20, cov=0, r=0

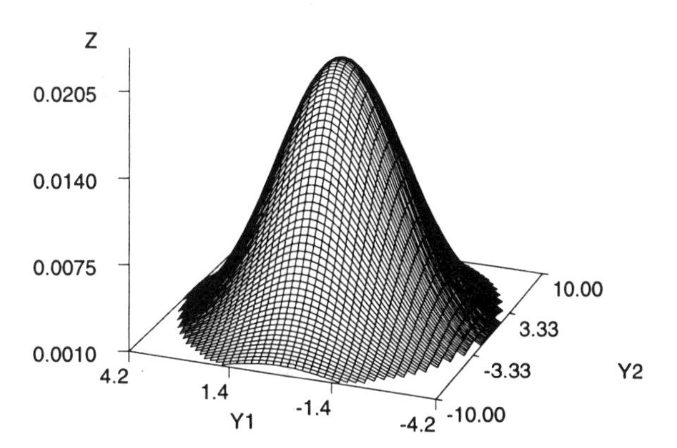

Notice that in the first plot, a cross-wise slice would result in an oval shape, whereas in the second plot, a circular shape would result. This is because in the first plot there is a covariance of one, but for the second plot the covariance is zero. See Khattree and Naik for many more graphical representations of multivariate data using SAS software.

1.10 Multivariate Skewness and Kurtosis

Another method for examining the assumption of multivariate normality is to compute the measures of skewness and kurtosis. If the data are multivariate normally distributed, these measures should be near zero. If the distribution is leptokurtic (has heavy tails), the measure of kurtosis will be large. If the distribution is platykurtic (has light tails), the kurtosis coefficient will be small.

Mardia (1970) defines the measures of multivariate skewness and kurtosis as

$$\begin{aligned} \beta_{1,p} &= E\left\{(\mathbf{x}-\mu)'\mathbf{\Sigma}^{-1}(\mathbf{y}-\mu)\right\}^3 \\ \beta_{2,p} &= E\left\{(\mathbf{y}-\mu)'\mathbf{\Sigma}^{-1}(\mathbf{y}-\mu)\right\}^2 \end{aligned} \tag{1.34}$$

where $\mathbf{x}$ and $\mathbf{y}$ are identically and independently distributed. Sample estimates of these quantities are

$$\begin{aligned} b_{1,p} &= \frac{1}{n^2}\sum_{i=1}^{n}\sum_{j=1}^{n}\left[(\mathbf{y}_i-\bar{\mathbf{y}})'\mathbf{S}^{-1}(\mathbf{y}_j-\bar{\mathbf{y}})\right]^3 \\ b_{2,p} &= \frac{1}{n}\sum_{i=1}^{n}\left[(\mathbf{y}_i-\bar{\mathbf{y}})'\mathbf{S}^{-1}(\mathbf{y}_i-\bar{\mathbf{y}})\right]^2 \end{aligned} \tag{1.35}$$

When $\mathbf{y} \sim N_p(\mu, \mathbf{\Sigma})$, then $\beta_{1,p} = 0$, $\beta_{2,p} = p(p+2)$, $\kappa_1 = nb_{1,p}/6$ has an asymptotic chi-square distribution with $p(p+1)(p+2)/6$ degrees of freedom, and $\kappa_2 = \left[b_{2,p} - p(p+2)\right]/\left[8p(p+2)/n\right]^{1/2}$ has an asymptotic standard normal distribution. Thus, when $n > 50$ one may develop large sample tests of multivariate normality. Mardia (1974) developed tables of approximate percentiles for $p = 2$ and $n \geq 10$ and alternative large sample approximations. In general, tests of hypotheses regarding means are sensitive to highly skewed multivariate distributions.

To illustrate the calculations to evaluate multivariate skewness and kurtosis, we again will use the Rhower data; however, given the small sample size of only $n = 32$ the significance level associated with the test may be in error. The SAS code to perform the calculations and to estimate p-values is given in Program 1_10.sas.

Program 1_10.sas

```
/* Program 1_10.sas                                          */
/* Program to calculate Mardia's measure of multivariate     */
/* skewness and kurtosis                                     */

options ls=80 ps=60 nodate nonumber;
title 'Output 1.10: Mardias Multivariate Skewness & Kurtosis';

data Rhower;
   infile 'c;\ 5_1.dat';
   input y1-y3 x0-x5;
proc print data=Rhower;

proc iml;
   use Rhower;
   v={y1 y2 y3};
   w={x0 x1 x2 x3 x4 x5};
   read all var v into y;
   read all var w into x;
   beta=inv(x`*x)*x`*y;
   n=nrow(y);
   p=ncol(y);
   k=ncol(x);
   s=(y`*y-y`*x*beta)/(n-k);
   s_inv=inv(s);
   q=i(n)-x*(inv(x`*x)*x`);
   d=q*y*s_inv*y`*q;

   b1=(sum(d#d))/(n*n);
   b2=trace(d#d)/n;

   kappa1= n*b1/6;
   kappa2=(b2-p*(p+2))/sqrt(8*p*(p+2)/n);

   dfchi=p*(p+1)*(p+2)/6;

   pvalskew=1-probchi(kappa1,dfchi);
   pvalkurt=2*(1-probnorm(abs(kappa2)));

   print s; print s_inv;
   print b1; print kappa1; print pvalskew;
   print b2; print kappa2; print pvalkurt;
quit;
```

First the three dependent and five independent variables of the Rhower data set are read within the DATA step and then printed. Next, using PROC IML, the matrix **Y** is defined as the 32×3 matrix of dependent variables, and the matrix **X** is defined as the 32×5 matrix of independent variables. The variance-covariance matrix is then computed and named s. Next, $b1$ and $b2$ of (1.35) are computed and then Kappa1 and Kappa2 and the corresponding p-values, pvalskew and pvalkurt, are printed.

Result and Interpretation 1_10.sas

Output 1.10: *Mardias Multivariate Skewness & Kurtosis*

Output 1.10: Mardias Multivariate Skewness & Kurtosis

OBS	Y1	Y2	Y3	X0	X1	X2	X3	X4	X5
1	68	15	24	1	0	10	8	21	22
2	82	11	8	1	7	3	21	28	21
3	82	13	88	1	7	9	17	31	30
4	91	18	82	1	6	11	16	27	25
5	82	13	90	1	20	7	21	28	16
6	100	15	77	1	4	11	18	32	29
7	100	13	58	1	6	7	17	26	23
8	96	12	14	1	5	2	11	22	23
9	63	10	1	1	3	5	14	24	20
10	91	18	98	1	16	12	16	27	30
11	87	10	8	1	5	3	17	25	24
12	105	21	88	1	2	11	10	26	22
13	87	14	4	1	1	4	14	25	19
14	76	16	14	1	11	5	18	27	22
15	66	14	38	1	0	0	3	16	11
16	74	15	4	1	5	8	11	12	15
17	68	13	64	1	1	6	10	28	23
18	98	16	88	1	1	9	12	30	18
19	63	15	14	1	0	13	13	19	16
20	94	16	99	1	4	6	14	27	19
21	82	18	50	1	4	5	16	21	24
22	89	15	36	1	1	6	15	23	28
23	80	19	88	1	5	8	14	25	24
24	61	11	14	1	4	5	11	16	22
25	102	20	24	1	5	7	17	26	15
26	71	12	24	1	0	4	8	16	14
27	102	16	24	1	4	17	21	27	31
28	96	13	50	1	5	8	20	28	26
29	55	16	8	1	4	7	19	20	13
30	96	18	98	1	4	7	10	23	19
31	74	15	98	1	2	6	14	25	17
32	78	19	50	1	5	10	18	27	26

```
      S
149.9613 10.822548 49.215424
10.822548 6.8088795 23.991269
49.215424 23.991269  659.0046

      S_INV
0.0075446 -0.011479 -0.000146
-0.011479 0.1859442 -0.005912
-0.000146 -0.005912 0.0017435

       B1
1.9804688

   KAPPA1
  10.5625

  PVALSKEW
0.3926023

       B2
8.8696232
```

Output 1.10 *(continued)*

```
KAPPA2
-3.165713

PVALKURT
0.001547
```

From Output 1.10, it appears that for the Rhower data, there is no evidence of multivariate skewness. The distribution may, however, be platykurtic since the Kurtosis coefficient is small (-3.166) and the p-value is significant ($p < .01$).

1.11 Box-Cox Transformations

Data that are not normally distributed can often be transformed to near normality. Count data can be made to be more nearly normal by taking the square root. Proportions are often transformed using the logit, $\text{logit}\,(p) = \frac{1}{2}\log\left(\frac{p}{1-p}\right)$, and correlations are transformed using Fisher's r to z transformation, $\frac{1}{2}\log\left(\frac{1+r}{1-r}\right)$, or the arc-sine square root. Alternately, the sample data can be used for finding an appropriate power transformation by using the Box-Cox family of power transformations

$$x^{(\lambda)} = \begin{cases} \dfrac{x^{\lambda}-1}{\lambda} & \lambda \neq 0 \quad x > 0 \\ \ln x & \lambda = 0 \end{cases} \tag{1.36}$$

where x is the observed data. The appropriate λ to use for the transformation is that value of λ that maximizes the expression

$$L(\lambda) = -\frac{n}{2}\ln\left[\frac{1}{n}\sum_{i=1}^{n}(x_i^{(\lambda)} - \overline{x^{(\lambda)}})^2\right] + (\lambda-1)\sum_{i=1}^{n}\ln x_i \tag{1.37}$$

$$\overline{x^{(\lambda)}} = \frac{1}{n}\sum_{i=1}^{n} x_i^{(\lambda)} = \frac{1}{n}\sum_{i=1}^{n}\left(\frac{(x_i)^{\lambda}-1}{\lambda}\right)$$

Program 1_11.sas computes and graphs the function L (λ) for values of λ: -1.0 (.1) 1.3. The appropriate value to use for transforming nonnormally distributed observations is obtained from the plot by locating the $\hat{\lambda}$ that maximizes the likelihood. When the data are normally distributed there will be no maximum value for the graph.

Program 1_11.sas

```
/* Program 1_11.sas                                             */
/* Program to compute Box Cox Transformations                   */
/* To run this program on your own dataset change               */
/* the name of the file in the file=____ statement              */
/* the number of rows in the n=____ statement and               */
/* the number of columns (variables) in the p=___ statement */
```

```
options ls=80 ps=60 nodate nonumber mprint;
%let file=c:\exp1_6.dat;
%let n=50;
%let p=1;

 /*macro to expand the string of variables that are processed */
%macro expand(cols);
   %do j=1 %to &cols;
      x&j
   %end;
%mend expand;

 /*macro to perform the Box-Cox transformation on the data matrix */
%macro loop(cols);
   %do i=1 %to &cols;
      proc iml;
         use matrix;
         read all var {x&i};
         in=i(&n);
         allh={-1.0, -0.9, -0.8, -0.7, -0.6, -0.5, -0.4, -0.3, -0.2, -0.1,
           0.1, 0.2, 0.3, 0.4, 0.5, 0.6, 0.7, 0.8, 0.9, 1, 1.1, 1.2, 1.3};
         one=j(&n,1,1);
         c=in-(one*(inv(one`*one))*one`);
         do k=1 to 23 by 1;
            h=allh[k,1];
            xh=x&i##h;
            hinv=1/h;
            vhinv=j(&n,1,hinv);
            y=(xh-one)#vhinv;
            my=(one`*y)/&n;
            ycy=y`*c*y;
            lnx=log(x&i);
            slnx=one`*lnx;
            ycyn=ycy/&n;
            if ycyn > 0 then lhp1=-(&n/2)*log(ycyn); else lhp1=.;
            lhp2=(h-1)*slnx;
            if ycyn > 0 then lh=lhp1+lhp2; else lh=.;
            lhs=lhs//lh;
         end;
         Lambda=allh||lhs;
         print, "Lambda and corresponding likelihood for variables x&i",
           lambda;
         call pgraf(lambda, '*', 'lambda', 'likelihood',
         "plot of lambda vs likelihood for variable x&i");
      quit;
  %end;
%mend loop;

 /*input the data and process the macro */
data matrix;
   infile "&file";
   input (%expand(&p)) (&p*:25.) ;
   title "Output 1.11: Box-Cox Transformation plots of &file";
proc print;
%loop(&p)
run;
```

First the name of the data file is specified in the statement `%let file = exp1_6.dat`, and the number of observations and the number of columns are specified in the following two statements, respectively. In the macro named loop, PROC IML computes the values of $L(\lambda)$ of (1.37) for each of the values of λ in the vector named **allh**. The likelihood value is named lh and at the end of each iteration through the values of **allh**, the value of lh is appended to the bottom of a vector named **lhs**. The vectors **allh** and **lhs** are then appended side by side and plotted using the PGRAF call.

Result and Interpretation 1_11.sas

Output 1.11 gives the results of using Program 1_11.sas for variable $y1$ of the data generated by Program 1_6.sas, but transformed by the exponential transformation $y_i^* = \exp(y_i)$.

Output 1.11: *Box-Cox Transformation plots of exp1_6.dat*

```
           Output 1.11: Box-Cox Transformation plots of exp1_6.dat

                             OBS        X1
                               1    29063.57
                               2   151698.83
                               3     3698.38
                               4     4473.25
                               5    22789.77
                               6   381699.28
                               7     2386.98
                               8    38226.50
                               9    24487.89
                              10   208847.22
                              11     6082.12
                              12   177373.85
                              13    17846.14
                              14   111387.69
                              15     8053.79
                              16    52212.82
                              17   264210.87
                              18    89527.14
                              19   136181.37
                              20    82360.05
                              21    87709.27
                              22   957392.80
                              23    14595.46
                              24     5590.57
                              25    81208.83
                              26   220662.85
                              27   451228.04
                              28     7171.21
                              29   204031.97
                              30     5078.52
                              31     6667.42
                              32     4376.54
                              33    13820.96
                              34    75634.08
                              35    33116.29
                              36   180406.47
                              37   204440.99
                              38    15697.88
```

Output 1.11 (*continued*)

```
39      18038.71
40       9306.02
41      77575.93
42       2583.53
43       7311.21
44       2659.12
45     118760.90
46      16748.52
47     116633.26
48       2973.87
49      25203.74
50     126357.47

LAMBDA
    -1 -587.4075
  -0.9 -580.1185
  -0.8 -573.3246
  -0.7 -567.0842
  -0.6 -561.4603
  -0.5 -556.5192
  -0.4 -552.3292
  -0.3 -548.9588
  -0.2 -546.4744
  -0.1 -544.9384
   0.1 -544.9296
   0.2 -546.5448
   0.3   -549.28
   0.4 -553.1478
   0.5 -558.1424
   0.6 -564.2375
   0.7 -571.3853
   0.8 -579.5187
   0.9 -588.5554
     1 -598.4042
   1.1 -608.9711
   1.2 -620.1652
   1.3 -631.9018
```

Output 1.11 (*continued*)

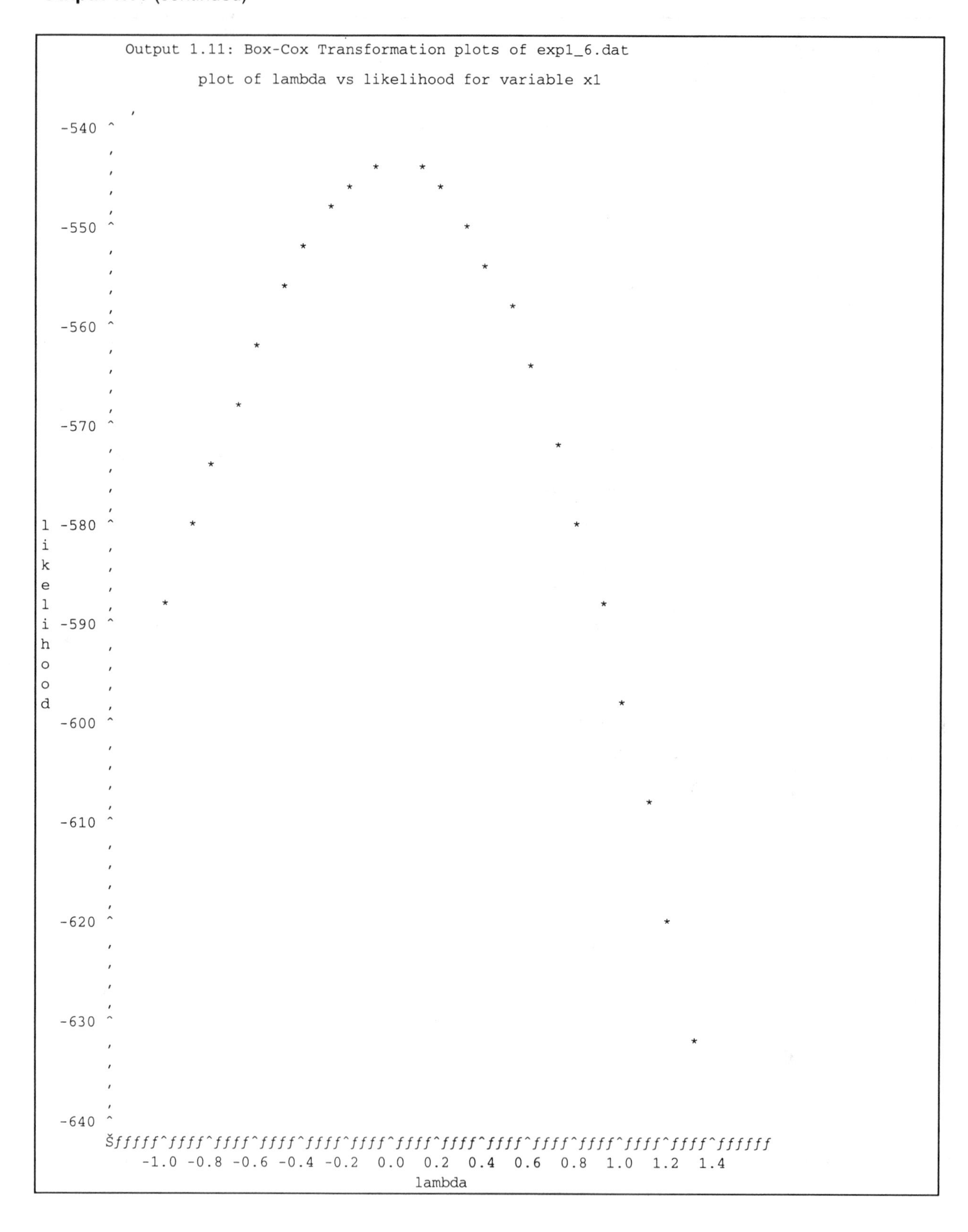

Notice that the value of the likelihood reaches the maximum at lambda equals 0. This suggests that the natural log transformation should be used, as we would have expected, since we transformed the normally distributed variable by exp(y).

When using the Box-Cox transformation, one must transform the data and reevaluate normality, a variable at a time. Alternatively, one may estimate the shape $(\hat{\eta})$ and scale $(\hat{\lambda})$ parameters of a gamma distribution using D_i^2 and construct gamma probability plots to assess multivariate normality (Gnandesikan, 1980). One may also construct multivariate Box-Cox transformations to improve joint multivariate normality (Andrews, 1971). However, such procedures are not readily available except through custom software (Gnandesikan, 1997). Only recently have regression graphics reached the desktop for the multiple regression case (Cook and Weisberg, 1994).

2. Unrestricted General Linear Models

Theory

Applications

2.1 Introduction

Unrestricted (univariate) linear models or Gauss-Markov models are linear models that specify a relationship between a set of random, independent, identically distributed (*iid*) dependent variables $(y_1, y_2, \ldots, y_n) = \mathbf{y}'$ and a matrix of fixed, nonrandom, independent variables $(\mathrm{x}_{ik}) = \mathbf{X}$ such that

$$E(y_i) = \mathrm{x}_{i1}\beta_1 + \mathrm{x}_{i2}\beta_2 + \ldots + \mathrm{x}_{ik}\beta_k \qquad (i = 1, 2, \ldots, n)$$

The variance of each y_i is constant (σ^2) or homogeneous, and the relationship is linear in the unknown, nonrandom parameters $(\beta_1, \beta_2, \ldots, \beta_k) = \beta'$. Special classes of such models are called multiple linear regression models, analysis of variance (ANOVA) models, and intraclass covariance models. In this chapter, we review both ordinary least squares (OLS) and maximum likelihood (ML) estimation of β, and also discuss general linear hypotheses and simultaneous inference for these models. We also introduce the general linear mixed model (GLMM); however, estimation and hypothesis testing for the GLMM are not discussed until Chapter 9. Applications discussed include a multiple linear regression analyses and the analysis of several analysis of variance designs.

2.2 Gauss-Markov Models

For multiple linear regression, ANOVA, and intraclass covariance models, we assume that the covariance structure for the *iid* random variables in $\mathbf{y}$ has the structure

$$\mathbf{\Omega} = \sigma^2\mathbf{I}_n \tag{2.1}$$

where $\mathbf{I}_n$ is an $n \times n$ identity matrix. The error structure for the observations is said to be homogeneous or spherical. Models of the form (1.1) with covariance structure (2.1) are called unrestricted (univariate) linear models or Gauss-Markov models.

To estimate β, the vector of unknown, nonrandom, fixed effect regression coefficients, the method of ordinary least squares (OLS) is commonly utilized. The least squares criterion requires minimizing the error sum of squares, $\sum_{i=1}^{n} e_i^2 = \text{Tr}(\mathbf{ee}')$, where $\text{Tr}(\cdot)$ is the trace operator. Minimizing the error sum of squares leads to the normal equations

$$(\mathbf{X}'\mathbf{X})\hat{\beta} = \mathbf{X}'\mathbf{y} \tag{2.2}$$

Because $\mathbf{X}$ has full column rank, $\text{rank}(\mathbf{X}) = k$, the ordinary least squares estimator (OLSE) of β is the unique solution to the normal equations (2.2),

$$\hat{\beta} = (\mathbf{X}'\mathbf{X})^{-1}\mathbf{X}'\mathbf{y} \tag{2.3}$$

This is also called the best linear unbiased estimator (BLUE) of β since among all parametric functions $\psi = \mathbf{c}'\beta$, $\hat{\psi} = \mathbf{c}'\hat{\beta}$ is unbiased for ψ and has smallest variance:

$$\begin{aligned} E(\hat{\psi}) &= \psi \\ \text{var}(\hat{\psi}) &= \sigma^2\mathbf{c}'(\mathbf{X}'\mathbf{X})^{-1}\mathbf{c} \end{aligned} \tag{2.4}$$

Using the responsewise version of the GLM, the form of $\hat{\beta}$ is given by $\hat{\beta} = (\mathbf{XX}')^{-1}\mathbf{Xy}$ and the $\text{var}(\hat{\psi}) = \sigma^2\mathbf{c}'(\mathbf{XX}')^{-1}\mathbf{c}$.

If the matrix $\mathbf{X}$ in (1.1) is not of full rank k, one may either reparameterize the model to full rank or use a generalized inverse for $\mathbf{X}'\mathbf{X}$ in (2.3) to solve the normal equations.

Definition 2.1 A generalized inverse of a real matrix $\mathbf{A}$ is any matrix $\mathbf{G}$ that satisfies the condition $\mathbf{AGA} = \mathbf{A}$. The generalized inverse of $\mathbf{A}$ is written as $\mathbf{G} = \mathbf{A}^{-}$.

Because $\mathbf{A}^{-}$ is not unique, (2.2) has no unique solution if $\mathbf{X}'\mathbf{X}$ is not of full rank k; however, linear combinations of β, $\psi = \mathbf{c}'\beta$ may be found that are unique even though a unique estimate of β is not available. Several SAS procedures use a full rank design matrix while others do not; more will be said about this when the applications are discussed. For a thorough discussion of the analysis of univariate linear models see Searle (1971) and Milliken and Johnson (1992). Following Searle (1987), and Timm and Carlson (1975), we will usually assume in our discussion in this chapter that the design matrix $\mathbf{X}$ is of full rank.

2.3 Hypothesis Testing

Once the parameters β have been estimated, the next step is usually to test hypotheses. To test hypotheses about β, we assume that $\mathbf{e}$ follows a spherical multivariate normal distribution:

$$\mathbf{e} \sim N_n(\mathbf{0}, \mathbf{\Omega} = \sigma^2\mathbf{I}_n) \tag{2.5}$$

so that

$$\mathbf{y} \sim N_n(\mu = \mathbf{X}\beta, \mathbf{\Omega} = \sigma^2\mathbf{I}_n) \tag{2.6}$$

The maximum likelihood (ML) estimator of the population parameters β and σ^2 assuming normality are

$$\begin{aligned}\hat{\beta}_{ML} &= (\mathbf{X}'\mathbf{X})^{-1}\mathbf{X}'\mathbf{y} \\ \hat{\sigma}^2_{ML} &= (\mathbf{y}-\mathbf{X}\hat{\beta})'(\mathbf{y}-\mathbf{X}\hat{\beta})/n \\ &= (\mathbf{y}'\mathbf{y}-n\bar{\mathbf{y}}^2)/n\end{aligned} \tag{2.7}$$

where the likelihood function using (1.17) has the form

$$L(\sigma^2, \beta|\mathbf{y}) = (2\pi\sigma^2)^{-n/2}\exp[-(\mathbf{y}-\mathbf{X}\beta)'(\mathbf{y}-\mathbf{X}\beta)/2\sigma^2] \tag{2.8}$$

We see that the OLS (2.3) and ML (2.7) estimate of β are identical.

To test the hypothesis of the form $H\colon \mathbf{C}\beta = \xi$ (1.8), we may perform a likelihood ratio test (1.27) which requires maximizing (2.8) with respect to β and σ^2 under the hypothesis $L(\hat{\omega})$ and over the entire parameter space $L(\hat{\mathbf{\Omega}}_o)$. Over the entire parameter space, $\mathbf{\Omega}_o$, the estimates of β and σ^2 are given in (2.7). The corresponding estimates under the hypothesis are

$$\begin{aligned}\hat{\beta}_\omega &= \hat{\beta} - (\mathbf{X}'\mathbf{X})^{-1}\mathbf{C}'[\mathbf{C}(\mathbf{X}'\mathbf{X})^{-1}\mathbf{C}']^{-1}(\mathbf{C}\hat{\beta}-\xi) \\ \hat{\sigma}^2_\omega &= (\mathbf{y}-\mathbf{X}\hat{\beta}_\omega)'(\mathbf{y}-\mathbf{X}\hat{\beta}_\omega)/n,\end{aligned} \tag{2.9}$$

(see Timm 1975, p. 178). Substituting the estimates under ω and $\mathbf{\Omega}_o$ into the likelihood function (2.8), the likelihood ratio defined in (1.27) becomes

$$\begin{aligned}\lambda &= \frac{L(\hat{\omega})}{L(\hat{\mathbf{\Omega}}_o)} = \frac{(2\pi\hat{\sigma}^2_\omega)^{-n/2}}{(2\pi\hat{\sigma}^2)^{-n/2}} \\ &= \left[\frac{(\mathbf{y}-\mathbf{X}\hat{\beta})'(\mathbf{y}-\mathbf{X}\hat{\beta})}{(\mathbf{y}-\mathbf{X}\hat{\beta}_\omega)'(\mathbf{y}-\mathbf{X}\hat{\beta}_\omega)}\right]^{n/2}\end{aligned} \tag{2.10}$$

so that

$$\Lambda = \lambda^{2/n} = \frac{\mathbf{E}}{\mathbf{E}+\mathbf{H}} \tag{2.11}$$

where

$$\begin{aligned}\mathbf{E} &= \mathbf{y}'\mathbf{y} - n\bar{\mathbf{y}}\bar{\mathbf{y}}' = \mathbf{y}'(\mathbf{I}-\mathbf{X}(\mathbf{X}'\mathbf{X})^{-1}\mathbf{X}')\mathbf{y} \\ \mathbf{H} &= (\mathbf{C}\hat{\beta}-\xi)'[\mathbf{C}(\mathbf{X}'\mathbf{X})^{-1}\mathbf{C}']^{-1}(\mathbf{C}\hat{\beta}-\xi)\end{aligned} \tag{2.12}$$

Details are provided in Timm (1975, 1993b, Chapter 3) and Searle (1987, Chapter 8).

The likelihood ratio test is to reject

$$H(\mathbf{C}\beta = \xi) \text{ if } \Lambda < \mathrm{c} \tag{2.13}$$

where c is determined such that the $P(\Lambda < c|H) = \alpha$. The statistic Λ is related to a beta distribution represented generally as U_{p,v_h,v_e} where p is the number of variables, v_h is the degrees of freedom for the hypothesis, $v_h = \text{rank}(\mathbf{C}) = g$, and v_e is the degrees of freedom for error, $v_e = n - \text{rank}(\mathbf{X})$.

Theorem 2.1 When $p = 1$,

$$U_{1,v_h,v_e} = \frac{1}{1+(v_h F_{v_h,v_e} / v_e)}$$

or

$$[v_e(1-U_{1,v_h,v_e})] = F_{v_h,v_e}$$

where F denotes the F-distribution and U the beta distribution.

From Theorem 2.1, we see that rejecting H for small values of U is equivalent to rejecting H for large values of F where

$$F = \frac{\mathbf{H} / v_h}{\mathbf{E} / v_e} = \frac{MS_h}{MS_e} \tag{2.14}$$

is the familiar F-statistic, where MS_h refers to the mean square for hypothesis and MS_e refers to the mean square for error.

2.4 Simultaneous Inference

While the parametric function that led to the rejection of H may or may not be of interest to the researcher, one can easily find the combination of the parameters β that led to rejection. To find the function $\psi_* = \mathbf{c}'_* \beta$, observe by (2.14) and (2.12) with $\hat{\psi} = \mathbf{C}\hat{\beta}$ that

$$(\hat{\psi} - \psi)'(\mathbf{C}(\mathbf{X}'\mathbf{X})^{-1}\mathbf{C}')^{-1}(\hat{\psi} - \psi) > gMS_e F^{\alpha} \tag{2.15}$$

where F^{α} is the upper α percentage value of the F-distribution. Using the Cauchy-Schwartz (C-S) inequality, $(\mathbf{x}'\mathbf{y})^2 \leq (\mathbf{x}'\mathbf{x})(\mathbf{y}'\mathbf{y})$, with $\mathbf{x} = \mathbf{F}\mathbf{a}$, $\mathbf{y} = \mathbf{F}^{-1}\mathbf{b}$ and $\mathbf{G} = \mathbf{F}'\mathbf{F}$, we have that $(\mathbf{a}'\mathbf{b})^2 \leq (\mathbf{a}'\mathbf{G}\mathbf{a})(\mathbf{b}'\mathbf{G}^{-1}\mathbf{b})$. Letting $\mathbf{b} = (\hat{\psi} - \psi)$ and $\mathbf{G} = \mathbf{C}(\mathbf{X}\mathbf{X})^{-1}\mathbf{C}'$, the

$$\sup_{\mathbf{a}} \frac{[\mathbf{a}'(\hat{\psi} - \psi)]^2}{\mathbf{a}'\mathbf{C}(\mathbf{X}'\mathbf{X})^{-1}\mathbf{C}'\mathbf{a}} \leq (\hat{\psi} - \psi)'(\mathbf{C}(\mathbf{X}'\mathbf{X})^{-1}\mathbf{C}')^{-1}(\psi - \psi)$$

For (2.15), the

$$\sup_{\mathbf{a}} \frac{[\mathbf{a}'(\hat{\psi} - \psi)]^2}{[\mathbf{a}'\mathbf{C}(\mathbf{X}'\mathbf{X})^{-1}\mathbf{C}')\mathbf{a}]^{1/2}} \geq (g\, MS_e\, F^{\alpha})^{1/2} \tag{2.16}$$

By again applying the C-S inequality $(\mathbf{a}'\mathbf{a})^2 \leq (\mathbf{a}'\mathbf{a})(\mathbf{a}'\mathbf{a})$ or $(\mathbf{a}'\mathbf{a}) \leq [(\mathbf{a}'\mathbf{a})(\mathbf{a}'\mathbf{a})]^{1/2}$, we have that

$$|\mathbf{a}'\hat{\psi} - \psi)|^2 \leq \mathbf{a}'(\hat{\psi} - \psi)(\hat{\psi} - \psi)'\mathbf{a}$$

Hence, for (2.16) the

$$\sup_{\mathbf{a}} \frac{\mathbf{a}'\mathbf{G}_1\mathbf{a}}{\mathbf{a}'\mathbf{G}_2\mathbf{a}} \geq (gMS_e F^{\alpha}) \tag{2.17}$$

where $\mathbf{G}_1 = (\hat{\psi} - \psi)(\hat{\psi} - \psi)'$ and $\mathbf{G}_2 = \mathbf{C}(\mathbf{X}'\mathbf{X})^{-1}\mathbf{C}'$. Recall, however, that the supremum of the ratio of two quadratic forms is the largest characteristic root of the determinantal equation $|\mathbf{G}_1 - \lambda \mathbf{G}_2| = 0$ with associated eigenvector $\mathbf{a} = \mathbf{a}_*$. Solving $|\mathbf{G}_2^{-1}\mathbf{G}_1 - \lambda\, \mathbf{I}| = 0$, we find that there exists a matrix $\mathbf{P}$ such that $\mathbf{G}_2^{-1}\mathbf{G}_1\mathbf{P} = \Lambda \mathbf{P}$ where Λ are the roots and $\mathbf{P}$ is the matrix of associated eigenvectors of the determinantal equation. For (2.17), we have that

$$
\begin{aligned}
\mathbf{G}_2^{-1}\mathbf{G}_1[\mathbf{G}_2^{-1}(\hat{\psi} - \psi)] &= \mathbf{G}_2^{-1}(\hat{\psi} - \psi)(\hat{\psi} - \psi)'\mathbf{G}_2^{-1}(\hat{\psi} - \psi) \\
&= [(\hat{\psi} - \psi)'\mathbf{G}_2^{-1}(\hat{\psi} - \psi)]\mathbf{G}_2^{-1}(\hat{\psi} - \psi) \\
&= \lambda\, \mathbf{G}_2^{-1}(\hat{\psi} - \psi)
\end{aligned}
$$

so that $\mathbf{a}_* = \mathbf{G}_2^{-1}(\hat{\psi} - \psi)$ is the eigenvector of $\mathbf{G}_2^{-1}\mathbf{G}_1$ for the maximum root of $|\mathbf{G}_2^{-1}\mathbf{G}_1 - \lambda\,\mathbf{I}| = 0$. Thus, the eigenvector $\mathbf{a}_*$ may be used to find the linear parametric function of β that is *most* significantly different from ξ. The function is

$$\psi'_* = (\mathbf{a}'_*\mathbf{C})\beta = \mathbf{c}'_*\beta \tag{2.18}$$

The Scheffé-type simultaneous confidence interval for $\psi = \mathbf{c}'\beta$ for all nonnull vectors $\mathbf{c}'$, such that the $\sum_i c_i = 0$ and the c_i are elements of the vector $\mathbf{c}$, is given by

$$\hat{\psi} - c_o\hat{\sigma}_{\hat{\psi}} \leq \psi \leq \hat{\psi} + c_o\hat{\sigma}_{\hat{\psi}} \tag{2.19}$$

where $\psi = \mathbf{c}'\beta$, $\hat{\psi} = \mathbf{c}'\hat{\beta}$, $\hat{\sigma}_{\hat{\psi}}$ is an estimate of the standard error of $\hat{\psi}$ given in (2.4) and

$$c_o^2 = (g)F^{\alpha}_{g,v_e} \tag{2.20}$$

where $g = v_h$ (see Scheffé, 1959, p. 69).

To construct a UI test of $H(\mathbf{C}\beta = \xi)$, one writes the null hypothesis as an intersection of hypotheses:

$$H = \bigcap_i H_i \text{ where } H_i = \mathbf{a}'\mathbf{C}\beta = \mathbf{a}'\xi \tag{2.21}$$

or

$$H = \bigcap_{\mathbf{a}} H_{\mathbf{a}} \text{ (say)} \tag{2.22}$$

where $\mathbf{a}$ is a nonnull g-dimensional vector. Hence, H is the intersection of a set of elementary tests $H_{\mathbf{a}}$. We reject H if we reject $H_{\mathbf{a}}$ for any $\mathbf{a}$. By the union-intersection principle, it follows that if we reject for any $\mathbf{a}$, we reject for the $\mathbf{a} = \mathbf{a}_*$ that maximizes (2.16); thus, the UI test for this situation is equivalent to the F-test or a likelihood ratio test. For additional details, see for example Casella and Berger (1990, Section 11.2.2).

2.5 Linear Mixed Models

The general fixed effects linear model was given in (1.1) as

$$\mathbf{y}_{n\times 1} = \mathbf{X}_{n\times k}\beta_{k\times 1} + \mathbf{e}_{n\times 1} \tag{2.23}$$

To establish a form for the general linear mixed model (GLMM), we add to the model a vector of random effects **b** and a corresponding known design matrix **Z**:

$$\mathbf{y}_{n\times 1} = \mathbf{X}_{n\times k}\ \beta_{k\times 1} + \mathbf{Z}_{n\times h}\ \mathbf{b}_{h\times 1} + \mathbf{e}_{n\times 1} \tag{2.24}$$

The vector **b** is often partitioned into a set of t subvectors to correspond to the model parameters:

$$\mathbf{b}' = (\mathbf{b}_1', \mathbf{b}_2', \dots, \mathbf{b}_t') \tag{2.25}$$

For example, for a two-way random model, $t = 3$ and $\mu_1 = \alpha$, $\mu_2 = \beta$ and $\mu_3 = \gamma$ if $y_{ijk} = \mu + \alpha_i + \beta_j + \gamma_{ij} + e_{ijk}$ and μ is a fixed effect. For the linear model in (2.24), we assume that

$$E(\mathbf{y}) = \mathbf{X}\beta \text{ and } E(\mathbf{y}|\mathbf{b}) = \mathbf{X}\beta + \mathbf{Z}\mathbf{b} \tag{2.26}$$

so that the random error vector **e** is defined as

$$\mathbf{e} = \mathbf{y} - E(\mathbf{y}|\mathbf{b}) \tag{2.27}$$

where

$$E(\mathbf{y}) = \mathbf{X}\beta,\ E(\mathbf{e}) = \mathbf{0} \text{ and } E(\mathbf{b}) = \mathbf{0} \tag{2.28}$$

The vector **e** has the usual structure, $\text{var}(\mathbf{e}) = \sigma^2\mathbf{I}_n$. However, because the $\mathbf{b}_i$ are random, we assume

$$\begin{aligned} \text{cov}(\mathbf{b}_i) &= \sigma_i^2\mathbf{I}_{q_i} && \text{for all } i \\ \text{cov}(\mathbf{b}_i, \mathbf{b}_j) &= \mathbf{0} && \text{for all } i \neq j \\ \text{cov}(\mathbf{b}, \mathbf{e}) &= \mathbf{0} \end{aligned} \tag{2.29}$$

where q_i represents the number of elements in $\mathbf{b}_i$. Thus the covariance matrix for **b** is

$$\mathbf{V} = \text{cov}(\mathbf{b}) = \begin{bmatrix} \sigma_1^2\mathbf{I}_{q_1} & \cdots & \cdots & \cdots & \cdots & \cdots \\ \vdots & \sigma_2^2\mathbf{I}_{q_2} & \cdots & \cdots & \cdots & \vdots \\ \vdots & \cdots & \ddots & \cdots & \cdots & \vdots \\ \vdots & \cdots & \cdots & \ddots & \cdots & \vdots \\ \vdots & \cdots & \cdots & \cdots & \ddots & \vdots \\ \cdots & \cdots & \cdots & \cdots & \cdots & \sigma_t^2\mathbf{I}_{q_t} \end{bmatrix} \tag{2.30}$$

Partitioning **Z** conformable with **b**, so that $\mathbf{Z} = (\mathbf{z}_1, \mathbf{z}_2, \dots, \mathbf{z}_t)$, yields

$$\begin{aligned} \mathbf{y} &= \mathbf{X}\beta + \mathbf{Z}\mathbf{b} + \mathbf{e} \\ &= \mathbf{X}\beta + \sum_{i=1}^{t} \mathbf{Z}_i\mathbf{b}_i + \mathbf{e} \end{aligned} \tag{2.31}$$

so that the covariance matrix for **y** becomes

$$\begin{aligned} \Omega &= \mathbf{Z}\mathbf{V}\mathbf{Z}' + \sigma^2\mathbf{I}_n \\ &= \sum_{i=1}^{t} \sigma_i^2\mathbf{Z}_i\mathbf{Z}_i' + \sigma^2\mathbf{I}_n \end{aligned} \tag{2.32}$$

Since $\mathbf{e}$ is a random vector just like $\mathbf{b}_i$, we may write (2.24) as

$$\mathbf{y} = \mathbf{X}\beta + \mathbf{Z}\mathbf{b} + \mathbf{b}_o \tag{2.33}$$

and (2.31) becomes

$$\mathbf{y} = \mathbf{X}\beta + \sum_{i=0}^{t} \mathbf{Z}_i \mathbf{b}_i \tag{2.34}$$

where $\mathbf{Z}_0 \equiv \mathbf{I}_n$ and $\sigma_0^2 = \sigma^2$. Thus, the covariance structure $\mathbf{\Omega}$ may be written as

$$\mathbf{\Omega} = \sum_{i=0}^{t} \sigma_i^2 \mathbf{Z}_i \mathbf{Z}_i' \tag{2.35}$$

As an example of the structure given in (2.32), assume $\mathbf{b}$ has one random effect where

$$y_{ij} = \mu + \alpha_i + e_{ij} \qquad i = 1, 2, 3, 4 \text{ and } j = 1, 2, 3$$

the α_i are random effects with mean 0 and variance σ_α^2, and the covariances among the α_i, and the e_{ij} and α_i are zero; then the

$$\text{cov}(y_{ij}, y_{i'j'}) = \begin{cases} \sigma_\alpha^2 + \sigma^2 & \text{for } i = i' \text{ and } j = j' \\ \sigma^2 & \text{for } i = i' \text{ and } j \neq j' \\ 0 & \text{otherwise} \end{cases}$$

or

$$\text{cov}\begin{pmatrix} y_{11} \\ y_{12} \\ y_{13} \end{pmatrix} = \begin{pmatrix} \sigma_\alpha^2 + \sigma^2 & \sigma_\alpha^2 & \sigma_\alpha^2 \\ \sigma_\alpha^2 & \sigma_\alpha^2 + \sigma^2 & \sigma_\alpha^2 \\ \sigma_\alpha^2 & \sigma_\alpha^2 & \sigma_\alpha^2 + \sigma^2 \end{pmatrix}$$

$$\begin{aligned} &= \sigma_\alpha^2 \mathbf{1}\mathbf{1}' + \sigma^2 \mathbf{I} \\ &= \sigma_\alpha^2 \mathbf{J} + \sigma^2 \mathbf{I} \\ &= \mathbf{\Sigma}_1 \end{aligned}$$

where $\mathbf{J} = \mathbf{1}\mathbf{1}'$. Hence, $\mathbf{\Omega} = \text{diag}\{\mathbf{\Sigma}_i\}$,

$$\mathbf{\Omega} = \begin{pmatrix} \mathbf{\Sigma}_1 & 0 & 0 & 0 \\ 0 & \mathbf{\Sigma}_2 & 0 & 0 \\ 0 & 0 & \mathbf{\Sigma}_3 & 0 \\ 0 & 0 & 0 & \mathbf{\Sigma}_4 \end{pmatrix}$$

where $\mathbf{\Sigma}_i = \sigma_\alpha^2 \mathbf{J} + \sigma^2 \mathbf{I}$ or using direct sum notation,

$$\mathbf{\Omega} = \bigoplus_{i=1}^{4} \mathbf{\Sigma}_i = \bigoplus_{i=1}^{4} (\sigma_\alpha^2 \mathbf{J} + \sigma^2 \mathbf{I})$$

For our simple example, we let $\mathbf{\Sigma} = \mathbf{\Sigma}_i = \sigma_\alpha^2 \mathbf{1}\mathbf{1}' + \sigma^2 \mathbf{I}$. The structure for $\mathbf{\Sigma}$ is commonly called the compound symmetry structure for $\mathbf{\Sigma}$.

One of the primary goals of mixed models is to estimate the variance components σ_i^2. Another is to test hypotheses regarding β; for this large sample theory is used unless $\mathbf{\Omega}$ is known or has special structure. Alternatively, one may employ a hierarchical modeling approach. The theory and applications of the general linear mixed model (GLMM) are discussed in Chapter 9.

2.6 Multiple Linear Regression

The general linear model without restrictions has wide applicability in educational and psychological research. Multiple linear regression, ANOVA, and intraclass covariance designs can all be expressed in the form of the general linear model without restrictions. In this section we illustrate how to express multiple linear regression in the form of (1.1), and we show how to use SAS software to perform the analysis.

In multiple linear regression, we are interested in predicting one dependent variable using a set of p independent variables, where the independent variables are assumed to be fixed. The researcher may be interested in testing hypotheses about the vector of regression coefficients β. Before performing the regression analysis and the testing of hypotheses, it is advisable to evaluate the normality of the observations on the dependent variable since the assumption of normality is made when testing hypotheses. It is also advisable to analyze the data for possible outliers since these can adversely affect model fit. Various tools for evaluating normality and detecting outliers were presented in Chapter 1; these included evaluation of skewness and kurtosis, and Q-Q plots. The use of the Box-Cox transformation for transforming data to normality was also illustrated in Chapter 1. Several additional methods for the detection of outliers and detection of influential data points will be introduced after the following brief discussion of the general linear model representation of the multiple linear regression model.

Multiple linear regression models can be expressed in the form of (1.1) where $\mathbf{y}_{n\times 1}$ is a vector of observations on the dependent variable and $\mathbf{X}_{n\times k}$ is a matrix of observations on p independent variables. Here $k = p + 1$, and the first column of $\mathbf{X}$ is a column of 1's and is used to obtain the regression intercept. $\beta_{k\times 1}$ is a vector of parameters to be estimated, and $\mathbf{e}_{n\times 1}$ is a vector of errors, or residuals. The vector of parameters β is estimated by the least squares estimator (2.3), and the predicted values of $\mathbf{y}$ are thus

$$\hat{\mathbf{y}} = \mathbf{X}\hat{\beta} = \mathbf{X}(\mathbf{X}'\mathbf{X})^{-1}\mathbf{X}'\mathbf{y} \tag{2.36}$$

where $\mathbf{X}(\mathbf{X}'\mathbf{X})^{-1}\mathbf{X}'$ is defined as the hat matrix. Recall that to estimate β we minimized $\sum_{i=1}^{n} e_i^2 = \sum_{i=1}^{n} (y_i - \hat{y}_i)^2$, the sum of squares due to error. For the regression analysis the sums of squares may be partitioned as follows:

$$\sum_i (y_i - \bar{y})^2 = \sum_i (\hat{y}_i - \bar{y})^2 + \sum_i (y_i - \hat{y}_i)^2$$

where the term on the left of the equality is the sum of squares total (SST), corrected for the mean, with $n - 1$ degrees of freedom. The next term is the sum of squares due to regression (SSR) with p degrees of freedom, and the last term is the sum of squares due to error (SSE), or residual, with $n - p - 1$ degrees of freedom. One may test the hypothesis $H{:}\,\beta = \mathbf{0}$ using the F ratio:

$$F = \frac{SSR / p}{SSE / n - p - 1} = \frac{MSR}{MSE} \tag{2.37}$$

where F has degrees of freedom $(p, n - p - 1)$ and is compared to the F^{α} critical value.

The multiple correlation coefficient is defined as

$$R^2 = \frac{SSR}{SST} \tag{2.38}$$

and can be interpreted as the proportion of variability in **y** that can be explained by the set of independent variables in the model.

The independent variables should be analyzed for possible multicollinearity; independent variables that are highly correlated with other independent variables cause the resulting parameter estimates to be unstable, with inflated variance. One may look at the intercorrelation matrix of the independent variables to identify highly correlated variables. Alternately, the variance inflation factor (VIF) is a measure of multicollinearity. Recall that

$$\begin{aligned} \text{var}(\hat{\beta}) &= (\mathbf{X'X})^{-1}\mathbf{X'}\ \text{var}(\mathbf{y})\mathbf{X}(\mathbf{X'X})^{-1} \\ &= (\sigma^2\mathbf{I})(\mathbf{X'X})^{-1} \\ &= \sigma^2(\mathbf{X'X})^{-1} \end{aligned} \tag{2.39}$$

$(\mathbf{X'X})^{-1}$ is called the VIF. Note that the inverse of a matrix can be expressed as follows:

$$(\mathbf{X'X})^{-1} = \frac{\text{adj}\ (\mathbf{X'X})}{|\mathbf{X'X}|} \tag{2.40}$$

and that if $|\mathbf{X'X}| = 0$ there is dependency among the x's. From (2.40) we can see that as the x's become nearly dependent, the VIF increases. The VIF for the kth variable is $(\text{VIF})_k = (1 - R_k^2)^{-1}$ for $k = 1, 2, \ldots, p$. We usually want the VIF for each variable to be near 1; any variable with a VIF of 10 or greater should be considered for exclusion or combined with other related variables.

Various methods are available for selecting the independent variables to include in the prediction equation. These include forward, backward, and stepwise methods, methods based on R^2, adjusted R^2, and C_p, among many others. Much has already been written on this topic. See, for example, Stevens (1992) and Kleinbaum, Kupper, and Muller (1988).

Plots of residuals should be made to check for the validity of model assumptions. The plot of the standardized residuals (r_s) versus the predicted values is useful. If the assumptions of a linear relationship and of homoscedasticity are accurate, these plotted values should scatter randomly around the line $r_s = 0$. Any fanning pattern is an indication of nonconstant variance, and any curvilinear pattern is evidence that the linearity assumption is not adequate (see Chapter 4 for an example of data with nonconstant variance).

The data should be examined for outliers and influential data points; if such observations are found they should be considered for possible exclusion from the analysis. There are a number of methods for detection of outliers and influential points. Outliers on the dependent variable can be easily detected by looking at the standardized residuals, r_s. Since these are essentially standard normally distributed, we would expect approximately 95% of the r_s to be $-2 \le r_s \le 2$, and approximately 99% to be $-3 \le r_s \le 3$. Thus, any observation with a $r_s \le -3$ or ≥ 3 should be examined as a possible outlier.

Outliers on the set of independent variables can be detected by utilizing the diagonal entries of the hat matrix, $\mathbf{X}(\mathbf{X}'\mathbf{X})^{-1}\mathbf{X}'$. These are also called leverages. The average leverage is *k*/*n*. Stevens (1992) suggests that any observation with leverage $\geq 3k/n$ be examined.

Mahalanobis' distances can be used to detect outliers on the independent variables x. Mahalanobis' distance for observation *i* is

$$D_i^2 = (\mathbf{x}_i - \overline{\mathbf{x}})'\mathbf{S}^{-1}(\mathbf{x}_i - \overline{\mathbf{x}}) \tag{2.41}$$

where $\mathbf{x}_i$ is the $p \times 1$ vector of data for the prediction of the ith observation and $\overline{\mathbf{x}}$ is the vector of means of the predictors. Any large D_i^2 in (2.41) indicates an observation vector of independent variables that lies far from the other $n - 1$ observations (see Stevens, 1992, for more details).

Influential data points are observations that have a large effect on the value of the estimated regression coefficients. Cook's distance is one measure for detection of influential data points. Cook's distance is for the ith observation:

$$CD_i = [(\hat{\beta} - \hat{\beta}_{(i)})'\mathbf{X}'\mathbf{X}(\hat{\beta} - \hat{\beta}_{(i)}) / (p+1)]MS_{\text{error}} \tag{2.42}$$

$\hat{\beta}_{(i)}$ is the vector of estimated parameters with the ith observation deleted, and MS_{error} is from the model with all observations included. A $CD_i > 1$ is considered large (Cook and Weisberg 1994, p. 118). Cook's distance can also be expressed as

$$CD_i = \frac{1}{p+1} r_s^2 \frac{h_{ii}}{1 - h_{ii}} \tag{2.43}$$

where r_s is the standardized residual and h_{ii} is the leverage of the ith observation.

DFBETAs are another useful measure for detecting influential data points. *DFBETAs* are used to determine how much each regression coefficient will change if the ith observation is deleted. *DFBETAs* are defined as

$$DFBETAs = \frac{b_j - b_{j(i)}}{s_{(i)}\sqrt{(\mathbf{X}'\mathbf{X})^{jj}}}$$

where $(\mathbf{X}'\mathbf{X})^{jj}$ is the jj^{th} element of $(\mathbf{X}'\mathbf{X})^{-1}$. The *DFBETA* is a measure of the number of standard errors the jth regression coefficient changes when the ith observation is deleted. A $DFBETA > 2$ is considered an influential observation (Belsley, Kuh and Welsch, 1980).

Another measure of influence that can be obtained from the REG procedure is *COVRATIO*. This measure is defined as

$$COVRATIO = \frac{\left|(s_{(i)}^2(\mathbf{X}_{(i)}{}'\mathbf{X}_{(i)})^{-1}\right|}{\left|(s^2(\mathbf{X}'\mathbf{X})^{-1}\right|} \tag{2.44}$$

and is a measure of change in the determinant of the covariance matrix of the estimated parameters from deleting the ith observation compared to the data set with all observations included. According to Belsley et al. (1980) an observation with $|COVRATIO - 1| \geq \frac{3k}{n}$ requires further investigation.

The problem and hypothetical data set given in Kleinbaum et al. (1988, p. 117-118) are used to illustrate a multiple regression analysis using PROC REG. Here, a sociologist is interested in predicting homicide rate per 100,000 city population (Y) with the predictors of city population size (x1), the percent of families with yearly income less than $5,000 (x2), and rate of unemployment (x3).

Using the diagnostic procedures in Chapter 1, we see that there do not appear to be any violations of the normality assumption. However, with a sample size of $n = 20$ it is difficult to assess normality.

The general linear model expressed in the form (1.1) for the regression model including all three independent variables and an intercept is

$$\begin{bmatrix} 11.2 \\ 13.4 \\ 40.7 \\ \vdots \\ 25.8 \\ 21.7 \\ 25.7 \end{bmatrix} = \begin{bmatrix} 1 & 587 & 16.5 & 6.2 \\ 1 & 643 & 20.5 & 6.4 \\ 1 & 635 & 26.3 & 9.3 \\ \vdots & \vdots & \vdots & \vdots \\ 1 & 921 & 22.4 & 8.6 \\ 1 & 595 & 20.2 & 8.4 \\ 1 & 3353 & 16.9 & 6.7 \end{bmatrix} \begin{bmatrix} \beta_o \\ \beta_1 \\ \beta_2 \\ \beta_3 \end{bmatrix} + \begin{bmatrix} \varepsilon_1 \\ \varepsilon_2 \\ \varepsilon_3 \\ \vdots \\ \varepsilon_{n-2} \\ \varepsilon_{n-1} \\ \varepsilon_n \end{bmatrix}. \tag{2.45}$$

Here, the first column of **X** is a vector of 1s which is included to estimate the intercept in the model.

The SAS code for analyzing these data is given in Program 2_6.sas.

Program 2_6.sas

```
/* Program 2_6.sas                                                          */
/* Program to perform Multiple Linear Regression Analysis of homicide data */
/* Data from exercise 4, p 117, Kleinbaum, Kupper, Muller                  */

options ls=80 ps=60 nodate nonumber;
title 'Output 2.6: Multiple Linear Regression Analysis of Homicide data';

data reg;
   infile 'c:\2_6.dat';
   input city y x1 x2 x3;
   label  y='homicide rate'
         x1='population size'
         x2='percent low income'
         x3='unemployment rate';
proc print;
proc univariate;
   var y;
proc corr;
   var y x1 x2 x3;
proc reg;
   model y = x1 x2 x3 /vif;
proc reg;
   model y = x1 x2 x3 /partial selection = backward;
   model y = x2 x3 /vif r collin influence;
   paint student. ge 2 or student. le -2 /symbol = 'o';
   plot r.*x2 r.*x3 /hplots=2;
   plot r.*p. student.*p.;
run;
```

First the data are read in, printed, and univariate statistics as well as the correlation matrix of the variables are computed. The first regression analysis requested using PROC REG forces all three independent variables in the model, and the VIF option is specified to obtain the variance inflation factor for each of the independent variables. Next, PROC REG is used with the SELECTION = BACKWARD option; thus the selection of independent variables is performed using the backward elimination procedure. For the last analysis, the model statement includes only those variables that were selected in the previous analysis. Here the R option is used to obtain information about the residuals, the COLLIN option to obtain some collinearity measures, and the INFLUENCE option to obtain measures to use for detecting influential data points. Plots of residuals versus each independent variable and versus predicted values are output.

Result and Interpretation 2_6.sas

Output 2.6: *Multiple Linear Regression Analysis of Homicide data*

```
      Output 2.6: Multiple Linear Regression Analysis of Homicide data

Model: MODEL1
Dependent Variable: Y          homicide rate

                              Analysis of Variance

                                 Sum of         Mean
        Source          DF      Squares       Square       F Value       Prob>F

        Model            3   1518.14494    506.04831        24.022       0.0001
        Error           16    337.05706     21.06607
        C Total         19   1855.20200

            Root MSE        4.58978     R-square        0.8183
            Dep Mean       20.57000     Adj R-sq        0.7843
            C.V.           22.31297

                              Parameter Estimates

                       Parameter        Standard     T for H0:
      Variable  DF      Estimate           Error    Parameter=0     Prob > |T|

      INTERCEP   1    -36.764925      7.01092577         -5.244         0.0001
      X1         1      0.000763      0.00063630          1.199         0.2480
      X2         1      1.192174      0.56165391          2.123         0.0497
      X3         1      4.719821      1.53047547          3.084         0.0071

                        Variance  Variable
      Variable  DF     Inflation    Label

      INTERCEP   1    0.00000000  Intercept
      X1         1    1.05997406  population size
      X2         1    2.99089922  percent low income
      X3         1    3.07837605  unemployment rate
```

Output 2.6 *(continued)*

Model: MODEL2
Dependent Variable: Y homicide rate

Analysis of Variance

Source	DF	Sum of Squares	Mean Square	F Value	Prob>F
Model	2	1487.85941	743.92970	34.428	0.0001
Error	17	367.34259	21.60839		
C Total	19	1855.20200			

Root MSE	4.64848	R-square	0.8020
Dep Mean	20.57000	Adj R-sq	0.7787
C.V.	22.59836		

Parameter Estimates

Variable	DF	Parameter Estimate	Standard Error	T for H0: Parameter=0	Prob > |T|
INTERCEP	1	-34.072533	6.72654510	-5.065	0.0001
X2	1	1.223931	0.56820473	2.154	0.0459
X3	1	4.398936	1.52616848	2.882	0.0103

Variable	DF	Variance Inflation	Variable Label
INTERCEP	1	0.00000000	Intercept
X2	1	2.98424836	percent low income
X3	1	2.98424836	unemployment rate

Collinearity Diagnostics

Number	Eigenvalue	Condition Index	Var Prop INTERCEP	Var Prop X2	Var Prop X3
1	2.97893	1.00000	0.0027	0.0009	0.0011
2	0.01621	13.55677	0.9783	0.0643	0.1224
3	0.00486	24.75442	0.0191	0.9347	0.8765

Obs	Dep Var Y	Predict Value	Std Err Predict	Residual	Std Err Residual	Student Residual
1	11.2000	13.3957	1.529	-2.1957	4.390	-0.500
2	13.4000	19.1712	1.592	-5.7712	4.367	-1.321
3	40.7000	39.0269	2.466	1.6731	3.941	0.425
4	5.3000	9.4367	1.791	-4.1367	4.289	-0.964
5	24.8000	21.5392	1.322	3.2608	4.457	0.732
6	12.7000	12.0760	1.487	0.6240	4.404	0.142
7	20.9000	18.8041	1.478	2.0959	4.407	0.476
8	35.7000	25.4291	1.196	10.2709	4.492	2.286
9	8.7000	8.5339	2.350	0.1661	4.011	0.041
10	9.6000	11.5829	2.670	-1.9829	3.805	-0.521
11	14.5000	14.4742	1.350	0.0258	4.448	0.006
12	26.9000	26.7524	1.746	0.1476	4.308	0.034
13	15.7000	14.8184	1.792	0.8816	4.289	0.206
14	36.2000	33.9894	1.954	2.2106	4.218	0.524
15	18.1000	17.2857	1.112	0.8143	4.513	0.180
16	28.9000	32.9145	2.021	-4.0145	4.186	-0.959

Output 2.6 (*continued*)

```
      17   14.9000   17.3087      1.294   -2.4087      4.465     -0.539
      18   25.8000   31.1744      1.883   -5.3744      4.250     -1.265
      19   21.7000   27.6019      2.271   -5.9019      4.056     -1.455
      20   25.7000   16.0848      1.685    9.6152      4.332      2.219

                                 Cook's            Hat Diag       Cov
     Obs     -2-1-0 1 2              D  Rstudent          H     Ratio     Dffits

       1   |      *|      |      0.010   -0.4889    0.1083    1.2866   -0.1703
       2   |     **|      |      0.077   -1.3533    0.1173    0.9816   -0.4932
       3   |       |      |      0.024    0.4141    0.2813    1.6165    0.2591
       4   |      *|      |      0.054   -0.9623    0.1485    1.1899   -0.4019
       5   |       |*     |      0.016    0.7213    0.0808    1.1855    0.2139
       6   |       |      |      0.001    0.1375    0.1023    1.3314    0.0464
       7   |       |      |      0.008    0.4645    0.1011    1.2818    0.1558
       8   |       |****  |      0.124    2.6657    0.0662    0.4265    0.7098
       9   |       |      |      0.000    0.0402    0.2556    1.6109    0.0236
      10   |      *|      |      0.045   -0.5097    0.3300    1.7058   -0.3577
      11   |       |      |      0.000    0.0056    0.0843    1.3099    0.0017
      12   |       |      |      0.000    0.0332    0.1411    1.3963    0.0135
      13   |       |      |      0.002    0.1997    0.1485    1.3982    0.0834
      14   |       |*     |      0.020    0.5126    0.1767    1.3874    0.2375
      15   |       |      |      0.001    0.1752    0.0573    1.2650    0.0432
      16   |      *|      |      0.071   -0.9566    0.1890    1.2517   -0.4618
      17   |      *|      |      0.008   -0.5279    0.0775    1.2345   -0.1530
      18   |     **|      |      0.105   -1.2889    0.1641    1.0669   -0.5711
      19   |     **|      |      0.221   -1.5089    0.2388    1.0572   -0.8450
      20   |       |****  |      0.248    2.5549    0.1314    0.4948    0.9937

          INTERCEP        X2        X3
     Obs   Dfbetas   Dfbetas   Dfbetas

       1   -0.1228    0.1019   -0.0412
       2   -0.0516   -0.3436    0.3650
       3   -0.2151    0.0836    0.0589
       4   -0.3136   -0.0460    0.2251
       5    0.0249   -0.1213    0.1291
       6    0.0375   -0.0169   -0.0027
       7    0.0244    0.0989   -0.1095
       8   -0.2382    0.0417    0.1679
       9    0.0138    0.0110   -0.0194
      10   -0.2029    0.3233   -0.2270
      11    0.0010    0.0003   -0.0009
      12   -0.0056    0.0104   -0.0066
      13    0.0311    0.0493   -0.0673
      14   -0.1786    0.0921    0.0284
      15    0.0212   -0.0037   -0.0056
      16    0.3231   -0.2844    0.0724
      17   -0.0749    0.0877   -0.0573
      18    0.2939    0.1666   -0.3941
      19    0.1580    0.5768   -0.7491
      20    0.5019   -0.7725    0.5590

Sum of Residuals                        0
Sum of Squared Residuals         367.3426
Predicted Resid SS (Press)       490.4840
```

Output 2.6 (*continued*)

```
     „ƒˆƒƒƒˆƒƒƒˆƒƒƒˆƒƒƒˆƒƒƒˆƒƒƒˆƒƒƒˆƒƒ†        „ˆƒƒƒƒˆƒƒƒƒˆƒƒƒƒˆƒƒƒƒˆƒƒƒƒˆƒƒƒƒˆƒ†
     ‚                                ‚        ‚                                ‚
     ‚                                ‚        ‚                                ‚
     ‚                                ‚        ‚                                ‚
     ‚                                ‚        ‚                                ‚
  12 ˆ                                ˆ     12 ˆ                                ˆ
     ‚                                ‚        ‚                                ‚
     ‚                                ‚        ‚                                ‚
     ‚                                ‚        ‚                                ‚
     ‚                o               ‚        ‚                  o             ‚
  10 ˆ                                ˆ     10 ˆ                                ˆ
     ‚       o                        ‚        ‚              o                 ‚
     ‚                                ‚        ‚                                ‚
     ‚                                ‚        ‚                                ‚
     ‚                                ‚        ‚                                ‚
   8 ˆ                                ˆ      8 ˆ                                ˆ
     ‚                                ‚        ‚                                ‚
     ‚                                ‚        ‚                                ‚
     ‚                                ‚        ‚                                ‚
     ‚                                ‚        ‚                                ‚
   6 ˆ                                ˆ      6 ˆ                                ˆ
     ‚                                ‚        ‚                                ‚
     ‚                                ‚        ‚                                ‚
R    ‚                                ‚ R      ‚                                ‚
E    ‚                                ‚ E      ‚                                ‚
S  4 ˆ                                ˆ S    4 ˆ                                ˆ
I    ‚                                ‚ I      ‚                                ‚
D    ‚           1                    ‚ D      ‚                 1              ‚
U    ‚                                ‚ U      ‚                                ‚
A    ‚                      1         ‚ A      ‚                       1        ‚
L  2 ˆ             1                  ˆ L    2 ˆ            1                   ˆ
     ‚                          1     ‚        ‚                           1    ‚
     ‚                                ‚        ‚                                ‚
     ‚      1   11                    ‚        ‚         11  1                  ‚
     ‚                                ‚        ‚                                ‚
   0 ˆ       1 1         1            ˆ      0 ˆ     1    1      1              ˆ
     ‚                                ‚        ‚                                ‚
     ‚                                ‚        ‚                                ‚
     ‚                                ‚        ‚                                ‚
     ‚                                ‚        ‚                                ‚
  -2 ˆ  1   1                         ˆ     -2 ˆ           11                   ˆ
     ‚         1                      ‚        ‚              1                 ‚
     ‚                                ‚        ‚                                ‚
     ‚                                ‚        ‚                                ‚
     ‚                                ‚        ‚                                ‚
  -4 ˆ      1                1        ˆ     -4 ˆ       1              1         ˆ
     ‚                                ‚        ‚                                ‚
     ‚                                ‚        ‚                                ‚
     ‚                  1             ‚        ‚                       1        ‚
     ‚              1                 ‚        ‚            1                   ‚
  -6 ˆ             1                  ˆ     -6 ˆ                      1         ˆ
     ‚                                ‚        ‚                                ‚
     ‚                                ‚        ‚                                ‚
     ‚                                ‚        ‚                                ‚
     ŠƒˆƒƒƒˆƒƒƒˆƒƒƒˆƒƒƒˆƒƒƒˆƒƒƒˆƒƒƒˆƒƒŒ        ŠˆƒƒƒƒˆƒƒƒƒˆƒƒƒƒˆƒƒƒƒˆƒƒƒƒˆƒƒƒƒˆƒŒ
      14  16  18  20  22  24  26  28           4    5    6    7    8    9    10

                  X2                                        X3
```

Multicollinearity does not appear to be a problem with these data; we therefore did not have to combine any of the independent variables into a composite. The VIF values 1.06, 2.99, and 3.08 are all much less than 10. In the matrix of intercorrelations the highest correlation among the independent variables is 0.815.

When looking at the standardized residuals, we would expect 95% to lie within ± 2. Thus, in this example with 20 observations, we would expect one observation to lie outside this range. We can see from Output 2.6 that two of the observations, 8 and 20, have values >2 in the column labeled `Student Residual`. These observations are painted with an O in the plots but were not removed from the data set. The plots of residuals versus the variables x2 and x3 do not reveal any model violations. The two observations painted with an O appear to be outliers, but we did not delete them, since they were not influential on the parameter estimates based on the influence statistics; however, if both were deleted simultaneously, the parameter estimates might change significantly.

The leverages appear in Output 2.6 in the column labeled `Hat Diagonal` *H*. None of the observations have leverages greater than $3k/n = (3)(4)/(20) = .6$. In addition, no observations have Cook's distance greater than 1 or *DFBETA* values greater than 2. For the *COVRATIO*, the criteria are $|COVRATIO - 1| \geq (3)(4)/(2) = .6$. Thus, a $COVRATIO \leq .4$ or ≥ 1.6 is suspect. Observation 3 has a *COVRATIO* value of 1.62, 9 a value of 1.61, and observation 10 a value of 1.71; however, we did not remove these observations from the data set.

The sum of squares regression was 1487.859 and sum of squares error was 367.343; R-squared is then equal to 0.8020. Thus, 80% of the variance in homicide rate can be explained by the two variables: percent of families with yearly income less than $5,000 (x2) and rate of unemployment (x3).

2.7 One-Way Analysis of Variance

For ANOVA designs without restrictions, model (1.1) is again assumed. ANOVA designs can be analyzed using a full rank model, or using the overparameterized, less than full rank model. The full rank model is easy to understand, whereas the overparameterized model requires knowledge of estimable functions, side conditions, reparameterization, and generalized inverses. For a one-way ANOVA design, the full rank model is

$$\begin{aligned} y_{ij} &= \beta_i + e_{ij} \qquad i = 1, \ldots, I; \quad j = 1, \ldots, J \\ e_{ij} &\sim N(0, \sigma^2) \end{aligned} \tag{2.46}$$

In this chapter we will discuss only the full rank model.

In model (2.46) in the form (1.1), $\mathbf{y}_{n\times 1}$ is a vector of observations, $\mathbf{X}_{n\times k}$ is the design matrix of rank $k \leq n$ where the k columns (one for each cell in the design) consist of zeros and ones representing cell membership, $\beta_{p\times 1}$ is the parameter vector of cell means, and $\mathbf{e}_{n\times 1}$ is a vector of random errors. The parameter vector is estimated by (2.3), which is equivalent to the vector of observed cell means, $\hat{\beta} = \bar{\mathbf{y}}$.

All linear parametric functions $\psi = \mathbf{c}'\beta$ are estimable and estimated by $\hat{\psi} = \mathbf{c}'\hat{\beta}$, with $\text{var}(\hat{\psi})$ given in (2.4) where σ^2 is estimated by the unbiased estimator

$$s^2 = (\mathbf{y} - \mathbf{X}\hat{\beta})'(\mathbf{y} - \mathbf{X}\hat{\beta}) / (n - k) \tag{2.47}$$

The null hypothesis to be tested can be written as $\mathbf{C}\beta = \xi$, where the hypothesis test matrix **C** must be of full row rank; thus, the largest number of independent rows in **C** is $(I - 1)$, where I is the number of cells in the design. **C** is not unique and one usually constructs this matrix based upon the specific contrasts of interest to the researcher. Note for the case of equal n, that the matrix **C** is an orthogonal contrast matrix if $\mathbf{CC}' = \mathbf{D}$, where **D** is a diagonal matrix. For unequal n, **C** is an orthogonal contrast matrix if $\mathbf{C}(\mathbf{X}'\mathbf{X})^{-1}\mathbf{C}' = \mathbf{CD}^{-1}\mathbf{C}'$, and **D** is a diagonal matrix with n_i as diagonal entries. For unequal n_i then, to construct an orthogonal **C**, we use contrasts that are data dependent, depending on the n_i. A nonorthogonal **C** can be constructed that is not data dependent in this case, but the rows of the matrix **C** will not be independent.

As a numerical example of the analysis of an unrestricted one-way design with unequal n, the data from Winer (1971, p. 213) are analyzed using SAS. The data were also analyzed by Timm and Carlson (1975, p. 51). The data set consists of measurements from four treatment conditions.

The full rank representation of this design using (1.1) is

$$
\begin{array}{ccccc}
\mathbf{y} & = & \mathbf{X} & \beta & + \quad \mathbf{e}
\end{array}
$$

$$
\begin{bmatrix} 3 \\ 2 \\ 4 \\ 3 \\ 1 \\ 5 \\ 7 \\ 8 \\ 4 \\ 10 \\ 6 \\ 3 \\ 2 \\ 1 \\ 2 \\ 4 \\ 2 \\ 3 \\ 1 \\ 10 \\ 12 \\ 8 \\ 5 \\ 12 \\ 10 \\ 9 \end{bmatrix} = \begin{bmatrix} 1 & 0 & 0 & 0 \\ 1 & 0 & 0 & 0 \\ 1 & 0 & 0 & 0 \\ 1 & 0 & 0 & 0 \\ 1 & 0 & 0 & 0 \\ 1 & 0 & 0 & 0 \\ 0 & 1 & 0 & 0 \\ 0 & 1 & 0 & 0 \\ 0 & 1 & 0 & 0 \\ 0 & 1 & 0 & 0 \\ 0 & 1 & 0 & 0 \\ 0 & 0 & 1 & 0 \\ 0 & 0 & 1 & 0 \\ 0 & 0 & 1 & 0 \\ 0 & 0 & 1 & 0 \\ 0 & 0 & 1 & 0 \\ 0 & 0 & 1 & 0 \\ 0 & 0 & 1 & 0 \\ 0 & 0 & 1 & 0 \\ 0 & 0 & 0 & 1 \\ 0 & 0 & 0 & 1 \\ 0 & 0 & 0 & 1 \\ 0 & 0 & 0 & 1 \\ 0 & 0 & 0 & 1 \\ 0 & 0 & 0 & 1 \\ 0 & 0 & 0 & 1 \end{bmatrix} \begin{bmatrix} \beta_1 \\ \beta_2 \\ \beta_3 \\ \beta_4 \end{bmatrix} + \begin{bmatrix} e_{11} \\ e_{12} \\ e_{13} \\ e_{14} \\ e_{15} \\ e_{16} \\ e_{21} \\ e_{22} \\ e_{23} \\ e_{24} \\ e_{25} \\ e_{31} \\ e_{32} \\ e_{33} \\ e_{34} \\ e_{35} \\ e_{36} \\ e_{37} \\ e_{38} \\ e_{41} \\ e_{42} \\ e_{43} \\ e_{44} \\ e_{45} \\ e_{46} \\ e_{47} \end{bmatrix} \tag{2.48}
$$

The hypothesis we wish to test is H: $\beta_1 = \beta_2 = \beta_3 = \beta_4$, i.e., that all four cell means are equivalent. This is the same as simultaneously testing

$$
\begin{aligned}
\beta_1 - \beta_4 &= 0 \\
\beta_2 - \beta_4 &= 0 \\
\beta_3 - \beta_4 &= 0
\end{aligned} \tag{2.49}
$$

Thus for testing this hypothesis, $\mathbf{C}\beta = \xi$ is expressed as follows:

$$
\begin{bmatrix} 1 & 0 & 0 & -1 \\ 0 & 1 & 0 & -1 \\ 0 & 0 & 1 & -1 \end{bmatrix} \begin{bmatrix} \beta_1 \\ \beta_2 \\ \beta_3 \\ \beta_4 \end{bmatrix} = \begin{bmatrix} 0 \\ 0 \\ 0 \end{bmatrix} \tag{2.50}
$$

Program 2_7.sas contains the SAS code for analyzing these data.

Program 2_7.sas

```
/* Program 2_7.sas                                          */
/* Program to perform analysis of a One-Way Design with     */
/* Unequal n, using Full Rank model                         */

options ls=80 ps=60 nodate nonumber;
filename treat'c:\2_7.dat';
title 'Output 2.7: One-Way Design with Unequal n, Full Rank Model';

data treat;
   infile treat;
   input treat y;
proc print;
proc glm;
   class treat;
   model y=treat /noint e xpx;
   contrast 'treatments'    treat 1 0 0 -1,
                            treat 0 1 0 -1,
                            treat 0 0 1 -1;
   estimate 'treat1-treat4' treat 1 0 0 -1;
   estimate 'treat2-treat4' treat 0 1 0 -1;
   estimate 'treat3-treat4' treat 0 0 1 -1;
run;
proc glm;
   class treat;
   model y=treat /noint e xpx;
   contrast 'treats123 vs 4' treat -1 -1 -1 3;
   estimate 'treats123 vs 4' treat -1 -1 -1 3;
run;
```

First the data set are read in and then printed. The first analysis is performed using PROC GLM. The variable TREAT represents a nominal variable and is thus specified in the CLASS statement. In the MODEL statement the NOINT option is used to omit the intercept from the model so that a full rank model is used. The hypothesis test matrix **C** of (2.50) is specified in the CONTRAST statement, and the ESTIMATE statements are then used to obtain the individual estimates of the contrasts and their standard errors. The second analysis using PROC GLM is performed to test the hypothesis in (2.51); note again that the matrix **C** is specified in the CONTRAST statement.

Result and Interpretation 2_7.sas

Output 2.7: *One-Way Design with Unequal n, Full Rank Model*

```
           Output 2.7: One-Way Design with Unequal n, Full Rank Model

                          General Linear Models Procedure

Dependent Variable: Y
                                       Sum of             Mean
Source                      DF        Squares           Square   F Value       Pr > F

Model                        4   961.78571429    240.44642857      72.25       0.0001

Error                       22    73.21428571      3.32792208

Uncorrected Total           26  1035.00000000

                  R-Square              C.V.         Root MSE                 Y Mean

                  0.766175          34.62098        1.8242593              5.2692308


Source                      DF      Type I SS     Mean Square   F Value       Pr > F

TREAT                        4   961.78571429    240.44642857      72.25       0.0001

Source                      DF    Type III SS     Mean Square   F Value       Pr > F

TREAT                        4   961.78571429    240.44642857      72.25       0.0001

Contrast                    DF    Contrast SS     Mean Square   F Value       Pr > F

treatments                   3   239.90109890     79.96703297      24.03       0.0001

                                           T for H0:     Pr > |T|    Std Error of
Parameter                   Estimate     Parameter=0                  Estimate

treat1-treat4            -6.42857143           -6.33       0.0001      1.01492419
treat2-treat4            -2.42857143           -2.27       0.0331      1.06817688
treat3-treat4            -7.17857143           -7.60       0.0001      0.94414390

Dependent Variable: Y
                                       Sum of             Mean
Source                      DF        Squares           Square   F Value       Pr > F

Model                        4   961.78571429    240.44642857      72.25       0.0001

Error                       22    73.21428571      3.32792208

Uncorrected Total           26  1035.00000000
```

Output 2.7 *(continued)*

```
                    R-Square              C.V.          Root MSE                Y Mean

                    0.766175          34.62098         1.8242593             5.2692308

Source                    DF          Type I SS      Mean Square   F Value      Pr > F

TREAT                      4       961.78571429     240.44642857     72.25      0.0001

Source                    DF        Type III SS      Mean Square   F Value      Pr > F

TREAT                      4       961.78571429     240.44642857     72.25      0.0001

Contrast                  DF        Contrast SS      Mean Square   F Value      Pr > F

treats123 vs 4             1       144.67586834     144.67586834     43.47      0.0001

                                         T for H0:      Pr > |T|     Std Error of
Parameter                    Estimate   Parameter=0                     Estimate

treats123 vs 4             16.0357143          6.59        0.0001       2.43207428
```

The ANOVA summary table for testing this hypothesis $H\colon \mathbf{C}\beta = \mathbf{0}$, with **C** as previously given, was taken from Output 2.7 and is as follows:

Table 2.1 *ANOVA Summary Table*

Source	SS	df	MS	F	p
Contrast	239.90	3	79.97	24.03	<.0001
Error	73.21	22	3.33		

Say that we had set our Type I error at .01; we would thus reject the null hypothesis that the four treatment means are equivalent in the population. The estimates of the individual contrasts $\hat{\psi} = \mathbf{C}\hat{\beta}$ from Output 2.7 are

Table 2.2 *Contrast Estimates and Standard Errors*

Contrast		$\hat{\psi}$	$\hat{\sigma}_{\hat{\psi}}$
Treatment	1 vs 4	−6.43	1.01
Treatment	2 vs 4	−2.43	1.07
Treatment	3 vs 4	−7.18	0.94

Suppose we had wanted to test the null hypothesis that treatments 1, 2, and 3 averaged together are equivalent to treatment 4. This corresponds to a test of the null hypothesis $\mathbf{C}\beta = \xi$ written as

$$H\text{:}\ \begin{bmatrix} -1 & -1 & -1 & 3 \end{bmatrix} \begin{bmatrix} \beta_1 \\ \beta_2 \\ \beta_3 \\ \beta_4 \end{bmatrix} = 0 \tag{2.51}$$

The ANOVA summary table for hypothesis (2.51) taken from Output 2.7 is as follows.

Table 2.3 *ANOVA Summary Table*

Source	SS	df	MS	F	p
Contrast	1	144.68	144.68	43.47	.0001
Error	73.21	22	3.33		

Thus, we reject the null hypothesis. The contrast is $\hat{\psi} = 16.04$ with a standard error of $\hat{\sigma}_{\hat{\psi}} = 2.43$. These results agree with those in Timm and Carlson (1975) with their FULRNK program.

2.8 Two-Way Nested Designs

In the next example of the general linear model without restrictions we discuss the two-way nested design. In educational research, data often have a nested structure; students are nested in classes, classes are nested in schools, schools are nested in school districts, and so on. We discuss the two-way nested design using the full rank model. Suppose the two factors are A and B, with B nested within A. There are $i = 1, \ldots, I$ levels of A, $j = 1, \ldots, J_i$ levels of B within $A_{(i)}$, and $k = 1, \ldots, n_{ij}$ observations within the cell $A_{(i)}B_{(j)}$. The full rank nested linear model can be written as

$$y_{ijk} = \beta_{ij} + e_{ijk} \tag{2.52}$$

As a numerical example of a two-way nested design, we use the data analyzed by Searle (1971, p. 249) and by Timm and Carlson (1975, p. 84). (Note there was a typographical error in the data set printed in Timm and Carlson, 1975: Course 1, Section 1 should have a value of 5, not 4). In this data set two factors are involved, A and B, where A is *course* and B is *section*. Nested within course 1 are two sections and nested within course 2 are three sections.

To represent the model (2.52) in the form of (1.1), we write the model as follows:

$$\begin{array}{ccccc} \mathbf{y} & = & \mathbf{X} & \beta & + \quad \mathbf{e} \end{array}$$

$$\begin{bmatrix} 5 \\ 8 \\ 10 \\ 9 \\ 8 \\ 10 \\ 6 \\ 2 \\ 1 \\ 3 \\ 3 \\ 7 \end{bmatrix} = \begin{bmatrix} 1 & 0 & 0 & 0 & 0 \\ 0 & 1 & 0 & 0 & 0 \\ 0 & 1 & 0 & 0 & 0 \\ 0 & 1 & 0 & 0 & 0 \\ 0 & 0 & 1 & 0 & 0 \\ 0 & 0 & 1 & 0 & 0 \\ 0 & 0 & 0 & 1 & 0 \\ 0 & 0 & 0 & 1 & 0 \\ 0 & 0 & 0 & 1 & 0 \\ 0 & 0 & 0 & 1 & 0 \\ 0 & 0 & 0 & 0 & 1 \\ 0 & 0 & 0 & 0 & 1 \end{bmatrix} \begin{bmatrix} \beta_{11} \\ \beta_{12} \\ \beta_{21} \\ \beta_{22} \\ \beta_{23} \end{bmatrix} + \begin{bmatrix} e_{111} \\ e_{121} \\ e_{122} \\ e_{123} \\ e_{211} \\ e_{212} \\ e_{221} \\ e_{222} \\ e_{223} \\ e_{224} \\ e_{231} \\ e_{232} \end{bmatrix} \tag{2.53}$$

Program 2_8.sas contains the SAS code to analyze this data set.

Program 2_8.sas

```
/* Program 2_8.sas                                            */
/* Program to perform analysis of a nested design with        */
/* Unequal n, using Full Rank model                           */

options ls=80 ps=60 nodate nonumber;
filename nested 'c:\2_8.dat';
title 'Output 2.8: Nested Design with Unequal n, Full Rank Model';

data nest;
   infile nested;
   input a b y;
   label a='course'
         b='section';
proc print;
proc glm;
   class a b;
   model y = a b(a) /noint e;

   contrast 'ha_1' a 1 -1;
   estimate 'ha_1' a 1 -1 /e;

   contrast 'hb_1'        b(a) 1 -1  0  0  0;
   estimate 'hb_1vs2'     b(a) 1 -1  0  0  0 /e;

   contrast 'hb_2'        b(a) 0  0  1  0 -1,
                          b(a) 0  0  0  1 -1;
   estimate 'hb_1vs3'     b(a) 0  0  1  0 -1 /e;
   estimate 'hb_2vs3'     b(a) 0  0  0  1 -1 /e;

   contrast 'hb_1and2'    b(a) 1 -1  0  0  0,
                          b(a) 0  0  1  0 -1,
                          b(a) 0  0  0  1 -1;
```

```
proc glm;
   class a b;
   model y = a b(a);
run;
```

To use a full rank model representation in PROC GLM, we must include the NOINT option in the MODEL statement so that no intercept is fit in the model. The MODEL statement is written as model $y = a\ b(a)\ e$; specifying that factor B is nested within factor A. The CONTRAST statements are used for testing specific hypotheses (discussed below along with the output) and ESTIMATE statements for obtaining the contrast estimates and standard errors. The last analysis of the program uses PROC GLM to perform the less than full rank analysis of these data.

Result and Interpretation 2_8.sas

Output 2.8: *Nested Design with Unequal n, Full Rank Model*

```
          Output 2.8: Nested Design with Unequal n, Full Rank Model

                         General Linear Models Procedure

Dependent Variable: Y
                                      Sum of              Mean
Source                   DF          Squares            Square   F Value      Pr > F

Model                     5     516.00000000      103.20000000     27.78      0.0002

Error                     7      26.00000000        3.71428571

Uncorrected Total        12     542.00000000

                   R-Square             C.V.          Root MSE                Y Mean

                   0.763636         32.12080         1.9272482             6.0000000

Source                   DF        Type I SS       Mean Square   F Value      Pr > F

A                         2     456.00000000      228.00000000     61.38      0.0001
B(A)                      3      60.00000000       20.00000000      5.38      0.0309

Source                   DF      Type III SS       Mean Square   F Value      Pr > F

A                         2     378.20000000      189.10000000     50.91      0.0001
B(A)                      3      60.00000000       20.00000000      5.38      0.0309

Contrast                 DF      Contrast SS       Mean Square   F Value      Pr > F

ha_1                      1       3.76470588        3.76470588      1.01      0.3476
hb_1                      1      12.00000000       12.00000000      3.23      0.1153
hb_2                      2      48.00000000       24.00000000      6.46      0.0257
hb_1and2                  3      60.00000000       20.00000000      5.38      0.0309

                                          T for H0:    Pr > |T|    Std Error of
Parameter                     Estimate   Parameter=0                 Estimate

ha_1                        1.33333333          1.01      0.3476      1.32437467
hb_1vs2                    -4.00000000         -1.80      0.1153      2.22539456
hb_1vs3                     4.00000000          2.08      0.0766      1.92724822
hb_2vs3                    -2.00000000         -1.20      0.2698      1.66904592
```

Output 2.8 (*continued*)

```
                              General Linear Models Procedure
                         Class Level Information
                         Class     Levels     Values

                       A              2     1 2

                       B              5     1 2 3 4 5
                 Number of observations in data set = 12

                       General Linear Models Procedure

Dependent Variable: Y
                                      Sum of             Mean
Source                     DF        Squares           Square   F Value      Pr > F

Model                       4    84.00000000      21.00000000      5.65      0.0236

Error                       7    26.00000000       3.71428571

Corrected Total            11   110.00000000

                  R-Square              C.V.         Root MSE                Y Mean

                  0.763636          32.12080        1.9272482             6.0000000

Source                     DF      Type I SS      Mean Square   F Value      Pr > F

A                           1    24.00000000      24.00000000      6.46      0.0386
B(A)                        3    60.00000000      20.00000000      5.38      0.0309

Source                     DF    Type III SS      Mean Square   F Value      Pr > F

A                           1     3.76470588       3.76470588      1.01      0.3476
B(A)                        3    60.00000000      20.00000000      5.38      0.0309
```

There are several hypotheses which may be of interest with regard to these data. For example, one may be interested in testing that the mean of course 1 equals the mean of course 2. This could be performed as an unweighted test or as a weighted test. The unweighted test is

$$H_A: \quad \frac{\beta_{11}+\beta_{12}}{2} = \frac{\beta_{21}+\beta_{22}+\beta_{23}}{3}$$

Here $\mathbf{C}\beta = \xi$ is as follows:

$$\begin{bmatrix} 1/2 & 1/2 & -1/3 & -1/3 & -1/3 \end{bmatrix} \quad \begin{bmatrix} \beta_{11} \\ \beta_{12} \\ \beta_{21} \\ \beta_{22} \\ \beta_{23} \end{bmatrix} = 0$$

This was tested in Program 2_8.sas using the CONTRAST statement `contrast 'ha_1' a 1 -1;` . The ANOVA summary table of this test, taken from the SAS output, is

Table 2.4 *ANOVA Summary Table*

Source	SS	df	MS	F	p
ha_1	3.76	1	3.76	1.01	.35
Error	26.00	7	3.71		

Letting the Type 1 error equal .05, we do not reject the null hypothesis H_A.

Alternatively, the hypothesis test that the mean of course 1 equals the mean of course 2 can be tested using a weighted analysis, by weighting the cell means by n_{ij}, the number of observations in the cell. This hypothesis is expressed as

$$H_{A^{\bullet}}: \quad \frac{(1)(\beta_{11})+(3)(\beta_{12})}{4} = \frac{(2)(\beta_{21})+(4)(\beta_{22})+(2)(\beta_{23})}{8}$$

This test was performed in Program 2_8.sas using PROC GLM without the NOINT option to obtain the less than full rank model. The results of this weighted test are found at the end of Output 2.8 under `Type I SS` for the *A* effect. The ANOVA summary table is

Table 2.5 *ANOVA Summary Table*

Source	SS	df	MS	F	p
A	24.00	1	24.00	6.46	0.04
Error	26.00	7	3.71		

Thus, if the Type I error is set at .05, we reject this weighted null hypothesis $H_{A^{\bullet}}$. It is important to note, however, that $H_{A^{\bullet}}$ is data dependent since the weights were determined from the number of observations in each cell. The contrast estimate, $\hat{\psi}_{\text{h\_a1}}$, is 1.33 with a standard error of 1.32.

Now suppose we wish to test the null hypothesis that the section population means are equal within course 1. This is $H_{\beta(1)}: \beta_{11} = \beta_{12}$. In addition, we wish to test the null hypothesis that the section population means are equal within course 2. This is $H_{\beta(2)}: \beta_{21} = \beta_{22} = \beta_{23}$. To test $H_{\beta(1)}$, $\mathbf{C} = (1 \quad -1 \quad 0 \quad 0 \quad 0)$ and to test $H_{\beta(2)}$, $\mathbf{C} = \begin{pmatrix} 0 & 0 & 1 & 0 & -1 \\ 0 & 0 & 0 & 1 & -1 \end{pmatrix}$. To simultaneously test $H_{\beta(1)}$ and $H_{\beta(2)}$, $\mathbf{C} = \begin{pmatrix} 1 & -1 & 0 & 0 & 0 \\ 0 & 0 & 1 & 0 & -1 \\ 0 & 0 & 0 & 1 & -1 \end{pmatrix}$

These contrasts are labeled in Output 2.8 as *hb_1*, *hb_2*, and *hb_1* and 2, respectively. The ANOVA summary table for these hypotheses is as follows:

Table 2.6 *ANOVA Summary Table*

Source	SS	df	MS	F	p
$H_{\beta(1)}$	12.00	1	12.00	3.23	.12
$H_{\beta(2)}$	48.00	2	24.00	6.46	.03
$H_{\beta(1) \text{ and } \beta(2)}$	60.00	3	20.00	5.38	.03
Error	26.00	7	3.71		

These results agree with those obtained by Timm and Carlson (1975, p. 85) with their FULRNK program.

2.9 Intraclass Covariance Models

In Section 2.7, we discussed a one-way ANOVA with four treatment groups. Suppose we have an additional variable in the experiment that is correlated with the dependent variable, but not related to the treatment group. This variable could be included in the analysis as a covariate. In the analysis of covariance (ANCOVA), the regression slopes of the dependent variable on the covariable are assumed to be equal across treatment groups (the assumption of parallelism). In this section we discuss a more general model that allows the slopes of the regression lines within the populations to differ, the general intraclass model. In Chapter 3 we will place restrictions on this general model and discuss ANCOVA more fully.

In the general intraclass model, it is assumed that each population has a distinct slope parameter. Considering a one-factor intraclass covariance model, we write the model as

$$\begin{aligned} y_{ij} &= \alpha_i + \beta_i z_{ij} + e_{ij} \qquad i = 1, \dots I, \; j = 1, \dots J_i \\ e_{ij} &\sim N(0, \sigma^2) \end{aligned} \tag{2.54}$$

where β_i and α_i represent the slope and intercept for the regression of y on z in the ith population. To express this model in the form of (1.1), we write

$$\mathbf{X} = \begin{bmatrix} \mathbf{1}_1 & \mathbf{0}_1 & \cdots & \cdots & \cdots & \mathbf{0}_1 & \mathbf{Z}_1 & \mathbf{0}_1 & \cdots & \cdots & \cdots & \mathbf{0}_1 \\ \mathbf{0}_2 & \mathbf{1}_2 & \cdots & \cdots & \cdots & \mathbf{0}_2 & \mathbf{0}_2 & \mathbf{Z}_2 & \cdots & \cdots & \cdots & \mathbf{0}_2 \\ \vdots & & \ddots & & & \vdots & \vdots & \vdots & \ddots & & & \vdots \\ \vdots & & & \ddots & & \vdots & \vdots & \vdots & & \ddots & & \vdots \\ \vdots & & & & \ddots & \vdots & \vdots & \vdots & & & \ddots & \vdots \\ \mathbf{0}_I & \mathbf{0}_I & \cdots & \cdots & \cdots & \mathbf{1}_I & \mathbf{0}_I & \mathbf{0}_I & \cdots & \cdots & \cdots & \mathbf{Z}_I \end{bmatrix}$$

$$\beta' = \begin{bmatrix} \alpha_1 & \alpha_2 & \cdots & \cdots & \cdots & \alpha_I & \beta_1 & \beta_2 & \cdots & \cdots & \cdots & \beta_I \end{bmatrix}$$

Where $\mathbf{Z}_i$ is a vector of J_i values on the covariate z in the ith population, and $\mathbf{1}_i$ and $\mathbf{0}_i$ are vectors of J_i ones and zeros, respectively. In order that $\mathbf{X'X}$ be of full rank so that the least squares estimate of β exists, there must be at least two observations in each group in the sample having different values on the covariate (if this is not the case, there are two or more columns of **X** that are linearly dependent, and thus **X** is not of full column rank).

One can test hypotheses about the slope and intercept parameters either separately or simultaneously. To test the hypothesis

$$H_1: \ \beta_1 = \beta_2 = \ldots = \beta_I \tag{2.55}$$

a test that the slopes are identical in all I populations, we can write $\mathbf{C}\beta = \xi$ as

$$\begin{bmatrix} 0 & 0 & \cdots & \cdots & \cdots & 0 & 1 & 0 & 0 & \cdots & \cdots & & -1 \\ 0 & 0 & \cdots & \cdots & \cdots & 0 & 0 & 1 & 0 & \cdots & \cdots & & -1 \\ \vdots & \vdots & \ddots & \cdots & \cdots & \vdots & \vdots & \vdots & \vdots & & & & \\ \vdots & & & \ddots & & \vdots & \vdots & \vdots & \vdots & & & & \\ \vdots & & & & \ddots & \vdots & \vdots & \vdots & \vdots & & & & \\ 0 & 0 & \cdots & \cdots & \cdots & 0 & 0 & 0 & \cdots & \cdots & & 1 & -1 \end{bmatrix} \begin{bmatrix} \alpha_1 \\ \alpha_2 \\ \vdots \\ \vdots \\ \vdots \\ \alpha_I \\ \cdots\cdots \\ \beta_1 \\ \beta_2 \\ \vdots \\ \vdots \\ \vdots \\ \beta_I \end{bmatrix} = \begin{bmatrix} 0 \\ 0 \\ \vdots \\ \vdots \\ \vdots \\ \vdots \\ \cdots\cdots \\ \vdots \\ \vdots \\ \vdots \\ \vdots \\ \vdots \\ 0 \end{bmatrix}$$

H_1 is identical to the assumption of equality of slopes (parallelism) that underlies ANCOVA. Thus, we are interested in performing this test whenever we use ANCOVA and wish to use the data to examine the assumption of parallelism.

One could also test that the slopes and intercepts, and thus the regression lines themselves, are identical in the I populations:

$$H_2: \begin{bmatrix} \alpha_1 = \alpha_2 = \ldots = \alpha_I \\ \beta_1 = \beta_2 = \ldots = \beta_I \end{bmatrix}$$

This hypothesis does not relate specifically to a hypothesis in the analysis of covariance model but is of interest when the research question is about whether the regression lines are identical in the I populations, the test of coincidence.

One could also test the hypothesis of equality of the intercepts in the different populations:

$$H_3: \ \alpha_1 = \alpha_2 = \cdots = \alpha_I$$

However, the usefulness of this test is questionable because of the relationship between the intercepts and the slopes.

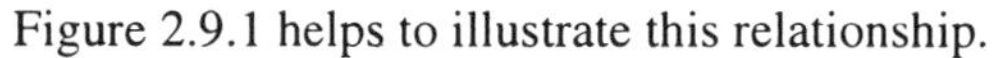

Figure 2.9.1 helps to illustrate this relationship.

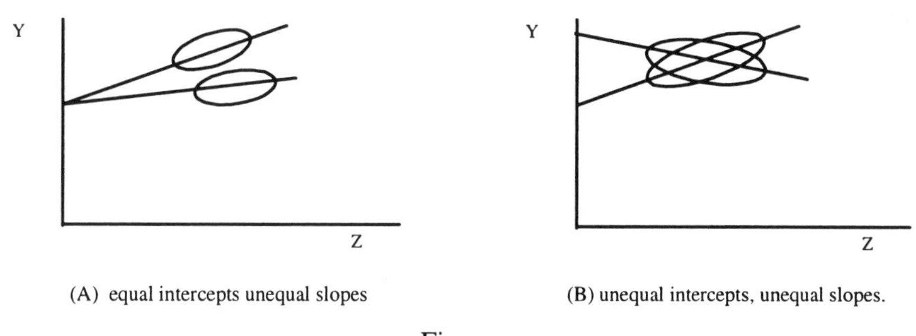

(A) equal intercepts unequal slopes (B) unequal intercepts, unequal slopes.

Figure 2.9.1

Examining the two cases illustrated in Figure 2.9.1 and considering other possible cases, one can see that a test of equality of the intercepts, without any assumption about the slopes of the regression lines in the different populations, is not useful.

As an example of the intraclass model, we analyzed the following data using SAS software:

A_1		A_2		A_3	
Y	Z	Y	Z	Y	Z
3	6	5	5	5	4
2	6	6	5	7	1
3	5	8	3	8	1
3	4	8	2	7	2
4	3	7	1		
5	3	9	1		
5	2				
6	2				

The SAS code to test hypothesis H_1 in (2.55) is given in Program 2_9.sas.

Program 2_9.sas

```
/* Program 2_9.sas                                                    */
/* Program to perform analysis of covariance with different slopes and */
/* Unequal n, using Full Rank model                                   */
/* the one-way Unbalanced Intraclass covariance model                 */

options ls=80 ps=60 nodate nonumber;
filename ex29 'c:\2_9.dat';
title 'Output 2.9: One-way Unbalanced Intraclass covariance model';
```

```
data cov;
   infile ex29;
   input y a z a1 a2 a3 z1 z2 z3;
proc print;

/* Using the full rank model */
proc reg;
   model y = a1 a2 a3 z1 z2 z3 /noint;
   parallel: mtest z1-z2=0,
                   z1-z3=0 /print details;
run;

/* Using the less than full rank model */
proc glm;
   class a;
   model y=a z a*z /e xpx solution;
   lsmeans a /stderr pdiff cov;
run;
```

The data are first analyzed using the full rank model by using the NOINT option in the MODEL statement of PROC REG. Note the way that the variables are specified in the MODEL statement: three variable are used to specify group membership, and three variables-one for each group-are used to specify the values of the covariate. The MTEST statement is used to specify the hypothesis, H_1 of (2.55), that we wish to test. Note that $\mathbf{C}$ in this example is

$$\mathbf{C} = \begin{bmatrix} 0 & 0 & 0 & 1 & -1 & 0 \\ 0 & 0 & 0 & 1 & 0 & -1 \end{bmatrix}$$

The analysis is also performed with PROC GLM using the less than full rank model, for purposes of comparing output.

Result and Interpretation 2_9.sas

Output 2.9: *One-way Unbalanced Intraclass covariance model*

Output 2.9: One-way Unbalanced Intraclass covariance model

OBS	Y	A	Z	A1	A2	A3	Z1	Z2	Z3
1	3	1	6	1	0	0	6	0	0
2	2	1	6	1	0	0	6	0	0
3	3	1	5	1	0	0	5	0	0
4	3	1	4	1	0	0	4	0	0
5	4	1	3	1	0	0	3	0	0
6	5	1	3	1	0	0	3	0	0
7	5	1	2	1	0	0	2	0	0
8	6	1	2	1	0	0	2	0	0
9	5	2	5	0	1	0	0	5	0
10	6	2	5	0	1	0	0	5	0
11	8	2	3	0	1	0	0	3	0
12	8	2	2	0	1	0	0	2	0
13	7	2	1	0	1	0	0	1	0
14	9	2	1	0	1	0	0	1	0
15	5	3	4	0	0	1	0	0	4
16	7	3	1	0	0	1	0	0	1
17	8	3	1	0	0	1	0	0	1
18	7	3	2	0	0	1	0	0	2

Output 2.9 *(continued)*

```
Model: MODEL1
NOTE: No intercept in model. R-square is redefined.
Dependent Variable: Y

                              Analysis of Variance

                                 Sum of          Mean
        Source          DF      Squares        Square        F Value         Prob>F

        Model            6    632.25064     105.37511        187.351         0.0001
        Error           12      6.74936       0.56245
        U Total         18    639.00000

            Root MSE        0.74996      R-square        0.9894
            Dep Mean        5.61111      Adj R-sq        0.9842
            C.V.           13.36570

                              Parameter Estimates

                       Parameter       Standard    T for H0:
      Variable  DF      Estimate          Error   Parameter=0     Prob > T|

      A1         1      6.774834     0.71954746         9.415        0.0001
      A2         1      8.990099     0.60163974        14.943        0.0001
      A3         1      8.416667     0.71803603        11.722        0.0001
      Z1         1     -0.748344     0.17262236        -4.335        0.0010
      Z2         1     -0.643564     0.18279127        -3.521        0.0042
      Z3         1     -0.833333     0.30617159        -2.722        0.0185

Multivariate Test: PARALLEL

                             L Ginv(X'X) L'    LB-cj

                    0.112386073       0.0529801325      -0.104780014
                   0.0529801325       0.2196467991      0.0849889625

                       Inv(L Ginv(X'X) L')     Inv()(LB-cj)

                   10.039460539       -2.421578422      -1.257742258
                   -2.421578422       5.1368631369      0.6903096903

                              E, the Error Matrix

                                  6.7493552335

                            H, the Hypothesis Matrix

                                  0.1904549563

                             T, the H + E Matrix

                                  6.9398101898
```

Output 2.9 (*continued*)

```
                              Eigenvectors

                                 0.3796000001

                              Eigenvalues

                                 0.0274438279

                   Multivariate Statistics and Exact F Statistics

                            S=1      M=0      N=5

   Statistic                    Value          F       Num DF    Den DF  Pr > F

   Wilks' Lambda              0.97255617     0.1693        2         12  0.8462
   Pillai's Trace             0.02744383     0.1693        2         12  0.8462
   Hotelling-Lawley Trace     0.02821824     0.1693        2         12  0.8462
   Roy's Greatest Root        0.02821824     0.1693        2         12  0.8462

 Dependent Variable: Y
                                         Sum of            Mean
 Source                     DF          Squares          Square   F Value      Pr > F

 Model                       5      65.52842254     13.10568451     23.30      0.0001

 Error                      12       6.74935523      0.56244627

 Corrected Total            17      72.27777778

                      R-Square            C.V.        Root MSE                 Y Mean

                      0.906619        13.36570       0.7499642              5.6111111

 Source                     DF        Type I SS     Mean Square   F Value      Pr > F

 A                           2      43.81944444     21.90972222     38.95      0.0001
 Z                           1      21.51852314     21.51852314     38.26      0.0001
 Z*A                         2       0.19045496      0.09522748      0.17      0.8462

 Source                     DF      Type III SS     Mean Square   F Value      Pr > F

 A                           2       3.22575628      1.61287814      2.87      0.0960
 Z                           1      17.74468236     17.74468236     31.55      0.0001
 Z*A                         2       0.19045496      0.09522748      0.17      0.8462

                                             T for H0:     Pr > |T|    Std Error of
 Parameter                  Estimate       Parameter=0                  Estimate

 INTERCEPT               8.416666667 B          11.72       0.0001      0.71803603
 A          1           -1.641832230 B          -1.62       0.1323      1.01652560
            2            0.573432343 B           0.61       0.5519      0.93677432
            3            0.000000000 B            .          .           .
 Z                      -0.833333333 B          -2.72       0.0185      0.30617159
 Z*A        1            0.084988962 B           0.24       0.8130      0.35148190
            2            0.189768977 B           0.53       0.6043      0.35658617
            3            0.000000000 B            .          .           .
```

Output 2.9 *(continued)*

```
NOTE: The X'X matrix has been found to be singular and a generalized inverse
      was used to solve the normal equations.   Estimates followed by the
      letter 'B' are biased, and are not unique estimators of the parameters.

                          General Linear Models Procedure
                                Least Squares Means

A            Y          Std Err      Pr > |T|     Pr > |T| H0: LSMEAN(i)=LSMEAN(j)
        LSMEAN           LSMEAN   H0:LSMEAN=0     i/j      1          2          3

1   4.44665195     0.29613169        0.0001     1   .          0.0001     0.0368
2   6.98789879     0.31035332        0.0001     2  0.0001       .         0.0737
3   5.82407407     0.50630154        0.0001     3  0.0368     0.0737       .

NOTE: To ensure overall protection level, only probabilities associated with
      pre-planned comparisons should be used.
```

The results of the test of parallelism, H_1, which we labeled the test of parallel in Output 2.9, are labeled `Multivariate Statistics and Exact F-Statistics`. The resulting $F = .1693$, with 2 and 12 degrees of freedom, with a p-value of 0.8426; thus, we do not reject the null hypothesis that the slopes of the three groups are equivalent.

Alternatively, the test of H_1 using PROC GLM and the less than full rank model was performed by including the term $z * a$ in the MODEL statement. The results of this test are found at the end of Output 2.9 in the row labeled $z * a$; note that the results are identical to those obtained using PROC REG.

3. Restricted General Linear Models

Theory

Applications

3.1 Introduction

In Chapter 2 we discussed of the general linear model without restrictions. There we assumed that the covariance structure of the GLM was $\mathbf{\Omega} = \sigma^2\mathbf{I}_n$, with no restrictions of the form (1.6) associated with the model. However, in many statistical applications it is necessary to place restrictions on the parameter vector β in (1.1). For example, in the analysis of a two-way factorial design without interaction one must place the restriction that the interactions are equal to zero. In this chapter we consider the situation where restrictions of the form $\mathbf{R}\beta = \theta$ are associated with the model and the rank of the restriction matrix $\mathbf{R}$ is s, $\text{rank}(\mathbf{R}) = s > 0$. This model is called the restricted general linear model (RGLM) or the restricted general Gauss-Markov model. We discuss how to obtain OLS and ML estimates of β given the restrictions, and we also test hypotheses with restrictions and obtain simultaneous confidence sets for linear parametric functions of β.

We then discuss several examples of the restricted general linear model using the full rank model. In the first example, we discuss the analysis of a two-way factorial design with unequal cell frequencies. We first discuss this model including the interactions (unrestricted model) and then excluding interactions (restricted model). In the second example, we discuss the analysis of a Latin square design. In the third example, we discuss the split-plot repeated measures design. Last, we discuss the analysis of covariance model, with one covariate and with two covariates; then we briefly discuss a nested design.

3.2 Estimation and Hypothesis Testing

In the unrestricted linear model, the fixed parameter vector β was unconstrained in the parameter space. However, when one wants to perform a regression analysis and constrain the intercept to equal zero, or constrain independent regression lines to be parallel, or create additive analysis of variance models, one constrains the elements of β by imposing restrictions on the GLM. When the elements of β are constrained, we obtain the restricted ordinary least squares (OLS) estimator of β by minimizing the error sum of squares with the estimable restrictions $\mathbf{R}\beta = \theta$ as part of the model. For this situation, the restriction estimator of β is

$$\hat{\beta}_r = \hat{\beta} - (\mathbf{X}'\mathbf{X})^{-1}\mathbf{R}'[\mathbf{R}(\mathbf{X}'\mathbf{X})^{-1}\mathbf{R}']^{-1}(\mathbf{R}\hat{\beta} - \theta) \tag{3.1}$$

which is identical to the estimator given in (2.9) since the matrix $\mathbf{R}$ takes the place of $\mathbf{C}$, with $\hat{\beta}$ defined by (2.3). The estimator in (3.1) is the restricted OLSE, which is again BLUE. Letting $\mathbf{G} = (\mathbf{X}'\mathbf{X})^{-1}$, we may rewrite (3.1) as

$$\begin{aligned}\hat{\beta}_r &= (\mathbf{I} - \mathbf{GR}'(\mathbf{RGR}')^{-1}\mathbf{R})\hat{\beta} + \mathbf{GR}'(\mathbf{RGR}')^{-1}\theta \\ &= \mathbf{F}\hat{\beta} + \mathbf{GR}'(\mathbf{RGR}')^{-1}\theta\end{aligned} \tag{3.2}$$

where $\mathbf{F}$ is idempotent, $\mathbf{F}^2 = \mathbf{F} = \mathbf{I} - \mathbf{GR}'(\mathbf{RGR}')^{-1}\mathbf{R}$, and symmetric so that $\mathbf{F}$ is a projection operator. When $\theta = \mathbf{0}$, the restrictions on the model project $\hat{\beta}$ into a smaller subspace of the parameter space.

To test hypotheses about β, we again assume spherical multivariate normality as given in (2.6), where the likelihood function in (2.8) is maximized with the added restriction that $\mathbf{R}\beta = \theta$. For this restricted case, the maximum likelihood (ML) estimators of β and σ^2 are

$$\begin{aligned}\hat{\beta}_{ML_r} &= \hat{\beta}_r \\ \hat{\sigma}^2_{ML_r} &= (\mathbf{y} - \mathbf{X}\hat{\beta}_r)'(\mathbf{y} - \mathbf{X}\hat{\beta}_r)/n \\ &= [\mathbf{E} + (\mathbf{R}\hat{\beta} - \theta)'(\mathbf{R}(\mathbf{X}'\mathbf{X})^{-1}\mathbf{R}')^{-1}(\mathbf{R}\hat{\beta} - \theta)]/n\end{aligned} \tag{3.3}$$

where $\mathbf{E}$ is defined in (2.12). To obtain an unbiased estimator for σ^2, we replace the value of n in (3.3) by $v_e = n - \text{rank}(\mathbf{X}) + \text{rank}(\mathbf{R}) = n - k + s$.

The likelihood ratio test of $H(\mathbf{C}\beta = \xi)$ in (1.8) again requires maximizing the likelihood function (2.8) under ω and $\mathbf{\Omega}_o$ which now includes the restrictions. If we let the matrix $\mathbf{Q} = \binom{\mathbf{R}}{\mathbf{C}}$ and $\eta = \binom{\theta}{\xi}$, the ML estimator of β and σ^2 are

$$\begin{aligned}\hat{\beta}_{r,\omega} &= \hat{\beta} - (\mathbf{X}'\mathbf{X})^{-1}\mathbf{Q}'[\mathbf{Q}(\mathbf{X}'\mathbf{X})^{-1}\mathbf{Q}']^{-1}(\mathbf{Q}\hat{\beta} - \eta) \\ \hat{\sigma}^2_{r,\omega} &= (\mathbf{y} - \mathbf{X}\hat{\beta}_{r,\omega})'(\mathbf{y} - \mathbf{X}\hat{\beta}_{r,\omega})/n.\end{aligned} \tag{3.4}$$

Forming the likelihood ratio,

$$\lambda = \frac{L(\hat{\omega})}{L(\hat{\mathbf{\Omega}}_o)} = \left[\frac{(\mathbf{y} - \mathbf{X}\hat{\beta}_r)'(\mathbf{y} - \mathbf{X}\hat{\beta}_r)}{(\mathbf{y} - \mathbf{X}\hat{\beta}_{r,\omega})'(\mathbf{y} - \mathbf{X}\hat{\beta}_{r,\omega})}\right]^{n/2} \tag{3.5}$$

so that

$$\Lambda = \lambda^{2/n} = \frac{\mathbf{E}_r}{\mathbf{E}_r + \mathbf{H}_r} \tag{3.6}$$

where

$$\begin{aligned}
\mathbf{E}_r &= \mathbf{E} + (\mathbf{R}\hat{\beta} - \theta)'(\mathbf{RGR}')^{-1}(\mathbf{R}\hat{\beta} - \theta) \\
\mathbf{H}_r &= (\mathbf{C}\hat{\beta}_r - \xi)'[\mathbf{CGC}' - \mathbf{CGR}'(\mathbf{RGR}')^{-1}\mathbf{RGC}']^{-1}(\mathbf{C}\hat{\beta}_r - \xi) \\
&= (\mathbf{C}\hat{\beta}_r - \xi)'[\mathbf{C}(\mathbf{FGF}')\mathbf{C}']^{-1}(\mathbf{C}\hat{\beta}_r - \xi) \\
&= (\mathbf{C}\hat{\beta}_r - \xi)'(\mathbf{CFGC}')^{-1}(\mathbf{C}\hat{\beta}_r - \xi)
\end{aligned} \tag{3.7}$$

on simplifying (3.5) using properties of partitioned matrices

$$\text{If } \mathbf{A} = \begin{pmatrix} \mathbf{A}_{11} & \mathbf{A}_{12} \\ \mathbf{A}_{21} & \mathbf{A}_{22} \end{pmatrix}. \text{ Then } \mathbf{A}^{-1} = \begin{pmatrix} \mathbf{A}_{11}^{-1} + \mathbf{A}_{11}^{-1}\mathbf{A}_{12}\mathbf{T}^{-1}\mathbf{A}_{21}\mathbf{A}_{11}^{-1} & -\mathbf{A}_{11}^{-1}\mathbf{A}_{12}^{-1}\mathbf{T}^{-1} \\ -\mathbf{T}^{-1}\mathbf{A}_{21}\mathbf{A}_{11} & \mathbf{T}^{-1} \end{pmatrix} \text{ where} \tag{3.8}$$

$\mathbf{T} = \mathbf{A}_{22} - \mathbf{A}_{21}\mathbf{A}_{11}^{-1}\mathbf{A}_{12}$, $\mathbf{G} = (\mathbf{X}'\mathbf{X})^{-1}$, $\mathbf{F} = \mathbf{I} - \mathbf{GR}'(\mathbf{RGR}')^{-1}\mathbf{R}$, $\mathbf{F}^2 = \mathbf{F}$, and $\mathbf{E}$ is defined in (2.12). For details, see Timm and Carlson (1975), or Searle (1987, p. 311).

Using the likelihood ratio statistic one can again test $H(\mathbf{C}\beta = \xi)$; however, the degrees of freedom for the hypothesis is the rank($\mathbf{C}$) = g. No row of $\mathbf{C}$ may be dependent on the rows of $\mathbf{R}$, i.e., the intersection of $\mathbf{R}$ and $\mathbf{C}$ must be null. Having determined $\mathbf{E}_r$ and $\mathbf{H}_r$, one may test hypotheses again using the F-distribution defined in (2.14) by replacing $\mathbf{H}$ with $\mathbf{H}_r$, $\mathbf{E}$ with $\mathbf{E}_r$, v_h = rank($\mathbf{C}$) = g, and $v_e = n$ − rank($\mathbf{X}$) + rank($\mathbf{R}$). The $E(\mathbf{E}_r) = (n - k + s)\sigma^2$. To obtain confidence intervals for the parametric functions $\psi = \mathbf{c}'\beta$, we have that

$$\begin{aligned}
\hat{\psi} &= \mathbf{c}'\hat{\beta}_r \\
E(\hat{\psi}) &= \psi \\
\text{var}(\hat{\psi}) &= \sigma^2\mathbf{c}'(\mathbf{FGF}')\mathbf{c} = \sigma^2\mathbf{c}'(\mathbf{FG})\mathbf{c}
\end{aligned} \tag{3.9}$$

where an unbiased estimate s^2 of σ^2 obtained from (3.3) is $s^2 = n\hat{\sigma}^2_{ML_r} / (n - k + s)$, $\mathbf{G} = (\mathbf{X}'\mathbf{X})^{-1}$ and $\mathbf{F} = \mathbf{I} - \mathbf{GR}'(\mathbf{RGR}')^{-1}\mathbf{R}$.

3.3 Two-Way Factorial Design without Interaction

In this section we discuss the two-way factorial design without interaction, a restricted model. We first discuss the two-way factorial design with interaction, an unrestricted model, and then discuss the restricted model. The two-way design with interaction using the full rank representation can be written as

$$\begin{aligned}
y_{ijk} &= \mu_{ij} + e_{ijk} \qquad i = 1, \ldots, I; \; j = 1, \ldots, J; \; k = 1, \ldots, K_{ij} \\
e_{ijk} &\sim N(0, \sigma^2)
\end{aligned} \tag{3.10}$$

where K_{ij} represents the number of observations in cell ij, and $K_{ij} \geq 1$.

We consider the analysis of the data from Timm & Carlson (1975, p. 58). The data are shown in Table 3.1.

Table 3.1 *Two-way Design, Timm, and Carlson Data*

		Factor B			
		b_1	b_2	b_3	b_4
Factor A	a_1	3, 6, 3	3, 4, 5, 4, 3	7, 8, 7	6, 7, 8, 9, 8
	a_2	1, 2, 2, 2	2, 3, 4, 3	5, 6, 5, 6	10, 10, 9, 11

If we associate with each cell in the table the means μ_{ij}, we can write the design in Table 3.2.

Table 3.2 *Cell Means*

		Factor B				
		b_1	b_2	b_3	b_4	
Factor A	a_1	μ_{11}	μ_{12}	μ_{13}	μ_{14}	$\mu_{1.}$
	a_2	μ_{21}	μ_{22}	μ_{23}	μ_{24}	$\mu_{2.}$
		$\mu_{.1}$	$\mu_{.2}$	$\mu_{.3}$	$\mu_{.4}$	

The full rank representation of these data in the form of (1.1) is as follows:

$$\mathbf{y} = \mathbf{X}\beta + \mathbf{e} \tag{3.11}$$

$$\begin{bmatrix} 3 \\ 6 \\ 3 \\ 3 \\ 4 \\ 5 \\ \vdots \\ \vdots \\ \vdots \\ 10 \\ 10 \\ 9 \\ 11 \end{bmatrix} = \begin{bmatrix} 1 & 0 & 0 & 0 & 0 & 0 & 0 & 0 \\ 1 & 0 & 0 & 0 & 0 & 0 & 0 & 0 \\ 1 & 0 & 0 & 0 & 0 & 0 & 0 & 0 \\ 0 & 1 & 0 & 0 & 0 & 0 & 0 & 0 \\ 0 & 1 & 0 & 0 & 0 & 0 & 0 & 0 \\ 0 & 1 & 0 & 0 & 0 & 0 & 0 & 0 \\ \vdots & \vdots & \vdots & \vdots & \vdots & \vdots & \vdots & \vdots \\ \vdots & \vdots & \vdots & \vdots & \vdots & \vdots & \vdots & \vdots \\ \vdots & \vdots & \vdots & \vdots & \vdots & \vdots & \vdots & \vdots \\ 0 & 0 & 0 & 0 & 0 & 0 & 0 & 1 \\ 0 & 0 & 0 & 0 & 0 & 0 & 0 & 1 \\ 0 & 0 & 0 & 0 & 0 & 0 & 0 & 1 \\ 0 & 0 & 0 & 0 & 0 & 0 & 0 & 1 \end{bmatrix} \begin{bmatrix} \mu_{11} \\ \mu_{12} \\ \mu_{13} \\ \mu_{14} \\ \mu_{21} \\ \mu_{22} \\ \mu_{23} \\ \mu_{24} \end{bmatrix} + \begin{bmatrix} e_{111} \\ e_{112} \\ e_{113} \\ e_{121} \\ e_{122} \\ e_{123} \\ \vdots \\ \vdots \\ \vdots \\ e_{241} \\ e_{242} \\ e_{243} \\ e_{244} \end{bmatrix}$$

Now suppose we wish to test the hypothesis of no interaction. To do this in the form of the full rank model, we write the hypothesis as

$$H_{AB}: \ \mu_{ij} - \mu_{i'j} - \mu_{ij'} + \mu_{i'j'} = 0 \quad \text{all } i, j, i', j' \tag{3.12}$$

If no interaction exists, the difference between the means of i and i' at level j should be the same as the difference at level j'.

To construct the hypothesis test matrix **C**, we construct linearly independent rows in **C** of the form specified in H_{AB} (3.12). There are $(I-1)(J-1)$ independent functions of this form. One possible construction of **C** that we can use to test (3.12) for the data in Table 3.1 is

$$\mathbf{C}\beta = \mathbf{0} \tag{3.13}$$

$$\begin{bmatrix} 1 & -1 & 0 & 0 & -1 & 1 & 0 & 0 \\ 0 & 1 & -1 & 0 & 0 & -1 & 1 & 0 \\ 0 & 0 & 1 & -1 & 0 & 0 & -1 & 1 \end{bmatrix} \begin{bmatrix} \mu_{11} \\ \mu_{12} \\ \mu_{13} \\ \mu_{14} \\ \mu_{21} \\ \mu_{22} \\ \mu_{23} \\ \mu_{24} \end{bmatrix} = \begin{bmatrix} 0 \\ 0 \\ 0 \end{bmatrix}$$

which corresponds to

$$\begin{aligned} \mu_{11} - \mu_{21} - \mu_{12} + \mu_{22} &= 0 \\ \mu_{12} - \mu_{22} - \mu_{13} + \mu_{23} &= 0 \\ \mu_{13} - \mu_{23} - \mu_{14} + \mu_{24} &= 0 \end{aligned}$$

Program 3_3.sas contains the SAS code for testing the hypothesis H_{AB} as well as the code for testing other hypotheses discussed later in this section.

Program 3_3.sas

```
/* Program 3_3.sas                                    */
/* Two-Way Factorial Design, Unequal Cell Freq        */
/* with and without interaction                       */
/* data from Timm and Carlson (1975, page 58)         */

options ls=80 ps=60 nodate nonumber;
title 'Output 3.3: 2-Way Factorial Design';

data two;
   infile 'c:\3_3.dat';
   input a b y1 x11 x12 x13 x14 x21 x22 x23 x24;
proc print data=two;
run;

proc reg;
   title2 'Unrestricted: 2-Way w/Interaction (univar)--Proc Reg';
   model y1 = x11 x12 x13 x14 x21 x22 x23 x24 /noint;
   AB: mtest x11-x12-x21+x22=0,
             x12-x13-x22+x23=0,
             x13-x14-x23+x24=0 /print details;
   A: mtest x11+x12+x13+x14-x21-x22-x23-x24=0 /print details;
   B: mtest x11+x21-x14-x24=0,
            x12+x22-x14-x24=0,
            x13+x23-x14-x24=0 /print details;
   wtA: mtest .1875*x11+.3125*x12+.1875*x13+.3125*x14-
              .2500*x21-.2500*x22-.2500*x23-.2500*x24=0 /print details;
   wtB: mtest .4286*x11+.5714*x21-.5556*x14-.4444*x24=0,
              .5556*x12+.4444*x22-.5556*x14-.4444*x24=0,
              .4286*x13+.5714*x23-.5556*x14-.4444*x24=0 /print details;
run;
```

```
proc glm;
   title2 'Unrestricted: 2-Way w/Interaction (univar)--Proc GLM';
   class a b;
   model y1 = a b a*b /ss1 ss2 ss3;
proc glm;
   class a b;
   model y1 = b a a*b /ss1 ss2 ss3;
run;
proc reg;
   title2 'Restricted: 2-Way w/o Interaction (univar)--Proc Reg';
   model y1 = x11 x12 x13 x14 x21 x22 x23 x24 /noint;
   restrict x11 - x12 - x21 + x22 = 0,
            x12 - x13 - x22 + x23 = 0,
            x13 - x14 - x23 + x24 = 0;
   A: mtest x11+x12+x13+x14-x21-x22-x23-x24=0 /print details;
   B: mtest x11+x21-x14-x24=0,
            x12+x22-x14-x24=0,
            x13+x23-x14-x24=0 /print details;
   wtA: mtest .1875*x11+.3125*x12+.1875*x13+.3125*x14-
              .2500*x21-.2500*x22-.2500*x23-.2500*x24=0 /print details;
   wtB: mtest .4286*x11+.5714*x21-.5556*x14-.4444*x24=0,
              .5556*x12+.4444*x22-.5556*x14-.4444*x24=0,
              .4286*x13+.5714*x23-.5556*x14-.4444*x24=0 /print details;
run;
proc glm;
   title2 'Restricted: 2-Way w/o Interaction (univar)--Proc GLM';
   class a b;
   model y1 = a b /ss1;
proc glm;
   class a b;
   model y1 = b a /ss1;
run;
```

The data set is first read and then analyzed using PROC REG with the NOINT option specified to omit the intercept term from the model to obtain the full rank model. Notice that the independent variables specified in the MODEL statement are the columns of the design matrix, **X**. The hypothesis (3.13) is tested by using the MTEST statement to specify the rows of **C**, and is labeled `AB`.

Result and Interpretation 3_3.sas

Output 3.3: *2-Way Factorial Design Unrestricted: 2-Way w/Interaction (univar)--Proc Reg*

```
                    Output 3.3: 2-Way Factorial Design
           Unrestricted: 2-Way w/Interaction (univar)--Proc Reg

Multivariate Test: AB

                          L Ginv(X'X) L'    LB-cj

          1.0333333333          -0.45                 0           1.45
                 -0.45   1.0333333333      -0.583333333   -1.033333333
                     0   -0.583333333      1.0333333333   4.2333333333

                     Inv(L Ginv(X'X) L')    Inv()(LB-cj)

          1.3410138249   0.8571428571      0.4838709677   3.1071428571
          0.8571428571   1.9682539683      1.1111111111   3.9126984127
          0.4838709677   1.1111111111      1.5949820789   6.3055555556
```

Output 3.3 (*continued*)

```
                         E, the Error Matrix

                              20.416666667

                       H, the Hypothesis Matrix

                              27.155753968

                         T, the H + E Matrix

                              47.572420635

                             Eigenvectors

                               0.144984767

                             Eigenvalues

                              0.5708297708


           Multivariate Statistics and Exact F Statistics

                         S=1     M=0.5     N=11

Statistic                     Value          F       Num DF    Den DF  Pr > F

Wilks' Lambda              0.42917023    10.6406        3        24  0.0001
Pillai's Trace             0.57082977    10.6406        3        24  0.0001
Hotelling-Lawley Trace     1.33007775    10.6406        3        24  0.0001
Roy's Greatest Root        1.33007775    10.6406        3        24  0.0001
```

The results of the hypothesis test are found in Output 3.3, labeled `Multivariate Test: AB`. The sum of squares for the hypothesis if found in the output labeled `H, the Hypothesis Matrix`, the sum of squares error labeled `E, the Error Matrix`, the degrees of freedom, value of the F statistic, and probability are under the label `Multivariate Statistics and Exact F Statistics`. The results of this hypothesis test are summarized in Table 3.3.

To test hypotheses regarding the row effects or column effects in this unrestricted two-way design with interaction and unequal cell sizes, we can use either the unweighted or the weighted analysis. To test H_A: $\mu_{1.} = \mu_{2.}$ using the unweighted test we construct **C** as $\mathbf{C}_A = (1/4 \quad 1/4 \quad 1/4 \quad 1/4 \quad -1/4 \quad -1/4 \quad -1/4 \quad -1/4)$.

Alternately, we can test the following weighted hypothesis:

$$H_{A^*}: \quad \frac{\sum_j K_{1j}\mu_{1j}}{K_{1+}} = \frac{\sum_j K_{2j}\mu_{2j}}{K_{2+}} \tag{3.14}$$

which weights each cell mean according to the number of observations in the cell. The hypothesis test matrix **C** to test H_{A^*} in the above data set is

$$\mathbf{C}_A = (3/16 \quad 5/16 \quad 3/16 \quad 5/16 \quad -4/16 \quad -4/16 \quad -4/16 \quad -4/16)$$

Note that the weights are data dependent.

Program 3_3.sas includes the SAS code to test the above hypotheses, both weighted and unweighted. The hypothesis test matrices are specified using the MTEST statements commands and labeled `A`, `B`, `WTA`, and `WTB`. The results are given in the continuation Output 3.3 that follows.

Output 3.3 (*continued*)

```
                     Output 3.3:  2-Way Factorial Design
              Unrestricted: 2-Way W/Interaction (univar)-Proc Reg

                              Multivariate Test: A

                              L Ginv(X'X) L'    LB-cj

                             2.0666666667      2.4833333333

                         Inv(L Ginv(X'X) L')     Inv()(LB-cj)

                             0.4838709677      1.2016129032

                              E, the Error Matrix

                                    20.416666667

                             H, the Hypothesis Matrix

                                    2.9840053763

                              T, the H + E Matrix

                                    23.400672043

                                 Eigenvectors

                                    0.2067215892

                                 Eigenvalues

                                     0.127517935

                 Multivariate Statistics and Exact F Statistics

                              S=1    M=-0.5    N=11

  Statistic                        Value          F      Num DF    Den DF  Pr > F

  Wilks' Lambda               0.87248207     3.5077          1        24  0.0733
  Pillai's Trace              0.12751793     3.5077          1        24  0.0733
  Hotelling-Lawley Trace      0.14615537     3.5077          1        24  0.0733
  Roy's Greatest Root         0.14615537     3.5077          1        24  0.0733

Multivariate Test: B

                              L Ginv(X'X) L'    LB-cj

         1.0333333333           0.45              0.45          -11.85
                 0.45            0.9              0.45           -10.8
                 0.45           0.45      1.0333333333    -4.766666667
```

Output 3.3 *(continued)*

```
                    Inv(L Ginv(X'X) L')     Inv()(LB-cj)

        1.3410138249      -0.483870968      -0.373271889        -8.8859447
        -0.483870968      1.5949820789      -0.483870968      -9.185483871
        -0.373271889      -0.483870968      1.3410138249      3.2569124424

                            E, the Error Matrix

                                 20.416666667

                          H, the Hypothesis Matrix

                                 188.97705453

                            T, the H + E Matrix

                                  209.3937212

                               Eigenvectors

                                 0.0691063846

                               Eigenvalues

                                 0.9024962805


                Multivariate Statistics and Exact F Statistics

                            S=1    M=0.5    N=11

  Statistic                    Value          F      Num DF    Den DF  Pr > F

  Wilks' Lambda              0.09750372   74.0482         3        24  0.0001
  Pillai's Trace             0.90249628   74.0482         3        24  0.0001
  Hotelling-Lawley Trace     9.25601900   74.0482         3        24  0.0001
  Roy's Greatest Root        9.25601900   74.0482         3        24  0.0001


Multivariate Test: WTA

                           L Ginv(X'X) L'   LB-cj

                                 0.125            0.625

                    Inv(L Ginv(X'X) L')     Inv()(LB-cj)

                                   8                 5

                            E, the Error Matrix

                                 20.416666667

                          H, the Hypothesis Matrix

                                        3.125

                            T, the H + E Matrix

                                 23.541666667
```

Output 3.3 (*continued*)

```
                                 Eigenvectors

                                      0.206101616

                                  Eigenvalues

                                     0.1327433628

Multivariate Statistics and Exact F Statistics

                             S=1    M=-0.5    N=11

  Statistic                     Value            F       Num DF    Den DF  Pr > F

  Wilks' Lambda              0.86725664     3.6735          1        24  0.0673
  Pillai's Trace             0.13274336     3.6735          1        24  0.0673
  Hotelling-Lawley Trace     0.15306122     3.6735          1        24  0.0673
  Roy's Greatest Root        0.15306122     3.6735          1        24  0.0673


Multivariate Test: WTB


                             L Ginv(X'X) L'   LB-cj

          0.2539682553       0.111111112       0.111111112          -5.95221
           0.111111112       0.222222224       0.111111112          -5.22208
           0.111111112       0.111111112      0.2539682553      -2.380793333

                      Inv(L Ginv(X'X) L')     Inv()(LB-cj)

          5.4687499778      -1.968749989      -1.531249999      -18.62458857
          -1.968749989      6.4687499534      -1.968749989      -17.37472953
          -1.531249999      -1.968749989      5.4687499778      6.3753280121

                              E, the Error Matrix

                                     20.416666667

                            H, the Hypothesis Matrix

                                     186.41135151

                              T, the H + E Matrix

                                     206.82801817

                                  Eigenvectors

                                      0.069533696

                                  Eigenvalues

                                     0.9012867461
```

Output 3.3 (*continued*)

```
              Multivariate Statistics and Exact F Statistics

                          S=1    M=0.5    N=11

Statistic                     Value            F      Num DF    Den DF  Pr > F

Wilks' Lambda             0.09871325     73.0428         3        24  0.0001
Pillai's Trace            0.90128675     73.0428         3        24  0.0001
Hotelling-Lawley Trace    9.13035191     73.0428         3        24  0.0001
Roy's Greatest Root       9.13035191     73.0428         3        24  0.0001
```

The results of the hypothesis tests are summarized in Table 3.3.

Table 3.3 *Hypotheses Test Results, Unrestricted*

Hypothesis	Reduction Notation	SS	df	MS	F	P
H_{AB}	$R(AB \mid \mu)$	27.16	3	9.05	10.64	<.0001
H_A	$R(A \mid \mu,B)$	2.98	1	2.98	3.51	.0733
H_B	$R(B \mid \mu,A)$	188.98	3	62.99	74.05	<.0001
H_{A^*}	$R(A \mid \mu)$	3.13	1	3.13	3.67	.0673
H_{B^*}	$R(B \mid \mu)$	186.42	3	62.14	73.04	<.0001
Error		20.42	24	.85		

The analysis could also be performed using PROC GLM and the same results obtained. The second section of Program 3_3.sas contains the code for this analysis. Note that one must run PROC GLM two times, once `with model` $y1 = a\ b\ a*b$ and then with `model` $y1 = b\ a\ a*b$ in order to obtain the weighted tests for both A and B. The results of this analysis are given in the continuation of Output 3.3 that follows. PROC GLM output is discussed in Chapter 5.

Output 3.3 (*continued*)

```
                    Output 3.3:  2-way Factorial Design
               Unrestricted:  2-way w/Interaction (univar)-PROC GLM
                      General Linear Models Procedure

Dependent Variable: Y1
                                         Sum of             Mean
Source                        DF        Squares           Square   F Value     Pr > F

Model                          7   215.08333333      30.72619048     36.12     0.0001

Error                         24    20.41666667       0.85069444

Corrected Total               31   235.50000000
```

Output 3.3 (*continued*)

```
                 R-Square            C.V.         Root MSE               Y1 Mean

                 0.913305        17.15965        0.9223310             5.3750000

Source                 DF      Type I SS      Mean Square   F Value      Pr > F

A                       1     3.12500000       3.12500000      3.67      0.0673
B                       3   184.80257937      61.60085979     72.41      0.0001
A*B                     3    27.15575397       9.05191799     10.64      0.0001

Source                 DF     Type II SS      Mean Square   F Value      Pr > F

A                       1     1.50694444       1.50694444      1.77      0.1957
B                       3   184.80257937      61.60085979     72.41      0.0001
A*B                     3    27.15575397       9.05191799     10.64      0.0001

Source                 DF    Type III SS      Mean Square   F Value      Pr > F

A                       1     2.98400538       2.98400538      3.51      0.0733
B                       3   188.97705453      62.99235151     74.05      0.0001
A*B                     3    27.15575397       9.05191799     10.64      0.0001

                       General Linear Models Procedure

Dependent Variable: Y1
                                     Sum of             Mean
Source                 DF           Squares           Square   F Value      Pr > F

Model                   7      215.08333333      30.72619048     36.12      0.0001

Error                  24       20.41666667       0.85069444

Corrected Total        31      235.50000000

                 R-Square            C.V.         Root MSE               Y1 Mean

                 0.913305        17.15965        0.9223310             5.3750000

Source                 DF      Type I SS      Mean Square   F Value      Pr > F

B                       3   186.42063492      62.14021164     73.05      0.0001
A                       1     1.50694444       1.50694444      1.77      0.1957
A*B                     3    27.15575397       9.05191799     10.64      0.0001

Source                 DF     Type II SS      Mean Square   F Value      Pr > F

B                       3   184.80257937      61.60085979     72.41      0.0001
A                       1     1.50694444       1.50694444      1.77      0.1957
A*B                     3    27.15575397       9.05191799     10.64      0.0001

Source                 DF    Type III SS      Mean Square   F Value      Pr > F

B                       3   188.97705453      62.99235151     74.05      0.0001
A                       1     2.98400538       2.98400538      3.51      0.0733
A*B                     3    27.15575397       9.05191799     10.64      0.0001
```

The results in Table 3.3 are given in the PROC GLM output, among the Type I and III sums of square results.

In the previous example the two-way design with interaction was an unrestricted design. Suppose now we wish to analyze the same data, but this time assume there is no interaction. This requires placing restrictions on the parameter vector where the restrictions $\mathbf{R}\beta = \mathbf{0}$ are now considered part of the model. The full rank specification of the two-way design without interaction is

$$
\begin{aligned}
&y_{ijk} = \mu_{ij} + e_{ijk} && i = 1,\ldots,I;\quad j = 1,\ldots,J;\quad k = 1,\ldots,K_{ij} \\
&\mu_{ij} - \mu_{i'j} - \mu_{ij'} + \mu_{i'j'} = 0 && \text{for a linearly independent set } (I-1)(J-1) \\
&e_{ijk} \sim N(0,\sigma^2)
\end{aligned}
\tag{3.15}
$$

To represent the restrictions in this design with the data in Table 3.1, we write $\mathbf{R}\beta = \theta$ where

$$
\mathbf{R} = \begin{bmatrix} 1 & -1 & 0 & 0 & -1 & 1 & 0 & 0 \\ 0 & 1 & -1 & 0 & 0 & -1 & 1 & 0 \\ 0 & 0 & 1 & -1 & 0 & 0 & -1 & 1 \end{bmatrix} \tag{3.16}
$$

$\beta = \mu$, and $\theta = \mathbf{0}$. Note that this restriction matrix $\mathbf{R}$ is identical to the hypothesis test matrix $\mathbf{C}$ of (3.13) to test H_{AB}.

The SAS code for the analysis of the data in Table 3.1 using the restricted model is given in Program 3_3.sas. Again PROC REG is used, but now the RESTRICT statement is added where the rows specified in this statement correspond to the rows in (3.16).

To test hypotheses with the restricted model, we construct hypothesis test matrices $\mathbf{C}$, of rank g, where the rows of $\mathbf{C}$ are not dependent on the rows of $\mathbf{R}$. Again, because of the unequal cell frequencies, we can use either a weighted or an unweighted hypothesis test of equal row means (or column means).

To test H_A: all $\mu_{i'.s}$ equal, which is the unweighted test, $\mathbf{Q} = \binom{\mathbf{R}}{\mathbf{C}}$, and η are as follows:

$$
\mathbf{Q} = \begin{bmatrix} \mathbf{R} \\ \\ \\ \cdots \\ \mathbf{C} \end{bmatrix} = \begin{bmatrix} 1 & -1 & 0 & 0 & -1 & 1 & 0 & 0 \\ 0 & 1 & -1 & 0 & 0 & -1 & 1 & 0 \\ 0 & 0 & 1 & -1 & 0 & 0 & -1 & 1 \\ \cdots & \cdots & \cdots & \cdots & \cdots & \cdots & \cdots & \cdots \\ 1/4 & 1/4 & 1/4 & 1/4 & -1/4 & -1/4 & -1/4 & -1/4 \end{bmatrix} \tag{3.17}
$$

$$
\eta = \begin{bmatrix} \theta \\ \\ \\ \cdots \\ \xi \end{bmatrix} = \begin{bmatrix} 0 \\ 0 \\ 0 \\ \cdots \\ 0 \end{bmatrix}
$$

Similarly, one can test H_B: all $\mu_{.j's}$ are equal.

Weighted tests, H_{A^*} and H_{B^*} can also be performed, where the weights are based upon cell frequencies. For the weighted test of A in the data set above, the weights are as follows:

$$
\mathbf{C} = (3/16 \quad 4/16 \quad 3/16 \quad 5/16 \quad -4/16 \quad -4/16 \quad -4/16 \quad -4/16)
$$

(see Timm and Carlson, 1975). The results of these hypotheses are given in Output 3.3, treated in the same manner as in the previous unrestricted analyses.

Output 3.3 (*continued*)

```
                  Output 3.3:  2-way Factorial Design
          Restricted:  2-way w/o Interaction (univar)--PROC GLM

Multivariate Test: A

                                L Ginv(X'X) L'   LB-cj

                         2.0322580645             1.75

                   Inv(L Ginv(X'X) L')     Inv()(LB-cj)

                         0.4920634921      0.8611111111

                          E, the Error Matrix

                                47.572420635

                         H, the Hypothesis Matrix

                                1.5069444444

                          T, the H + E Matrix

                                49.079365079

                            Eigenvectors

                                0.1427415907

                            Eigenvalues

                                0.0307042367


            Multivariate Statistics and Exact F Statistics

                         S=1    M=-0.5    N=12.5

  Statistic                    Value          F      Num DF    Den DF  Pr > F

  Wilks' Lambda              0.96929576    0.8553         1        27  0.3633
  Pillai's Trace             0.03070424    0.8553         1        27  0.3633
  Hotelling-Lawley Trace     0.03167685    0.8553         1        27  0.3633
  Roy's Greatest Root        0.03167685    0.8553         1        27  0.3633

Multivariate Test: B

                         L Ginv(X'X) L'   LB-cj

        1.0240655402     0.4444444444     0.4526369688     -11.79365079
        0.4444444444     0.8888888889     0.4444444444     -10.44444444
        0.4526369688     0.4444444444     1.0240655402     -4.650793651

                   Inv(L Ginv(X'X) L')     Inv()(LB-cj)

            1.359375        -0.484375        -0.390625          -9.15625
           -0.484375         1.609375        -0.484375          -8.84375
           -0.390625        -0.484375         1.359375           3.34375
```

Output 3.3 (*continued*)

```
                          E, the Error Matrix

                                47.572420635

H, the Hypothesis Matrix

                                184.80257937

                          T, the H + E Matrix

                                     232.375

                              Eigenvectors

                                0.0656002204

                              Eigenvalues

                                0.7952773722

                 Multivariate Statistics and Exact F Statistics

                         S=1    M=0.5    N=12.5

  Statistic                   Value          F       Num DF     Den DF   Pr > F

  Wilks' Lambda             0.20472263    34.9619         3         27   0.0001
  Pillai's Trace            0.79527737    34.9619         3         27   0.0001
  Hotelling-Lawley Trace    3.88465789    34.9619         3         27   0.0001
  Roy's Greatest Root       3.88465789    34.9619         3         27   0.0001


Multivariate Test: WTA

                         L Ginv(X'X) L'    LB-cj

                              0.125              0.625

                    Inv(L Ginv(X'X) L')     Inv()(LB-cj)

                                 8                  5

                          E, the Error Matrix

                                47.572420635

                        H, the Hypothesis Matrix

                                       3.125

                             H + E Matrix

                                50.697420635

                              Eigenvectors

                                 0.140445254
```

Output 3.3 (*continued*)

```
                                  Eigenvalues

                                      0.0616402168

Multivariate Statistics and Exact F Statistics

                           S=1    M=-0.5    N=12.5

  Statistic                    Value          F      Num DF    Den DF  Pr > F

  Wilks' Lambda             0.93835978     1.7736         1        27  0.1941
  Pillai's Trace            0.06164022     1.7736         1        27  0.1941
  Hotelling-Lawley Trace    0.06568932     1.7736         1        27  0.1941
  Roy's Greatest Root       0.06568932     1.7736         1        27  0.1941

Multivariate Test: WTB

                            L Ginv(X'X) L'   LB-cj

          0.253968254       0.1111111111       0.1111111111       -5.952387897
         0.1111111111       0.2222222222       0.1111111111       -5.222222222
         0.1111111111       0.1111111111        0.253968254       -2.380959325

                     Inv(L Ginv(X'X) L')     Inv()(LB-cj)

         5.4687499995           -1.96875           -1.53125       -18.62502734
             -1.96875       6.4687499995           -1.96875       -17.37497266
             -1.53125           -1.96875       5.4687499995       6.3749726578

                             E, the Error Matrix

                                  47.572420635

                           H, the Hypothesis Matrix

                                  186.42080506

                             T, the H + E Matrix

                                  233.99322569

                                Eigenvectors

                                  0.0653729913

                                Eigenvalues

                                  0.7966931714


                Multivariate Statistics and Exact F Statistics

                           S=1    M=0.5    N=12.5

  Statistic                    Value          F      Num DF    Den DF  Pr > F

  Wilks' Lambda             0.20330683    35.2681         3        27  0.0001
  Pillai's Trace            0.79669317    35.2681         3        27  0.0001
  Hotelling-Lawley Trace    3.91867394    35.2681         3        27  0.0001
  Roy's Greatest Root       3.91867394    35.2681         3        27  0.0001
```

The results are summarized in Table 3.4.

Table 3.4 *Hypotheses Test Results, Restricted*

Hypothesis	Reduction Notation	SS	df	MS	F	P
H_A	$R(A \mid \mu B)$	1.51	1	1.51	.86	.3633
H_B	$R(B \mid \mu A)$	184.80	3	61.60	34.96	<.0001
H_{A^*}	$R(A \mid \mu)$	3.125	1	3.125	1.77	.1941
H_{B^*}	$R(B \mid \mu)$	186.42	3	62.14	35.27	<.0001
Error		47.57	27	1.76		

Again, these hypotheses could all be tested using PROC GLM, as was the case above for the unrestricted model. Program 3_3.sas contains the SAS code to perform the PROC GLM analysis. The results are given in the continuation of Output 3.3 that follows.

Output 3.3 (*continued*)

```
                  Output 3.3:  2-way Factorial Design
           Restricted:  2-way w/o Interaction (univar)--PROC GLM

                     General Linear Models Procedure

Dependent Variable: Y1
                                      Sum of           Mean
Source                  DF           Squares         Square   F Value     Pr > F

Model                    4      187.92757937     46.98189484     26.66     0.0001

Error                   27       47.57242063      1.76194150

Corrected Total         31      235.50000000

                  R-Square              C.V.        Root MSE             Y1 Mean

                  0.797994          24.69547       1.3273814           5.3750000


Source                  DF         Type I SS     Mean Square   F Value     Pr > F

A                        1        3.12500000      3.12500000      1.77     0.1941
B                        3      184.80257937     61.60085979     34.96     0.0001

                     General Linear Models Procedure

Dependent Variable: Y1
                                      Sum of           Mean
Source                  DF           Squares         Square   F Value     Pr > F

Model                    4      187.92757937     46.98189484     26.66     0.0001

Error                   27       47.57242063      1.76194150

Corrected Total         31      235.50000000
```

Output 3.3 (*continued*)

	R-Square	C.V.	Root MSE		Y1 Mean
	0.797994	24.69547	1.3273814		5.3750000
Source	DF	Type I SS	Mean Square	F Value	Pr > F
B	3	186.42063492	62.14021164	35.27	0.0001
A	1	1.50694444	1.50694444	0.86	0.3633

The results obtained using PROC GLM are identical to those obtained using PROC REG.

Note that in the analysis of these data a test for interactions was included which was in fact significant. Thus, we would not have, in practice, analyzed these data with the restriction of no interaction.

3.4 Latin Square Designs

In the previous example we discussed the two-way factorial design. In that design, both the row and the column effects were of equal interest. At times it is the case that only one of the variables, say A, is of interest and the other variable, say B, is a blocking variable included in the design to help explain variation and thus reduce the error variance. Analysis of a completely randomized block design is the same as for the previous factorial design.

When the researcher wishes to include two blocking variables in the design and the number of levels of the treatment (C) and the two blocking variables (A and B) are all equivalent, say m, the $m \times m$ Latin square design is one possible design that can be utilized. In the Latin square arrangement, each row (A) and each column (B) receives each treatment (each level of C) only once (Box, Hunter, and Hunter, 1978). Levels of factors A and B also occur together exactly once and thus the design is completely balanced. In the usual situation for Latin square designs there is only one observation in each cell, only one replication, as Table 3.3 illustrates.

Table 3.5 *3x3 Latin Square Design*

	b_1	b_2	b_3
a_1	C_2	C_3	C_1
a_2	C_3	C_1	C_2
a_3	C_1	C_2	C_3

There are 11 other ways of arranging three treatments in a 3 by 3 square and the researcher randomly selects one of the 12 possible arrangements.

There are larger Latin square designs, for example, 4 by 4 and 5 by 5, but there is no 2 by 2 design with one observation per cell. This is so because there would be no way of estimating the error variance in such a design. This can be seen from examination of the degrees of freedom. With one experimental unit per cell in an $m \times m$ Latin square design, the total number of degrees of freedom is $m^2 - 1$. With three factors of m levels each, the number of degrees of freedom for the main effects is $3(m-1)$. Thus, the number of degrees of freedom remaining for the error term is $(m^2-1)-3(m-1)=(m-1)(m-2)$. Note that this equals zero when $m=2$. If there were more than one observation per cell, the within cell variation could can be used to estimate the error variation and the $(m-1)(m-2)$ degrees of freedom could can be used as a partial test of the interactions. With one observation per cell, however, we must be able to assume no interaction, additivity, and we must use the restricted full rank model with $(m-1)(m-2)$ restrictions in most instances. Strictly speaking, some degrees of freedom could be used in a partial test of additivity, but it is not possible to test all interactions without using a complete design with multiple observations per cell.

The full rank representation of the $m \times m$ Latin square design is

$$\begin{aligned} & y_{ijk} = \mu_{ijk} + e_{ijk} && \text{for the } m^2 \text{ means defined by the arrangement} \\ & \text{restriction:} && \text{no interactions} \end{aligned} \tag{3.18}$$

We assume that the effects are additive and thus we place the restriction of no interaction on the model. The analysis of this design is carried out identically to a three-way factorial design with no interaction and empty cells (see Timm, 1975, for a discussion and illustration of a Latin square using the less than full rank model).

As an example of the analysis of a 3×3 Latin square design using SAS, and the full rank model, we analyze the data in Table 3.6.

Table 3.6 3×3 *Latin Square Data*

	b_1	b_2	b_3
a_1	C_2 7	C_3 5	C_1 4
a_2	C_3 5	C_1 6	C_2 11
a_3	C_1 8	C_2 8	C_3 9

Here, factors *A* and *B* are blocking factors, and *C* is a treatment factor. The numbers within the cells are the values obtained on the dependent variable.

Representing this design in terms of (1.1) we have

$$\mathbf{y} = \mathbf{X}\beta + \mathbf{e} \tag{3.19}$$

$$\begin{bmatrix} 7 \\ 5 \\ 4 \\ 5 \\ 6 \\ 11 \\ 8 \\ 8 \\ 9 \end{bmatrix} = \begin{bmatrix} 1 & 0 & 0 & 0 & 0 & 0 & 0 & 0 & 0 \\ 0 & 1 & 0 & 0 & 0 & 0 & 0 & 0 & 0 \\ 0 & 0 & 1 & 0 & 0 & 0 & 0 & 0 & 0 \\ 0 & 0 & 0 & 1 & 0 & 0 & 0 & 0 & 0 \\ 0 & 0 & 0 & 0 & 1 & 0 & 0 & 0 & 0 \\ 0 & 0 & 0 & 0 & 0 & 1 & 0 & 0 & 0 \\ 0 & 0 & 0 & 0 & 0 & 0 & 1 & 0 & 0 \\ 0 & 0 & 0 & 0 & 0 & 0 & 0 & 1 & 0 \\ 0 & 0 & 0 & 0 & 0 & 0 & 0 & 0 & 1 \end{bmatrix} \begin{bmatrix} \mu_{112} \\ \mu_{123} \\ \mu_{131} \\ \mu_{213} \\ \mu_{221} \\ \mu_{232} \\ \mu_{311} \\ \mu_{322} \\ \mu_{333} \end{bmatrix} + \begin{bmatrix} e_{112} \\ e_{123} \\ e_{131} \\ e_{213} \\ e_{221} \\ e_{232} \\ e_{311} \\ e_{322} \\ e_{333} \end{bmatrix}$$

A matrix of restrictions, **R**, that can be used for the analysis of this design is

$$\mathbf{R} = \begin{bmatrix} 1 & 0 & -1 & -1 & 1 & 0 & 0 & -1 & 1 \\ -1 & 1 & 0 & 0 & -1 & 1 & 1 & 0 & -1 \end{bmatrix} \quad (3.20)$$

The number of degrees of freedom for the residual is $(m-1)(m-2) = (3-1)(3-2) = 2$; this is the rank of **R**. Again, the restriction matrix represents a set of $(m-1)(m-2)$ linearly independent equations representing the assumption of no interaction.

There are three main effect hypotheses to be tested in a Latin square design:

$$\begin{aligned} H_A&: \text{ all } \mu_{i..'s} \text{ are equal} \\ H_B&: \text{ all } \mu_{.j.'s} \text{ are equal} \\ H_C&: \text{ all } \mu_{..k's} \text{ are equal} \end{aligned}$$

where the $\mu_{i..'s}$, $\mu_{.j.'s}$, and $\mu_{..k's}$ are simple averages, for example, $\mu_{1..} = (\mu_{112} + \mu_{123} + \mu_{131})/3$ in the previous design. Because the Latin square design is balanced, these hypotheses are orthogonal. To test the above main effect hypotheses, we use the following hypothesis test matrices:

$$\mathbf{C}_A = \begin{bmatrix} 1/3 & 1/3 & 1/3 & 0 & 0 & 0 & -1/3 & -1/3 & -1/3 \\ 0 & 0 & 0 & 1/3 & 1/3 & 1/3 & -1/3 & -1/3 & -1/3 \end{bmatrix} \quad (3.21)$$

$$\mathbf{C}_B = \begin{bmatrix} 1/3 & 0 & -1/3 & 1/3 & 0 & -1/3 & 1/3 & 0 & -1/3 \\ 0 & 1/3 & -1/3 & 0 & 1/3 & -1/3 & 0 & 1/3 & -1/3 \end{bmatrix} \quad (3.22)$$

$$\mathbf{C}_C = \begin{bmatrix} 0 & -1/3 & 1/3 & -1/3 & 1/3 & 0 & 1/3 & 0 & -1/3 \\ 1/3 & -1/3 & 0 & -1/3 & 0 & 1/3 & 0 & 1/3 & -1/3 \end{bmatrix} \quad (3.23)$$

Program 3_4.sas contains the SAS code to analyze these data using PROC REG.

Program 3_4.sas

```
/* Program 3_4.sas                      */
/* Latin Square Design, Full Rank Model */

options ls=80 ps=60 nodate nonumber;
title 'Output 3.4: Latin Square Design, Full Rank Model';

data Latin;
   infile 'c:\3_4.dat';
   input a b c y1 x112 x123 x131 x213 x221 x232 x311 x322 x333;
proc print data=latin;
run;

proc reg;
  model y1 = x112 x123 x131 x213 x221 x232 x311 x322 x333 /noint;
   restrict  x112 - x131 - x213 + x221 - x322 + x333 = 0,
            -x112 + x123 - x221 + x232 + x311 - x333 = 0;
   A: mtest  x112 + x123 + x131 - x311 - x322 - x333 = 0,
             x213 + x221 + x232 - x311 - x322 - x333 = 0 /print details;
   B: mtest  x112 - x131 + x213 - x232 + x311 - x333 = 0,
             x123 - x131 + x221 - x232 + x322 - x333 = 0 /print details;
```

```
C: mtest -x123 + x131 - x213 + x221 + x311 - x333 = 0,
             x112 - x123 - x213 + x232 + x322 - x333 = 0 /print details;
run;
```

First the data are read and then PROC REG is used with the NOINT option to omit the intercept from the model so that the full rank model is used. The columns of the design matrix **X** of (3.19) are included in the MODEL statement as the independent variables. The RESTRICT and MTEST statements are used as in previous examples to specify the rows of the restriction matrix, **R**, and hypothesis test matrix, **C**, respectively. The results are given in Output 3.4.

Result and Interpretation 3_4.sas

Output 3.4: *Latin Square Design, Full Rank Model*

```
                 Output 3.4: Latin Square Design, Full Rank Model

Multivariate Test: A

                             L Ginv(X'X) L'    LB-cj

                              6                 3                -9
                              3                 6                -3

                      Inv(L Ginv(X'X) L')     Inv()(LB-cj)

                    0.2222222222       -0.111111111       -1.666666667
                    -0.111111111       0.2222222222       0.3333333333

                              E, the Error Matrix

                                     8.6666666667

                            H, the Hypothesis Matrix

                                               14

                              T, the H + E Matrix

                                     22.666666667

                                  Eigenvectors

                                     0.2100420126

                                  Eigenvalues

                                     0.6176470588

                 Multivariate Statistics and Exact F Statistics

                               S=1     M=0     N=0

  Statistic                     Value            F       Num DF     Den DF   Pr > F

  Wilks' Lambda              0.38235294     1.6154           2          2   0.3824
  Pillai's Trace             0.61764706     1.6154           2          2   0.3824
  Hotelling-Lawley Trace     1.61538462     1.6154           2          2   0.3824
  Roy's Greatest Root        1.61538462     1.6154           2          2   0.3824
```

Output 3.4 *(continued)*

```
Multivariate Test: B

                        L Ginv(X'X) L'    LB-cj


                        6                3               -4
                        3                6               -5

                    Inv(L Ginv(X'X) L')      Inv()(LB-cj)

                  0.2222222222      -0.111111111      -0.333333333
                  -0.111111111      0.2222222222      -0.666666667

                         E, the Error Matrix


                              8.6666666667


                        H, the Hypothesis Matrix


                              4.6666666667


                         T, the H + E Matrix


                              13.333333333

                            Eigenvectors

                              0.2738612788

                            Eigenvalues

                                      0.35

                Multivariate Statistics and Exact F Statistics

                          S=1      M=0      N=0
```

Statistic	Value	F	Num DF	Den DF	Pr > F
Wilks' Lambda	0.65000000	0.5385	2	2	0.6500
Pillai's Trace	0.35000000	0.5385	2	2	0.6500
Hotelling-Lawley Trace	0.53846154	0.5385	2	2	0.6500
Roy's Greatest Root	0.53846154	0.5385	2	2	0.6500

Output 3.4 *(continued)*

```
Multivariate Test: C

                         L Ginv(X'X) L'    LB-cj

                           6                3                 -1
                           3                6                  7


                  Inv(L Ginv(X'X) L')     Inv()(LB-cj)

                0.2222222222       -0.111111111                -1
                -0.111111111       0.2222222222       1.6666666667

                          E, the Error Matrix

                                8.6666666667

                         H, the Hypothesis Matrix

                                12.666666667

                          T, the H + E Matrix

                                21.333333333

                              Eigenvectors

                                0.2165063509

                              Eigenvalues

                                     0.59375

              Multivariate Statistics and Exact F Statistics

                          S=1      M=0      N=0

 Statistic                     Value           F        Num DF     Den DF   Pr > F

 Wilks' Lambda              0.40625000      1.4615           2          2  0.4063
 Pillai's Trace             0.59375000      1.4615           2          2  0.4063
 Hotelling-Lawley Trace     1.46153846      1.4615           2          2  0.4063
```

The values of the sums of squares, df, F, and p are given in the output as they were in the analyses of Section 3.3. Table 3.7 summarizes the results of the tests.

Table 3.7 *Hypothesis Test Results, Latin Square*

Hypothesis	SS	df	MS	F	p
H_A	14.00	2	7.0	1.62	.3824
H_B	4.67	2	2.34	.54	.6500
H_C	12.67	2	6.34	1.46	.4063
Error	8.67	2	4.34		

3.5 Split Plot Repeated Measures Designs

In educational and behavioral research, differences among subjects may be so large relative to differences in treatment effects that the treatment effects are obscured by subject heterogeneity. One way to reduce this variability that is due to individual differences is to divide subjects into blocks on the basis of a concomitant variable correlated with the dependent variable and then to randomly assign the subjects within blocks to treatments. This reduces the size of the error variance and makes it easier to detect differences in treatments. To further separate true error variance from variability that is due to individual differences, each subject may serve as its own control if we take repeated measurements on a subject over several levels of an independent treatment variable where the order of presentation is randomized independently for each subject. Such designs are called one-sample repeated measures designs.

Repeated measurement designs have increased precision over completely randomized and randomized block designs and require fewer subjects. There are many experimental situations in educational and behavioral research for which this design is appropriate. For example, it is a natural choice when one is concerned with analyzing performance over time.

3.5.1 Univariate Mixed-Model ANOVA, Full Rank Representation

Extending the notion of repeated measures to experiments having more than one treatment leads us to a split-plot repeated measures design. In the next example of the general linear model with restrictions, we discuss the univariate split-plot repeated measures design, also known in the literature as a two-sample profile analysis (Timm, 1975) or a mixed within-subjects factorial design (Keppel, 1991). The name split-plot comes from agricultural experiments where each level of a treatment is applied to a large plot of land, and all levels of a second treatment are applied each to a subplot within the large plot (Winer, 1971).

A simple example of such a design involves randomly assigning k subjects to I treatments, labeled factor A, and observing the subjects repeatedly over J levels of a second factor B as illustrated in Table 3.8.

Table 3.8 *Simple Split-Plot Design*

	a_1		
Subjects	b_1	b_2	b_3
S_1	Y_{111}	Y_{121}	Y_{131}
S_2	Y_{112}	Y_{121}	Y_{132}

	a_2		
Subjects	b_1	b_2	b_3
S_3	Y_{213}	Y_{223}	Y_{233}
S_4	Y_{214}	Y_{224}	Y_{234}

In this chapter we analyze the split-plot repeated measures design as a univariate model. This can be done if we can assume circularity of the variance covariance matrix (Section 3.5.3). At the end of this chapter we discuss ways of testing assumptions of covariance structures. In Chapter 5 we present the multivariate analysis of the repeated measures design, which does not require the assumption of circularity; in Chapter 9 the design is analyzed as a general linear mixed model.

The full rank model with restrictions for the split-plot design in Table 3.8 is

$$y_{ijk} = \mu_{ij} + e_{ijk} \quad i = 1,\ldots,I;\ j = 1,\ldots,J;\ k = 1,\ldots,K_{ij} \tag{3.24}$$

restrictions: no interaction between S and B within each level of A.

The parameters of the design are as follows:

Table 3.9 *Simple Split-Plot Cell Means*

	a_1			
Subject	b_1	b_2	b_3	
S_1	μ_{111}	μ_{121}	μ_{131}	$\mu_{1.1}$
S_2	μ_{112}	μ_{122}	μ_{132}	$\mu_{1.2}$
	$\mu_{11.}$	$\mu_{12.}$	$\mu_{13.}$	$\mu_{1..}$

	a_2			
Subject	b_1	b_2	b_3	
S_3	μ_{213}	μ_{223}	μ_{233}	$\mu_{2.3}$
S_4	μ_{214}	μ_{224}	μ_{234}	$\mu_{2.4}$
	$\mu_{21.}$	$\mu_{22.}$	$\mu_{23.}$	$\mu_{2..}$
	$\mu_{.1.}$	$\mu_{.2.}$	$\mu_{.3.}$	$\mu_{...}$

For the model that is given in (3. 24) and represented in Table 3.9, there are several hypotheses of interest:

$$\begin{array}{ll} H_A:\ \mu_{1..} = \mu_{2..} & \text{test of group difference} \\ H_{(A)S}\mu_{1.1} = \mu_{1.2} & \\ \quad \mu_{2.1} = \mu_{2.2} & \text{test for subject differences within treatment} \\ H_B:\ \mu_{.1.} = \mu_{.2.} = \mu_{.3.} & \text{test for differences among conditions} \\ H_{AB}: \mu_{11.} - \mu_{21.} - \mu_{12.} + \mu_{22.} = 0 & \\ \quad \mu_{12.} - \mu_{22.} - \mu_{13.} + \mu_{23.} = 0 & \text{test for parallelism} \end{array} \tag{3.25}$$

These hypotheses must be tested subject to the restriction that there is no interaction between S and B within each level of A. The following represents this restriction

$$\begin{aligned} \mu_{111} - \mu_{112} - \mu_{121} + \mu_{122} &= 0 \\ \mu_{121} - \mu_{122} - \mu_{131} + \mu_{132} &= 0 \\ \mu_{211} - \mu_{212} - \mu_{221} + \mu_{222} &= 0 \\ \mu_{221} - \mu_{222} - \mu_{213} + \mu_{232} &= 0 \end{aligned} \tag{3.26}$$

To test the hypotheses in (3.25) using the full rank model, we associate a hypothesis test matrix **C** with each hypothesis. However, for the split-plot design the factor associated with subjects is random rather than fixed. To determine the appropriate F-ratios we now assume

$$\mu_{ijk} = \mu_{ij} + s_{(i)k} \tag{3.27}$$

where the μ_{ij} are constants and the $s_{(i)k}$ are random components jointly independent of the $e_{(i)jk}$. In addition, we assume that the $s_{(i)k} \sim IN(0, \rho\sigma^2)$ and $e_{(i)jk} \sim IN(0,(1-\rho)\sigma^2)$. Letting

$$
\begin{aligned}
y_{i..} &= \mu_{i.} + s_{i.} + e_{i..} \\
y_{.j.} &= \mu_{.j} + s_{..} + e_{.j.} \\
y_{..k} &= \mu_{..} + s_{.k} + e_{..k} \\
y_{ij.} &= \mu_{.j} + s_{..} + e_{.j.} \\
y_{i.k} &= \mu_{ij} + s_{i.} + e_{ij.} \\
y_{...} &= \mu_{..} + s_{..} + e_{...}
\end{aligned}
$$

and substituting these into expressions for the sums of squares for each hypothesis and error, we summarize the expected mean squares for the hypotheses given in (3.25) in Table 3.10.

Table 3.10 *Expected Mean Squares for the Split-Plot Repeated Measures Design*

Hypotheses	df	E(MS)	F ratios
H_A [1]	I -1	$\sigma_e^2 + J\sigma_s^2 + \sigma_A^2/(I-1)$	[1]/[2]
$H_{(A)S}$ [2]	I(k-1)	$\sigma_e^2 + J\sigma_s^2$	
H_B [3]	(J-1)	$\sigma_e^2 + \sigma_B^2/(J-1)$	[3]/[5]
H_{AB} [4]	(I-1)(J-1)	$\sigma_e^2 + \sigma_{AB}^2/(I-1)(J-1)$	[4]/[5]
Error [5]	I(J-1)(k-1)	σ_e^2	

From Table 3.10 we see that the usual error term for the fixed effect model is used only to test H_B and H_{AB} and that the hypothesis mean square for $H_{(A)S}$ is the error term for testing H_A.

To illustrate how one would analyze a split-plot repeated measures design using SAS, we use the data in Table 3.11.

a) data

Table 3.11 *Split-Plot (Data Set 1)*

			B		
			b_1	b_2	b_3
A	A_1	S_1	3	4	3
		S_2	2	2	1
	A_2	S_1'	3	7	7
		S_2'	5	4	6
	A_3	S_1''	3	4	6
		S_2''	2	3	5

b) <u>means</u>

			B			
			b_1	b_2	b_3	(means)
	A_1	S_1	μ_{111}	μ_{121}	μ_{131}	$\mu_{1.1}$
		S_2	μ_{112}	μ_{122}	μ_{132}	$\mu_{1.2}$
		(means)	$\mu_{11.}$	$\mu_{12.}$	$\mu_{13.}$	$\mu_{1..}$
A	A_2	S_1'	μ_{211}	μ_{221}	μ_{231}	$\mu_{2.1}$
		S_2'	μ_{212}	μ_{222}	μ_{232}	$\mu_{2.2}$
		(means)	$\mu_{21.}$	$\mu_{22.}$	$\mu_{23.}$	$\mu_{2..}$
	A_3	S_1''	μ_{311}	μ_{321}	μ_{331}	$\mu_{3.1}$
		S_2''	μ_{312}	μ_{322}	μ_{332}	$\mu_{3.2}$
		(means)	$\mu_{31.}$	$\mu_{32.}$	$\mu_{33.}$	$\mu_{3..}$
		means	$\mu_{..1}$	$\mu_{..2}$	$\mu_{..3}$	$\mu_{...}$

With $\mu' = (\mu_{111}, \mu_{121}, \mu_{131}, \mu_{112}, \mu_{122}, \mu_{132}, \mu_{211}, \mu_{221}, \mu_{231}, \mu_{212}, \mu_{222}, \mu_{232}, \mu_{311}, \mu_{321}, \mu_{331}, \mu_{312}, \mu_{322}, \mu_{332})$, hypothesis test matrices associated with each hypothesis of (3.25) are as follows:

$$\mathbf{C}_A = \begin{bmatrix} 1 & 1 & 1 & 1 & 1 & 1 & -1 & -1 & -1 & -1 & -1 & -1 & 0 & 0 & 0 & 0 & 0 & 0 \\ 0 & 0 & 0 & 0 & 0 & 0 & 1 & 1 & 1 & 1 & 1 & 1 & -1 & -1 & -1 & -1 & -1 & -1 \end{bmatrix}$$

$$\mathbf{C}_B = \begin{bmatrix} 1 & 0 & -1 & 1 & 0 & -1 & 1 & 0 & -1 & 1 & 0 & -1 & 1 & 0 & -1 & 0 & -1 & 1 \\ 0 & 1 & -1 & 0 & 1 & -1 & 0 & 1 & -1 & 0 & 1 & -1 & 0 & 1 & -1 & 1 & -1 & 0 \end{bmatrix}$$

$$\mathbf{C}_{(A)S} = \begin{bmatrix} 1 & 1 & 1 & -1 & -1 & -1 & 0 & 0 & 0 & 0 & 0 & 0 & 0 & 0 & 0 & 0 & 0 & 0 \\ 0 & 0 & 0 & 0 & 0 & 0 & 1 & 1 & 1 & -1 & -1 & -1 & 0 & 0 & 0 & 0 & 0 & 0 \\ 0 & 0 & 0 & 0 & 0 & 0 & 0 & 0 & 0 & 0 & 0 & 0 & 1 & 1 & 1 & -1 & -1 & -1 \end{bmatrix}$$

$$\mathbf{C}_{AB} = \begin{bmatrix} 1 & -1 & 0 & 1 & -1 & 0 & -1 & 1 & 0 & -1 & 1 & 0 & 0 & 0 & 0 & 0 & 0 & 0 \\ 0 & 1 & -1 & 0 & 1 & -1 & 0 & -1 & 1 & 0 & -1 & 1 & 0 & 0 & 0 & 0 & 0 & 0 \\ 0 & 0 & 0 & 0 & 0 & 0 & 1 & -1 & 0 & 1 & -1 & 0 & -1 & 1 & 0 & -1 & 1 & 0 \\ 0 & 0 & 0 & 0 & 0 & 0 & 0 & 1 & -1 & 0 & 1 & -1 & 0 & -1 & 1 & 0 & -1 & 1 \end{bmatrix}$$

The restriction matrix for the example is

$$\mathbf{R} = \begin{bmatrix} 1 & -1 & 0 & -1 & 1 & 0 & 0 & 0 & 0 & 0 & 0 & 0 & 0 & 0 & 0 & 0 & 0 & 0 \\ 0 & 1 & -1 & 0 & -1 & 1 & 0 & 0 & 0 & 0 & 0 & 0 & 0 & 0 & 0 & 0 & 0 & 0 \\ 0 & 0 & 0 & 0 & 0 & 0 & 1 & -1 & 0 & -1 & 1 & 0 & 0 & 0 & 0 & 0 & 0 & 0 \\ 0 & 0 & 0 & 0 & 0 & 0 & 0 & 1 & -1 & 0 & -1 & 1 & 0 & 0 & 0 & 0 & 0 & 0 \\ 0 & 0 & 0 & 0 & 0 & 0 & 0 & 0 & 0 & 0 & 0 & 0 & 1 & -1 & 0 & -1 & 1 & 0 \\ 0 & 0 & 0 & 0 & 0 & 0 & 0 & 0 & 0 & 0 & 0 & 0 & 0 & 1 & -1 & 0 & -1 & 1 \end{bmatrix}$$

The SAS code to analyze these data is given in Program 3_5_1.sas.

Program 3_5_1.sas

```
/* Program 3_5_1.sas                    */
/* Split-Plot (Two-Group Profile) Design */

options ls=80 ps=60 nodate nonumber;
title 'Output 3.5.1: Split-Plot Repeated Measures Design';

data split;
   infile 'c:\3_5_1.dat';
   input y x111 x121 x131 x112 x122 x132
           x211 x221 x231 x212 x222 x232
           x311 x321 x331 x312 x322 x332;
proc print data=split;
run;

/* Using full rank model */
proc reg;
   title2 'Full Rank Model';
   model y = x111 x121 x131 x112 x122 x132 x211 x221 x231 x212
             x222 x232 x311 x321 x331 x312 x322 x332 /noint;
   restrict x111 - x121 - x112 + x122 = 0,
            x121 - x131 - x122 + x132 = 0,
            x211 - x221 - x212 + x222 = 0,
            x221 - x231 - x222 + x232 = 0,
            x311 - x321 - x312 + x322 = 0,
            x321 - x331 - x322 + x332 = 0;
   A: mtest x111+x121+x131+x112+x122+x132-x211-x221-x231-x212-x222-x232=0,
            x211+x221+x231+x212+x222+x232-x311-x321-x331-x312-x322-x332=0
            /print details;
   B: mtest x111-x131+x112-x132+x211-x231+x212-x232+x311-x331+x312-x332=0,
            x121-x131+x122-x132+x221-x231+x222-x232+x321-x331+x322-x332=0
            /print details;
   AB: mtest x111-x121+x112-x122-x221+x231-x222+x232=0,
             x121-x131+x122-x132-x221+x231-x222+x232=0,
             x211-x221+x212-x222-x311+x321-x312+x322=0,
             x221-x231+x222-x232-x321+x331-x322+x332=0 /print details;
   AS: mtest x111+x121+x131-x112-x122-x132=0,
             x211+x221+x231-x212-x222-x232=0,
             x311+x321+x331-x312-x322-x332=0 /print details;
run;

/* Using less than full rank model */
data split2;
   input a b1 b2 b3;
   cards;
   1 3 4 3
   1 2 2 1
   2 3 7 7
   2 5 4 6
```

```
   3 3 4 6
   3 2 3 5
   ;
run;
proc glm;
   title2 'Less than Full Rank Model';
   class a;
   model b1 b2 b3 = a;
   repeated b 3 profile /nom summary;
run;
```

PROC REG is used with NOINT option to obtain the full rank model. Note that to analyze these data using the full rank model in this way, we must enter the columns of the design matrix as the independent variables in the model, which becomes cumbersome for larger designs. The restriction matrix **R** is specified with the RESTRICT statement, and the hypothesis test matrices are specified in the MTEST statements similar to the previous examples in this chapter. The results are given in Output 3.5.1.

Result and Interpretation 3_5_1.sas

Output 3.5.1: *Split-Plot Repeated Measures Design Full Rank Model*

```
              Output 3.5.1: Split-Plot Repeated Measures Design
                                Full Rank Model

Multivariate Test: A

                           L Ginv(X'X) L'   LB-cj

                           12                -6                 -17
                           -6                12                   9

                    Inv(L Ginv(X'X) L')     Inv()(LB-cj)

                 0.1111111111       0.0555555556       -1.388888889
                 0.0555555556       0.1111111111       0.0555555556

                            E, the Error Matrix

                                  6.6666666667

                          H, the Hypothesis Matrix

                                  24.111111111

                            T, the H + E Matrix

                                  30.777777778

                               Eigenvectors

                                  0.1802525304

                               Eigenvalues

                                  0.7833935018
```

Output 3.5.1 (*continued*)

```
                Multivariate Statistics and Exact F Statistics

                              S=1     M=0     N=2

  Statistic                     Value           F      Num DF    Den DF   Pr > F

  Wilks' Lambda              0.21660650     10.8500         2         6   0.0102
  Pillai's Trace             0.78339350     10.8500         2         6   0.0102
  Hotelling-Lawley Trace     3.61666667     10.8500         2         6   0.0102
  Roy's Greatest Root        3.61666667     10.8500         2         6   0.0102

Multivariate Test: B

                             L Ginv(X'X) L'    LB-cj

                             12               6              -10
                              6              12               -4

                      Inv(L Ginv(X'X) L')     Inv()(LB-cj)

                    0.1111111111      -0.055555556      -0.888888889
                    -0.055555556      0.1111111111      0.1111111111

                              E, the Error Matrix

                                    6.6666666667

                          H, the Hypothesis Matrix

                                    8.4444444444

                              T, the H + E Matrix

                                    15.111111111

                                 Eigenvectors

                                    0.2572478777

                                 Eigenvalues

                                    0.5588235294

                Multivariate Statistics and Exact F Statistics

                              S=1     M=0     N=2

  Statistic                     Value           F      Num DF    Den DF   Pr > F

  Wilks' Lambda              0.44117647      3.8000         2         6   0.0859
  Pillai's Trace             0.55882353      3.8000         2         6   0.0859
  Hotelling-Lawley Trace     1.26666667      3.8000         2         6   0.0859
  Roy's Greatest Root        1.26666667      3.8000         2         6   0.0859
```

Output 3.5.1 *(continued)*

```
Multivariate Test: AB

                       L Ginv(X'X) L'    LB-cj

             8              2              2              -4              1
             2              8              2              -4              4
             2              2              8              -4             -1
            -4             -4             -4               8              2

                   Inv(L Ginv(X'X) L')      Inv()(LB-cj)

  0.1666666667   4.240247E-18    3.0832E-18   0.0833333333   0.3333333333
  4.240247E-18   0.1666666667    3.0832E-18   0.0833333333   0.8333333333
    3.0832E-18     3.0832E-18  0.1666666667   0.0833333333   -1.54228E-17
  0.0833333333   0.0833333333  0.0833333333           0.25   0.8333333333

                        E, the Error Matrix

                             6.6666666667

                       H, the Hypothesis Matrix

                             5.3333333333

                        T, the H + E Matrix

                                       12

                           Eigenvectors

                             0.2886751346

                           Eigenvalues

                             0.4444444444

                                 Full Rank Model

              Multivariate Statistics and Exact F Statistics

                          S=1      M=1      N=2

  Statistic                    Value            F      Num DF    Den DF   Pr > F

  Wilks' Lambda              0.55555556      1.2000         4         6   0.4001
  Pillai's Trace             0.44444444      1.2000         4         6   0.4001
  Hotelling-Lawley Trace     0.80000000      1.2000         4         6   0.4001
  Roy's Greatest Root        0.80000000      1.2000         4         6   0.4001

Multivariate Test: AS

                        L Ginv(X'X) L'    LB-cj

                   6              0              0              5
                   0              6              0              2
                   0              0              6              3
```

Output 3.5.1 (*continued*)

```
                 Inv(L Ginv(X'X) L')      Inv()(LB-cj)

      0.1666666667             0                 0       0.8333333333
                 0      0.1666666667             0       0.3333333333
                 0             0      0.1666666667                0.5

                          E, the Error Matrix

                              6.6666666667

                        H, the Hypothesis Matrix

                              6.3333333333

                          T, the H + E Matrix

                                        13

                            Eigenvectors

                              0.2773500981

                            Eigenvalues

                              0.4871794872

            Multivariate Statistics and Exact F Statistics

                          S=1    M=0.5    N=2

 Statistic                     Value          F     Num DF    Den DF  Pr > F

 Wilks' Lambda            0.51282051     1.9000          3         6  0.2307
 Pillai's Trace           0.48717949     1.9000          3         6  0.2307
 Hotelling-Lawley Trace   0.95000000     1.9000          3         6  0.2307
 Roy's Greatest Root      0.95000000     1.9000          3         6  0.2307
```

Table 3.12 summarizes the test results.

Table 3.12 *Hypotheses Test Results, Split Plot*

Hypotheses	SS	df	MS	F	P
H_A	24.11	2	12.06	5.71	.0949
$H_{(A)S}$	6.33	3	2.11		
H_B	8.44	2	4.22	3.80	.0859
H_{AB}	8.22	4	2.06	1.85	.2385
Error	6.67	6	1.11		

Program 3_5_1.sas contains the SAS code to analyze these data using PROC GLM with the REPEATED statement. The results follow.

Output 3.5.1 *(continued)*

```
              Output 3.5.1: Split-Plot Repeated Measures Design
                            Less than Full Rank Model

                          General Linear Models Procedure
                       Repeated Measures Analysis of Variance
                  Tests of Hypotheses for Between Subjects Effects

Source                  DF      Type III SS     Mean Square   F Value     Pr > F

A                        2      24.11111111     12.05555556      5.71     0.0949

Error                    3       6.33333333      2.11111111

General Linear Models Procedure
                       Repeated Measures Analysis of Variance
              Univariate Tests of Hypotheses for Within Subject Effects

Source: B
                                                                   Adj  Pr > F
     DF        Type III SS       Mean Square   F Value   Pr > F    G - G    H - F
      2         8.44444444        4.22222222      3.80   0.0859   0.1451   0.0859

Source: B*A
                                                                   Adj  Pr > F
     DF        Type III SS       Mean Square   F Value   Pr > F    G - G    H - F
      4         8.22222222        2.05555556      1.85   0.2385   0.2985   0.2385

Source: Error(B)

     DF        Type III SS       Mean Square
      6         6.66666667        1.11111111

                           Greenhouse-Geisser Epsilon = 0.5076
                                  Huynh-Feldt Epsilon = 1.0307
```

The results of the test for *A* are labeled `Tests of Hypotheses for Between Subjects Effects` and the test for *B* and *A*B* are labeled `Univariable Tests of Hypotheses for Within Subjects Effects`. The Greenhouse-Geisser (G-G) and Huhyn-Feldt (H-F) adjusted *p*-values are given, as well as the values of these test statistics, epsilon. These are used in situations of mild violations of the circularity assumption. In cases of extreme violation of the assumption, the use of multivariate methods is advised.

3.5.2 Univariate Mixed-Model ANOVA, Less Than Full Rank Represention

In a second example of how to analyze a split-plot repeated measures design (this time with a larger data set), the data from Timm (1975, p. 244) are utilized. Data were collected from subjects of low short-term memory capacity (GROUP I) and from subjects with high short-term memory capacity (GROUP II). The data are presented in Table 3.13.

Table 3.13 *Split-Plot (Timm Data)*

		Probe-Word Position				
		1	2	3	4	5
GROUP I	S_1	20	21	42	32	32
	S_2	67	29	56	39	41
	S_3	37	25	28	31	34
	S_4	42	38	36	19	35
	S_5	57	32	21	30	29
	S_6	39	38	54	31	28
	S_7	43	20	46	42	31
	S_8	35	34	43	35	42
	S_9	41	23	51	27	30
	S_{10}	39	24	35	26	32
MEAN		42.0	28.4	41.2	31.2	33.4
GROUP II	S_{11}	47	25	36	21	27
	S_{12}	53	32	48	46	54
	S_{13}	38	33	42	48	49
	S_{14}	60	41	67	53	50
	S_{15}	37	35	45	34	46
	S_{16}	59	37	52	36	52
	S_{17}	67	33	61	31	50
	S_{18}	43	27	36	33	32
	S_{19}	64	53	62	40	43
	S_{20}	41	34	47	37	46
MEAN		50.9	35.0	49.6	37.9	44.9

To analyze these data using the full rank representation would be very cumbersome. The vector μ is a 100×1 vector as there are 100 cells in the design. We therefore analyze these data utilizing PROC GLM and the less than full rank representation. (This same data set is later analyzed using multivariate methods in Chapter 5). Program 3_5_2.sas contains the SAS code to perform the analysis.

Program 3_5_2.sas

```
/* Program 3_5_2.sas                                         */
/* Split-Plot (Two-Group Profile) Design, Full Rank Model */
/* Timm (1975, p244) data                                    */

options ls=80 ps=60 nodate nonumber;
title 'Output 3_5_2: Two Group Profile Analysis';

data splitb;
   infile 'c:\3_5_2.dat';
   input grp p1 p2 p3 p4 p5;
proc print data=splitb;
run;

proc glm;
   class grp;
   model p1 p2 p3 p4 p5 = grp;
   repeated position 5 profile /nom summary;
run;
```

The results are given in Output 3_5_2.

Result and Interpretation 3_5_2.sas

Output 3_5_2: *Two Group Profile Analysis*

```
                    Output 3_5_2: Two Group Profile Analysis
                        General Linear Models Procedure
                      Repeated Measures Analysis of Variance
                 Tests of Hypotheses for Between Subjects Effects

Source                     DF       Type III SS     Mean Square   F Value     Pr > F

GRP                         1       1772.410000     1772.410000      8.90     0.0080

Error                      18       3583.140000      199.063333

                        General Linear Models Procedure
                      Repeated Measures Analysis of Variance


           Univariate Tests of Hypotheses for Within Subject Effects

Source: POSITION
                                                                   Adj  Pr > F
     DF        Type III SS       Mean Square  F Value   Pr > F    G - G    H - F
      4     3371.30000000     842.82500000     14.48   0.0001   0.0001   0.0001

Source: POSITION*GRP
                                                                   Adj  Pr > F
     DF        Type III SS       Mean Square  F Value   Pr > F    G - G    H - F
      4       79.94000000      19.98500000      0.34   0.8479   0.8068   0.8479

Source: Error(POSITION)

     DF        Type III SS       Mean Square
     72     4191.96000000      58.22166667

                      Greenhouse-Geisser Epsilon = 0.8009
                             Huynh-Feldt Epsilon = 1.0487
```

The results of the hypotheses tested are summarized in Table 3.14.

Table 3.14 *Hypothesis Test Results, Split-Plot (Timm Data)*

Hypotheses	SS	df	MS	F	P
Group	1772.41	1	1772.41	8.90	.0080
	3583.14	18	199.06		
Position	3371.30	4	842.825	14.48	<.0001
Group*Position	79.94	4	19.99	0.34	.8479
Error	4191.96	72	58.22		

3.5.3 Test for Equal Variance-Covariance Matrices and Test for Circularity

There are certain covariance structures modeling the dependence in the data that would permit the univariate mixed-model ANOVA F-test that was discussed earlier to be used. The assumption of circularity is a necessary and sufficient condition by which the mixed-model F can be used. The circularity assumption is

$$\mathbf{A}^{*\prime}\Sigma\mathbf{A}^{*} = \sigma^2\mathbf{I}$$

where $\mathbf{A}^{*\prime}$ is any orthonormal hypothesis test matrix of order $(p-1)\times p$, Σ is the $p\times p$ variance-covariance matrix, σ^2 is a scalar and $\mathbf{I}$ is an identity matrix of order $(p-1)\times(p-1)$. When there is more than one group in the design, then the multisample circularity assumption must be satisfied in order to validly utilize the within-subject mixed-model F-tests. The multisample circularity assumption is

$$\mathbf{A}^{*\prime}\Sigma_1\mathbf{A}^{*} = \mathbf{A}^{*\prime}\Sigma_2\mathbf{A}^{*} = \ldots = \mathbf{A}^{*\prime}\Sigma_g\mathbf{A}' = \sigma^2\mathbf{I}$$

where the $\mathbf{A}^{*\prime}$ is as previously shown, and Σ_j is the population variance-covariance matrix for each of the $j = 1, \ldots, g$ groups.

Before assuming any covariance structure, a formal test can be performed. Several tests follow (from Timm, 1975, and Kirk, 1995) along with SAS code to perform the tests. Examples of the tests are given using the data of Table 3.13 from Timm (1975, p. 244). Note that the following tests assume multivariate normality.

To test the null hypothesis

$$H\colon\ \mathbf{A}^{*\prime}\Sigma_1\mathbf{A}^{*} = \mathbf{A}^{*\prime}\Sigma_2\mathbf{A}^{*} = \ldots = \mathbf{A}^{*\prime}\Sigma_g\mathbf{A}^{*}$$

we use the methods discussed in Timm (1975, p. 251-252) and Kirk (1995, p. 524-530). The test statistic had also been described by Box (1950). We form the statistic

$$M = (N-g)\ln|\mathbf{A}^{*\prime}\mathbf{S}_{pooled}\mathbf{A}^{*\prime}| - \sum_{i=1}^{g} v_i \ln|\mathbf{A}^{*\prime}\mathbf{S}_i\mathbf{A}|$$

where g is the number of groups, $N = \Sigma n_i, i = 1, 2, \ldots, g$, $\mathbf{S}_{pooled}$ is the pooled within estimate of the variance-covariance matrix defined as $\mathbf{S}_{pooled} = \frac{1}{N-g}\sum_{i=1}^{g} v_i\mathbf{S}_i$, and $\mathbf{S}_i$ is the unbiased estimate of Σ_i for the ith group and $v_i = n_i - 1$.

The quantity $X_B^2 = (1-C)M$ approximates a chi-square distribution with $v = q(q+1)(g-1)/2$ when H is true and $N \to \infty$, where

$$C = \frac{2q^2+3q-1}{6(q+1)(g-1)}\left(\sum_{i=1}^{g}\frac{1}{v_i} - \frac{1}{(N-g)}\right)$$

and q = the number of rows of $\mathbf{A}^{*\prime}$. We reject H: $\mathbf{A}^{*\prime}\Sigma_1\mathbf{A}^* = \ldots = \mathbf{A}^{*\prime}\Sigma_g\mathbf{A}^*$ at level α if $X_B^2 > \chi_v^2(\alpha)$. This approximation works reasonably well when each $n_i > 20$ and $q < 6$ and $g < 6$.

If n_i are small or q or g are larger than 6, the F distribution can be used. We calculate

$$C_o = \frac{(q-1)(q+2)}{6(g-1)}\left(\sum_{i=1}^{g}\frac{1}{v_i^2} - \frac{1}{(N-g)^2}\right)$$

$$v_o = \frac{v+2}{C_o - C^2}$$

$$F = \frac{1-C-v/v_o}{v}M$$

where $F \sim F(v, v_o)$, under the null hypothesis. The hypothesis of homogeneity is rejected at level α if $F > F_{(v,v_o)}^{(\alpha)}$.

The first sections of Program 3_5_3.sas contain the PROC IML code to perform the above hypothesis test on the data in Table 3.13.

Program 3_5_3.sas

```
/* Program 3_5_3.sas              */
/* Tests of Covariance Structures */

options ls=80 ps=60 nodate nonumber mprint;
title 'Output 3.5.3: Tests of Covariance Structures';

data struct;
   infile 'c:\3_5_2.dat';
   input grp p1 p2 p3 p4 p5;
proc print data=struct;
run;

proc iml;
   use struct;
   read all var {p1 p2 p3 p4 p5} where (grp=1) into y1;
   read all var {p1 p2 p3 p4 p5} where (grp=2) into y2;
   g=2;
   n1=nrow(y1);
   n2=nrow(y2);
   n=n1+n2;
   p=ncol(y1);
   df1=n1-1;
   df2=n2-1;
   s1=(y1`*(i(n1)-(1/n1)*j(n1,n1))*y1)/df1;
   s2=(y2`*(i(n2)-(1/n2)*j(n2,n2))*y2)/df2;
   s=(1/(n-g))*(df1*s1+df2*s2);

   /* the hypothesis test matrix a` */
   ap={1  -1   0   0   0,
       0   1  -1   0   0,
       0   0   1  -1   0,
       0   0   0   1  -1};
```

```
    a=ap`;
    q=nrow(ap);
    call gsorth(as,t,lindep,a);
    asp=as`;
    as1a=as`*s1*as;
    as2a=as`*s2*as;
    asa=(1/(n-g))*(df1*as1a+df2*as2a);
    print "A` and A*`", ap asp;
    print s1 s2;
    print "Reduced Covariance Matrices", as1a as2a;
    print "Reduced Pooled Covaraince Matrix", asa;

    /* Step 1 : Test of Equality of the Reduced Covariance Matrices */
    m1=(n-g)*(log(det(asa)));
    m2=df1*(log(det(as1a)))+df2*(log(det(as2a)));
    m=m1-m2;
    print m;
    e1=(2*q*q+3*q-1)/(6*(q+1)*(g-1));
    e2=(1/df1)+(1/df2)-(1/(n-g));
    e=e1*e2;
    print e;
    xb=(1-e)*m;
    v1=(q*(q+1)*(g-1))/2;
    probxb=1-probchi(xb,v1);
    print "Test of Equality of Reduced Covariance Matrices", xb v1;
    print "Probability of Xb with df=v1", probxb;
    print "The above test works well for ni>20 and q<6 and g<6";
    eo1=(1/(df1*df1))+(1/(df2*df2))-(1/((n-g)*(n-g)));
    eo2=(q-1)*(q+2)/(6*g-6);
    eo=eo2*eo1;
    vo=(v1+2)/(eo-(e*e));
    fb=((1-e-(v1/vo))/v1)*m;
    probfb=1-probf(fb,v1,vo);
    print "Test of Equality of Reduced Covariance Matrices", fb v1 vo;
    print "Probability of Fb with df=v1,vo", probfb;
    print "The above test can be used for small ni and/or q and/or g >6";

    /* Step 2 : Test of Circularity */
    const=-( (n-1)-((2*q*q+q+2)/(6*q)) );
    lds=log(det(asa));
    qlt=q*(log((trace(asa))/q));
    print const lds qlt;
    xs=const*(lds-qlt);
    dfxs=((q*q+q)/2)-1;
    probxs=1-probchi(xs,dfxs);
    print "Tests of Circularity (Mauchly)", xs dfxs;
    print "Probability of Xs with df=dfxs", probxs;
    sasa=asa*asa;
    v=trace(sasa)/((trace(asa))*(trace(asa)));
    vs=((q*q*n)/2)*(v-(1/q));
    probvs=1-probchi(vs,dfxs);
    print "Test of Circularity (L.B.I.)", vs, dfxs;
    print "Probability of Vs with df=dfxs", probvs;
quit;
run;
```

The results of the test are given in Output 3.5.3.

Result and Interpretation 3_5_3.sas

Output 3.5.3: *Tests of Covariance Structures*

```
              Output 3.5.3: Tests of Covariance Structures

                             A` and A*`

                  AP
                   1        -1         0         0         0
                   0         1        -1         0         0
                   0         0         1        -1         0
                   0         0         0         1        -1

                 ASP
           0.7071068 -0.707107         0         0         0
           0.4082483 0.4082483 -0.816497         0         0
           0.2886751 0.2886751 0.2886751 -0.866025         0
           0.2236068 0.2236068 0.2236068 0.2236068 -0.894427

                  S1
           158.66667 21.222222 8.6666667 18.333333 12.555556
           21.222222 46.044444 -0.866667 -16.42222 5.9333333
           8.6666667 -0.866667 128.17778 28.177778 10.466667
           18.333333 -16.42222 28.177778 43.066667 8.1333333
           12.555556 5.9333333 10.466667 8.1333333 22.711111

                  S2
           124.32222 45.111111 96.511111 10.322222 30.322222
           45.111111 60.666667 64.111111 33.666667 27.333333
           96.511111 64.111111 116.71111 48.066667 56.177778
           10.322222 33.666667 48.066667 86.322222 55.544444
           30.322222 27.333333 56.177778 55.544444 77.211111

                       Reduced Covariance Matrices

     AS1A                                   AS2A
81.133333 27.007163 3.6515542 20.621564 47.383333 -0.330373 33.902752 9.6062968
27.007163 121.44444 17.245549 -5.629371 -0.330373 16.594444 4.4312025 4.3655516
3.6515542 17.245549 49.833333 11.834116 33.902752 4.4312025 78.144444 31.067064
20.621564 -5.629371 11.834116 28.042222 9.6062968 4.3655516 31.067064 43.197778

                     Reduced Pooled Covaraince Matrix

                       ASA
                64.258333 13.338395 18.777153  15.11393
                13.338395 69.019444 10.838376  -0.63191
                18.777153 10.838376 63.988889  21.45059
                 15.11393  -0.63191  21.45059     35.62

                                  M
                          13.379623

                                  E
                          0.2388889

          Test of Equality of Reduced Covariance Matrices

                            XB         V1

                        10.18338         10

                       Probability of Xb with df=v1
```

Output 3.5.3 *(continued)*

```
                              PROBXB
                           0.4245548

       The above test works well for ni>20 and q<6 and g<6

         Test of Equality of Reduced Covariance Matrices

                       FB           V1         VO
                1.0097004           10   1549.004

                  Probability of Fb with df=v1,vo

                              PROBFB
                            0.432616

The above test can be used for small ni and/or q and/or g >6
```

The value of $X_B = 10.18$, with 10 degrees of freedom. The χ^2 probability is $p = 0.42$; thus, we do not reject H. The F-test for small samples (as we have in this example) gives $F_B = 1.0097$ with 10 and 1549 degrees of freedom. The F probability is 0.43; thus, we do not reject the null hypothesis that the reduced covariance matrices are equivalent.

The null hypothesis of circularity is

$$H\text{: } \mathbf{A}^{*\prime}\boldsymbol{\Sigma}\mathbf{A}^{*} = \sigma^2\mathbf{I}$$

where $\mathbf{A}^{*\prime}$ is an orthonormal hypothesis test matrix. Mauchley (1940) developed a chi-square statistic to test this hypothesis

$$X_s^2 = -\left[(N-1) - \frac{2q^2+q+2}{6q}\right]\left[\ln|\mathbf{A}^{*\prime}\mathbf{S}_{pooled}\mathbf{A}^{*}| - q\ln\left(\frac{Tr(\mathbf{A}^{*\prime}\mathbf{S}_{pooled}\mathbf{A}^{*})}{q}\right)\right]$$

where X_s^2 is distributed approximately chi-square with $v = [q(q-1)/2] - 1$ degree of freedom when H is true. We reject H when $X_s^2 > \chi_v^2(\alpha)$.

Another test for circularity which has been found to be more powerful than the Mauchley test, at least for small sample sizes, is the locally best invariant test statistic

$$V = \frac{\text{Tr}[\mathbf{A}^{*\prime}\text{S}_{\text{pooled}}\,\mathbf{A}^{*}]}{[\text{Tr}(\mathbf{A}^{*\prime}\text{S}_{\text{pooled}}\,\mathbf{A}^{*})]^2}$$

where

$$V^{*} = \frac{q^2 n}{2}\{V - q^{-1}\}$$

Under H, V^{*} is asymptotically distributed as a χ^2 with $df = (q(q+1)/2) - 1$, (Cornell, Young, Seaman, and Kirk, 1992).

Program 3_5_3. sas contains the PROC IML code to perform these tests on these data.

Output 3.5.3 (*continued*)

```
    CONST         LDS         QLT
-17.41667  15.693554  16.25703

Tests of Circularity (Mauchly)

             XS        DFXS
      9.8138804           9

Probability of Xs with df=dfxs

              PROBXS
           0.3657618

 Test of Circularity (L.B.I.)

                  VS
           9.9445755

                DFXS
                   9

Probability of Vs with df=dfxs

              PROBVS
           0.3549908
```

The value of $X_s^2 = 9.81$ with 9 degrees of freedom. The χ^2 probability is 0.37; thus, we do not reject the null hypothesis of circularity using Mauchley's test. The value of $V^* = 9.94$ with 9 degrees of freedom. The χ^2 probability is 0.35. Thus, we do not reject the null hypothesis of circularity using the locally best invariant test.

3.6 Analysis of Covariance

In Chapter 2 we discussed the general intraclass model, an unrestricted model that allows the slopes of the regression lines within the different populations to differ. In this section we now place restrictions on this general model to develop the analysis of covariance (ANCOVA). In the one-factor analysis of covariance design we require that the slopes of the regression lines within the different populations be identical. The ANCOVA model is a special case of the intraclass model, in which the slopes of the regression lines are assumed to be equal and thus the statement, $\beta_1 = \beta_2 = \ldots = \beta_I$ becomes a restriction on the model parameters. As Figure 2.9.1 shows, unambiguous interpretation of differences among the intercepts is not possible if the slopes are unequal. It does not make sense, of course, to assume equal slopes, and thus to use the ANCOVA model when the slopes are not truly equal in the populations. In the absence of information about equality of the slopes, therefore, it makes sense to test the hypothesis H: $\beta_1 = \beta_2 = \ldots = \beta_I$, as we did in Chapter 2, as a formal test of the assumption underlying the analysis of covariance. If the sample data support the assumption, we can be more comfortable about the use of the

ANCOVA model than if the data do not support the assumption. However, we must always consider the possibility of making a Type I or Type II error on any hypothesis test.

Since the analysis of covariance model is like the intraclass covariance model with the addition of a set of restrictions, it may be written as

$$\mathbf{y} = \mathbf{X}\beta^* + \mathbf{e}$$
$$R\text{: } \beta_1 = \beta_2 = \ldots = \beta_I = \beta_p$$

where β_p is used to represent the common (pooled) slope of the I populations. The design matrix is identical to that in the intraclass model, but the parameter vector is

$$\beta^{*\prime} = \left[\mu_1^*,\ \mu_2^* \ldots \mu_I^*,\ \ \beta_p,\ \beta_p \ldots \beta_p\right] \tag{3.28}$$

The common slope, β_p, appears *I* times in the parameter vector as a consequence of our method of forming the model, but there is only one slope parameter. The intercepts are denoted μ_i^* in order to emphasize that they are different from the α_i of the intraclass covariance model and the μ_i of the analysis of variance model.

The classical analysis of covariance F-test is a test of the hypothesis

$$H\text{: } \mu_1^* = \mu_2^* = \ldots = \mu_I^* \tag{3.29}$$

which can be thought of as a test of equality of intercepts in the presence of equal slopes. Using the classical less than full rank model, many writers discuss this test as a test of the equality of *adjusted means*. The adjusted means are the values in units of the dependent variable where the I regression lines intercept a line perpendicular to the covariate axis at the overall mean on the covariate, $\bar{z}_{..}$, as shown in Figure 3.6.1 for three populations.

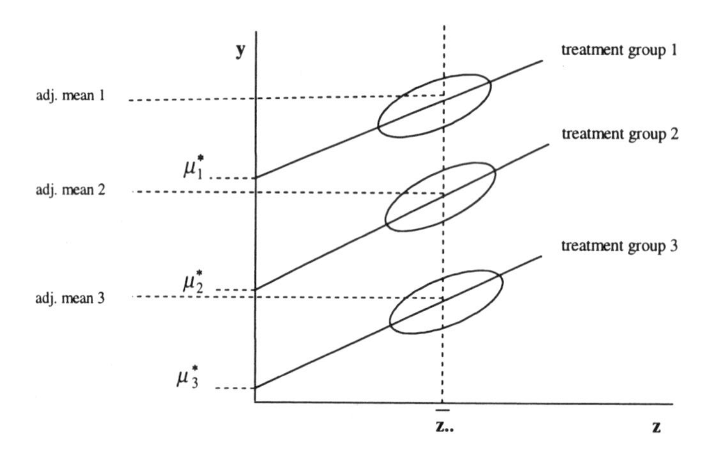

Figure 3.6.1 Analysis of Covariance

Since we are assuming equal slopes, the regression lines are parallel and it is immaterial whether we consider the analysis of covariance test to be a test of homogeneity of intercepts or of adjusted means. This test is perhaps best thought of as a test of coincidence of the regression lines, assuming parallelism. The error variance estimate for this model is based on variation about lines with equal slopes fitted to each sample. The degrees of freedom for error are

$$n - \begin{bmatrix} \text{No. Columns in} \\ \text{Design Matrix} \end{bmatrix} + \text{No. Restrictions}$$
$$= n - 2\text{I} + (\text{I} - 1)$$
$$= n - \text{I} - 1$$

for the restricted model of this section.

We may also test the hypothesis that the common slope is zero

$$H_{reg}: \quad \beta_p = 0. \tag{3.30}$$

This test may be thought of as a test of the effectiveness of the covariate much like the test of block means in a blocking design. The assumption in using a covariate is, of course, that this hypothesis is false.

There is no requirement that the regression of Y and Z be linear in the different populations, as long as the regressions are identical and can be expressed in linear form. Thus, if theory or previous research suggests a regression such as

$$y = \log z$$
$$y = z^2$$
$$y = (z)^{1/2}$$

or some similar form, we can use the techniques described in this section but using the logs, squares, square roots, etc., of the values of z in the design matrix rather than the raw values, z_{ij} .

If the regression is nonlinear but involves more than two parameters, however, there are some slight modifications which must be made. If, for example, we wish to assume a model of the form

$$y_{ij} = \mu_i^* + \beta_{1p}\, z_{ij} + \beta_{2p}\, z_{ij}^2 \tag{3.31}$$

where β_{1p} and β_{2p} represent pooled parameters for the linear and quadratic terms of the regression, there are two regression parameters (as well as the I intercepts) that must be estimated from the sample data. Using the intraclass model in the presence of such a regression, we would have 3I columns in the design matrix, 2I columns like those of X for the intraclass made in Chapter 2, and I additional columns containing the z_{ij}^2 values in vectors like those containing the z_{ij} values. Expressing the regression parameters of this intraclass model as $\beta_{11}, \beta_{12}, \ldots \beta_{1I}$, and $\beta_{21}, \beta_{22}, \ldots \beta_{2I}$, the restrictions for the analysis of covariance are

$$R: \quad \begin{bmatrix} \beta_{11} = \beta_{12} = \ldots = \beta_{1I} = \beta_{1p} \\ \beta_{21} = \beta_{22} = \ldots = \beta_{2I} = \beta_{2p} \end{bmatrix} \tag{3.32}$$

and the error degrees of freedom for the restricted full rank model are

$$n - 3\text{I} + 2(\text{I} - 1)$$
$$= n - \text{I} - 2$$

Using the intraclass model we can, of course, test the assumption of equality of the linear and quadratic regression coefficients as stated in the previous restrictions. To test the effectiveness of the covariate, we could also test that the two regression parameters in the restricted model are simultaneously equal to zero. A test that β_{2p} is zero could be used to assess the effectiveness of adding the quadratic term to the regression using only the linear term.

Using basically the same procedures as we do for the quadratic regression model, we can incorporate two or more covariates into a design. Replacing z_{ij} and z_{ij}^2 by z_{1ij} and z_{2ij}, the values on the two covariates, and letting β_{1p} and β_{2p} be the parameters associated with the two covariates, we use a model similar to (3.31) as a model having two covariates. Such a model is easily extended to more than two covariates as long as the design matrix is of full column rank so that $(\mathbf{X}'\mathbf{X})^{-1}$ exists. It is not good practice, however, to add several covariates to a design without some evidence that each accounts for some unique variation in the dependent variable. One degree of freedom is lost from the error term for each added covariate, and the loss of error degrees of freedom without a concomitant decrease in error variation results in a loss in precision.

3.6.1 ANCOVA with One Covariate

In order to illustrate the ANCOVA discussed above, we analyze the data in Table 3.15 using SAS.

Table 3.15 *Analysis of Covariance Data*

A_1			A_2			A_3		
Y	Z_1	Z_2	Y	Z_1	Z_2	Y	Z_1	Z_2
3	6	2	5	5	2	5	4	3
2	6	2	6	5	1	7	1	1
3	5	2	8	3	2	8	1	6
3	4	4	8	2	7	7	2	6
4	3	5	7	1	8			
5	3	5	9	1	7			
5	2	6						
6	2	5						

First, we assume that we wish to use only Z_1 as a covariate. Recall from the intraclass model example in Chapter 2, where we used these same data, that the hypothesis of equality of slopes is not rejected for these data. Thus, there is no evidence that the homogeneity of regression assumption underlying the analysis of covariance F-test has been violated.

To perform the ANCOVA, we use $\mathbf{R} = \begin{bmatrix} 0 & 0 & 0 & 1 & 0 & -1 \\ 0 & 0 & 0 & 0 & 1 & -1 \end{bmatrix}$ as a restriction matrix, (note this is equivalent to $\mathbf{C}$ of the intraclass model in Chapter 2) and $\beta^{*\prime} = (\mu_1^*, \mu_2^*, \mu_3^*, \beta_p, \beta_p, \beta_p)$ as the parameter vector. Note that β_p is stated three times in this vector because of the way in which we have formulated the model, but there is only one β_p parameter and only one estimate of it, based on pooling the data from all three groups. Program 3_6_1.sas contains the SAS code to perform the analysis.

Program 3_6_1.sas

```
/* Program 3_6_1.sas                                    */
/* Program to perform ANCOVA with one covariate and     */
/* Unequal n, using Full Rank model                     */

options ls=80 ps=60 nodate nonumber;
filename examp4 'c:\2_9.dat';
title 'Output 3.6.1: ANCOVA, One Covariate, Unequal n';

data onecov;
   infile examp4;
   input y a z a1 a2 a3 z1 z2 z3;
proc print;
run;

/* Using the full rank model */
proc reg;
   title2 'Full Rank Model';
   model y = a1 a2 a3 z1 z2 z3 /noint;
   restrict  z1-z2=0,
             z1-z3=0;
   A: mtest a1-a3=0,
            a2-a3=0 /print details;
   REG: mtest z3=0 /print details;
run;

/* Using the less than full rank model */
proc glm;
   title2 'Less than Full Rank Model';
   class a;
   model y=a z /e xpx solution;
   lsmeans a /stderr pdiff cov;
run;
```

The analysis is first performed using PROC REG with the NOINT option to obtain the full rank model. The restrictions are specified in the RESTRICT statement. The hypothesis test matrices

$$\mathbf{C}_A = \begin{pmatrix} 1 & 0 & -1 & 0 & 0 & 0 \\ 0 & 1 & -1 & 0 & 0 & 0 \end{pmatrix}$$

$$\mathbf{C}_{reg} = \begin{pmatrix} 0 & 0 & 0 & 0 & 0 & 1 \end{pmatrix}$$

are specified using the MTEST statements. The ANCOVA is also performed using PROC GLM and the less than full rank model. The results of both analyses are given in Output 3.6.1.

Result and Interpretation 3_6_1.sas

Output 3.6.1: *ANCOVA, One Covariate, Unequal n Full Rank Model*

```
                 Output 3.6.1: ANCOVA, One Covariate, Unequal n
                               Full Rank Model

Multivariate Test: A

                              L Ginv(X'X) L'    LB-cj

                  0.4592907093      0.2874625375      -1.528221778
                  0.2874625375      0.4333166833      1.0152347652

                     Inv(L Ginv(X'X) L')     Inv()(LB-cj)

                  3.7231759657      -2.469957082      -8.197424893
                  -2.469957082      3.9463519313      7.7811158798

                               E, the Error Matrix

                                    6.9398101898

                             H, the Hypothesis Matrix

                                     20.4271426

                               T, the H + E Matrix

                                     27.36695279

                                   Eigenvectors

                                    0.1911554913

                                   Eigenvalues

                                     0.746416408

                 Multivariate Statistics and Exact F Statistics

                               S=1     M=0     N=6

  Statistic                      Value           F       Num DF     Den DF  Pr > F

  Wilks' Lambda               0.25358359    20.6043          2         14  0.0001
  Pillai's Trace              0.74641641    20.6043          2         14  0.0001
  Hotelling-Lawley Trace      2.94347281    20.6043          2         14  0.0001
  Roy's Greatest Root         2.94347281    20.6043          2         14  0.0001

Multivariate Test: REG

                              L Ginv(X'X) L'   LB-cj

                              0.023976024      -0.718281718

                       Inv(L Ginv(X'X) L')     Inv()(LB-cj)

                             41.708333333      -29.95833333

                               E, the Error Matrix

                                    6.9398101898
```

Output 3.6.1 (*continued*)

```
                          H, the Hypothesis Matrix

                                21.518523144

                            T, the H + E Matrix

                                28.458333333

                                Eigenvectors

                                0.1874542404

                                Eigenvalues

                                0.7561413696

              Multivariate Statistics and Exact F Statistics

                            S=1    M=-0.5    N=6

  Statistic                     Value          F      Num DF    Den DF  Pr > F

  Wilks' Lambda               0.24385863    43.4103        1        14  0.0001
  Pillai's Trace              0.75614137    43.4103        1        14  0.0001
  Hotelling-Lawley Trace      3.10073656    43.4103        1        14  0.0001
  Roy's Greatest Root         3.10073656    43.4103        1        14  0.0001

                               Less than Full Rank Model
                            General Linear Models Procedure

Dependent Variable: Y
                                       Sum of            Mean
Source                   DF           Squares          Square   F Value       Pr > F

Model                     3       65.33796759     21.77932253     43.94       0.0001

Error                    14        6.93981019      0.49570073

Corrected Total          17       72.27777778

                 R-Square              C.V.        Root MSE                 Y Mean

                 0.903984          12.54761       0.7040602              5.6111111


Source                   DF         Type I SS     Mean Square   F Value       Pr > F

A                         2       43.81944444     21.90972222     44.20       0.0001
Z                         1       21.51852314     21.51852314     43.41       0.0001

Source                   DF       Type III SS     Mean Square   F Value       Pr > F

A                         2       20.42714260     10.21357130     20.60       0.0001
Z                         1       21.51852314     21.51852314     43.41       0.0001
```

Output 3.6.1 *(continued)*

```
                                      T for H0:     Pr > |T|    Std Error of
Parameter                 Estimate    Parameter=0                 Estimate

INTERCEPT             8.186563437 B        19.77      0.0001      0.41408322
A          1         -1.528221778 B        -3.20      0.0064      0.47714855
           2          1.015234765 B         2.19      0.0459      0.46346024
           3          0.000000000 B          .          .           .
Z                    -0.718281718          -6.59      0.0001      0.10901804

NOTE: The X'X matrix has been found to be singular and a generalized inverse
      was used to solve the normal equations.    Estimates followed by the
      letter 'B' are biased, and are not unique estimators of the parameters.


                      General Linear Models Procedure
                              Least Squares Means

A          Y          Std Err      Pr > |T|     Pr > |T| H0: LSMEAN(i)=LSMEAN(j)
        LSMEAN        LSMEAN     H0:LSMEAN=0    i/j      1        2        3

1    4.42368742    0.26248383        0.0001     1    .        0.0001   0.0064
2    6.96714397    0.28902221        0.0001     2  0.0001      .       0.0459
3    5.95190920    0.37228744        0.0001     3  0.0064   0.0459      .

NOTE: To ensure overall protection level, only probabilities associated with
      pre-planned comparisons should be used.
```

The results are summarized in Table 3.16.

Table 3.16 *Hypothesis Test Results, ANCOVA 1*

Source of Variation	SS	DF	MS	F	p-value
Regression	21.52	1	21.52	43.41	<.0001
A (adjusted treatment)	20.43	2	10.21	20.60	<.0001
Error	6.94	14	0.50		

Note that in our example the test of the regression is significant, indicating that the covariate Z_1 accounts for a significant portion of the variation in the dependent variate. If this test is nonsignificant one might question whether the covariate should appear in the model.

Since the analysis of covariance F-test is significant at the .05 level, we wish to compute the 95% confidence intervals for the contrasts used in forming $\mathbf{C}_A$. These are

$$-2.833 \le \mu_1^* - \mu_3^* \le -0.223$$

$$-0.252 \le \mu_2^* - \mu_3^* \le \ \ 2.283.$$

Thus, we conclude that there is a difference between the regression intercepts for treatments 1 and 3 but not between those for treatments 2 and 3. Since the distance between the lines for treatments 1 and 2 is estimated to be

$$\hat{\mu}_1^* - \hat{\mu}_2^* = 6.658 - 9.202 = -2.544$$

which is larger than the other differences, we might expect that it is also significant. We must, however, compute the standard error of this contrast before we can reach any conclusion. From the estimated variance-covariance matrix of restricted least squares estimates in the Output 3.6.1, we calculate

$$\begin{aligned}\hat{\sigma}_{\hat{\mu}_1^*-\hat{\mu}_2^*} &= \sqrt{\hat{\sigma}^2_{\hat{\mu}_1^*}+\hat{\sigma}^2_{\hat{\mu}_2^*}-2\hat{\sigma}_{\hat{\mu}_1^*\hat{\mu}_2^*}} \\ &= \sqrt{.240+.178-2(.130)} \\ &= 0.398.\end{aligned}$$

We also calculate

$$c_o = \sqrt{(v_h)F_{(v_h,v_e)}} = \sqrt{2F_{(2,14)}} = \sqrt{2(3.74)} = 2.73$$

so that the limits of the 95% confidence interval are

$$(\hat{\mu}_1^* - \hat{\mu}_2^*) - c_o\hat{\sigma}_{\hat{\mu}_1^*-\hat{\mu}_2^*} = -2.544 - 2.73(.398) = -3.631$$

$$(\hat{\mu}_1^* - \hat{\mu}_2^*) + c_o\hat{\sigma}_{\hat{\mu}_1^*-\hat{\mu}_2^*} = -2.544 + 2.73(.398) = -1.457$$

Thus, we also conclude that the regression lines for treatments 1 and 2 have different intercepts.

3.6.2 ANCOVA with Two Covariates

If we wish to use both Z_1 and Z_2 as covariates in analyses of the data in Table 3.15 using an intraclass model, the design matrix and parameter vector are

$$\mathbf{X} = \begin{bmatrix} 1 & 0 & 0 & 6 & 0 & 0 & 2 & 0 & 0 \\ 1 & 0 & 0 & 6 & 0 & 0 & 2 & 0 & 0 \\ 1 & 0 & 0 & 5 & 0 & 0 & 2 & 0 & 0 \\ 1 & 0 & 0 & 4 & 0 & 0 & 4 & 0 & 0 \\ 1 & 0 & 0 & 3 & 0 & 0 & 5 & 0 & 0 \\ 1 & 0 & 0 & 3 & 0 & 0 & 5 & 0 & 0 \\ 1 & 0 & 0 & 2 & 0 & 0 & 6 & 0 & 0 \\ 1 & 0 & 0 & 2 & 0 & 0 & 5 & 0 & 0 \\ 0 & 1 & 0 & 0 & 5 & 0 & 0 & 2 & 0 \\ 0 & 1 & 0 & 0 & 5 & 0 & 0 & 1 & 0 \\ 0 & 1 & 0 & 0 & 3 & 0 & 0 & 2 & 0 \\ 0 & 1 & 0 & 0 & 2 & 0 & 0 & 7 & 0 \\ 0 & 1 & 0 & 0 & 1 & 0 & 0 & 8 & 0 \\ 0 & 1 & 0 & 0 & 1 & 0 & 0 & 7 & 0 \\ 0 & 0 & 1 & 0 & 0 & 4 & 0 & 0 & 3 \\ 0 & 0 & 1 & 0 & 0 & 1 & 0 & 0 & 1 \\ 0 & 0 & 1 & 0 & 0 & 1 & 0 & 0 & 6 \\ 0 & 0 & 1 & 0 & 0 & 2 & 0 & 0 & 6 \end{bmatrix} \qquad \beta = \begin{bmatrix} \mu_1^* \\ \mu_2^* \\ \mu_3^* \\ \beta_{11} \\ \beta_{12} \\ \beta_{13} \\ \beta_{21} \\ \beta_{22} \\ \beta_{23} \end{bmatrix}$$

To test the hypothesis of parallelism of the regression planes (note that with two covariates we are fitting regression *planes* to the data rather than regression *lines*) we use, for example,

$$\mathbf{C}_{par} = \begin{bmatrix} 0 & 0 & 0 & 1 & 0 & -1 & 0 & 0 & 0 \\ 0 & 0 & 0 & 0 & 1 & -1 & 0 & 0 & 0 \\ 0 & 0 & 0 & 0 & 0 & 0 & 1 & 0 & -1 \\ 0 & 0 & 0 & 0 & 0 & 0 & 0 & 1 & -1 \end{bmatrix}$$

The SAS code to perform this analysis is found in Program 3_6_2.sas.

Program 3_6_2.sas

```
/* Program 3_6_2.sas                                        */
/* Program to perform ANCOVA with two covariates and        */
/* Unequal n, using Full Rank model                         */

options ls=80 ps=60 nodate nonumber;
title 'Output 3.6.2: ANCOVA, Two Covariates, Unequal n';

data twocov;
   infile 'c:\ 3_6_2.dat';
   input y a z1 z2 a1 a2 a3 z11 z12 z13 z21 z22 z23;
proc print;
run;

/* Using the full rank model */
proc reg;
   title2 'Full Rank Model';
   model y = a1 a2 a3 z11 z12 z13 z21 z22 z23 /noint;
   parallel: mtest z11-z13=0,
                   z12-z13=0,
                   z21-z23=0,
                   z22-z23=0 /print details;
run;
proc reg;
   model y = a1 a2 a3 z11 z12 z13 z21 z22 z23 /noint;
   restrict  z11-z13=0,
             z12-z13=0,
             z21-z23=0,
             z22-z23=0;
   A: mtest a1-a2=0,
            a2-a3=0 /print details;
   REG: mtest z13=0,
              z23=0 /print details;
run;

/* Using the less than full rank model */
proc glm;
   title2 'Less than full rank model';
   class a;
   model y=a z1 z2 a*z1 a*z2 /e solution;
run;
proc glm;
   class a;
   model y=a z1 z2 /e xpx solution;
   lsmeans a /stderr pdiff cov;
run;
```

Result and Interpretation 3_6_2.sas

Output 3.6.2: *ANCOVA, Two Covariates, Unequal n Full Rank Model*

```
                    Output 3.6.2: ANCOVA, Two Covariates, Unequal n
                                  Full Rank Model

Multivariate Test: PARALLEL

                             L Ginv(X'X) L'   LB-cj

    0.8484044792    0.1682242991    0.6624989475    0.0093457944      -0.39445988
    0.1682242991    0.5460868945    0.0093457944    0.2116358707     -0.558857102
    0.6624989475    0.0093457944    0.7362549465    0.0560747664     -0.646038562
    0.0093457944    0.2116358707    0.0560747664    0.1845734941     -0.634313096

                          Inv(L Ginv(X'X) L')     Inv()(LB-cj)

    10.158676237    -7.058544737     -9.85672018    10.573639801     -0.401649225
    -7.058544737     8.246479645     7.104076489     -11.2564517     0.7263163662
     -9.85672018     7.104076489     10.97364528    -10.98046682     -0.206427045
    10.573639801     -11.2564517    -10.98046682    21.125335598     -4.186400745

                              E, the Error Matrix

                                     4.2409235054

                            H, the Hypothesis Matrix

                                     2.5413760947

                              T, the H + E Matrix

                                        6.7822996

                                  Eigenvectors

                                     0.3839825734

                                  Eigenvalues

                                     0.3747071413


                 Multivariate Statistics and Exact F Statistics

                              S=1     M=1     N=3.5

  Statistic                    Value          F       Num DF     Den DF   Pr > F

  Wilks' Lambda             0.62529286     1.3483          4          9   0.3247
  Pillai's Trace            0.37470714     1.3483          4          9   0.3247
  Hotelling-Lawley Trace    0.59925063     1.3483          4          9   0.3247
  Roy's Greatest Root       0.59925063     1.3483          4          9   0.3247
```

The test of parallelism is summarized in Table 3.17.

Table 3.17 *Test of Parallelism*

Source of Variation	SS	DF	MS	F	p-value
Parallelism	2.54	4	0.64	1.35	.3247
Error	4.24	9	0.47		

Using $\mathbf{C}_{\text{par}}$ as a restriction and the hypothesis test matrices

$$\mathbf{C}_A = \begin{bmatrix} 1 & -1 & 0 & 0 & 0 & 0 & 0 & 0 & 0 \\ 0 & 1 & -1 & 0 & 0 & 0 & 0 & 0 & 0 \end{bmatrix}$$

$$\mathbf{C}_{reg} = \begin{bmatrix} 0 & 0 & 0 & 0 & 0 & 1 & 0 & 0 & 0 \\ 0 & 0 & 0 & 0 & 0 & 0 & 0 & 0 & 1 \end{bmatrix}$$

we perform the analysis of covariance F-test and the test of the significance of the regression on the two covariates. The results of these tests are given in the continuation of Output 3.6.2 that follows.

Output 3.6.2 (*continued*)

```
Model: MODEL1
NOTE: Restrictions have been applied to parameter estimates.
NOTE: No intercept in model. R-square is redefined.
Dependent Variable: Y

                         Analysis of Variance

                              Sum of         Mean
        Source        DF     Squares       Square      F Value       Prob>F

        Model          5   632.21770    126.44354      242.361       0.0001
        Error         13     6.78230      0.52172
        U Total       18   639.00000

            Root MSE       0.72230     R-square       0.9894
            Dep Mean       5.61111     Adj R-sq       0.9853
            C.V.          12.87265

                         Parameter Estimates

                      Parameter      Standard     T for H0:
      Variable  DF     Estimate         Error   Parameter=0    Prob > |T|

      A1         1     7.191606    1.09313702         6.579        0.0001
      A2         1     9.701249    1.00678496         9.636        0.0001
      A3         1     8.593241    0.85338595        10.070        0.0001
      Z11        1    -0.790176    0.17213062        -4.591        0.0005
      Z12        1    -0.790176    0.17213062        -4.591        0.0005
      Z13        1    -0.790176    0.17213062        -4.591        0.0005
      Z21        1    -0.065722    0.11961221        -0.549        0.5920
      Z22        1    -0.065722    0.11961221        -0.549        0.5920
      Z23        1    -0.065722    0.11961221        -0.549        0.5920
      RESTRICT  -1    -0.401649    2.30215928        -0.174        0.8642
      RESTRICT  -1     0.726316    2.07420227         0.350        0.7318
      RESTRICT  -1    -0.206427    2.39272214        -0.086        0.9326
      RESTRICT  -1    -4.186401    3.31985119        -1.261        0.2295
```

Output 3.6.2 (*continued*)

```
Multivariate Test: A

                    L Ginv(X'X) L'    LB-cj

          0.3249410991      -0.125938305      -2.509643307
          -0.125938305      0.4879595639      1.1080077804

             Inv(L Ginv(X'X) L')     Inv()(LB-cj)

          3.4195344372      0.8825533957      -7.603935685
          0.8825533957      2.2771298296      0.3081833453

                       E, the Error Matrix

                            6.7822996

                    H, the Hypothesis Matrix

                          19.424635845

                      T, the H + E Matrix

                          26.206935445

                         Eigenvectors

                          0.1953403136

                         Eigenvalues

                          0.7412021099

         Multivariate Statistics and Exact F Statistics

                      S=1    M=0    N=5.5

 Statistic                  Value          F      Num DF    Den DF  Pr > F

 Wilks' Lambda            0.25879789    18.6161        2        13  0.0002
 Pillai's Trace           0.74120211    18.6161        2        13  0.0002
 Hotelling-Lawley Trace   2.86401914    18.6161        2        13  0.0002
 Roy's Greatest Root      2.86401914    18.6161        2        13  0.0002


        Output 3.6.2: ANCOVA, Two Covariates, Unequal n
                        Full Rank Model

Multivariate Test: REG

                    L Ginv(X'X) L'    LB-cj

          0.0567914087      0.0299983563      -0.790175881
          0.0299983563      0.0274231549      -0.065722426

             Inv(L Ginv(X'X) L')     Inv()(LB-cj)

          41.708333333          -45.625       -29.95833333
               -45.625           86.375             30.375
```

Output 3.6.2 *(continued)*

```
                              E, the Error Matrix

                                     6.7822996

H, the Hypothesis Matrix

                                   21.676033733

                              T, the H + E Matrix

                                   28.458333333

                                 Eigenvectors

                                   0.1874542404

                                 Eigenvalues

                                   0.7616761488

           Multivariate Statistics and Exact F Statistics

                           S=1     M=0     N=5.5

  Statistic                     Value           F      Num DF    Den DF  Pr > F

  Wilks' Lambda              0.23832385    20.7738          2        13  0.0001
  Pillai's Trace             0.76167615    20.7738          2        13  0.0001
  Hotelling-Lawley Trace     3.19597113    20.7738          2        13  0.0001
  Roy's Greatest Root        3.19597113    20.7738          2        13  0.0001
```

The results of these tests are summarized in Table 3.18.

Table 3.18 *Hypothesis Test Results, ANCOVA 2*

Source of Variation	SS	DF	MS	F	p-value
Regression	21.68	2	10.84	20.77	<.0001
A (adjusted treatments)	19.42	2	9.71	18.62	.0002
Error	6.78	13	0.52		

Note that the error variance estimate of 0.52 for this analysis is larger, by a small amount, than the error variance estimate of 0.50 that we got when we used only one covariate. This is an indication that the addition of the second covariate is not effective in accounting for variation in the dependent variable. With samples as small as those of our example it is highly unlikely that two covariates will both be effective in accounting for variation in a dependent variable.

Program 3_6_2.sas also contains the code to perform the less than full rank ANCOVA. The results are found in the continuation of Output 3.6.2 that follows.

Output 3.6.2 (*continued*)

```
              Output 3.6.2: ANCOVA, Two Covariates, Unequal n
                        Less than full rank model

                      General Linear Models Procedure

Dependent Variable: Y
                                      Sum of            Mean
Source                  DF           Squares          Square   F Value     Pr > F

Model                    8       68.03685427      8.50460678     18.05     0.0001

Error                    9        4.24092351      0.47121372

Corrected Total         17       72.27777778

                  R-Square              C.V.        Root MSE               Y Mean

                  0.941325          12.23376       0.6864501            5.6111111

Source                  DF         Type I SS     Mean Square   F Value     Pr > F

A                        2       43.81944444     21.90972222     46.50     0.0001
Z1                       1       21.51852314     21.51852314     45.67     0.0001
Z2                       1        0.15751059      0.15751059      0.33     0.5773
Z1*A                     2        0.07573143      0.03786571      0.08     0.9234
Z2*A                     2        2.46564467      1.23282233      2.62     0.1272

Source                  DF       Type III SS     Mean Square   F Value     Pr > F

A                        2        2.36746801      1.18373401      2.51     0.1359
Z1                       1        9.23132928      9.23132928     19.59     0.0017
Z2                       1        0.64638942      0.64638942      1.37     0.2716
Z1*A                     2        0.63396351      0.31698176      0.67     0.5342
Z2*A                     2        2.46564467      1.23282233      2.62     0.1272

                                        T for H0:    Pr > |T|   Std Error of
Parameter                Estimate     Parameter=0                Estimate

INTERCEPT            7.647196262 B           7.94      0.0001      0.96338719
A          1         2.686137072 B           0.60      0.5615      4.45591735
           2         5.435500939 B           2.20      0.0552      2.46837620
           3         0.000000000 B            .          .          .
Z1                  -0.803738318 B          -2.85      0.0189      0.28154857
Z2                   0.177570093 B           1.09      0.3030      0.16255214
Z1*A       1        -0.394459880 B          -0.62      0.5482      0.63228145
           2        -0.558857102 B          -1.10      0.2992      0.50727077
           3         0.000000000 B            .          .          .
Z2*A       1        -0.646038562 B          -1.10      0.3012      0.58901056
           2        -0.634313096 B          -2.15      0.0600      0.29491281
           3         0.000000000 B            .          .          .

NOTE: The X'X matrix has been found to be singular and a generalized inverse
      was used to solve the normal equations.   Estimates followed by the
      letter 'B' are biased, and are not unique estimators of the parameters.
```

Output 3.6.2 (*continued*)

```
                          General Linear Models Procedure
                              Class Level Information

                             Class     Levels     Values

                             A              3     1 2 3

Number of observations in data set = 18
                         General Linear Models Procedure

                                   The X'X Matrix

                 INTERCEPT             A 1           A 2           A 3             Z1

INTERCEPT               18               8             6             4             56
A 1                      8               8             0             0             31
A 2                      6               0             6             0             17
A 3                      4               0             0             4              8
Z1                      56              31            17             8            226
Z2                      74              31            27            16            183
Y                      101              31            43            27            266

                        Z2               Y

INTERCEPT               74             101
A 1                     31              31
A 2                     27              43
A 3                     16              27
Z1                     183             266
Z2                     392             452
Y                      452             639


                          General Linear Models Procedure
                        General Form of Estimable Functions

Effect         Coefficients

INTERCEPT      L1

A          1   L2
           2   L3
           3   L1-L2-L3

Z1             L5

Z2             L6

                          General Linear Models Procedure

Dependent Variable: Y
                                      Sum of            Mean
Source                      DF       Squares          Square    F Value      Pr > F

Model                        4   65.49547818     16.37386954      31.38      0.0001

Error                       13    6.78229960      0.52171535

Corrected Total             17   72.27777778
```

Output 3.6.2 *(continued)*

```
                   R-Square            C.V.          Root MSE              Y Mean

                   0.906163         12.87265        0.7222987           5.6111111

Source                   DF         Type I SS      Mean Square   F Value      Pr > F

A                         2       43.81944444      21.90972222     42.00      0.0001
Z1                        1       21.51852314      21.51852314     41.25      0.0001
Z2                        1        0.15751059       0.15751059      0.30      0.5920

Source                   DF       Type III SS      Mean Square   F Value      Pr > F

A                         2       19.42463585       9.71231792     18.62      0.0002
Z1                        1       10.99423200      10.99423200     21.07      0.0005
Z2                        1        0.15751059       0.15751059      0.30      0.5920

                                          T for H0:     Pr > |T|    Std Error of
Parameter                 Estimate       Parameter=0                  Estimate

INTERCEPT             8.593241466 B           10.07       0.0001      0.85338595
A          1         -1.401635527 B           -2.59       0.0224      0.54101281
           2          1.108007780 B            2.20       0.0468      0.50455525
           3          0.000000000 B             .          .           .
Z1                   -0.790175881             -4.59       0.0005      0.17213062
Z2                   -0.065722426             -0.55       0.5920      0.11961221

NOTE: The X'X matrix has been found to be singular and a generalized inverse
      was used to solve the normal equations.   Estimates followed by the
      letter 'B' are biased, and are not unique estimators of the parameters.


                         General Linear Models Procedure
                               Least Squares Means

A            Y          Std Err       Pr > |T|     Pr > |T| H0: LSMEAN(i)=LSMEAN(j)
          LSMEAN         LSMEAN     H0:LSMEAN=0    i/j      1        2        3

1     4.46308878     0.27866776          0.0001    1     .        0.0001   0.0224
2     6.97273209     0.29668361          0.0001    2   0.0001     .        0.0468
3     5.86472431     0.41358042          0.0001    3   0.0224   0.0468      .

NOTE: To ensure overall protection level, only probabilities associated with
      pre-planned comparisons should be used.
```

3.6.3 ANCOVA In Nested Designs

In order to further illustrate the principles involved in constructing full rank models for designs having covariates, we now discuss methods used in a nested design.

Consider the design

Table 3.19 *ANCOVA Nested Design*

<table>
<tr><td colspan="4">A_1</td><td colspan="6">A_2</td></tr>
<tr><td colspan="2">$B_{(1)1}$</td><td colspan="2">$B_{(1)2}$</td><td colspan="2">$B_{(2)1}$</td><td colspan="2">$B_{(2)2}$</td><td colspan="2">$B_{(2)3}$</td></tr>
<tr><td>α_{11}</td><td>β_{11}</td><td>α_{12}</td><td>β_{12}</td><td>α_{21}</td><td>β_{21}</td><td>α_{22}</td><td>β_{22}</td><td>α_{23}</td><td>β_{23}</td></tr>
</table>

The intraclass model for this design is

$$\begin{aligned} y_{ijk} &= \alpha_{ij} + \beta_{ij}\, z_{ijk} + e_{ijk} & i &= 1, 2, \ldots, I \\ e_{ijk} &\sim IN(0, \sigma^2) & j &= 1, 2, \ldots, J_i \\ & & k &= 1, 2, \ldots, K_{ij} \end{aligned} \qquad (3.33)$$

Since there are ten parameters in this model the design matrix has ten columns.

We may test the equality of all six slope parameters

$$H: \text{ all } \beta_{ij} \text{ are equal} \qquad (3.34)$$

using the same procedures as we used with other designs. If all β_{ij} are assumed to be equal, we have the nested analysis of covariance design having the form

$$y_{ijk} = \mu^*_{ij} + \beta_p\, z_{ijk} + e_{ijk} \qquad (3.35)$$

which is just (3.33) with the statement in (3.34) used as a restriction rather than as hypothesis. We then proceed to test hypotheses about the μ^*_{ij} using procedures analogous to those presented in Chapter 2 for nested designs having no covariates.

Suppose, however, that theory or previous research suggests that (3.34) is not true but that the slopes for levels of B within A are equal. That is, there is reason to believe that

$$\begin{aligned} \beta_{11} &= \beta_{12} = \beta_1 \\ \beta_{21} &= \beta_{22} = \beta_{23} = \beta_2 \end{aligned}$$

Then we may formulate the model as

$$y_{ijk} = \alpha_{ij} + \beta_i z_{ijk} + e_{ijk} \qquad (3.36)$$

and test hypotheses about B within A. It would not make sense, however, to test hypotheses about factor A using such a model since we would then be faced with problems associated with testing for differences between intercepts when the slopes were known to be different. These problems were discussed in Chapter 2.

To illustrate the procedures for the design in Table 3.19 with the intraclass parameter vector

$$\beta' = [\alpha_{11}, \alpha_{12}, \alpha_{21}, \alpha_{22}, \alpha_{23}, \beta_{11}, \beta_{12}, \beta_{21}, \beta_{22}, \beta_{23}]$$

we may perform the test of (3.36) using

$$\mathbf{C}_{\text{hom}} = \begin{bmatrix} 0 & 0 & 0 & 0 & 0 & 1 & 0 & 0 & 0 & -1 \\ 0 & 0 & 0 & 0 & 0 & 0 & 1 & 0 & 0 & -1 \\ 0 & 0 & 0 & 0 & 0 & 0 & 0 & 1 & 0 & -1 \\ 0 & 0 & 0 & 0 & 0 & 0 & 0 & 0 & 1 & -1 \end{bmatrix} \qquad (3.37)$$

The analysis of covariance (assuming equality of all five slopes) can be performed by using the matrix $\mathbf{C}_{\text{hom}}$ in (3.36) as a restriction and

$$\mathbf{C}_A = (1/2 \quad 1/2 \quad -1/3 \quad -1/3 \quad -1/3 \quad 0 \quad 0 \quad 0 \quad 0 \quad 0)$$

to test factor A differences, and

$$\mathbf{C}_{BWA} = \begin{bmatrix} 1 & -1 & 0 & 0 & 0 & 0 & 0 & 0 & 0 & 0 \\ 0 & 0 & 1 & 0 & -1 & 0 & 0 & 0 & 0 & 0 \\ 0 & 0 & 0 & 1 & -1 & 0 & 0 & 0 & 0 & 0 \end{bmatrix} \tag{3.38}$$

to test for differences among the levels of B within A.

If, however, we assume only that the slopes of the regressions for levels of B within A are equivalent, we use the restriction matrix

$$\mathbf{R} = \begin{bmatrix} 0 & 0 & 0 & 0 & 0 & 1 & -1 & 0 & 0 & 0 \\ 0 & 0 & 0 & 0 & 0 & 0 & 0 & 1 & 0 & -1 \\ 0 & 0 & 0 & 0 & 0 & 0 & 0 & 0 & 1 & -1 \end{bmatrix}$$

and test only the hypothesis about the levels of B within A, using (3.38) as a hypothesis test matrix. In this case we are using (3.36) as a model rather than as the full nested analysis of covariance model (3.35). The partitioning of variation in the analysis of these two models is shown in Table 3.20.

Table 3.20 *Partitioning of Variation*

Full Analysis of Covariance (3.6.3)		Analysis of B with A (3.6.4)	
Source of Variation	DF	Source of Variation	DF
A	1	B within A	3
B within A	3	Covariate	2
Covariate	1	Error	$n-7$
Error	$n-6$		

There is one fewer degree of freedom for error obtained in using model (3.36) than in using model (3.35) because we have one more parameter to estimate in the former. Also we do not account for all $n-1$ degrees of freedom between persons in using (3.36) because there is one degree of freedom associated with A, and we do not test the hypothesis about A.

4. Weighted General Linear Models

Theory

Applications

4.1 Introduction

In Chapters 2 and 3 we discussed the GLM assuming a simple structure for $\mathbf{\Omega}$, namely $\mathbf{\Omega} = \sigma^2 \mathbf{I}_n$, where the observations are independent with common variance σ^2. There we used ordinary least squares estimation (OLSE). Two additional structures that $\mathbf{\Omega}$ may have are

$$\mathbf{\Omega} = \sigma^2 \mathbf{V} \text{ } (\mathbf{V} \text{ known and nonsingular}) \tag{4.1}$$

$$\mathbf{\Omega} = \mathbf{\Sigma} \text{ } (\mathbf{\Sigma} \text{ unknown}) \tag{4.2}$$

When $\mathbf{V}$ is unknown, (4.1) is equivalent to (4.2). Alternatively when $\mathbf{\Sigma}$ is known, (4.2) reduces to (4.1) by letting $\sigma^2 = 1$ and $\mathbf{V} \equiv \mathbf{\Sigma}$.

When $\mathbf{V}$ is known, case (4.1) is often called generalized least squares (GLS), weighted least squares estimation (WLSE) or the Gauss-Markov (GM) model. In WLSE, $\mathbf{V}$ is often a diagonal weight matrix or a correlation matrix, $\mathbf{V} = \mathbf{R} = (r_{ij})$. With a correlation matrix $\mathbf{V}$, a common structure for $\mathbf{\Omega}$ is

$$\begin{aligned}\mathbf{\Omega} = \sigma^2 \mathbf{V} &= \sigma^2[(1-\rho)\mathbf{I} + \rho\mathbf{J}] \\ &= \sigma^2 \begin{pmatrix} 1 & \cdots & \cdots & \cdots & \cdots & \rho \\ \vdots & \ddots & \cdots & \cdots & \cdots & \vdots \\ \vdots & \cdots & \ddots & \cdots & \cdots & \vdots \\ \vdots & \cdots & \cdots & \ddots & \cdots & \vdots \\ \vdots & \cdots & \cdots & \cdots & \ddots & \vdots \\ \rho & \cdots & \cdots & \cdots & \cdots & 1 \end{pmatrix}\end{aligned} \tag{4.3}$$

where $\mathbf{J}$ is a matrix of 1's, $\mathbf{I}$ is an identity matrix, and ρ is the population correlation coefficient. The structure of $\mathbf{\Omega}$ in (4.3) is known as compound symmetry, common in repeated measurement designs which we discussed in Chapter 3.

Since $\mathbf{V}$ is known and nonsingular it may be factored such that $\mathbf{V} = \mathbf{FF}'$, so that $\mathbf{\Omega}$ may be reduced to spherical structure:

$$(\mathbf{F}'\mathbf{F})^{-1}\mathbf{F}'\mathbf{\Omega}\mathbf{F}(\mathbf{F}'\mathbf{F})^{-1} = \mathbf{A}'\mathbf{\Omega}\mathbf{A} = \sigma^2\mathbf{I} \tag{4.4}$$

When (4.1) can be reduced to (4.4), the matrix $\mathbf{\Omega}$ is said to be a type H matrix or Hankel matrix; (4.4) is the circularity assumption in repeated measurement designs discussed in Section 3.5.3.

Case (4.2), when $\mathbf{\Sigma}$ is unknown with no special structure and thus must be estimated by $\hat{\mathbf{\Sigma}}$, is often called feasible generalized least squares (FGLS) estimation or the general Gauss-Markov (GGM) model. If σ^2 and $\mathbf{V}$ are both unknown, then this is a special case of (4.2) since $\mathbf{\Sigma}$ has structure $\mathbf{\Sigma} = \sigma^2\mathbf{V}$.

In this chapter, we show how to estimate the parameter vector β given the more general structure of the covariance matrix shown in (4.1) and (4.2). With both known and unknown covariance structures, we then show how to test linear hypotheses. OLSE and FGLS estimates are compared. In addition, basic asymptotic theory and Fisher's information matrix are developed to establish large sample tests. Examples in this chapter include weighted regression analysis and categorical data analysis.

4.2 Estimation and Hypothesis Testing

Assume for the moment that $\mathbf{\Omega} = \mathbf{\Sigma}$ and that $\mathbf{\Sigma}$ is known. Let

$$\mathbf{y}_o = \mathbf{\Sigma}^{-1/2}\mathbf{y} \quad \text{and} \quad \mathbf{X}_o = \mathbf{\Sigma}^{-1/2}\mathbf{X} \tag{4.5}$$

with $\mathbf{\Sigma}^{-1/2} = \mathbf{P}\mathbf{\Lambda}^{1/2}$, where $\mathbf{P}$ is an orthogonal matrix ($\mathbf{P}'\mathbf{P} = \mathbf{PP}' = \mathbf{I}$) such that $\mathbf{P}'\mathbf{\Sigma}\mathbf{P} = \mathbf{I}$ (by the spectral decomposition theorem, Theorem 7.3). Then, substituting $\mathbf{y}_o$ for $\mathbf{y}$ in the GLM (1.1), the structure for $\mathbf{y}_o$ becomes $\sigma^2\mathbf{I}_n$. Thus, the OLSE = BLUE for β is

$$\hat{\beta} = (\mathbf{X}_o'\mathbf{X}_o)^{-1}\mathbf{X}_o'\mathbf{y}_o = (\mathbf{X}'\mathbf{\Sigma}^{-1}\mathbf{X})^{-1}\mathbf{X}'\mathbf{\Sigma}^{-1}\mathbf{y} \tag{4.6}$$

(4.6) is called the generalized least squares (GLS) estimator of β or the Gauss-Markov estimator of β. The covariance matrix of $\hat{\beta}$ is

$$\operatorname{cov}\hat{\beta} = (\mathbf{X}'\mathbf{\Sigma}^{-1}\mathbf{X})^{-1} \tag{4.7}$$

If $\mathbf{\Omega} = \sigma^2\mathbf{V}$ and $\mathbf{V}$ is known and nonsingular, the weighted least squares estimator (WLSE) of β is

$$\hat{\beta} = (\mathbf{X}_o'\mathbf{X}_o)^{-1}\mathbf{X}_o'\mathbf{y}_o = (\mathbf{X}'\mathbf{V}^{-1}\mathbf{X})^{-1}\mathbf{X}'\mathbf{V}^{-1}\mathbf{y} \tag{4.8}$$

and does not involve the unknown common variance σ^2. Alternatively, the estimators given in (4.6) and (4.8) may be obtained by minimizing the more general least squares criterion

$$(\mathbf{y} - \mathbf{X}\beta)'\mathbf{\Sigma}^{-1}(\mathbf{y} - \mathbf{X}\beta) = \operatorname{Tr}(\mathbf{\Sigma}^{-1}\mathbf{ee}') \tag{4.9}$$

with regard to β, which requires $\mathbf{\Sigma}$ to be nonsingular. For a more general treatment, see Rao (1973, pp. 297-302). When $\mathbf{\Omega} = \sigma^2\mathbf{V}$, and $\mathbf{V}$ is known and nonsingular, the Gauss-Markov theory applies so that a GLSE of β is obtained without making distributional assumptions regarding $\mathbf{y}$. However, to test the hypothesis $H(\mathbf{C}\beta = \xi)$, we make the distributional assumption $\mathbf{y} \sim \mathbf{N}_n(\mathbf{X}\beta, \boldsymbol{\Omega} = \sigma^2\mathbf{V})$ so that $\mathbf{y}_o = \mathbf{V}^{-1/2}\mathbf{y}$ is normal with mean $\mathbf{V}^{-1/2}\mathbf{X}\beta$ and covariance matrix $\boldsymbol{\Omega} = \sigma^2\mathbf{I}_n$. Hence, the general hypothesis testing theory of Chapters 2 and 3 may be used for GLS.

When $\boldsymbol{\Omega} = \boldsymbol{\Sigma}$ and $\boldsymbol{\Sigma}$ is known and Gramian (nonnegative definite, represented as $\boldsymbol{\Sigma} \geq 0$), one can apply the general Gauss-Markov theory to estimate β without making distributional assumptions. However, to test hypotheses, we must make distributional assumptions. Usually we assume that $\mathbf{y} \sim \mathbf{N}_n(\mathbf{X}\beta, \boldsymbol{\Omega} = \boldsymbol{\Sigma})$. The maximum likelihood (ML) estimate of β is

$$\hat{\beta}_{ML} = (\mathbf{X}'\boldsymbol{\Sigma}^{-1}\mathbf{X})^{-1}\mathbf{X}'\boldsymbol{\Sigma}^{-1}\mathbf{y} \tag{4.10}$$

Comparing (4.9) with the density of the multivariate normal distribution (1.14), observe that (4.9) is included in the exponent of the distribution indicating that the OLSE = BLUE = MLE of β under multivariate normality and known $\boldsymbol{\Sigma}$. With known $\boldsymbol{\Sigma}$, the statistic

$$X^2 = (\mathbf{C}\hat{\beta}_{ML} - \xi)'(\mathbf{C}(\mathbf{X}'\boldsymbol{\Sigma}^{-1}\mathbf{X})^{-1}\mathbf{C}')^{-1}(\mathbf{C}\hat{\beta}_{ML} - \xi) \tag{4.11}$$

has a noncentral chi-square distribution. The likelihood ratio test of $H(\mathbf{C}\beta = \xi)$ is to reject H if the $P(X^2 > c|H) = \alpha$ where X^2 has a central chi-square distribution when H is true and degrees of freedom $v = \text{rank}(\mathbf{C}) = g$.

When $\boldsymbol{\Sigma}$ is unknown, estimation of β and hypothesis testing is more complicated since it requires a consistent estimator of $\boldsymbol{\Sigma}$. To see how we might proceed, we know that the ML estimate of β in (4.10) is normally distributed since $\mathbf{y}$ is normally distributed. $\boldsymbol{\Sigma}$ is unknown, but suppose we can find a (weakly) consistent estimator $\hat{\boldsymbol{\Sigma}}$ of $\boldsymbol{\Sigma}$ that converges in probability to $\hat{\boldsymbol{\Sigma}}$, plim $\hat{\boldsymbol{\Sigma}} = \boldsymbol{\Sigma}$. Let the feasible generalized least squares (FGLS) estimator of β be $\hat{\hat{\beta}}_{\text{FGLS}}$ or

$$\hat{\hat{\beta}}_{\text{FGLS}} = \hat{\hat{\beta}}_{CAN} = (\mathbf{X}'\hat{\boldsymbol{\Sigma}}^{-1}\mathbf{X})^{-1}\mathbf{X}'\hat{\boldsymbol{\Sigma}}^{-1}\mathbf{y} \tag{4.12}$$

so that $\hat{\hat{\beta}}_{CAN}$ is a consistent asymptotically normal (CAN) estimator of β since the $\text{plim}\hat{\hat{\beta}}_{\text{CAN}} = \beta$. If $\hat{\Sigma}$ is the ML estimate of Σ, then by maximum likelihood estimation theory, the estimator in (4.12) is the best asymptotically normal (BAN) estimator of β.

Definition 4.1 A CAN estimator $\hat{\hat{\beta}}_k$ of order k based on a sample of size n is said to be a BAN estimator for β if $n^{1/2}(\hat{\hat{\beta}}_k - \beta)$ converges in distribution to a multivariate normal distribution,

$$n^{1/2}(\hat{\hat{\beta}}_k - \beta) \xrightarrow{d} N_k(\mathbf{0}, \mathbf{F}^{-1}(\beta))$$

where $\mathbf{F}^{-1}(\beta)$ is the inverse of the $k \times k$ Fisher's information matrix (the negative matrix of expectations of second order partial derivatives of the ln of the likelihood of β evaluated at its true value times $1/n$).

In general, maximum likelihood estimates are BAN provided the density of the estimator satisfies certain general regularity conditions. For a discussion of these conditions, see Stuart and Ord (1991, vol. II, pp. 649-682). The information matrix for $\hat{\hat{\beta}}$ in (4.12) is given by

$$\mathbf{F}_{n,\beta} = \mathbf{X}'\mathbf{\Sigma}^{-1}\mathbf{X} / n \tag{4.13}$$

where Σ is unknown. The asymptotic covariance matrix for $\hat{\hat{\beta}}$ is then

$$\frac{1}{n}\mathbf{F}_{n,\beta}^{-1} = (\mathbf{X}'\Sigma^{-1}\mathbf{X})^{-1} \tag{4.14}$$

since $\mathbf{X}$ is of full rank and Σ is nonsingular.

To test the hypothesis $H(\mathbf{C}\beta = \xi)$ where $\mathbf{\Sigma}$ is unknown, we may use the test statistic according to Wald (1943). Wald showed that

$$W = (\mathbf{C}\hat{\hat{\beta}}_{\text{FGLS}} - \xi)'(\mathbf{C}(\mathbf{X}'\hat{\mathbf{\Sigma}}^{-1}\mathbf{X})^{-1}\mathbf{C}')^{-1}(\mathbf{C}\hat{\hat{\beta}}_{\text{FGLS}} - \xi) \tag{4.15}$$

converges asymptotically to a central chi-square distribution with degrees of freedom $g = \text{rank}(\mathbf{C})$, when H is true. The null hypothesis H is rejected at the nominal level α if $W > c$ where c is the critical value of the central chi-square distribution such that

$$P(W > c | H) = \alpha \tag{4.16}$$

When n is small, W/g may be approximated by an F-distribution with degrees of freedom $g = \text{rank}(\mathbf{C})$ and $v_e = n - \text{rank}(\mathbf{X})$, (Theil, 1971, pp. 402-403). However, the accuracy of the approximation is unknown. An estimate of Σ is not usually available with univariate data; thus, the procedure outlined is more useful with multivariate data where one has multiple observations for $\mathbf{y}$. The estimator for Σ depends on the observed data.

4.3 OLSE Versus FGLS

To estimate β and to test hypotheses of the form $H(\mathbf{C}\beta = \xi)$ using the GLM (1.1) having unknown, nonsingular covariance structure $\Omega = \Sigma$, we obtained the GM estimator of β and substituted a consistent estimator for $\mathbf{\Sigma}$ to obtain the FGLS estimate in (4.12). Using asymptotic theory, we are able to estimate β optimally and to test hypotheses. The FGLS is used when $\mathbf{\Sigma}$ is unknown and we have no knowledge about the structure of $\mathbf{\Sigma}$. When we do have knowledge of the structure of $\mathbf{\Sigma}$, $\mathbf{\Sigma} = \sigma^2\mathbf{I}_n$, or $\mathbf{\Sigma} = \sigma^2\mathbf{V}$ and $\mathbf{V}$ is known, using (4.5) we can use the OLSE, and it is the BLUE.

Now suppose we let $\Sigma = \sigma^2\mathbf{V}$ where $\mathbf{V}$ is not known. The WLSE of β given in (4.8) required $\mathbf{V}$ to be known. For the WLSE of β to be the BLUE, the matrix $\mathbf{V}$ must be eliminated from the WLSE or

$$(\mathbf{X}'\mathbf{V}^{-1}\mathbf{X})^{-1}\mathbf{X}'\mathbf{V}^{-1} = (\mathbf{X}'\mathbf{X})^{-1}\mathbf{X}' \tag{4.17}$$

McElroy (1967) showed that a necessary and sufficient condition for the OLSE of β to be the BLUE is that $\mathbf{V}$ must have the structure of compound symmetry given in (4.3) with $0 \le \rho < 1$ which ensures the existence of $\mathbf{V}^{-1}$ for all values of n. For a given n, the correlation ρ can be replaced by $-1/(n-1) < \rho < 1$; for more detail, see Arnold (1981, pp. 232-238). He calls this situation the exchangeable linear model. The optimal structure of compound symmetry for Σ is a special case of the more general class of covariance matrices of the form

$$\mathbf{\Sigma} = \mathbf{ZVZ}' + \sigma^2\mathbf{I} \tag{4.18}$$

which arises for the mixed model given in (2.24)

$$\mathbf{y} = \mathbf{X}\beta + \mathbf{Zb} + \mathbf{e} \tag{4.19}$$

where $\mathbf{b}$ is a random vector, $\mathbf{b} \sim N(\mathbf{0}, \mathbf{V} = \mathbf{I}_n \otimes \mathbf{D})$, $\mathbf{e} \sim N(\mathbf{0}, \sigma^2\mathbf{I})$, and $\text{cov}(\mathbf{b}, \mathbf{e}) = \mathbf{0}$.

Rao (1967) compared the OLSE with the FGLS estimate under the assumption that Σ has the general form given in (4.18). Rao shows that

$$\text{cov}[(\mathbf{X}'\mathbf{X})^{-1}\mathbf{X}'\mathbf{y}] = \text{cov}(\hat{\beta}_{\text{OLSE}}) = (\mathbf{X}'\mathbf{\Sigma}^{-1}\mathbf{X})^{-1} \tag{4.20}$$

$$\text{cov}(\mathbf{X}'\hat{\mathbf{\Sigma}}^{-1}\mathbf{X})^{-1}\mathbf{X}'\hat{\mathbf{\Sigma}}^{-1}\mathbf{y} = \text{cov}(\hat{\hat{\beta}}_{\text{FGLS}}) = \frac{n-1}{n-k-1}(\mathbf{X}'\mathbf{\Sigma}^{-1}\mathbf{X})^{-1} \tag{4.21}$$

so that the effect of using a maximum likelihood estimate, $\hat{\mathbf{\Sigma}}$ of $\mathbf{\Sigma}$ when $\mathbf{\Sigma}$ is a member of the class of matrices defined by (4.18), is to reduce the estimator's efficiency; thus OLSE is better than the FGLS. Rao further showed that if Σ is not a member of the class, then the

$$\text{cov}(\hat{\beta}_{OLSE}) = (\mathbf{X}'\mathbf{X})^{-1}\mathbf{X}'\mathbf{\Sigma}\mathbf{X}(\mathbf{X}'\mathbf{X})^{-1} \tag{4.22}$$

and that the $\text{cov}(\hat{\beta}_{\text{FGLS}})$ is smaller than this only if

$$|(\mathbf{X}'\mathbf{X})^{-1}(\mathbf{X}'\mathbf{\Sigma}\mathbf{X})(\mathbf{X}'\mathbf{X})^{-1}| > |(\mathbf{X}'\mathbf{\Sigma}^{-1}\mathbf{X})^{-1}| \tag{4.23}$$

thus compensating for the multiplication factor in (4.21). Note that the $\text{cov}(\hat{\beta}_{\text{FGLS}})$ is given by (4.21) whether or not Σ is a member of the class in (4.18).

More generally Rao (1967) also found the class of covariance matrices, Σ, necessary and sufficient for the FGLS estimate to equal the OLSE. That is, to determine Σ such that

$$(\mathbf{X}'\mathbf{\Sigma}^{-1}\mathbf{X})^{-1}\mathbf{X}'\mathbf{\Sigma}^{-1} = (\mathbf{X}'\mathbf{X})^{-1}\mathbf{X}' \tag{4.24}$$

Theorem 4.1 Given the linear model $\mathbf{y} = \mathbf{X}\beta + \mathbf{e}$ where $E(\mathbf{e}) = \mathbf{0}$ and $\text{cov}(\mathbf{e}) = \mathbf{\Sigma}$, the necessary and sufficient condition for the OLSE (Gauss-Markov estimator) of this model to be equivalent to that obtained for the model with $\text{cov}(\mathbf{e}) = \sigma^2\mathbf{I}_n$ is that Σ has the structure

$$\mathbf{\Sigma} = \mathbf{X\Delta X}' + \mathbf{Z\Gamma Z}' + \sigma^2\mathbf{I}_n$$

where $\mathbf{X}_{n\times k}$, $\text{rank}(\mathbf{X}) = k$, $\mathbf{Z}'\mathbf{X} = \mathbf{0}$ and $\mathbf{\Delta}, \mathbf{\Gamma}$ and σ^2 are unknown.

This theorem implies that $\mathbf{y}$ has the mixed model form

$$\mathbf{y} = \mathbf{X}\beta + \mathbf{X}\delta + \mathbf{Z}\gamma + \mathbf{e} \tag{4.25}$$

where δ, γ, and $\mathbf{e}$ are uncorrelated random vectors with covariance matrices $\mathbf{\Delta}, \mathbf{\Gamma}$, and $\sigma^2\mathbf{I}_n$ and means $\mathbf{0}$.

4.4 Maximum Likelihood Estimation and Fisher's Information Matrix

The principle of maximum likelihood estimation states that the best value to assign to an unknown parameter is the value that maximizes the likelihood of the observed sample. That is, the value of the parameter that is most likely, given the data.

To obtain maximum likelihood estimates, let us first consider the random vector $\mathbf{y}$ with mean zero and density $f(\mathbf{y}|\theta)$ and the joint density $P(\mathbf{Y}|\theta)$ where

$$P(\mathbf{Y}|\theta) = P(\mathbf{y}_1, \mathbf{y}_2, \ldots, \mathbf{y}_n|\theta) = \prod_{i=1}^{n} f(\mathbf{y}_i|\theta) \tag{4.26}$$

The likelihood function of θ is defined as the joint density $P(\mathbf{Y}|\theta)$ where the observations are given and θ is a vector of parameters. Thus the likelihood is a function of θ:

$$L(\theta|\mathbf{y}) \propto P(\mathbf{Y}|\theta) \tag{4.27}$$

The principle of ML estimation consists of finding $\hat{\theta}$ as an estimate of θ, where

$$L(\hat{\theta}|\mathbf{y}) = \max_{\theta} L(\theta|\mathbf{y}) \tag{4.28}$$

If $\hat{\theta}$ is the ML estimator and $\hat{\delta}$ is any other estimator, the following inequality holds:

$$L(\hat{\theta}|\mathbf{y}) > L(\hat{\delta}|\mathbf{y}) \tag{4.29}$$

If $\theta_{m\times 1}$ is continuous and $L(\theta|\mathbf{y})$ is differentiable for all values of θ, the maximum of $L(\theta|\mathbf{y})$ can be obtained by equating the partial derivatives to zero:

$$\frac{\partial L(\theta|\mathbf{y})}{\partial \theta_j} = 0 \qquad j = 1, 2, \ldots, m \tag{4.30}$$

Since the $\ln L(\theta|\mathbf{y})$ is a monotone increasing function of $L(\theta|\mathbf{y})$, both attain their maximum value at the same value of θ. Therefore, we may write (4.30) as follows:

$$\begin{aligned} &\frac{\partial \ln L(\theta|\mathbf{y})}{\partial \theta_j} = 0 \qquad j = 1, 2, \ldots, m \\ \text{or} \qquad &\sum_{i=1}^{n} \frac{\partial \ln f(\mathbf{y}_i|\theta)}{\partial \theta_j} = 0 \qquad j = 1, 2, \ldots, m \end{aligned} \tag{4.31}$$

The system of equations in (4.31) is called the ML equations or the normal equations, and a solution to the system of equations yields the ML estimate, $\hat{\theta}$.

If $\hat{\theta}$ is the ML estimator of θ, then $\hat{\theta}$ is invariant under single-valued transformations of the parameters so that if $T(\theta) = \delta$, then the ML estimate of δ is $\hat{\delta} = T(\theta)$. More important, ML estimates are asymptotically efficient (BAN) under general regularity conditions on the likelihood (Stuart and Ord, 1991).

Expanding (4.31) in a Taylor series about $\theta = \theta_o$, we have that the

$$\frac{\partial \ln L(\theta_o|\mathbf{y})}{\partial \theta_j} + (\hat{\theta} - \theta_o)\left[\frac{\partial^2 \ln L(\theta_o|\mathbf{y})}{\partial \theta_j \partial \theta_k}\right] + \ldots = 0 \qquad j, k = 1, 2, \ldots, m \tag{4.32}$$

Neglecting the higher order terms, observe that

$$(\hat{\theta}-\theta_o)=-\left[\sum_{i=1}^{n}\frac{\partial^2 \ln f(\mathbf{y}_i|\theta)}{\partial\theta_j\partial\theta_k}\right]^{-1}\left[\sum_{i=1}^{n}\frac{\partial^2 \ln f(\mathbf{y}_i|\theta)}{\partial\theta_j}\right] \tag{4.33}$$

Multiplying (4.33) by the $n^{1/2}$, we have that

$$n^{1/2}(\hat{\theta}-\theta_o)=-\left[\frac{1}{n}\sum_{i=1}^{n}\frac{\partial^2 \ln f(\mathbf{y}_i|\theta)}{\partial\theta_j\partial\theta_k}\right]^{-1}\left[\frac{1}{n^{1/2}}\sum_{i=1}^{n}\frac{\partial^2 \ln f(\mathbf{y}_i|\theta)}{\partial\theta_j}\right] \tag{4.34}$$

The second term to the right of the equal sign in (4.34) is the sum of n independent and identically distributed random vectors. The random vector $\partial \ln f(\mathbf{y}_i|\theta)/\partial\theta_j$ has zero mean and covariance matrix

$$\text{var}\left(\frac{\partial \ln f(\mathbf{y}_i|\theta)}{\partial\theta_j}\right)=E\left[\frac{\partial \ln f(\mathbf{y}_i|\theta)}{\partial\theta_j}\frac{\partial \ln f(\mathbf{y}_i|\theta)}{\partial\theta_k}\right] \quad \text{for all } j \text{ and } k \tag{4.35}$$

since the $E[\partial \ln f(\mathbf{y}_i|\theta)/\partial\theta_j]=0$.

Taking the expected value of the matrix of second derivatives results in the following

$$-E\left[\frac{\partial^2 \ln f(\mathbf{y}_i|\theta)}{\partial\theta_j\partial\theta_k}\right]=\text{var}\left(\frac{\partial \ln f(\mathbf{y}_i|\theta)}{\partial\theta_j}\right) \tag{4.36}$$

which is defined as the information matrix $\mathbf{F}_\theta$, (Casella and Berger, 1990, pp. 308-312) and

$$\mathbf{F}_\theta=\text{var}\left(\frac{\partial \ln f(\mathbf{y}_i|\theta)}{\partial\theta_j}\right)=-\left[E\left(\frac{\partial^2 \ln f(\mathbf{y}_i|\theta)}{\partial\theta_j\partial\theta_k}\right)\right] \tag{4.37}$$

We now consider the two terms to the right of the equality in (4.34). By the central limit theorem, the random vector $n^{1/2}\sum_{i=1}^{n}\partial \ln f(\mathbf{y}_i|\theta)/\partial\theta_j$ as $n\to\infty$ has a multivariate normal distribution with mean $\mathbf{0}$ and covariance matrix $\mathbf{F}_\theta$. Furthermore, by the weak law of large numbers, the matrix

$$n^{-1}\left[\sum_{i=1}^{n}\frac{\partial^2 \ln f(\mathbf{y}_i|\theta)}{\partial\theta_j\partial\theta_k}\right]$$

is asymptotically equal (converges in probability) to $-\mathbf{F}_\theta$ so that

$$-\left[\frac{1}{n}\sum_{i=1}^{n}\frac{\partial \ln f(\mathbf{y}_i|\theta)}{\partial\theta_j\partial\theta_k}\right]^{-1}\approx\mathbf{F}_\theta^{-1} \tag{4.38}$$

Hence for large n, the quantity $\text{n}^{1/2}(\hat{\theta}-\theta_o)$ has a multivariate normal distribution with mean $\mathbf{0}$ and covariance matrix $\mathbf{F}_\theta^{-1}$ so that the

$$\text{var}(\hat{\theta}-\theta)=\text{var}(\hat{\theta})\approx\mathbf{F}_\theta^{-1}/n$$

which for large n means that the covariance matrix of the ML estimator attains the Cramer-Rao lower bound and is hence BAN, which establishes Definition 4.1. This result implies that an estimate of the asymptotic covariance matrix of the estimators can be derived by evaluating the negative inverse of Fisher's information matrix at the point estimate corresponding to the maximum of the likelihood function and then by multiplying by the factor $1/n$.

4.5 WLSE for Data Heteroscedasticity

The application of WLSE is to stabilize the variance of the dependent variable in the GLM when $\mathbf{\Omega} \neq \sigma^2\mathbf{I}$. For the classical model, we had assumed that the variance of the dependent variable Y_i was constant and that $\mathbf{\Omega} = \sigma^2\mathbf{I}$. When this is not the case, we have heteroscedasticity

$$\text{var}(Y_i) = \sigma_i^2 \tag{4.39}$$

To remove heteroscedasticity, we have two options: either transform the data to remove the effect of lack of homogeneity or use WLSE where the weights are, for example, $w_i = 1/\sigma_i^2$. The weights in the diagonal weight matrix are selected proportional to the reciprocal of the variances, thereby discounting unreliability that is due to large variability.

Assuming a simple linear regression model $(Y_i = \alpha + \beta X_i + e_i)$, we estimate the parameters with estimators that minimize the weighted sum of squares

$$\sum_i \frac{1}{\sigma_i^2}(y_i - \alpha - \beta x_i)^2 = \sum_i w_i (y_i - \alpha - \beta x_i)^2 \tag{4.40}$$

to obtain WLS estimates of the parameters. For example, suppose the variance of Y_i is proportional to X_i^2 so that σ_i is proportional to X_i:

$$\text{var}(\mathbf{Y}_i) = \sigma_i^2 = c^2 X_i^2$$
$$\sigma_i = c X_i$$

When one fits a simple linear regression model using OLSE and plots the residuals $(\hat{e}_i = \mathbf{Y}_i - \hat{\mathbf{Y}}_i)$ versus the fitted values $(\hat{\mathbf{Y}}_i)$ a fan pattern will result, suggesting that σ_i may be proportional to X_i. To remove this dependence of σ_i^2 on X_i, one may transform the original data by dividing Y_i by X_i and then perform a simple linear regression using OLSE with this transformed variable

$$Y_i^* = Y_i/x_i = \alpha(1/x_i) + \beta + (e_i/x_i)$$

so that the $\text{var}(Y_i^*) = c^2$, a constant which permits OLSE. Alternatively, we can minimize (4.40) directly using WLSE. Using matrix notation, the weight matrix is

$$\mathbf{W} = \begin{pmatrix} w_1 & & & 0 \\ & w_2 & & \\ & & \ddots & \\ 0 & & & w_n \end{pmatrix} = \begin{pmatrix} 1/\sigma_1^2 & & & 0 \\ & 1/\sigma_2^2 & & \\ & & \ddots & \\ 0 & & & 1/\sigma_n^2 \end{pmatrix} = \mathbf{V}^{-1} \tag{4.41}$$

For our simple linear regression example, the weight matrix for $c = 1$ is

$$\mathbf{W} = \begin{pmatrix} 1/s_1 & & & 0 \\ & 1/s_2 & & \\ & & \ddots & \\ & 0 & & 1/s_n \end{pmatrix} = \mathbf{V}^{-1} \tag{4.42}$$

In applications of multiple linear regression, one often finds that the dependent variable is related in a known way to one of several independent variables, for example X_1, and then $\sigma_i = cX_{1i}$ as described above. Reiterating, one may solve this heteroscedasticity problem using variable transformations in which all independent variables and the dependent variable are divided by $\mathbf{X}_{1i}$, or one may use the WLSE procedure where **W** in (4.42) contains values of s_i. Other patterns of heteroscedasticity include

$$\text{var}(\mathbf{Y}_i) = \sigma_i^2 = c^2\mathbf{X}_{1i}$$
$$\text{var}(\mathbf{Y}_i) = \sigma_i^2 = c^2\sqrt{\mathbf{X}_{1i}}$$

To illustrate the WLSE procedure using SAS, we analyze a problem given in Neter, Wasserman, and Kutner (1990, p. 430, problem 11.22) using OLSE and WLSE with the REG procedure. For a comprehensive overview of how to use the SAS System for regression see Littell, Freund, and Spector (1991). In this application, a health educator is interested in studying the relation between diastolic blood pressure (BLPRESS) and age among healthy adult women aged 20 to 60 years (AGE), using a random sample of 54 women. The SAS code for the analysis is provided in Program 4_5.sas.

Program 4_5.sas

```
/* Program 4_5.sas                                        */
/* WLSE for data with heteroscedasticity                  */
/* Data are from Neter, Wasserman, and Kutner, 1990, p. 421 */

options ls=78 ps=60 nodate nonumber;
title1 'Output 4.5: Data with Heteroscedasticity';

data heter;
   infile 'c:\4_5.dat';
   input age blpress;
   if age lt 30 then agegrp=1;
   else if age lt 40 then agegrp=2;
   else if age lt 50 then agegrp=3;
   else agegrp=4;
run;
proc sort;
   by agegrp;
run;
proc means mean std var n;
   var blpress;
   by agegrp;
run;

/* Ordinary Least Square Regression */
proc reg data=heter;
   title2 'Ordinary Least Squares Regression ';
   model blpress=age/ p cli;
   output out=resid1 r=olsresid;
run;
```

```
/* Plot of OLS residuals */
filename out1 'c:\4_5_1.cgm';
goptions device=cgmmwwc gsfname=out1 gsfmode=replace
   colors=(black) vsize=5in hsize=4in htitle=1;
proc gplot data=resid1;
   title1 'OLS Residuals';
   title2;
   plot olsresid*age /frame;
   symbol1 v=;
run;

/* Add weight vector to the SAS dataset */
data wlse;
   set heter;
   if agegrp=1 then w=(1/sqrt(24.9231));
   else if agegrp=2 then w=(1/sqrt(50.5256));
   else if agegrp=3 then w=(1/sqrt(96.3524));
   else if agegrp=4 then w=(1/sqrt(133.1410));
run;

/* Weighted Least Squares Regression */
proc reg data=wlse;
   title1 'Output 4.5: Data with Heteroscedasticity';
   title2 'Weighted Least Squares Regression ';
   model blpress=age/ p cli;
   weight w;
   output out=resid2 r=wlsresid;
run;
data resid3;
   set resid2;
   wwlsresi=w*wlsresid;
   label wwlsresi = 'weighted residual';
run;

/* Plot of weighted WLS residuals */
filename out2 'c:\4_5_2.cgm';
goptions device=cgmmwwc gsfname=out2 gsfmode=replace
   colors=(black) vsize=5in hsize=4in htitle=1;
proc gplot data=resid3;
   title1 'Weighted WLS Residuals';
   title2;
   plot wwlsresi*age /frame;
   symbol1 v=;
run;
```

The data are read from the file 4_5.dat and then the mean and standard deviation of blpress are computed for each age group. To obtain the OLSE estimates for the simple linear regression equation $E(Y) = \alpha + \beta x$, the MODEL statement blpress = age is used. The P and CLI options are used to output the predicted values $\hat{Y}$ and the confidence bounds for the predicted values. The output statement is used to read the predicted values and residuals to an output file we named resid. The procedure GPLOT is then used to plot these residuals versus the predicted values. The results are given in Output 4.5.

Result and Interpretation 4_5.sas

Output 4.5: *Data with Heteroscedasticity*

```
                Output 4.5: Data with Heteroscedasticity

                     Analysis Variable : BLPRESS


----------------------------------- AGEGRP=1 --------------------------------

                     Mean        Std Dev       Variance   N
               ---------------------------------------------
                 70.6153846     4.9923018     24.9230769  13
               ---------------------------------------------

----------------------------------- AGEGRP=2 --------------------------------

                     Mean        Std Dev       Variance   N
               ---------------------------------------------
                 76.2307692     7.1081391     50.5256410  13
               ---------------------------------------------

----------------------------------- AGEGRP=3 --------------------------------

                     Mean        Std Dev       Variance   N
               ---------------------------------------------
                 82.2666667     9.8159249     96.3523810  15
               ---------------------------------------------

----------------------------------- AGEGRP=4 --------------------------------

                     Mean        Std Dev       Variance   N
               ---------------------------------------------
                 86.8461538    11.5386752    133.1410256  13
               ---------------------------------------------



Model: MODEL1
Dependent Variable: BLPRESS

                          Analysis of Variance

                             Sum of         Mean
       Source          DF       Squares        Square       F Value       Prob>F

       Model            1    2374.96833    2374.96833        35.793       0.0001
       Error           52    3450.36501      66.35317
       C Total         53    5825.33333

           Root MSE       8.14575     R-square        0.4077
           Dep Mean      79.11111     Adj R-sq        0.3963
           C.V.          10.29659

                          Parameter Estimates

                       Parameter      Standard    T for H0:
     Variable  DF       Estimate         Error   Parameter=0    Prob > |T|

     INTERCEP   1      56.156929    3.99367376        14.061        0.0001
     AGE        1       0.580031    0.09695116         5.983        0.0001
```

Output 4.5 *(continued)*

Obs	Dep Var BLPRESS	Predict Value	Std Err Predict	Lower95% Predict	Upper95% Predict	Residual
1	73.0000	71.8178	1.648	55.1411	88.4944	1.1822
2	66.0000	68.3376	2.115	51.4502	85.2250	-2.3376
3	63.0000	68.9176	2.033	52.0707	85.7645	-5.9176
4	79.0000	71.2377	1.721	54.5314	87.9441	7.7623
5	68.0000	70.6577	1.796	53.9195	87.3959	-2.6577
6	67.0000	72.3978	1.577	55.7485	89.0470	-5.3978
7	75.0000	70.0777	1.873	53.3054	86.8499	4.9223
8	71.0000	70.6577	1.796	53.9195	87.3959	0.3423
9	70.0000	69.4976	1.952	52.6892	86.3061	0.5024
10	65.0000	67.7575	2.198	50.8274	84.6877	-2.7575
11	79.0000	72.9778	1.510	56.3538	89.6019	6.0222
12	72.0000	70.0777	1.873	53.3054	86.8499	1.9223
13	70.0000	67.7575	2.198	50.8274	84.6877	2.2425
14	91.0000	78.1981	1.119	61.6990	94.6972	12.8019
15	76.0000	74.7179	1.330	58.1559	91.2799	1.2821
16	69.0000	75.2979	1.279	58.7522	91.8437	-6.2979
17	66.0000	74.1379	1.386	57.5575	90.7183	-8.1379
18	73.0000	75.8780	1.233	59.3461	92.4099	-2.8780
19	78.0000	77.6181	1.136	61.1142	94.1220	0.3819
20	87.0000	78.1981	1.119	61.6990	94.6972	8.8019
21	76.0000	75.2979	1.279	58.7522	91.8437	0.7021
22	79.0000	76.4580	1.194	59.9377	92.9783	2.5420
23	73.0000	73.5579	1.446	56.9567	90.1590	-0.5579
24	68.0000	77.6181	1.136	61.1142	94.1220	-9.6181
25	80.0000	74.1379	1.386	57.5575	90.7183	5.8621
26	75.0000	78.7781	1.110	62.2815	95.2748	-3.7781
27	89.0000	82.8383	1.272	66.2948	99.3819	6.1617
28	101.0	84.5784	1.437	67.9805	101.2	16.4216
29	70.0000	79.3582	1.109	62.8617	95.8547	-9.3582
30	72.0000	80.5182	1.133	64.0152	97.0213	-8.5182
31	80.0000	81.0983	1.157	64.5885	97.6080	-1.0983
32	83.0000	82.8383	1.272	66.2948	99.3819	0.1617
33	75.0000	81.0983	1.157	64.5885	97.6080	-6.0983
34	80.0000	84.5784	1.437	67.9805	101.2	-4.5784
35	90.0000	79.3582	1.109	62.8617	95.8547	10.6418
36	70.0000	83.9984	1.377	67.4209	100.6	-13.9984
37	85.0000	80.5182	1.133	64.0152	97.0213	4.4818
38	71.0000	81.6783	1.189	65.1595	98.1970	-10.6783
39	80.0000	82.8383	1.272	66.2948	99.3819	-2.8383
40	96.0000	83.4184	1.322	66.8589	99.9778	12.5816
41	92.0000	82.2583	1.227	65.7283	98.7883	9.7417
42	76.0000	88.0586	1.862	71.2916	104.8	-12.0586
43	71.0000	87.4786	1.785	70.7453	104.2	-16.4786
44	99.0000	89.2187	2.021	72.3776	106.1	9.7813
45	86.0000	86.3185	1.637	69.6461	103.0	-0.3185
46	79.0000	86.8986	1.710	70.1968	103.6	-7.8986
47	92.0000	88.6387	1.940	71.8357	105.4	3.3613
48	85.0000	86.3185	1.637	69.6461	103.0	-1.3185
49	109.0	89.2187	2.021	72.3776	106.1	19.7813
50	71.0000	85.1585	1.500	68.5379	101.8	-14.1585
51	90.0000	90.3787	2.185	73.4551	107.3	-0.3787
52	91.0000	85.1585	1.500	68.5379	101.8	5.8415
53	100.0	86.3185	1.637	69.6461	103.0	13.6815
54	80.0000	89.7987	2.102	72.9174	106.7	-9.7987

Output 4.5: *Data with Heteroscedasticity Ordinary Least Squares Regression*

```
             Output 4.5: Data with Heteroscedasticity
                       Ordinary Least Squares Regression

Sum of Residuals                        0
Sum of Squared Residuals        3450.3650
Predicted Resid SS (Press)      3724.5142
```

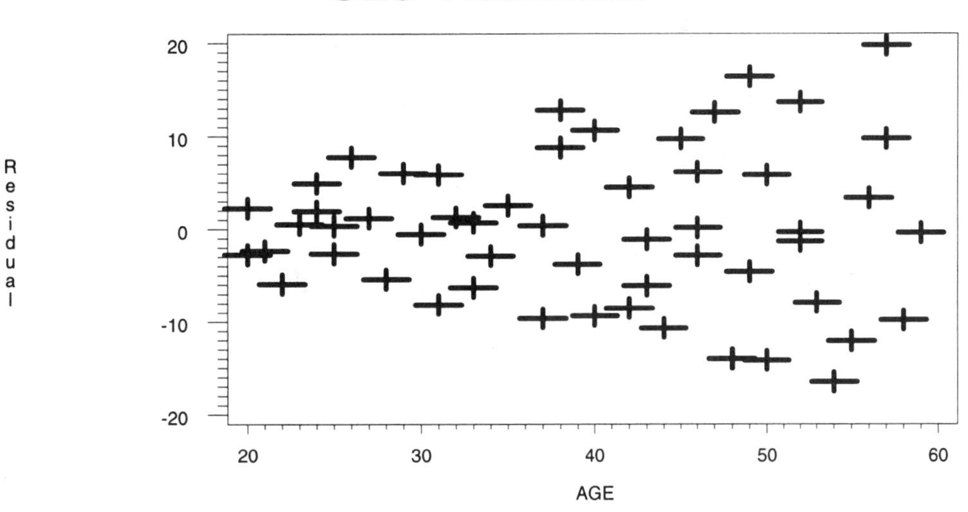

The OLS equation is

$$\hat{Y} = 56.156929 + 0.580031X$$

where X = age and $\hat{Y}$ = predicted blood pressure.

The plot of the OLS residuals versus the fitted values provides an evaluation of the homogeneity of variance assumption. From the plot in Output 4.5, we can see that there is clearly a fan shape, indicating a lack of homogeneity of variance. Upon examining the four subgroups of age groups < 30, < 40, < 50 and 55+ respectively, one observes that the standard deviations are 4.99, 7.11, 9.82, and 11.54, respectively. Therefore, we now use WLSE with weights $1/s_j$.

To perform WLSE on these data, we use the REG procedure using a vector of weights that are constructed in the DATA step. The weight vector is calculated as $1/s_j$, where s_j is the standard deviation of the age group. The WEIGHT statement in PROC REG causes the SAS System to perform a WLS analysis. The WLSE are found in the continuation of Output 4.5 that follows, along with a plot of the weighted WLS residuals versus the predicted values.

Output 4.5 (*continued*)

Model: MODEL1
Dependent Variable: BLPRESS

Analysis of Variance

Source	DF	Sum of Squares	Mean Square	F Value	Prob>F
Model	1	310.27305	310.27305	43.666	0.0001
Error	52	369.48755	7.10553		
C Total	53	679.76061			

Root MSE	2.66562	R-square	0.4564
Dep Mean	77.15645	Adj R-sq	0.4460
C.V.	3.45482		

Parameter Estimates

Variable	DF	Parameter Estimate	Standard Error	T for H0: Parameter=0	Prob > |T|
INTERCEP	1	56.208449	3.32443339	16.908	0.0001
AGE	1	0.581610	0.08801531	6.608	0.0001

Obs	Weight	Dep Var BLPRESS	Predict Value	Std Err Predict	Lower95% Predict	Upper95% Predict	Residual
1	0.2003	73.0000	71.9119	1.278	59.6886	84.1352	1.0881
2	0.2003	66.0000	68.4223	1.658	56.0163	80.8282	-2.4223
3	0.2003	63.0000	69.0039	1.589	56.6345	81.3733	-6.0039
4	0.2003	79.0000	71.3303	1.334	59.0827	83.5779	7.6697
5	0.2003	68.0000	70.7487	1.394	58.4744	83.0230	-2.7487
6	0.2003	67.0000	72.4935	1.225	60.2920	84.6951	-5.4935
7	0.2003	75.0000	70.1671	1.456	57.8635	82.4707	4.8329
8	0.2003	71.0000	70.7487	1.394	58.4744	83.0230	0.2513
9	0.2003	70.0000	69.5855	1.522	57.2502	81.9208	0.4145
10	0.2003	65.0000	67.8407	1.729	55.3957	80.2856	-2.8407
11	0.2003	79.0000	73.0751	1.176	60.8928	85.2575	5.9249
12	0.2003	72.0000	70.1671	1.456	57.8635	82.4707	1.8329
13	0.2003	70.0000	67.8407	1.729	55.3957	80.2856	2.1593
14	0.1407	91.0000	78.3096	1.016	63.9036	92.7156	12.6904
15	0.1407	76.0000	74.8200	1.062	60.4008	89.2392	1.1800
16	0.1407	69.0000	75.4016	1.036	60.9900	89.8132	-6.4016
17	0.1407	66.0000	74.2384	1.094	59.8094	88.6673	-8.2384
18	0.1407	73.0000	75.9832	1.017	61.5770	90.3893	-2.9832
19	0.1407	78.0000	77.7280	1.005	63.3252	92.1308	0.2720
20	0.1407	87.0000	78.3096	1.016	63.9036	92.7156	8.6904
21	0.1407	76.0000	75.4016	1.036	60.9900	89.8132	0.5984
22	0.1407	79.0000	76.5648	1.005	62.1619	90.9677	2.4352
23	0.1407	73.0000	73.6568	1.133	59.2159	88.0977	-0.6568
24	0.1407	68.0000	77.7280	1.005	63.3252	92.1308	-9.7280
25	0.1407	80.0000	74.2384	1.094	59.8094	88.6673	5.7616
26	0.1407	75.0000	78.8912	1.035	64.4799	93.3026	-3.8912
27	0.1019	89.0000	82.9625	1.332	65.9922	99.9329	6.0375
28	0.1019	101.0	84.7073	1.519	67.6738	101.7	16.2927
29	0.1019	70.0000	79.4729	1.061	62.5797	96.3660	-9.4729
30	0.1019	72.0000	80.6361	1.131	63.7245	97.5476	-8.6361
31	0.1019	80.0000	81.2177	1.175	64.2942	98.1412	-1.2177
32	0.1019	83.0000	82.9625	1.332	65.9922	99.9329	0.0375
33	0.1019	75.0000	81.2177	1.175	64.2942	98.1412	-6.2177
34	0.1019	80.0000	84.7073	1.519	67.6738	101.7	-4.7073

Output 4.5 *(continued)*

35	0.1019	90.0000	79.4729	1.061	62.5797	96.3660	10.5271
36	0.1019	70.0000	84.1257	1.454	67.1151	101.1	-14.1257
37	0.1019	85.0000	80.6361	1.131	63.7245	97.5476	4.3639
38	0.1019	71.0000	81.7993	1.223	64.8620	98.7366	-10.7993
39	0.1019	80.0000	82.9625	1.332	65.9922	99.9329	-2.9625
40	0.1019	96.0000	83.5441	1.392	66.5545	100.5	12.4559
41	0.1019	92.0000	82.3809	1.276	65.4280	99.3338	9.6191
42	0.0867	76.0000	88.1970	1.948	69.6117	106.8	-12.1970
43	0.0867	71.0000	87.6154	1.873	69.0611	106.2	-16.6154
44	0.0867	99.0000	89.3602	2.101	70.7080	108.0	9.6398
45	0.0867	86.0000	86.4522	1.727	67.9551	104.9	-0.4522
46	0.0867	79.0000	87.0338	1.799	68.5089	105.6	-8.0338
47	0.0867	92.0000	88.7786	2.024	70.1606	107.4	3.2214
48	0.0867	85.0000	86.4522	1.727	67.9551	104.9	-1.4522
49	0.0867	109.0	89.3602	2.101	70.7080	108.0	19.6398
50	0.0867	71.0000	85.2890	1.587	66.8425	103.7	-14.2890
51	0.0867	90.0000	90.5234	2.257	71.7978	109.2	-0.5234
52	0.0867	91.0000	85.2890	1.587	66.8425	103.7	5.7110
53	0.0867	100.0	86.4522	1.727	67.9551	104.9	13.5478
54	0.0867	80.0000	89.9418	2.179	71.2537	108.6	-9.9418

```
Sum of Residuals                      0
Sum of Squared Residuals       369.4876
Predicted Resid SS (Press)     396.9641
NOTE: The above statistics use observation weights or frequencies.
```

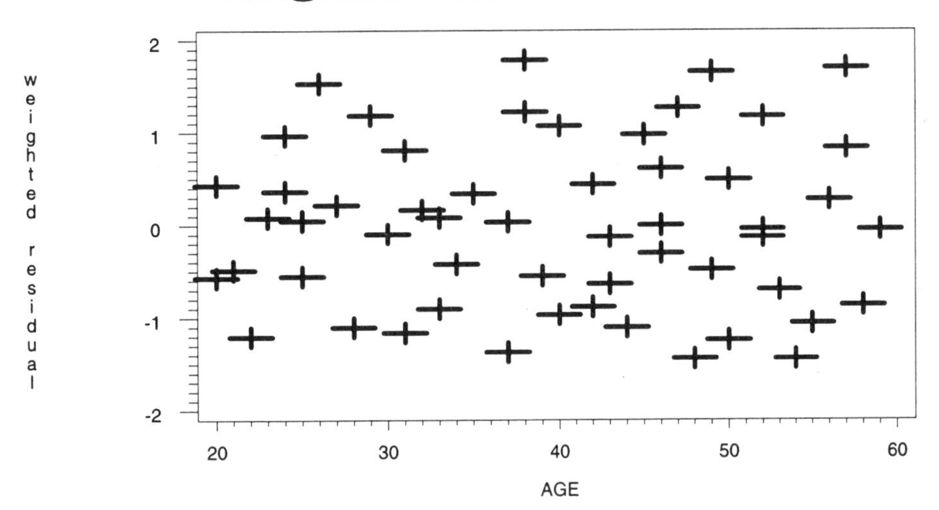

Comparing the WLS estimates with the OLS estimates,

$$\text{WLSE } \hat{Y} = 56.208449 + 0.581610\text{X}$$

$$\text{OLSE } \hat{Y} = 56.156929 + 0.580031\text{X}$$

we see little differences in the model equations. However, because the standard errors of the regression coefficients are smaller for the WLS, it provides a better fit to the data; the WLSE procedure provides tighter bounds on the fitted values. Note that the fan shape pattern has been removed from the WLS residual plot.

4.6 WLSE for Correlated Errors

WLSE was used to modify a regression equation when the variance of the dependent variable increased with the independent variable resulting in heteroscedasticity. The GLSE procedure may also be used to modify the regression equation in the study of longitudinal change where the errors are correlated over time. This occurs frequently in the behavioral sciences--in particular in studies of learning, maturation and development.

To illustrate, here is the multiple regression model for time-series data

$$\begin{aligned} \mathbf{Y}_t &= \mathbf{x}_t \beta + e_t \\ e_t &= \rho\, e_{t-1} + v_t \qquad t = \ldots -2, -1, 0, 1, 2 \ldots \end{aligned} \tag{4.43}$$

where $v_t \sim \mathbf{N}(0, \sigma^2)$ and $|\rho| < 1$, the so-called Markov chain model. Markov models have no "memory" of past states of the process. The probability of the system being in state $\mathbf{Y}_t$ depends only on the state of the system at time $t-1$ and no prior times. The error terms v_t are assumed to satisfy the usual assumptions, $\mathrm{E}(v) = \mathbf{0}$ and $\mathrm{cov}(v) = \sigma^2 \mathbf{I}$. These assumptions regarding v ensure that $\mathrm{E}(\mathbf{e}) = \mathbf{0}$. Writing e_t

$$\begin{aligned} e_t &= \rho\, e_{t-1} + v_t \\ &= \rho(\rho e_{t-2} + v_{t-1}) + v_t \\ &= \rho\left[\rho(\rho e_{t-3} + v_{t-2}) + v_{t-1}\right] + v_t \\ &= v_t + \rho v_{t-1} + \rho^2 v_{t-2} + \rho^3 v_{t-3} + \ldots \\ &= \sum_{s=o}^{\infty} \rho^s v_{t-s} \end{aligned} \tag{4.44}$$

so that $\mathrm{E}(e_t) = 0$, for all t. From (4.43), the

$$\mathrm{var}(e_t) = \sigma^2 \sum_{s=o}^{\infty} \rho^{2s} \rho^{2s} = \sigma^2 / (1 - \rho^2) \tag{4.45}$$

Provided the $|\rho| < 1$, e_t is stationary, we may sum the infinite geometric series. To obtain the covariance between e_t and e_{t+1}, observe that the

$$\begin{aligned} \mathrm{cov}(e_t, e_{t+1}) = \mathrm{E}(e_t e_{t+1}) &= \mathrm{E}\left[e_t(\rho e_t + v_{t+1})\right] \\ &= \mathrm{E}(\rho e_t^2) + \mathrm{E}(e_t v_{t+1}) \\ &= \rho\, \mathrm{var}(e_t) + 0 \end{aligned}$$

since the $\mathrm{E}(v_t v_{t+1}) = 0$ for all t. Furthermore,

$$\begin{aligned} e_{t+2} &= \rho e_{t+1} + v_{t+2} \\ &= \rho(\rho e_t + v_{t+1}) + v_{t+2} \end{aligned}$$

so that the

$$\begin{aligned} \mathrm{cov}(e_t, e_{t+2}) &= \mathrm{E}(e_t e_{t+2}) \\ &= \mathrm{E}\left[e_t(\rho^2 e_t + \rho v_{t+1} + v_{t+2})\right] \\ &= \rho^2 \mathrm{E}(e_t^2) + \rho \mathrm{E}(e_t v_{t+1}) + \mathrm{E}(e_t v_{t+2}) \\ &= \rho^2\, \mathrm{var}(e_t) + 0 + 0 \end{aligned}$$

More generally, the

$$\operatorname{cov}(e_t, e_{t+s}) = \rho^s \operatorname{var}(e_t) = \rho^s \left(\frac{\sigma^2}{1-\rho^2} \right) \qquad t = 0, 1, 2, \ldots \tag{4.46}$$

so that the autocorrelation parameter ρ is the correlation between two successive error terms:

$$\rho(e_t, e_{t+1}) = \frac{\rho(\sigma^2/(1-\rho^2))}{\sqrt{\sigma^2/(1-\rho^2)}\sqrt{\sigma^2/(1-\rho^2)}} = \rho$$

The correlation between e_t and e_{t+s} is then ρ^s.

Returning to model (4.43), we see that the covariance matrix of the errors is

$$\boldsymbol{\Sigma} = \frac{\sigma^2}{1-\rho^2} \begin{pmatrix} 1 & \rho & \rho^2 & \cdots & \cdots & \\ \rho & 1 & \rho & \cdots & \cdots & \\ \rho^2 & \rho & 1 & \cdots & \cdots & \\ \vdots & \vdots & \vdots & \ddots & & \\ \vdots & \vdots & \vdots & & \ddots & \\ \vdots & \vdots & \vdots & & & 1 \end{pmatrix}$$

where

$$\boldsymbol{\Sigma}^{-1} = \frac{1}{\sigma^2} \begin{pmatrix} 1 & -\rho & & & & \\ -\rho & 1+\rho^2 & -\rho & & 0 & \\ & -\rho & 1+\rho^2 & -\rho & & \\ & & \ddots & \ddots & \ddots & \\ & 0 & & -\rho & 1+\rho^2 & -\rho \\ & & & & -\rho & 1 \end{pmatrix}$$

The GLSE of β is

$$\hat{\beta} = (\mathbf{X}'\boldsymbol{\Sigma}^{-1}\mathbf{X})^{-1}\mathbf{X}'\boldsymbol{\Sigma}^{-1}\mathbf{y} \tag{4.47}$$

Alternatively, we can transform the observation vector **y** so that the errors are uncorrelated, which in the analysis of serially correlated time series data is called the analysis of the "first differences." It involves calculating the generalized differences

$$\mathrm{Y}_t = \mathrm{Y}_t - \rho \mathrm{Y}_{t-1}$$
$$\mathbf{x}_t - \rho \mathbf{x}_{t-1}$$

and

$$\mathrm{Y}_1 = \mathrm{Y}_1 \sqrt{1-\rho^2}$$
$$\mathbf{x}_1 = \mathbf{x}_1 \sqrt{1-\rho^2}$$

and performing OLSE. The matrix of differences for **Dy** and **DX** is defined using

$$\mathbf{D} = \begin{pmatrix} \sqrt{1-\rho^2} & & & 0 \\ -\rho & 1 & & \\ & \ddots & \ddots & \\ 0 & & -\rho & 1 \end{pmatrix}$$

motivated by the observation that $-\rho e_{t-1} + e_t = v_t$. In forming the matrix product $\mathbf{D'D}$, observe that $\mathbf{D'D} = \mathbf{\Sigma}^{-1}/\sigma^2$ and that the constant σ^2 cancels using (4.47) so that transforming the observations by the matrix $\mathbf{D}$ is equivalent to GLSE. While this example illustrates the GLS procedure, it is seldom used since ρ is unknown. When ρ is unknown, iterative methods are employed (Neter et al., 1990, p. 496).

4.7 FGLS for Categorical Data

We have shown how WLSE is used to improve estimates when the variance of Y depends on X and when the errors of a linear model have known serial correlation. For illustrations of the application of FGLS and the Wald statistic we now discuss the analysis of categorical data. In particular we present first a brief overview of the theory, and then tests of marginal homogeneity, homogeneity of proportions, and independence.

4.7.1 Overview of the Categorical Data Model

Assume that the categorical data to be analyzed are samples of size $n_{i+}, i = 1,2,\ldots,s$ from s independent multinomial populations of r mutually exclusive and exhaustive categories. Letting N_{ij} be the number of observations falling into cell ij for population i, the multinomial distribution of N_{ij} is

$$P(N_{i1} = n_{i1}, N_{i2} = n_{i2}, \ldots, N_{ir} = n_{ir}) = \frac{n_{i+}!}{\prod_{j=1}^{r} n_{ij}!} \prod_{j=1}^{r} \pi_{ij}^{n_{ij}} \tag{4.48}$$

where π_{ij} is the probability of a single observation falling into cell j, the $\sum_{j=1}^{r} \pi_{ij} = 1$, and $\sum_{j=1}^{r} n_{ij} = n_{i+}$. Given i, the probability of an observation falling into cell j follows a binomial distribution so the $\mathrm{E}(N_{ij}) = n_{i+}\pi_{ij}$ and the $\mathrm{var}(N_{ij}) = n_{i+}\, \pi_{ij}(1-\pi_{ij})$. Furthermore, we can think of the N_{ij} as an experiment of n_{i+} independent Bernoulli trials so that $N_{ij} = \sum_{u=1}^{n_{i+}} Y_u$ and $N_{ij'} = \sum_{v=1}^{n_{i+}} X_v$ where $Y_u = 1$ if trial t results in class j, and $X_v = 1$ if trial t results in class j'. Hence, the

$$\begin{aligned}
\mathrm{cov}(N_{ij}, N_{ij'}) &= \sum_{u=1}^{n_{i+}} \sum_{v=1}^{n_{i+}} \mathrm{cov}(Y_u, X_v) \\
&= \sum_{u=1}^{n_{i+}} \mathrm{cov}(Y_u, X_v) + 2 \sum_{u<1} \mathrm{cov}(Y_u, X_v) \\
&= \sum_{u=1}^{n_{i+}} (-\pi_{iu}\pi_{iv}) \\
&= -n_{i+}\, \pi_{iu}\, \pi_{iv}
\end{aligned} \tag{4.49}$$

so that the N_{ij} are not independent. Since n_{ij} represents the observed number of observations falling into each cell, $p_{ij} = n_{ij}/n_{i+}$ is an estimate of the population proportion π_{ij} and $\mathrm{E}(p_{ij}) = \pi_{ij}$. Grizzle, Starmer, and Koch (1969) proposed analyzing the population proportions as a linear model.

For a s×r contingency table, we let

$$\underset{1\times r}{\pi_i} = (\pi_{i1}, \pi_{i2}, \ldots \pi_{ir})$$
$$\underset{1\times rs}{\pi'} = (\pi_1', \pi_2', \ldots, \pi_s')$$
$$\underset{1\times r}{\mathbf{p}_i'} = (p_{i1}', p_{i2}', \ldots, p_{ir}')$$
$$\underset{1\times rs}{\mathbf{p}'} = (\mathbf{p}_1', \mathbf{p}_2', \ldots, \mathbf{p}_s')$$

Since the N_{ij} are not independent, by (4.49) the covariance matrix for the proportions in the ith population is

$$\text{var}(\mathbf{p}_i) = \underset{r\times r}{\boldsymbol{\Sigma}} = \frac{1}{n_{i+}} \begin{pmatrix} \pi_{i1}(1-\pi_{i1}) & -\pi_{i1}\pi_{i2} & \cdots & -\pi_{i1}\pi_{ir} \\ -\pi_{i2}\pi_{i1} & \pi_{i2}(1-\pi_{i2}) & \cdots & -\pi_{i2}\pi_{ir} \\ \vdots & \vdots & & \vdots \\ -\pi_{ir}\pi_{i1} & -\pi_{ir}\pi_{i2} & \cdots & \pi_{ir}(1-\pi_{ir}) \end{pmatrix} \tag{4.50}$$

and the $\text{var}(\mathbf{p}) = \boldsymbol{\Omega} = \mathbf{I}_s \otimes \boldsymbol{\Sigma}$. Furthermore, let

$f_m(\pi) =$ any function(s) of the elements of π that have partial derivatives up to second order with respect to π_{ij}, $m = 1, 2, \ldots, q < (r-1)s$

$f_m(\mathbf{p}) = f_m(\pi)$ evaluated at $\pi = \mathbf{p}$

$$[\text{F}(\pi)]' = [f_1(\pi), f_2(\pi), \ldots, f_q(\pi)]$$

$$\hat{\mathbf{F}}' = [\text{F}(\mathbf{p})]' = [f_1(\mathbf{p}), f_2(\mathbf{p}), \ldots, f_q(\mathbf{p})]$$

$$\mathbf{H}_{q\times rs} = \left[\frac{\partial f_m(\pi)}{\partial \pi_{ij}} \bigg| \pi_{ij} = p_{ij} \right]$$

By the multivariate δ-method, $\hat{\mathbf{F}}$ converges in distribution to a multivariate normal with covariance matrix $\boldsymbol{\Sigma}_{\text{F}} = \mathbf{H}\boldsymbol{\Omega}\mathbf{H}'$ where $\boldsymbol{\Omega} = \mathbf{I}_s \otimes \boldsymbol{\Sigma}_{r\times r}$. The approximate large sample estimate of Σ_{F} is

$$\mathbf{S}_{q\times q} = \mathbf{H}\hat{\boldsymbol{\Omega}}\mathbf{H}', \quad \hat{\boldsymbol{\Omega}} = \mathbf{I}_s \otimes \hat{\boldsymbol{\Sigma}}_{r\times r} \tag{4.51}$$

where $\hat{\boldsymbol{\Sigma}}_{r\times r}$ is the sample estimate of $\boldsymbol{\Sigma}$ in (4.50) with sample proportions substituted for π_{ij}. See Agresti (1990, p. 424) or Bishop, Fienberg, and Holland (1975, p. 493). When $f_m(\mathbf{p})$ are linear functions of $\boldsymbol{p}$, **S** is the exact covariance matrix. Finally, assume functions are chosen such that they are jointly independent of each other and of the population constraint $\sum_{j=1}^{r} \pi_{ij} = 1$, for $i = 1, 2, \ldots, s$, then, **H** and $\mathbf{H}\hat{\boldsymbol{\Omega}}\mathbf{H}'$ are of full rank q.

With these preliminaries, Grizzle et al. (1969) proposed a linear model for functions of π

$$\mathbf{F}_{q\times 1}(\pi) = \mathbf{X}_{q\times k}\, \beta_{k\times 1} \tag{4.52}$$

where β is a vector of parameters, and **X** is a design matrix of rank $k \le q$. For the parameter vector β in (4.52), the FGLS estimator of β is

$$\hat{\hat{\beta}} = (\mathbf{X}'\mathbf{S}^{-1}\mathbf{X})^{-1}\mathbf{X}'\mathbf{S}^{-1}\hat{\mathbf{F}} \tag{4.53}$$

This estimate minimizes the quadratic form

$$Q = (\hat{\mathbf{F}} - \mathbf{X}\beta)'\,\mathbf{S}^{-1}(\hat{\mathbf{F}} - \mathbf{X}\beta) \tag{4.54}$$

The FGLS in (4.53) has an asymptotic multivariate normal distribution with estimated covariance matrix

$$\operatorname{cov}(\hat{\hat{\beta}}) = (\mathbf{X}'\mathbf{S}^{-1}\mathbf{X})^{-1}$$

To evaluate the fit of the model, one examines the residual sum of squares

$$\begin{aligned} Q &= (\hat{\mathbf{F}} - \mathbf{X}\hat{\hat{\beta}})' \mathbf{S}^{-1}(\hat{\mathbf{F}} - \mathbf{X}\hat{\hat{\beta}}) \\ &= \hat{\mathbf{F}}'\mathbf{S}^{-1}\hat{\mathbf{F}} - \hat{\hat{\beta}}'(\mathbf{X}'\mathbf{S}^{-1}\mathbf{X})\hat{\hat{\beta}} \end{aligned} \tag{4.55}$$

sometimes called a lack of fit or goodness of fit test (Timm, 1975, p. 291). Under the hypothesis $H\colon \mathrm{F}(\pi) - \mathbf{X}\beta = \mathbf{0}$, Q has an asymptotic chi-square distribution with degrees of freedom q-k, the difference between the number of functions and the number of estimated parameters in β. Given the model, the test of the hypothesis $H(\mathbf{C}\beta = \xi)$ may be tested by the Wald statistic given in (4.15),

$$W = (\mathbf{C}\hat{\hat{\beta}} - \xi)' (\mathbf{C}(\mathbf{X}'\mathbf{S}^{-1}\mathbf{X})^{-1}\mathbf{C}')^{-1} (\mathbf{C}\hat{\hat{\beta}} - \xi) \tag{4.56}$$

which has an asymptotic chi-square distribution with degrees of freedom g = rank ($\mathbf{C}$), when H is true.

In applying the general theory presented above, two classes of functions cover most of the problems of interest in categorical data analysis. The first class of functions are linear functions of the cell probabilities. In this case,

$$\mathbf{F}(\pi_{q\times 1}) = \mathbf{A}_{q\times rs}\,\pi_{rs\times 1} = \mathbf{X}_{q\times k}\,\beta_{k\times 1} \tag{4.57}$$

where the rank $\mathbf{A} = \mathrm{q} \le (\mathrm{r}-1)\mathrm{s}$. The elements of $\mathbf{A}$ are arbitrary constants, a_{ijk} where $i = 1, 2, \ldots, q$; $j = 1, 2, \ldots, \mathrm{s}$ and $k = 1, 2, \ldots, r$. For the linear model in (4.57), $\mathbf{H} = \partial\mathbf{F}/\partial\pi = \mathbf{A}$, and $\mathbf{S} = \mathbf{A}\hat{\mathbf{\Omega}}\mathbf{A}'$ where $\hat{\mathbf{\Omega}}$ is a block diagonal $(rs \times rs)$ matrix and

$$\mathbf{A} = \begin{pmatrix} \mathrm{a}_{111} & \mathrm{a}_{112} & \cdots & \mathrm{a}_{11\mathrm{r}}; & \mathrm{a}_{121} & \mathrm{a}_{122} & \cdots & \mathrm{a}_{12\mathrm{r}} & ;\cdots; & \mathrm{a}_{1\mathrm{s}1} & \mathrm{a}_{1\mathrm{s}2} & \cdots & \mathrm{a}_{1\mathrm{sr}} \\ \mathrm{a}_{211} & \mathrm{a}_{212} & \cdots & \mathrm{a}_{21\mathrm{r}}; & \mathrm{a}_{221} & \mathrm{a}_{222} & \cdots & \mathrm{a}_{22\mathrm{r}} & ;\cdots; & \mathrm{a}_{2\mathrm{s}1} & \mathrm{a}_{2\mathrm{s}1} & \cdots & \mathrm{a}_{2\mathrm{sr}} \\ \vdots & \vdots & \vdots & \vdots & \vdots & \vdots & \vdots & \vdots & \cdots & \vdots & \vdots & \vdots & \vdots \\ \vdots & \vdots & \vdots & \vdots & \vdots & \vdots & \vdots & \vdots & \cdots & \vdots & \vdots & \vdots & \vdots \\ \vdots & \vdots & \vdots & \vdots & \vdots & \vdots & \vdots & \vdots & \cdots & \vdots & \vdots & \vdots & \vdots \\ \mathrm{a}_{\mathrm{u}11} & \mathrm{a}_{\mathrm{u}12} & \cdots & \mathrm{a}_{\mathrm{u}1\mathrm{r}}; & \mathrm{a}_{\mathrm{u}21} & \mathrm{a}_{\mathrm{u}22} & \cdots & \mathrm{a}_{\mathrm{u}2\mathrm{r}} & ;\cdots; & \mathrm{a}_{\mathrm{us}1} & \mathrm{a}_{\mathrm{us}2} & \cdots & \mathrm{a}_{\mathrm{usr}} \end{pmatrix} \tag{4.58}$$

The second class of functions include logarithmic functions of the cell probabilities. For this situation,

$$\mathbf{F}(\pi_{t\times 1}) = \mathbf{K}_{t\times q}\,\ln(\mathbf{A}_{q\times rs}\,\pi_{rs\times 1}) = \mathbf{X}_{q\times k}\,\beta_{k\times 1} \tag{4.59}$$

For this model, $\mathbf{H} = \partial\mathbf{F}/\partial\pi = \mathbf{K}\mathbf{D}^{-1}\mathbf{A}$, and $\mathbf{S} = \mathbf{K}\mathbf{D}^{-1}\mathbf{A}\hat{\mathbf{\Omega}}\mathbf{A}'\mathbf{D}^{-1}\mathbf{K}'$ where $\mathbf{A}$ is given in (4.58) and

$$\mathbf{D} = \begin{pmatrix} \mathbf{a}_1'\mathbf{p} & \mathbf{0} & \cdots & \mathbf{0} \\ \mathbf{0} & \mathbf{a}_2'\mathbf{p} & \cdots & \mathbf{0} \\ \vdots & \vdots & \cdots & \vdots \\ \vdots & \vdots & \cdots & \vdots \\ \vdots & \vdots & \cdots & \vdots \\ \mathbf{0} & \mathbf{0} & \cdots & \mathbf{a}_\mathrm{q}'\mathbf{p} \end{pmatrix} \tag{4.60}$$

where $\mathbf{a}_i'$ is the ith row of $\mathbf{A}$. The model given in (4.52) for categorical data is a special case of a more general class of models called generalized linear models (Agresti, 1990; Dobson, 1990; and McCullagh and Nelder, 1989).

The FGLS procedure and the Wald statistic are also used in the analysis of longitudinal data when observations are correlated, missing and normally distributed with unknown covariance structure, as discussed in Chapter 10.

However, when normality cannot be assumed as with binary data, we must develop additional statistical theory. Liang and Zeger (1986) extended the estimating equations for the generalized linear model to the generalized estimating equations (GEE) to analyze binary longitudinal data. An elementary introduction is provided by Dunlop (1994). These procedures are beyond the scope of this book.

4.7.2 Marginal Homogeneity

A common problem in the analysis of categorical data in the behavioral and social sciences is to test the equality of two dependent proportions for longitudinal data consisting of two time points, before and after. Using a contingency table, this is the test of marginal homogeneity (McNemar's test for change for 2×2 tables or Stuart's test for I×I tables). For a 3×3 table, the null hypothesis of marginal homogeneity is

$$H\colon \pi_{i+} = \pi_{+j} \qquad i = j = 1, 2, 3 \tag{4.61}$$

where $\pi_{i+} = \sum_j \pi_{ij}$ and $\pi_{+j} = \sum_i \pi_{ij}$. The sample estimates of the population proportions are $\hat{\pi}_{i+} = n_{i+}/n_{++}$ and $\hat{\pi}_{+j} = n_{+j}/n_{++}$. To represent (4.61) as a linear model, observe that

$$\begin{aligned}
\pi_{1+} = \pi_{+1} \quad &\Rightarrow \quad \pi_{12} + \pi_{13} - \pi_{21} - \pi_{31} = 0 \\
\pi_{2+} = \pi_{+2} \quad &\Rightarrow \quad \pi_{21} + \pi_{23} - \pi_{12} - \pi_{32} = 0 \\
\pi_{3+} = \pi_{+3} \quad &\Rightarrow \quad \pi_{31} + \pi_{32} - \pi_{13} - \pi_{23} = 0
\end{aligned}$$

Thus, H is a linear function of the probabilities. Letting

$$\pi'_{1\times rs} = (\pi_{11}, \pi_{12}, \pi_{13}, \pi_{21}, \pi_{22}, \pi_{23}, \pi_{31}, \pi_{32}, \pi_{33})$$

the null hypothesis has the form

$$\mathbf{F}(\pi) = \mathbf{A}\,\pi = \begin{pmatrix} 0 & 1 & 1 & -1 & 0 & 0 & -1 & 0 & 0 \\ 0 & -1 & 0 & 1 & 0 & 1 & 0 & -1 & 0 \\ 0 & 0 & -1 & 0 & 0 & -1 & 1 & 1 & 0 \end{pmatrix} \pi \tag{4.62}$$

The matrix $\mathbf{A}$ is singular since (-1) times the first row plus (-1) times the second row yields the third row. The matrix $\mathbf{A}$ must be of full rank for $\mathbf{S}$ to be of full rank. Since the chi-square statistic is invariant to the row basis of $\mathbf{A}$, we may delete any row of $\mathbf{A}$ to obtain a full rank matrix. A suitable matrix is

$$\mathbf{A}^* = \begin{pmatrix} 0 & 1 & 1 & -1 & 0 & 0 & -1 & 0 & 0 \\ 0 & -1 & 0 & 1 & 0 & 1 & 0 & -1 & 0 \end{pmatrix} \tag{4.63}$$

Using the matrix $\mathbf{A}^*$ in (4.63), the equivalent hypothesis becomes

$$H\colon\ \mathbf{F}^* = \mathbf{A}^*\pi = \mathbf{X}\beta = \mathbf{0} \tag{4.64}$$

Since $\mathbf{X}\beta$ is zero, if $\hat{\mathbf{p}} = \hat{\pi}$ is an estimate of π, the test statistic for testing (4.64), from (4.55), is

$$\begin{aligned}
Q &= \hat{\mathbf{F}}'\mathbf{S}^{-1}\hat{\mathbf{F}} \\
&= \hat{\mathbf{F}}'^*(\mathbf{A}^*\hat{\Omega}\mathbf{A}^{*\prime})^{-1}\hat{\mathbf{F}}^* \\
&= (\mathbf{A}^*\mathbf{p})'(\mathbf{A}^*\hat{\Omega}\mathbf{A}^{*\prime})^{-1}\mathbf{A}^*\mathbf{p}
\end{aligned} \tag{4.65}$$

which has a chi-square distribution with degrees of freedom $g = \text{Rank}(\mathbf{A}^*) = q - 1 = 2$. The estimate of the covariance matrix for **p** is

$$\text{var}(\mathbf{p}) = \frac{1}{n_{++}}\left[\mathbf{D}(\mathbf{p}) - \mathbf{p}\mathbf{p}'\right] = \hat{\Omega} \tag{4.66}$$

where $\mathbf{D}(\mathbf{p})$ is a diagonal matrix of proportions since we have only one population.

As an example of the application of testing for of marginal homogeneity, we use data from Marascuilo and McSweeney (1977, p. 174). The data are from a study of school desegregation in which the racial composition of a city's schools in 1955 is compared to racial composition in 1965. The racial composition of a sample of 124 schools is given in Table 4.1.

Table 4.1 *Racial Composition of Schools*

1955 / 1965	Segregated 90-100% Black	Desegregated 11-89% Black/White	Segregated 90-100% White	Total
Segregated Black	50	9	10	69
Desegregated	2	6	29	37
Segregated White	0	0	18	18
Total	52	15	57	124

To analyze the categorical data in Table 4.1 for the equality of marginal proportions, we use the CATMOD procedure and also show how the IML procedure may be used to perform the analysis. Other SAS procedures for the analysis of categorical data are the LOGISTIC and GENMOD procedures. PROC LOGISTIC is a procedure used to analyze logistic regression models, while PROC GENMOD is used in the analysis of generalized linear models (Stokes, Davis, and Koch, 1995). The SAS code for the analysis of the example is provided in Program 4_7_2.sas.

Program 4_7_2.sas

```
/* Program 4_7_2.sas                                        */
/* Data are from Marascuilo and McSweeney, 1977, p.174      */
/* Using PROC CATMOD to evaluate Marginal Homogeniety       */

options ls=78 ps=60 nodate nonumber;
title1 'Output 4.7.2: Test of Marginal Homogeniety';

/* Using PROC IML */
title2 'Using PROC IML';
proc iml;
   start cat1;
```

```
      /* Input variables and data */
      n=124;
      s=3;
      r=3;
      freq={50 9 10,
             2 6 29,
             0 0 18};
      print freq;

      /* Form proportions and shape p */
      prop=freq/n;
      p=shape(prop,r*s,1);
      print p;

      /* Construct Covariance matrix and test matrix */
      omega=(diag(p)-p*p`)/n;
      qr=2;
      A={0  1 1 -1 0 0 -1  0 0,
         0 -1 0  1 0 1 0  -1 0};
      S=A*Omega*A`;
      print S;
      InvS=Inv(S);
      print InvS;
      F=A*p;
      print F;

      /* Calculate Wald statistic and p-value */
      Q=F`*InvS*F; pv=1-probchi(Q,qr);
      print 'Chi-square Test of Marginal Homogeniety',Q, 'p-value',pv;
   finish cat1;
run cat1;
quit;
run;

/* Using PROC CATMOD */
data school;
   input t55 t65 count @@;
   cards;
   1 1 50 1 2 9 1 3 10
   2 1  2 2 2 6 2 3 29
   3 1  0 3 2 0 3 3 18
   ;
run;

proc catmod;
title2 'Using PROC CATMOD';
   weight count;
   response marginals;
   model t55*t65=_response_;
   repeated year 2;
run;
```

The analysis of the school desegregation data using PROC IML is straightforward. We have only to evaluate the matrix expression given in (4.65) with $\hat{\mathbf{\Omega}}$ defined in (4.66) and $\mathbf{A}^*$ defined in (4.63). The IML code first reads the data into a matrix named **freq**, prints **freq**, calculates the covariance matrix, **S,** and prints it, and calculates the Wald statistic, W, and the associated *p*-value, and prints these.

Result and Interpretation 4_7_2.sas

Output 4.7.2: *Test of Marginal Homogeneity Using PROC IML*

```
            Output 4.7.2: Test of Marginal Homogeneity
                         Using PROC IML

                        FREQ
                          50         9         10
                           2         6         29
                           0         0         18

                                     P
                             0.4032258
                             0.0725806
                             0.0806452
                              0.016129
                             0.0483871
                              0.233871
                                     0
                                     0
                             0.1451613

                                S
                        0.0012142 -0.000912
                        -0.000912 0.0023476

                             INVS
                        1162.4717 451.37977
                        451.37977  601.2337

                                     F
                             0.1370968
                             0.1774194

            Chi-square Test of Marginal Homogeneity

                                     Q
                             62.733059

                              p-value

                                    PV
                             2.387E-14
```

Since the critical value of the chi-square distribution with 2 *df* is 5.99, the hypothesis of marginal homogeneity is rejected at the nominal α =.05 level of significance. The racial composition of schools in 1965 differs from that in 1955.

Program 4_7_2.sas also contains the code to perform the analysis using PROC CATMOD. First the data are input; this is done with the INPUT statement. The variables input are T55 (category of desegregation in 1955), T65 (category of desegregation in 1965), and COUNT (the number of schools). The @@ symbol allows for the input of more than one cell frequency in a line.

The four statements used in PROC CATMOD are WEIGHT, RESPONSE, MODEL, and REPEATED. The WEIGHT statement includes the variable that contains the cell frequencies. The RESPONSE statement defines the response function **F**, where the MARGINAL function specifies that the response functions are marginal proportions. Only the independent marginal proportions $\hat{\pi}_{1+}$, $\hat{\pi}_{2+}$, $\hat{\pi}_{+1}$, and $\hat{\pi}_{+2}$ are calculated and associated with the response function $\hat{\mathbf{F}}$, as seen in the following output using PROC CATMOD. The RESPONSE statement directs the CATMOD procedure to build a full rank factorial design on the response function with an intercept and slope parameter representing the year factor:

$$\begin{pmatrix} p_{1+} \\ p_{2+} \\ p_{+1} \\ p_{+2} \end{pmatrix} = \begin{pmatrix} 1 & 0 & 1 & 0 \\ 0 & 1 & 0 & 1 \\ 1 & 0 & -1 & 0 \\ 0 & 1 & 0 & -1 \end{pmatrix} \begin{pmatrix} \beta_1 \\ \beta_2 \\ \beta_3 \\ \beta_4 \end{pmatrix} \tag{4.67}$$

$$\mathbf{f} \quad = \quad \mathbf{X} \quad \beta$$

PROC CATMOD generates a full rank design matrix for each population of the form:

$$\mathbf{X}_o = \begin{pmatrix} 1 & 1 \\ 1 & -1 \end{pmatrix}$$

so that the design matrix for the repeated measures follows:

Sample	Response Function	Design Matrix: Intercept		A	
1	1	1	0	1	0
1	2	0	1	0	1
2	1	1	0	−1	0
2	2	0	1	0	−1

Letting $\hat{\pi} = p$, we define a linear model for the response function **f** using the MODEL and REPEATED statements. Because we have two repeated factors, the REPEATED statement identifies the factor and the number of independent levels. The MODEL statement says that the response variables are crossed.

Output 4.7.2 (*continued*)

```
                 Output 4.7.2: Test of Marginal Homogeniety
                            Using PROC CATMOD

                             CATMOD PROCEDURE

     Response: T55*T65                      Response Levels (R)=     7
     Weight Variable: COUNT                 Populations      (S)=     1
     Data Set: SCHOOL                       Total Frequency (N)=   124
     Frequency Missing: 0                   Observations   (Obs)=     7


                                      Sample
                             Sample     Size
                             ƒƒƒƒƒƒƒƒƒƒƒƒƒƒƒƒ
                                 1        124
```

Output 4.7.2 *(continued)*

```
                          RESPONSE PROFILES

                          Response  T55  T65
                          ƒƒƒƒƒƒƒƒƒƒƒƒƒƒƒƒƒƒ
                              1       1    1
                              2       1    2
                              3       1    3
                              4       2    1
                              5       2    2
                              6       2    3
                              7       3    3


          Function     Response           DESIGN MATRIX
  Sample   Number      Function       1        2        3        4
  ƒƒƒƒƒƒƒƒƒƒƒƒƒƒƒƒƒƒƒƒƒƒƒƒƒƒƒƒƒƒƒƒƒƒƒƒƒƒƒƒƒƒƒƒƒƒƒƒƒƒƒƒƒƒƒƒƒƒƒƒƒƒƒ
     1        1        0.55645        1        0        1        0
              2        0.29839        0        1        0        1
              3        0.41935        1        0       -1        0
              4        0.12097        0        1        0       -1


                    ANALYSIS-OF-VARIANCE TABLE

       Source                    DF    Chi-Square      Prob
       --------------------------------------------------
       INTERCEPT                  2        505.23    0.0000
       YEAR                       2         62.73    0.0000

       RESIDUAL                   0          .         .
```

For the marginal response function (4.67), the estimate of β is calculated and an ANOVA table formed. The ANOVA table for the factor YEARS tests that the two contrasts $\psi_1 = \hat{\pi}_{+1} - \hat{\pi}_{1+}$ and $\psi_2 = \hat{\pi}_{+2} - \hat{\pi}_{2+}$ are simultaneously zero. The test statistic for YEAR is again 62.73 as calculated using PROC IML.

4.7.3 Homogeneity of Proportions

Another common problem in the analysis of contingency tables is to determine whether proportions are homogeneous across independent populations, a one-way ANOVA. This example, taken from Grizzle, Starmer and Koch (1969), deals with a study of the effects of four different procedures on the dumping syndrome, an undesirable aftermath of duodenal ulcer. The data are provided in Table 4.2.

Table 4.2 *Severity of the Dumping Syndrome*

	Hospital															
Clinical Evaluation of Dumping	1 Surgical Procedure				2 Surgical Procedure				3 Surgical Procedure				4 Surgical Procedure			
	A	B	C	D	A	B	C	D	A	B	C	D	A	B	C	D
None	23	23	20	24	18	18	13	9	8	12	11	7	12	18	14	13
Slight	7	10	13	10	6	6	13	15	6	4	6	7	9	3	8	6

Moderate	2	5	5	6	1	2	2	2	3	4	2	4	1	2	3	4
Total N	32	38	38	40	25	26	28	26	17	20	19	18	22	20	25	23
Average Score	1.3	1.5	1.6	1.6	1.3	1.4	1.6	1.7	1.7	1.6	1.5	1.8	1.5	1.4	1.6	1.6

There are several hypotheses of interest in this study. The primary analysis is concerned with fitting an additive linear model to the observed cell probabilities for the four hospitals and different surgical procedures A, B, C, and D. If the model fits, one can then test hypotheses concerning the model parameters. Finally, a test of linear trend is performed. The four surgical procedures consist of operations which remove differing amounts of the stomach: A = none, B = 1/4, C = 1/2, and D = 3/4. It is of interest to determine if the severity of the dumping syndrome is linearly related to the amount of the stomach removed.

For each independent group,

$$\mathbf{p}_i' = (p_{i1}, p_{i2}, p_{i3})$$

so that

$$\begin{aligned} \mathbf{p}_1' &= (23, 7, 2)/32 \\ \mathbf{p}_2' &= (23, 10, 5)/38 \\ &\vdots \\ \mathbf{p}_{16}' &= (13, 16, 4)/23 \end{aligned}$$

where the

$$\text{var}(\mathbf{p}_i) = \hat{\boldsymbol{\Sigma}}_{3\times 3} = \frac{1}{n_{i+}} \begin{pmatrix} p_{i1}(1-p_{i1}) & -p_{i1}\,p_{i2} & -p_{i1}\,p_{i3} \\ -p_{i1}\,p_{i2} & p_{i2}(1-p_{i2}) & -p_{i2}\,p_{i3} \\ -p_{i3}\,p_{i1} & p_{i3}\,p_{i2} & p_{i3}(1-p_{i3}) \end{pmatrix}$$

so that

$$\begin{pmatrix} \text{var}(\mathbf{p}_1) & & & 0 \\ & \text{var}(\mathbf{p}_2) & & \\ & & \ddots & \\ 0 & & & \text{var}(\mathbf{p}_{16}) \end{pmatrix} = \hat{\boldsymbol{\Omega}}_{48\times 48} = \mathbf{I} \otimes \hat{\boldsymbol{\Sigma}}$$

For each of the categories of response, (none, slight, and moderate), the scores of 1, 2, and 3 are assigned. Hence, the mean score of the 16 populations is calculated as $1\text{p}_{i1} + 2\text{p}_{i2} + 3\text{p}_{i3}$. It is this mean score within each treatment combination that is used in the analysis. To calculate these means, the matrix $\mathbf{A}$ is postmultiplied by $\mathbf{p}$ where $\mathbf{A}$ is

$$\mathbf{A} = \begin{pmatrix} 123 & 000 & 000 & \cdots & 000 \\ 000 & 123 & 000 & \cdots & 000 \\ \cdots & \cdots & \cdots & \cdots & \cdots \\ \cdots & \cdots & \cdots & \cdots & \cdots \\ 000 & 000 & 000 & \cdots & 123 \end{pmatrix}$$

The function of the population proportion of interest for the study is

$$\mathbf{F}(\pi) = \mathbf{A}\mathbf{p} = \mathbf{X}\beta \tag{4.68}$$

an addition linear model. The design matrix and parameter vector take the following form:

$$\mathbf{X}_{16\times 7} = \begin{pmatrix} 1 & 1 & 0 & 0 & 1 & 0 & 0 \\ 1 & 1 & 0 & 0 & 0 & 1 & 0 \\ 1 & 1 & 0 & 0 & 0 & 0 & 1 \\ 1 & 1 & 0 & 0 & -1 & -1 & -1 \\ 1 & 0 & 1 & 0 & 1 & 0 & 0 \\ 1 & 0 & 1 & 0 & 0 & 1 & 0 \\ 1 & 0 & 1 & 0 & 0 & 0 & 1 \\ 1 & 0 & 1 & 0 & -1 & -1 & -1 \\ 1 & 0 & 0 & 1 & 1 & 0 & 0 \\ 1 & 0 & 0 & 1 & 0 & 1 & 0 \\ 1 & 0 & 0 & 1 & 0 & 0 & 1 \\ 1 & 0 & 0 & 1 & -1 & -1 & -1 \\ 1 & -1 & -1 & -1 & 1 & 0 & 0 \\ 1 & -1 & -1 & -1 & 0 & 1 & 0 \\ 1 & -1 & -1 & -1 & 0 & 0 & 1 \\ 1 & -1 & -1 & -1 & -1 & -1 & -1 \end{pmatrix} \qquad \beta = \begin{pmatrix} \mu \\ \alpha_1 \\ \alpha_2 \\ \alpha_3 \\ \lambda_1 \\ \lambda_2 \\ \lambda_3 \end{pmatrix}$$

where: μ = overall mean

α_k = effect of kth hospital, $k = 1, 2, 3$

λ_m = effect of mth treatment, $m = 1, 2, 3$

Because **X** is of full rank, the last effect is estimated from the prior effects: $\hat{\alpha}_4 = -\sum_k \hat{\alpha}_k$ and $\hat{\lambda}_4 = -\sum_m \hat{\lambda}_m$. For the model,

$$\mathbf{S} = \mathbf{A}\hat{\Omega}\mathbf{A}'$$

$$\hat{\hat{\beta}} = (\mathbf{X}'\mathbf{S}^{-1}\mathbf{X})^{-1}\mathbf{X}'\mathbf{S}^{-1}\mathbf{F}$$

$$Q = \mathbf{F}'\mathbf{S}^{-1}\mathbf{F} - \hat{\hat{\beta}}'(\mathbf{X}'\mathbf{S}^{-1}\mathbf{X})\hat{\hat{\beta}}$$

Program 4_7_3.sas

```
/* Program 4_7_3.sas                           */
/* Data are from  Grizzle et al.,1969          */
/* Testing for Homogeniety of Proportions   */

options ls=78 ps=60 nodate nonumber;
title1 'Output 4.7.3: Testing Homogeniety of Proportions, One-Way ANOVA';

/* Using PROC IML */
title2 'Using PROC IML';
proc iml;
   start cat2;
      freq={23  7 2, 23 10 5, 20 13 5, 24 10 6,
            18  6 1, 18  6 2, 13 13 2,  9 15 2,
             8  6 3, 12  4 4, 11  6 2,  7  7 4,
            12  9 1, 15  3 2, 14  8 3, 13  6 4};
```

```
      s=nrow(freq);
      r=ncol(freq);
      wt={1 2 3};
      rowsum=freq[,+];
      prob=freq/(rowsum*repeat(1,1,r));
      p=shape(prob[,1:r],0,1);
      A=I(16)@(repeat(1,1,r)#wt);
      D=diag(rowsum);
      V=(diag(p)-p*p`)#(Inv(D)@repeat(1,r,r));
      S=A*V*A`;
      InvS=Inv(S);
      F=A*p; print F;
      X={1  1  0  0  1  0  0,
         1  1  0  0  0  1  0,
         1  1  0  0  0  0  1,
         1  1  0  0 -1 -1 -1,
         1  0  1  0  1  0  0,
         1  0  1  0  0  1  0,
         1  0  1  0  0  0  1,
         1  0  1  0 -1 -1 -1,
         1  0  0  1  1  0  0,
         1  0  0  1  0  1  0,
         1  0  0  1  0  0  1,
         1  0  0  1 -1 -1 -1,
         1 -1 -1 -1  1  0  0,
         1 -1 -1 -1  0  1  0,
         1 -1 -1 -1  0  0  1,
         1 -1 -1 -1 -1 -1 -1};
      Beta=Inv(X`*InvS*X)*X`*InvS*F; print Beta;
      Q=F`*InvS*F-Beta`*(X`*InvS*X)*Beta;
      dfq=9;
      pvq=1-probchi(Q,dfq);
      print 'Chi-square Test of Fit',Q,'p-value for fit',pvq;
      CT={ 0 0 0 0 1 0 0,
           0 0 0 0 0 1 0,
           0 0 0 0 0 0 1};
      dfct=3;
      CH={ 0 1 0 0 0 0 0,
           0 0 1 0 0 0 0,
           0 0 0 1 0 0 0};
      dfch=3;
      L={0 0 0 0 3 2 1};
      dfL=1;
      Wh=(CH*Beta)`*Inv(CH*Inv(X`*InvS*X)*CH`)*(CH*Beta);
      pvWh=1-probchi(Wh,dfch);
      Wt=(CT*Beta)`*Inv(CT*Inv(X`*InvS*X)*CT`)*(CT*Beta);
      pvWt=1-probchi(Wt,dfct);
      W=(L*Beta)`*Inv(L*Inv(X`*InvS*X)*L`)*(L*Beta);
      pvL=1-probchi(W,dfL);
      print 'Hosp Chi-Square Test',WH, 'p-value Hosp',pvWh,
            'Treat Chi-Square Test',WT,'p-value Treat',pvWt,
            'Linear Trend',W,'p-value Linear',pvL;
   finish cat2;
run cat2;
quit;
run;

/* Using PROC CATMOD */
data operate;
   input treat hosp $ severity $ wt @@;
   cards;
   1 a N 23 1 a S  7 1 a M 2
   1 b N 18 1 b S  6 1 b M 1
   1 c N  8 1 c S  6 1 c M 3
```

```
   1 d N 12 1 d S  9 1 d M 1
   2 a N 23 2 a S 10 2 a M 5
   2 b N 18 2 b S  6 2 b M 2
   2 c N 12 2 c S  4 2 c M 4
   2 d N 15 2 d S  3 2 d M 2
   3 a N 20 3 a S 13 3 a M 5
   3 b N 13 3 b S 13 3 b M 2
   3 c N 11 3 c S  6 3 c M 2
   3 d N 14 3 d S  8 3 d M 3
   4 a N 24 4 a S 10 4 a M 6
   4 b N  9 4 b S 15 4 b M 2
   4 c N  7 4 c S  7 4 c M 4
   4 d N 13 4 d S  6 4 d M 4
   ;
run;

proc catmod order=data;
   title2 'Using PROC CATMOD' ;
   weight wt;
   response 1 2 3;
   model severity = hosp treat/ freq oneway;
   contrast 'Linear Trend - Treatment' treat 3 2 1;
run;
```

Result and Interpretation 4_7_3.sas

Output 4.7.3: *Testing Homogeneity of Proportions, One-Way ANOVA Using PROC IML*

```
        Output 4.7.3: Testing Homogeniety of Proportions, One-Way ANOVA
                                Using PROC IML

                                          F
                                    1.34375
                                  1.5263158
                                  1.6052632
                                       1.55
                                       1.32
                                  1.3846154
                                  1.6071429
                                  1.7307692
                                  1.7058824
                                        1.6
                                  1.5263158
                                  1.8333333
                                        1.5
                                       1.35
                                       1.56
                                  1.6086957


                                       BETA
                                  1.5448619
                                  -0.040816
                                  -0.035645
                                  0.1061082
                                  -0.110471
                                  -0.072985
                                  0.0496295
```

Output 4.7.3 (*continued*)

```
                    Chi-square Test of Fit

                              Q
                        6.3262338

                      p-value for fit

                            PVQ
                        0.7068753

                     Hosp Chi-Square Test

                             WH
                        2.3317129

                        p-value Hosp

                           PVWH
                        0.5064728

                    Treat Chi-Square Test

                             WT
                        8.8971439

                       p-value Treat

                           PVWT
                        0.0306901

                        Linear Trend

                              W
                        8.7418996

                       p-value Linear

                            PVL
                        0.0031098
```

From Output 4.7.2 we see that the estimate of β is

$$\hat{\mu} = 1.54, \quad \hat{\alpha}_1 = -0.04, \quad \hat{\alpha}_2 = -0.04, \quad \hat{\alpha}_3 = 0.11, \quad \hat{\alpha}_4 = -0.03$$
$$\hat{\lambda}_1 = -0.11, \quad \hat{\lambda}_2 = -0.07, \quad \hat{\lambda}_3 = 0.05, \quad \hat{\lambda}_4 = 0.13$$

Because the value of $Q = 6.32$, the additive model fits the data, and further hypotheses concerning the parameters may be tested using the Wald statistic (4.54). To test for hospital and treatment effects, the matrices

$$\mathbf{C}_H = \begin{pmatrix} 0 & 1 & 0 & 0 & 0 & 0 & 0 \\ 0 & 0 & 1 & 0 & 0 & 0 & 0 \\ 0 & 0 & 0 & 1 & 0 & 0 & 0 \end{pmatrix}$$

$$\mathbf{C}_T = \begin{pmatrix} 0 & 0 & 0 & 0 & 1 & 0 & 0 \\ 0 & 0 & 0 & 0 & 0 & 1 & 0 \\ 0 & 0 & 0 & 0 & 0 & 0 & 1 \end{pmatrix}$$

are used, resulting in the ANOVA table shown in Table 4.3.

Table 4.3 *ANOVA Table*

Source	Chi-Square	d.f	p-value
Hospitals	2.33	3	0.5065
Treatments	8.90	3	0.0307
Error	6.33	9	

The degrees of freedom in the ANOVA table for the main effect tests are the number of independent rows of the hypothesis test matrix **C**. The degrees of freedom for the "fit" are $q - k = 16 - 7 = 9$, the difference between the number of functions and the number of independent parameters in β. Comparing the test statistics with the critical values, we find that this treatment effect is significant with a p-value less than $\alpha = 0.05$. As usual, Scheffé-type confidence intervals may be obtained for contrasts $\psi = \mathbf{c}'\beta$

$$\mathbf{c}'\hat{\beta} \pm \left[\chi^2_{\upsilon}\mathbf{c}'(\mathbf{X}'\mathbf{S}^{-1}\mathbf{X})^{-1}\mathbf{c}\right]^{1/2}$$

where v is the rank of the matrix **C** for the overall test.

The hypothesis that severity of syndrome is linearly related to the amount of stomach removed may be tested by the orthogonal linear polynomial

$$\mathrm{H_L}{:}-3\lambda_1 - \lambda_2 + \lambda_3 + 3\lambda_4 = 0$$

However, for the reparameterized model, this becomes $-3\lambda_1 - \lambda_2 + \lambda_3 + 3(-\lambda_1 - \lambda_2 - \lambda_3) = -6\lambda_1 - 4\lambda_2 - 2\lambda_3 = 0$ or $3\lambda_1 + 2\lambda_2 + \lambda_3 = 0$. Hence,

$$\mathbf{C}_L = (0 \quad 0 \quad 0 \quad 0 \quad 3 \quad 2 \quad 1)$$

in terms of model (4.68). The Wald statistic for linear trend is W = 8.74, which when compared to a chi-square value of $\chi^2_{3,0.95} = 7.815$ (p-value = .0031), is significant.

The SAS code for the CATMOD procedure is also included in Program 4_7_3.sas and the continuation of that output follows.

Output 4.7.3 *(continued)*

ANALYSIS-OF-VARIANCE TABLE

Source	DF	Chi-Square	Prob
INTERCEPT	1	1999.88	0.0000
HOSP	3	2.33	0.5065
TREAT	3	8.90	0.0307
RESIDUAL	9	6.33	0.7069

ANALYSIS OF WEIGHTED-LEAST-SQUARES ESTIMATES

Effect	Parameter	Estimate	Standard Error	Chi-Square	Prob
INTERCEPT	1	1.5449	0.0345	1999.88	0.0000
HOSP	2	-0.0408	0.0527	0.60	0.4388
	3	-0.0356	0.0535	0.44	0.5055
	4	0.1061	0.0703	2.28	0.1312
TREAT	5	-0.1105	0.0541	4.17	0.0411
	6	-0.0730	0.0579	1.59	0.2073
	7	0.0496	0.0560	0.78	0.3757

ANALYSIS OF CONTRASTS

Contrast	DF	Chi-Square	Prob
Linear Trend - Treatment	1	8.74	0.0031

4.7.4 Independence

The examples of the linear model approach to the analysis of categorical data have assumed only linear function of π. We now consider the hypothesis of independence which utilizes the logarithmic function. The standard form of the test of independence for a two-way table is

$$H:\quad \pi_{ij} = \pi_{i+}\pi_{+j} \tag{4.69}$$

for all i and j. To write this in the form $\mathbf{F}(\pi) = \mathbf{0}$, we let $\mu_{ij} = \ln \pi_{ij}$ and write μ_{ij} as a linear model:

$$\mu_{ij} = \mu + \alpha_i + \beta_j + \gamma_{ij}$$

Using dot notation to represent averages, one observes, with the usual side conditions on the parameters, that

$$\alpha_i = \mu_{i.} - \mu_{..}, \quad \beta_j = \mu_{.j} - \mu_{..}, \quad \gamma_{ij} = \mu_{ij} - \mu_{i.} - \mu_{.j} + \mu_{..}$$

For a 2×2 table,

$$\alpha_1 = (\mu_{11} + \mu_{12} - \mu_{21} - \mu_{22})/4 = \sum_j \ln(\pi_{1j}/\pi_{2j})$$

$$\beta_1 = (\mu_{11} - \mu_{12} + \mu_{21} - \mu_{22})/4 = \sum_i \ln(\pi_{i1}/\pi_{i2})$$

$$\gamma_{11} = (\mu_{11} - \mu_{12} - \mu_{21} + \mu_{22})/4 = \ln\left[(\pi_{11}\pi_{22})/(\pi_{12}\pi_{21})\right]/4$$

Thus, the interaction term is proportional to the logarithm of the cross-product ratio in the 2×2 table, which is also seen to be an odds-ratio: $(\pi_{11}/\pi_{21})/(\pi_{12}/\pi_{22})$. Hence, for a 2×2 table, if $\lambda_{11} > 0$ the expected frequency in cell

(1,1) is greater than $n\,\pi_{1+}\pi_{+1}$ corresponding to independence. More generally, the test of independence has the general form:

$$\ln\left(\frac{\pi_{ij}\pi_{i'j'}}{\pi_{ij'}\pi_{i'j}}\right) = 0 \qquad i \neq i',\ j \neq j' \tag{4.70}$$

For a 2×2 table, the test of independence is written as

$$\begin{aligned}
\mathbf{H}\colon \mathbf{F}(\pi) &= \mathbf{K}\ln \mathbf{A}(\pi) = \mathbf{0} \\
&= \ln\left(\frac{\pi_{11}\pi_{22}}{\pi_{12}\pi_{21}}\right) = 0 \\
&= \ln \pi_{11} - \ln \pi_{12} - \ln \pi_{21} + \ln \pi_{22} = 0
\end{aligned}$$

To obtain the test statistic using (4.59), observe that

$$\mathbf{A} = \mathbf{I}_4$$
$$\mathbf{K} = (1, -1, -1, 1)$$

$$\hat{\boldsymbol{\Omega}} = \frac{1}{n}\begin{pmatrix} p_{11}(1-p_{11}) & -p_{11}p_{12} & -p_{11}p_{21} & -p_{11}p_{22} \\ -p_{12}p_{11} & p_{12}(1-p_{12}) & -p_{12}p_{21} & -p_{12}p_{22} \\ -p_{21}p_{11} & -p_{21}p_{12} & p_{21}(1-p_{21}) & -p_{21}p_{22} \\ -p_{22}p_{11} & -p_{22}p_{12} & -p_{22}p_{21} & p_{22}(1-p_{22}) \end{pmatrix}$$

$$\mathbf{D} = \begin{pmatrix} p_{11} & 0 & 0 & 0 \\ 0 & p_{12} & 0 & 0 \\ 0 & 0 & p_{13} & 0 \\ 0 & 0 & 0 & p_{14} \end{pmatrix}$$

so that

$$\mathbf{K}\mathbf{D}^{-1}\mathbf{A} = \left(\frac{1}{p_{11}},\ \frac{-1}{p_{12}},\ \frac{-1}{p_{21}},\ \frac{1}{p_{22}},\right)$$

$$\mathbf{K}\mathbf{D}^{-1}\mathbf{A}\hat{\boldsymbol{\Omega}} = \frac{1}{n}(1,\ \ -1,\ \ -1,\ \ 1)$$

$$\mathbf{S} = \mathbf{K}\mathbf{D}^{-1}\mathbf{A}\hat{\boldsymbol{\Omega}}\mathbf{A}'\mathbf{D}^{-1}\mathbf{K}' = \frac{1}{n}\left(\frac{1}{p_{11}} + \frac{1}{p_{12}} + \frac{1}{p_{21}} + \frac{1}{p_{22}}\right)$$

Since $\mathbf{X}\beta = \mathbf{0}$ for the test of independence, the test statistic is

$$\begin{aligned}
Q &= \left[\mathbf{K}\ln(\mathbf{Ap})\right]'\mathbf{S}^{-1}\left[\mathbf{K}\ln(\mathbf{Ap})\right] \\
&= \frac{n(\ln p_{11} - \ln p_{12} - \ln p_{21} + \ln p_{22})^2}{\dfrac{1}{p_{11}} + \dfrac{1}{p_{12}} + \dfrac{1}{p_{21}} + \dfrac{1}{p_{22}}}
\end{aligned} \tag{4.71}$$

which has a chi-square distribution with 1 degree of freedom. For a general I×J table, Bhapkar (1966) showed that the Wald statistic is equal to Neyman's (1949) modified chi-square statistic

$$X_N^2 = \sum_i \sum_j (n_{ij} - n\hat{\pi}_{ij})^2 \big/ n_{ij}$$

In general, this is not equal to Karl Pearson's chi-square statistic

$$X_{KP}^2 = \sum_i \sum_j (n_{ij} - n\hat{\pi}_{ij})^2 / n\hat{\pi}_{ij}$$

which is also generally not the same as the Neyman-Pearson likelihood ratio statistic

$$X_{NP}^2 = -2 \sum_i \sum_j n_{ij} \ln\left(\frac{n\hat{\pi}_{ij}}{n_{ij}}\right)$$

These statistics are calculated with PROC FREQ to test for independence.

Our example of the test of independence is taken from Marascuilo and McSweeney (1977, p. 205). The study is designed to determine if there is any association between how a teacher performs and whether the teacher held an elected office while a student. The data are shown in Table 4. 4.

Table 4.4 *Frequencies of Teacher Performance and Whether Office Was Held*

	Teacher Performance		
Held Office	Successful	Unsuccessful	Total
Yes	50	20	70
No	10	20	30
Total	60	40	100

Program 4_7_4.sas includes the SAS code to perform the test of independence using PROC CATMOD.

Program 4_7_4.sas

```
/* Program 4_7_4.sas         */
/* Testing for Independence */

options ls=78 ps=60 nodate nonumber;
title 'Output 4.7.4: Test of Independence ' ;

data independ;
   input office $ perform $ count @@;
   cards;
   Y Success 50  Y Usuccess 20
   N Success 10  N Usuccess 20
   ;
run;

proc catmod;
   weight count;
   model office*perform=_response_/ freq wls;
   loglin office perform;
run;
```

Result and Interpretation 4_7_4.sas

Output 4.7.4: *Test of Independence*

```
              Output 4.7.4: Test of Independence

                  ANALYSIS-OF-VARIANCE TABLE

       Source                  DF   Chi-Square       Prob
       --------------------------------------------------
       OFFICE                   1         9.81     0.0017
       PERFORM                  1         3.42     0.0643

       RESIDUAL                 1        11.77     0.0006
```

Using formula (4.71), we see that the value of the Wald chi-square statistic is

$$Q = \frac{100(\ln(.5) - \ln(.2) - \ln(.1) + \ln(.2))^2}{\frac{1}{.5} + \frac{1}{.2} + \frac{1}{.1} + \frac{1}{.2}}$$

$$= \frac{100(1.609)^2}{22}$$

$$= 11.77$$

Hence, there appears to be an association between teacher performance and having held an elected office as a student.

The test of independence from the PROC CATMOD Output 4.7.4 is the chi-square goodness-of-fit statistic $Q = 11.77$ with 1 *df* and *p*-value 0.0006.

5. Multivariate General Linear Models

Theory

Applications

5.1 Introduction

The multivariate general linear model without restrictions has wide applicability in regression analysis, analysis of variance, analysis of covariance, analysis of repeated measurements, and, more broadly defined, experimental designs involving several dependent variables. Timm (1975, 1993b) provides numerous examples of applications in education and psychology.

In this chapter we develop the multivariate general linear model, estimate the model parameters and test general linear hypotheses. The multivariate normal distribution is extended to the matrix normal distribution and the general multivariate linear mixed model (GMLMM) is introduced. A special case of the GMLMM, known as Scheffé's mixed model, is used to analyze repeated measurement data. We then illustrate the general theory using examples of multivariate regression, MANOVA, and MANCOVA. We show the relationship between univariate and multivariate analysis of repeated measurement data and discuss how one may analyze extended linear hypotheses.

5.2 Developing the Model

The univariate GLM was obtained from (1.1) by assuming the covariance structure $\mathbf{\Omega} = \sigma^2 \mathbf{I}_n$ where $\mathbf{y}$ was a vector of n independent observations for one dependent variable. To extend the model to p-variables, we write a linear model for each variable

$$\begin{aligned} \mathbf{y}_i &= \mathbf{X}\beta_i + \mathbf{e}_i \qquad i = 1, 2, \ldots, p \\ \operatorname{var}(\mathbf{y}_i) &= \sigma_{ii}\mathbf{I}_n \\ \operatorname{cov}(\mathbf{y}_i, \mathbf{y}_j) &= \sigma_{ij}\mathbf{I}_n \qquad \text{for } i \neq j \end{aligned} \tag{5.1}$$

In (5.1), we again assume $E(\mathbf{e}_i) = \mathbf{0}$. The p n-variate observation vectors $\mathbf{y}_i(n \times 1)$ are related to the p k-variate parameter vectors β_i by a common design matrix $\mathbf{X}_{n \times k}$ of rank k. The p-separate univariate models are related by the $p(p-1)/2$ covariances among the pairs of variables.

The classical matrix representation for (5.1) is to let

$$\mathbf{Y}_{n\times p} = (\mathbf{y}_1, \mathbf{y}_2, \ldots, \mathbf{y}_p), \quad \mathbf{B}_{k\times p} = (\beta_1, \beta_2, \ldots, \beta_p) \tag{5.2}$$

and $\mathbf{U}_{n\times p}$ be the matrix of errors appropriately partitioned. Then (5.1) is written as

$$\begin{gathered} \mathbf{Y}_{n\times p} = \mathbf{X}_{n\times k}\mathbf{B}_{k\times p} + \mathbf{U}_{n\times p} \\ \mathbf{\Omega} = \text{cov}(\mathbf{Y}) = \mathbf{I}_n \otimes \mathbf{\Sigma} \end{gathered} \tag{5.3}$$

Here $\mathbf{\Sigma} = (\sigma_{ij})$ is the common $(p \times p)$ covariance matrix and $\otimes$ is a Kronecker product defined as $\mathbf{A} \otimes \mathbf{B} = (\mathbf{a}_{ij}\mathbf{B})$, so that each row of the data matrix $\mathbf{Y}$ has a common covariance structure $\mathbf{\Sigma}$.

Stacking the p n-vectors $\mathbf{y}_i$ into a single $(np \times 1)$ vector $\mathbf{y}$, the p k-vectors β_i into a $(kp \times 1)$ vector β, and the p n-vectors $\mathbf{e}_i$ into a $(np \times 1)$ vector $\mathbf{e}$, we may write (5.1) in vector form as follows:

$$\begin{gathered} \mathbf{y} = \begin{pmatrix} \mathbf{y}_1 \\ \mathbf{y}_2 \\ \vdots \\ \vdots \\ \vdots \\ \mathbf{y}_p \end{pmatrix} = \mathbf{D} \begin{pmatrix} \beta_1 \\ \beta_2 \\ \vdots \\ \vdots \\ \vdots \\ \beta_p \end{pmatrix} + \begin{pmatrix} \mathbf{e}_1 \\ \mathbf{e}_2 \\ \vdots \\ \vdots \\ \vdots \\ \mathbf{e}_p \end{pmatrix} \\ \mathbf{y} \quad = \mathbf{D} \quad \beta \quad + \quad \mathbf{e} \end{gathered} \tag{5.4}$$

where $\mathbf{D} = (\mathbf{I}_p \otimes \mathbf{X})$ and $\mathbf{\Omega} = \mathbf{\Sigma} \otimes \mathbf{I}_{\text{n}}$. Thus, (5.4) has the GLM structure (1.1). Expression (5.4) may also be obtained from (5.3) using the vec(.) operator. The vec(.) operator creates a vector from a matrix by stacking the columns of a matrix sequentially; hence, $\text{vec}(\mathbf{Y}) = \mathbf{y}$ and $\text{vec}(\mathbf{B}) = \beta$. Using this notation, (5.4) becomes

$$\begin{gathered} E[\text{vec}(\mathbf{Y})] = \mathbf{D}\ \text{vec}(\mathbf{B}) = (\mathbf{I}_p \otimes \mathbf{X})\text{vec}(\mathbf{B}) \\ \text{cov}[\text{vec}(\mathbf{Y})] = \mathbf{\Omega} = \mathbf{\Sigma} \otimes \mathbf{I}_{\text{n}} \end{gathered} \tag{5.5}$$

Alternatively, the ith row $\mathbf{y}_i$ $(p \times 1)$ of $\mathbf{Y}$ has the structure

$$\mathbf{y}_i = (\mathbf{I}_p \otimes \mathbf{x}_i')\ \text{vec}(\mathbf{B}) + \mathbf{e}_i \tag{5.6}$$

where $\mathbf{x}_i'$ and $\mathbf{e}_i'$ are the ith rows of $\mathbf{X}$ and $\mathbf{U}$, respectively. The GLM includes the multivariate regression model, MANOVA, the multivariate analysis of covariance (MANCOVA) model, and numerous other multivariate linear models.

Instead of using the classical matrix form (5.3), we may use the responsewise orientation. Then (5.3) is written as

$$\mathbf{Y}_{p\times n} = (\mathbf{y}_1, \mathbf{y}_2, \ldots, \mathbf{y}_n) = \mathbf{B}_{p\times k}\mathbf{X}_{k\times n} + \mathbf{U}_{p\times n} \tag{5.7}$$

where now each column of $\mathbf{Y}$ or $\mathbf{U}$ in (5.7) has common covariance structure $\mathbf{\Sigma}$ for $j = 1, 2, \ldots, n$.

5.3 Estimation Theory and Hypothesis Testing

To estimate $\mathbf{B}$ or $\beta = \text{vec}(\mathbf{B})$, we have from (5.4) and the generalized least squares result (4.10) that the BLUE of β is

$$\hat{\beta} = (\mathbf{D}'\mathbf{\Omega}^{-1}\mathbf{D})^{-1}\mathbf{D}'\mathbf{\Omega}^{-1}\mathbf{y} \tag{5.8}$$

where $\mathbf{\Omega} = \mathbf{\Sigma} \otimes \mathbf{I}_n$ is known and $\mathbf{D} = (\mathbf{I}_p \otimes \mathbf{X})$. However, even when $\mathbf{\Omega}$ is unknown, (5.8) may be used because the estimator of β does not depend on $\mathbf{\Sigma}$; $\mathbf{\Sigma}$ drops out of (5.8) because the design matrix $\mathbf{X}$ is the same for each variable. To see this, observe that on substituting $\mathbf{\Omega}$ and $\mathbf{D}$ into (5.8) that

$$\begin{aligned} \text{vec}(\hat{\mathbf{B}}) = \hat{\beta} &= [(\mathbf{I}_p \otimes \mathbf{X})'(\mathbf{\Sigma} \otimes \mathbf{I}_n)^{-1}(\mathbf{I}_p \otimes \mathbf{X})]^{-1}(\mathbf{I}_p \otimes \mathbf{X})'(\mathbf{\Sigma} \otimes \mathbf{I}_n)^{-1}\mathbf{y} \\ &= (\mathbf{\Sigma}^{-1} \otimes \mathbf{X}'\mathbf{X})^{-1}(\mathbf{\Sigma}^{-1} \otimes \mathbf{X}')\mathbf{y} \\ &= (\mathbf{I}_p \otimes (\mathbf{X}'\mathbf{X})^{-1}\mathbf{X}')\mathbf{y} \\ &= (\mathbf{I}_p \otimes (\mathbf{X}'\mathbf{X})^{-1}\mathbf{X}')\text{vec}(\mathbf{Y}) \\ &= \text{vec}[(\mathbf{X}'\mathbf{X})^{-1}\mathbf{X}'\mathbf{Y}] \end{aligned} \tag{5.9}$$

where the last step in (5.9) uses the identity that the $\text{vec}(\mathbf{ABC}) = (\mathbf{C}' \otimes \mathbf{A})\ \text{vec}(\mathbf{B})$ with $\mathbf{A} = (\mathbf{X}'\mathbf{X})^{-1}\mathbf{X}'$ and $\mathbf{C}' = \mathbf{I}_p$. From (5.9), we have for the classical model (5.3) that

$$\hat{\mathbf{B}} = (\mathbf{X}'\mathbf{X})^{-1}\mathbf{X}'\mathbf{Y} \tag{5.10}$$

Partitioning $\mathbf{B}$ and $\mathbf{Y}$ and equating columns, the BLUE of β_i is the OLSE, $\hat{\beta}_i = (\mathbf{X}'\mathbf{X})^{-1}\mathbf{X}'\mathbf{y}_i$. To obtain the covariance matrix for $\hat{\beta}$, formula (4.7) is used

$$\begin{aligned} \text{cov}(\hat{\beta}) &= (\mathbf{D}'\mathbf{\Omega}^{-1}\mathbf{D})^{-1} \\ &= [(\mathbf{I}_p \otimes \mathbf{X})'(\mathbf{\Sigma}^{-1} \otimes \mathbf{I}_n)(\mathbf{I}_p \otimes \mathbf{X})]^{-1} \\ &= \mathbf{\Sigma} \otimes (\mathbf{X}'\mathbf{X})^{-1} \end{aligned} \tag{5.11}$$

We have obtained the BLUE of $\mathbf{B}$ using the vector version of the GLM with covariance structure $\mathbf{\Omega} = \mathbf{\Sigma} \otimes \mathbf{I}_n$. To directly obtain the OLSE = BLUE of $\mathbf{B}$ one may minimize the matrix error sum of squares using the classical model (5.3),

$$\text{Tr}(\mathbf{U}'\mathbf{U}) = \text{Tr}[(\mathbf{Y} - \mathbf{XB})'(\mathbf{Y} - \mathbf{XB})] \tag{5.12}$$

to obtain the matrix normal equations

$$(\mathbf{X}'\mathbf{X})\hat{\mathbf{B}} = \mathbf{X}'\mathbf{Y} \tag{5.13}$$

and directly obtain result (5.10).

To test hypotheses regarding the parameter matrix $\mathbf{B}$ or the vector $\beta = \text{vec}(\mathbf{B})$, we again make distributional assumptions. Using (5.3), we assume that the n rows $\mathbf{e}_1', \mathbf{e}_2', \ldots, \mathbf{e}_n'$ of $\mathbf{U}$ are independently, normally distributed

$$\begin{aligned} \mathbf{e}_i' &\sim IN_p(\mathbf{0}, \mathbf{\Sigma}) \\ \mathbf{e} = \text{vec}(\mathbf{U}) &\sim N_{np}(\mathbf{0}, \mathbf{\Omega} = \mathbf{\Sigma} \otimes \mathbf{I}_n) \end{aligned} \tag{5.14}$$

Alternatively using the observation vector **y**, we assume that

$$\mathbf{y} = \text{vec}(\mathbf{Y}) \sim N_{np}(\mathbf{D}\beta, \boldsymbol{\Omega} = \boldsymbol{\Sigma} \otimes \mathbf{I}_n) \tag{5.15}$$

where the design matrix $\mathbf{D} = \mathbf{I}_p \otimes \mathbf{X}$. Letting $N = np$, by (1.14) the density for **y** is

$$(2\pi)^{-N/2} |\boldsymbol{\Omega}|^{-1/2} \exp\left[-\frac{1}{2}(\mathbf{y} - \mathbf{D}\beta)'\boldsymbol{\Omega}^{-1}(\mathbf{y} - \mathbf{D}\beta)\right] \tag{5.16}$$

While (5.16) may be used to find the ML estimate of β and $\boldsymbol{\Omega}$ and hence $\boldsymbol{\Sigma}$, it is convenient to use the matrix normal distribution for the data matrix $\mathbf{Y}_{n \times p}$. By simplification of (5.16), observe first that the

$$|\boldsymbol{\Omega}|^{-1/2} = |\boldsymbol{\Sigma}_p \otimes \mathbf{I}_n|^{-1/2} = |\boldsymbol{\Sigma}|^{-n/2} |\mathbf{I}_n|^{-p/2} = |\boldsymbol{\Sigma}|^{-n/2}$$

using the identity that the $|\mathbf{A}_m \otimes \mathbf{B}_n| = |\mathbf{A}|^n |\mathbf{B}|^m$. Next we use the results that the $\text{vec}(\mathbf{ABC}) = (\mathbf{C}' \otimes \mathbf{A})\ \text{vec}(\mathbf{B})$ and that the $\text{Tr}(\mathbf{A}'\mathbf{B}) = (\text{vec}\,\mathbf{A})'\text{vec}\,\mathbf{B}$ so that

$$\begin{aligned}
(\mathbf{y} - \mathbf{D}\beta)'\boldsymbol{\Omega}^{-1}(\mathbf{y} - \mathbf{D}\beta) &= [\text{vec}(\mathbf{Y} - \mathbf{XB})]'(\boldsymbol{\Sigma} \otimes \mathbf{I}_n)^{-1}\text{vec}(\mathbf{Y} - \mathbf{XB}) \\
&= [\text{vec}(\mathbf{Y} - \mathbf{XB})]'\text{vec}[(\mathbf{Y} - \mathbf{XB})\boldsymbol{\Sigma}^{-1})] \\
&= \text{Tr}[(\mathbf{Y} - \mathbf{XB})'(\mathbf{Y} - \mathbf{XB})\boldsymbol{\Sigma}^{-1}] \\
&= \text{Tr}[\boldsymbol{\Sigma}^{-1}(\mathbf{Y} - \mathbf{XB})'(\mathbf{Y} - \mathbf{XB})]
\end{aligned}$$

The matrix normal distribution for $\mathbf{Y}_{n \times p}$ is

$$(2\pi)^{-np/2} |\boldsymbol{\Sigma}|^{-n/2}\, \text{etr}\left[-\frac{1}{2}\boldsymbol{\Sigma}^{-1}(\mathbf{Y} - \mathbf{XB})'(\mathbf{Y} - \mathbf{XB})\right] \tag{5.17}$$

Using (5.17), we immediately observe from the exponent that the ML estimate of **B** is equal to the BLUE. Furthermore, the maximum likelihood estimate of $\boldsymbol{\Sigma}$ is

$$\begin{aligned}
\hat{\boldsymbol{\Sigma}} &= \mathbf{E}/n = \mathbf{Y}'(\mathbf{I} - \mathbf{X}(\mathbf{X}'\mathbf{X})^{-1}\mathbf{X}')\mathbf{Y}/n \\
&= (\mathbf{Y} - \mathbf{X}\hat{\mathbf{B}})'(\mathbf{Y} - \mathbf{X}\hat{\mathbf{B}})/n
\end{aligned} \tag{5.18}$$

Result (5.17) is a special case of the following definition (5.1) with $\mathbf{W} \equiv \mathbf{I}$.

Definition 5.1 The data matrix $\mathbf{Y}_{n \times p}$ has a matrix normal distribution with parameters μ, $\boldsymbol{\Sigma}$ and **W** written as $\mathbf{Y} \sim \mathbf{N}_{n,p}(\mu, \boldsymbol{\Sigma}, \mathbf{W})$ if the density for **Y** is

$$(2\pi)^{-np/2} |\boldsymbol{\Sigma}|^{-n/2} |\mathbf{W}|^{-p/2}\, \text{etr}\left[-\frac{1}{2}\boldsymbol{\Sigma}^{-1}(\mathbf{Y} - \mu)'\mathbf{W}^{-1}(\mathbf{Y} - \mu)\right]$$

or $\text{vec}(\mathbf{Y}) \sim N_{np}(\text{vec}\,\mu, \boldsymbol{\Sigma} \otimes \mathbf{W})$.

From the matrix normal distribution, the ML of μ and $\boldsymbol{\Sigma}$ is

$$\begin{aligned}
\hat{\mu} &= (\mathbf{X}'\mathbf{W}^{-1}\mathbf{X})^{-1}\mathbf{X}'\mathbf{W}^{-1}\mathbf{Y} \\
\hat{\boldsymbol{\Sigma}} &= (\mathbf{Y}'\mathbf{W}^{-1}\mathbf{Y} - \mathbf{Y}'\mathbf{X}(\mathbf{X}'\mathbf{W}^{-1}\mathbf{X})^{-1}\mathbf{X}'\mathbf{Y})/n \\
&= (\mathbf{Y} - \mathbf{X}\hat{\mu})'(\mathbf{Y} - \mathbf{X}\hat{\mu})/n
\end{aligned} \tag{5.19}$$

To test hypotheses using (5.3), we are usually interested in bilinear parametric functions of the form $\psi = \mathbf{c}'\mathbf{B}\mathbf{a}$ or linear sets of functions $\mathbf{\Psi} = \mathbf{CBA}$ where $\mathbf{C}_{g\times k}$ and $\mathbf{A}_{p\times u}$ are known matrices, $\mathbf{c}_{k\times 1}$ is a row in $\mathbf{C}$, and $\mathbf{a}_{i(p\times 1)}$ is a column of $\mathbf{A}$. Letting

$$\beta' = (\beta_1', \beta_2', \ldots, \beta_p') = \text{vec}(\mathbf{B})$$

where β_i is the ith column of $\mathbf{B}$, $\mathbf{a}' = (\mathbf{a}_1, \mathbf{a}_2, \ldots, \mathbf{a}_p)$, and $\mathbf{c}_i' = \mathbf{a}_i\mathbf{C}'$ with $i = 1, 2, \ldots, p$, observe that

$$\psi = \mathbf{c}'\mathbf{B}\mathbf{a} = \sum_{i=1}^{p} \mathbf{c}_i'\beta_i = \mathbf{c}_*'\beta \tag{5.20}$$

where $\mathbf{c}_*' = (\mathbf{c}_1', \mathbf{c}_2', \ldots, \mathbf{c}_p')$. Hence, we may consider the more general problem of estimating $\mathbf{c}_*'\beta$ where $\mathbf{c}_*'$ is arbitrary and apply the result to the special case $\psi = \mathbf{c}'\mathbf{B}\mathbf{a}$. In a similar manner, instead of estimating $\mathbf{\Psi} = \mathbf{CBA}$, we may investigate the more general case of estimating $\mathbf{\Psi} = \mathbf{C}_*\beta$, where $\mathbf{C}_*' = (\mathbf{C}_1', \mathbf{C}_2', \ldots, \mathbf{C}_q')$ is a known matrix. To apply the general result to the special case $\mathbf{\Psi} = \mathbf{CBA}$, we let $\mathbf{C}_{i\,(gu\times k)}' = \mathbf{a}_i \otimes \mathbf{C}\,(i = 1, 2, \ldots, p)$, $\mathbf{a}_i$ equal to the ith row of $\mathbf{A}$, and $\mathbf{C}_* = \mathbf{A}_{u\times p}' \otimes \mathbf{C}_{g\times k}$. Using the classical form of the model, hypotheses of the form

$$H{:}\,\mathbf{CBA} = \mathbf{\Gamma} \Leftrightarrow (\mathbf{A}' \otimes \mathbf{C})\text{vec}(\mathbf{B}) = \text{vec}(\mathbf{\Gamma}) \;\Leftrightarrow (\mathbf{A}' \otimes \mathbf{C})\beta = \gamma \tag{5.21}$$

are of general interest, where a hypothesis is testable if and only if $\mathbf{B}$ is estimable. Alternatively, general hypotheses H become $H{:}\,\mathbf{C}_*\beta = \gamma$. Under multivariate normality of $\hat{\beta}$, the ML estimate of $\gamma = \text{vec}(\mathbf{\Gamma})$ is $\hat{\gamma} = \mathbf{C}_*\hat{\beta}$ where $\hat{\beta} = \text{vec}(\hat{\mathbf{B}})$ and $\hat{\mathbf{B}}$ is defined in (5.10), the ML estimate of $\mathbf{B}$. Then, the distribution of $\hat{\gamma}$ is multivariate normal:

$$\hat{\gamma} \sim \mathbf{N}_{gu}[\gamma, (\mathbf{C}_*(\mathbf{D}'\mathbf{\Omega}^{-1}\mathbf{D})^{-1}\mathbf{C}_*')] \tag{5.22}$$

However, one can simplify the structure of the covariance matrix:

$$\begin{aligned}(\mathbf{C}_*(\mathbf{D}'\mathbf{\Omega}^{-1}\mathbf{D})^{-1}\mathbf{C}_*') &= [(\mathbf{A}' \otimes \mathbf{C})[(\mathbf{I} \otimes \mathbf{X}')(\mathbf{\Sigma} \otimes \mathbf{I})^{-1}(\mathbf{I} \otimes \mathbf{X})]^{-1}(\mathbf{A} \otimes \mathbf{C}')]\\ &= [(\mathbf{A}' \otimes \mathbf{C})(\mathbf{\Sigma} \otimes (\mathbf{X}'\mathbf{X})^{-1})(\mathbf{A} \otimes \mathbf{C}')]\\ &= \mathbf{A}'\mathbf{\Sigma}\mathbf{A} \otimes \mathbf{C}(\mathbf{X}'\mathbf{X})^{-1}\mathbf{C}'\end{aligned}$$

When $\mathbf{A}'\mathbf{\Sigma}\mathbf{A}$ is known, following (4.11), the likelihood ratio test of

$$H{:}\,\mathbf{CBA} = \mathbf{\Gamma} \Leftrightarrow H{:}\,\gamma = \gamma_o \tag{5.23}$$

where $\mathbf{C}$, $\mathbf{A}$, and $\mathbf{\Gamma}$ are known and where $\hat{\mathbf{B}}$ is the ML estimate of $\mathbf{B}$ is to reject H if $X^2 > c$ where c is chosen such that the $P(X^2 > c \mid H \text{ true}) = \alpha$, and X^2 is defined in (5.24)

$$\begin{aligned}X^2 &= (\hat{\gamma} - \gamma_o)'[(\mathbf{A}'\mathbf{\Sigma}\mathbf{A}) \otimes (\mathbf{C}(\mathbf{X}'\mathbf{X})^{-1}\mathbf{C}')]^{-1}(\hat{\gamma} - \gamma_o)\\ &= \text{Tr}[(\mathbf{C}\hat{\mathbf{B}}\mathbf{A} - \mathbf{\Gamma})'(\mathbf{C}(\mathbf{X}'\mathbf{X})^{-1}\mathbf{C}')^{-1}(\mathbf{C}\hat{\mathbf{B}}\mathbf{A} - \mathbf{\Gamma})(\mathbf{A}'\mathbf{\Sigma}\mathbf{A})^{-1}]\end{aligned} \tag{5.24}$$

using the identities $\text{vec}(\mathbf{ABC}) = (\mathbf{C}' \otimes \mathbf{A})\,\text{vec}\,\mathbf{B}$ and $\text{Tr}(\mathbf{A}'\mathbf{B}) = (\text{vec}\,\mathbf{A})'\,\text{vec}\,\mathbf{B}$. X^2 has a central chi-square distribution with $v = gu$ degrees of freedom when H is true.

To develop the likelihood ratio test of $H(\mathbf{CBA} = \mathbf{\Gamma})$ with unknown $\mathbf{\Sigma}$, we let $\mu = \mathbf{XB}$ in the matrix normal distribution and obtain the maximum likelihood estimates of $\mathbf{B}$ and $\mathbf{\Sigma}$ under $\mathbf{\Omega}_o$ and ω. With $\mu = \mathbf{XB}$, the ML estimate of $\mathbf{B}$ and $\mathbf{\Sigma}$ are given in (5.19) (Seber, 1984, p. 406). Following (2.9), the corresponding ML estimates under ω are

$$\begin{aligned}\hat{\mathbf{B}}_\omega &= \hat{\mathbf{B}} - (\mathbf{X}'\mathbf{X})^{-1}\mathbf{C}'[\mathbf{C}(\mathbf{X}'\mathbf{X})^{-1}\mathbf{C}']^{-1}(\mathbf{C}\hat{\mathbf{B}}\mathbf{A} - \mathbf{\Gamma}) \\ \hat{\mathbf{\Sigma}}_\omega &= (\mathbf{Y} - \mathbf{X}\hat{\mathbf{B}}_\omega)'(\mathbf{Y} - \mathbf{X}\hat{\mathbf{B}}_\omega)/n\end{aligned} \tag{5.25}$$

Following the univariate linear model procedure,

$$\begin{aligned}\lambda &= |\hat{\mathbf{\Sigma}}_\omega|^{-n/2} / |\hat{\mathbf{\Sigma}}|^{-n/2} = \left[|\hat{\mathbf{\Sigma}}|/|\hat{\mathbf{\Sigma}}_\omega|\right]^{n/2} \\ &= \left[\frac{|(\mathbf{Y} - \mathbf{X}\hat{\mathbf{B}})'(\mathbf{Y} - \mathbf{X}\hat{\mathbf{B}})|}{|(\mathbf{Y} - \mathbf{X}\hat{\mathbf{B}}_\omega)'(\mathbf{Y} - \mathbf{X}\hat{\mathbf{B}}_\omega)|}\right]^{n/2}\end{aligned} \tag{5.26}$$

so that

$$\Lambda = \lambda^{2/n} = \frac{|\mathbf{E}|}{|\mathbf{E} + \mathbf{H}|}$$

where

$$\begin{aligned}\mathbf{E} &= \mathbf{A}'\mathbf{Y}'(\mathbf{I} - \mathbf{X}(\mathbf{X}'\mathbf{X})^{-1}\mathbf{X}')\mathbf{Y}\mathbf{A} \\ \mathbf{H} &= (\mathbf{C}\hat{\mathbf{B}}\mathbf{A} - \mathbf{\Gamma})'(\mathbf{C}(\mathbf{X}'\mathbf{X})^{-1}\mathbf{C}')^{-1}(\mathbf{C}\hat{\mathbf{B}}\mathbf{A} - \mathbf{\Gamma}) \\ &= (\hat{\mathbf{\Gamma}} - \mathbf{\Gamma}_o)'(\mathbf{C}(\mathbf{X}'\mathbf{X})^{-1}\mathbf{C}')^{-1}(\hat{\mathbf{\Gamma}} - \mathbf{\Gamma}_o)\end{aligned} \tag{5.27}$$

$\mathbf{E}$ and $\mathbf{H}$ have independent Wishart distributions

$$\begin{aligned}\mathbf{E} &\sim \mathbf{W}_u(v_e = n - k,\ \mathbf{A}'\mathbf{\Sigma}\mathbf{A},\ \mathbf{\Delta} = \mathbf{0}) \\ \mathbf{H} &\sim \mathbf{W}_u(v_h = g,\ \mathbf{A}'\mathbf{\Sigma}\mathbf{A},\ (\mathbf{A}'\mathbf{\Sigma}\mathbf{A})^{-1}\mathbf{\Delta})\end{aligned}$$

where the noncentrality matrix

$$\mathbf{\Delta} = (\mathbf{\Gamma} - \mathbf{\Gamma}_o)'(\mathbf{C}(\mathbf{X}'\mathbf{X})^{-1}\mathbf{C}')^{-1}(\mathbf{\Gamma} - \mathbf{\Gamma}_o) \tag{5.28}$$

The likelihood ratio test is to reject $H(\mathbf{CBA} = \mathbf{\Gamma})$ if

$$\Lambda = \frac{|\mathbf{E}|}{|\mathbf{E} + \mathbf{H}|} = \prod_{i=1}^{s}(1 + \lambda_i) \le U^{\alpha}_{u,g,n-k}$$

where $U^{\alpha}_{u,g,n-k}$ is the upper α 100% of the Λ-statistic, λ_i are the roots of $|\mathbf{H} - \lambda\mathbf{E}| = 0$, $s = \min(g, u)$, $g = \text{rank}(\mathbf{C})$, $u = \text{rank}(\mathbf{A})$, and $k = \text{rank}(\mathbf{X})$.

Theorem 5.1 The null distribution of the likelihood ratio criterion or, more generally, the U-statistic

$$U = \frac{|\mathbf{E}|}{|\mathbf{E} + \mathbf{H}|} = \frac{1}{|\mathbf{I} + \mathbf{H}\mathbf{E}^{-1}|}$$

when $v_e \ge u$, $v_h \ge u$, $\mathbf{H} \sim W_u(v_h, \mathbf{\Sigma})$, $\mathbf{E} \sim W_u(v_e, \mathbf{\Sigma})$ and $\mathbf{H}$ and $\mathbf{E}$ are independent is the distribution of the product of u independent beta random variables V_i, $\prod_{i=1}^{u} V_i$, where V_i is beta $[(v_e - i + 1)/2, v_h/2]$.

Theorem 5.2 When $\nu_e \geq u > \nu_h$, the null distribution of U_{ν_h,u,ν_e} is the same as U_{u,ν_h,ν_e}.

Theorem 5.3 The asymptotic distribution of U is given by

$$X^2 = -[\nu_e - (u - \nu_h + 1)/2] \ln U \,\dot{\sim}\, X^2(u\nu_h)$$

a chi-square distribution with $\nu = u\nu_h$ degrees of freedom, as the sample size n tends to infinity.

Box's asymptotic expansion may be used, to obtain p-values for the U-statistic (Anderson, 1984, p. 317). Using *Oh* notation, $O(n^{-\gamma})$ represents terms in the expansion that are bounded when divided by $n^{-\gamma}$.

Theorem 5.4 The

$$P\{\rho n \ln U_{u,g,n-k} > c\} = P(X_\nu^2 > c) + \gamma[P(X_{\nu+4}^2 > c) - P(X_\nu^2 > c)] + O[(\rho n)^{-3}]$$

where $\nu = ug$, $\gamma = \nu(u^2 + g^2 - 5)/48(\rho n)^2$, $\rho = 1 - n^{-1}[k - g + (u + g + 1)/2]$ and X_ν^2 represents a chi-square random variable with ν degrees of freedom.

Anderson states that the expression in Theorem 5.4 is accurate to three decimal places if $u^2 + g^2 \leq \rho n/3$. An alternative to the chi-square approximation for obtaining p-values is to use Rao's F-approximation (Seber, 1984, p. 41).

Theorem 5.5 The

$$P[(U^{-a} - 1)(\nu_2/\nu_1)] = P(F_{(\nu_1,\nu_2)} > c)$$

where

$$a = [(u^2 + \nu_h^2 - 5)/(\nu_h u)^2 - 4]^{1/2}, \nu_1 = \nu_h u$$
$$\nu_2 = a^{-1}[\nu_e - (u - \nu_h + 1)/2] - (\nu_h u - 2)/2$$

and $u = \text{rank}(\mathbf{A})$, $\nu_h = \text{rank}(\mathbf{C})$ and $\nu_e = n - \text{rank}(\mathbf{X})$. If u or ν_h is 1 or 2, the approximation is exact.

The Wishart distribution is a multivariate distribution that generalizes the chi-square distribution. A distribution that generalizes the F-distribution is the multivariate beta distribution.

Theorem 5.6 Let **H** and **E** be independently distributed Wishart distributions: $\mathbf{H} \sim W_u(\nu_h, \mathbf{\Sigma})$ and $\mathbf{E} \sim W_u(\nu_e, \mathbf{\Sigma})$ with $\nu_h, \nu_e \geq u$. Then

$$\mathbf{F} = (\mathbf{H} + \mathbf{E})^{-1/2}\mathbf{H}(\mathbf{H} + \mathbf{E})^{-1/2}$$

where $(\mathbf{H} + \mathbf{E})^{1/2}$ is any nonsingular factorization of $(\mathbf{H} + \mathbf{E})$. Then the *pdf* of **F** is a multivariate beta Type I distribution:

$$c\,|\mathbf{F}|^{(\nu_h - u - 1)/2}\,|\mathbf{I}_u - \mathbf{F}|^{(\nu_e - u - 1)2}$$

or

$$\mathbf{F} \sim \mathbf{M}\beta_{I(u,\,\nu_h/2,\,\nu_e/2)}$$

The constant c is chosen such that the total probability is equal to one in the sample space.

Inspection of the likelihood ratio criterion suggests that the null hypothesis should be rejected if the roots of $|\mathbf{H}-\lambda\mathbf{E}|=0$ are large. Several proposed criteria are dependent on the roots of the following determinantal equations:

$$\begin{aligned} &|\mathbf{H}-\lambda\mathbf{E}|=0 \\ &|\mathbf{H}-\mathrm{v}(\mathbf{H}+\mathbf{E})|=0 \\ &|\mathbf{H}-\theta(\mathbf{H}+\mathbf{E})|=0 \end{aligned} \tag{5.29}$$

Using the roots in (5.29), Wilks' likelihood ratio criterion (Wilks, 1932) is

$$\Lambda=\frac{|\mathbf{E}|}{|\mathbf{E}+\mathbf{H}|}=\prod_{i=1}^{s}\mathrm{v}_i=\prod_{i=1}^{s}(1+\lambda_i)^{-1}=\prod_{i=1}^{s}(1-\theta_i)=|\mathbf{I}_u-\mathbf{F}| \tag{5.30}$$

where $s=\min(v_h,u)$, $v_h=\mathrm{rank}(\mathbf{C})$, and $u=\mathrm{rank}(\mathbf{A})$. H is rejected if $\mathbf{\Lambda}>U^{\alpha}_{u,v_h,v_e}$. Bartlett-Nanda-Pillai suggested the trace criterion (Bartlett, 1939; Nanda, 1950; Pillai, 1955)

$$V=\mathrm{Tr}(\mathbf{H}(\mathbf{H}+\mathbf{E})^{-1})=\sum_{i=1}^{s}\theta_i=\sum_{i=1}^{s}\frac{\lambda_i}{1+\lambda_i}=\sum_{i=1}^{s}(1-\mathrm{v}_i) \tag{5.31}$$

where H is rejected if $V>V^{\alpha}_{s,M,N}$ where $M=(|v_h-u|-1)/2$, and $N=(v_e-u-1)/2$. Employing his Union-Intersection principle, Roy (1953) recommended the largest root statistic

$$\theta_1=\frac{\lambda_1}{1+\lambda_1}=1-\mathrm{v}_1 \tag{5.32}$$

where H is rejected if $\theta_1>\theta^{\alpha}_{s,M,N}$. Another trace criterion was developed by Bartlett-Lawley-Hotelling (Bartlett, 1939; Lawley, 1938; Hotelling, 1931)

$$T_o^2=v_e\mathrm{Tr}(\mathbf{HE}^{-1})=v_e\sum_{i=1}^{s}\lambda_i=v_e\sum_{i=1}^{s}\left(\frac{1-\mathrm{v}_i}{\mathrm{v}_i}\right)=v_e\sum_{i=1}^{s}\left(\frac{\theta_i}{1-\theta_i}\right) \tag{5.33}$$

where H is rejected if T_o^2 is larger than some constant k^* chosen to attain a predetermined Type I error level. The statistic is also called Hotelling's generalized T_o^2. An alternative form for T_o^2 is

$$\begin{aligned} T_o^2&=(\hat{\gamma}-\gamma_o)'[(\mathbf{A}'\hat{\mathbf{\Sigma}}\mathbf{A})\otimes\mathbf{C}(\mathbf{X}'\mathbf{X})^{-1}\mathbf{C}']^{-1}(\hat{\gamma}-\gamma_o) \\ &=\mathrm{Tr}[(\mathbf{C}\hat{\mathbf{B}}\mathbf{A}-\mathbf{\Gamma})'(\mathbf{C}(\mathbf{X}'\mathbf{X})^{-1}\mathbf{C}')^{-1}(\mathbf{C}\hat{\mathbf{B}}\mathbf{A}-\mathbf{\Gamma})(\mathbf{A}'\hat{\mathbf{\Sigma}}\mathbf{A})^{-1}] \end{aligned} \tag{5.34}$$

where $\hat{\mathbf{\Sigma}}=\mathbf{E}/n$ is the ML estimator of $\mathbf{\Sigma}$ which is identical to the LR statistic given in (5.24) with $\mathbf{\Sigma}=\hat{\mathbf{\Sigma}}$. To derive (5.34), observe that the

$$\begin{aligned} \mathrm{vec}(\mathbf{ABC})&=(\mathbf{C}'\otimes\mathbf{A})\,\mathrm{vec}\,\mathbf{B} \\ \mathrm{Tr}(\mathbf{AB})&=(\mathrm{vec}\,\mathbf{A}')'\mathrm{vec}\,\mathbf{B} \end{aligned} \tag{5.35}$$

implies that the

$$\begin{aligned} \mathrm{Tr}(\mathbf{AZ}'\mathbf{BZC})&=\mathrm{Tr}(\mathbf{Z}'\mathbf{BZCA}) \\ &=(\mathrm{vec}\,\mathbf{Z})'(\mathbf{CA}'\otimes\mathbf{B})(\mathrm{vec}\,\mathbf{Z}) \\ &=(\mathrm{vec}\,\mathbf{Z})'(\mathbf{A}'\mathbf{C}'\otimes\mathbf{B})\mathrm{vec}\,\mathbf{Z} \end{aligned} \tag{5.36}$$

Now, letting $\mathrm{vec}\,\mathbf{Z}=\hat{\gamma}-\gamma_o$, $\mathbf{B}=(\mathbf{C}(\mathbf{X}'\mathbf{X})^{-1}\mathbf{C}')^{-1}$, $\mathbf{A}=\mathbf{I}$ and $\mathbf{C}=(\mathbf{A}'\hat{\mathbf{\Sigma}}\mathbf{A})^{-1}$ in (5.36), the first expression in (5.34) follows. Reapplication of (5.35) allows one to express T_o^2 in matrix form.

Having rejected an overall test of $H(\mathbf{CBA} = \mathbf{\Gamma})$, one may obtain $1-\alpha$ simultaneous confidence intervals for bilinear parametric functions of the form $\psi = \mathbf{c'Ba}$ by evaluating the expression

$$\hat{\psi} - c_o\hat{\sigma}_{\hat{\psi}} \leq \psi \leq \hat{\psi} + c_o\hat{\sigma}_{\hat{\psi}} \tag{5.37}$$

where $\hat{\psi} = \mathbf{c}'\hat{\mathbf{B}}\mathbf{a}$, $\hat{\sigma}^2_{\hat{\psi}} = (\mathbf{a'Sa})(\mathbf{c}'(\mathbf{X'X})^{-1}\mathbf{c})$, $E(\mathbf{S}) = \mathbf{\Sigma}$ and c_o:

$$\begin{aligned}
c_o^2 &= v_e\left(\frac{1-U^\alpha}{U^\alpha}\right) && \text{Wilks}\\
c_o^2 &= v_e\left(\frac{\theta^\alpha}{1-\theta^\alpha}\right) && \text{Roy}\\
c_o^2 &= v_e U_o^\alpha = T^2_{o,\alpha} && \text{Bartlett - Lawley - Hotelling}\\
c_o^2 &= v_e\left(\frac{V^\alpha}{1-V^\alpha}\right) && \text{Bartlett - Nanda - Pillai}
\end{aligned} \tag{5.38}$$

The critical values U^α, θ^α, U_o^α, V^α correspond to those procured in testing $H(\mathbf{CBA} = \mathbf{\Gamma})$. Tables for each are provided in Timm (1975, 1993b).

The power of a multivariate test for a specific noncentrality matrix in (5.28) is readily calculated using F-distribution approximations to the noncentral Wishart distributions (Muller et al., 1992).

5.4 Multivariate Mixed Models and the Analysis of Repeated Measurements

When employing a mixed model, the number of independent parameters in $\mathbf{\Sigma}$ is reduced by assuming a linear structure for $\mathbf{\Phi}$. For the classical MANOVA model no assumptions were made about $\mathbf{\Phi} = \mathbf{A'\Sigma A}$, other than positive definiteness. The general multivariate linear mixed model corresponding to the multivariate general linear model is

$$\begin{aligned}
\mathbf{Y}_{n\times t} &= \mathbf{X}^*{}_{n\times k}\,\mathbf{B}^*{}_{k\times t} + \mathbf{Z}^*_{n\times r}\,\mathbf{\Lambda}_{r\times t} + \mathbf{E}_{o(n\times t)}\\
&= \mathbf{X}\,\mathbf{B} + \mathbf{U}
\end{aligned} \tag{5.39}$$

where $\mathbf{\Lambda}$ and $\mathbf{E}_o$ are random matrices, and

$$\text{vec}(\mathbf{Y}) \sim N[(\mathbf{I}_t \otimes \mathbf{X})\,\text{vec}(\mathbf{B}), \mathbf{Z}^*\mathbf{D}^*\mathbf{Z}^{*\prime} + \mathbf{\Psi}] \tag{5.40}$$

where $\mathbf{D}^*$ is a positive definite block diagonal matrix, $\mathbf{Z}^*$ is a matrix of fixed covariates, and $\mathbf{\Psi} = \mathbf{\Sigma}_e \otimes \mathbf{I}_n$. Alternatively, from the general linear model $\mathbf{Y} = \mathbf{XB} + \mathbf{U}$, the ith row of $\mathbf{Y}$ by (5.6) is

$$\mathbf{y}_i = (\mathbf{I}_t \otimes \mathbf{x}_i')\,\text{vec}(\mathbf{B}) + \mathbf{u}_i \tag{5.41}$$

Then, a simple multivariate linear mixed model is constructed by modeling $\mathbf{u}_i$

$$\mathbf{u}_i = \mathbf{Z}_i\theta_i + \mathbf{e}_i \qquad i = 1, 2, \ldots, n \tag{5.42}$$

where $\mathbf{Z}_i$ is a $(t \times h)$ within-subject design matrix for random effects, the $\text{rank}(\mathbf{Z}_i) = h$,

$$\begin{aligned}
\theta_i &\sim N_h(\mathbf{0}, \mathbf{D}_h)\\
\mathbf{e}_i &\sim N(\mathbf{0}, \sigma^2\mathbf{I}_t)
\end{aligned} \tag{5.43}$$

and θ_i and $\mathbf{e}_i$ are independent. Thus,

$$\mathbf{y}_i \sim N_t[(\mathbf{I}_t \otimes \mathbf{x}_i')\text{vec}(\mathbf{B}), \mathbf{\Sigma}_i] \qquad i = 1, 2, \ldots, n \tag{5.44}$$

where

$$\mathbf{\Sigma}_i = \mathbf{Z}_i\mathbf{D}_h\mathbf{Z}_i' + \sigma^2\mathbf{I}_t \tag{5.45}$$

which is of the same form as (2.32), the univariate linear mixed model. Comparing (5.41) and (5.45) with (5.6), we see that the simple multivariate linear mixed model is identical to the classical multivariate linear model if we rearrange the data matrix, assume homogeneity of $\mathbf{\Sigma} = \mathbf{\Sigma}_i$ for $i = 1, 2, \ldots, n$, and impose Hankel structure on $\mathbf{\Phi} = \mathbf{A}'\mathbf{\Sigma}\mathbf{A}$.

Using (5.41), the ML estimator of $\mathbf{\Gamma}$ is $\hat{\mathbf{\Gamma}} = \mathbf{C}\hat{\mathbf{B}}\mathbf{A}$ or

$$\text{vec}(\hat{\mathbf{\Gamma}}) = (\mathbf{A}' \otimes \mathbf{C})\left[\sum_{i=1}^{n} (\hat{\mathbf{\Sigma}}_i^{-1} \otimes \mathbf{x}_i\mathbf{x}_i')\right]^{-1} \sum_{i=1}^{n} (\hat{\mathbf{\Sigma}}_i^{-1} \otimes \mathbf{x}_i)\mathbf{y}_i \tag{5.46}$$

where $\hat{\mathbf{\Sigma}}_i = \mathbf{Z}_i\hat{\mathbf{D}}_h\mathbf{Z}_i' + \hat{\sigma}^2\mathbf{I}_t$, $\hat{\mathbf{D}}_h$ is the ML estimate of $\mathbf{D}_h$ and $\hat{\sigma}^2$ is the ML estimate of σ^2. To test hypotheses about $\mathbf{\Gamma}$, one usually employs large sample theory, except if $\mathbf{\Sigma}_i$ has special structure. If the $\mathbf{\Sigma}_i$ are not allowed to vary among subjects, (5.45) becomes

$$\mathbf{\Sigma}_i = \mathbf{\Sigma} = \mathbf{Z}\mathbf{D}_h\mathbf{Z}' + \sigma^2\mathbf{I}_t \tag{5.47}$$

The simplest general linear mixed model, attributed to Scheffé (1959), is obtained by associating $\mathbf{Z}$ with a vector of t 1's, $\mathbf{Z} = \mathbf{1}_t$ so that $h = 1$, then

$$\mathbf{\Sigma} = \sigma_s^2\mathbf{1}\mathbf{1}' + \sigma^2\mathbf{I}_t \tag{5.48}$$

Postmultipication of (5.39) by $\mathbf{A}$ yields

$$\begin{aligned} \mathbf{YA} &= \mathbf{XBA} + \mathbf{UA} \\ \mathbf{Z}_{n\times u} &= \mathbf{X}_{n\times k}\mathbf{\Theta}_{k\times u} + \xi_{n\times u} \end{aligned} \tag{5.49}$$

The distribution of $\text{vec}(\xi)$ is normal and $\text{vec}(\xi) \sim N(\mathbf{0}, \sigma^2\mathbf{I})$ if $\mathbf{\Phi} = \mathbf{A}'\mathbf{\Sigma}\mathbf{A} = \sigma^2\mathbf{I}_u$ has Hankel structure. This is also known as circularity or sphericity which Huynh and Feldt (1970) showed was a necessary and sufficient condition for a univariate mixed model analysis to be valid. Hence, one should be able to recover univariate mixed model F-statistics from the multivariate linear model. To see this, consider the model in Scheffé (1959, Chapter 8). The model with side conditions is

$$\begin{aligned} & y_{ij} = \mu + \alpha_j + e_{ij} \quad i = 1, 2, \ldots, n \text{ and } j = 1, 2, \ldots, t \\ & \sum_j \alpha_j = 0 \\ & E(e_{ij}) = 0 \\ & \text{cov}(e_{ij}, e_{i'j'}) = \delta_{ii'}\sigma_{jj'} \end{aligned} \tag{5.50}$$

where $\delta_{ii'} = 1$ if $i = i'$ and 0 otherwise. Thus, for $n = 3$ and $t = 3$:

$$\begin{pmatrix} y_{11} \\ y_{12} \\ y_{13} \\ y_{21} \\ y_{22} \\ y_{23} \\ y_{31} \\ y_{32} \\ y_{33} \end{pmatrix} = \begin{pmatrix} 1 & 1 & 0 \\ 1 & 0 & 1 \\ 1 & -1 & -1 \\ 1 & 1 & 0 \\ 1 & 0 & 1 \\ 1 & -1 & 1 \\ 1 & 1 & 0 \\ 1 & 0 & 1 \\ 1 & -1 & -1 \end{pmatrix} \begin{pmatrix} \mu \\ \alpha_1 \\ \alpha_2 \end{pmatrix} + \begin{pmatrix} e_{11} \\ e_{12} \\ e_{13} \\ e_{21} \\ e_{22} \\ e_{23} \\ e_{31} \\ e_{32} \\ e_{33} \end{pmatrix} \tag{5.51}$$

$$\mathbf{y} \quad = \quad \mathbf{X} \quad \beta \quad + \quad \mathbf{e}$$

where $E(\mathbf{y}) = \mathbf{0}$ and $\text{cov}(\mathbf{y}) = \mathbf{I}_n \otimes \mathbf{\Sigma}$. The random effects are contained within the covariance matrix $\mathbf{\Sigma}$. The problem with specification (5.51) is that the pattern of the elements in $\mathbf{\Sigma}$ is unspecified. The univariate model assumes a certain sphericity pattern for $\mathbf{\Sigma}$. We shall show how a univariate repeated measures design may be represented as a multivariate linear model and, with additional restrictions on $\mathbf{\Sigma}$, how to recover univariate tests.

We can rearrange the univariate vector into a data matrix with t occasions and n subjects:

$$\begin{pmatrix} y_{11} & y_{21} & y_{31} \\ y_{12} & y_{22} & y_{32} \\ y_{13} & y_{23} & y_{33} \end{pmatrix} = \begin{pmatrix} 1 & 1 & 0 \\ 1 & 0 & 1 \\ 1 & -1 & -1 \end{pmatrix} \begin{pmatrix} \mu \\ \alpha_1 \\ \alpha_2 \end{pmatrix} \begin{pmatrix} 1 & 1 & 1 \end{pmatrix} + \begin{pmatrix} e_{11} & e_{21} & e_{31} \\ e_{12} & e_{22} & e_{32} \\ e_{13} & e_{23} & e_{33} \end{pmatrix} \tag{5.52}$$

$$\mathbf{Y}' \quad = \quad \mathbf{X}_W \quad \mathbf{B}' \quad \mathbf{X}'_B \quad + \quad \mathbf{U}'$$

Letting

$$\mathbf{A}' = \begin{pmatrix} 1/3 & 1/3 & 1/3 \\ 2/3 & -1/3 & -1/3 \\ -1/3 & 2/3 & -1/3 \end{pmatrix}$$

notice that $\mathbf{A}'\mathbf{X}_W = \mathbf{I}$. Hence, multiplying (5.52) on the left by $\mathbf{A}'$ reduces (5.52) to the transpose of a standard multivariate linear model

$$\begin{aligned} \mathbf{A}'\mathbf{Y}' &= \mathbf{B}'\mathbf{X}'_B + \mathbf{A}'\mathbf{U}' \\ \mathbf{Z}' &= \mathbf{B}'\mathbf{X}'_B + \xi' \end{aligned} \tag{5.53}$$

or taking transposes, the standard multivariate linear model results

$$\begin{aligned} \mathbf{YA} &= \mathbf{X}_B\mathbf{B} + \mathbf{UA} \\ \mathbf{Z} &= \mathbf{X}_B\mathbf{B} + \xi \end{aligned} \tag{5.54}$$

where $\mathbf{A} = (\mathbf{X}'_W)^{-1}$ and $\mathbf{Z} = \mathbf{Y}(\mathbf{X}'_W)^{-1}$. Finally, observe that $\mathbf{X} = \mathbf{X}_B \otimes \mathbf{X}_W$ or $\mathbf{X}' = \mathbf{X}'_B \otimes \mathbf{X}'_W$. The condition that the univariate design matrix $\mathbf{X}$ can be represented as the product $\mathbf{X}_B \otimes \mathbf{X}_W$ is called the condition of separability. Using (5.54), the BLUE of $\mathbf{B}$ is

$$\begin{aligned} \hat{\mathbf{B}} &= (\mathbf{X}'_B\mathbf{X}_B)^{-1}\mathbf{X}'_B\mathbf{Z} \\ &= (\mathbf{X}'_B\mathbf{X}_B)^{-1}\mathbf{X}'_B\mathbf{Y}(\mathbf{X}'_B)^{-1} \end{aligned} \tag{5.55}$$

However, recall that $\text{vec}(\mathbf{ABC}) = (\mathbf{C}' \otimes \mathbf{A})\,\text{vec}(\mathbf{B})$. Hence, by applying the vec(.) operator to both sides of (5.55),

$$\hat{\beta}' = \text{vec}(\hat{\mathbf{B}}')[(\mathbf{X}_B'\mathbf{X}_B)^{-1}\mathbf{X}_B' \otimes \mathbf{X}_W^{-1}]\text{vec}(\mathbf{Y}') \tag{5.56}$$

Next observe that the general form of the univariate repeated measures model (5.51) is

$$E(\mathbf{y}) = \mathbf{X}\beta \text{ and } \text{cov}(\mathbf{y}) = \mathbf{I}_n \otimes \mathbf{\Sigma} = \mathbf{\Omega} \tag{5.57}$$

Further, assume that $\mathbf{X} = \mathbf{X}_B \otimes \mathbf{X}_W$ so that the design matrix is separable. Then, by (4.6)

$$\begin{aligned}\hat{\beta} &= (\mathbf{X}'\mathbf{\Omega}^{-1}\mathbf{X})^{-1}\mathbf{X}'\mathbf{\Omega}^{-1}\mathbf{y} \\ &= [(\mathbf{X}_B \otimes \mathbf{X}_W)'(\mathbf{I}_n \otimes \mathbf{\Sigma})^{-1}(\mathbf{X}_B \otimes \mathbf{X}_W)]^{-1}(\mathbf{X}_B \otimes \mathbf{X}_W)'(\mathbf{I} \otimes \mathbf{\Sigma})^{-1}\mathbf{y} \\ &= [(\mathbf{X}_B'\mathbf{X}_B)^{-1}\mathbf{X}_B' \otimes (\mathbf{X}_W'\mathbf{\Sigma}^{-1}\mathbf{X}_W)^{-1}\mathbf{X}_W'\mathbf{\Sigma}^{-1}]\mathbf{y}\end{aligned} \tag{5.58}$$

Comparing (5.58) with (5.56), we have equivalence of the univariate and multivariate estimates if and only if

$$\mathbf{X}_W^{-1} = (\mathbf{X}_W'\mathbf{\Sigma}^{-1}\mathbf{X}_W)^{-1}\mathbf{X}_W'\mathbf{\Sigma}^{-1} \tag{5.59}$$

so that $\mathbf{X}_W^{-1}$ must exist or the within-design matrix must be of full rank, $\text{rank}(\mathbf{X}_W) = t$, or saturated. Hence, the univariate and multivariate estimates are identical if and only if the univariate design matrix is separable ($\mathbf{X} = \mathbf{X}_B \otimes \mathbf{X}_W$) and the within-design matrix is saturated, ($\text{rank}(\mathbf{X}_W) = t$ or $\mathbf{X}_W^{-1}$ exists). In conclusion, the two models are really one model, arranged in two ways, Davidson (1988).

To demonstrate the equivalence of the two models for hypothesis testing is more complicated. Using the reduced form of model (5.49), $\text{vec}(\xi) \sim N(\mathbf{0}, \mathbf{\Phi} \otimes \mathbf{I}_n)$ where $\mathbf{\Phi} = \mathbf{A}'\mathbf{\Sigma}\mathbf{A} = \sigma^2\mathbf{I}_u$, the maximum likelihood estimator of σ^2 is

$$\hat{\sigma}^2 = \hat{\xi}'\hat{\xi}/n = \text{Tr}(\hat{\xi}\hat{\xi}')/n \tag{5.60}$$

so that the ML estimate of $\mathbf{\Phi}$, assuming sphericity, is

$$\tilde{\mathbf{\Phi}} = \mathbf{I}_u\text{Tr}(\mathbf{E})/un \tag{5.61}$$

where $u = \text{rank}(\mathbf{A})$. The usual unbiased estimator of $\mathbf{\Phi}$ is

$$\hat{\mathbf{\Phi}} = \mathbf{I}_u\text{Tr}(\mathbf{E})/uv_e \tag{5.62}$$

where $v_e = n - \text{rank}(\mathbf{X}) = n - k$ and $\mathbf{E}$ is the SSP matrix, $\mathbf{E} = \mathbf{Y}'(\mathbf{I} - \mathbf{X}(\mathbf{X}'\mathbf{X})^{-1}\mathbf{X}')\mathbf{Y}$. Substituting $[nuv_h/(n-k)]\hat{\mathbf{\Phi}}$ into (5.24) for $\mathbf{\Phi}$ results in the likelihood ratio statistic for testing $H{:}\,\mathbf{C\Theta} = \mathbf{0}$:

$$F = \frac{(v_e/v_h)[(\text{vec}\,\mathbf{\Theta})'\{\mathbf{I}_u \otimes \mathbf{C}'(\mathbf{C}(\mathbf{X}'\mathbf{X})^{-1}\mathbf{C}')^{-1}\mathbf{C}\}(\text{vec}\,\mathbf{\Theta})]}{\text{Tr}(\mathbf{E})} \tag{5.63}$$

which under sphericity has an F-distribution with degrees of freedom uv_h and uv_e, and noncentrality parameter

$$\delta = \theta'\{\mathbf{I}_u \otimes \mathbf{C}'[\mathbf{C}(\mathbf{X}'\mathbf{X})^{-1}\mathbf{C}']^{-1}\mathbf{C}^{-1}\mathbf{C}\}\theta/\sigma^2 \tag{5.64}$$

where $\theta' = \text{vec}(\mathbf{\Theta})$ and $\mathbf{\Theta} = \mathbf{BA}$.

An alternative approach to establish the equivalence of the models for hypothesis testing is to observe that the multivariate hypothesis is equivalent to the univariate hypothesis,

$$H{:}\,\mathbf{C}_B\mathbf{B}\mathbf{A}_W = \mathbf{0} \Leftrightarrow \mathbf{C}\beta = \mathbf{0} \tag{5.65}$$

if and only if $\mathbf{C} = \mathbf{C}'_B \otimes \mathbf{A}_W$ where $\text{vec}(\mathbf{B}') = \beta$ so that the univariate hypothesis is separable. If both the univariate design matrix **X** and hypothesis test matrix **C** are separable, and $\mathbf{\Phi} = \mathbf{A}'_W \mathbf{\Sigma} \mathbf{A}_W = \sigma^2 \mathbf{I}$ has spherical structure, the univariate hypothesis and error sum of squares defined in (2.12) have the following representation

$$SS_H = \text{Tr}(\mathbf{H}) \text{ and } SS_E = \text{Tr}(\mathbf{E}) \tag{5.66}$$

where

$$\begin{aligned}\mathbf{H} &= (\mathbf{C}_B \hat{\mathbf{B}} \mathbf{A}_W)'(\mathbf{C}_B(\mathbf{X}'_B\mathbf{X}_B)^{-1}\mathbf{C}'_B)^{-1}(\mathbf{C}_B\hat{\mathbf{B}}\mathbf{A}_W) \\ \mathbf{E} &= \mathbf{A}'_W\mathbf{Y}'(\mathbf{I}-\mathbf{X}_B(\mathbf{X}'_B\mathbf{X}_B)^{-1}\mathbf{X}'_B)\mathbf{Y}\mathbf{A}_W\end{aligned} \tag{5.67}$$

The F-statistic for testing $H\!:\!\mathbf{C}\beta = \mathbf{0}$ is

$$F = \frac{SS_H / v_h^*}{SS_E / v_e^*} = \frac{MS_h}{MS_e} \tag{5.68}$$

where $v_e^* = \text{rank}(\mathbf{E})\,\text{rank}(\mathbf{A}_W) = v_e u$, and $v_h^* = \text{rank}(\mathbf{H})\,\text{rank}(\mathbf{A}_W) = v_h u$ as discussed by Boik (1988), Davidson (1988), and Timm (1980a).

5.5 Extended Linear Hypotheses

In our discussion of the multivariate GLM model, we demonstrated how one tests standard hypotheses of the form $H(\mathbf{CBA} = \mathbf{\Gamma})$ and obtains simultaneous confidence intervals for bilinear parametric functions $\psi = \mathbf{c}'\mathbf{Ba}$. However, not all potential hypotheses may be represented as $H\!:\!\mathbf{CBA} = \mathbf{\Gamma}$. To illustrate, suppose we consider a simple repeated measures design for two groups (I and II), two conditions (1 and 2), and two stimuli (*A* and *B*) as shown in Figure 5.1.

Figure 5.1 *2x2 Repeated Measures Design*

		Conditions 1	Conditions 2
Groups	I	A μ_{11}	B μ_{12}
	II	B μ_{21}	A μ_{22}

For the design in Figure 5.1, the parameter matrix **B** is

$$\mathbf{B} = \begin{pmatrix} \mu_{11} & \mu_{12} \\ \mu_{21} & \mu_{22} \end{pmatrix} \tag{5.69}$$

The overall test for differences in conditions is to select the matrices **C** and **A** such that

$$\mathbf{C}_o = \begin{pmatrix} 1 & 0 \\ 0 & 1 \end{pmatrix} \text{ and } \mathbf{A}_o = \begin{pmatrix} 1 \\ -1 \end{pmatrix} \tag{5.70}$$

so that the test becomes

$$H\!: \begin{pmatrix} \mu_{11} \\ \mu_{21} \end{pmatrix} = \begin{pmatrix} \mu_{12} \\ \mu_{22} \end{pmatrix} \tag{5.71}$$

Having rejected the overall test of H using one of the multivariate test criteria, one investigates parametric functions $\psi = \mathbf{c'Ba}$ for significance. For this example, one may be interested in the contrasts

$$\psi_1 : \mu_{11} - \mu_{22} = 0 \qquad \psi_2 : \mu_{21} - \mu_{12} = 0 \tag{5.72}$$

that the differences in conditions are due either to stimulus A or stimulus B. However, these contrasts are not of the standard bilinear form, $\psi = \mathbf{c'Ba}$, since the pattern of significance is not the same for each variable.

Using the more general form for ψ_1 and ψ_2, $H_\psi(\mathbf{C}_* \beta = \gamma)$, the contrasts in (5.72) become

$$\begin{pmatrix} 1 & 0 & 0 & -1 \\ 0 & 1 & -1 & 0 \end{pmatrix} \begin{pmatrix} \mu_{11} \\ \mu_{21} \\ \mu_{12} \\ \mu_{22} \end{pmatrix} = \begin{pmatrix} 0 \\ 0 \end{pmatrix} \tag{5.73}$$

Because $\mathbf{C}_* = \mathbf{A}_o' \otimes \mathbf{C}_o$ for the matrices $\mathbf{C}_o$ and $\mathbf{A}_o$ in (5.70), the overall test and the contrasts may be combined and written

$$\mathbf{C}_* \beta = \begin{pmatrix} 1 & 0 & 0 & -1 \\ 0 & 1 & -1 & 0 \\ 1 & 0 & -1 & 0 \\ 0 & 1 & 0 & -1 \end{pmatrix} \begin{pmatrix} \mu_{11} \\ \mu_{21} \\ \mu_{12} \\ \mu_{22} \end{pmatrix} = \begin{pmatrix} 0 \\ 0 \\ 0 \\ 0 \end{pmatrix} \tag{5.74}$$

where the first two rows of $\mathbf{C}_*$ in (5.74) are the contrasts associated with ψ_1 and ψ_2 and the last two rows represent the overall test H. The LR test for testing (5.74) is to use the chi-square statistic given in (5.24)

$$X_v^2 = (\gamma - \gamma_o)'(\mathbf{C}_*(\mathbf{D}'\mathbf{\Omega}^{-1}\mathbf{D})^{-1}\mathbf{C}_*')^{-1}(\hat{\gamma} - \gamma_o) \tag{5.75}$$

where $v = \text{rank}(\mathbf{C}_*)$, $\mathbf{\Omega} = \mathbf{\Sigma} \otimes \mathbf{I}_n$, and $\mathbf{D} = \mathbf{I}_p \otimes \mathbf{X}$. Because $\mathbf{\Sigma}$ is unknown, we must replace $\mathbf{\Sigma}$ by a consistent estimator so that (5.75) is the Wald statistic given in (4.15). When $\mathbf{C}_* = \mathbf{A}' \otimes \mathbf{C}$, (5.75) is T_o^2 by (5.34).

Alternatively, observe that (5.74) may be written as the joint test of

$$\begin{aligned}
&\text{Tr}\left[\begin{pmatrix} 1 & 0 \\ 0 & -1 \end{pmatrix}\begin{pmatrix} \mu_{11} & \mu_{12} \\ \mu_{21} & \mu_{22} \end{pmatrix}\right] = 0 \\
&\text{Tr}\left[\begin{pmatrix} 0 & 1 \\ -1 & 0 \end{pmatrix}\begin{pmatrix} \mu_{11} & \mu_{12} \\ \mu_{21} & \mu_{22} \end{pmatrix}\right] = 0 \\
&\text{Tr}\left[\begin{pmatrix} 1 & 0 \\ -1 & 0 \end{pmatrix}\begin{pmatrix} \mu_{11} & \mu_{12} \\ \mu_{21} & \mu_{22} \end{pmatrix}\right] = 0 \\
&\text{Tr}\left[\begin{pmatrix} 0 & 1 \\ 0 & -1 \end{pmatrix}\begin{pmatrix} \mu_{11} & \mu_{12} \\ \mu_{21} & \mu_{22} \end{pmatrix}\right] = 0
\end{aligned} \tag{5.76}$$

where the last two expressions in (5.76) are constructed from the fact that $\psi = \mathbf{c'Ba} = \text{Tr}(\mathbf{ac'B}) = \text{Tr}(\mathbf{MB})$. More generally, this suggests a union-intersection representation for the multivariate hypothesis ω_H:

$$\omega_H = \bigcap_{M \varepsilon M_o} \{H(\mathbf{M}) : \text{Tr}(\mathbf{MB}) = \text{Tr}(\mathbf{M\Gamma})\} \tag{5.77}$$

where $\mathbf{M}_o$ is some set of matrices of order $p \times k$. Letting $\mathbf{M}_o$ be defined as the family of matrices

$$\mathbf{M}_o = \left\{ \mathbf{M} | \mathbf{M} = \sum_{j=1}^{q} \theta_j \mathbf{C}'_j, \theta_j\text{' s are real} \right\} \tag{5.78}$$

the set of q linearly independent matrices forms a matrix decomposition of the multivariate hypothesis called the extended linear hypothesis of order q by Mudholkar et al. (1974) who use Roy's UI principle and symmetric gauge functions to develop test statistics for testing (5.77), a generalized T_o^2 statistic.

As we saw in our example, the family ω_H includes the general linear hypotheses implied by the maximal hypothesis H so that a test procedure constructed for ω_H includes H and each minimal hypothesis ω_M: $\text{Tr}(\mathbf{MB}) = \text{Tr}(\mathbf{M\Gamma})$. By the UI principle, suppose we can construct a test statistic $t_\psi(\mathbf{M})$ for each minimal hypothesis $\omega_M \subseteq \omega_H$ where the maximum hypothesis H: $\mathbf{C}_o\mathbf{BA}_o = \mathbf{\Gamma}$ is a subfamily of ω_H so that a test procedure for ω_M will include H. Constructing a test statistic $\omega_M \subseteq \omega_H$ independent of $\mathbf{M} \varepsilon \mathbf{M}_o$, the extended linear hypothesis ω_H is rejected if

$$t(\mathbf{M}) = \sup_{\mathbf{M}\varepsilon\mathbf{M}_o} t_\psi(\mathbf{M}) \geq c_\psi(\alpha) \tag{5.79}$$

is significant for some minimal hypothesis ω_M where the critical value $c_\psi(\alpha)$ is chosen such that the $P(t(\mathbf{M}) \leq c_\psi(\alpha) | \omega_H) = 1 - \alpha$.

To construct a test of ω_H, Mudholkar et al. (1974) relate $t_\psi(\mathbf{M})$ to symmetric gauge functions (*sgf*) to generate a class of invariant tests containing both Roy's largest root test and the Bartlett-Lawley-Hotelling trace test statistics. Informally, an *sgf* is a mapping from a vector space defined by $\mathbf{M}$ of order $p \times k$, $p \leq k$ and $p = \text{rank}(\mathbf{M})$ to the real numbers. The mapping is invariant under nonsingular transformations of $\mathbf{M}$ and generates a matrix norm $\|\mathbf{M}\|_\psi = \psi(\lambda_1^{1/2}, \ldots, \lambda_k^{1/2})$ where $\lambda_1 > \lambda_2 > \ldots > \lambda_k$ are the eigenvalues of $\mathbf{MM}'$. Formally, an *sgf* is a mapping from a normed vector space (a vector space on which a norm is defined) to the real numbers (a Banach space or complete normed vector space). The mapping is invariant to permutations and arbitrary sign changes of the coordinates. For a normed vector space of order *r*, the *sgf* is

$$t_{\psi_r}(\mathbf{M}) = \left\{ \sum_i \sum_j |\mathbf{m}_{ij}|^r \right\}^{1/r} \tag{5.80}$$

For $r = 2$, the *sgf* defines the familiar Euclidean (Frobenius) matrix norm

$$t_{\psi_2}(\mathbf{M}) = \left\{ \sum_i \sum_j m_{ij}^2 \right\}^{1/2}, \|\mathbf{M}\|_{\psi_2} = \{\text{Tr}(\mathbf{MM}')\}^{1/2} = \left(\sum_{i=1}^{p} \lambda_i \right)^{1/2} \tag{5.81}$$

Associated with an *sgf* is a normed conjugate *sgf*. For $t_{\psi_r}(\mathbf{M})$, the associated conjugate *sgf* is

$$t_{\psi_s}(\mathbf{M}) = \left\{ \sum_i \sum_j |\mathbf{m}_{ij}|^s \right\}^{1/s} \qquad (1/r) + (1/s) = 1 \tag{5.82}$$

Hence for $r = s = 2$, the Euclidean norm is self-conjugate. For $r = 1$, $t_{\psi_1}(\mathbf{M}) = \sum_i \sum_j |m_{ij}|$, the matrix norm for ψ_1 is $\|\mathbf{M}\|_{\psi_1} = \sum_{i=1}^{p} \lambda_i^{1/2}$. The corresponding conjugate *sgf* is $t_{\psi_\infty}(\mathbf{M}) = \max(m_{ij})$ with matrix norm $\|\mathbf{M}\|_{\psi_\infty} = \lambda^{1/2}$, also called the spectral norm. To construct a test statistic for ω_H, a result due to Mudholkar that relates two conjugate norms ψ and ϕ is required.

Theorem 5.7 For real matrices **A** and **B** of order $p \times k$, $p \leq k$, the (5.83)

$$\|\mathbf{B}\|_\psi = \sup_{\mathbf{A} \neq \mathbf{0}} \operatorname{Tr} \mathbf{AB}' / \|\mathbf{A}\|_\phi$$

With (5.79) and (5.83), we can develop a general class of tests based on *sgf* ψ to test ω_H. We consider first the subfamily of maximal hypotheses $H{:}\mathbf{C}_o\mathbf{B}\mathbf{A}_o = \boldsymbol{\Gamma}$; from (5.27) let

$$\begin{aligned}
\mathbf{E}_o &= \mathbf{A}_o'\mathbf{Y}'(\mathbf{I}-\mathbf{X}(\mathbf{X}'\mathbf{X})^{-1}\mathbf{X}')\mathbf{Y}\mathbf{A}_o \\
\mathbf{H}_o &= (\hat{\mathbf{B}}_o - \boldsymbol{\Gamma})'\mathbf{W}_o^{-1}(\hat{\mathbf{B}}_o - \boldsymbol{\Gamma}) \\
\mathbf{W}_o &= \mathbf{C}_o(\mathbf{X}'\mathbf{X})^{-1}\mathbf{C}_o' \\
\hat{\mathbf{B}}_o &= \mathbf{C}_o\hat{\mathbf{B}}\mathbf{A}_o,\ \hat{\mathbf{B}} = (\mathbf{X}'\mathbf{X})^{-1}\mathbf{X}'\mathbf{Y}
\end{aligned} \tag{5.84}$$

for testing *H*. Then, if we associate matrices **A** and **B** in (5.83) with the matrices in (5.84)

$$\mathbf{A} = \mathbf{E}_o^{1/2}\mathbf{M}\mathbf{W}_o^{1/2} \text{ and } \mathbf{B} = \mathbf{E}_o^{-1/2}\hat{\mathbf{B}}_o'\mathbf{W}_o^{-1/2} \tag{5.85}$$

the $\operatorname{Tr}(\mathbf{AB}') = \operatorname{Tr}(\mathbf{M}\hat{\mathbf{B}}_o)$, **M** is of order $(u \times g)$ and the $\|\mathbf{B}\|_\psi$ depends only on the eigenvalues of $\mathbf{H}_o\mathbf{E}_o^{-1}$. Comparing (5.83) with (5.79), we find a general test statistic for the minimal hypothesis ω_M: $\operatorname{Tr}(\mathbf{MB}_o) = \operatorname{Tr}(\mathbf{M}\boldsymbol{\Gamma})$ is

$$|t_\psi(\mathbf{M})| = \operatorname{Tr}[\mathbf{M}(\hat{\mathbf{B}}_o - \boldsymbol{\Gamma})] / \|\mathbf{W}_o^{1/2}\mathbf{M}'\mathbf{E}_o^{1/2}\|_\phi \tag{5.86}$$

with the critical value $c_\psi(\alpha)$ of $\psi(\lambda_1^{1/2}, \dots, \lambda_s^{1/2})$, $s = \min(g, u)$, $g = \operatorname{rank}(\mathbf{C}_o)$, $u = \operatorname{rank}(\mathbf{A}_o)$, λ_i are the roots of $\mathbf{H}_o\mathbf{E}_o^{-1}$, and there ϕ and ψ are conjugate *sgf*'s. Since the denominator in (5.86) is a matrix norm,

$$\|\mathbf{W}_o^{1/2}\mathbf{M}'\mathbf{E}_o^{1/2}\|_\phi = \phi(\lambda_1^{1/2}, \dots, \lambda_s^{1/2}) \text{ where } \lambda_i = \lambda_i(\mathbf{M}\mathbf{W}_o\mathbf{M}'\mathbf{E}_o) \tag{5.87}$$

represent the characteristic roots of $\mathbf{M}\mathbf{W}_o\mathbf{M}'\mathbf{E}_o$. When $\psi = \psi_2$ is the self-conjugate *sgf*, the test statistic for ω_M: $\operatorname{Tr}(\mathbf{MB}_o) = \operatorname{Tr}(\mathbf{M}\boldsymbol{\Gamma})$ becomes

$$t_\psi(\mathbf{M}) = \frac{|\operatorname{Tr}[\mathbf{M}(\hat{\mathbf{B}}_o - \boldsymbol{\Gamma})]|}{\{\operatorname{Tr}(\mathbf{M}\mathbf{B}_o\mathbf{M}'\mathbf{E}_o)\}^{1/2}} \tag{5.88}$$

a generalized T_o^2 statistic. The hypothesis ω_H is rejected if the supremum of $t_\psi(\mathbf{M})$, $t(\mathbf{M})$, exceeds the critical constant $c_\psi(\alpha)$. To compute the supremum, let

$$\mathbf{M} = \theta_1\mathbf{M}_1 + \theta_2\mathbf{M}_2 + \dots + \theta_q\mathbf{M}_q$$

for a matrix hypothesis ω_H. Furthermore, let

$$\tau_i = \operatorname{Tr}[\mathbf{M}_i(\hat{\mathbf{B}}_o - \boldsymbol{\Gamma})] \text{ and } \mathbf{T} = (t_{ij}),\ t_{ij} = \operatorname{Tr}(\mathbf{M}_i\mathbf{W}_o\mathbf{M}_j'\mathbf{E}_o) \tag{5.89}$$

then by (5.49), $t(\mathbf{M}) = \{t_\psi(\mathbf{M})\}^2 = (\theta'\tau)^2 / \theta'\mathbf{T}\theta$ is the supremum over all vectors θ. To evaluate this supremum, recall from Timm (1975, p. 107) that the sup $\theta'\mathbf{A}\theta / \theta'\mathbf{B}\theta$ is the largest root of the determinantal equation $|\mathbf{A} - \lambda\mathbf{B}| = 0$

for symmetric matrices $\mathbf{A}$ and $\mathbf{B}$, $\lambda_1 = \lambda_1(\mathbf{AB}^{-1})$, the largest root of $\mathbf{AB}^{-1}$. Letting $\mathbf{A} = \tau\tau'$ and $\mathbf{B} = \mathbf{T}$, $\lambda_1 = \lambda_1(\tau\tau'\mathbf{T}^{-1}) = \tau'\mathbf{T}^{-1}\tau$, a quadratic form. Hence,

$$t(\mathbf{M}) = \tau' \, \mathbf{T}^{-1} \, \tau \tag{5.90}$$

and ω_H is rejected if $t(\mathbf{M})$ exceeds $c_\psi^2(\alpha)$. More important, $1-\alpha$ simultaneous confidence intervals may be constructed for contrasts of the parametric functions $\psi = \text{Tr}(\mathbf{MB}_o) = \sum_i \sum_j \beta_{oij}$ estimated by $\hat{\psi} = \text{Tr}(\mathbf{M\hat{B}}_o)$. The simultaneous confidence intervals again have the general form

$$\hat{\psi} - c_o\hat{\sigma}_{\hat{\psi}} \le \psi \le \hat{\psi} + c_o\hat{\sigma}_{\hat{\psi}} \tag{5.91}$$

where

$$\hat{\sigma}_{\hat{\psi}}^2 = \text{Tr}(\mathbf{MW}_o\mathbf{M}'\mathbf{A}_o'\mathbf{SA}_o) \tag{5.92}$$

$E(\mathbf{S}) = \mathbf{\Sigma}$ and the critical constant is $c_o^2 = v_e U_{o(s,M,N)}^\alpha$ for the overall hypothesis H. Letting the *sgf* be $t_{\psi_\infty}(\mathbf{M})$, one obtains Roy's extended largest root statistic. For this criterion, the $1-\alpha$ simultaneous confidence intervals in (5.91) are defined with

$$\begin{aligned} c_o^2 &= \frac{\theta_{s,M,N}^\alpha}{1-\theta_{S,m,n}^\alpha} \\ \hat{\sigma}_{\hat{\psi}} &= \sum_{i=1}^{s} \lambda_i^{1/2}(\mathbf{MW}_o\mathbf{M}'\mathbf{E}_o) \end{aligned} \tag{5.93}$$

A computer program for tests of extended linear hypotheses and the establishment of confidence intervals is included in Krishnaiah et al. (1980).

In (5.75) we showed how the Wald statistic X_v^2 may be used to test the general hypotheses $H(\mathbf{C}_*\beta = \gamma_o)$ by substituting for the unknown covariance matrix $\mathbf{\Sigma}$ any consistent estimate. Hecker (1987) derived a test for H in (5.74) for situations in which a Wishart-distributed estimator of $\mathbf{\Sigma}$ is available and independent of X_v^2 in (5.75). To test H, let $\hat{\mathbf{\Sigma}} \sim W_p(v_e, \mathbf{\Sigma})$ independent of $X_v^2(v = \text{rank } \mathbf{C}^*)$ where

$$\begin{aligned} \lambda_1 &= \text{Tr}(\hat{\mathbf{\Sigma}})/pv_e \\ \lambda_2 &= |\hat{\mathbf{\Sigma}}|^{v_e/2}/(\text{Tr}\hat{\mathbf{\Sigma}}/p)^{pv_e/2} \\ \lambda &= \lambda_1\lambda_2. \end{aligned} \tag{5.94}$$

The hypothesis $H(\mathbf{C}_*\beta = \gamma_o)$ is rejected if the

$$\text{P}(\text{X}_v^2/\lambda \;>\; \text{k}) \;=\; \alpha \tag{5.95}$$

for some constant k. To find the constant k, the expression

$$C(v, v_e, p) = X_v^2 / \lambda_1\lambda_2 \tag{5.96}$$

must be evaluated where X_v^2 and λ_1 have chi-square distributions with v and pv_e degree of freedom, respectively, and

$$k^* = \lambda_2^{2/v_e} \tag{5.97}$$

is distributed as a product of $p-1$ independent beta random variables V_i with $(v_e - i)/2$ and $i\left(\frac{1}{2}+p^{-1}\right)$ degrees of freedom for $i = 1, 2, \ldots, p-1$, (Srivastava and Khatri, 1979, p. 209). The distribution of the test statistic X_v^2 / λ proposed by Hecker is unknown; however, its value is between the distribution of two known distributions

$$\begin{aligned} F = X_v^2 / \lambda_1 &\sim F_{(v, pv_e)} \\ X_v^2 &\sim \chi^2_{(v)} \end{aligned} \tag{5.98}$$

when $\hat{\Sigma}$ is the pooled within-covariance matrix for the standard MANOVA model that is independent of the mean.

Hence, we may test extended linear hypotheses using the generalized T_o^2 statistic defined in (5.88), Roy's extended largest root procedure using (5.93), Wald's statistic defined in (5.75) with Σ replaced by a consistent estimator $\hat{\Sigma}$, or Hecker's (1987) statistic. While all the proposed methods are asymptotically equivalent, the performance of each is unknown for small sample sizes.

5.6 Multivariate Regression

The first example that we illustrate of the general multivariate linear model as presented in (5.3) is the multivariate linear regression model. In multivariate regression, we are provided with k fixed independent variables and are interested in predicting simultaneously p dependent variables where the design matrix $\mathbf{X}$ is common to each of the p-variables. Having fit a model to the data, we may be interested in testing hypotheses about the matrix of regression coefficients to help in the evaluation of model fit. The multivariate regression model is limited in its use because of the restriction that the design matrix is common to all dependent variables. When the design matrix is not common, it is often better to analyze the data as p multiple regression models.

In multivariate regression, as in multiple regression, we first want to evaluate whether the matrix $\mathbf{Y}$ follows a multivariate normal distribution and determine whether there are any outliers in the data. In Chapter 1, we illustrated the use of chi-square plots of the squared Mahalanobis distances $D_i^2 = (\mathbf{y}_i - \bar{\mathbf{y}})'\mathbf{S}^{-1}(\mathbf{y}_i - \bar{\mathbf{y}})$ and calculated multivariate skewness and kurtosis measures to evaluate multivariate normality and to graphically locate multivariate outliers. We also showed how to use Box-Cox univariate transformations to transform data to marginal normality.

Having examined the data for multivariate normality and identified observations that are potential outliers relative to either X or Y, we next need to determine whether these outlying observations are influential on the parameter estimates. Multivariate extensions of some univariate measures have been proposed by Barrett and Ling (1992); these include Cook's distance, DFFITS, COVRATIO for a single observation or a subset of observations, and others. If we let $\hat{\mathbf{B}}$ be the OLSE of $\mathbf{B}$ under (5.3), and $\hat{\mathbf{B}}_{(I)}$ be the estimate of $\mathbf{B}$ with I observations in the data matrix $\mathbf{Y}$ excluded (the complement of $\mathbf{B}_{(I)}$ is $\mathbf{B}_I$) then the multivariate extension of Cook's distance measure which is used to evaluate the combined effect of I observations on the estimation of $\mathbf{B}$ is given by

$$D_I^2 = [\mathrm{vec}(\hat{\mathbf{B}} - \hat{\mathbf{B}}_{(I)})]'[\mathbf{S}^{-1} \otimes (\mathbf{X}'\mathbf{X})]\,\mathrm{vec}(\hat{\mathbf{B}} - \hat{\mathbf{B}}_{(I)})/k \tag{5.99}$$

where $\mathbf{S} = \mathbf{E}/(n-k)$ is the unbiased estimate of $\boldsymbol{\Sigma}$ and $\hat{\mathbf{B}}_{(I)} = (\mathbf{X}'_{(I)}\mathbf{X}_{(I)})^{-1}\mathbf{X}'_{(I)}\mathbf{Y}_{(I)}$. Letting $\mathbf{V} = k\mathbf{S}$ and $\mathbf{M} = \mathbf{X}'\mathbf{X}$, (5.99) has the general quadratic form

$$(\hat{\gamma} - \hat{\gamma}_o)'(\mathbf{V} \otimes \mathbf{M})^{-1}(\hat{\gamma} - \gamma_o)$$

which is similar to (5.24). While D_I^2 does not have a chi-square distribution, like the univariate case, one may relate the value of D_I^2 to the corresponding chi-square distribution with $v = kp$ degrees of freedom. If the percentile value is larger than 50%, the distance between $\hat{\mathbf{B}}$ and $\hat{\mathbf{B}}_{(I)}$ is large so that the I observations affect the fit. Since the 50th percentile for large data sets is near v, if $D_I^2 > v$ the I cases are considered influential and some remedial action may have to be taken.

Barrett and Ling (1992) show that D_I^2 may be written as a function of the leverage or "hat" matrix

$$\mathbf{H}_I = \mathbf{X}_I(\mathbf{X}'\mathbf{X})^{-1}\mathbf{X}'_I$$

and the residual matrix

$$\mathbf{Q}_{(I)} = \mathbf{U}_{(I)}(\mathbf{U}'\mathbf{U})^{-1}\mathbf{U}'_{(I)}$$

where $\mathbf{U}_{(I)} = \mathbf{Y}_{(I)} - \mathbf{X}_{(I)}\hat{\mathbf{B}}_{(I)}$ and $\mathbf{Y}_{(I)}, \hat{\mathbf{B}}_{(I)}$ and $\mathbf{X}_{(I)}$ are the complements of $\mathbf{Y}_I$, $\hat{\mathbf{B}}_I$ and $\mathbf{X}_I$, respectively, with the I indexed observations included. After some manipulation, Cook's distance measure for the multivariate regression model may be expressed

$$D_I^2 = ((n-k)k)\, Tr[(\mathbf{I} - \mathbf{H}_I)^{-2}\mathbf{H}_I\mathbf{Q}_I] \tag{5.100}$$

From (5.100), we see that D_I^2 has the general form $f(\cdot)\, Tr(\mathbf{L}_I\mathbf{R}_I)$ where

$$\mathbf{L}_I = \mathbf{H}_I(\mathbf{I} - \mathbf{H}_I)^{-1}$$
$$\mathbf{R}_I = (\mathbf{I} - \mathbf{H}_I)^{-1/2}\, \mathbf{Q}_I(\mathbf{I} - \mathbf{H}_I)^{-1/2}$$

and $\mathbf{L}_I$ and $\mathbf{R}_I$ represent general leverage and residual matrices, respectively. From vector geometry, the $\|\mathbf{x}\|\,\|\mathbf{y}\|\cos\theta = \mathbf{x}'\mathbf{y} = Tr(\mathbf{y}\mathbf{x}')$ where $0^o \leq \theta \leq 180^o$ is the angle between $\mathbf{x}$ and $\mathbf{y}$, so that the

$$\begin{aligned} Tr(\mathbf{L}_R\mathbf{R}_I) &= \|\text{vec}(\mathbf{L}_I)\|\ \|\text{vec}(\mathbf{R}_I)\|\cos\theta_I \\ &= \|\mathbf{L}_I\|(\cos\theta_I)^{1/2}\ \|\mathbf{R}_I\|(\cos\theta_I)^{1/2} \\ &= \mathbf{L}_I^*\mathbf{R}_I^* \end{aligned}$$

where $\mathbf{L}_I^*$ and $\mathbf{R}_I^*$ represent the relative contribution of leverage and residual to Cook's total influence measure. Barrett and Ling (1992) recommend plotting the logarithms of $\mathbf{L}_I^*$ and $\mathbf{R}_I^*$, $\ln \mathbf{L}_I^*$ versus $\ln \mathbf{R}_I^*$, for various values of I to assess the joint influence of multiple observations on the estimation of $\mathbf{B}$ for the multivariate regression model. While other multivariate criteria have been proposed by Barrett and Ling (1992) to evaluate the influence of multivariate observations on fitted values and model parameters, considerable work has to be done in this area to provide tests or graphical methods to guide researchers.

Note that the squared form of the ith univariate internally standardized residual, $e_i(\mathbf{e}'\mathbf{e})^{-1/2}$, is generalized by $\mathbf{Q}_I$. The squared form of the univariate externally standardized residual $e_i(\mathbf{e}'_{(i)}\mathbf{e}_{(i)})^{-1/2}$ is generalized by

$\mathbf{U}_I(\mathbf{U}'_{(I)}\mathbf{U}_{(I)})^{-1}\mathbf{U}'_I = \mathbf{Q}_I(\mathbf{I}-\mathbf{H}_I-\mathbf{Q}_I)^{-1}(\mathbf{I}-\mathbf{H}_I)$. One may also Studentize the residuals. The generalization of the univariate internally Studentized residual, $e_i / [MSE_{(i)}(1-h_{ii})]^{1/2}$, (sometimes called the Studentized residual) is generalized to $(\mathbf{I}-\mathbf{H}_I)^{-1/2}\ \mathbf{Q}_I(\mathbf{I}-\mathbf{H}_I)^{-1/2}$ for the multivariate model.

The multivariate regression model using (5.4) has the simple form

$$\mathbf{y}_i = \beta_{oi} + \beta_{1i}\mathbf{x}_1 + \beta_{2i}\mathbf{x}_2 + \ldots + \beta_{qi}\mathbf{x}_q + \mathbf{e}_i$$

where $\mathbf{y}_i$, $\mathbf{e}_i$ and $\mathbf{x}_1$ to $\mathbf{x}_q$ are $n\times 1$ vectors, $q + 1 = k$, and $i = 1, 2, \ldots, p$. Using (5.3), the matrix representation for the model is

$$\begin{aligned} \mathbf{Y}_{n\times p} &= (\mathbf{y}_1, \mathbf{y}_2, \ldots, \mathbf{y}_p) \\ \mathbf{X}_{n\times k} &= (\mathbf{1}_n, \mathbf{x}_1, \mathbf{x}_2, \ldots, \mathbf{x}_q) \\ \mathbf{B}_{k\times p} &= \begin{pmatrix} \beta_{01} & \beta_{02} & \cdots & \beta_{0p} \\ \beta_{11} & \beta_{12} & \cdots & \beta_{1p} \\ \cdots & \cdots & \cdots & \cdots \\ \beta_{q1} & \beta_{q2} & \cdots & \beta_{qp} \end{pmatrix} = \begin{pmatrix} \beta'_o \\ \Gamma \end{pmatrix} \end{aligned} \tag{5.101}$$

where $k = q + 1$, $\mathbf{Y}$ is the data matrix of dependent variables, and $\mathbf{X}$ is the design matrix of q independent variables and the constant term. The vector $\mathbf{1}_n$ of n 1's is used to account for the constant term in the model.

While one may test general hypotheses of the form (5.23) for the regression model, one is usually only interested in whether all rows of $\mathbf{B}$ are zero or that some subset of the rows is zero. Thus, the hypothesis tests take the general form $H\text{:}\,\mathbf{CB} = \mathbf{0}$ since the postmatrix $\mathbf{A}$ in (5.23) equals $\mathbf{I}$. If some row of $\mathbf{B}$ is zero, this indicates that the variable should not be included in the model.

To analyze multivariate regression models using SAS, one may use either PROC REG or PROC GLM. While we will illustrate both, PROC REG is designed for regression models, whereas PROC GLM is designed for experimental designs, designs where the independent variables have values of zero and one.

While PROC REG computes all of the test statistics for testing the multivariate hypotheses, Roy's largest root statistic is calculated as λ_1 and not θ_1 as given in (5.32). PROC REG uses an F approximation to obtain the p-values of the test criteria which by Theorem 5.5 is exact for the Wilks' Λ criterion. The relationships between the criteria and the F-distribution are provided in the *SAS/STAT User's Guide* (1990, pp. 17-19).

To illustrate the general theory in Section 5.3 for the multivariate regression model, we again consider the Rhower data given in Timm (1975, p. 281 and p. 345). Rhower was interested in predicting the performance on three standardized tests (Peabody Picture Vocabulary (y_1), Student Achievement Test (y_2), and the Raven Progressive Matrices Test (y_3)) given five paired-associate learning-proficiency tasks (named (x_1), still (x_2), named still (x_3), named action (x_4) and sentence still (x_5)) for $n = 32$ randomly selected school children in an upper-class, white, residential school. These data were used in Section 1.7.4 to illustrate *Q-Q* plots for the dependent variables and to identify outliers in the independent variables. Barrett and Ling (1992) used the data to analyze the single and multiple-case effect of the observations on the estimation of $\mathbf{B}$. The SAS code to analyze the Rhower data is provided in Program 5_6.sas.

Program 5_6.sas

```
/* Program 5_6.sas                          */
/* Rhower data from Timm(1975), pp.281, 345 */

options ls = 80 ps = 60 nodate nonumber;
title1 'Output 5.6: Multivariate Regression';

data rhower;
   infile 'c:\5_6.dat';
   input y1-y3 x0-x5;
proc print data=rhower;
run;

/* Calculations for Regression and Multivariate Cooks Distance */
proc iml;
   title2 'Using PROC IML, including Cooks Distance';
   use rhower;
   v={y1 y2 y3};
   w={x0 x1 x2 x3 x4 x5};
   read all var v into y;
   read all var w into x;
   beta=inv(x`*x)*x`*y;
   print 'Regression Coefficients' beta;
   n=nrow(y);
   p=ncol(y);
   k=ncol(x);
   S=(y`*y-y`*x*beta)/(n-k);
   print 'Estimated Covarianc Matrix' S;

   v=inv(k*s)@(x`*x);
   b=shape(beta`,p*k,1);
   m=n-1;
   x1=x[1:m,1:k];
   y1=y[1:m,1:p];
   beta1=inv(x1`*x1)*x1`*y1;
   b1=shape(beta1`,p*k,1);
   d1=(b-b1)`*v*(b-b1);
   obs=32;
   index=obs;
   do i=2 to 31;
      j=n-i;
      x2=x[1:j,1:k]; x1=x[j+2:n,1:k];
      x2=x2//x1;
      y2=y[1:j,1:p]; y1=y[j+2:n,1:p];
      y2=y2//y1;
      beta2=inv(x2`*x2)*x2`*y2;
      b2=shape(beta2`,p*k,1);
      d2=(b-b2)`*v*(b-b2);
      d1=d1//d2;
      obs=obs-1;
      index=index//obs;
   end;
   x1=x[2:n,1:k];
   y1=y[2:n,1:p];
   beta1=inv(x1`*x1)*x1`*y1;
   b1=shape (beta1`,p*k,1);
   d2=(b-b1)`*v*(b-b1);
   d1=d1//d2;
   obs=1;
   index=index//obs;
   D=index||d1;
   print 'Influence Obs # and Multivariate Cooks Distance', D;
quit;
```

```
/* Multivariate Regression using PROC REG */
proc reg data=rhower;
   title2 'Using PROC REG';

/* Full Model Analysis */
   model y1-y3 = x1-x5;
   Gammaa:mtest x1-x5/ print;   /* test that all coefficients equal zero */
   Gamma1:mtest x1;             /* test that x1 equals zero */
   Gamma2:mtest x2;             /* test that x2 equals zero */
   Gamma3:mtest x3;             /* test that x3 equals zero */
   Gamma4:mtest x4;             /* test that x4 equals zero */
   Gamma5:mtest x5;             /* test that x5 equals zero */
   Beta:mtest intercept,x1-x5;  /* test that int and all coeff are zero */

/* Reduced Model Analysis*/
   model y1-y3 = x2-x4;
   Gamma:mtest x2-x4;   /* test that gamma=0 for reduced model */
   Gamma2:mtest x2;
   Gamma3:mtest x3;
   Gamma4:mtest x4;
run;

/* Calculation of Multivariate Eta Squared for Full and Reduced Models */
proc iml;
   title2 'Multivariate Eta Squared for Full and Reduced Models';
   n=32;
   g1=6;
   g2=4;
   p=3;
   LmdaF=.81206193;
   LmdaR=.70761904;
   Full=1-n*LmdaF/(n-g1+LmdaF);
   Reduced=1-n*LmdaR/(n-g2+LmdaR);
   print 'Eta Squared full model' Full, 'Eta Squared reduced model' Reduced;
quit;

/* Multivariate Regression using PROC GLM and Full Rank Model */
proc glm data=rhower;
   title2 'Using PROC GLM';
   model y1-y3=x2-x4/ nouni;
   manova h=x2-x4/ printe printh;
run;
```

First the DATA step reads the data and then prints it. Note that x0 is a column of 1s. PROC IML then performs the multivariate regression analysis. Next, PROC REG is used for the analysis and PROC IML calculates multivariate eta squared. Lastly, PROC GLM is used for the analysis.

Result and Interpretation 5_6.sas

The results from PROC IML are shown in Output 5.6. The matrix of estimated regression coefficients and the estimated covariance matrix are output. Using (5.99), Cook's distance was computed to evaluate the influence of each observation (thus $I = 1$).

Output 5.6: *Multivariate Regression Using PROC IML, including Cook's Distance*

```
                 Output 5.6: Multivariate Regression
              Using PROC IML, including Cook's Distance

                                        BETA
        Regression Coefficients  39.69709 13.243836 -28.46747
                                0.0672825 0.0593474 3.2571323
                                0.3699843  0.492444 2.9965832
                                 -0.37438 -0.164022 -5.859063
                                 1.523009 0.1189805 5.6662226
                                0.4101567 -0.121156 -0.622653

                                             S
       Estimated Covarianc Matrix  149.9613 10.822548 49.215424
                                  10.822548 6.8088795 23.991269
                                  49.215424 23.991269  659.0046

           Influence Obs # and Multivariate Cooks Distance

                                  D
                                32 0.0550321
                                31 0.0975839
                                30  0.045052
                                29 0.3025963
                                28  0.034224
                                27 0.3386616
                                26  0.042609
                                25 0.2600835
                                24 0.0729437
                                23 0.0303567
                                22 0.0402451
                                21 0.1516428
                                20 0.0373328
                                19 0.1732064
                                18 0.0567071
                                17 0.1444825
                                16 0.1183235
                                15 0.0151908
                                14 0.1642736
                                13 0.0426712
                                12 0.1162936
                                11 0.0456802
                                10  0.063392
                                 9 0.0404034
                                 8 0.1476762
                                 7 0.0252955
                                 6 0.0145846
                                 5 0.8467173
                                 4 0.0064545
                                 3 0.0741106
                                 2  0.035759
                                 1 0.1106688
```

Comparing the calculated values of D^2 with the value of $v = kp = (6)(3) = 18$, we see that none of the observed D^2 values even come close to the value of 18. Note, however, that we have a small data set ($n = 32$) and it is for large data sets that the 50th percentile of the chi-square distribution is near v. Observation {5} has the largest influence value (0.8467); this is consistent with the findings reported by Barrett and Ling (1992) who found case {5} to have the largest influence and the largest leverage component. Barrett and Ling (1992) suggest investigating subsets of observation that have large joint-leverage or joint-residual components, but small influence. They found subsets {5, 14} and {5, 25} to have the largest joint-leverages, although not sufficiently large to suggest data modification.

PROC REG computes the OLSE of **B** using the normal equations (5.13), a variable at a time. See the continuation of Output 5.6 that follows. Comparing the estimated regression coefficients for each dependent variable, y_1, y_2, and y_3 from the following PROC REG output with those in the matrix **B** of the previous PROC IML output, we see that they are equivalent.

Output 5.6: *(continued)*

```
                     Output 5.6: Multivariate Regression
                               Using PROC REG

Model: MODEL1
Dependent Variable: Y1

                              Analysis of Variance

                               Sum of         Mean
        Source          DF     Squares        Square      F Value       Prob>F

        Model            5   2133.72496    426.74499        2.846       0.0353
        Error           26   3898.99379    149.96130
        C Total         31   6032.71875

            Root MSE      12.24587     R-square       0.3537
            Dep Mean      83.09375     Adj R-sq       0.2294
            C.V.          14.73741

                             Parameter Estimates

                        Parameter       Standard     T for H0:
      Variable   DF      Estimate          Error   Parameter=0    Prob > |T|

      INTERCEP    1     39.697090    12.26873491         3.236        0.0033
      X1          1      0.067283     0.61812654         0.109        0.9142
      X2          1      0.369984     0.71554864         0.517        0.6095
      X3          1     -0.374380     0.73699367        -0.508        0.6157
      X4          1      1.523009     0.63846633         2.385        0.0246
      X5          1      0.410157     0.54409367         0.754        0.4577


Dependent Variable: Y2

                            Analysis of Variance

                               Sum of         Mean
        Source          DF     Squares        Square      F Value       Prob>F

        Model            5     78.96913     15.79383        2.320       0.0720
        Error           26    177.03087      6.80888
        C Total         31    256.00000

            Root MSE       2.60938     R-square       0.3085
            Dep Mean      15.00000     Adj R-sq       0.1755
            C.V.          17.39589
```

Output 5.6 *(continued)*

```
                            Parameter Estimates

                         Parameter        Standard     T for H0:
        Variable  DF      Estimate           Error    Parameter=0     Prob > |T|

        INTERCEP   1     13.243836      2.61425537          5.066         0.0001
        X1         1      0.059347      0.13171208          0.451         0.6560
        X2         1      0.492444      0.15247105          3.230         0.0033
        X3         1     -0.164022      0.15704061         -1.044         0.3059
        X4         1      0.118980      0.13604614          0.875         0.3898
        X5         1     -0.121156      0.11593696         -1.045         0.3056


Dependent Variable: Y3

                            Analysis of Variance

                                    Sum of          Mean
          Source          DF       Squares        Square       F Value        Prob>F

          Model            5   21545.09904    4309.01981         6.539        0.0005
          Error           26   17134.11971     659.00460
          C Total         31   38679.21875

              Root MSE       25.67108      R-square        0.5570
              Dep Mean       47.65625      Adj R-sq        0.4718
              C.V.           53.86719

                            Parameter Estimates

                         Parameter        Standard     T for H0:
        Variable  DF      Estimate           Error    Parameter=0     Prob > |T|

        INTERCEP   1    -28.467471     25.71901959         -1.107         0.2785
        X1         1      3.257132      1.29578221          2.514         0.0185
        X2         1      2.996583      1.50000873          1.998         0.0563
        X3         1     -5.859063      1.54496406         -3.792         0.0008
        X4         1      5.666223      1.33842065          4.234         0.0003
        X5         1     -0.622653      1.14058670         -0.546         0.5898
```

While SAS provides a measure of the strength of relationship using $R^2 (= SSR / SST)$ a variable at a time, no multivariate measure is provided. The set of independent variables appears to simultaneously predict variable y_3 best, with over 50% of total variation accounted for by the independent variables.

To test hypotheses regarding **B** using PROC REG, one uses the MTEST statement with the independent variables specified. For example, to test $H\colon \mathbf{\Gamma} = \mathbf{0}$, the MTEST statement is

```
mtest x1-x5/print;
```

To include the intercept in the test, the statement to use is

```
mtest intercept, x1-x5/print;
```

A statement with only one independent variable included is testing $H{:}\gamma_i = \mathbf{0}$. Thus,

```
mtest x5/print;
```

tests $H{:}\gamma_5 = \mathbf{0}$. The PRINT option prints the hypothesis test matrix **H** and the error matrix **E**. The output of the tests follows.

Output 5.6 (*continued*)

```
                    Output 5.6: Multivariate Regression
                              Using PROC REG

Multivariate Test: GAMMAA

                              E, the Error Matrix

                 3898.9937916       281.38623868       1279.6010132
                 281.38623868        177.0308674       623.77300322
                 1279.6010132       623.77300322       17134.119711

                            H, the Hypothesis Matrix

                  26.47691243       -13.93859057       -299.5989233
                 -13.93859057       7.3378762553        157.7218166
                 -299.5989233        157.7218166        3390.105061

               Multivariate Statistics and Exact F Statistics
                            S=1      M=0.5     N=11

  Statistic                        Value          F      Num DF    Den DF   Pr > F
  Wilks' Lambda               0.81206193     1.8515           3        24   0.1648
  Pillai's Trace              0.18793807     1.8515           3        24   0.1648
  Hotelling-Lawley Trace      0.23143318     1.8515           3        24   0.1648
  Roy's Greatest Root         0.23143318     1.8515           3        24   0.1648

Multivariate Test: GAMMA 1

               Multivariate Statistics and Exact F Statistics
                            S=1      M=0.5     N=11
  Statistic                        Value          F      Num DF    Den DF   Pr > F
  Wilks' Lambda               0.79822682     2.0222           3        24   0.1376
  Pillai's Trace              0.20177318     2.0222           3        24   0.1376
  Hotelling-Lawley Trace      0.25277675     2.0222           3        24   0.1376
  Roy's Greatest Root         0.25277675     2.0222           3        24   0.1376


Multivariate Test: GAMMA2

               Multivariate Statistics and Exact F Statistics
                            S=1      M=0.5     N=11
  Statistic                        Value          F      Num DF    Den DF   Pr > F
  Wilks' Lambda               0.69030111     3.5891           3        24   0.0284
  Pillai's Trace              0.30969889     3.5891           3        24   0.0284
  Hotelling-Lawley Trace      0.44864320     3.5891           3        24   0.0284
  Roy's Greatest Root         0.44864320     3.5891           3        24   0.0284
```

Output 5.6 *(continued)*

```
Multivariate Test: GAMMA 3

                 Multivariate Statistics and Exact F Statistics

                              S=1    M=0.5    N=11
  Statistic                    Value           F      Num DF    Den DF  Pr > F
  Wilks' Lambda              0.64205049     4.4601         3        24  0.0126
  Pillai's Trace             0.35794951     4.4601         3        24  0.0126
  Hotelling-Lawley Trace     0.55750991     4.4601         3        24  0.0126
  Roy's Greatest Root        0.55750991     4.4601         3        24  0.0126

Multivariate Test: GAMMA 4

                 Multivariate Statistics and Exact F Statistics
                              S=1    M=0.5    N=11
  Statistic                    Value           F      Num DF    Den DF  Pr > F
  Wilks' Lambda              0.53484385     6.9576         3        24  0.0016
  Pillai's Trace             0.46515615     6.9576         3        24  0.0016
  Hotelling-Lawley Trace     0.86970457     6.9576         3        24  0.0016
  Roy's Greatest Root        0.86970457     6.9576         3        24  0.0016

Multivariate Test: GAMMA 5

                 Multivariate Statistics and Exact F Statistics
                              S=1    M=0.5    N=11
  Statistic                    Value           F      Num DF    Den DF  Pr > F
  Wilks' Lambda              0.91126676     0.7790         3        24  0.5173
  Pillai's Trace             0.08873324     0.7790         3        24  0.5173
  Hotelling-Lawley Trace     0.09737351     0.7790         3        24  0.5173
  Roy's Greatest Root        0.09737351     0.7790         3        24  0.5173

Multivariate Test: BETA

                  Multivariate Statistics and F Approximations
                              S=2     M=0     N=11
Statistic                    Value           F      Num DF    Den DF  Pr > F
  Wilks' Lambda              0.30183585     6.5614         6        48  0.0001
  Pillai's Trace             0.80828405     5.6521         6        50  0.0002
  Hotelling-Lawley Trace     1.94822532     7.4682         6        46  0.0001
  Roy's Greatest Root        1.73835198    14.4863         3        25  0.0001

          NOTE: F Statistic for Roy's Greatest Root is an upper bound.
                NOTE: F Statistic for Wilks' Lambda is exact.
```

The following output is for the model excluding the independent variables X_1 and X_5. PROC REG is used to test the hypotheses that the rows of **B** are equal to zero.

Output 5.6 (*continued*)

```
Model: MODEL 2
Dependent Variable: Y1

                          Analysis of Variance

                              Sum of          Mean
        Source          DF      Squares        Square      F Value        Prob>F

        Model            3   2047.14999     682.38333        4.794        0.0081
        Error           28   3985.56876     142.34174
        C Total         31   6032.71875

            Root MSE       11.93071     R-square        0.3393
            Dep Mean       83.09375     Adj R-sq        0.2686
            C.V.           14.35813

                          Parameter Estimates

                       Parameter       Standard    T for H0:
      Variable  DF      Estimate          Error   Parameter=0    Prob > |T|

      INTERCEP   1     41.695185    11.38237810         3.663        0.0010
      X2         1      0.546450     0.65326733         0.836        0.4100
      X3         1     -0.286485     0.62589608        -0.458        0.6507
      X4         1      1.710677     0.57349467         2.983        0.0059

Dependent Variable: Y2

                          Analysis of Variance

                              Sum of          Mean
        Source          DF      Squares        Square      F Value        Prob>F

        Model            3     70.03116      23.34372        3.515        0.0280
        Error           28    185.96884       6.64174
        C Total         31    256.00000

            Root MSE        2.57716     R-square        0.2736
            Dep Mean       15.00000     Adj R-sq        0.1957
            C.V.           17.18105

                          Parameter Estimates

                       Parameter       Standard    T for H0:
      Variable  DF      Estimate          Error   Parameter=0    Prob > |T|

      INTERCEP   1     12.356643     2.45871363         5.026        0.0001
      X2         1      0.432342     0.14111263         3.064        0.0048
      X3         1     -0.144616     0.13520015        -1.070        0.2939
      X4         1      0.066049     0.12388089         0.533        0.5981
```

Output 5.6 *(continued)*

```
Dependent Variable: Y3

                              Analysis of Variance

                                   Sum of          Mean
        Source          DF        Squares        Square       F Value       Prob>F

        Model            3    17150.42901    5716.80967         7.435       0.0008
        Error           28    21528.78974     768.88535
        C Total         31    38679.21875

            Root MSE       27.72878     R-square        0.4434
            Dep Mean       47.65625     Adj R-sq        0.3838
            C.V.           58.18499

                              Parameter Estimates

                        Parameter       Standard     T for H0:
      Variable  DF       Estimate          Error    Parameter=0    Prob > |T|

      INTERCEP   1     -44.093409    26.45438459         -1.667        0.1067
      X2         1       2.390491     1.51829302          1.574        0.1266
      X3         1      -4.068687     1.45467806         -2.797        0.0092
      X4         1       5.487488     1.33288918          4.117        0.0003

Multivariate Test: GAMMA

                 Multivariate Statistics and Exact F Statistics

                              S=1      M=0.5      N=12

  Statistic                        Value           F       Num DF     Den DF   Pr > F

  Wilks' Lambda               0.70761904      3.5810            3         26   0.0273
  Pillai's Trace              0.29238096      3.5810            3         26   0.0273
  Hotelling-Lawley Trace      0.41318979      3.5810            3         26   0.0273
  Roy's Greatest Root         0.41318979      3.5810            3         26   0.0273

Multivariate Test: GAMMA 2

                 Multivariate Statistics and Exact F Statistics

                              S=1      M=0.5      N=12

  Statistic                        Value           F       Num DF     Den DF   Pr > F

  Wilks' Lambda               0.74429094      2.9775            3         26   0.0499
  Pillai's Trace              0.25570906      2.9775            3         26   0.0499
  Hotelling-Lawley Trace      0.34356063      2.9775            3         26   0.0499
  Roy's Greatest Root         0.34356063      2.9775            3         26   0.0499
```

Output 5.6 *(continued)*

```
Multivariate Test: GAMMA 3

              Multivariate Statistics and Exact F Statistics

                         S=1     M=0.5     N=12

 Statistic                     Value          F     Num DF    Den DF   Pr > F

 Wilks' Lambda            0.78144647     2.4239          3        26   0.0885
 Pillai's Trace           0.21855353     2.4239          3        26   0.0885
 Hotelling-Lawley Trace   0.27967818     2.4239          3        26   0.0885
 Roy's Greatest Root      0.27967818     2.4239          3        26   0.0885

Multivariate Test: GAMMA 4

              Multivariate Statistics and Exact F Statistics

                         S=1     M=0.5     N=12

 Statistic                     Value          F     Num DF    Den DF   Pr > F

 Wilks' Lambda            0.51605757     8.1273          3        26   0.0006
 Pillai's Trace           0.48394243     8.1273          3        26   0.0006
 Hotelling-Lawley Trace   0.93776830     8.1273          3        26   0.0006
 Roy's Greatest Root      0.93776830     8.1273          3        26   0.0006
```

Reviewing the output, we see that the multivariate hypothesis tests suggest that the independent variables X_1 (see GAMMA1) and X_5 (see GAMMA5), with p-values 0.1376 and 0.5173 should be excluded from the full model. MODEL2 in the output includes only variables X_2, X_3 and X_4, (excluding variables X_1 and X_5) a reduced model.

While the value of R^2 or adjusted R^2 may be compared for the full model versus the reduced model, these measures do not evaluate the overall fit for the set of dependent variables simultaneously. To evaluate the overall fit of the models, we need a generalization of R^2. For the multivariate regression model, an extension of R^2 is given by $1-\Lambda = (|SSR|/|SST|)$ when testing $H\colon \boldsymbol{\Gamma} = \mathbf{0}$. However, the measure of association is positively biased. A less biased estimator recommended by Jobson (1992, p. 218) is

$$\eta^2 = 1 - n\Lambda / (n - g + \Lambda) \tag{5.102}$$

where $g = q + 1 = k$ is the rank of $\mathbf{X}$, and $\boldsymbol{\Lambda}$ is the Wilks criterion associated with testing $H\colon \boldsymbol{\Gamma} = \mathbf{0}$. The PROC IML code to perform these calculations is included in Program 5_6.sas. The Λ values were obtained from the PROC REG output for the tests $H\colon \boldsymbol{\Gamma} = \mathbf{0}$ for each model. The results are given in the continuation of Output 5.6 that follows.

Output 5.6 *(continued)*

```
              Output 5.6: Multivariate Regression
        Multivariate Eta Squared for Full and Reduced Models

                                                   FULL
                     Eta Squared full model    0.03081


                                                   REDUCED
                   Eta Squared reduced model 0.2112265
```

We see that MODEL2, the reduced model, is much better than the full model with 21% of the variation in the Y's accounted for versus only 3% for the full model.

Lastly, PROC GLM is used to perform the multivariate regression analysis and to test hypotheses (see the continuation of Output 5.6 that follows). PROC GLM may be used to test that any variable of **B** is zero. Currently, PROC GLM may not be used to test that more than one row of the matrix **B** is simultaneously zero. Comparing the output from PROC REG with the following output from PROC GLM for the reduced model (x2 – x4), we see that the multivariate tests of $H\!:\boldsymbol{\gamma}_i = \mathbf{0}$ agree for the two procedures. The PRINTE and PRINTH options are used in the MANOVA statement of PROC GLM to output the matrices **E** and **H** for the individual tests.

Output 5.6 *(continued)*

```
                     Output 5.6: Multivariate Regression
                                Using PROC GLM

                       General Linear Models Procedure

                  Number of observations in data set = 32

E = Error SS&CP Matrix

                                 Y1                Y2                Y3

          Y1        3985.5687624      257.64138293        1225.710055
          Y2        257.64138293      185.96884254       741.40238546
          Y3         1225.710055      741.40238546       21528.789739

                       General Linear Models Procedure
                      Multivariate Analysis of Variance

   Partial Correlation Coefficients from the Error SS&CP Matrix / Prob > |r|

                  DF = 28            Y1          Y2          Y3

                  Y1           1.000000    0.299261    0.132322
                                 0.0001      0.1148      0.4938

                  Y2           0.299261    1.000000    0.370531
                                 0.1148      0.0001      0.0479

                  Y3           0.132322    0.370531    1.000000
                                 0.4938      0.0479      0.0001
```

Output 5.6 (*continued*)

```
                    General Linear Models Procedure
                   Multivariate Analysis of Variance

                   H = Type III SS&CP Matrix for X2

                         Y1                  Y2                  Y3

          Y1       99.598008187       78.800297482       435.70003948
          Y2       78.800297482       62.345492608       344.71866806
          Y3       435.70003948       344.71866806       1906.0072371

          Characteristic Roots and Vectors of: E Inverse * H, where
          H = Type III SS&CP Matrix for X2    E = Error SS&CP Matrix

Characteristic    Percent              Characteristic Vector  V'EV=1
     Root
                                         Y1              Y2              Y3

   0.34356063    100.00          -0.00051248      0.06866505      0.00112423
   0.00000000      0.00          -0.00032702     -0.03969090      0.00725321
   0.00000000      0.00           0.01659451     -0.02097429      0.00000000

               Manova Test Criteria and Exact F Statistics for
                  the Hypothesis of no Overall X2 Effect
          H = Type III SS&CP Matrix for X2    E = Error SS&CP Matrix

                          S=1     M=0.5     N=12

 Statistic                       Value          F       Num DF    Den DF  Pr > F

 Wilks' Lambda              0.74429094     2.9775          3        26  0.0499
 Pillai's Trace             0.25570906     2.9775          3        26  0.0499
 Hotelling-Lawley Trace     0.34356063     2.9775          3        26  0.0499
 Roy's Greatest Root        0.34356063     2.9775          3        26  0.0499

                   H = Type III SS&CP Matrix for X3

                         Y1                  Y2                  Y3

          Y1       29.821715119       15.053799291       423.53026109
          Y2       15.053799291       7.5990556609       213.79519986
          Y3       423.53026109       213.79519986        6015.008907

          Characteristic Roots and Vectors of: E Inverse * H, where
          H = Type III SS&CP Matrix for X3    E = Error SS&CP Matrix

Characteristic    Percent              Characteristic Vector  V'EV=1
     Root
                                         Y1              Y2              Y3

   0.27967818    100.00           0.00048501      0.00035962      0.00677192
   0.00000000      0.00           0.01597581     -0.00005962     -0.00112277
```

Output 5.6 *(continued)*

```
                    General Linear Models Procedure
                   Multivariate Analysis of Variance

        Characteristic Roots and Vectors of: E Inverse * H, where
        H = Type III SS&CP Matrix for X3    E = Error SS&CP Matrix

Characteristic   Percent              Characteristic Vector  V'EV=1
     Root
                                        Y1          Y2            Y3

   0.00000000      0.00          -0.00450394     0.08203685    -0.00259875

               Manova Test Criteria and Exact F Statistics for
                   the Hypothesis of no Overall X3 Effect
        H = Type III SS&CP Matrix for X3    E = Error SS&CP Matrix

                           S=1     M=0.5     N=12

 Statistic                     Value          F      Num DF     Den DF  Pr > F

 Wilks' Lambda              0.78144647     2.4239         3         26  0.0885
 Pillai's Trace             0.21855353     2.4239         3         26  0.0885
 Hotelling-Lawley Trace     0.27967818     2.4239         3         26  0.0885
 Roy's Greatest Root        0.27967818     2.4239         3         26  0.0885

                    H = Type III SS&CP Matrix for X4

                           Y1               Y2               Y3

        Y1       1266.513119      48.899620009      4062.7040889
        Y2      48.899620009      1.8879968956       156.8595565
        Y3      4062.7040889       156.8595565      13032.288624

        Characteristic Roots and Vectors of: E Inverse * H, where
        H = Type III SS&CP Matrix for X4    E = Error SS&CP Matrix

Characteristic   Percent              Characteristic Vector  V'EV=1
     Root
                                        Y1          Y2            Y3

   0.93776830    100.00           0.00926079    -0.02889542     0.00594358
   0.00000000      0.00          -0.00397312     0.07532180     0.00033200
   0.00000000      0.00          -0.01319846    -0.01489490     0.00429378

                    General Linear Models Procedure
                   Multivariate Analysis of Variance

               Manova Test Criteria and Exact F Statistics for
                   the Hypothesis of no Overall X4 Effect
        H = Type III SS&CP Matrix for X4    E = Error SS&CP Matrix

                           S=1     M=0.5     N=12

 Statistic                     Value          F      Num DF     Den DF  Pr > F

 Wilks' Lambda              0.51605757     8.1273         3         26  0.0006
 Pillai's Trace             0.48394243     8.1273         3         26  0.0006
 Hotelling-Lawley Trace     0.93776830     8.1273         3         26  0.0006
 Roy's Greatest Root        0.93776830     8.1273         3         26  0.0006
```

If we use (5.37), the last calculations one may want to perform for these data are the construction of simultaneous confidence intervals for the individual elements of **B**, but since we have at most *kp* planned comparisons of interest, procedure (5.37) is really too conservative (Timm, 1975, p. 317). An alternative is to utilize adjusted values of the Student t-distribution to control the familywise error rate at the nominal level α or to use finite intersection tests (FIT) (Timm, 1995).

In our discussion of the GLM theory and in our example of multivariate linear regression, we have assumed that the design matrix **X** has full rank or may be reparameterized to a matrix of full rank. Unfortunately, when this is not the case, PROC GLM does not reparameterize the design matrix to full rank, but instead, solves the normal equations (5.13) using a generalized inverse of $\mathbf{X'X}$ so that a solution to (5.13) is

$$\hat{\mathbf{B}} = (\mathbf{X'X})^{-}\ \mathbf{X'Y} \tag{5.103}$$

which is not unique. Even though the matrix $\hat{\mathbf{B}}$ is not unique, certain linear functions of $\hat{\mathbf{B}}$ are unique independent of $\hat{\mathbf{B}}$. These functions are called estimable functions.

Theorem 5.8 A parametric function $\psi = \mathbf{c}'B\mathbf{a}$ is estimable if and only if $\mathbf{c'H} = \mathbf{c}'$ where $\mathbf{H} = (\mathbf{X'X})^{-}\ (\mathbf{X'X})$.

For a proof see Timm (1975, p. 174). A linear hypothesis is said to be testable if and only if it is estimable.

5.7 Nonorthogonal MANOVA Designs

Consider first a completely randomized orthogonal factorial design with $n_{ij} > 0$ observations per cell, factors *A* and *B*, and all interactions in the model. The overparameterized parametric form of the two-factor design is

$$\begin{aligned} \mathbf{y}_{ijk} &= \mu + \alpha_i + \beta_j + \gamma_{ij} + \mathbf{e}_{ijk} \\ \mathbf{e}_{ijk} &\sim IN_p(\mathbf{0}, \boldsymbol{\Sigma}) \end{aligned} \tag{5.104}$$

where $\mathbf{y}_{ijk}$ is a *p*-variate random vector, $i = 1, 2, \ldots, a$; $j = 1, 2, \ldots, b$; $k = 1, 2, \ldots, n_{ij} > 0$. The subscripts *i* and *j* in the model are associated with the levels of factors *A* and *B*. The parameters α_i and β_j are the main effects, and γ_{ij} is the vector of interactions. Letting $n_{ij} = 2$, $a = 3$ and $b = 2$, we present the following design matrix $\mathbf{X}$ and parameter matrix $\mathbf{B}$ in (5.3) for an orthogonal design.

$$\mathbf{B} = \begin{pmatrix} \mu_{11} & \mu_{12} & \cdots & \mu_{1p} \\ \alpha_{11} & \alpha_{12} & \cdots & \alpha_{1p} \\ \alpha_{21} & \alpha_{22} & \cdots & \alpha_{2p} \\ \alpha_{31} & \alpha_{32} & \cdots & \alpha_{3p} \\ \beta_{11} & \beta_{12} & \cdots & \beta_{1p} \\ \beta_{21} & \beta_{22} & \cdots & \beta_{2p} \\ \gamma_{111} & \gamma_{112} & \cdots & \gamma_{11p} \\ \gamma_{121} & \gamma_{122} & \cdots & \gamma_{12p} \\ \cdots & \cdots & \cdots & \cdots \\ \gamma_{321} & \gamma_{322} & \cdots & \gamma_{32p} \end{pmatrix}$$

$$\mathbf{X} = \begin{pmatrix} 1 & 1 & 0 & 0 & 1 & 0 & 1 & 0 & 0 & 0 & 0 & 0 \\ 1 & 1 & 0 & 0 & 1 & 0 & 1 & 0 & 0 & 0 & 0 & 0 \\ 1 & 1 & 0 & 0 & 0 & 1 & 0 & 1 & 0 & 0 & 0 & 0 \\ 1 & 1 & 0 & 0 & 0 & 1 & 0 & 1 & 0 & 0 & 0 & 0 \\ 1 & 0 & 1 & 0 & 1 & 0 & 0 & 0 & 1 & 0 & 0 & 0 \\ 1 & 0 & 1 & 0 & 1 & 0 & 0 & 0 & 1 & 0 & 0 & 0 \\ 1 & 0 & 1 & 0 & 0 & 1 & 0 & 0 & 0 & 1 & 0 & 0 \\ 1 & 0 & 1 & 0 & 0 & 1 & 0 & 0 & 0 & 1 & 0 & 0 \\ 1 & 0 & 0 & 1 & 1 & 0 & 0 & 0 & 0 & 0 & 1 & 0 \\ 1 & 0 & 0 & 1 & 1 & 0 & 0 & 0 & 0 & 0 & 1 & 0 \\ 1 & 0 & 0 & 1 & 0 & 1 & 0 & 0 & 0 & 0 & 0 & 1 \\ 1 & 0 & 0 & 1 & 0 & 1 & 0 & 0 & 0 & 0 & 0 & 1 \end{pmatrix}$$

Thus, we see that each column of $\mathbf{B}$ is associated with a row of the design matrix. Unfortunately, this design matrix $\mathbf{X}$ is not of full column rank, so a unique inverse does not exist; the $\text{rank}(\mathbf{X}) = ab$. While one may reparameterize the model to full rank, this is not the approach taken in PROC GLM; instead, PROC GLM obtains a generalized inverse and provides a general form of estimable functions using Theorem 5.8. For the two-factor factorial design, the general form of parametric functions that are estimable and hence testable are given in Timm (1975, p. 404)

$$\begin{aligned} \psi &= \mathbf{c'Ba} = \mathbf{a'}\left[\sum_{j=1}^{a}\sum_{j=1}^{b}(\mu+\alpha_i+\beta_j+\gamma_{ij})\right] \\ \hat{\psi} &= \mathbf{c'}\hat{\mathbf{B}}\mathbf{a} = \mathbf{a'}\left(\sum_{j=1}^{a}\sum_{j=1}^{b} t_{ij}\,\mathbf{y}_{ij.}\right) \end{aligned} \tag{5.105}$$

where $\mathbf{y}_{ij}.$ is a vector of cell means and $\mathbf{t'} = (t_{ij})$ is a vector of arbitrary constants.

As in a univariate analysis, ψ's involving the individual effects of α and β are not directly estimable. One can see this by using (5.105) and noting that no set of t_{ij}'s may be found to estimate the individual effects. However, if the $\sum_j t_{ij} = \sum_j t_{i'j} = 1$, then

$$\psi = \alpha_{is} - \alpha_{i's} + \sum_{j=1}^{b} t_{ij}(\beta_{js}+\gamma_{ijs}) - \sum_{j=1}^{b} t_{i'j}(\beta_{js}+\gamma_{i'js})$$

for $s = 1, 2, \cdots, p$ is estimated by $\hat{\psi} = \sum_j t_{ij}\, y_{ij.}^{(s)} - \sum_j t_{i'j}\, y_{i'j}^{(s)}.$ where $y_{ij.}^{(s)}$ is a cell mean for variable s. A similar expression exists for the β's. Thus, the hypothesis for main effects becomes

$$\begin{aligned} H_{A^*}&: \alpha_i + \sum_j t_{ij}(\beta_j+\gamma_{ij}) \text{ are equal} \\ H_{B^*}&: \beta_i + \sum_i t_{ij}(\alpha_i+\gamma_{ij}) \text{ are equal} \end{aligned} \tag{5.106}$$

for arbitrary t_{ij} that sum to unity. For example, letting $t_{ij} = 1/b$ for H_{A^*} and $t_{ij} = 1/a$ for H_{B^*}, (5.106) becomes

$$\begin{aligned} H_A&: \alpha_i + \sum_j \gamma_{ij}/b \text{ are equal} \\ H_B&: \beta_j + \sum_i \gamma_{ij}/a \text{ are equal} \end{aligned} \tag{5.107}$$

Hence, tests of main effects are confounded by interaction, whether the design is orthogonal or nonorthogonal.

To test for interactions we have to obtain functions of the γ_{ij} that are estimable. Using (5.105), observe that

$$\psi = \gamma_{ij} - \gamma_{i'j} - \gamma_{ij'} + \gamma_{i'j'} \tag{5.108}$$

is estimable and estimated by

$$\hat{\psi} = \mathbf{y}_{ij.} - \mathbf{y}_{i'j.} - \mathbf{y}_{ij'.} + \mathbf{y}_{i'j'.} \tag{5.109}$$

The parametric function in (5.108) is called a tetrad. The interaction hypothesis tests that all tetrads are zero; when they are zero, the parameter γ_{ij} may be removed from the model and tests of main effects are no longer confounded by interaction.

When the factorial design is nonorthogonal, the coefficients t_{ij} are design dependent. For example, letting the t_{ij} depend on the cell frequencies n_{ij}, we have the weighted tests H_{A^*} and H_{B^*}:

$$\begin{aligned} H_{A^*}&: \text{all } \alpha_i + \sum_j n_{ij}(\beta_j+\gamma_{ij})/n_{i+} \text{ are equal} \\ H_{B^*}&: \text{all } \beta_j + \sum_i n_{ij}(\alpha_i+\gamma_{ij})/n_{+j} \text{ are equal} \end{aligned} \tag{5.110}$$

where n_{ij} is a cell frequency, n_{i+} is a sum over column frequencies, and n_{+j} is a sum over row frequencies. Choosing weights proportional to the number of levels in the design, we have the unweighted tests

$$\begin{aligned} H_A&: \text{all } \alpha_i + \sum_j \gamma_{ij} / b \text{ are equal} \\ H_B&: \text{all } \beta_j + \sum_i \gamma_{ij} / a \text{ are equal} \end{aligned} \tag{5.111}$$

Other, more complicated functions are of course estimable and testable. One function tested in PROC GLM for a factorial design is

$$\psi_i = \mathbf{a}' \left\{ \sum_{j=1}^{b} [n_{ij}(\mu + \alpha_i + \beta_j + \gamma_{ij}) - \sum_{k=1}^{a} \frac{n_{ij}n_{kj}}{n_{+j}}(\mu + \alpha_i + \beta_j + \gamma_{ij})] \right\} \tag{5.112}$$

The inclusion or exclusion of the interaction parameter also determines which functions of the model parameters are estimable and their estimates. For additional detail see Timm (1975, p. 511-526) for the multivariate case and Milliken and Johnson (1992, p. 138) for the univariate case.

The orthogonal decomposition of the hypothesis test space is not unique for a nonorthogonal design (Timm and Carlson, 1975). PROC GLM evaluates the reduction in the error sum of squares for the parameters in the model for some hierarchical sequence of the parameters. For example, one may fit α in the model

$$\begin{gathered} \text{I: } \alpha \text{ after } \mu \\ \text{II: } \alpha \text{ after } \mu \text{ and } \beta \\ \text{III: } \alpha \text{ after } \mu, \beta, \gamma \end{gathered}$$

These are termed the Type I, Type II, and Type III sum of squares, respectively. The Type I sum of squares evaluates the effect of the parameter added to the model in the order defined by the MODEL statement. The Type II sum of squares evaluates the effect after adjusting for all other terms in the model that do not include it. The Type III sum of squares evaluates the effect of the last term given that all others terms are in the model. For nonorthogonal designs, the Type I , Type II, and Type III sum of squares do not add to the total sum of squares about the mean. This was illustrated in Section 3.3.

For a two-factor design that is orthogonal, the Type II sum of squares equals the Type III sum of squares and adds to the total sum of squares. SAS also computes a Type IV sum of squares for use with designs having empty cells; for a discussion of Type IV sum of squares see Milliken and Johnson (1992, p. 186). In general, the various sum of squares for a nonorthogonal design are testing very different hypotheses. For a nonorthogonal two-factor design, the Type I sum of squares is a weighted test of A, $H_{A^\bullet}$ defined in (5.110). The Type III sum of squares is an unweighted test of A, H_A defined in (5.111). The Type II sum of squares is testing (5.112) that all ψ_i are equal. For most nonorthogonal designs, the most appropriate test is that which uses the Type III sum of squares (Timm and Carlson, 1975; and Searle, 1994). Additional advice on the utilization of the various sum of squares is provided by Milliken and Johnson (1992, p. 150) for the univariate model.

To illustrate the general theory of Chapter 5 for an experimental design and to extend the discussion of nonorthogonality to a more complex design, we will analyze a completely randomized fixed effect nonorthogonal three-factor factorial design with $n_{ijk} > 0$ observations per cell. The overparameterized model for the design is

$$\mathbf{y}_{ijkm} = \mu + \alpha_i + \beta_j + \lambda_k + (\alpha\beta)_{ij} + (\alpha\lambda)_{ik} + (\beta\lambda)_{jk} + (\alpha\beta\lambda)_{ijk} + \mathbf{e}_{ijkm} \tag{5.113}$$

where $i = 1, 2, \cdots, a$; $j = 1, 2, \cdots, b$; $k = 1, 2, \cdots, c$; and $m = 1, 2, \cdots, n_{ijk} > 1$. The subscripts i, j, and k are associated with the levels of factors *A, B,* and *C* and all interactions are included in the model. The data we use are hypothetical with factor *A* having three levels, factor *B* having three levels, and factor *C* having two levels.

Before analyzing the data for a nonorthogonal design, one must determine the appropriate sum of squares to test both main effects and interactions. For a two-factor design, the test of the two-factor interaction was invariant to the design weights; this is also the case for the highest interaction (three-way) in the three-factor design. However, both the main effect and two-factor interaction tests now depend on the design weights, the sum of squares in PROC GLM. If the unequal cell frequencies are the result of treatment effects so that upon replication of the study one would expect the same proportion of cell frequencies, then the most appropriate design weights are the cell frequencies, yielding weighted tests or Type I sum of squares. If, however, the nonorthogonal nature of the design is not related to treatment and the fixed treatments represent all levels of the study so that upon replication one would have the same number of fixed levels in the design, then the most appropriate design weights are the levels of the design, yielding unweighted tests or Type III sum of squares. For most experimental designs, the sequential Type II sum of squares is not appropriate for hypothesis testing; however, they may be used in exploratory studies to evaluate the contribution of the parameters in the model, adjusting for the parameters already in the model. The SAS code for the analysis of the data are provided in Program 5_7.sas.

Program 5_7.sas

```
/* Program 5_7.sas              */
/* Nonorthogonal MANOVA Design */

options ls=80 ps=60 nodate nonumber;
title1 'Output 5.7: Nonorthogonal Three Factor MANOVA';

data three;
   infile 'c:\5_7.dat';
   input a b c y1 y2;
proc print data=three;
run;

/* Unweighted MANOVA Analysis */
proc glm;
   title2 'Unweighted Analysis';
   class a b c;
   model y1 y2 = a b c a*b a*c b*c a*b*c/ ss3 e;
   contrast 'a1 vs a3' a 1 0 -1;
   estimate 'a1 vs a3' a 1 0 -1/ divisor=3 e;
   contrast 'c1 vs c2' c .5 -.5;
   estimate 'c1 vs c2' c .5 -.5/ e;
   lsmeans a b c a*b a*c b*c a*b*c;
   manova h=a|b|c;
run;
```

```
/* Full Rank Model for MANOVA Design */
proc glm;
   title2 'Full Rank Model with Unweighted Contrast';
   class a b c;
   model y1 y2 = a*b*c/ noint;
   contrast 'a*b at c1' a*b*c 1 0 -1 0  0 0 -1 0  1 0  0 0  0 0  0 0 0 0,
                        a*b*c 0 0  1 0 -1 0  0 0 -1 0  1 0  0 0  0 0 0 0,
                        a*b*c 0 0  0 0  0 0  1 0 -1 0  0 0 -1 0  1 0 0 0,
                        a*b*c 0 0  0 0  0 0  0 0  1 0 -1 0  0 0 -1 0 1 0/ e;
   manova h=a*b*c/ printe printh;
run;

/* Weighted MANOVA Analysis */
proc glm data=three;
   title2 'Weighted Analysis';
   class a b c;
   model y1 y2 = a b a*b c a*c b*c a*b*c/ ss1;
   means a b c a*b a*c b*c a*b*c;
   manova h=a a*b a*b*c;
run;

proc glm;
   class a b c;
   model y1 y2 = b a a*b c c*b c*a a*b*c/ ss1;
   manova h=b;
run;

proc glm;
   class a b c;
   model y1 y2 = c b b*c a a*c a*b a*b*c/ ss1;
   manova h= c c*b;
run;

proc glm;
   class a b c;
   model y1 y2 = a c a*c b a*b b*c a*b*c/ ss1;
   manova h=a*c;
run;

/* Cell Means Model for MANOVA Design              */
proc glm;
   title2 'Full Rank Model with Weighted Contrast';
   class a b c;
   model y1 y2 = a*b*c/ noint nouni;
   contrast 'a*b at c1'a*b*c 1 0 -1 0  0 0 -1 0  1 0  0 0  0 0  0 0 0 0,
                       a*b*c 0 0  1 0 -1 0  0 0 -1 0  1 0  0 0  0 0 0 0,
                       a*b*c 0 0  0 0  0 0  1 0 -1 0  0 0 -1 0  1 0 0 0,
                       a*b*c 0 0  0 0  0 0  0 0  1 0 -1 0  0 0 -1 0 1 0/ e;
   contrast 'a1 vs a3'a*b*c .2143 .1429 .1429 .2143 .1429 .1429 0 0 0 0 0 0
                           -.1333 -.0667 -.200 -.200 -.200 -.200/ e;
   contrast 'c1 vs c2'a*b*c .1429 -.1053 .0952  -.1579 .0952 -.1053 .0952
                           -.0526  .0476 -.1053 .1429 -.1053 .0952 -.0526
                            .1429 -.1579 .1429 -.1579/ e;
   manova h=a*b*c/printe printh;
run;
```

First the DATA step reads in the data. Next PROC GLM, with the less-than-full-rank model, performs an unweighted analysis of the data. Then an unweighted analysis using the full-rank model is performed. Lastly weighted analyses are performed, first using the less-than-full-rank model and then using the full-rank model.

Result and Interpretation 5_7.sas

To analyze the data for this example of a three-factor design, we first assume that the unequal sample sizes are not due to treatment and that the levels for the design exhaust the categories of interest so that the Type III sum of squares is most appropriate, unweighted tests. Unweighted tests are obtained by specifying the option SS3 in the MODEL statement. The ESTIMATE statements generate contrasts $\hat{\psi}$ while the CONSTRAST statements perform the tests that $\psi = \mathbf{0}$. To illustrate that differences in main effects are dependent on other model parameters we include an ESTIMATE statement that compares the average of $A1$ with the average of $A3$; the average is obtained with the option DIVISOR = 3. The option E in the ESTIMATE statement causes the coefficients of the estimable function to be output. We also include the contrast and estimate of the average of $C1$ vs $C2$ which is the same as the overall test. The contrast for $A1$ vs $A3$ performs the multivariate test of the null hypothesis that the contrast is significantly different from zero. If one included all linearly independent rows in the CONTRAST statement:

```
contrast 'Test of A' a 1 0 -1, a 0 1 -1;
```

the overall test of the contrast would be identical to the overall test of A. The MANOVA statement can be used to perform all unweighted tests of the model parameters. The statements associated with the unweighted MANOVA analysis in Program 5_7.sas generate all univariate Type III F-tests and the Type III multivariate tests for all factors in the MODEL statement. Because the design is nonorthogonal, the sum of squares do not add to the total sum of squares about the mean.

The unweighted test of the interaction AB is testing that all

$$\psi = (\alpha\beta)_{ij} - (\alpha\beta)_{i'j} - (\alpha\beta)_{ij'} + (\alpha\beta)_{i'j'} = \mathbf{0} \tag{5.114}$$

where

$$\hat{\psi} = \mathbf{y}_{ij.} - \mathbf{y}_{i'j.} - \mathbf{y}_{ij'.} + \mathbf{y}_{i'j'.} \tag{5.115}$$

and $\mathbf{y}_{ij.}$ is a simple average of the cell means. The three-factor test of interaction is testing that

$$[(\alpha\beta\gamma)_{ijk} - (\alpha\beta\gamma)_{i'jk} - (\alpha\beta\gamma)_{ij'k} + (\alpha\beta\gamma)_{i'j'k}] - [(\alpha\beta\gamma)_{ijk'} - (\alpha\beta\gamma)_{i'jk'} - (\alpha\beta\gamma)_{ij'k'} + (\alpha\beta\gamma)_{i'j'k'}] = \mathbf{0}$$

for all indices. When analyzing a three-factor design, one analyzes the factors in a hierarchical manner, highest order to lowest order. The process is similar to the univariate strategy suggested by Milliken and Johnson (1992, p. 198). The univariate output for each dependent variable and the multivariate test for the test of the three-way interaction follow.

Output 5.7: *Nonorthogonal Three-Factor MANOVA Unweighted Analysis*

```
                Output 5.7: Nonorthogonal Three Factor MANOVA
                            Unweighted Analysis
                     General Linear Models Procedure

Dependent Variable: Y1
                                     Sum of            Mean
Source                  DF          Squares          Square    F Value      Pr > F

Model                   17      86.10000000      5.06470588       5.71      0.0001

Error                   22      19.50000000      0.88636364

Corrected Total         39     105.60000000

                  R-Square             C.V.        Root MSE                Y1 Mean

                  0.815341         27.69026       0.9414689              3.4000000


Source                  DF      Type III SS     Mean Square    F Value      Pr > F

A                        2       1.75024248      0.87512124       0.99      0.3885
B                        2       6.35714286      3.17857143       3.59      0.0449
C                        1       0.24107143      0.24107143       0.27      0.6072
A*B                      4      32.01598819      8.00399705       9.03      0.0002
A*C                      2       3.89185257      1.94592629       2.20      0.1351
B*C                      2       5.16763848      2.58381924       2.92      0.0753
A*B*C                    4      25.87630849      6.46907712       7.30      0.0007

Contrast                DF      Contrast SS     Mean Square    F Value      Pr > F

a1 vs a3                 1       1.13636364      1.13636364       1.28      0.2697
c1 vs c3                 1       0.24107143      0.24107143       0.27      0.6072

                                           T for H0:    Pr > |T|   Std Error of
Parameter                   Estimate     Parameter=0                 Estimate

a1 vs a3                 -0.13888889           -1.13      0.2697     0.12266335
c1 vs c2                 -0.08333333           -0.52      0.6072     0.15979083

Dependent Variable: Y2
                                     Sum of            Mean
Source                  DF          Squares          Square    F Value      Pr > F

Model                   17     9857.2750000     579.8397059       6.35      0.0001

Error                   22     2008.5000000      91.2954545

Corrected Total         39    11865.7750000

                  R-Square             C.V.        Root MSE                Y2 Mean

                  0.830732         24.14369       9.5548655              39.575000


Source                  DF      Type III SS     Mean Square    F Value      Pr > F

A                        2     1253.0761397     626.5380698       6.86      0.0048
B                        2        1.4857467       0.7428733       0.01      0.9919
C                        1     1573.0029762    1573.0029762      17.23      0.0004
A*B                      4     2964.3869136     741.0967284       8.12      0.0004
```

Output 5.7 *(continued)*

```
A*C                          2       806.9616877      403.4808438        4.42       0.0244
B*C                          2       246.9823453      123.4911727        1.35       0.2792
A*B*C                        4      1995.8981718      498.9745430        5.47       0.0033

Contrast                    DF       Contrast SS      Mean Square    F Value      Pr > F

a1 vs a3                     1      1242.5050505     1242.5050505      13.61       0.0013
c1 vs c2                     1      1573.0029762     1573.0029762      17.23       0.0004

                                            T for H0:     Pr > |T|    Std Error of
Parameter                      Estimate    Parameter=0                  Estimate

a1 vs a3                    -4.59259259          -3.69      0.0013      1.24489699
c1 vs c2                    -6.73148148          -4.15      0.0004      1.62169979

                       Manova Test Criteria and F Approximations for
                          the Hypothesis of no Overall A*B*C Effect
                  H = Type III SS&CP Matrix for A*B*C   E = Error SS&CP Matrix

                                 S=2     M=0.5     N=9.5

  Statistic                         Value           F        Num DF    Den DF   Pr > F

  Wilks' Lambda                0.24745566      5.3038             8        42   0.0001
  Pillai's Trace               0.99325167      5.4263             8        44   0.0001
  Hotelling-Lawley Trace       2.06839890      5.1710             8        40   0.0002
  Roy's Greatest Root          1.34538972      7.3996             4        22   0.0006

                  NOTE: F Statistic for Roy's Greatest Root is an upper bound.
                         NOTE: F Statistic for Wilks' Lambda is exact.

                   Characteristic Roots and Vectors of: E Inverse * H, where
                 H = Contrast SS&CP Matrix for a1 vs a3    E = Error SS&CP Matrix

                Characteristic    Percent         Characteristic Vector  V'EV=1
                     Root
                                                                Y1             Y2

                    0.65730970     100.00          -0.06433914     0.02494617
                    0.00000000       0.00           0.25728216    -0.00778071
```

Because there is a significant three-way interaction, one would analyze the two-way treatment effects (say *AB*) at selected levels of the other factor, usually the factor of least interest (say *C*). While one may use the overparameterized model for this analysis it would involve finding the constants using the general form of estimable functions as discussed by Milliken and Johnson (1992, p. 142). It is more convenient to use the cell means (μ_{ijk}) model, the full-rank model. Using the full-rank model, one merely has to specify the cell means to construct tetrads of the form

$$\psi = \mu_{ij1} - \mu_{i'j1} - \mu_{ij'1} + \mu_{i'j'1} \tag{5.116}$$

In Program 5_7.sas we we use PROC GLM with the NOINT option in the MODEL statement to obtain the less-than-full-rank model. The CONTRAST statement is used to test for the *AB* interaction at the first level of *C*. The output follows.

Output 5.7 (*continued*)

```
              Output 5.7: Nonorthogonal Three Factor MANOVA
                  Full Rank Model with Unweighted Contrast

               Manova Test Criteria and F Approximations for
               the Hypothesis of no Overall a*b at c1 Effect
       H = Contrast SS&CP Matrix for a*b at c1    E = Error SS&CP Matrix

                          S=2     M=0.5     N=9.5

Statistic                     Value           F     Num DF     Den DF  Pr > F

Wilks' Lambda            0.42748067      2.7797          8         42  0.0146
Pillai's Trace           0.60141358      2.3651          8         44  0.0327
Hotelling-Lawley Trace   1.27169514      3.1792          8         40  0.0070
Roy's Greatest Root      1.21611491      6.6886          4         22  0.0011

        NOTE: F Statistic for Roy's Greatest Root is an upper bound.
               NOTE: F Statistic for Wilks' Lambda is exact.
```

The p-value for the multivariate test of AB at $C1$ is $p = 0.0146$ using Wilks' Λ criterion; thus AB is significant at $C1$. One may also test AB at $C2$. Because the test of interaction is significant, one would not investigate main effects since they are confounded by interaction. If all interactions are nonsignificant, contrasts of main effects may be tested. Any number of contrasts may be included in SAS. Depending on the number of interest, one would use either an overall multivariate criterion to determine significance or a test procedure that distributes α for the test over the specific number of planned comparisons. PROC MULTTEST, a procedure for addressing the multiple testing problem, can be used to obtain adjusted p-values.

To obtain weighted tests using PROC GLM one must run PROC GLM several times, reordering the sequence of the effects to obtain the correct Type I sum of squares for each main effect and interaction. While we have not stopped the printing of the univariate tests with the NOUNI option, one must be careful when using the output since all of the tests are printed; one may control the output of the multivariate tests with the MANOVA statement.

The output for the dependent variable Y1 from the weighted analysis using the less-than-full-rank model follows.

Output 5.7 (*continued*)

```
                  Output 5.7: Nonorthogonal Three Factor MANOVA
                               Weighted Analysis

                        General Linear Models Procedure

Dependent Variable: Y1
                                        Sum of            Mean
Source                  DF             Squares          Square    F Value      Pr > F

Model                   17         86.10000000      5.06470588       5.71      0.0001

Error                   22         19.50000000      0.88636364

Corrected Total         39        105.60000000
```

Output 5.7 (*continued*)

R-Square	C.V.	Root MSE	Y1 Mean
0.815341	27.69026	0.9414689	3.4000000

Source	DF	Type I SS	Mean Square	F Value	Pr > F
A	2	4.88961039	2.44480519	2.76	0.0853
B	2	7.93898370	3.96949185	4.48	0.0234
A*B	4	33.23807257	8.30951814	9.37	0.0001
C	1	1.66666667	1.66666667	1.88	0.1841
A*C	2	4.24653345	2.12326672	2.40	0.1145
B*C	2	8.24382473	4.12191237	4.65	0.0207
A*B*C	4	25.87630849	6.46907712	7.30	0.0007

General Linear Models Procedure

Dependent Variable: Y1

Source	DF	Sum of Squares	Mean Square	F Value	Pr > F
Model	17	86.10000000	5.06470588	5.71	0.0001
Error	22	19.50000000	0.88636364		
Corrected Total	39	105.60000000			

R-Square	C.V.	Root MSE	Y1 Mean
0.815341	27.69026	0.9414689	3.4000000

Source	DF	Type I SS	Mean Square	F Value	Pr > F
B	2	9.09740260	4.54870130	5.13	0.0148
A	2	3.73119150	1.86559575	2.10	0.1457
A*B	4	33.23807257	8.30951814	9.37	0.0001
C	1	1.66666667	1.66666667	1.88	0.1841
B*C	2	6.54979743	3.27489871	3.69	0.0414
A*C	2	5.94056075	2.97028037	3.35	0.0537
A*B*C	4	25.87630849	6.46907712	7.30	0.0007

General Linear Models Procedure

Dependent Variable: Y1

Source	DF	Sum of Squares	Mean Square	F Value	Pr > F
Model	17	86.10000000	5.06470588	5.71	0.0001
Error	22	19.50000000	0.88636364		
Corrected Total	39	105.60000000			

R-Square	C.V.	Root MSE	Y1 Mean
0.815341	27.69026	0.9414689	3.4000000

Output 5.7 *(continued)*

```
Source                     DF          Type I SS      Mean Square    F Value       Pr > F

C                           1         1.94085213       1.94085213       2.19       0.1531
B                           2         8.06451712       4.03225856       4.55       0.0222
B*C                         2         5.86248790       2.93124395       3.31       0.0555
A                           2         4.09467760       2.04733880       2.31       0.1229
A*C                         2        10.48829938       5.24414969       5.92       0.0088
A*B                         4        29.77285738       7.44321435       8.40       0.0003
A*B*C                       4        25.87630849       6.46907712       7.30       0.0007

General Linear Models Procedure

Dependent Variable: Y1
                                          Sum of             Mean
Source                     DF            Squares           Square    F Value       Pr > F

Model                      17        86.10000000       5.06470588       5.71       0.0001

Error                      22        19.50000000       0.88636364

Corrected Total            39       105.60000000

                     R-Square               C.V.         Root MSE                 Y1 Mean

                     0.815341           27.69026        0.9414689               3.4000000

Source                     DF          Type I SS      Mean Square    F Value       Pr > F

A                           2         4.88961039       2.44480519       2.76       0.0853
C                           1         2.18355229       2.18355229       2.46       0.1308
A*C                         2         8.25064685       4.12532342       4.65       0.0206
B                           2         7.08096265       3.54048133       3.99       0.0331
A*B                         4        29.57509460       7.39377365       8.34       0.0003
B*C                         2         8.24382473       4.12191237       4.65       0.0207
A*B*C                       4        25.87630849       6.46907712       7.30       0.0007
```

Table 5.1 summarizes the weighted tests for dependent variable Y1.

Table 5.1 *Nonorthogonal ANOVA for Variable Y1*

Hypotheses	df	SS	MS	F	p-value
H_{A^*}	2	4.89	2.44	2.76	0.0853
H_{B^*}	2	9.10	4.55	5.13	0.0148
H_{C^*}	1	1.94	1.94	2.19	0.1531
H_{AB^*}	4	33.24	8.31	9.37	0.0001
H_{AC^*}	2	8.25	4.13	4.56	0.0206
H_{BC^*}	2	5.86	2.93	3.91	0.0555
H_{ABC^*}	4	19.50	6.47	7.30	0.0007
Error	22	19.50	0.89		

A similar table may be constructed for the dependent variable Y2.

For the run with first sequence of factors, we requested MANOVA weighted tests of A, AB, and ABC. Only the weighted test of ABC is identical to the unweighted MANOVA test. The weighted test of AB is not equal to the unweighted test, nor are the other two-way tests of interaction. This is because the parameter ψ in (5.114) is not the same for the two models. The estimate of ψ for AB is now estimated

$$\hat{\psi}_{AB} = \overline{\mathbf{y}}_{ij.} - \overline{\mathbf{y}}_{i'j.} - \overline{\mathbf{y}}_{ij'.} + \overline{\mathbf{y}}_{i'j'.} \quad (5.117)$$

where $\overline{\mathbf{y}}_{ij.}$ is a weighted average of cell means. Similarly, the weighted test of A is not the same for the two models. The MANOVA statements in the other runs of PROC GLM are used to generate tests of B, C, CB, and AC.

There is no LSMEANS statement included with the Type I analysis since we are interested in weighted averages of cell means to construct marginal means which is accomplished using the MEANS statement. The ESTIMATE statement in PROC GLM uses LSMEANS to obtain estimates of marginal means which are unweighted estimates of cell means. To obtain weighted estimates, it is convenient to use the cell means model, the full-rank model. The code for the contrasts of the average of $A1$ vs $A3$ and $C1$ vs $C2$ is included in Program 5_7.sas. The coefficients are constructed from the cell frequencies. For example, the coefficients for $C1$ vs $C2$ are

$$(3/21\ -2/19\ 2/21\ -3/19\ 2/21\ -2/19\ 2/21\ -1/19\ 1/21\ -2/19\ 3/21\ -2/19\ 2/21\ -1/19\ 3/21\ -3/19\ 3/21\ -3/19) \quad (5.118)$$

The test of AB at $C1$ for the weighted analysis is identical to the unweighted analysis, since at fixed levels of C, the test of AB is invariant to the design weights. This is also the case for two-way interactions at each level of the other factors. The output with the results of the multivariate tests of these contrasts follows.

Output 5.7 (*continued*)

```
                 Output 5.7: Nonorthogonal Three Factor MANOVA
                      Full Rank Model with Weighted Contrast

                  Manova Test Criteria and F Approximations for
                  the Hypothesis of no Overall a*b at c1 Effect
         H = Contrast SS&CP Matrix for a*b at c1    E = Error SS&CP Matrix

                             S=2     M=0.5     N=9.5

Statistic                       Value          F       Num DF    Den DF   Pr > F

Wilks' Lambda              0.42748067     2.7797          8        42   0.0146
Pillai's Trace             0.60141358     2.3651          8        44   0.0327
Hotelling-Lawley Trace     1.27169514     3.1792          8        40   0.0070
Roy's Greatest Root        1.21611491     6.6886          4        22   0.0011

        NOTE: F Statistic for Roy's Greatest Root is an upper bound.
               NOTE: F Statistic for Wilks' Lambda is exact.
```

Output 5.7 *(continued)*

```
              Manova Test Criteria and Exact F Statistics for
                 the Hypothesis of no Overall a1 vs a3 Effect
        H = Contrast SS&CP Matrix for a1 vs a3    E = Error SS&CP Matrix

                              S=1     M=0     N=9.5

  Statistic                     Value           F       Num DF    Den DF  Pr > F

  Wilks' Lambda              0.68637959     4.7977          2        21  0.0192
  Pillai's Trace             0.31362041     4.7977          2        21  0.0192
  Hotelling-Lawley Trace     0.45691977     4.7977          2        21  0.0192
  Roy's Greatest Root        0.45691977     4.7977          2        21  0.0192


Manova Test Criteria and Exact F Statistics for
                 the Hypothesis of no Overall c1 vs c2 Effect
        H = Contrast SS&CP Matrix for c1 vs c2    E = Error SS&CP Matrix

                              S=1     M=0     N=9.5

  Statistic                     Value           F       Num DF    Den DF  Pr > F

  Wilks' Lambda              0.53998843     8.9449          2        21  0.0015
  Pillai's Trace             0.46001157     8.9449          2        21  0.0015
  Hotelling-Lawley Trace     0.85189153     8.9449          2        21  0.0015
  Roy's Greatest Root        0.85189153     8.9449          2        21  0.0015
```

We have illustrated the analysis of the three-factor design using both the overparameterized model and a cell means model using PROC GLM. When all of the parameters are not included in the model, PROC GLM may not be used to generate a cell means model since currently restrictions cannot be imposed on the model parameters. With this being the case we may use either the overparameterized GLM procedure or the REG procedure, as discussed in Chapter 3. The multivariate restricted model analysis is discussed in Chapter 7.

5.8 MANCOVA Designs

We have illustrated how one may apply the general theory of Chapter 5 to analyze a multivariate regression model and to analyze a MANOVA design. Combining regression with experimental design into a single model is called multivariate analysis of covariance (MANCOVA). When one is analyzing MANCOVA designs, $\mathbf{X}$ contains values for the fixed, categorical design variables and $\mathbf{Z}$ the fixed regression variables (also called the covariates or concomitant variables). The multivariate linear model for the design is

$$\mathbf{Y} = \mathbf{XB} + \mathbf{Z\Gamma} + \mathbf{U} \tag{5.119}$$

where $\mathbf{Y}$ is the data matrix, $\mathbf{B}$ and $\mathbf{\Gamma}$ are unknown matrices of coefficients, and $\mathbf{U}$ is the matrix of random errors. Letting $\mathbf{X}_o = \binom{\mathbf{X}}{\mathbf{Z}}$ and $\mathbf{B}_o = \binom{\mathbf{B}}{\mathbf{\Gamma}}$, we see that (5.119) has the general form (5.3) or $\mathbf{Y} = \mathbf{X}_o\mathbf{B}_o + \mathbf{U}$, a multivariate linear model.

The normal equations for the MANCOVA design are

$$\begin{pmatrix} \mathbf{X'X} & \mathbf{X'Z} \\ \mathbf{Z'X} & \mathbf{Z'Z} \end{pmatrix} \begin{pmatrix} \mathbf{B} \\ \mathbf{\Gamma} \end{pmatrix} = \begin{pmatrix} \mathbf{X'Y} \\ \mathbf{Z'Y} \end{pmatrix} \tag{5.120}$$

Solving these equations, we have from (7.40) that

$$\begin{aligned} \hat{\mathbf{B}} &= (\mathbf{X}'\mathbf{X})^{-1}\mathbf{X}'\mathbf{Y} - (\mathbf{X}'\mathbf{X})^{-1}\mathbf{X}'\mathbf{Z}\hat{\mathbf{\Gamma}} \\ \hat{\mathbf{\Gamma}} &= (\mathbf{Z}'\mathbf{Q}\mathbf{Z})^{-1}\mathbf{Z}'\mathbf{Q}\mathbf{Y} = \mathbf{E}_{ZZ}^{-1}\mathbf{E}_{ZY} \end{aligned} \tag{5.121}$$

where $\mathbf{Q} = \mathbf{I} - \mathbf{X}(\mathbf{X}'\mathbf{X})^{-1}\mathbf{X}'$, a projection matrix.

To test the general hypothesis $H{:}\,\mathbf{C}\binom{\mathbf{B}}{\mathbf{\Gamma}}\mathbf{A} = \mathbf{0}$, one may select $\mathbf{C} = \mathbf{I}$ and $\mathbf{A} = \binom{\mathbf{0}}{\mathbf{I}}$ to test the null hypothesis $H{:}\,\mathbf{\Gamma} = \mathbf{0}$. Alternatively, to test hypotheses regarding the elements of $\mathbf{B}$, we select $\mathbf{C} = (\mathbf{C}_1, \mathbf{0})$. Using the general theory in Chapter 5, the residual sum of squares for the MANCOVA model is

$$\begin{aligned} \mathbf{E} &= (\mathbf{Y} - \mathbf{X}_o\hat{\mathbf{B}}_o)'(\mathbf{Y} - \mathbf{X}_o\hat{\mathbf{B}}_o) \\ \mathbf{E} &= \mathbf{Y}'\mathbf{Q}\mathbf{Y} - \hat{\mathbf{\Gamma}}'(\mathbf{Z}\mathbf{Q}\mathbf{Z})\hat{\mathbf{\Gamma}} \\ &= \mathbf{E}_{YY} - \mathbf{E}_{YZ}\mathbf{E}_{ZZ}^{-1}\mathbf{E}_{ZY} \end{aligned} \tag{5.122}$$

with degrees of freedom $v_e = n - R(\mathbf{X}) - R(\mathbf{Z})$. To test the hypothesis $H{:}\,\mathbf{\Gamma} = \mathbf{\Gamma}_o$, that the covariates have no effect on the dependent variables, the hypothesis sum of squares is

$$\mathbf{H}_{\Gamma} = (\hat{\mathbf{\Gamma}} - \mathbf{\Gamma}_o)'(\mathbf{Z}'\mathbf{Q}\mathbf{Z})(\hat{\mathbf{\Gamma}} - \mathbf{\Gamma}_o) \tag{5.123}$$

with degrees of freedom $v_h = R(\mathbf{Z})$. The MANCOVA model, like the ANCOVA model, assumes that the regression matrix $\mathbf{\Gamma}$ is common across groups, the assumption of parallelism. A test of multivariate parallelism is given in Timm (1975, p. 344).

To test hypotheses regarding the matrix $\mathbf{B}$, $H{:}\,\mathbf{B} = \mathbf{B}_o$, we construct the hypothesis test matrix directly using properties of partitioned matrices. The hypothesis matrix is

$$\mathbf{H} = (\mathbf{C}_1\hat{\mathbf{B}}\mathbf{A} - \mathbf{C}_1\mathbf{B}_o\mathbf{A})'[\mathbf{C}_1(\mathbf{X}'\mathbf{X})^{-}\mathbf{C}_1' + \mathbf{C}_1(\mathbf{X}'\mathbf{X})^{-}\mathbf{X}'\mathbf{Z}(\mathbf{Z}'\mathbf{Q}\mathbf{Z})^{-1}\mathbf{Z}'\mathbf{X}(\mathbf{X}'\mathbf{X})^{-}\mathbf{C}_1]^{-1}(\mathbf{C}_1\hat{\mathbf{B}}\mathbf{A} - \mathbf{C}_1\mathbf{B}_o\mathbf{A}) \tag{5.124}$$

where $v_h = R(\mathbf{C}_1)$. An alternative to the determination of $\mathbf{H}$ is to perform a two-step least square procedure (Seber, 1984, p. 465):

(1) Find $\hat{\mathbf{B}} = (\mathbf{X}'\mathbf{X})^{-1}\mathbf{X}'\mathbf{Y}$ and $\mathbf{Y}'\mathbf{Q}\mathbf{Y} (= \mathbf{E}_{YY})$ for the MANOVA model $\mathbf{E}(\mathbf{Y}) = \mathbf{X}\mathbf{B}$.

(2) $\hat{\mathbf{\Gamma}} = (\mathbf{Z}'\mathbf{Q}\mathbf{Z})^{-1}\mathbf{Z}'\mathbf{Q}\mathbf{Y} = \mathbf{E}_{ZZ}^{-1}\mathbf{E}_{ZY}$

(3) $$\begin{aligned} \mathbf{E} &= (\mathbf{Y} - \mathbf{Z}\hat{\mathbf{\Gamma}})'\mathbf{Q}(\mathbf{Y} - \mathbf{Z}\hat{\mathbf{\Gamma}}) \\ &= \mathbf{Y}'\mathbf{Q}\mathbf{Y} - \hat{\mathbf{\Gamma}}'(\mathbf{Z}'\mathbf{Q}\mathbf{Z})\hat{\mathbf{\Gamma}} \\ &= \mathbf{E}_{YY} - \mathbf{E}_{YZ}\mathbf{E}_{ZZ}^{-1}\mathbf{E}_{ZY} \end{aligned}$$

(4) Replace $\mathbf{Y}$ by $\mathbf{Y} - \mathbf{Z}\hat{\mathbf{\Gamma}}$ in $\hat{\mathbf{B}}$ to obtain $\hat{\mathbf{B}}_R = (\mathbf{X}'\mathbf{X})^{-1}\mathbf{X}'(\mathbf{Y} - \mathbf{Z}\hat{\mathbf{\Gamma}})$

(5) To obtain $\mathbf{E}_{\omega}$ where $\omega{:}\,\mathbf{C}_1\mathbf{B} = \mathbf{0}$, obtain $\mathbf{Y}'\mathbf{Q}_{\omega}\mathbf{Y} (= \mathbf{E}_{HYY}$, say) for the model $\omega{:}\,E(\mathbf{Y}) = \mathbf{X}\mathbf{B}$ and $\omega{:}\,E(\mathbf{Y}) = \mathbf{X}\mathbf{B}$ and $\mathbf{C}_1\mathbf{B} = \mathbf{0}$, and steps (2) and (3) are repeated. From (3),

$$\begin{aligned} \mathbf{E}_{\omega} &= \mathbf{Y}'\mathbf{Q}_{\omega}\mathbf{Y} - \mathbf{Y}'\mathbf{Q}_{\omega}\mathbf{Z}(\mathbf{Z}'\mathbf{Q}_{\omega}\mathbf{Z})^{-1}\mathbf{Z}'\mathbf{Q}_{\omega}\mathbf{Y} \\ &= \mathbf{E}_{HYY} - \mathbf{E}_{HYY}\mathbf{E}_{HZZ}^{-1}\mathbf{E}_{HZY} \end{aligned}$$

and

$$\Lambda = \frac{|\mathbf{E}|}{|\mathbf{E}_\omega|} = \frac{|\mathbf{E}_{YY} - \mathbf{E}_{YZ}\mathbf{E}_{ZZ}^{-1}\mathbf{E}_{ZY}|}{|\mathbf{E}_{HYY} - \mathbf{E}_{HYZ}\mathbf{E}_{HZZ}^{-1}\mathbf{E}_{HZY}|} \sim U_{p, v_h, v_e} \tag{5.125}$$

A test closely associated with the MANCOVA design is Rao's test for additional information, (Rao, 1973, p. 551). In many MANOVA or MANCOVA designs, one collects data on p response variables and one is interested in determining whether the additional information provided by the last $(p - s)$ variables, independent of the first s variables, is significant. To develop a test procedure of this hypothesis, we begin with the linear model $\mathbf{\Omega}_o$: $\mathbf{Y} = \mathbf{XB} + \mathbf{U}$ where the usual hypothesis is H: $\mathbf{CB} = \mathbf{0}$. Partitioning the data matrix $\mathbf{Y} = (\mathbf{Y}_1, \mathbf{Y}_2)$ and $\mathbf{B} = (\mathbf{B}_1, \mathbf{B}_2)$, we consider the alternative model

$$\begin{aligned} &\mathbf{\Omega}_1: \mathbf{Y}_1 = \mathbf{XB}_1 + \mathbf{U}_1 \\ &H_{01}: \mathbf{CB}_1 = \mathbf{0} \end{aligned} \tag{5.126}$$

where

$$\begin{aligned} E(\mathbf{Y}_2|\mathbf{Y}_1) &= \mathbf{XB}_2 + (\mathbf{Y}_1 - \mathbf{XB}_1)\mathbf{\Sigma}_{11}^{-1}\mathbf{\Sigma}_{12} \\ &= \mathbf{X}(\mathbf{B}_2 - \mathbf{B}_1\mathbf{\Sigma}_{11}^{-1}\mathbf{\Sigma}_{12}) + \mathbf{Y}_1\mathbf{\Sigma}_{11}^{-1}\mathbf{\Sigma}_{12} \\ &= \mathbf{X\Theta} + \mathbf{Y}_1\mathbf{\Gamma} \\ \mathbf{\Sigma}_{22.1} &= \mathbf{\Sigma}_{22} - \mathbf{\Sigma}_{21}\mathbf{\Sigma}_{11}^{-1}\mathbf{\Sigma}_{12} \end{aligned}$$

following Timm (1975, p. 125). Thus, the conditional model is

$$\mathbf{\Omega}_2: E(\mathbf{Y}_2|\mathbf{Y}_1) = \mathbf{X\Theta} + \mathbf{Y}_1\mathbf{\Gamma} \tag{5.127}$$

the MANCOVA model. Under $\mathbf{\Omega}_2$, testing

$$H_{02}: \mathbf{C}(\mathbf{B}_2 - \mathbf{B}_1\mathbf{\Gamma}) = \mathbf{0} \tag{5.128}$$

corresponds to testing H_{02}: $\mathbf{C\Theta} = \mathbf{0}$. If $\mathbf{C} = \mathbf{I}_p$, so that $\mathbf{\Theta} = \mathbf{0}$, then the conditional distribution of $\mathbf{Y}_2|\mathbf{Y}_1$ depends only on Γ and does not involve $\mathbf{B}_1$; thus $\mathbf{Y}_2$ provides no additional information on $\mathbf{B}_1$. Because $\mathbf{\Omega}_2$ is the standard MANCOVA model with $\mathbf{Y} \equiv \mathbf{Y}_2$ and $\mathbf{Z} \equiv \mathbf{Y}_1$, we may test H_{02} using (5.125)

$$\Lambda_{2.1} = \frac{|\mathbf{E}_{22} - \mathbf{E}_{21}\mathbf{E}_{11}^{-1}\mathbf{E}_{12}|}{\mathbf{E}_{H22} - \mathbf{E}_{H21}\mathbf{E}_{H11}^{-1}\mathbf{E}_{H12}} \sim U_{p-s, v_h, v_e} \tag{5.129}$$

where $v_e = n - p - s$ and $v_h = R(\mathbf{C})$. Because $H(\mathbf{CB} = \mathbf{0})$ is true if and only if H_{01} and H_{02} are true, we may partition Λ as $\Lambda = \Lambda_1\Lambda_{2.1}$ where Λ_1 is from the test of H_{01}; this results in a stepdown test of H (Seber, 1984, p. 472).

To illustrate the analysis of a MANCOVA design, we consider a study to examine the effect of four drugs on the apgar scores for twins at birth, adjusting for the weight in kilograms of the mother. Subjects were randomly assigned to the four treatment drug conditions with an equal number of subjects per cell. The SAS code to perform the analysis is provided in Program 5_8.sas.

Program 5_8.sas

```
/* Program 5_8.sas */
/* MANCOVA Designs */
```

```
options ls=80 ps=60 nodate nonumber;
title 'Output 5.8: Multivariate MANCOVA';
data mancova;
   infile 'c:\5_8.dat';
   input drugs $ apgar1 apgar2 x;
proc print;
run;

proc glm;
   title2 'Test of Parallelism';
   class drugs;
   model Apgar1 Apgar2 = drugs x drugs*x;
   manova h= drugs*x;
run;

proc glm;
   title2 'MANCOVA Tests';
   class drugs;
   model apgar1 apgar2 = drugs x/ solution;
   contrast '1 vs. 4' drugs 1 0 0 -1;
   manova h=drugs x;
   means drugs;
   lsmeans drugs/ stderr pdiff tdiff cov out=adjmeans;
run;

proc print data=adjmeans;
   title2 'Adjusted Means';
run;
```

Result and Interpretation 5_8.sas

Having ensured that the data are normally distributed with equal conditional covariance matrices, we test that the regression lines are parallel. This is accomplished using PROC GLM by testing for the significance of an interaction between the independent variable, drugs, and the covariate x, the weight of the mother in kilograms.

The results of the test are found in Output 5.8.

Output 5.8: *Multivariate MANCOVA Test of Parallelism*

```
                    Output 5.8: Multivariate MANCOVA
                           Test of Parallelism

                     General Linear Models Procedure

Dependent Variable: APGAR1
                                      Sum of            Mean
Source                   DF          Squares          Square   F Value      Pr > F

Model                     7     228.88478151     32.69782593    118.63      0.0001

Error                    24       6.61521849      0.27563410

Corrected Total          31     235.50000000

                   R-Square             C.V.        Root MSE          APGAR1 Mean

                   0.971910         9.767603       0.5250087            5.3750000

Source                   DF          Type I SS     Mean Square   F Value      Pr > F

DRUGS                     3     194.50000000     64.83333333    235.22      0.0001
X                         1      33.95315418     33.95315418    123.18      0.0001
X*DRUGS                   3       0.43162733      0.14387578      0.52      0.6713
```

Output 5.8 (*continued*)

Source	DF	Type III SS	Mean Square	F Value	Pr > F
DRUGS	3	0.35507226	0.11835742	0.43	0.7338
X	1	25.84884937	25.84884937	93.78	0.0001
X*DRUGS	3	0.43162733	0.14387578	0.52	0.6713

Dependent Variable: APGAR2

Source	DF	Sum of Squares	Mean Square	F Value	Pr > F
Model	7	202.40574686	28.91510669	18.71	0.0001
Error	24	37.09425314	1.54559388		
Corrected Total	31	239.50000000			

R-Square	C.V.	Root MSE	APGAR2 Mean
0.845118	23.12966	1.2432192	5.3750000

Source	DF	Type I SS	Mean Square	F Value	Pr > F
DRUGS	3	194.50000000	64.83333333	41.95	0.0001
X	1	7.53065989	7.53065989	4.87	0.0371
X*DRUGS	3	0.37508697	0.12502899	0.08	0.9698

Source	DF	Type III SS	Mean Square	F Value	Pr > F
DRUGS	3	0.14655011	0.04885004	0.03	0.9922
X	1	5.72220737	5.72220737	3.70	0.0663
X*DRUGS	3	0.37508697	0.12502899	0.08	0.9698

Manova Test Criteria and F Approximations for
the Hypothesis of no Overall X*DRUGS Effect
H = Type III SS&CP Matrix for X*DRUGS E = Error SS&CP Matrix

S=2 M=0 N=10.5

Statistic	Value	F	Num DF	Den DF	Pr > F
Wilks' Lambda	0.93101088	0.2790	6	46	0.9440
Pillai's Trace	0.06927413	0.2870	6	48	0.9403
Hotelling-Lawley Trace	0.07379516	0.2706	6	44	0.9478
Roy's Greatest Root	0.06938293	0.5551	3	24	0.6497

NOTE: F Statistic for Roy's Greatest Root is an upper bound.
NOTE: F Statistic for Wilks' Lambda is exact.

From Output 5.8, we see that the p-values for the test for variable Apgar 1 and Apgar 2 are $p = 0.6713$ and $p = 0.9698$, respectively. In addition, the overall multivariate test of interaction is nonsignificant, $p = 0.9440$. Thus, the assumption of parallelism appears tenable.

Given parallelism, we next want to test for differences in drugs. Although we have an equal number of observations per cell, a balanced design, PROC GLM generates both Type I and Type III sum of squares for the design since the covariate is changing from mother to mother. For this design, the Type III sum of squares is the appropriate sum of squares for the tests. The results of the hypothesis test for a drug effect are found in the continuation of Output 5.8 that follows.

Output 5.8 (*continued*)

```
                        Output 5.8: Multivariate MANCOVA
                                 MANCOVA Tests

                         General Linear Models Procedure

Dependent Variable: APGAR1
                                        Sum of            Mean
Source                    DF           Squares           Square   F Value      Pr > F

Model                      4      228.45315418      57.11328855    218.83      0.0001

Error                     27        7.04684582       0.26099429

Corrected Total           31      235.50000000

                    R-Square              C.V.         Root MSE          APGAR1 Mean

                    0.970077          9.504670        0.5108760            5.3750000

Source                    DF         Type I SS      Mean Square   F Value      Pr > F

DRUGS                      3      194.50000000      64.83333333    248.41      0.0001
X                          1       33.95315418      33.95315418    130.09      0.0001

Source                    DF       Type III SS      Mean Square   F Value      Pr > F

DRUGS                      3        1.79283521       0.59761174      2.29      0.1010
X                          1       33.95315418      33.95315418    130.09      0.0001

Contrast                  DF       Contrast SS      Mean Square   F Value      Pr > F

1 vs. 4                    1        0.03072558       0.03072558      0.12      0.7342

                                          T for H0:      Pr > |T|    Std Error of
Parameter                  Estimate      Parameter=0                   Estimate

INTERCEPT              -4.136387217 B          -3.55       0.0014      1.16580898
DRUGS       G1          0.213270388 B           0.34       0.7342      0.62157850
            G2          0.228807844 B           0.41       0.6879      0.56349497
            G3          0.670496991 B           1.70       0.1002      0.39393361
            G4          0.000000000 B            .          .           .
X                       0.167877153            11.41       0.0001      0.01471862

NOTE: The X'X matrix has been found to be singular and a generalized inverse
      was used to solve the normal equations.   Estimates followed by the
      letter 'B' are biased, and are not unique estimators of the parameters.

Dependent Variable: APGAR2
                                        Sum of            Mean
Source                    DF           Squares           Square   F Value      Pr > F

Model                      4      202.03065989      50.50766497     36.40      0.0001

Error                     27       37.46934011       1.38775334

Corrected Total           31      239.50000000
```

Output 5.8 (*continued*)

```
               R-Square              C.V.          Root MSE            APGAR2 Mean

               0.843552          21.91683         1.1780294              5.3750000

Source                DF          Type I SS      Mean Square   F Value      Pr > F

DRUGS                  3       194.50000000      64.83333333     46.72      0.0001
X                      1         7.53065989       7.53065989      5.43      0.0276

Source                DF        Type III SS      Mean Square   F Value      Pr > F

DRUGS                  3         9.15646327       3.05215442      2.20      0.1112
X                      1         7.53065989       7.53065989      5.43      0.0276

Contrast              DF        Contrast SS      Mean Square   F Value      Pr > F

1 vs. 4                1         6.94378412       6.94378412      5.00      0.0338

                                           T for H0:    Pr > |T|   Std Error of
Parameter                  Estimate      Parameter=0                 Estimate

INTERCEPT              2.813394895 B          1.05       0.3046      2.68823996
DRUGS     G1          -3.206111226 B         -2.24       0.0338      1.43329841
          G2          -2.802007678 B         -2.16       0.0401      1.29936355
          G3          -1.139110811 B         -1.25       0.2206      0.90837186
          G4           0.000000000 B           .           .             .
X                      0.079062046            2.33       0.0276      0.03393967

NOTE: The X'X matrix has been found to be singular and a generalized inverse
      was used to solve the normal equations.   Estimates followed by the
      letter 'B' are biased, and are not unique estimators of the parameters.


                       General Linear Models Procedure
                      Multivariate Analysis of Variance

          Characteristic Roots and Vectors of: E Inverse * H, where
         H = Type III SS&CP Matrix for DRUGS   E = Error SS&CP Matrix

        Characteristic   Percent        Characteristic Vector  V'EV=1
             Root
                                                APGAR1         APGAR2

           0.31635172     64.57            0.26800234     0.10781061
           0.17358815     35.43           -0.26576729     0.12316171

                Manova Test Criteria and F Approximations for
                  the Hypothesis of no Overall DRUGS Effect
         H = Type III SS&CP Matrix for DRUGS   E = Error SS&CP Matrix

                              S=2    M=0    N=12

  Statistic                    Value           F       Num DF    Den DF  Pr > F

  Wilks' Lambda              0.64731004     2.1053          6        52  0.0683
  Pillai's Trace             0.38823694     2.1679          6        54  0.0604
  Hotelling-Lawley Trace     0.48993987     2.0414          6        50  0.0773
  Roy's Greatest Root        0.31635172     2.8472          3        27  0.0562

         NOTE: F Statistic for Roy's Greatest Root is an upper bound.
                NOTE: F Statistic for Wilks' Lambda is exact.
```

The p-values for the APGAR1 and APGAR2 dependent variables are $p = 0.1010$ and $p = 0.1112$, respectively. The overall multivariate test using Wilks' lambda criterion is $p = 0.0683$. Thus, there appears to be no difference in the apgar scores. If one did not take into account the weights of the mothers, the difference in apgar scores would have been judged as significant.

When one performs an analysis of covariance, PROC GLM can compute adjusted means and perform pairwise comparisons, a variable at a time, if one uses the LSMEANS statement. The option OUT = ADJMEANS in the LSMEANS statement creates a SAS data set with the adjusted means for each variable. To compute the adjusted means for each variable in the jth group, PROC GLM uses the formula $\overline{Y}_{adj,j} = \overline{Y}_j - \hat{\beta}_w(\overline{x}_j - \overline{x})$ where $\overline{x}_j$ is the mean of the covariate for the jth group, $\overline{x}$ is the overall mean for the covariates, $\overline{y}$ is the mean for the jth on the dependent variable and $\hat{\beta}_w$ is the common estimate of the slope. The common estimate for the slope of the regression equation is output by using the SOLUTION option in the MODEL statement in PROC GLM. The values can be found in the univariate ANCOVA test output in the column labeled `ESTIMATE` and the row for variable X. For group 1 and Apgar 1, we have that

$$\overline{Y}_{adj.1} = 2.75 - 0.167877153\,(39.75 - 55) = 5.31012658$$

The standard error of the adjusted mean is

$$\begin{aligned}\hat{\sigma}_{adj.1} &= \sqrt{MS_e}\sqrt{\frac{1}{n_1} + (\overline{x}_1 - \overline{x})^2 / \sum_i (x_{ij} - \overline{x})^2} \\ &= 0.510876\sqrt{\frac{1}{8} + (39.75 - 55)^2 / 1204.75} \\ &= 0.28810778\end{aligned}$$

as found in the continuation of Output 5.8 that follows. The calculations for the other values are computed similarly.

Output 5.8 (*continued*)

```
                      Output 5.8: Multivariate MANCOVA
                               Adjusted Means

OBS _NAME_ DRUGS  LSMEAN  STDERR NUMBER   COV1      COV2      COV3      COV4

 1  APGAR1  G1   5.31013 0.28811   1    0.08301  0.03593 -0.00950 -0.07681
 2  APGAR1  G2   5.32566 0.24134   2    0.03593  0.05825 -0.00677 -0.05478
 3  APGAR1  G3   5.76735 0.18551   3   -0.00950 -0.00677  0.03441  0.01448
 4  APGAR1  G4   5.09686 0.38695   4   -0.07681 -0.05478  0.01448  0.14973
 5  APGAR2  G1   3.95570 0.66435   1    0.44136  0.19104 -0.05050 -0.40842
 6  APGAR2  G2   4.35980 0.55651   2    0.19104  0.30970 -0.03601 -0.29125
 7  APGAR2  G3   6.02270 0.42777   3   -0.05050 -0.03601  0.18299  0.07700
 8  APGAR2  G4   7.16181 0.89227   4   -0.40842 -0.29125  0.07700  0.79614
```

In Program 5_8.sas, the test of the hypothesis of no difference in drug 1 versus drug 4 is performed using CONTRAST statements.

5.9 Stepdown Analysis

In this section we illustrate the test of additional information, a stepdown analysis. We illustrate with a model that involves two covariates; data reproduced in Seber (1984, p. 469) from Smith et al. (1962) are utilized. The data consist of urine samples from men in four weight categories; the 11 variables and two covariates are

$y_1 = pH,$
$y_2 =$ modified createnine coefficient,
$y_3 =$ pegment createnine,
$y_4 =$ phosphate (mg / ml),
$y_5 =$ calcium (mg / ml),
$y_6 =$ phosphours (mg / ml),
$y_7 =$ createnine (mg / ml),
$y_8 =$ chloride (mg / ml),
$y_9 =$ bacon $(\mu g / ml)$,
$y_{10} =$ choline $(\mu g / ml)$,
$y_{11} =$ copper $(\mu g / ml)$,
$x_1 =$ volume (ml),
$x_2 =$ (specific gravity $-$ 1) x10^3.

The model considered by Smith et al. (1962) is a one-way MANCOVA model with two covariates

$$\mathbf{y}_{ij} = \boldsymbol{\mu} + \boldsymbol{\alpha}_i + \boldsymbol{\gamma}_1 \mathrm{x}_{ij1} + \boldsymbol{\gamma}_2 \mathrm{x}_{ij2} + \mathbf{e}_{ij} \tag{5.130}$$

with the unweighted Type III restriction that the $\sum_i \boldsymbol{\alpha}_i = \mathbf{0}$. The hypotheses considered by the Smith et al. (1962) were

$$\begin{aligned} H_{01}&: \boldsymbol{\mu} = \mathbf{0} \\ H_{02}&: \boldsymbol{\alpha}_1 = \boldsymbol{\alpha}_2 = \boldsymbol{\alpha}_3 = \boldsymbol{\alpha}_4 = \mathbf{0} \\ H_{03}&: \boldsymbol{\gamma}_1 = \mathbf{0} \\ H_{04}&: \boldsymbol{\gamma}_2 = \mathbf{0} \end{aligned} \tag{5.131}$$

where H_{01} is the null hypothesis that the overall mean is zero, H_{02} is the null hypothesis of no difference between the weight groups, and H_{03} and H_{04} are testing that the covariates are zero. The SAS code to perform these tests and to evaluate parallelism for x_1 and x_2 is provided in Program 5_9.sas.

Program 5_9.sas

```
/* Program 5_9.sas                                      */
/* Data from Smith, Gnanadesikan and Hughes (1962) */
/* Stepdown Analysis--MANCOVA                           */

options ls=80 ps=60 nodate nonumber;
title1 'Output 5.9: Multivariate MANCOVA with Stepdown Analysis';

data mancova;
   infile 'c:\5_9.dat';
   input group $ y1 y2 y3 y4 y5 y6 y7 y8 y9 y10 y11 x1 x2;
proc print data=mancova;
run;

/* Test of Parallelism */
proc glm data=mancova;
   title2 'Test of Parallelism';
   class group;
   model y1-y11 = group x1 x2 group*x1 group*x2/ nouni ss3;
   manova h = group *x1 group*x2;
run;
```

```
/* MANCOVA Tests */
proc glm data=mancova;
   title2 'MANCOVA Tests';
   class group;
   model y1-y11 = x1 x2 group/ e nouni ss3;
   contrast 'Ov Mean' intercept 1/ e;
   manova h=group x1 x2/ printh;
   means group;
   lsmeans group/ stderr pdiff tdiff cov out=adjmeans;
run;

proc print data=adjmeans;
   title2 'Adjusted Means';
run;

/* Mancova stepdown analysis for additional information  */
proc glm data=mancova;
   class group;
   title2 'Test of Gamma 2.1';
   model y2 y3 y4 y6 y9-y11 = group y1 y5 y7 y8 x1 x2/ nouni ss3;
   manova h = group y1 y5 y7 y8 x1 x2;
run;

proc glm data=mancova;
   class group;
   title2 'Test of Gamma1';
   model y1 y5 y7 y8 = group x1 x2/ nouni ss3;
   manova h = group x1 x2;
run;
```

Result and Interpretation 5_9.sas

The output with the results of the test of parallelism and each of the hypotheses given in (5.131) follows.

Test of parallelism:

Output 5.9: *Multivariate MANCOVA with Stepdown Analysis Test of Parallelism*

```
        Output 5.9: Multivariate MANCOVA with Stepdown Analysis
                             Test of Parallelism
               the Hypothesis of no Overall X1*GROUP Effect
      H = Type III SS&CP Matrix for X1*GROUP   E = Error SS&CP Matrix

                            S=3     M=3.5     N=10.5

Statistic                     Value           F      Num DF    Den DF   Pr > F

Wilks' Lambda              0.19697346      1.5265        33  68.46624   0.0708
Pillai's Trace             1.18924880      1.4927        33        75   0.0779
Hotelling-Lawley Trace     2.35299226      1.5449        33        65   0.0677
Roy's Greatest Root        1.42789610      3.2452        11        25   0.0071

        NOTE: F Statistic for Roy's Greatest Root is an upper bound.
```

Output 5.9 *(continued)*

```
              Manova Test Criteria and F Approximations for
                the Hypothesis of no Overall X2*GROUP Effect
      H = Type III SS&CP Matrix for X2*GROUP    E = Error SS&CP Matrix

                              S=3     M=3.5     N=10.5

Statistic                       Value           F       Num DF    Den DF   Pr > F

Wilks' Lambda                0.21977061     1.3951          33  68.46624   0.1231
Pillai's Trace               1.10605882     1.3273          33        75   0.1566
Hotelling-Lawley Trace       2.20021973     1.4446          33        65   0.1030
Roy's Greatest Root          1.39647511     3.1738          11        25   0.0081

        NOTE: F Statistic for Roy's Greatest Root is an upper bound.
```

Test of H_{02}:

Output 5.9 *(continued)*

```
              Manova Test Criteria and F Approximations for
                the Hypothesis of no Overall GROUP Effect
        H = Type III SS&CP Matrix for GROUP    E = Error SS&CP Matrix

                              S=3     M=3.5     N=13.5

Statistic                       Value           F       Num DF    Den DF   Pr > F

Wilks' Lambda                0.06011042     4.1685          33  86.14335   0.0001
Pillai's Trace               1.50417898     2.8339          33        93   0.0001
Hotelling-Lawley Trace       7.65378673     6.4168          33        83   0.0001
Roy's Greatest Root          6.70400149    18.8931          11        31   0.0001

        NOTE: F Statistic for Roy's Greatest Root is an upper bound.
```

Test of H_{03}:

Output 5.9 *(continued)*

```
              Manova Test Criteria and Exact F Statistics for
                  the Hypothesis of no Overall X1 Effect
         H = Type III SS&CP Matrix for X1    E = Error SS&CP Matrix

                              S=1     M=4.5     N=13.5

Statistic                       Value           F       Num DF    Den DF   Pr > F

Wilks' Lambda                0.37745713     4.3482          11        29   0.0007
Pillai's Trace               0.62254287     4.3482          11        29   0.0007
Hotelling-Lawley Trace       1.64930750     4.3482          11        29   0.0007
```

Test of H_{04}:

Output 5.9 (*continued*)

```
              Manova Test Criteria and Exact F Statistics for
                  the Hypothesis of no Overall X2 Effect
        H = Type III SS&CP Matrix for X2    E = Error SS&CP Matrix

                            S=1     M=4.5     N=13.5

Statistic                     Value            F      Num DF    Den DF  Pr > F

Wilks' Lambda              0.25385207     7.7491         11        29  0.0001
Pillai's Trace             0.74614793     7.7491         11        29  0.0001
Hotelling-Lawley Trace     2.93930220     7.7491         11        29  0.0001
Roy's Greatest Root        2.93930220     7.7491         11        29  0.0001
```

Test of H_{01}:

Output 5.9 (*continued*)

```
              Manova Test Criteria and Exact F Statistics for
                the Hypothesis of no Overall Ov Mean Effect
      H = Contrast SS&CP Matrix for Ov Mean    E = Error SS&CP Matrix

                            S=1     M=4.5     N=13.5

Statistic                     Value            F      Num DF    Den DF  Pr > F

Wilks' Lambda              0.06920279    35.4598         11        29  0.0001
Pillai's Trace             0.93079721    35.4598         11        29  0.0001
Hotelling-Lawley Trace    13.45028460    35.4598         11        29  0.0001
Roy's Greatest Root       13.45028460    35.4598         11        29  0.0001
```

We see that the tests of parallelism are nonsignificant and that each test in (5.131) is significant. To summarize,

Hypothesis	Wilks	*p*-value	F
02	0.06011	<0.0001	4.17
03	0.37746	0.0007	4.35
04	0.25385	0.0001	7.75
01	0.06920	0.0001	35.46

These values do not agree with those reported by Smith et al. (1962). Their Λ values were 0.0734, 0.4718, 0.2657, and 0.0828 for H_{02}, H_{03}, H_{04}, and H_{01} respectively. However, the conclusions remain valid.

Given significance between weight groups, Smith et al. (1962) were interested in evaluating whether variables $y_2, y_3, y_4, y_6, y_9, y_{10}$, and y_{11} (set 2) add additional information to the analysis of weight difference over and above that provided by variables y_1, y_5, y_7, and y_8 (set 1). To evaluate the contribution of set 2 given set 1, we calculate $\Lambda_{2.1}$ using PROC GLM. The output follows.

Output 5.9 (*continued*)

```
          Output 5.9: Multivariate MANCOVA with Stepdown Analysis
                             Test of Gamma 2.1

               Manova Test Criteria and F Approximations for
                 the Hypothesis of no Overall GROUP Effect
        H = Type III SS&CP Matrix for GROUP    E = Error SS&CP Matrix

                           S=3     M=1.5     N=13.5

Statistic                      Value          F     Num DF    Den DF   Pr > F

Wilks' Lambda             0.16614119     3.4663        21  83.82245   0.0001
Pillai's Trace            1.08299153     2.5019        21        93   0.0014
Hotelling-Lawley Trace    3.62094249     4.7704        21        83   0.0001
Roy's Greatest Root       3.22916434    14.3006         7        31   0.0001

        NOTE: F Statistic for Roy's Greatest Root is an upper bound.
```

The test of additional information has a p-value < 0.0001 since $\Lambda_{2.1} = 0.16614119$. We also calculate Λ_1 to verify that $\Lambda = \Lambda_1 \Lambda_{2.1}$ where Λ is Wilks' criterion for the MANCOVA test H_{02} in (5.131). From the following output we see that $\Lambda_1 = 0.36180325$.

Output 5.9 (*continued*)

```
          Output 5.9: Multivariate MANCOVA with Stepdown Analysis
                              Test of Gamma1

                      General Linear Models Procedure
                     Multivariate Analysis of Variance

             Manova Test Criteria and Exact F Statistics for
                  the Hypothesis of no Overall X2 Effect
        H = Type III SS&CP Matrix for X2    E = Error SS&CP Matrix

                           S=1     M=1     N=17

Statistic                      Value          F     Num DF    Den DF   Pr > F

Wilks' Lambda             0.36319830    15.7799         4        36   0.0001
Pillai's Trace            0.63680170    15.7799         4        36   0.0001
Hotelling-Lawley Trace    1.75331682    15.7799         4        36   0.0001
Roy's Greatest Root       1.75331682    15.7799         4        36   0.0001
```

With the MULTTEST procedure, stepdown p-value adjustments can easily be obtained. The stepdown option includes STEPBON for stepdown Bonferroni, STEPBOOT for stepdown for bootstrap resampling, STEPPERM for stepdown permutation resampling, and STEPSID for stepdown Sidak adjustments (see *SAS/STAT Software: Changes and Enhancements for Release 6.12* for details).

5.10 Repeated Measurement Analysis

As our next example of the GLM, we illustrate how to use the model to analyze repeated measurement data. Using the data in Timm (1975, p. 244), we showed in Chapter 3 how to analyze repeated measurement data using a restricted univariate linear model which required the covariance matrix $\mathbf{\Sigma}$ within each group to have a Hankel structure (Kirk, 1995, p. 525). To analyze repeated measurement data using the multivariate linear model, we allow $\mathbf{\Sigma}$ to be arbitrary but homogeneous across groups where the number of subjects within each group is at least equal to the number of repeated measurements acquired on each subject.

For a repeated measurement experiment for I groups and a p–variate reponse vector of observations, we assume that

$$\mathbf{y}'_{ij} = (y_{ij1}, y_{ij2}, \ldots, y_{ijp}) \sim IN_p(\mu_i, \mathbf{\Sigma})$$

where $i = 1, 2, \ldots, I$; $j = 1, 2, \ldots, n_i$ and $n = \sum_{i=1}^{I} n_i$. If we employ a general linear model, the data matrix is $\mathbf{Y}_{n\times p}$ and the parameter matrix $\mathbf{B}_{I\times p}$ is

$$\mathbf{B} = \begin{pmatrix} \mu_{11} & \mu_{12} & \cdots & \mu_{1p} \\ \mu_{21} & \mu_{22} & \cdots & \mu_{2p} \\ \vdots & \vdots & \cdots & \vdots \\ \mu_{I1} & \mu_{I2} & \cdots & \mu_{Ip} \end{pmatrix} \tag{5.132}$$

where the ith row is a $1{\times}p$ vector of means, $\mu'_i = (\mu_{i1}, \mu_{i2}, \ldots, \mu_{ip})$. The design matrix $\mathbf{X}_{n\times I}$ has the general form

$$\underset{n\times I}{\mathbf{X}} = \begin{pmatrix} \mathbf{1}_{n_1} & \cdots & \cdots & 0 \\ 0 & \mathbf{1}_{n_2} & \cdots & 0 \\ \vdots & \vdots & \vdots & \vdots \\ 0 & \cdots & \cdots & \mathbf{1}_{n_I} \end{pmatrix} = \mathbf{I}_n \otimes \mathbf{1}_{n_i}$$

To create the multivariate repeated measures model with PROC GLM, we may use two MODEL statements

```
model y5 = group/intercept nouni;
model y5 = group/noint nouni;
```

In the first statement, SAS generates a less-than-full-rank model of the form

$$E(\mathbf{y}_{ij}) = \mu + \alpha_i \qquad i = 1, 2, \ldots, I$$

and prints hypotheses associated with the intercept. In the second statement, SAS generates a full-rank model of the form

$$E(\mathbf{y}_{ij}) = \mu_i \qquad i = 1, 2, \ldots, I$$

Either model may be used to test hypotheses regarding the parameters in $\mathbf{B}$. The multivariate model corresponds to a univariate split-plot design where the subjects are random, the I groups are fixed, and the p measurements represent p levels of the fixed univariate within (condition) factor.

When one analyzes a repeated measures design, three hypotheses are of primary interest:

H_{01} : Are the profiles for the I groups parallel?

H_{02}: Are there differences among the p conditions?

H_{03}: Are there differences among the I groups?

If the profiles are parallel, or equivalently when there is no group by condition interaction, one may alter H_{02} and H_{03} to test for differences in groups or conditions given parallelism. Furthermore, if there is an unequal number of subjects in each group that is due to the treatment, one may test for weighted differences in group means given parallelism. In terms of the model parameters, the tests are

$$H_{01}: \begin{pmatrix} \mu_{11}-\mu_{12} \\ \mu_{12}-\mu_{13} \\ \vdots \\ \mu_{1(p-1)}-\mu_{1p} \end{pmatrix} = \cdots = \begin{pmatrix} \mu_{I1}-\mu_{I2} \\ \mu_{I2}-\mu_{I3} \\ \vdots \\ \mu_{I(p-1)}-\mu_{Ip} \end{pmatrix}$$

$$H_{02}: \begin{pmatrix} \mu_{11} \\ \mu_{21} \\ \vdots \\ \mu_{I1} \end{pmatrix} = \cdots = \begin{pmatrix} \mu_{1p} \\ \mu_{2p} \\ \vdots \\ \mu_{Ip} \end{pmatrix} \qquad (5.133)$$

$$H_{03}: \quad \mu_1 = \cdots = \mu_I$$

$$H_{02}^{*}: \sum_{i=1}^{I} \mu_{i1} / I = \cdots = \sum_{i=1}^{I} \mu_{ip} / I$$

$$H_{03}^{*}: \sum_{j=1}^{p} \mu_{1j} / p = \cdots = \sum_{j=1}^{p} \mu_{Ij} / p$$

$$H_{03}^{**}: \sum_{i=1}^{I} n_i \mu_{i1} / n = \cdots = \sum_{i=1}^{I} n_i \mu_{ip} / n$$

where H_{02}^{*} and H_{03}^{*} are tests for finding differences in conditions and groups given parallelism (no group x condition interaction), respectively, and H_{03}^{**} is the weighted test of group differences. The SAS code to perform the tests is provided in Program 5_10.sas.

Program 5_10.sas

```
/* Program 5_10.sas                    */
/* Data from Timm(1975, page 244) */
/* Repeated Measures Analysis     */

options ls=80 ps=60 nodate nonumber;
title1 'Output 5.10: Repeated Measurement Analysis';

data timm;
   infile 'c:\5_10.dat';
   input group y1 y2 y3 y4 y5;
proc print data=timm;
run;
```

```
proc glm;
   title2 'Multivariate Tests';
   class group;
   model y1-y5 = group/ intercept nouni;
   means group;
   manova h = group/ printe printh;   /*test of group diffs for means */
   manova h =_all_ m=(1 -1 0 0 0,
                      0 1 -1 0 0,
                      0 0 1 -1 0,
                      0 0 0 1 -1)
      prefix = diff/ printe printh;   /* test for parallel profiles */

proc glm;
   class group;
   model y1-y5= group/noint nouni;
   contrast 'Mult Cond' group 1 0,
                        group 0 1;
   manova m=(1 -1 0 0 0,
             0 1 -1 0 0,
             0 0 1 -1 0,
             0 0 0 1 -1)
      prefix = diff/ printe printh; /* test of condtions as vectors */
run;

proc glm;
   title2 'Tests given Parallelism of Profiles';
   class group;
   model y1 - y5 = group/ noint nouni;
   contrast 'Univ Gr' group 1 -1;
   manova m=(.2 .2 .2 .2 .2) prefix=Gr/
      printe printh;  /* test of group means given parallel profiles */
run;

/* Univariate Test given Parallelism and Sphericity */
proc glm;
   title2 'Univariate Test given Parallelism and Sphericity';
   class group;
   model y1 - y5 = group/ nouni;
   repeated cond 5 (1 2 3 4 5) profile/printm summary;
   manova h=group m=(1 -1  0  0  0,
                     0  1 -1  0  0,
                     0  0  1 -1  0,
                     0  0  0  1 -1) prefix = diff / printe printh;
run;
```

Result and Interpretation 5_10.sas

To interpret how SAS is used to analyze repeated measurement data, it is convenient to express the hypotheses in (5.133) using the general matrix product form $\mathbf{CBA} = \mathbf{0}$ where $\mathbf{B}$ is the matrix defined in (5.132). In this example $I = 2$ and $p = 5$ so that

$$\mathbf{B} = \begin{pmatrix} \mu_{11} & \mu_{12} & \mu_{13} & \mu_{14} & \mu_{15} \\ \mu_{21} & \mu_{22} & \mu_{23} & \mu_{24} & \mu_{25} \end{pmatrix} \tag{5.134}$$

To test H_{03}, differences between groups, we set $\mathbf{C} = (1, -1)$ to obtain the difference in group vectors and $\mathbf{A} = \mathbf{I}_5$. The within-matrix $\mathbf{A}$ is equal to the identity since we are evaluating the equivalence of the means across the p– variables simultaneously. In PROC GLM, this test is performed with the statement

```
manova  h = group/printe printh;
```

where the options PRINTE and PRINTH are used to print $\mathbf{H}$ and $\mathbf{E}$ for hypothesis testing.

To test H_{02}, differences among treatments, the matrices

$$\mathbf{C} = \mathbf{I}_2 \text{ and } \mathbf{A} = \begin{pmatrix} 1 & 0 & 0 & 0 \\ -1 & 1 & 0 & 0 \\ 0 & -1 & 1 & 0 \\ 0 & 0 & -1 & 1 \\ 0 & 0 & 0 & -1 \end{pmatrix}$$

are used. The matrix $\mathbf{A}$ forms differences within variables, the within-subject dimension, and $\mathbf{C}$ is the identity since we are evaluating vectors across groups, simultaneously. To test this hypothesis using PROC GLM, one uses the CONTRAST statement, the full rank model (NOINT option in the MODEL statement) and the MANOVA statement

```
contrast 'Mult Cond' group 1 0
                     group 0 1;
```

$$\texttt{manova m} = \begin{pmatrix} 1 & -1 & 0 & 0 & 0, \\ 0 & 1 & -1 & 0 & 0, \\ 0 & 0 & 1 & -1 & 0, \\ 0 & 0 & 0 & 1 & -1 \end{pmatrix} \texttt{ prefix = diff/printe printh;}$$

where $m = \mathbf{A}'$ and the group matrix is the identity matrix $\mathbf{I}_2$.

To test for parallelism of profiles, H_{01}, we use the matrices

$$\mathbf{C} = (1, -1) \text{ and } \mathbf{A} = \begin{pmatrix} 1 & 0 & 0 & 0 \\ -1 & 1 & 0 & 0 \\ 0 & -1 & 1 & 0 \\ 0 & 0 & -1 & 1 \\ 0 & 0 & 0 & -1 \end{pmatrix}$$

The matrix $\mathbf{A}$ again forms differences across variables while $\mathbf{C}$ creates differences among groups. For more than two groups, say four, $\mathbf{C}$ would take the form

$$\mathbf{C} = \begin{pmatrix} 1 & 0 & 0 & -1 \\ 0 & 1 & 0 & -1 \\ 0 & 0 & 1 & -1 \end{pmatrix}$$

which compares each group with the last and the rank $(\mathbf{C}) = 3 = I - 1$. Of course the matrices **C** and **A** are not unique since other differences could be specified. The tests are invariant to the specific form provided they span the same test space. To test H_{02}^{*}, differences in conditions given parallelism, we use the matrices

$$\mathbf{C} = (1/2, 1/2) \text{ and } \mathbf{A} = \begin{pmatrix} 1 & 0 & 0 & 0 \\ -1 & 1 & 0 & 0 \\ 0 & -1 & 1 & 0 \\ 0 & 0 & -1 & 1 \\ 0 & 0 & 0 & -1 \end{pmatrix}$$

To test these hypotheses using PROC GLM, we use the statement

```
                                  (1  -1   0   0  0,)
                                  (0   1  -1   0  0,)
manova h =_all_          m =      (0   0   1  -1  0,)
                                  (0   0   0   1 -1)  prefix = diff/printe printh;
```

In this statement, $\text{h} = \begin{pmatrix} 1 & 1 \\ 1 & -1 \end{pmatrix}$ and the test of "intercept" is the test of equal condition means (summed over groups) given parallelism of profiles using the less-than-full-rank model. Because the multivariate tests are invariant to scale, it does not matter whether we use $\mathbf{C} = (1,1)$ or $\mathbf{C} = (1/2, 1/2)$.

The final test of interest for these data is H_{03}^{*}, differences in groups given parallelism, sometimes called the test of coincidence. For this hypothesis we select the matrices

$$\mathbf{C} = (1, -1) \text{ and } \mathbf{A} = \begin{pmatrix} 1/5 \\ 1/5 \\ 1/5 \\ 1/5 \\ 1/5 \end{pmatrix}$$

To test this hypothesis using PROC GLM, it is again convenient to form a contrast using either the full-rank or less-than-full-rank model:

```
contrast   'Univ Gr' group    1  -1;
manova     m = (.2 .2 .2 .2 .2) prefix = Gr/printe, printh;
```

where the matrix $\mathbf{A}' = m$ in SAS. For more than two groups, one would add rows to the contrast statement. For example, for three groups we would use the statement

```
contrast         'Univ Gr' group    1    0   -1,     0    1    -1;
```

The results of Program 5_10.sas are found in the following Output 5.10.

Output 5.10: *Repeated Measurement Analysis Multivariate Tests*

```
                 Output 5.10: Repeated Measurement Analysis
                             Multivariate Tests

                       General Linear Models Procedure
              Manova Test Criteria and Exact F Statistics for
                 the Hypothesis of no Overall GROUP Effect
         H = Type III SS&CP Matrix for GROUP   E = Error SS&CP Matrix

                            S=1    M=1.5    N=6

  Statistic                    Value          F      Num DF     Den DF   Pr > F

  Wilks' Lambda             0.55607569     2.2353         5         14   0.1083
  Pillai's Trace            0.44392431     2.2353         5         14   0.1083
  Hotelling-Lawley Trace    0.79831634     2.2353         5         14   0.1083
  Roy's Greatest Root       0.79831634     2.2353         5         14   0.1083

General Linear Models Procedure
                        Multivariate Analysis of Variance
              Manova Test Criteria and Exact F Statistics for
                 the Hypothesis of no Overall GROUP Effect
           on the variables defined by the M Matrix Transformation
         H = Type III SS&CP Matrix for GROUP   E = Error SS&CP Matrix

                            S=1    M=1    N=6.5

  Statistic                    Value          F      Num DF     Den DF   Pr > F

  Wilks' Lambda             0.83905852     0.7193         4         15   0.5919
  Pillai's Trace            0.16094148     0.7193         4         15   0.5919
  Hotelling-Lawley Trace    0.19181198     0.7193         4         15   0.5919
  Roy's Greatest Root       0.19181198     0.7193         4         15   0.5919



                       General Linear Models Procedure
               Manova Test Criteria and F Approximations for
                the Hypothesis of no Overall Mult Cond Effect
           on the variables defined by the M Matrix Transformation
       H = Contrast SS&CP Matrix for Mult Cond   E = Error SS&CP Matrix

                           S=2    M=0.5    N=6.5

  Statistic                    Value          F      Num DF     Den DF   Pr > F

  Wilks' Lambda             0.19454894     4.7519         8         30   0.0008
  Pillai's Trace            0.88327644     3.1638         8         32   0.0094
  Hotelling-Lawley Trace    3.74006487     6.5451         8         28   0.0001
  Roy's Greatest Root       3.62985954    14.5194         4         16   0.0001

         NOTE: F Statistic for Roy's Greatest Root is an upper bound.
                NOTE: F Statistic for Wilks' Lambda is exact.
```

Tests given Parallelism of Profiles:

Output 5.10 (*continued*)

```
                     General Linear Models Procedure
              Manova Test Criteria and Exact F Statistics for
                the Hypothesis of no Overall Univ Gr Effect
          on the variables defined by the M Matrix Transformation

       H = Contrast SS&CP Matrix for Univ Gr    E = Error SS&CP Matrix

                             S=1    M=-0.5    N=8

Statistic                     Value          F      Num DF    Den DF  Pr > F

Wilks' Lambda              0.66905173     8.9037         1        18  0.0080
Pillai's Trace             0.33094827     8.9037         1        18  0.0080
Hotelling-Lawley Trace     0.49465273     8.9037         1        18  0.0080
Roy's Greatest Root        0.49465273     8.9037         1        18  0.0080

                    General Linear Model Procedure
              Manova Test Criteria and Exact F Statistics
                  for the Hypothesis of no COND Effect
        H = Type III SS&CP Matrix for COND    E = Error SS&CP Matrix

                             S=1    M=1    N=6.5

Statistic                     Value          F      Num DF    Den DF  Pr > F

Wilks' Lambda              0.21986464    13.3059         4        15  0.0001
Pillai's Trace             0.78013536    13.3059         4        15  0.0001
Hotelling-Lawley Trace     3.54825288    13.3059         4        15  0.0001
Roy's Greatest Root        3.54825288    13.3059         4        15  0.0001


                    General Linear Models Procedure
              Manova Test Criteria and Exact F Statistics for
                 the Hypothesis of no COND*GROUP Effect
      H = Type III SS&CP Matrix for COND*GROUP    E = Error SS&CP Matrix

                             S=1    M=1    N=6.5

Statistic                     Value          F      Num DF    Den DF  Pr > F

Wilks' Lambda              0.83905852     0.7193         4        15  0.5919
Pillai's Trace             0.16094148     0.7193         4        15  0.5919
Hotelling-Lawley Trace     0.19181198     0.7193         4        15  0.5919
Roy's Greatest Root        0.19181198     0.7193         4        15  0.5919
```

Following Timm (1980a), it is convenient to create two MANOVA tables, Multivariate Analysis I and II. The Multivariate Analysis I table is used if the test of parallelism is significant, i.e., we reject the null hypothesis of parallelism. Given parallelism, one would use the results in the Multivariate Analysis II table. Tables 5.2 and 5.3 summarize the results.

Table 5.2 *Multivariate Analysis I*

Hypotheses	*H* Matrix	df	Λ	*p*-value
H_{03} (Groups)	$\begin{pmatrix} 396.05 & & & & (\text{Sym}) \\ 293.7 & 217.8 & & & \\ 373.8 & 277.2 & 352.8 & & \\ 298.15 & 221.1 & 281.4 & 224.45 & \\ 511.75 & 379.5 & 483 & 385.25 & 661.25 \end{pmatrix}$	1	0.556	0.1083
H_{02} (Conditions)	$\begin{pmatrix} 4377.7 & & & (\text{Sym}) \\ -4062.2 & & & \\ 3229.3 & -2988.2 & 2368.9 & \\ -1412.2 & 1303.6 & -1039 & 538.4 \end{pmatrix}$	2	0.195	0.0008
H_{01} (Parallelism)	$\begin{pmatrix} 26.45 & & & (\text{Sym}) \\ -20.7 & 16.2 & & \\ 19.55 & -15.3 & 14.45 & \\ -55.2 & 43.2 & -40.80 & 115.2 \end{pmatrix}$	1	0.839	0.5919

Table 5.3 *Multivariate Analysis II*

Hypotheses	*H* Matrix	df	Λ	*p*-value
H_{03}^{*} (Groups$\mid$Parallelism)	354.482	1	0.669	0.0080
H_{02}^{*} (Conditions$\mid$Parallelism)	$\begin{pmatrix} 4351.25 & & & (\text{Sym}) \\ -4041.5 & 3753.8 & & \\ 3200.75 & -2972.9 & 2354.45 & \\ -1357 & 1260.4 & -998.2 & 423.2 \end{pmatrix}$	1	0.220	0.0001
H_{01} (Parallelism)	$\begin{pmatrix} 26.45 & & & (\text{Sym}) \\ -20.7 & 16.2 & & \\ 19.55 & -15.3 & 14.45 & \\ -55.2 & 43.2 & -40.8 & 115.2 \end{pmatrix}$	1	0.839	0.5919

Finally, if one can assume a Hankel structure for $\mathbf{\Sigma}$ or the sphericity of the matrix $\mathbf{A}'\mathbf{\Sigma}\mathbf{A}$ where $\mathbf{A}$ is the post profile matrix given in general by

$$\underset{(p-1)xp}{\mathbf{A}'} = \begin{pmatrix} 1 & -1 & 0 & \cdots & \cdots & \cdots & 0 & 0 \\ 0 & 1 & -1 & \cdots & \cdots & \cdots & 0 & 0 \\ \cdots & \cdots & \cdots & \cdots & \cdots & \cdots & \cdots & \cdots \\ 0 & 0 & 0 & \cdots & \cdots & \cdots & 1 & -1 \end{pmatrix}$$

then a univariate analysis is most appropriate. If in addition one chooses $\mathbf{A}$ such that $\mathbf{A}'\mathbf{A} = \mathbf{I}$, so that $\mathbf{A}$ is semi-orthogonal, one may obtain the mixed model univariate F-ratios for the split-plot design as illustrated in Chapter 3. To accomplish this using PROC GLM, one employs the REPEATED statement as shown in Program 5_10.sas. The output follows.

Output 5.10 (*continued*)

```
                 Output 5.10: Repeated Measurement Analysis

              Univariate Test given Parallelism and Sphericity

                        General Linear Models Procedure
                     Repeated Measures Analysis of Variance
                 Tests of Hypotheses for Between Subjects Effects

Source                  DF      Type III SS     Mean Square    F Value      Pr > F

GROUP                    1      1772.410000     1772.410000       8.90      0.0080

Error                   18      3583.140000      199.063333


                        General Linear Models Procedure
                     Repeated Measures Analysis of Variance
            Univariate Tests of Hypotheses for Within Subject Effects

Source: COND
                                                                      Adj  Pr > F
      DF       Type III SS        Mean Square    F Value    Pr > F    G - G    H - F
       4     3371.30000000      842.82500000      14.48    0.0001    0.0001   0.0001

Source: COND*GROUP
                                                                      Adj  Pr > F
      DF       Type III SS        Mean Square    F Value    Pr > F    G - G    H - F
       4       79.94000000       19.98500000       0.34    0.8479    0.8068   0.8479

Source: Error(COND)

      DF       Type III SS        Mean Square
      72     4191.96000000       58.22166667

                      Greenhouse-Geisser Epsilon = 0.8009
                             Huynh-Feldt Epsilon = 1.0487
```

Because the profile matrix $\mathbf{A}$ used in Multivariate Analysis II was not normalized on input, one may not directly obtain the univariate tests by averaging diagonal elements. However, PROC GLM performs the normalization by the REPEATED statement and the results of the three tests found in Output 5.10 are summarized in Table 5.4.

Table 5.4 *Test Results*

F-ratios	p-values
$F_g = \dfrac{354.482/1}{716.629/18} = 8.90 \sim F(1,18)$	0.0080
$F_c = 14.48 \sim F(4,72)$	< 0.0001
$F_{cg} = 0.34 \sim F(4,72)$	0.8479

Both the multivariate (Multivariate Analysis I and II) and univariate test results show that there are differences among conditions. The Multivariate Analysis I failed to detect group differences whereas with the univariate analysis group differences were detected. This is due to the fact that the univariate result examines only one contrast of the means while the multivariate tests an infinite number of contrasts; hence, the multivariate test is more conservative.

Finally, we observe that PROC GLM calculates both the Greenhouse-Geisser and Huynh-Feldt Epsilon adjusted tests. However, as we discussed in Chapter 3, these should not be used if the multivariate tests can be used since these tests are only approximate and have no power advantage over the exact multivariate tests.

Having found a significant overall multivariate effect, one may want to investigate contrasts in the means of the general form $\psi = \mathbf{c'Ba}$ where $\mathbf{B}$ is defined in (5.133). The vectors $\mathbf{c}$ and $\mathbf{a}$ are selected depending upon whether one is interested in differences among groups, differences in conditions, or an interaction. Because there is not a significant group x condition interaction in our example, the *differences* in conditions or groups, averaging over the other factor, may be of interest. Following (5.37), the confidence interval for ψ is

$$\hat{\psi} - c_o\hat{\sigma}_{\hat{\psi}} \leq \psi \leq \hat{\psi} + c_o\hat{\sigma}_{\hat{\psi}}$$

where $\hat{\sigma}^2_{\hat{\psi}} = (\mathbf{a'Sa})\mathbf{c'(X'X)}^{-1}\mathbf{c}$ and c_o is established by the overall criterion used to test the hypothesis. For our example, one finds $\mathbf{S}$ from the PROC GLM output, by dividing the Error *SS & CP* matrix by v_e. The Error *SS & CP* matrix is

```
E = Error SS&CP Matrix

              Y1            Y2            Y3            Y4            Y5

Y1        2546.9           597         946.6         257.9         385.9
Y2           597         960.4         569.2         155.2         299.4
Y3         946.6         569.2          2204         686.2         599.8
Y4         257.9         155.2         686.2        1164.5         573.1
Y5         385.9         299.4         599.8         573.1         899.3
```

So that $\mathbf{S} = \mathbf{E}/18$:

$$\mathbf{S} = \begin{pmatrix} 141.49 & & & & (\text{Sym}) \\ 33.17 & 53.36 & & & \\ 52.59 & 31.62 & 122.44 & & \\ 14.33 & 8.62 & 38.12 & 64.69 & \\ 21.44 & 16.63 & 33.22 & 31.83 & 49.96 \end{pmatrix}$$

The Bartlett-Lawley-Hotelling criterion resulted in $c_o^2 = 14.667$ when testing for differences in conditions given parallelism for $\alpha = 0.05$, (Timm, 1975, p. 605). Using formula (5.134), with $c_o^2 = 14.667$, we see that confidence intervals for the differences in conditions one versus two, and two versus five are significant

$$5.04 \le \mu_1 - \mu_2 \le 24.46$$
$$-14.62 \le \mu_2 - \mu_5 \le -0.28$$

where

$$\hat{\sigma}^2_{1 \text{ vs } 2} = \frac{141.49 + 53.36 - 2(33.17)}{20} = 6.426$$
$$\hat{\sigma}^2_{2 \text{ vs } 5} = 3.503$$

and $\mathbf{c}' = (1, 1)$, $\mathbf{a}_1' = (1, -1, 0, 0, 0)$ and $\mathbf{a}_2' = (0, 1, 0, 0, -1)$ for the two comparisons. Comparing each of level one, three and four with level five, we find each comparison to be nonsignificant (Timm, 1975, p. 249). For additional examples, see Khattree and Naik (1995).

5.11 Extended Linear Hypotheses

When comparing means in any MANOVA design, but especially in repeated measurement designs, one frequently encounters contrasts in the means that are of interest, but are not expressible as a bilinear form, $\psi = \mathbf{c}'B\mathbf{a}$. When this type of contrast occurs, one may use the extended linear hypothesis as defined in (5.77) to test hypotheses and evaluate contrasts in cell means. To illustrate, suppose one has a one-way MANOVA design involving three groups and three variables, or three repeated measures, so that the parameter matrix is

$$\mathbf{B} = \begin{pmatrix} \mu_{11} & \mu_{12} & \mu_{13} \\ \mu_{21} & \mu_{22} & \mu_{23} \\ \mu_{31} & \mu_{32} & \mu_{33} \end{pmatrix}$$

Suppose that the primary hypothesis of interest is the equality of group means

$$H_g\text{:} \quad \begin{pmatrix} \mu_{11} \\ \mu_{12} \\ \mu_{13} \end{pmatrix} = \begin{pmatrix} \mu_{21} \\ \mu_{22} \\ \mu_{23} \end{pmatrix} = \begin{pmatrix} \mu_{31} \\ \mu_{32} \\ \mu_{33} \end{pmatrix}$$

To test H_g, one may select $\mathbf{C} \equiv \mathbf{C}_o$ and $\mathbf{A} \equiv \mathbf{A}_o$ to test the overall hypothesis where for example,

$$\mathbf{C}_o = \begin{pmatrix} 1 & -1 & 0 \\ 0 & 1 & -1 \end{pmatrix} \text{ and } \mathbf{A}_o = \mathbf{I}_3 \tag{5.135}$$

so that we may represent H_g in the general form $\mathbf{C}_o\mathbf{B}\mathbf{A}_o = \mathbf{0}$.

Upon rejecting H_g, suppose we are interested in the contrast

$$\psi = \mu_{11} + \mu_{22} + \mu_{33} - (\mu_{12} + \mu_{21})/2 - (\mu_{13} + \mu_{31})/2 - (\mu_{23} + \mu_{32})/2 \tag{5.136}$$

which compares the diagonal means with the average of the off diagonals, a common situation in a repeated measures design. Contrast (5.136) may not be expressed as a bilinear form. Using the vec(·) operator on $\mathbf{B}$, $\psi = \mathbf{C}_*\text{vec}(\mathbf{B})$ where

$$\mathbf{C}_* = (1, -.5, -.5, -.5, 1, -.5, -.5, -.5, 1)$$

We may test that $\psi = 0$ using a chi-square statistic with $v = \text{rank}(\mathbf{C}_*) = 1$ by (5.75). Alternatively, letting

$$\mathbf{M} = \begin{pmatrix} 1 & .5 \\ -.5 & .5 \\ -.5 & 1 \end{pmatrix}$$

observe that ψ in (5.136) may be expressed as an extended linear hypothesis since

$$\psi = Tr(\mathbf{MB}_o) = Tr(\mathbf{MC}_o\mathbf{BA}_o) = Tr(\mathbf{A}_o\mathbf{MC}_o\mathbf{B}) = Tr(\mathbf{GB})$$

where $\mathbf{C}_o$ and $\mathbf{A}_o$ are defined in (5.135), $\mathbf{B}_o = \mathbf{C}_o\mathbf{BA}_o$, and $\mathbf{G} = \mathbf{A}_o\mathbf{MC}_o$. Thus, we have that

$$\mathbf{G} = \begin{pmatrix} 1 & -.5 & -.5 \\ -.5 & 1 & -.5 \\ -.5 & -.5 & 1 \end{pmatrix}$$

where the coefficients in each row and column of $\mathbf{G}$ sum to one. A contrast of this type has been termed a generalized contrast by Bradu and Gabriel (1974).

Testing that the contrast in (5.136) is zero is achieved using a single matrix $\mathbf{M}$. Alternatively, suppose we are interested in the following hypothesis

$$\begin{aligned} H_o\colon\ & \mu_{11} = \mu_{21} \\ & \mu_{12} = \mu_{22} = \mu_{32} \\ & \mu_{23} = \mu_{33} \end{aligned} \tag{5.137}$$

which is not a general linear hypothesis of the form $\mathbf{C}_o\mathbf{BA}_o = \mathbf{0}$, but an extended linear hypothesis. To see this, observe that hypothesis (5.137) is equivalent to testing

$$Tr(\mathbf{M}_i\mathbf{B}_o) = Tr(\mathbf{M}_i\mathbf{C}_o\mathbf{BA}_o) = Tr(\mathbf{A}_o\mathbf{M}_i\mathbf{C}_o\mathbf{B}) = Tr(\mathbf{G}_i\mathbf{B})$$

for $i = 1, 2, 3, 4$, where

$$\mathbf{M}_1 = \begin{pmatrix} 1 & 0 \\ 0 & 0 \\ 0 & 0 \end{pmatrix}, \quad \mathbf{M}_2 = \begin{pmatrix} 0 & 0 \\ 1 & 0 \\ 0 & 0 \end{pmatrix}, \quad \mathbf{M}_3 = \begin{pmatrix} 0 & 0 \\ 0 & 1 \\ 0 & 0 \end{pmatrix}, \quad \mathbf{M}_4 = \begin{pmatrix} 0 & 0 \\ 0 & 0 \\ 0 & 1 \end{pmatrix}$$

To test (5.137) requires obtaining the supremum in (5.88) as outlined in (5.89).

The examples we have illustrated have assumed $\mathbf{A}_o = \mathbf{I}$. We now consider the test of parallelism for a repeated measures design

$$\begin{pmatrix} 1 & -1 & 0 \\ 0 & 1 & -1 \end{pmatrix} \begin{pmatrix} \mu_{11} & \mu_{12} & \mu_{13} \\ \mu_{21} & \mu_{22} & \mu_{23} \\ \mu_{31} & \mu_{32} & \mu_{33} \end{pmatrix} \begin{pmatrix} 1 & 0 \\ -1 & 1 \\ 0 & -1 \end{pmatrix} = \mathbf{0}$$

$$\mathbf{C}_o\mathbf{B}\mathbf{A}_o = \mathbf{0}$$

Following the overall test, suppose we were interested in the sum of the following tetrads

$$\psi = (\mu_{21} + \mu_{12} - \mu_{31} - \mu_{22}) + (\mu_{32} + \mu_{23} - \mu_{13} - \mu_{22}) \tag{5.138}$$

Such a contrast may not be expressed using the simple bilinear form, $\psi = \mathbf{c}'\mathbf{B}\mathbf{a}$. However, selecting

$$\mathbf{M} = \begin{pmatrix} 0 & 1 \\ 1 & 0 \end{pmatrix}$$

we see that the contrast

$$\psi = Tr(\mathbf{M}\mathbf{B}_o) = Tr(\mathbf{M}\mathbf{C}_o\mathbf{B}\mathbf{A}_o) = Tr(\mathbf{A}_o\mathbf{M}\mathbf{C}_o\mathbf{B}) = Tr(\mathbf{G}\mathbf{B})$$

where

$$\mathbf{G} = \begin{pmatrix} 0 & 1 & -1 \\ 1 & -2 & 1 \\ -1 & 1 & 0 \end{pmatrix}$$

is a generalized contrast matrix. Now suppose we wish to test for equality of conditions for the repeated measures design

$$H_c\text{:} \quad \begin{pmatrix} \mu_{11} \\ \mu_{21} \\ \mu_{31} \end{pmatrix} = \begin{pmatrix} \mu_{12} \\ \mu_{22} \\ \mu_{32} \end{pmatrix} = \begin{pmatrix} \mu_{13} \\ \mu_{23} \\ \mu_{33} \end{pmatrix}$$

Expressing the hypothesis in the linear form, we have that $\mathbf{C}_o \equiv \mathbf{I}_3$ and

$$\mathbf{A}_o = \begin{pmatrix} 1 & 0 \\ -1 & 1 \\ 0 & -1 \end{pmatrix}$$

Following the overall test, suppose we are interested in the contrast

$$\psi = (\mu_{11} - \mu_{12}) + (\mu_{22} - \mu_{23}) + (\mu_{31} - \mu_{33}) \tag{5.139}$$

This contrast is again an extended linear hypothesis and tested with

$$\mathbf{M} = \begin{pmatrix} 1 & 0 & 1 \\ 0 & 1 & 1 \end{pmatrix} \quad \text{or} \quad \mathbf{G} = \begin{pmatrix} 1 & 0 & 1 \\ -1 & 1 & 0 \\ 0 & -1 & -1 \end{pmatrix}$$

in the expression $\psi = Tr(\mathbf{M}\mathbf{B}_o) = Tr(\mathbf{M}\mathbf{C}_o\mathbf{B}\mathbf{A}_o) = Tr(\mathbf{G}\mathbf{B})$.

To obtain a point estimate for $\psi = Tr(\mathbf{M}\mathbf{B}_o) = Tr(\mathbf{G}\mathbf{B})$, we use $\hat{\psi} = Tr(\mathbf{M}\hat{\mathbf{B}}_o) = Tr(\mathbf{G}\hat{\mathbf{B}})$. A test of $\psi = \mathbf{0}$ is rejected if $\hat{\psi} / \hat{\sigma}_{\hat{\psi}} > c_o$ where c_o is the critical value of the root and trace tests. To construct an approximate $1-\alpha$

simultaneous confidence interval, we utilize the F–approximation discussed in the *SAS/STAT User's Guide* (1990, p. 18). For Roy's root test, $\lambda_1 / (1+\lambda_1) = \theta$ so that $\theta / (1-\theta) = \lambda_1$. Letting $s = \min(v_h, u)$, $M = (|v_h - u| - 1)/2$, $N = (v_e - u - 1)/2$ where $v_h = \text{rank}(C_o)$ and $u = \text{rank}(A_o)$, we find that the critical value is

$$c_o^2 \doteq \frac{r}{v_e - r + v_h} F^{1-\alpha}_{(r,\, v_e - r + v_h)} \tag{5.140}$$

where $r = \max(v_h, u)$ and

$$\hat{\sigma}_{\hat{\psi}} = \sum_{i=1}^{s} \lambda_i^{1/2}(\mathbf{MW}_o\mathbf{M'E}_o)$$

Alternatively, using the Bartlett-Lawley-Hotelling trace criterion, also called T_o^2, we find that the critical value c_o is

$$c_o^2 \doteq \frac{s^2(2M+s+1)}{2(sN+1)} F^{1-\alpha}_{(s(2M+s+1),2(sN+1))} \tag{5.141}$$

and $\hat{\sigma}_{\hat{\psi}} = [Tr(\mathbf{MW}_o\mathbf{M'E}_o)]^{1/2}$. Thus, using (5.91) and (5.93) we may establish approximate $1-\alpha$ simultaneous confidence intervals for $\psi = Tr(\mathbf{MB}_o) = Tr(\mathbf{MC}_o\mathbf{BA}_o) = Tr(\mathbf{A}_o\mathbf{MC}_o\mathbf{B}) = Tr(\mathbf{GB})$ for arbitrary matrices $\mathbf{M}$ as given in (5.91) and (5.93). One may also use the approximations for c_o to construct $1-\alpha$ simultaneous confidence intervals for bilinear forms $\psi = \mathbf{c'Ba}$.

While Krishnaiah et al. (1980) have developed the "Roots" program to test extended linear hypotheses involving a single matrix **M**, we illustrate the procedure using PROC IML and data from Timm (1975, p. 454) involving three groups and three repeated measurements. The SAS code for the analysis is provided in Program 5_11.sas.

Program 5_11.sas

```
/* Program 5_11.sas                  */
/* Data from Timm(1975, page 454) */
/* Extended Linear Hypotheses        */

options ls=80 ps=60 nodate nonumber;
title1 'Output 5.11: Extended Linear Hypotheses';

data timm;
   infile 'c:\5_11.dat';
   input group y1 y2 y3 x1 x2 x3;
proc print data=timm;
run;

proc glm data=timm;
   title2 'Multivariate Test of Group--Using PROC GLM';
   class group;
   model y1-y3 = group/nouni;
   means group;
   manova h=group/printh printe;
run;

/* IML procedure for Extended Linear Hypothesis */
title2 'Multivariate Test of Group--Using PROC IML';
proc iml;
   use timm;
   a={x1 x2 x3};
   b={y1 y2 y3};
   read all var a into x;
```

```
read all var b into y;
beta=inv(x`*x)*x`*y;
print beta;
n=nrow(y);
p=ncol(y);
k=ncol(x);
nu_h=2; u=3; nu_e=n-k; s0=min(nu_h,u);
r=max(nu_h,u); alpha=.05;
denr=(nu_e-r+nu_h);
roy_2=(r/denr)*finv(1-alpha,r,denr);
rvalue=sqrt(roy_2);
m0=(abs(nu_h-u)-1)/2; n0=(nu_e-u-1)/2;
num=s0**2*(2*m0+s0+1); dent=2*(s0*n0+1);df=s0*(2*m0+s0+1);
t0_2=(num/dent)*finv(1-alpha,df,dent);
tovalue=sqrt(t0_2);
print s0 m0 n0;
e=(y`*y-y`*x*beta);
co={1 -1 0, 0 1 -1};
ao=i(3);
eo=ao`*e*ao;
bo=co*beta*ao;
wo=co*inv(x`*x)*co`;
ho=bo`*inv(wo)*bo;
print,"Overall Error Matrix", eo ,
      "Overall Hypothesis Test Matrix",ho;

/* c`c=eo where c is upper triangle Cholesky matrix */
c=root(eo);
f=inv(c`)*ho*inv(c);
eig=Eigval(round(f,.0001));
vec=inv(c)*eigvec(round(f,.0001));
print,"Eigenvalues & Eigenvectors of Overall Test of Ho (Groups)",
   eig vec;

/* Extended Linear Hypothesis following Overall Group Test */
m={1 .5, -.5 .5, -.5 -1};
g=ao*m*co;
print, "Extended Linear Hypothesis Test Matrix",m g;
psi=m*bo;
psi_hat=trace(psi);
tr_psi=abs(psi_hat);
h=m*wo*m`;
einv=inv(eo);
c=root(einv);
f=inv(c`)*h*inv(c);
xeig=Eigval(round(f,.0001));
print, "Eigenvalues of Extended Linear Hypothesis", xeig;
denrt=sum(sqrt(xeig));
dentr=sqrt(sum(xeig));
to_2=tr_psi/dent; print, "Extended To**2 Statistic", to_2;
print, "Extended To**2 Critical Value", tovalue;
root=tr_psi/denr; print, "Extended Largest Root Statistic", root;
print, "Extended Largest Root Critical Value", rvalue;
print psi_hat alpha;
ru=psi_hat+rvalue*denrt;
rl=psi_hat-rvalue*denrt;
vu=psi_hat+tovalue*dentr;
vl=psi_hat-tovalue*dentr;
print 'Approximate Simultaneous Confidence Intervals';
print 'Contrast Significant if interval does not contain zero';
print 'Extended Root interval:  ('rl ',' ru ')';
print 'Extended Trace interval: ('vl ',' vu ')';
```

```
   /* Multiple Extended Linear Hypothesis using To**2  */
   m1={1 0,0 0,0 0}; m2={0 0,1 0,0 0}; m3={0 0,0 1,0 0}; m4={0 0,0 0,0 1};
   print,"Multiple Extended Linear Hypothesis Test Matrices", m1,m2,m3,m4;

   g1=ao*m1*co; g2=ao*m2*co; g3=ao*m3*co; g4=ao*m4*co;
   t1=trace(m1*bo); t2=trace(m2*bo); t3=trace(m3*bo); t4=trace(m4*bo);
   tau=t1//t2//t3//t4;
   t11=trace(m1*wo*m1`*eo); t21=trace(m2*wo*m1`*eo);
   t22=trace(m2*wo*m2`*eo); t31=trace(m3*wo*m1`*eo);
   t32=trace(m3*wo*m2`*eo); t33=trace(m3*wo*m3`*eo);
   t41=trace(m4*wo*m1`*eo); t42=trace(m4*wo*m2`*eo);
   t43=trace(m4*wo*m3`*eo); t44=trace(m4*wo*m4`*eo);
   r1=t11||t21||t31||t41; r2=t21||t22||t32||t42;
   r3=t31||t32||t33||t43; r4=t41||t42||t43||t44;
   t=r1//r2//r3//r4;
   print tau,t;
   to_4=tau`*inv(t)*tau;
   print, "Extended Linear Hypothesis Criterion To**2 Squared", to_4;
   print, "Extended To**2 Critical Value", t0_2;
quit;

/* Multivariate test of Parallelism */
proc glm data=timm;
title2 'Multivariate Test of Parallelism--Using PROC GLM';
   class group;
   model y1-y3 = group/nouni;
   manova h = group m = ( 1 -1  0,
                          0  1 -1) prefix = diff/ printe printh;
run;

title2 'Multivariate Test of Parallelism--Using PROC IML';
proc iml;
   use timm;
   a={x1 x2 x3};
   b={y1 y2 y3};
   read all var a into x;
   read all var b into y;
   beta=inv(x`*x)*x`*y;
   n=nrow(y);
   p=ncol(y);
   k=ncol(x);
   nu_h=2; u=2; nu_e=n-k; s0=min(nu_h,u); r=max(nu_h,u); alpha=.05;
   denr=(nu_e-r+nu_h);
   roy_2=(r/denr)*finv(1-alpha,r,denr);
   rvalue=sqrt(roy_2);
   m0=(abs(nu_h-u)-1)/2; n0=(nu_e-u-1)/2;
   num=s0**2*(2*m0+s0+1); dent=2*(s0*n0+1); df=s0*(2*m0+s0+1);
   t0_2=(num/dent)*finv(1-alpha,df,dent);
   tovalue=sqrt(t0_2);
   print s0 m0 n0;
   e=(y`*y-y`*x*beta);
   co={1 -1 0, 0 1 -1};
   ao={1 0, -1 1, 0 -1};
   eo=ao`*e*ao;
   bo=co*beta *ao;
   wo=co*inv(x`*x)*co`;
   ho=bo`*inv(wo)*bo;
   c=root(eo);
   f=inv(c`)*ho*inv(c);
   eig=eigval(round(f,.0001));
   vec=inv(c)*eigvec(round(f,.0001));
   print,"Eigenvalues & Eigenvectors of Overall test of Ho (Parallelism)",
      eig vec;
```

```
   /* Extended Linear Hypothesis following overall Parallelism test */
   m={0 1,1 0};
   g=ao*m*co;
   print, "Extended Linear Hypothesis Test Matrix", m g;

   psi=m*bo;
   psi_hat=trace(psi);
   tr_psi=abs(psi_hat);
   h=m*wo*m`;
   einv=inv(eo);
   c=root(einv);
   f=inv(c`)*h*inv(c);
   xeig=eigval(round(f,.0001));
   print, "Eigenvalues of Extended Linear Hypothesis", xeig;

   to_2=tr_psi/sqrt(sum(xeig)); print, "Extended To**2 Statistic", to_2;
   print,"Extended To**2 Critical Value", tovalue;
   root=tr_psi/sum(sqrt(xeig)); print, "Extended Largest Root Statistic", root;
   print, "Extended Largest Root Critical Value", rvalue;

   print psi_hat alpha;
   ru=psi_hat+rvalue*sum(sqrt(xeig));
   rl=psi_hat-rvalue*sum(sqrt(xeig));
   vu=psi_hat+tovalue*sqrt(sum(xeig));
   vl=psi_hat-tovalue*sqrt(sum(xeig));
   print 'Approximate Simultaneous Confidence Intervals';
   print 'Contrast Significant if interval does not contain zero';
   print 'Extended Root  Interval: ('rl ',' ru ')';
   print 'Extended Trace Interval: ('vl '.' vu ')';
quit;

/* Multivariate test of Conditions as vectors   */
proc glm data=timm;
title2 'Multivariate Test of Conditions--Using PROC GLM';
   class group;
   model y1-y3 = group/noint nouni;
   contrast 'Mult Cond' group 1 0 0,
                        group 0 1 0,
                        group 0 0 1;
   manova m=(1 -1  0,
             0  1 -1) prefix = diff/ printe printh;
run;

title2 'Multivariate Test of Conditions--Using PROC IML';
proc iml;
   use timm;
   a={x1 x2 x3};
   b={y1 y2 y3};
   read all var a into x;
   read all var b into y;
   beta=inv(x`*x)*x`*y;
   n=nrow(y);
   p=ncol(y);
   k=ncol(x);
   nu_h=3; u=2; nu_e=n-k; s0=min(nu_h,u); r=max(nu_h,u); alpha=.05;
   denr=(nu_e-r+nu_h);
   roy_2=(r/denr)*finv(1-alpha,r,denr);
   rvalue=sqrt(roy_2);
   m0=(abs(nu_h-u)-1)/2; n0=(nu_e-u-1)/2;
   num=s0**2*(2*m0+s0+1); dent=2*(s0*n0+1); df=s0*(2*m0+s0+1);
   t0_2=(num/dent)*finv(1-alpha,df,dent);
   tovalue=sqrt(t0_2);
   print s0 m0 n0;
   e=(y`*y-y`*x*beta);
```

```
    co=i(3);
    ao={1 0, -1 1, 0 -1};
    eo=ao`*e*ao;
    bo=co*beta*ao;
    wo=co*inv(x`*x)*co`;
    ho=bo`*inv(wo)*bo;
    c=root(eo);
    f=inv(c`)*ho*inv(c);
    eig=eigval(round(f,.0001));
    vec=inv(c)*eigvec(round(f,.0001));
    print, "Eigenvalues & Eigenvectors of Overall test of Ho (Conditions)",
       eig vec;

    m={1 0 1, 0 1 1};
    g=ao*m*co;
    print, "Extended Linear Hypothesis Test Matrix", m g;

    psi=m*bo;
    psi_hat=trace(psi);
    tr_psi=abs(psi_hat);
    h=m*wo*m`;
    einv=inv(eo);
    c=root(eo);
    f=inv(c`)*h*inv(c);
    xeig=eigval(round(f,.0001));
    print, "Eigenvalues of Extended Linear Hypothesis", xeig;

    to_2=tr_psi/sqrt(sum(xeig)); print, "Extended To**2 Statistic", to_2;
    print, "Extended T0**2 Critical Value", tovalue;
    root= tr_psi/sum(sqrt(xeig)); print, "Extended Largest Root Statistic",
       root;
    print, "Extended Largest Root Critical Value", rvalue;

    print psi_hat alpha;
    ru=psi_hat+rvalue*sum(sqrt(xeig));
    rl=psi_hat-rvalue*sum(sqrt(xeig));
    vu=psi_hat+tovalue*sqrt(sum(xeig));
    vl=psi_hat-tovalue*sqrt(sum(xeig));
    print 'Approximate Simultaneous Confidence Intervals';
    print 'Contrast Significant if interval does not contain zero';
    print 'Extended Root  Interval: ('rl ',' ru ')';
    print 'Extended Trace Interval: ('vl ',' vu ')';
quit;
```

In Program 5_11.sas the test for a group effect, for parallelism, and for conditions are all run both with PROC GLM and with PROC IML to verify the calculation of eigenvalues and eigenvectors for each overall hypothesis. PROC IML is used to test the significance of several extended linear hypotheses, both simple (those involving a single matrix $\mathbf{M}$) and complex hypotheses (hypotheses that use several $\mathbf{M}_i$).

Result and Interpretation 5_11.sas

First PROC IML is used to test the overall hypothesis of group differences. Finding the difference to be significant, suppose we tested the extended linear hypothesis $\psi = \mathbf{0}$ where ψ is given in (5.136). PROC IML is used to verify the calculations of the overall test and estimate

$$\begin{aligned}\hat{\psi} &= Tr(\mathbf{M}\hat{\mathbf{B}}_o) = Tr(\mathbf{MC}_o\mathbf{BA}_o) = Tr(\mathbf{A}_o\mathbf{MC}_o\mathbf{B}) \\ &= Tr(\mathbf{G}\hat{\mathbf{B}}) = -2.5\end{aligned}$$

with the matrices **M**, **G** and $\hat{\mathbf{B}}$ defined

$$\mathbf{M} = \begin{pmatrix} 1 & .5 \\ -.5 & .5 \\ -.5 & 1 \end{pmatrix}, \quad \mathbf{G} = \begin{pmatrix} 1 & -.5 & -.5 \\ -.5 & 1 & -.5 \\ -.5 & -.5 & 1 \end{pmatrix} \quad \text{and} \quad \hat{\mathbf{B}} = \begin{pmatrix} 4 & 7 & 10 \\ 7 & 8 & 11 \\ 5 & 7 & 9 \end{pmatrix}.$$

Solving $\left|\mathbf{H} - \lambda \mathbf{E}_o^{-1}\right| = 0$ with $\mathbf{H} = \mathbf{MW}_o\mathbf{M}'$, $\mathbf{W}_o = \mathbf{C}_o(\mathbf{X}'\mathbf{X})^{-1}\mathbf{C}_o'$ and $\mathbf{E}_o = \mathbf{A}_o'\mathbf{EA}_o = \mathbf{A}_o'\mathbf{Y}'(\mathbf{I} - \mathbf{X}(\mathbf{X}'\mathbf{X})^{-1}\mathbf{X}')\mathbf{YA}_o$, we solve the eigenequation using the EIGVAL function for a symmetric matrix. The last eigenvalue is essentially zero and, $\lambda_1 = 6.881$ and $\lambda_2 = 2.119$. By (5.87), let

$$\sigma_{Trace} = \{Tr(\mathbf{MW}_o\mathbf{M}'\mathbf{E}_o)\}^{1/2} = (\sum_i \lambda_i)^{1/2}$$

$$\sigma_{Root} = \sum_{i=1}^{s} \lambda_i^{1/2}$$

for the trace and root criterion. Since $|\hat{\psi}| = 2.5$, the extended trace and root statistics for testing $\psi = 0$ are

$$|\hat{\psi}| / \sigma_{Trace} = 0.8333$$

$$|\hat{\psi}| / \sigma_{Root} = 0.6123$$

Evaluating (5.140) and (5.141), we find that approximate critical values for the criteria are 1.3312 and 0.9891, respectively for the trace and root criteria using $\alpha = 0.05$. Thus, we fail to reject the null hypothesis $H\colon \psi = 0$. Approximate confidence intervals may also be calculated for ψ as illustrated in Program 5_11. The PROC IML output for the extended linear hypothesis test of Group differences and confidence intervals follows.

Output 5.11 *(continued)*

```
              Output 5.11: Extended Linear Hypotheses
             Multivariate Test of Group--Using PROC IML

                          BETA
                             4            7           10
                             7            8           11
                             5            7            9

                            S0           M0           N0
                             2            0            4

                          Overall Error Matrix

                            EO
                            74           65           35
                            65           68           38
                            35           38           26

                     Overall Hypothesis Test Matrix

                            HO
                     23.333333    8.3333333           10
                     8.3333333    3.3333333            5
                            10            5           10
```

Output 5.11 (*continued*)

```
          Eigenvalues & Eigenvectors of Overall Test of Ho (Groups)

                       EIG       VEC
                 1.6707755 0.1982804 -0.212929 -0.047984
                 0.5810043 -0.384715 0.0522413 0.1920039
                 0.0000202 0.3854583 0.2563231 -0.048027

                  Extended Linear Hypothesis Test Matrix

                    M                   G
                    1       0.5         1        -0.5       -0.5
                 -0.5       0.5      -0.5           1       -0.5
                 -0.5        -1      -0.5        -0.5          1

                Eigenvalues of Extended Linear Hypothesis

                                     XEIG
                                6.8811895
                                2.1187947
                                0.0000158

                       Extended To**2 Statistic

                                     TO_2
                                0.1388889

                     Extended To**2 Critical Value


                 Output 5.11: Extended Linear Hypotheses
                Multivariate Test of Group--Using PROC IML

                                  TOVALUE
                                1.3319921

                    Extended Largest Root Statistic

                                     ROOT
                                0.2272727

                   Extended Largest Root Critical Value

                                   RVALUE
                                0.9891365

                              PSI_HAT     ALPHA
                                 -2.5      0.05

              Approximate Simultaneous Confidence Intervals

          Contrast Significant if interval does not contain zero


                                                 RL          RU
           Extended Root interval:  ( -6.538431 , 1.5384307 )

                                                 VL          VU
           Extended Trace interval: ( -6.495976 , 1.4959764 )
```

Next, we illustrate testing (5.137) using PROC IML by evaluating the supremum in (5.88) as outlined in (5.89). With $\lambda_i = Tr(\mathbf{M}_i'\hat{\mathbf{B}}_o)$ and $t_{ij} = Tr(\mathbf{M}_i\mathbf{W}_o\mathbf{M}_j'\mathbf{E}_o)$, we have in Output 5.11 that

$$\lambda' = (-3, -1, 1, 2)$$

$$\mathbf{T} = \begin{pmatrix} 29.6 & 26 & -13 & -7 \\ 26 & 27.2 & -13.6 & -7.6 \\ -13 & -13.6 & 27.2 & 15.2 \\ 7 & -7.6 & 15.2 & 10.4 \end{pmatrix}$$

so that $\lambda'\mathbf{T}^{-1}\lambda = 2.2081151$. Comparing this with the T_o^2 critical value $(1.3319921)^2 = 1.774203$, we see that the test is significant.

Output 5.11 (*continued*)

```
                Output 5.11: Extended Linear Hypotheses
               Multivariate Test of Group--Using PROC IML

         Multiple Extended Linear Hypothesis Test Matrices

                              M1
                               1           0
                               0           0
                               0           0

                              M2
                               0           0
                               1           0
                               0           0

                              M3
                               0           0
                               0           1
                               0           0

                              M4
                               0           0
                               0           0
                               0           1

                                   TAU
                                    -3
                                    -1
                                     1
                                     2

                    T
                 29.6         26        -13          -7
                   26       27.2      -13.6        -7.6
                  -13      -13.6       27.2        15.2
                   -7       -7.6       15.2        10.4

         Extended Linear Hypothesis Criterion To**2 Squared

                                  TO_4
                             2.2081151

                   Extended To**2 Critical Value

                                  TO_2
                              1.774203
```

Finally, Program 5_11.sas contains the code to test (5.138) and (5.139), following tests of parallelism and conditions, respectively, for the repeated measurement design. The PROC IML output follows:

Output 5.11 (*continued*)

```
              Output 5.11: Extended Linear Hypotheses
          Multivariate Test of Parallelism--Using PROC IML

                         S0         M0         N0
                          2       -0.5        4.5

     Eigenvalues & Eigenvectors of Overall test of Ho (Parallelism)

                        EIG        VEC
                     0.8333 0.2886751          0
                     0.1852         0  0.2357023


               Extended Linear Hypothesis Test Matrix

                 M                   G
                 0          1        0          1         -1
                 1          0        1         -2          1
                                    -1          1          0

             Eigenvalues of Extended Linear Hypothesis

                                  XEIG
                              9.174913
                              2.825087

                      Extended To**2 Statistic

                                  TO_2
                             0.2886751

                   Extended To**2 Critical Value

                               TOVALUE
                             1.0707159

                  Extended Largest Root Statistic

                                  ROOT
                             0.2123227

                Extended Largest Root Critical Value

                                RVALUE
                             0.8047043

                        PSI_HAT     ALPHA
                              1      0.05

           Approximate Simultaneous Confidence Intervals

       Contrast Significant if interval does not contain zero

                                                RL          RU
        Extended Root   Interval: ( -2.790005 , 4.7900054 )
```

Output 5.11 (*continued*)

```
                                    VL          VU
        Extended Trace Interval: ( -2.709069 . 4.7090687 )

         Multivariate Test of Conditions--Using PROC IML

                          S0         M0        N0
                           2          0       4.5


  Eigenvalues & Eigenvectors of Overall test of Ho (Conditions)

                         EIG        VEC
                   11.417272 0.2015038 0.2067113
                   0.5271281 0.1687791 -0.164527

              Extended Linear Hypothesis Test Matrix

           M                             G
           1         0         1         1         0         1
           0         1         1        -1         1         0
                                         0        -1        -1

            Eigenvalues of Extended Linear Hypothesis

                              XEIG
                         0.0424389
                         0.0130611

                    Extended To**2 Statistic

                              TO_2
                         42.447636

                 Extended T0**2 Critical Value

                           TOVALUE
                          1.248754

                Extended Largest Root Statistic

                              ROOT
                         31.221482

              Extended Largest Root Critical Value

                            RVALUE
                         0.9341165

                      PSI_HAT    ALPHA
                          -10     0.05

          Approximate Simultaneous Confidence Intervals

      Contrast Significant if interval does not contain zero

         Multivariate Test of Conditions--Using PROC IML

                                    RL          RU
        Extended Root  Interval: ( -10.29919 ,  -9.70081 )

                                    VL          VU
        Extended Trace Interval: ( -10.29419 , -9.705813 )
```

5.12 Power Analysis

The last topic to be illustrated in this section is the approximate calculation of the power of a fixed effects hypothesis associated with a MANOVA design recommended by Muller et al. (1992). Using noncentral F approximations to the noncentral Wishart distribution, Muller et al. have written a program named POWER.PRO to calculate power (see Appendix 1, "POWER.PRO Program").

To compute the approximate power of the multivariate GLM, one must specify the following population matrices

$$\mathbf{X}, \mathbf{B}, \mathbf{C}, \mathbf{A}, \mathbf{\Gamma} \text{ and } \mathbf{\Sigma},$$

and the size of the test, α. To illustrate the setup for the program, we use data from Timm (1975, p. 264) for a simple two-group MANOVA design. The technique allows for the power analysis of any fixed effects ANOVA/MANOVA model and repeated measures mixed models. The associated theory for the procedure is provided in Muller et al. (1992). In general, to perform a power analysis for a multivariate design requires substantial calculations since one wants to vary $\mathbf{\Sigma}$ and the multivariate hypotheses. The example considered here is very straightforward; we test that

$$H_o: \quad \mu_1 - \mu_2 = \begin{pmatrix} 20.38 \\ -38.33 \end{pmatrix}$$

$$\mathbf{\Sigma} = \begin{pmatrix} 307.08 & \\ 280.83 & 421.67 \end{pmatrix}$$

and $\alpha = 0.01$ with a sample size of $n = 10$.

Program 5_12.sas

```
/* Program 5_12.sas                                                */
/* Data are from Timm(1975, p.264)                                 */
/* The SAS routine power.pro was supplied by Muller, La Vange      */
/* Ramey, and Ramey (1992) as shown in their JASA (1992) article   */
/* "Power calculations for general linear models including repeated */
/* measures applications", pp 1209-1226                            */

options ls=80 ps=60 nodate nonumber;
title 'Output 5.12: Power Analysis for MANOVA design';

proc iml symsize=1000;
   %include 'c:\power.pro';
   alpha = .01;
   c={1 -1};
   u=i(2);
   beta={20.38 -38.33, 0 0};
   sigma={307.08 280.83, 280.83 421.67};
   essencex={1 0, 1 0, 1 0, 1 0, 1 0,
             0 1, 0 1, 0 1, 0 1, 0 1};
   run power;
quit;
```

Result and Interpretation 5_12.sas

Output 5.12: *Power Analysis for MANOVA design*

```
                    Output 5.12: Power Analysis for MANOVA design

                                   BETA
                                  20.38     -38.33
                                      0          0

                                  SIGMA
                                 307.08     280.83
                                 280.83     421.67

                                  _RHO_
                                      1  0.7804254
                              0.7804254          1

                                      C
                                      1         -1

                                      U
                                      1          0
                                      0          1

                                         CBETAU
                         C*BETA*U =       20.38     -38.33


                    Output 5.12: Power Analysis for MANOVA design

_HOLDPOW       CASE      ALPHA    SIGSCAL    RHOSCAL   BETASCAL    TOTAL_N    EPSILON
:          GG_EXEPS     GG_PWR      POWER

                  1       0.01          1          1          1         10      0.618
:             0.626      0.376      0.977
```

The power for the example is calculated as $1-\beta = 0.977$, which is found in the column labeled `POWER` in Output 5.12. For this example, the power calculation is exact. Numerous examples of the program power.pro are illustrated in O'Brien and Muller (1993).

6. Double Multivariate Linear Models

Theory

Applications

6.1 Introduction

In this chapter, we extend the multivariate general linear model to the double multivariate linear model (DMLM) which allows one to analyze vector-valued repeated measurements or panel data. Estimation and hypothesis testing theory for the classical and the responsewise forms of the model are presented. The double multivariate linear mixed model is also reviewed. Finally, data are analyzed using both the DMLM and mixed model approaches.

6.2 Classical Model Development

The multivariate general linear model is extended to a double multivariate linear model by obtaining p-variates on n subjects at each of t occasions. If we let the number of variables $p = 1$, the repeated measurement design over t occasions results. In this chapter, we assume that all vector observations are complete so that the matrices of responses $\mathbf{R}_i$ $(t \times p)$ for $i = 1, \dots, n$ have no missing values. Missing data problems are considered in Chapter 10.

The classical double multivariate linear model (CDMLM) for the p-variate repeated measurements over t occasions is

$$\mathbf{Y}_{n\times pt} = \mathbf{X}_{n\times k}\mathbf{B}_{k\times pt} + \mathbf{U}_{n\times pt} \tag{6.1}$$

where $\mathbf{Y}_{n\times pt}$ is the data matrix for a sample of n subjects with $pt \le n-k$ responses, $\mathbf{X}_{n\times k}$ is the between-subjects design matrix of full rank k, $\mathbf{B}_{k\times pt}$ is a matrix of fixed effects, and $\mathbf{U}_{n\times pt}$ is a matrix of random errors. For hypothesis testing we further assume that the n rows of $\mathbf{Y}$ are distributed independently multivariate normal

$$\mathbf{y}_i \sim N(\mathbf{0}, \mathbf{\Sigma}_i) \tag{6.2}$$

Assuming homogeneity of $\mathbf{\Sigma}_i$ so that $\mathbf{\Sigma}_i = \mathbf{\Sigma}$ for all $i = 1, 2, \dots, n$, we write (6.1) as

$$\begin{aligned} E(\mathbf{Y}) &= \mathbf{XB} \\ \text{cov}(\mathbf{Y}) &= \mathbf{I}_n \otimes \mathbf{\Sigma}_{pt\times pt} \end{aligned} \tag{6.3}$$

Letting $\mathbf{u}_i$ denote the columns of $\mathbf{U}$ and $\mathbf{\Sigma}_i = \mathbf{\Sigma}$, we have $\mathbf{u}_i \sim IN(\mathbf{0}, \mathbf{\Sigma})$ or

$$\text{vec}(\mathbf{U}) \sim N_{npt}(\mathbf{0}, \mathbf{\Sigma} \otimes \mathbf{I}_n) \tag{6.4}$$

We can then write (6.1) as

$$\text{vec}(\mathbf{Y}) = (\mathbf{I}_{pt} \otimes \mathbf{X})\ \text{vec}(\mathbf{B}) + \text{vec}(\mathbf{U}) \tag{6.5}$$

which is again a general linear model with unknown $\mathbf{\Sigma}$. The model for the ith subject is

$$\begin{aligned} \mathbf{y}_i &= (\mathbf{I}_{pt} \otimes \mathbf{x}_i')\text{vec}(\mathbf{B}) + \mathbf{u}_i \\ \mathbf{u}_i &\sim IN(\mathbf{0}, \mathbf{\Sigma}) \end{aligned} \tag{6.6}$$

Alternatively, the ith row of $\mathbf{Y}$ is related to the ith response matrix $\mathbf{R}_i$ $(t \times p)$:

$$\underset{pt \times 1}{\mathbf{y}_i} = \text{vec}(\mathbf{R}_i) = (\mathbf{I}_{pt} \otimes \mathbf{x}_i')\ \text{vec}(\mathbf{B}) + \mathbf{u}_i \tag{6.7}$$

To analyze contrasts among the t time periods, we use a postmatrix $\mathbf{P}$ of the form

$$\mathbf{P} = (\mathbf{I}_p \otimes \mathbf{A}) \tag{6.8}$$

where the matrix $\mathbf{A}$ again satisfies the semiorthogonal condition that $\mathbf{A}'\mathbf{A} = \mathbf{I}_u$, and the $\text{rank}(\mathbf{A}) = u$. Using $\mathbf{P}$, (6.1) has the reduced form

$$\mathbf{Z} = \mathbf{X}\mathbf{\Theta} + \xi \tag{6.9}$$

where $\mathbf{Z} = \mathbf{YP}$, $\mathbf{\Theta} = \mathbf{BP}$, $\xi = \mathbf{UP}$, and

$$\text{vec}(\xi) \sim N(\mathbf{0}, \tilde{\mathbf{\Phi}} \otimes \mathbf{I}_n) \tag{6.10}$$

where

$$\tilde{\mathbf{\Phi}} = (\mathbf{I}_p \otimes \mathbf{A})'\mathbf{\Sigma}(\mathbf{I}_p \otimes \mathbf{A}) \tag{6.11}$$

This is a GLM with $\mathbf{\Omega} = \tilde{\mathbf{\Phi}} \otimes \mathbf{I}_n$. The matrix $\tilde{\mathbf{\Phi}}$ may have Kronecker or direct product structure, $\tilde{\mathbf{\Phi}} = \mathbf{\Sigma}_e \otimes \mathbf{\Sigma}_t$ (Galecki, 1994).

The primary objective of the analysis is to test hypotheses about the coefficients of $\mathbf{B}$. A convenient form of the hypothesis H is

$$H{:}\, \mathbf{CB}(\mathbf{I}_p \otimes \mathbf{A}) = \mathbf{\Gamma}_o \quad \text{or} \quad H{:}\, \mathbf{\Gamma} = \mathbf{\Gamma}_o \tag{6.12}$$

where $\mathbf{C}_{v_h \times k}$ has rank v_h, and $\mathbf{A}_{t \times u}$ has rank u. Without loss of generality, we may assume $\mathbf{A}'\mathbf{A} = \mathbf{I}_u$. Assuming a CDMLM, the BLUE estimator of $\mathbf{\Gamma}$ is

$$\hat{\mathbf{\Gamma}} = \mathbf{C}\hat{\mathbf{B}}(\mathbf{I}_p \otimes \mathbf{A}) \text{ where } \hat{\mathbf{B}} = (\mathbf{X}'\mathbf{X})^{-1}\mathbf{X}'\mathbf{Y} \tag{6.13}$$

$\hat{\mathbf{\Gamma}}$ is also the ML estimate of $\mathbf{\Gamma}$. The estimator $\hat{\mathbf{\Gamma}}$ is the minimum variance unbiased estimator of $\mathbf{\Gamma}$ and does not depend on $\mathbf{\Sigma}$; that is, the maximum likelihood estimator is invariant to the assumed structure of $\mathbf{\Sigma}$. Under normality,

$$\hat{\gamma} = \text{vec}(\hat{\mathbf{\Gamma}}) \sim N[\text{vec}(\mathbf{\Gamma}), \tilde{\mathbf{\Phi}} \otimes \mathbf{C}(\mathbf{X}'\mathbf{X})^{-1}\mathbf{C}'] \tag{6.14}$$

where $\tilde{\mathbf{\Phi}} = (\mathbf{I}_p \otimes \mathbf{A}')\mathbf{\Sigma}(\mathbf{I}_p \otimes \mathbf{A})$. Thus, the distribution of $\hat{\mathbf{\Gamma}}$ depends on $\mathbf{\Sigma}$ so that inferences about $\mathbf{\Gamma}$ depend on the structure of $\mathbf{\Sigma}$. Again if $p = 1$, the standard multivariate model results, or

$$H{:}\, \mathbf{CBA} = \mathbf{\Gamma}_o \text{ or } H{:}\, \mathbf{\Gamma} = \mathbf{\Gamma}_o \tag{6.15}$$

where $\hat{\mathbf{\Gamma}} = \mathbf{C}\hat{\mathbf{B}}\mathbf{A}$, $\hat{\mathbf{B}} = (\mathbf{X}'\mathbf{X})^{-1}\mathbf{X}'\mathbf{Y}$, and $\mathbf{\Phi} = \mathbf{A}'\mathbf{\Sigma}\mathbf{A}$. Letting $\psi = \mathbf{c}'\mathbf{B}\mathbf{a}$, the $\text{var}(\hat{\psi}) = \mathbf{a}'\mathbf{\Sigma}\mathbf{a}[\mathbf{c}'(\mathbf{X}'\mathbf{X})^{-1}\mathbf{c}]$.

To test hypotheses about $\boldsymbol{\Gamma}$ in (6.12), the test statistics again depend on the characteristic roots λ_i of $|\mathbf{H} - \lambda\mathbf{E}| = 0$ where

$$\begin{aligned} \mathbf{E} &= (\mathbf{I}_p \otimes \mathbf{A})'\mathbf{Y}'(\mathbf{I}_n - \mathbf{X}(\mathbf{X}'\mathbf{X})^{-1}\mathbf{X}']\mathbf{Y}(\mathbf{I}_p \otimes \mathbf{A}) \\ \mathbf{H} &= (\hat{\boldsymbol{\Gamma}} - \boldsymbol{\Gamma}_o)'[\mathbf{C}(\mathbf{X}'\mathbf{X})^{-1}\mathbf{C}']^{-1}(\hat{\boldsymbol{\Gamma}} - \boldsymbol{\Gamma}_o) \end{aligned} \tag{6.16}$$

The matrices $\mathbf{E}$ and $\mathbf{H}$ have independent Wishart distributions

$$\mathbf{E} \sim W_{pu}(v_e, \tilde{\boldsymbol{\Phi}}, \mathbf{0}) \;\; \text{and} \;\; \mathbf{H} \sim W_{pu}(v_h, \tilde{\boldsymbol{\Phi}}, \tilde{\boldsymbol{\Phi}}^{-1}\boldsymbol{\Delta}) \tag{6.17}$$

where $v_e = n - k$ and the noncentrality matrix is

$$\boldsymbol{\Delta} = (\boldsymbol{\Gamma} - \boldsymbol{\Gamma}_o)'[\mathbf{C}(\mathbf{X}'\mathbf{X})^{-1}\mathbf{C}']^{-1}(\boldsymbol{\Gamma} - \boldsymbol{\Gamma}_o) \tag{6.18}$$

When $p = 1$, we have the following simplification

$$\begin{aligned} \mathbf{E} &= \mathbf{A}'\mathbf{Y}'[\mathbf{I}_n - \mathbf{X}(\mathbf{X}'\mathbf{X})^{-1}\mathbf{X}']\mathbf{Y}\mathbf{A} \\ \mathbf{H} &= (\mathbf{C}\hat{\mathbf{B}}\mathbf{A} - \boldsymbol{\Gamma}_o)'[\mathbf{C}(\mathbf{X}'\mathbf{X})^{-1}\mathbf{C}']^{-1}(\mathbf{C}\hat{\mathbf{B}}\mathbf{A} - \boldsymbol{\Gamma}_o) \end{aligned} \tag{6.19}$$

To test H in (6.12), we use several available test statistics that are functions of the roots of $|\mathbf{H} - \lambda\mathbf{E}| = 0$ for $\mathbf{H}$ and $\mathbf{E}$ defined in (6.16).

<u>Wilks' Λ:</u>

$$\Lambda = \frac{|\mathbf{E}|}{|\mathbf{E} + \mathbf{H}|} = \prod_{i=1}^{s} (1 + \lambda_i)^{-1} \sim U^{\alpha}_{g, v_h, v_e} \tag{6.20}$$

where $v_e = n - k$, $v_h = \text{rank}(\mathbf{C})$, and $g = \text{rank}(\mathbf{I}_p \otimes \mathbf{A}) = pu$.

<u>Roy's Largest Root:</u>

$$\theta_1 = \frac{\lambda_1}{1 + \lambda_1} \sim \theta^{\alpha}_{s,M,N} \;\; \text{where} \;\; s = \min(v_h, g), \;\; M = \frac{|g - v_h| - 1}{2}, \;\; N = \frac{v_e - g - 1}{2} \tag{6.21}$$

<u>Bartlett-Nanda-Pillai Trace:</u>

$$V = \text{Tr}[\mathbf{H}(\mathbf{H} + \mathbf{E})^{-1}] = \sum_{i=1}^{s} \frac{\lambda_i}{1 + \lambda_i} \sim V^{\alpha}_{s,M,N} \tag{6.22}$$

<u>Bartlett-Lawley-Hotelling Trace:</u>

$$T^2 = T_o^2 / v_e = \text{Tr}(\mathbf{H}\mathbf{E}^{-1}) / v_e = \sum_{i=1}^{s} \lambda_i / v_e \sim U^{\alpha}_{o(s,M,N)} \tag{6.23}$$

An alternative expression for T_o^2 is

$$T_o^2 = [\text{vec}(\hat{\boldsymbol{\Gamma}} - \boldsymbol{\Gamma}_o)]'\{\hat{\tilde{\boldsymbol{\Phi}}}^{-1} \otimes [\mathbf{C}(\mathbf{X}'\mathbf{X})^{-1}\mathbf{C}']^{-1}\}\text{vec}(\hat{\boldsymbol{\Gamma}} - \boldsymbol{\Gamma}_o)] \tag{6.24}$$

where

$$\hat{\tilde{\boldsymbol{\Phi}}} = \mathbf{E} / v_e = (\mathbf{I}_p \otimes \mathbf{A})'\mathbf{Y}'[\mathbf{I}_n \otimes \mathbf{X}(\mathbf{X}'\mathbf{X})^{-1}\mathbf{X}']\mathbf{Y}(\mathbf{I}_p \otimes \mathbf{A}) / v_e \tag{6.25}$$

6.3 Responsewise Model Development

Our presentation of the DMLM so far has followed the classical approach. In previous chapters we saw that when $p = 1$ there were problems in representing the model for mixed model analysis of the data. To reduce the rearrangement of data problems, we represent (6.1) using the transpose of the data matrix **Y**, following Boik (1988). The responsewise double multivariate linear model (RDMLM) is represented as

$$\mathbf{Y}_{pt\times n} = \mathbf{B}_{pt\times k}\,\mathbf{X}_{k\times n} + \mathbf{U}_{pt\times n} \tag{6.26}$$

which is no more than the transpose of the corresponding matrices in (6.1). Using representation (6.26) the $pt \times 1$ vector of random errors for the jth subject is again assumed to be normally distributed:

$$\mathbf{u}_j \sim IN(\mathbf{0}, \boldsymbol{\Sigma}) \tag{6.27}$$

Now, however,

$$\text{vec}(\mathbf{U}) \sim \mathbf{N}_{npt}(\mathbf{0}, \mathbf{I}_n \otimes \boldsymbol{\Sigma}) \tag{6.28}$$

which has a convenient block diagonal structure. Because the model has changed, the format for testing hypotheses must change and is the transpose of (6.12):

$$H\text{: } (\mathbf{A}' \otimes \mathbf{I}_p)\mathbf{B}\mathbf{C}' = \boldsymbol{\Gamma}_o \text{ or } H\text{: } \boldsymbol{\Gamma} = \boldsymbol{\Gamma}_o \tag{6.29}$$

To reduce (6.26) to perform a multivariate repeated measures analysis, we use the following reduced multivariate linear model

$$(\mathbf{A}' \otimes \mathbf{I}_p)\mathbf{Y} = (\mathbf{A}' \otimes \mathbf{I}_p)\mathbf{B}\mathbf{X} + (\mathbf{A}' \otimes \mathbf{I}_p)\mathbf{U} \tag{6.30}$$

However, because of the alternative model representation, the form of **E** and **H** also change:

$$\begin{aligned} \tilde{\mathbf{E}} &= (\mathbf{A} \otimes \mathbf{I}_p)'\mathbf{Y}(\mathbf{I}_n - \mathbf{X}'(\mathbf{X}\mathbf{X}')^{-1}\mathbf{X}]\mathbf{Y}'(\mathbf{A} \otimes \mathbf{I}_P) \\ \tilde{\mathbf{H}} &= (\hat{\boldsymbol{\Gamma}} - \boldsymbol{\Gamma}_o)[\mathbf{C}(\mathbf{X}\mathbf{X}')^{-1}\mathbf{C}']^{-1}(\hat{\boldsymbol{\Gamma}} - \boldsymbol{\Gamma}_o)' \end{aligned} \tag{6.31}$$

where

$$\hat{\boldsymbol{\Gamma}} = (\mathbf{A}' \otimes \mathbf{I}_p)\hat{\mathbf{B}}\mathbf{C}' \text{ and } \hat{\mathbf{B}}' = (\mathbf{X}\mathbf{X}')^{-1}\mathbf{X}\mathbf{Y}' \tag{6.32}$$

We again have that $\tilde{\mathbf{E}}$ and $\tilde{\mathbf{H}}$ follow Wishart distributions as defined by (6.17); however,

$$\tilde{\boldsymbol{\Phi}} \equiv \tilde{\boldsymbol{\Phi}}^* = (\mathbf{A}' \otimes \mathbf{I}_p)\boldsymbol{\Sigma}(\mathbf{A} \otimes \mathbf{I}_p) \tag{6.33}$$

and the noncentrality parameter becomes

$$\Delta \equiv \tilde{\Delta}^* = (\boldsymbol{\Gamma} - \boldsymbol{\Gamma}_o)(\mathbf{C}(\mathbf{X}\mathbf{X}')^{-1}\mathbf{C}')^{-1}(\boldsymbol{\Gamma} - \boldsymbol{\Gamma}_o)' \tag{6.34}$$

with $\boldsymbol{\Gamma}$ defined in (6.29).

6.4 The Multivariate Mixed Model

Returning to the classical form of the DMLM and $\mathbf{Y}_{n\times pt}$ as defined in (6.1), the ith row of **Y** is

$$\mathbf{y}_i = \text{vec}(\mathbf{R}_i) = (\mathbf{I}_{pt} \otimes \mathbf{x}_i')\text{vec}(\mathbf{B}) + \mathbf{u}_i \tag{6.35}$$

where $\mathbf{x}_i'$ and $\mathbf{u}_i'$ are the ith rows of **X** and **U,** respectively. One can construct a multivariate mixed model (MMM) by modeling $\mathbf{u}_i$:

$$\mathbf{u}_i = \mathbf{Z}_i\theta_i + \mathbf{e}_i \qquad i = 1, 2, \ldots, n \tag{6.36}$$

where $\mathbf{Z}_i$ is a $(pt \times h)$ within-subject design matrix for the random effects θ_i, the rank$(\mathbf{Z}_i) = h$, and the vectors θ_i and $\mathbf{e}_i$ are distributed independent multivariate normal:

$$\begin{aligned} \theta_i &\sim N_h(\mathbf{0}, \mathbf{D}_h) \\ \mathbf{e}_i &\sim N_{pt}[\mathbf{0}, \mathbf{\Sigma}_e \otimes \mathbf{I}_t] \end{aligned} \tag{6.37}$$

as discussed by Boik (1991). Hence,

$$\mathbf{y}_i \sim N[(\mathbf{I}_{pt} \otimes \mathbf{x}_i')\text{vec}(\mathbf{B}), \mathbf{\Sigma}_i] \tag{6.38}$$

where

$$\mathbf{\Sigma}_i = \mathbf{Z}_i \mathbf{D}_h \mathbf{Z}_i' + (\mathbf{\Sigma}_e \otimes \mathbf{I}_t) \qquad i = 1, 2, \ldots, n \tag{6.39}$$

The mixed model ML estimate of $\mathbf{\Gamma}$ is $\hat{\mathbf{\Gamma}} = \mathbf{C}\hat{\mathbf{B}}(\mathbf{I}_p \otimes \mathbf{A})$ and

$$\hat{\gamma} = \text{vec}(\hat{\mathbf{\Gamma}}) = [(\mathbf{I}_p \otimes \mathbf{A}) \otimes \mathbf{C}']'[\sum_{i=1}^{n} \hat{\mathbf{\Sigma}}_i^{-1} \otimes \mathbf{x}_i \mathbf{x}_i')]^{-1} \sum_{i=1}^{n} \hat{\mathbf{\Sigma}}_i^{-1} \otimes \mathbf{x}_i)\mathbf{y}_i \tag{6.40}$$

where $\hat{\mathbf{\Sigma}}_i = \mathbf{Z}_i \hat{\mathbf{D}}_h \mathbf{Z}_i' + (\hat{\mathbf{\Sigma}}_e \otimes \mathbf{I}_t)$, $\hat{\mathbf{D}}_h$ is the ML estimate of $\hat{\mathbf{D}}_h$ and $\hat{\mathbf{\Sigma}}_e$ is the ML estimate of $\mathbf{\Sigma}_e$. By the multivariate central limit theorem, the

$$\lim_{n \to \infty} n^{1/2} \text{vec}(\hat{\mathbf{\Gamma}} - \mathbf{\Gamma}) \sim N(\mathbf{0}, \mathbf{\Sigma}_{\hat{\Gamma}}) \tag{6.41}$$

where

$$\mathbf{\Sigma}_{\hat{\Gamma}} = \lim_{n \to \infty} [n(\mathbf{I}_p \otimes \mathbf{A}) \otimes \mathbf{C}']' \left[\sum_{i=1}^{n} (\mathbf{\Sigma}_i^{-1} \otimes \mathbf{x}_i \mathbf{x}_i') \right]^{-1} [(\mathbf{I}_p \otimes \mathbf{A}) \otimes \mathbf{C}'] \tag{6.42}$$

provided $\mathbf{\Sigma}_{\hat{\Gamma}}$ exists. Substituting the ML estimate $\hat{\mathbf{\Sigma}}_i = \mathbf{E}/n$ for $\mathbf{\Sigma}_i$, inference on $\mathbf{\Gamma}$ may be obtained using (6.41) in large samples since $(\hat{\gamma} - \gamma_o)' \hat{\mathbf{\Sigma}}_{\hat{\Gamma}}^{-1} (\hat{\gamma} - \gamma_o) \sim \chi_v^2$ and $v =$ rank$(\mathbf{C})$; the Wald statistic defined in (4.15) with $(\mathbf{X}'\hat{\mathbf{\Sigma}}^{-1}\mathbf{X})^{-1} \equiv \hat{\mathbf{\Sigma}}_{\hat{\Gamma}}$ may be used.

The p-variate generalization of Scheffé's mixed model under sphericity is obtained by associating $\mathbf{Z}_i$ in (6.39) with the matrix $\mathbf{I}_p \otimes \mathbf{1}_t$ so that $h = p$ and

$$\mathbf{\Sigma}_i = \mathbf{\Sigma} = (\mathbf{D}_p \otimes \mathbf{1}_t \mathbf{1}_t') + (\mathbf{\Sigma}_e \otimes \mathbf{I}_t) \qquad \text{for } i = 1, 2, \ldots, n \tag{6.43}$$

This is called the Scheffé MMM and satisfies the multivariate sphericity condition that

$$\tilde{\mathbf{\Phi}} = (\mathbf{\Sigma}_e \otimes \mathbf{I}_u) \tag{6.44}$$

as discussed by Reinsel (1982, 1984). It is a special case of (6.39).

Showing the equivalence of the Scheffé Multivariate Mixed Model (MMM) and the standard double multivariate model is again complicated because of the orientation of the data matrix $\mathbf{Y}_{n \times pt}$. To solve this problem, Boik (1988) formulates the Scheffé MMM by rearranging the elements of $\mathbf{Y}_{n \times pt}$ to $\mathbf{Y}_{pt \times n}$ as in (6.26). Using the RDMLM (6.26), the jth column $\mathbf{y}_j$ of $\mathbf{Y}_{pt \times 1}$ is the vector of responses for the jth subject. Boik shaped the vector of responses for the jth subject into a $(p \times t)$ matrix $\mathbf{Y}_j^*$ where the columns of $\mathbf{Y}_j^*$ are the t occasions and the rows are the p dependent variables. The rearranged matrix is represented as

$$\mathbf{Y}^* = (\mathbf{Y}_1^*, \mathbf{Y}_2^*, \ldots, \mathbf{Y}_n^*) \tag{6.45}$$

The $(pt \times 1)$ response vector for the jth subject is related

$$\mathbf{y}_j = \text{vec}(\mathbf{Y}_j^*) \tag{6.46}$$

which is a rearrangement of $\mathbf{R}_i$ $(t \times p)$ for the CDMLM. In a similar manner, the elements in $\mathbf{B}_{pt \times k}$ and $\mathbf{U}_{pt \times n}$ are rearranged so that

$$\mathbf{B}^* = (\mathbf{B}_1^*, \ldots, \mathbf{B}_k^*) \text{ and } \mathbf{U}^* = (\mathbf{U}_1^*, \ldots, \mathbf{U}_n^*) \tag{6.47}$$

where $\mathbf{B}_i^*$ and $\mathbf{U}_j^*$ are each $(p \times t)$ matrices, the ith column of $\mathbf{B}$ satisfies $\mathbf{B}_i = \text{vec}(\mathbf{B}_i^*)$ and the jth column of $\mathbf{U}$ satisfies $\mathbf{u}_j = \text{vec}(\mathbf{U}_j^*)$. Finally, $\text{vec}(\mathbf{Y}^*) = \text{vec}(\mathbf{Y})$, $\text{vec}(\mathbf{B}^*) = \text{vec}(\mathbf{B})$, and $\text{vec}(\mathbf{U}^*) = \text{vec}(\mathbf{U})$.

This rearrangement allows one to rewrite the linear model (6.26) and the reduced linear model (6.30):

$$\mathbf{Y}^* = \mathbf{B}^*(\mathbf{X} \otimes \mathbf{I}_t) + \mathbf{U}^* \tag{6.48}$$

$$\mathbf{Y}^*(\mathbf{I}_n \otimes \mathbf{A}) = \mathbf{B}^*(\mathbf{X} \otimes \mathbf{A}) + \mathbf{U}^*(\mathbf{I}_n \otimes \mathbf{A}) \tag{6.49}$$

Assuming $\tilde{\mathbf{\Phi}}^*$ in (6.33) satisfies the condition of multivariate sphericity

$$\tilde{\mathbf{\Phi}}^* = \mathbf{I}_u \otimes \mathbf{\Sigma}_e \tag{6.50}$$

an MMM analysis for testing

$$H\text{: } \mathbf{B}^*(\mathbf{C}' \otimes \mathbf{A}) = \mathbf{\Gamma}_o^* \quad \text{or} \quad H\text{: } \mathbf{\Gamma}^* = \mathbf{\Gamma}_o^* \tag{6.51}$$

is obtained by performing a multivariate analysis using the reduced model specified by (6.49).

To test (6.51), one uses the standard multivariate test statistics obtained from the reduced form of the model with the assumption that multivariate sphericity is satisfied. In testing (6.51), observe that the design matrix is separable, that it has the form $\mathbf{X} \otimes \mathbf{A}$, and that the hypothesis test matrix is also separable in that it takes the form $\mathbf{C}' \otimes \mathbf{A}$ as required using the standard multivariate linear model and the univariate mixed model. Establishing an expression for the error matrix, say $\mathbf{E}^*$, is complicated by the fact that the design matrix involves a Kronecker product. For the rearranged model (6.48),

$$\begin{aligned} \mathbf{B}^{*\prime} &= [(\mathbf{X} \otimes \mathbf{I}_t)(\mathbf{X} \otimes \mathbf{I}_t)']^{-1}(\mathbf{X} \otimes \mathbf{I}_p)\mathbf{Y}^{*\prime} \\ &= ((\mathbf{X}\mathbf{X}')^{-1}\mathbf{X} \otimes \mathbf{I}_t)\mathbf{Y}^{*\prime} \end{aligned} \tag{6.52}$$

For the reduced model (6.49),

$$\begin{aligned} \hat{\mathbf{B}}^{*\prime} &= [(\mathbf{X} \otimes \mathbf{A})(\mathbf{X} \otimes \mathbf{A})']^{-1}(\mathbf{X} \otimes \mathbf{A})(\mathbf{I}_n \otimes)'\mathbf{Y}^{*\prime} \\ & [\mathbf{X}\mathbf{X}' \otimes \mathbf{A}\mathbf{A}']^{-1}(\mathbf{X} \otimes \mathbf{A}\mathbf{A}')\mathbf{Y}^{*\prime} \\ &= ((\mathbf{X}\mathbf{X}')^{-1}\mathbf{X} \otimes \mathbf{I}_t)\mathbf{Y}^{*\prime} \end{aligned} \tag{6.53}$$

To find $\mathbf{E}^*$, recall that in general

$$\mathbf{E} = \mathbf{Z}(\mathbf{I} - \mathbf{M}'(\mathbf{M}\mathbf{M}')^{-1}\mathbf{M}')\mathbf{Z}' \tag{6.54}$$

where **Z** and **M** are column forms of the data matrix and design matrix, respectively. Letting $\mathbf{Z} = \mathbf{Y}^*(\mathbf{I}_n \otimes \mathbf{A})$ and $\mathbf{M} = \mathbf{X} \otimes \mathbf{A}$, then

$$
\begin{aligned}
\mathbf{E}^* &= \mathbf{Y}^*(\mathbf{I}_n \otimes \mathbf{A})\{(\mathbf{I} - \mathbf{X} \otimes \mathbf{A})'[(\mathbf{X} \otimes \mathbf{A})(\mathbf{X} \otimes \mathbf{A}')]^{-1}(\mathbf{X} \otimes \mathbf{A})\}(\mathbf{I}_n \otimes \mathbf{A})'\mathbf{Y}^{*'} \\
&= \mathbf{Y}^*(\mathbf{I}_n \otimes \mathbf{AA}')\mathbf{Y}^* - \hat{\mathbf{B}}^*(\mathbf{X} \otimes \mathbf{A})(\mathbf{X} \otimes \mathbf{A})'\hat{\mathbf{B}}^* \\
&= \mathbf{Y}^*(\mathbf{I}_n \otimes \mathbf{AA}')\mathbf{Y}^{*'} - \mathbf{Y}^*(\mathbf{X}'(\mathbf{XX}')^{-1}\mathbf{X} \otimes \mathbf{AA})\mathbf{Y}^{*'} \\
&= \mathbf{Y}^*\{[\mathbf{I}_n - \mathbf{X}'(\mathbf{XX}')^{-1}\mathbf{X}] \otimes \mathbf{AA}'\}\mathbf{Y}^{*'}
\end{aligned}
\tag{6.55}
$$

If we set $\hat{\mathbf{\Gamma}}^* = \hat{\mathbf{B}}^*(\mathbf{C}' \otimes \mathbf{A})$ where $\hat{\mathbf{B}}^{*'}$ is defined by (6.52), the hypothesis matrix becomes

$$
\mathbf{H}^* = (\hat{\mathbf{\Gamma}}^* - \mathbf{\Gamma}_o)(\mathbf{C}(\mathbf{XX}')^{-1}\mathbf{C}')^{-1}(\hat{\mathbf{\Gamma}}^* - \mathbf{\Gamma}_o)' \tag{6.56}
$$

However, because one usually tests $H\colon \mathbf{B}^*(\mathbf{C}' \otimes \mathbf{A}) = \mathbf{0}$, (6.56) reduces to

$$
\mathbf{H}^* = \mathbf{Y}^*\{\mathbf{X}'(\mathbf{XX}')^{-1}\mathbf{C}'[\mathbf{C}(\mathbf{XX}')^{-1}\mathbf{C}']^{-1}\mathbf{C}(\mathbf{XX}')^{-1}\mathbf{X} \otimes \mathbf{AA}'\}\mathbf{Y}^{*'} \tag{6.57}
$$

Assuming $H\colon \mathbf{\Gamma}^* = \mathbf{0}$, we can write $\mathbf{E}^*$ and $\mathbf{H}^*$ using **E** and **H** defined by the CDMLM, by using the generalized trace operator proposed by Thompson (1973). If we let **W** be $(pu \times pu)$ and $\mathbf{W}_{ij}$ be the ij^{th} $(u \times u)$ submatrix for $i, j = 1, 2, \dots, p$, the generalized trace operator $T_p(\mathbf{W})$ is a $(p \times p)$ matrix defined

$$
T_p(\mathbf{W}) \equiv [\mathrm{Tr}(\mathbf{W}_{ij})] \tag{6.58}
$$

Partitioning **E**, **H**, $\tilde{\mathbf{\Phi}}$, and $\mathbf{\Delta}$ into $(p \times p)$ submatrices: $\mathbf{E} = \{\mathbf{E}_{ij}\}, \mathbf{H} = \{\mathbf{H}_{ij}\}, \tilde{\mathbf{\Phi}} = \{\mathbf{\Phi}_{ij}\}$, and $\mathbf{\Delta} = \{\mathbf{\Delta}_{ij}\}$ $i = j, 2, \dots, u$ we define

$$
\begin{aligned}
\mathbf{E}^* &= T_p(\mathbf{E}) \\
\mathbf{H}^* &= T_p(\mathbf{H}) \\
\mathbf{\Phi}^* &= T_p(\tilde{\mathbf{\Phi}}) \\
\mathbf{\Delta}^* &= T_p(\mathbf{\Delta})
\end{aligned}
\tag{6.59}
$$

and assuming multivariate sphericity

$$
\mathbf{E}^* \sim W_p(uv_e, \mathbf{\Sigma}_e, \mathbf{0}) \text{ and } \mathbf{H}^* \sim W_p[uv_h, \mathbf{\Sigma}_e, \mathbf{\Sigma}_e^{-1}T_p(\mathbf{\Delta})] \tag{6.60}
$$

where $\mathbf{\Sigma}_e = u^{-1}\mathbf{T}_p(\tilde{\mathbf{\Phi}})$.

To test $H\colon \mathbf{B}^*(\mathbf{C}' \otimes \mathbf{A}) = \mathbf{0}$, we again use the multivariate criteria defined by formulas (6.20) through (6.23) substituting for **E** and **H** with $\mathbf{E}^*$ and $\mathbf{H}^*$. Again the formula simplifies; for example, T_o^2 (6.24) is T_o^2 with $\hat{\tilde{\mathbf{\Phi}}}$ replaced with $\hat{\tilde{\mathbf{\Phi}}} = \left(\dfrac{\mathbf{E}^*}{uv_e} \otimes \mathbf{I}_u\right)$ which is an unbiased estimator of $\tilde{\mathbf{\Phi}}$. If $p = 1$ then

$$
T_o^2 = \frac{\mathrm{Tr}(\mathbf{H}^*)/v_h^*}{\mathrm{Tr}(\mathbf{E}^*)/v_e^*} \sim F[uv_h, uv_e, \delta = \mathrm{Tr}(\mathbf{\Delta})/\mathrm{Tr}(\mathbf{\Sigma}_e)] \tag{6.61}
$$

a noncentral F-distribution with $v_h^* = uv_h$ and $v_e^* = uv_e$ degrees of freedom (Davidson, 1988, and Timm, 1980a).

6.5 Double Multivariate and Mixed Model Analysis of Repeated Measurement Data

In Chapter 2 we illustrated the analysis of a univariate split-plot repeated measures design when the covariance matrices across groups (plots) are homogeneous and satisfy the circularity condition or have spherical structure when transformed by the within-subjects matrix. When spherical structure does not hold, as tested using perhaps Mauchley's chi-square statistic, the appropriate analysis is to use a multivariate model which allows for arbitrary covariance structure as illustrated in Chapter 5. When, the number of subjects is smaller than the number of repeated measures the multivariate analysis cannot be performed; thus, one must use an ε-adjusted F-test. It is possible to extract the univariate mixed model results from the multivariate analysis hypothesis and error sum of matrices. For the situation where the repeated measures are vectors, if we can assume a block spherical structure for the transformed covariance matrix (multivariate sphericity) the standard MMM should be used to analyze the data. The MMM results may be obtained from the DMLM analysis. When the number of subjects is less than the number of total repeated measures, an ε-adjusted MMM test is used.

To illustrate the analysis of a DMLM, we use data from Timm (1980a). The data were provided by Dr. Thomas Zullo in the School of Dental Medicine, University of Pittsburgh. Nine subjects were assigned to two orthopedic treatments to study the effectiveness of three activator conditions on three variables associated with the adjustment of the mandible.

For this problem, we have $n = 18$ subjects for which we obtain $p = 3$ variables over $t = 3$ conditions (occasions) for $k = 2$ groups. To analyze these data, we may organize the p-variables at t time points so that the jth subject has a $p \times t$ data matrix: $\mathbf{Y}_j^*$. For the MMM, we let

$$\mathbf{Y}^* = (\mathbf{Y}_1^*, \mathbf{Y}_2^*, \ldots, \mathbf{Y}_n^*) \tag{6.62}$$

For this arrangement of the data, $\mathbf{y}_j = \text{vec}(\mathbf{Y}_j^*)$, and variables are nested within time. Thus, for p-variables and two time (occasions) points,

$$\mathbf{\Sigma} = \begin{pmatrix} \mathbf{\Sigma}_{11} & \mathbf{\Sigma}_{12} \\ \mathbf{\Sigma}_{21} & \mathbf{\Sigma}_{22} \end{pmatrix} \tag{6.63}$$

has a convenient block structure. Alternatively, one may nest the time (occasions) dimension within each variable. For this data arrangement, we have for our example the following:

$$\begin{matrix} & \text{v}_1 & & & \text{v}_2 & & & \text{v}_3 & \\ t_1 & t_2 & t_3 & t_1 & t_2 & t_3 & t_1 & t_2 & t_3 \end{matrix} \tag{6.64}$$

so that when the number of variables $p = 1$, the design reduces to a standard repeated measures model with t_1 , t_2 , and t_3 occasions. Most standard computer packages require organization (6.64); however, (6.62) is more useful for recovering mixed model results. To convert (6.64) to (6.62), it is convenient to order the postmultiplication matrix by variables and thus create a block diagonal matrix as in (6.63).

Program 6_5.sas contains the SAS code to analyze the Zullo dental data using both a DMLM and an MMM. In addition, we explain how to obtain from a normalized DMLM the MMM results, how to test for multivariate sphericity, and how to perform ε-adjusted MMM tests.

Program 6_5.sas

```
/* Program 6_5.sas                                                    */
/* Example from Timm(1980a) and Boik(1988,1991) - Zullo Dental data */

options ls=80 ps=60 nodate nonumber;
title 'Output 6.5: Double Multivariate Linear Model';

data dmlm;
   infile 'c:\mmm.dat';
   input group y1 - y9;
proc print data=dmlm;
run;

proc glm;
   title2 ' Double Multivariate Model Analysis';
   class group;
   model y1 - y9 = group/ nouni;
   means group;
   manova  h=group/ printh printe;
run;

/* Multivariate test of Parallelism */
proc glm;
   class group;
   model y1 - y9 = group / nouni;
   contrast 'Parallel'
             group 1 -1;
   manova m=(.7071 0 -.7071 0 0 0 0 0 0,
             0 0 0 -.408 .816 -.408 0 0 0,
             0 0 0 0 0 0 .7071 0 -.7071,
             -.408 .816 -.408 0 0 0 0 0 0,
             0 0 0 .7071 0 -.7071 0 0 0,
             0 0 0 0 0 0 -.408 .816 -.408) prefix = parl/ printh printe;
run;

/* Multivariate test of Conditions as vectors */
proc glm;
   class group;
   model y1 - y9 = group/ noint nouni;
   contrast 'Mult Cond' group 1 0,
                        group 0 1;
   manova m=(1 -1 0 0 0 0 0 0 0,
             0 1 -1 0 0 0 0 0 0,
             0 0 0 1 -1 0 0 0 0,
             0 0 0 0 1 -1 0 0 0,
             0 0 0 0 0 0 1 -1 0,
             0 0 0 0 0 0 0 1 -1) prefix = diff/ printh printe;
run;
```

```
/* Multivariate test of Conditions given Parallelism */
proc glm;
   class group;
   model y1- y9 = group/noint nouni;
   contrast 'Cond|Parl' group .5 .5;
   manova m=(.7071 0 -.7071 0 0 0 0 0 0,
             0 0 0 -.408 .816 -.408 0 0 0,
             0 0 0 0 0 0 .7071 0 -.7071,
             -.408 .816 -.408 0 0 0 0 0 0,
             0 0 0 .7071 0 -.7071 0 0 0,
             0 0 0 0 0 0 -.408 .816 -.408) prefix=cond/ printh printe;
run;

proc glm;
   class group;
   model y1 - y9 = group/noint nouni;
   contrast 'Group|Parl' group 1 -1;
   manova m=(.577 .577 .577 0 0 0 0 0 0,
             0 0 0 .577 .577 .577 0 0 0,
             0 0 0 0 0 0 .577 .577 .577) prefix=ovall/ printh printe;

/* Multivariate Mixed Model Analysis */
data mix;
   infile 'c:\mixed.dat';
   input group subj cond y1 y2 y3;
proc print data=mix;
run;

proc glm;
   title2 'Multivaraiate Mixed Model Analysis';
   class group subj cond;
   model y1 - y3 = group subj(group) cond cond*group;
   random subj(group);
   contrast 'Group' group 1 -1/ e=subj(group);
   manova h = cond group*cond/ printh printe;
run;

/* Test for Multivariate Spericity and calculation of Epsilon for MMM*/

proc iml;
   title2 'Multivariate Sphericity Test';
   print 'Test of Multivariate Sphericity UsingChi-Square and Adjusted Chi-
         Square Statistics';
   e={  9.6944  7.3056  -6.7972 -4.4264 -0.6736   3.7255,
        7.3056  8.8889  -4.4583 -3.1915 -3.2396   2.9268,
       -6.7972 -4.4583  18.6156  2.5772  0.8837 -10.1363,
       -4.4264 -3.1915   2.5772  5.3981  1.4259  -1.8546,
       -0.6736 -3.2396   0.8837  1.4259 18.3704   -.7769,
        3.7255  2.9268 -10.1363 -1.8546 -0.7769   6.1274};
   print e;
   n=18;
   p=3;
   t=3;
   k=2;
   u=6;
   q=u/p;
   nu_e=n-k;
   nu_h=1;
   e11=e[1:3,1:3];print e11;
   e22=e[4:6,4:6];print e22;
   dn=(e11+e22)/2;
   b=eigval(dn); print b;
   a=eigval(e); print a;
   b=log(b);
```

```
    a=log(a);
    chi_2=n#(q#sum(b)-sum(a));
    df=p#(q-1)#(p#q+p+1)/2;
    pvalue=1-probchi(chi_2,df);
    print chi_2 df pvalue;
    c1=p/(12#q#nu_e#df);
    rho= 1-c1#(2#p##2#(q##4-1)+3#p#(q##3-1)-(q##2-1));
    ro_chi_2=(rho#nu_e/n)#chi_2; print rho;
    c2=1/(2#rho##2);
    c3=((p#q-1)#p#q#(p#q+1)#(p#q+2))/(24#nu_e##2);
    c4=((p-1)#p#(p+1)#(p+2))/(24#q##2#nu_e##2);
    c5=df#(1-rho)##2/2;
    omega=c2#(c3-c4-c5); print omega;
    p1=1-probchi(ro_chi_2,df);
    p2=1-probchi(ro_chi_2,df+4);
    cpvalue=(1-omega)#p1+omega#p2;
    print ro_chi_2 cpvalue;

    s=e/nu_e;
    s11=s[1:3,1:3]; s12=s[1:3,4:6];
    s21=s[4:6,1:3]; s22=s[4:6,4:6];
    enum=trace((s11+s22)*(s11+s22))+(trace(s11+s22))##2;
    eden=q#( trace(s11)##2+trace(s11*s11)+trace(s12)##2+trace(s12*s12)+
             trace(s21)##2+trace(s21*s21)+trace(s22)##2+trace(s22*s22));
    epsilon=enum/eden;

    nu_h=nu_h#q; nu_e=nu_e#q; s0=min(nu_h,p);
    Mnu_h=nu_h#epsilon; Mnu_e=nu_e#epsilon; ms0=min(mnu_h,p);
    m0=(abs(mnu_h-p)-1)/2; n0=(mnu_e-p-1)/2;

    denom=s0##2#(2#m0+ms0+1); numer=2#(ms0#n0+1);
    df1=ms0#(2#m0+ms0+1); df2=2#(ms0#n0+1);

    print 'Epsison adjusted F-Statistics for cond and group X cond MMM tests';
    print epsilon;
    f_cond = df2#13.75139851/(ms0#df1);
    f_gXc  = df2#0.19070696/(ms0#df1);

    print f_cond f_gXc df1 df2;

    p_cond=1-probf(f_cond,df1,df2);
    p_gXc=1-probf(f_gXc,df1,df2);

    print 'Epsilon adjusted pvalues for MMM tests using T0**2 Criterion';

    print p_cond p_gXc;
quit;
```

Using organization (6.64), we present Zullo's dental experiment data in the continuation of Output 6.5 that follows. The three variables are individual measurements of the vertical position of the mandible at three activator treatment conditions and two treatment groups. The organization of the means for the analysis is presented in Table 6.1.

Output 6.5: *Double Multivariate Linear Model*

```
                  Output 6.5: Double Multivariate Linear Model

OBS   GROUP     Y1      Y2      Y3      Y4     Y5     Y6     Y7     Y8     Y9

  1     1     117.0   117.5   118.5   59.0   59.0   60.0   10.5   16.5   16.5
  2     1     109.0   110.5   111.0   60.0   61.5   61.5   30.5   30.5   30.5
  3     1     117.0   120.0   120.5   60.0   61.5   62.0   23.5   23.5   23.5
  4     1     122.0   126.0   127.0   67.5   70.5   71.5   33.0   32.0   32.5
  5     1     116.0   118.5   119.5   61.5   62.5   63.5   24.5   24.5   24.5
  6     1     123.0   126.0   127.0   65.5   61.5   67.5   22.0   22.0   22.0
  7     1     130.5   132.0   134.5   68.5   69.5   71.0   33.0   32.5   32.0
  8     1     126.5   128.5   130.5   69.0   71.0   73.0   20.0   20.0   20.0
  9     1     113.0   116.5   118.0   58.0   59.0   60.5   25.0   25.0   24.5
 10     2     128.0   129.0   131.5   67.0   67.5   69.0   24.0   24.0   24.0
 11     2     116.5   120.0   121.5   63.5   65.0   66.0   28.5   29.5   29.5
 12     2     121.5   125.5   127.0   64.5   67.5   69.0   26.5   27.0   27.0
 13     2     109.5   112.0   114.0   54.0   55.5   57.0   18.0   18.5   19.0
 14     2     133.0   136.0   137.5   72.0   73.5   75.5   34.5   34.5   34.5
 15     2     120.0   124.5   126.0   62.5   65.0   66.0   26.0   26.0   26.0
 16     2     129.5   133.5   134.5   65.0   68.0   69.0   18.5   18.5   18.5
 17     2     122.0   124.0   125.5   64.5   65.5   66.0   18.5   18.5   18.5
 18     2     125.0   127.0   128.0   65.5   66.5   67.0   21.5   21.5   21.6
```

Table 6.1 *DMLM Organization of Means*

	So$_r$ - Me (MM)			ANS - Me (MM)			Pal-MP Angle (Degrees)		
Variables	1			2			3		
Conditions	C_1	C_2	C_3	C_1	C_2	C_3	C_1	C_2	C_3
G_1 (Treatment)	μ_{11}	μ_{12}	μ_{13}	μ_{14}	μ_{15}	μ_{16}	μ_{17}	μ_{18}	μ_{19}
G_2 (Treatment)	μ_{21}	μ_{22}	μ_{23}	μ_{24}	μ_{25}	μ_{26}	μ_{27}	μ_{28}	μ_{29}

Result and Interpretation 6_5.sas

The first hypothesis test of interest for these data is to test whether the profiles for the two treatment groups are parallel, by testing for an interaction between conditions and groups. The hypothesis is

$$\begin{aligned} H_{GC}:\ & (\mu_{11}-\mu_{13}, \mu_{12}-\mu_{13}, \ldots, \mu_{17}-\mu_{18}, \mu_{18}-\mu_{19}) \\ & = (\mu_{21}-\mu_{23}, \mu_{22}-\mu_{23}, \ldots, \mu_{27}-\mu_{28}, \mu_{28}-\mu_{29}) \end{aligned} \tag{6.65}$$

The matrices $\mathbf{C}$ and $\mathbf{P} = (\mathbf{I}_3 \otimes \mathbf{A})$ of the form $\mathbf{CBP} = \mathbf{0}$ to test H_{GC} are

$$\mathbf{C} = (1,\ -1), \mathbf{P} = \begin{pmatrix} \mathbf{A} & \mathbf{0} & \mathbf{0} \\ \mathbf{0} & \mathbf{A} & \mathbf{0} \\ \mathbf{0} & \mathbf{0} & \mathbf{A} \end{pmatrix} \quad \text{and} \quad \mathbf{A} = \begin{pmatrix} 1 & 0 \\ 0 & 1 \\ -1 & -1 \end{pmatrix} \tag{6.66}$$

For the DMLM analysis in SAS, the matrix $\mathbf{P}$ is defined as $m = \mathbf{P}'$ so that the columns of $\mathbf{P}$ become the rows of m. This organization of the data and the structure for $\mathbf{P}$ do not result in a convenient block structure for $\mathbf{H}$ and $\mathbf{E}$. Instead, as suggested by Boik (1988), one may use the responsewise representation for the model. Alternatively, one may use the multivariate mixed model for the analysis which is convenient in SAS as illustrated in Program 6_5.sas. Note that when $\mathbf{A}$ is normalized, the MMM tests may be obtained from the DMLM results as illustrated in Timm (1980a).

The overall hypothesis to test for differences in Groups, $H_{G^\bullet}$, is

$$H_{G^\bullet}: \mu_1 = \mu_2 \tag{6.67}$$

and the matrices $\mathbf{C}$ and $\mathbf{P}$ to test this hypothesis are

$$\mathbf{C}_{G^\bullet} = (1, -1) \text{ and } \mathbf{P} = \mathbf{I}_2 \tag{6.68}$$

The hypothesis to test for vector differences in conditions becomes

$$H_{C^\bullet}: \begin{pmatrix} \mu_{11} \\ \mu_{21} \\ \mu_{14} \\ \mu_{24} \\ \mu_{17} \\ \mu_{27} \end{pmatrix} = \begin{pmatrix} \mu_{12} \\ \mu_{22} \\ \mu_{15} \\ \mu_{25} \\ \mu_{18} \\ \mu_{28} \end{pmatrix} = \begin{pmatrix} \mu_{13} \\ \mu_{23} \\ \mu_{16} \\ \mu_{26} \\ \mu_{19} \\ \mu_{29} \end{pmatrix} \tag{6.69}$$

The test matrices to test (6.69) are

$$\mathbf{C}_{C^\bullet} = \mathbf{I}_2 \text{ and } \mathbf{P} = \begin{pmatrix} 1 & 0 & 0 & 0 & 0 & 0 \\ -1 & 1 & 0 & 0 & 0 & 0 \\ 0 & -1 & 0 & 0 & 0 & 0 \\ 0 & 0 & 1 & 0 & 0 & 0 \\ 0 & 0 & -1 & 1 & 0 & 0 \\ 0 & 0 & 0 & -1 & 0 & 0 \\ 0 & 0 & 0 & 0 & 1 & 0 \\ 0 & 0 & 0 & 0 & -1 & 1 \\ 0 & 0 & 0 & 0 & 0 & -1 \end{pmatrix} \tag{6.70}$$

Tests (6.67) and (6.69) do not require parallelism of profiles. However, given parallelism, tests for differences between groups and among conditions are written

$$H_G: \begin{pmatrix} \sum_{j=1}^{3} \mu_{1j} / 3 \\ \sum_{j=4}^{6} \mu_{1j} / 3 \\ \sum_{j=4}^{9} \mu_{1j} / 3 \end{pmatrix} = \begin{pmatrix} \sum_{j=1}^{3} \mu_{2j} / 3 \\ \sum_{j=4}^{6} \mu_{2j} / 3 \\ \sum_{j=7}^{9} \mu_{2j} / 3 \end{pmatrix} \tag{6.71}$$

and

$$H_C\colon \begin{pmatrix} \sum_{i=1}^{2}\mu_{i1}/2 \\ \sum_{i=1}^{2}\mu_{i4}/2 \\ \sum_{i=1}^{2}\mu_{i7}/2 \end{pmatrix} = \begin{pmatrix} \sum_{i=1}^{2}\mu_{i2}/2 \\ \sum_{i=1}^{2}\mu_{i5}/2 \\ \sum_{i=1}^{2}\mu_{i8}/2 \end{pmatrix} = \begin{pmatrix} \sum_{i=1}^{2}\mu_{i3}/2 \\ \sum_{i=1}^{2}\mu_{i6}/2 \\ \sum_{i=1}^{2}\mu_{i9}/2 \end{pmatrix} \tag{6.72}$$

respectively. The hypothesis test matrices to test H_G and H_C are

$$\mathbf{C}_G = (1, -1) \text{ and } \mathbf{P}' = \begin{pmatrix} 1 & 1 & 1 & 0 & 0 & 0 & 0 & 0 & 0 \\ 0 & 0 & 0 & 1 & 1 & 1 & 0 & 0 & 0 \\ 0 & 0 & 0 & 0 & 0 & 0 & 1 & 1 & 1 \end{pmatrix} \tag{6.73}$$

$$\mathbf{C}_C = (1/2, 1/2) \text{ and } \mathbf{P} = \begin{pmatrix} 1 & 0 & 0 & 0 & 0 & 0 \\ 0 & 0 & 0 & 1 & 0 & 0 \\ -1 & 0 & 0 & -1 & 0 & 0 \\ 0 & 1 & 0 & 0 & 0 & 0 \\ 0 & 0 & 0 & 0 & 1 & 0 \\ 0 & -1 & 0 & 0 & -1 & 0 \\ 0 & 0 & 0 & 0 & 0 & 0 \\ 0 & 0 & 1 & 0 & 0 & 1 \\ 0 & 0 & -1 & 0 & 0 & -1 \end{pmatrix} \tag{6.74}$$

Normalization of the matrix **P** in (6.73) and (6.74) within each variable allows one to again obtain MMM results from the multivariate tests, given parallelism. This is not the case for the multivariate tests of $H_{G^\bullet}$ and $H_{C^\bullet}$ that do not assume parallelism of profiles. Some standard statistical packages for the analysis of this design, like SPSS, do not test $H_{G^\bullet}$ or $H_{C^\bullet}$ but assume the parallelism condition, and hence are testing H_G and H_C.

The SAS code to perform these hypothesis tests is provided in Program 6_5.sas. The results of the tests are given in the continuation of Output 6.5 that follows.

Output 6.5: *(continued)*

```
                          Output 6.5: Double Multivariate Linear Model
                                 Double Multivariate Model Analysis

                        General Linear Models Procedure
                       Multivariate Analysis of Variance

                 Manova Test Criteria and Exact F Statistics for
                    the Hypothesis of no Overall GROUP Effect
            H = Type III SS&CP Matrix for GROUP   E = Error SS&CP Matrix

                              S=1    M=3.5    N=3

Statistic                         Value          F       Num DF    Den DF  Pr > F

Wilks' Lambda                0.42223394     1.2163            9         8  0.3965
Pillai's Trace               0.57776606     1.2163            9         8  0.3965
Hotelling-Lawley Trace       1.36835531     1.2163            9         8  0.3965
Roy's Greatest Root          1.36835531     1.2163            9         8  0.3965

                 Manova Test Criteria and Exact F Statistics for
                   the Hypothesis of no Overall Parallel Effect
              on the variables defined by the M Matrix Transformation
          H = Contrast SS&CP Matrix for Parallel   E = Error SS&CP Matrix


                        General Linear Models Procedure
                       Multivariate Analysis of Variance

                              S=1    M=2    N=4.5

Statistic                         Value          F       Num DF    Den DF  Pr > F

Wilks' Lambda                0.58298973     1.3114            6        11  0.3292
Pillai's Trace               0.41701027     1.3114            6        11  0.3292
Hotelling-Lawley Trace       0.71529608     1.3114            6        11  0.3292
Roy's Greatest Root          0.71529608     1.3114            6        11  0.3292


                 Manova Test Criteria and F Approximations for
                  the Hypothesis of no Overall Mult Cond Effect
              on the variables defined by the M Matrix Transformation
          H = Contrast SS&CP Matrix for Mult Cond   E = Error SS&CP Matrix

                             S=2    M=1.5    N=4.5

Statistic                         Value          F       Num DF    Den DF  Pr > F

Wilks' Lambda                0.02635582     9.4595           12        22  0.0001
Pillai's Trace               1.17460543     2.8462           12        24  0.0141
Hotelling-Lawley Trace      29.31734905    24.4311           12        20  0.0001
Roy's Greatest Root         29.05491748    58.1098            6        12  0.0001

            NOTE: F Statistic for Roy's Greatest Root is an upper bound.
                   NOTE: F Statistic for Wilks' Lambda is exact.


                 Manova Test Criteria and Exact F Statistics for
                  the Hypothesis of no Overall Cond|Parl Effect
              on the variables defined by the M Matrix Transformation
          H = Contrast SS&CP Matrix for Cond|Parl   E = Error SS&CP Matrix
```

Output 6.5 (*continued*)

```
                       General Linear Models Procedure
                      Multivariate Analysis of Variance

                           S=1    M=2    N=4.5

Statistic                    Value          F      Num DF    Den DF  Pr > F

Wilks' Lambda              0.03378144    52.4371        6        11  0.0001
Pillai's Trace             0.96621856    52.4371        6        11  0.0001
Hotelling-Lawley Trace    28.60205297    52.4371        6        11  0.0001
Roy's Greatest Root       28.60205297    52.4371        6        11  0.0001

              Manova Test Criteria and Exact F Statistics for
               the Hypothesis of no Overall Group|Parl Effect
           on the variables defined by the M Matrix Transformation
       H = Contrast SS&CP Matrix for Group|Parl    E = Error SS&CP Matrix

                           S=1    M=0.5    N=6

Statistic                    Value          F      Num DF    Den DF  Pr > F

Wilks' Lambda              0.88386089     0.6132        3        14  0.6176
Pillai's Trace             0.11613911     0.6132        3        14  0.6176
Hotelling-Lawley Trace     0.13139977     0.6132        3        14  0.6176
Roy's Greatest Root        0.13139977     0.6132        3        14  0.6176
```

Table 6.2 summarizes the results of the hypothesis tests including Wilks' Λ criterion and F-ratios.

Table 6.2 *DMLM Analysis, Zullo Data*

Hypothesis	Wilks' Λ	df	F	df	p-value
GC	0.5830	(6, 1, 16)	1.3114	(6, 11)	0.3292
G^*	0.4222	(9, 1, 16)	1.216	(9, 8)	0.3965
C^*	0.0264	(6, 2, 16)	9.4595	(12, 22)	<0.0001
G	0.8839	(3, 1, 16)	0.6132	(3, 14)	0.6176
C	0.0338	6, 1, 16)	52.4371	(6, 11)	<0.0001

Because the test of parallelism is not significant, tests of conditions and group differences given parallelism are appropriate. From these tests one may derive MMM tests of H_C and H_{GC} given multivariate sphericity. The MMM test of group differences is identical to the test of H_G using the DMLM and does not depend on the sphericity assumption.

Thomas (1983) derived the likelihood ratio tests of multivariate sphericity and Boik (1988) showed that it was the necessary and sufficient condition for the MMM tests to be valid. Boik (1988) also developed ε-adjusted multivariate tests of H_C and H_{GC} when multivariate sphericity is not satisfied. Recall that for the simple split-plot design F-ratios are exact if and only if

$$\mathbf{A}'\mathbf{\Sigma}\mathbf{A} = \sigma^2\mathbf{I}$$

across groups. For the multivariate case, using the CDMLM, this becomes

$$(\mathbf{I}_p \otimes \mathbf{A})' \boldsymbol{\Sigma}(\mathbf{I}_p \otimes \mathbf{A}) = (\boldsymbol{\Sigma}_e \otimes \mathbf{I}_u) \tag{6.75}$$

where $\mathbf{A}$ is the submatrix in $\mathbf{P}$ of the DMLM appropriately normalized so that $\mathbf{A'A} = \mathbf{I}$. Alternatively, using the RDMLM, (6.75) becomes

$$(\mathbf{A} \otimes \mathbf{I}_p)' \boldsymbol{\Sigma}(\mathbf{A} \otimes \mathbf{I}_p) = (\mathbf{I}_u \otimes \boldsymbol{\Sigma}_e) \tag{6.76}$$

To test (6.76), Thomas (1983) showed that the LR test statistic is

$$\lambda = |\mathbf{E}|^{n/2} / \left| q^{-1} \sum_{i=1}^{q} \mathbf{E}_{ii} \right|^{nq/2} \tag{6.77}$$

where $u \equiv q = R(\mathbf{A})$. The asymptotic null distribution of $-2 \ln \lambda$ is a central chi-square distribution with degrees of freedom

$$f = p(q-1)(pq+p+1)/2 \tag{6.78}$$

If we let $\alpha_i (i = 1, \ldots, pq)$ be the eigenvalues of $\mathbf{E}$, and $\beta_i (i = 1, \ldots, p)$ the eigenvalues of $\sum_{i=1}^{q} \mathbf{E}_{ii}$, a simple form of (6.77) is

$$U = -2 \ln \lambda = n\left(q \sum_{i=1}^{p} \ln(\beta_i) - \sum_{i=1}^{pq} \ln(\alpha_i) \right) \tag{6.79}$$

When p or q are large relative to n, the asymptotic approximation U may be poor. To correct for this, Boik (1988) using Box's correction factor for the distribution of U showed that the

$$P(U \le U_o) = P(\rho^* U \le \rho^* U_o) = (1-\omega)P(X_f^2 \le \rho^* U_o) + \omega P(X_{f+4}^2 \le \rho^* U_o) + O(v_e^{-3}) \tag{6.80}$$

where f is defined in (6.78), and

$$\begin{aligned} \rho &= 1 - \left(\frac{p}{12qfv_e} \right) [2p^2(q^4-1) + 3p(q^3-1) - (q^2-1)] \\ \rho^* &= \rho v_e / n \\ \omega &= (2\rho^2)^{-1} \left\{ \left[\frac{(pq-1)pq(pq+1)(pq+2)}{24v_e^2} \right] - \left[\frac{(p-1)p(p+1)(p+2)}{24q^2v_e^2} \right] \left[\frac{f(1-\rho)^2}{2} \right] \right\} \end{aligned} \tag{6.81}$$

and $v_e = n - R(\mathbf{X})$. Hence, the p-value for the test of multivariate sphericity using Box's correction becomes

$$P(\rho^* U \ge U_o) = (1-\omega)P(X_f^2 \ge U_o) + \omega P(X_{f+4}^2 \ge U_o) + O(v_e^{-3}) \tag{6.82}$$

Using (6.78) and (6.81), PROC IML code is included at the end of Program 6_5.sas for the multivariate test of sphericity. The results of the test are given in the continuation of Output 6.5 that follows.

Output 6.5 (*continued*)

```
              Output 6.5: Double Multivariate Linear Model
                       Multivariate Sphericity Test

      Test of Multivariate Sphericity UsingChi-Square and Adjusted Chi-
                            Square Statistics

                                    E
   9.6944     7.3056    -6.7972    -4.4264    -0.6736     3.7255
   7.3056     8.8889    -4.4583    -3.1915    -3.2396     2.9268
  -6.7972    -4.4583    18.6156     2.5772     0.8837   -10.1363
  -4.4264    -3.1915     2.5772     5.3981     1.4259    -1.8546
  -0.6736    -3.2396     0.8837     1.4259    18.3704    -0.7769
   3.7255     2.9268   -10.1363    -1.8546    -0.7769     6.1274

                            E11
                         9.6944     7.3056    -6.7972
                         7.3056     8.8889    -4.4583
                        -6.7972    -4.4583    18.6156

                            E22
                         5.3981     1.4259    -1.8546
                         1.4259    18.3704    -0.7769
                        -1.8546    -0.7769     6.1274

                                    B
                               18.972307
                               10.357742
                               4.2173517

                                    A
                                32.29187
                               18.552285
                               10.952687
                               3.4969238
                               1.4751884
                               0.3258457

                         CHI_2          DF     PVALUE
                     74.367228          15 7.365E-10


                                  RHO
                                0.828125

                                  OMEGA
                               0.0342649

                           RO_CHI_2    CPVALUE
                          54.742543 2.7772E-6
```

Since $U = -2\ln\lambda = 74.367228$ with df $= 15$ and p-value for the test equal to 7.365×10^{-10} or using Box's correction, $\rho^{*}U = 54.742543$ with the p-value $= 2.7772 \times 10^{-6}$, we reject the null hypothesis of multivariate sphericity. The MMM assumption of sphericity is not satisfied with these data; thus, one should use the DMLM.

Assuming that the test of multivariate sphericity was not significant, we show how to analyze these data using the MMM. In addition, we illustrate the construction of the ε-adjusted MMM test statistics developed by Boik (1988) which extends the Greenhouse-Geisser type test procedure to the MMM.

While one may directly derive the MMM test matrices from the normalized DMLM tests, as illustrated in Timm (1980a), it is more convenient to use PROC GLM to obtain the test statistics and hypothesis test matrices using the mixed model. For the MMM analyses, the data must be reorganized; subject is a random factor nested within the fixed group factor, and groups are crossed with the fixed condition factor. The reorganized data follow in the continuation of Output 6.5.

Output 6.5 (*continued*)

Output 6.5: Double Multivariate Linear Model
Double Multivariate Model Analysis

OBS	GROUP	SUBJ	COND	Y1	Y2	Y3
1	1	1	1	117.0	59.0	10.5
2	1	1	2	117.5	59.0	16.5
3	1	1	3	118.5	60.0	16.5
4	1	2	1	109.0	60.0	30.5
5	1	2	2	110.5	61.5	30.5
6	1	2	3	111.0	61.5	30.5
7	1	3	1	117.0	60.0	23.5
8	1	3	2	120.0	61.5	23.5
9	1	3	3	120.5	62.0	23.5
10	1	4	1	122.0	67.5	33.0
11	1	4	2	126.0	70.5	32.0
12	1	4	3	127.0	71.5	32.5
13	1	5	1	116.0	61.5	24.5
14	1	5	2	118.5	62.5	24.5
15	1	5	3	119.5	63.5	24.5
16	1	6	1	123.0	65.5	22.0
17	1	6	2	126.0	61.5	22.0
18	1	6	3	127.0	67.5	22.0
19	1	7	1	130.5	68.5	33.0
20	1	7	2	132.0	69.5	32.5
21	1	7	3	134.5	71.0	32.0
22	1	8	1	126.5	69.0	20.0
23	1	8	2	128.5	71.0	20.0
24	1	8	3	130.5	73.0	20.0
25	1	9	1	113.0	58.0	25.0
26	1	9	2	116.5	59.0	25.0
27	1	9	3	118.0	60.5	24.5
28	2	1	1	128.0	67.0	24.0
29	2	1	2	129.0	67.5	24.0
30	2	1	3	131.5	69.0	24.0
31	2	2	1	116.5	63.5	28.5
32	2	2	2	120.0	65.0	29.5
33	2	2	3	121.5	66.0	29.5
34	2	3	1	121.5	64.5	26.5
35	2	3	2	125.5	67.5	27.0
36	2	3	3	127.0	69.0	27.0
37	2	4	1	109.5	54.0	18.0
38	2	4	2	112.0	55.5	18.5
39	2	4	3	114.0	57.0	19.0
40	2	5	1	133.0	72.0	34.5
41	2	5	2	136.0	73.5	34.5
42	2	5	3	137.5	75.5	34.5
43	2	6	1	120.0	62.5	26.0
44	2	6	2	124.5	65.0	26.0
45	2	6	3	126.0	66.0	26.0
46	2	7	1	129.5	65.0	18.5
47	2	7	2	133.5	68.0	18.5

Output 6.5 (*continued*)

```
         50      2      8      2     124.0     65.5     18.5
         48      2      7      3     134.5     69.0     18.5
         49      2      8      1     122.0     64.5     18.5
         51      2      8      3     125.5     66.0     18.5
         52      2      9      1     125.0     65.5     21.5
         53      2      9      2     127.0     66.5     21.5
         54      2      9      3     128.0     67.0     21.6
```

To perform this mixed model analysis using PROC GLM, the RANDOM statement is used to calculate the expected means squares to form the appropriate error sum of squares matrices for the multivariate tests. The Type III expected means squares are found in the continuation of Output 6.5 that follows.

Output 6.5 (*continued*)

```
        Output 6.5: Double Multivariate Linear Model

               Multivaraiate Mixed Model Analysis

                General Linear Models Procedure

Source         Type III Expected Mean Square
GROUP          Var(Error) + 3 Var(SUBJ(GROUP)) + Q(GROUP,GROUP*COND)
SUBJ(GROUP)    Var(Error) + 3 Var(SUBJ(GROUP))
COND           Var(Error) + Q(COND,GROUP*COND)
GROUP*COND     Var(Error) + Q(GROUP*COND)
```

From the expected mean square output, we see that the subject within-group effect is the error term to test for group differences given parallelism. The PROC GLM statements for the analysis from Program 6_5.sas are

```
model        y1 - y3 = group subj (group) card cond*group;
random       subj (group);
contrast     'Group' group 1 - 1 / e = subj (group);
manova       h = cond group*cond/printh printe;
```

The E option in the CONTRAST statement defines the error matrix for the test of groups. The "overall" error matrix is being used both to test for conditions given parallelism and for parallel profiles. The results of this analysis are found in the continuation of Output 6.5 that follows.

Output 6.5 (*continued*)

```
                    Output 6.5: Double Multivariate Linear Model
                         Multivaraiate Mixed Model Analysis

                      General Linear Models Procedure
                     Multivariate Analysis of Variance

              Manova Test Criteria and F Approximations for
                the Hypothesis of no Overall COND Effect
         H = Type III SS&CP Matrix for COND   E = Error SS&CP Matrix

                          S=2     M=0     N=14

 Statistic                     Value          F       Num DF     Den DF  Pr > F

 Wilks' Lambda            0.06053416    30.6443            6         60  0.0001
 Pillai's Trace           1.04650228    11.3413            6         62  0.0001
 Hotelling-Lawley Trace  13.75139851    66.4651            6         58  0.0001
 Roy's Greatest Root     13.62158994   140.7564            3         31  0.0001

         NOTE: F Statistic for Roy's Greatest Root is an upper bound.
                NOTE: F Statistic for Wilks' Lambda is exact.

                      General Linear Models Procedure

             Manova Test Criteria and F Approximations for
              the Hypothesis of no Overall GROUP*COND Effect
      H = Type III SS&CP Matrix for GROUP*COND   E = Error SS&CP Matrix

                          S=2     M=0     N=14

 Statistic                     Value          F       Num DF     Den DF  Pr > F

 Wilks' Lambda            0.83447804     0.9469            6         60  0.4687
 Pillai's Trace           0.17190315     0.9717            6         62  0.4519
 Hotelling-Lawley Trace   0.19070696     0.9218            6         58  0.4862
 Roy's Greatest Root      0.13337134     1.3782            3         31  0.2678

         NOTE: F Statistic for Roy's Greatest Root is an upper bound.
                NOTE: F Statistic for Wilks' Lambda is exact.


                      General Linear Models Procedure
                     Multivariate Analysis of Variance

             Manova Test Criteria and Exact F Statistics for
                the Hypothesis of no Overall Group Effect
H = Contrast SS&CP Matrix for Group   E = Type III SS&CP Matrix for SUBJ(GROUP)

                          S=1     M=0.5     N=6

 Statistic                     Value          F       Num DF     Den DF  Pr > F

 Wilks' Lambda            0.88386089     0.6132            3         14  0.6176
 Pillai's Trace           0.11613911     0.6132            3         14  0.6176
 Hotelling-Lawley Trace   0.13139977     0.6132            3         14  0.6176
 Roy's Greatest Root      0.13139977     0.6132            3         14  0.6176
```

The results of the mixed model analysis are summarized in Table 6.3.

Table 6.3 *MMM Analysis, Zullo Data*

Hypothesis	Wilks Λ	df	F	df	*p*-value
Group | Parallel	0.8839	(3, 1, 16)	0.6132	(3, 14)	0.6176
Cond | Parallel	0.0605	(3, 2, 32)	30.6443	(6, 60)	<0.0001
Group x Cond	0.8345	(3, 2, 32)	0.9469	(6, 60)	0.4687

To see that the MMM results may be obtained from the DMLM results, observe that for the test of conditions the hypothesis test matrix

$$\mathbf{H}^* = \begin{pmatrix} 152.93 & & (sym) \\ 95.79 & 62.73 & \\ 14.75 & 8.56 & 1.59 \end{pmatrix}$$

is the average of $\mathbf{H}_{11}$ and $\mathbf{H}_{22}$ where

$$\mathbf{H}_{11} = \begin{pmatrix} 148.02 & & (sym) \\ 96.32 & 62.67 & \\ 13.38 & 8.71 & 1.21 \end{pmatrix}, \quad \mathbf{H}_{22} = \begin{pmatrix} 4.89 & & (sym) \\ -0.53 & 0.06 & \\ 1.36 & -0.15 & 0.39 \end{pmatrix}$$

$\mathbf{H}_{11}$ and $\mathbf{H}_{22}$ the submatrices of $\mathbf{H}$ for the test of conditions given parallelism. The degrees of freedom for the mixed model analysis are $v_h^* = v_h\, u = v_h\ \text{Rank}(\mathbf{P}) / p = v_h \text{ x Rank}(\mathbf{A}) = 1 \text{ x } 2 = 2$ and $v_e^* = v_e\ \text{Rank}(\mathbf{P}) / p = v_e$ $\text{Rank}(\mathbf{A}) = 16 \text{ x } 6/3 = 32$. The results for the test of condxgroup are given similarly in the continuation of Output 6.5 that follows.

From the MMM, one may obtain the univariate split-plot univariate F-ratios, one variable at a time.

Output 6.5 (*continued*)

```
                  Output 6.5: Double Multivariate Linear Model
                        Multivaraiate Mixed Model Analysis

                         General Linear Models Procedure

Dependent Variable: Y1
                                      Sum of              Mean
Source                    DF         Squares            Square    F Value       Pr > F

Model                     21    2715.1666667       129.2936508     274.13       0.0001

Error                     32      15.0925926         0.4716435

Corrected Total           53    2730.2592593

                  R-Square              C.V.          Root MSE                 Y1 Mean

                  0.994472          0.557002         0.6867631               123.29630
```

Output 6.5 (*continued*)

```
Source                  DF        Type I SS     Mean Square   F Value      Pr > F

GROUP                    1      208.0740741     208.0740741    441.17      0.0001
SUBJ(GROUP)             16     2352.3518519     147.0219907    311.72      0.0001
COND                     2      152.9259259      76.4629630    162.12      0.0001
GROUP*COND               2        1.8148148       0.9074074      1.92      0.1626

Source                  DF      Type III SS     Mean Square   F Value      Pr > F

GROUP                    1      208.0740741     208.0740741    441.17      0.0001
SUBJ(GROUP)             16     2352.3518519     147.0219907    311.72      0.0001
COND                     2      152.9259259      76.4629630    162.12      0.0001
GROUP*COND               2        1.8148148       0.9074074      1.92      0.1626

                        General Linear Models Procedure

Dependent Variable: Y2
                                     Sum of            Mean
Source                  DF          Squares          Square   F Value      Pr > F

Model                   21     1172.3750000      55.8273810     65.54      0.0001

Error                   32       27.2592593       0.8518519

Corrected Total         53     1199.6342593

                  R-Square             C.V.        Root MSE                Y2 Mean

                  0.977277         1.418925       0.9229582              65.046296


Source                  DF        Type I SS     Mean Square   F Value      Pr > F

GROUP                    1       31.8935185      31.8935185     37.44      0.0001
SUBJ(GROUP)             16     1075.7407407      67.2337963     78.93      0.0001
COND                     2       62.7314815      31.3657407     36.82      0.0001
GROUP*COND               2        2.0092593       1.0046296      1.18      0.3205

Source                  DF      Type III SS     Mean Square   F Value      Pr > F

GROUP                    1       31.8935185      31.8935185     37.44      0.0001
SUBJ(GROUP)             16     1075.7407407      67.2337963     78.93      0.0001
COND                     2       62.7314815      31.3657407     36.82      0.0001
GROUP*COND               2        2.0092593       1.0046296      1.18      0.3205

                        General Linear Models Procedure

Dependent Variable: Y3
                                     Sum of            Mean
Source                  DF          Squares          Square   F Value      Pr > F

Model                   21     1615.8557407      76.9455115     99.51      0.0001

Error                   32       24.7429630       0.7732176

Corrected Total         53     1640.5987037

                  R-Square             C.V.        Root MSE                Y3 Mean

                  0.984918         3.578005       0.8793279              24.575926
```

Output 6.5 (*continued*)

Source	DF	Type I SS	Mean Square	F Value	Pr > F
GROUP	1	8.8816667	8.8816667	11.49	0.0019
SUBJ(GROUP)	16	1605.2103704	100.3256481	129.75	0.0001
COND	2	1.5892593	0.7946296	1.03	0.3693
GROUP*COND	2	0.1744444	0.0872222	0.11	0.8937

Source	DF	Type III SS	Mean Square	F Value	Pr > F
GROUP	1	8.8816667	8.8816667	11.49	0.0019
SUBJ(GROUP)	16	1605.2103704	100.3256481	129.75	0.0001
COND	2	1.5892593	0.7946296	1.03	0.3693
GROUP*COND	2	0.1744444	0.0872222	0.11	0.8937

The univariate F-ratios for the test of conditions from Output 6.5 are summarized in Table 6.4.

Table 6.4 *Summary of Univariate Output*

Variable	F-value	*p*-value
SOr (Y_1)	76.463/0.472 = 162.1	<0.001
ANS (Y_2)	31.366/0.852 = 36.82	<0.001
Pal (Y_3)	0.795/0.773 = 1.03	0.3694

These tests assume a spherical or Hankel structure across the three variables.

When multivariate sphericity is not satisfied or one cannot perform a DMLM analysis, Boik (1988) developed an ε-adjustment for the tests of conditions and conditions x group interaction following the original work of Box (1954). The multivariate tests may be approximated by a Wishart distribution when the null hypothesis is true:

$$\mathbf{E}^* \,\dot{\sim}\, W_p(v_e^* = \varepsilon\, q\, v_e, \boldsymbol{\Sigma}_e, 0)$$
$$\mathbf{H}^* \,\dot{\sim}\, W_p(v_h^* = \varepsilon\, q\, v_h, \boldsymbol{\Sigma}_e, 0)$$

Boik showed how one may use an estimate of ε to adjust the approximate F-tests for the multivariate criteria where

$$\hat{\varepsilon} = \frac{Tr[(\sum_{i=1}^{q} \mathbf{S}_{ii})^2] + [Tr(\sum_{i=1}^{q} \mathbf{S}_{ii})]^2}{q\{\sum_{i=1}^{q} \sum_{j=1}^{q} [(Tr\, \mathbf{S}_{ij})^2 + Tr(\mathbf{S}_{ij}^2)]} \tag{6.83}$$

$q = R(\mathbf{A})$ and v_h and v_e are the hypothesis and error degrees of freedom for the DMLM tests.

Using Wilks' Λ criterion, Boik (1988) shows how one may make the $\hat{\varepsilon}$-adjustment to the associated F-statistic. The result applies equally to the other criteria using Rao's F-approximation. In Program 6_5.sas, we use PROC IML code with the Bartlett-Lawley-Hotelling trace criterion, T_o^2. Then,

$$F = 2(sN+1)\, T_o^2 \,/\, s^2(2M+s+1) \,\dot{\sim}\, F[s(2M+s+1),\, 2(sN+1)] \tag{6.84}$$

where

$$s = \min(v_h^*, p),\ \ v_e^* = \hat{\varepsilon} q v_e,\ \ v_h^* = \hat{\varepsilon} q v_h$$
$$M = (\,|v_h^* - p| - 1\,)/2, \quad N = (v_e^* - p - 1)/2$$

The results are given in Output 6.5.

Output 6.5 (*continued*)

```
              Output 6.5: Double Multivariate Linear Model

                               EPSILON
                             0.7305055


          F_COND      F_GXC        DF1        DF2
       65.085984 0.9026246 4.3830331 30.308807

   Epsilon adjusted pvalues for MMM tests using T0**2 Criterion


                            P_COND      P_GXC
                         8.549E-15 0.4819311
```

From the output, we see that $\hat{\varepsilon} = 0.7305055$. The $\hat{\varepsilon}$-adjusted p-values are summarized in Table 6.5.

Table 6.5 $v_e\, T_o^2$ $\hat{\varepsilon}$-Adjusted *Analysis, Zullo Data*

Hypothesis	$v_e\, T_o^2$	F	df	p-value
Cond | Parallel	13.7514	65.090	(4.38, 30.31)	<0.0001
Group x Cond	0.1907	0.903	(4.38, 30.31)	0.482

The p-value for the unadjusted MMM was 0.4862 for the tests of parallelism using T_o^2. Using Wilks' Λ criterion, the p-values are 0.463 and 0.4687, for the adjusted MMM and the unadjusted test, respectively. The nominal p-values for the two adjusted tests are approximately equal 0.482 versus 0.463, for T_o^2 and Λ respectively. For this problem, the adjustment employing Wilks' Λ is better than that using T_o^2 since the Λ F-statistic is exact.

It appears that we have three competing strategies for the analysis of designs with vectors of repeated measures: the DMLM, the MMM, and the $\hat{\varepsilon}$-adjusted MMM. Given multivariate sphericity, the MMM is the most powerful. When multivariate sphericity does not hold, neither the adjusted MMMn or the DMLM analysis is the most powerful. Boik (1991) recommends using the DMLM. The choice between the two procedures depends on the ratio of the traces of the noncentrality matrices of the associated Wishart distributions which is seldom known in practice.

In this chapter we have illustrated the analysis of the DMLM with arbitrary covariance structure. When the covariance structure satisfies the multivariate sphericity condition, the Scheffé MMM analysis is optimal. However, in many repeated measurement experiments the structure of the covariance matrix $\tilde{\mathbf{\Phi}}$ may have Kronecker structure:

$$\tilde{\mathbf{\Phi}} = \mathbf{\Sigma}_e \otimes \mathbf{\Sigma}_t \tag{6.85}$$

where $\mathbf{\Sigma}_e$ is $p \times p$ and $\mathbf{\Sigma}_t$ is $t \times t$, and both matrices are positive definite, a structure commonly found in three-mode factor analysis (Benter and Lee, 1978). Under Release 6.12 of the SAS System, the MIXED procedure can be used to test hypotheses for the DMLM with Kronecker or direct product structure. For an illustration, see SAS Institute Inc. (1996, Chapter 5).

To test the null hypothesis that $\tilde{\mathbf{\Phi}}$ has the structure given in (6.85) versus the alternative that the rows $\mathbf{y}_i$ of $\mathbf{Y}$ have a general structure, under multivariate normality, we obtain a likelihood ratio test using the normal likelihood. The likelihood ratio statistic for testing the covariance structure is given by

$$\lambda = \frac{|\hat{\mathbf{\Sigma}}_{\mathbf{\Omega}_o}|^{n/2}}{|\hat{\mathbf{\Sigma}}_{\omega}|^{n/2}} = \frac{|\hat{\mathbf{\Sigma}}|^{n/2}}{|\mathbf{A}'\hat{\mathbf{\Sigma}}_e\mathbf{A}|^{np/2}\,|\hat{\mathbf{\Sigma}}_t|^{n(t-1)/2}} \tag{6.86}$$

where $\hat{\mathbf{\Sigma}}_e$ and $\hat{\mathbf{\Sigma}}_t$ are the maximum likelihood (ML) estimates of $\mathbf{\Sigma}_e$ and $\mathbf{\Sigma}_t$ under the null hypothesis, $\hat{\mathbf{\Sigma}} = \sum_{i=1}^{n} (\mathbf{y}_i - \overline{\mathbf{y}})(\mathbf{y}_i - \overline{\mathbf{y}})' / n$ is the ML estimate of the covariance matrix, and $\overline{\mathbf{y}} = \sum_{i=1}^{n} \mathbf{y}_i / n$.

To test for the structure in (6.85), the statistic -2lnλ is asymptotically distributed as a chi-square distribution with degrees of freedom $v = pt(pt+1)/2 - [p(p+1) + t(t+1) - 1]/2 = (p-1)(t-1)[(p+1)(t+1)+1]/2$. However, to solve the likelihood equations to obtain $\hat{\mathbf{\Sigma}}_e$ and $\hat{\mathbf{\Sigma}}_t$ involves an iterative process as outlined by Krishnaiah and Lee (1980), Boik (1991), and Naik and Rao (1996). Naik and Rao provide a computer program using the IML procedure to obtain the ML estimates. Under Release 6.12 of the SAS System, ML estimates may also be obtained using PROC MIXED.

Given that $\tilde{\mathbf{\Phi}}$ satisfies (6.85) and that $\mathbf{\Sigma}_e$ and $\mathbf{\Sigma}_t$ are estimated by $\hat{\mathbf{\Sigma}}_e$ and $\hat{\mathbf{\Sigma}}_t$, we develop a test of the extended linear hypothesis

$$\mathrm{H}: \psi = Tr(\mathbf{G}\mathbf{\Gamma}) = 0. \tag{6.87}$$

Following Mudholkar et al. (1974), a test statistic for testing (6.87) is

$$X^2 = \{U(\mathbf{G})\}^2 = [Tr(\mathbf{G}\hat{\mathbf{\Gamma}})]^2 / Tr[\mathbf{GC}(\mathbf{X}'\mathbf{X})^{-1}\mathbf{C}'\mathbf{G}(\hat{\mathbf{\Sigma}}_e \otimes \hat{\mathbf{\Sigma}}_t)] \tag{6.88}$$

since

$$\hat{\psi} \sim N(\psi, \hat{\sigma}^2_{\hat{\psi}} = Tr[\mathbf{GC}(\mathbf{X}'\mathbf{X})^{-1}\mathbf{C}'\mathbf{G}(\hat{\mathbf{\Sigma}}_e \otimes \hat{\mathbf{\Sigma}}_t)] \tag{6.89}$$

for fixed $\mathbf{G}$. The statistic X^2 converges to a chi-square distribution with one degree of freedom.

The statistic in (6.89) is an alternative to Boik's (1991) $\hat{\varepsilon}$-adjusted multivariate test procedure for the hypothesis $H: \mathbf{CB}(\mathbf{I}_p \otimes \mathbf{A}) = 0$ or $H: \mathbf{\Gamma} = 0$ when multivariate sphericity is not satisfied and $\tilde{\mathbf{\Phi}}$ has Kronecker structure. Naik and Rao (1996) have developed an alternative Satterthwaite-type approximate test for the MANOVA model when multivariate sphericity is not satisfied. The DMLM may also be analyzed using PROC MIXED as illustrated by Littell et al. (1996).

7. The Restricted Multivariate General Linear Model and the Growth Curve Model

Theory

Applications

7.1 Introduction

Repeated measurement data analysis is popular in the social and behavioral sciences and in other disciplines. A model that is useful for analyzing experiments involving repeated measurements is the restricted multivariate general linear model (RMGLM). Special cases of the RMGLM include the seemingly unrelated regression (SUR) model and the growth curve model of Potthoff and Roy (1964), also called the generalized MANOVA (GMANOVA) model. In this chapter we review the RMGLM and GMANOVA models. The canonical form of the GMANOVA model is also developed. While the SUR model is introduced in this chapter, it is discussed in detail in Chapter 8 along with the restricted GMANOVA model. A comprehensive discussion of linear and nonlinear growth curve models can be found in Vonesh and Chinchilli (1997).

7.2 The Restricted Multivariate General Linear Model

Following the classical matrix approach, we present the restricted multivariate general linear model (RMGLM):

$$\begin{aligned} \mathbf{Y}_{n\times p} &= \mathbf{X}_{n\times k}\mathbf{B}_{k\times p} + \mathbf{U}_{n\times p} \\ \operatorname{cov}(\mathbf{Y}) &= \mathbf{I}_n \otimes \boldsymbol{\Sigma} \\ \mathbf{R}_1\mathbf{B}\mathbf{R}_2 &= \boldsymbol{\Theta} \end{aligned} \tag{7.1}$$

where $\mathbf{R}_1(r_1 \times k)$ of rank r_1, $\mathbf{R}_2(p \times r_2)$ of rank r_2, and $\Theta_{r_1 \times r_2}$ are known matrices restricting the elements of the parameter matrix $\mathbf{B}$. From (7.1), we see that $\mathbf{R}_1$, $\mathbf{R}_2$ and Θ are similar to the hypothesis test matrices $\mathbf{C}$, $\mathbf{A}$ and $\mathbf{\Gamma}$ in the MANOVA model. Employing the vector version of the restricted model, we write (7.1) following (5.4) as follows:

$$
\begin{aligned}
\text{vec}(\mathbf{Y}) &= \mathbf{D}\ \text{vec}(\mathbf{B}) + \text{vec}(\mathbf{U}) \\
\mathbf{y} &= \mathbf{D}\beta + \mathbf{e} \\
\mathbf{R}\beta &= \theta
\end{aligned}
\tag{7.2}
$$

where $\mathbf{D} = \mathbf{I}_p \otimes \mathbf{X}, \mathbf{\Omega} = \mathbf{\Sigma} \otimes \mathbf{I}_n, \mathbf{R} = \mathbf{R}_2' \otimes \mathbf{R}_1$ and $\theta = \text{vec}(\Theta)$, a GLM. To estimate β in (7.2), recall that the estimator of β without restrictions from (5.8) is

$$
\begin{aligned}
\hat{\beta} &= (\mathbf{D}'\mathbf{\Omega}^{-1}\mathbf{D})^{-1}\mathbf{D}'\mathbf{\Omega}^{-1}\mathbf{y} = (\mathbf{I}_p \otimes (\mathbf{X}'\mathbf{X})^{-1}\mathbf{X}')\mathbf{y} \\
&= \text{vec}[(\mathbf{X}'\mathbf{X})^{-1}\mathbf{X}'\mathbf{Y}]
\end{aligned}
\tag{7.3}
$$

However, with $\mathbf{R} = \mathbf{R}_2' \otimes \mathbf{R}_1$, the restricted least squares (LS) estimator for β using model (7.2) is given in (3.1). Substituting $\hat{\beta}$ defined in (7.3) into (3.1) and simplifying, we find that the restricted LS estimator for β is

$$
\begin{aligned}
\hat{\beta}_r &= \hat{\beta} - (\mathbf{D}'\mathbf{D})^{-1}\mathbf{R}'[\mathbf{R}(\mathbf{D}'\mathbf{D})^{-1}\mathbf{R}']^{-1}(\mathbf{R}\hat{\beta} - \theta) \\
\hat{\mathbf{B}}_r &= \hat{\mathbf{B}} - (\mathbf{X}'\mathbf{X})^{-1}\mathbf{R}_1'(\mathbf{R}_1(\mathbf{X}'\mathbf{X})^{-1}\mathbf{R}_1')^{-1}(\mathbf{R}_1\hat{\mathbf{B}}\mathbf{R}_2 - \Theta)(\mathbf{R}_2'\mathbf{R}_2)^{-1}\mathbf{R}_2'
\end{aligned}
\tag{7.4}
$$

where $\hat{\mathbf{B}} = (\mathbf{X}'\mathbf{X})^{-1}\mathbf{X}\mathbf{Y}$. See Timm (1993a, p. 158) and Timm (1980b) for details on how the LS estimate $\hat{\mathbf{B}}_r$ is obtained by minimizing the least squares criterion $\text{Tr}[(\mathbf{Y} - \mathbf{X}\mathbf{B})'(\mathbf{Y} - \mathbf{X}\mathbf{B})]$. If one assumes matrix multivariate normality, the restricted LS estimate of $\mathbf{B}$ is not the ML estimate. To obtain the ML estimate of $\mathbf{B}$ given the restriction $\mathbf{R}_1\mathbf{B}\mathbf{R}_2 = \Theta$, one uses the matrix normal distribution given in Definition 5.1. As shown by Tubbs et al. (1975), the restricted ML estimate of $\mathbf{B}$ is

$$
\begin{aligned}
\hat{\mathbf{B}}_{ML} &= \hat{\mathbf{B}} - (\mathbf{X}'\mathbf{W}^{-1}\mathbf{X})^{-1}\mathbf{R}_1'(\mathbf{R}_1(\mathbf{X}'\mathbf{W}^{-1}\mathbf{X})^{-1}\mathbf{R}_1')^{-1}(\mathbf{R}_1\hat{\mathbf{B}}\mathbf{R}_2 - \Theta)(\mathbf{R}_2'\mathbf{\Sigma}\mathbf{R}_2)^{-1}\mathbf{R}_2'\mathbf{\Sigma} \\
&= \hat{\mathbf{B}} - (\mathbf{X}'\mathbf{X})^{-1}\mathbf{R}_1'(\mathbf{R}_1(\mathbf{X}'\mathbf{X})^{-1}\mathbf{R}_1')^{-1}(\mathbf{R}_1\hat{\mathbf{B}}\mathbf{R}_2 - \Theta)(\mathbf{R}_2'\mathbf{\Sigma}\mathbf{R}_2)^{-1}\mathbf{R}_2'\mathbf{\Sigma}
\end{aligned}
\tag{7.5}
$$

where $\mathbf{W} = \mathbf{I}$ and $\hat{\mathbf{B}} = (\mathbf{X}'\mathbf{X})^{-1}\mathbf{X}'\mathbf{Y}$. Hence, the restricted ML estimate of $\mathbf{B}$ involves the unknown matrix $\mathbf{\Sigma}$. However, from (7.4) observe that the ML and LS estimate of parametric functions of the form

$$
\Psi = \mathbf{C}\hat{\mathbf{B}}_r\mathbf{R}_2 = \mathbf{C}\hat{\mathbf{B}}\mathbf{R}_2 - (\mathbf{X}'\mathbf{X})^{-1}\mathbf{R}_1'(\mathbf{R}_1(\mathbf{X}'\mathbf{X})^{-1}\mathbf{R}_1')^{-1}(\mathbf{R}_1\hat{\mathbf{B}}\mathbf{R}_2 - \Theta)
\tag{7.6}
$$

do not contain the unknown covariance matrix $\mathbf{\Sigma}$ and are equal. Thus, we should be able to develop a test of $H(\mathbf{CBA} = \mathbf{\Gamma})$, if the restriction takes the form $\mathbf{R}_1\mathbf{B}\mathbf{R}_2 = \mathbf{R}_1\mathbf{B}\mathbf{A} = \Theta$, that is independent of the unknown value for $\mathbf{\Sigma}$ in the restricted ML estimate of $\mathbf{B}$. To derive a test statistic, we need an unbiased estimator of $\mathbf{\Sigma}$ under the restriction $\mathbf{R}_1\mathbf{B}\mathbf{R}_2 = \Theta$. The estimated restricted error sums of squares and products (SSP) matrix is

$$
\begin{aligned}
\tilde{\mathbf{E}} &= (\mathbf{Y} - \mathbf{X}\hat{\mathbf{B}}_r)'(\mathbf{Y} - \mathbf{X}\hat{\mathbf{B}}_r) \\
&= (\mathbf{Y} - \mathbf{X}\hat{\mathbf{B}})'(\mathbf{Y} - \mathbf{X}\hat{\mathbf{B}}) + (\hat{\mathbf{B}} - \hat{\mathbf{B}}_r)'(\mathbf{X}'\mathbf{X})(\hat{\mathbf{B}} - \hat{\mathbf{B}}_r) \\
&= \mathbf{E} + \mathbf{R}_2(\mathbf{R}_2'\mathbf{R}_2)^{-1}(\mathbf{R}_1\hat{\mathbf{B}}\mathbf{R}_2 - \Theta)'(\mathbf{R}_1(\mathbf{X}'\mathbf{X})^{-1}\mathbf{R}_1')^{-1}(\mathbf{R}_2'\mathbf{R}_2')^{-1}\mathbf{R}_2'
\end{aligned}
\tag{7.7}
$$

where $\mathbf{E} = \mathbf{Y}'(\mathbf{I} - \mathbf{X}(\mathbf{X}'\mathbf{X})^{-1}\mathbf{X}')\mathbf{Y}$. The matrix $\mathbf{S} = \tilde{\mathbf{E}}/(n - k + r_1)$ is an unbiased estimate of $\mathbf{\Sigma}$ given the restriction since $E(\tilde{\mathbf{E}}) = (n-k)\mathbf{\Sigma} + r_1\mathbf{\Sigma} = (n - k + r_1)\mathbf{\Sigma}$. Under multivariate normality, $\mathbf{R}_2'\tilde{\mathbf{E}}\mathbf{R}_2 \sim W_{r_2}(n - k + r_1, \mathbf{R}_2'\mathbf{\Sigma}\mathbf{R}_2)$.

To test hypotheses of the form $H(\mathbf{CBA}=\mathbf{\Gamma})$ assuming (7.1) is complicated by the restrictions. We consider two situations. The first situation is to assume that either $\mathbf{R}_2=\mathbf{A}$ and $\mathbf{A}$ is arbitrary or that $\mathbf{R}_2=\mathbf{A}=\mathbf{I}$ where no row of $\mathbf{C}$ is equal to or dependent on any row of $\mathbf{R}_1$ or other rows of $\mathbf{C}$. Following (3.4) for the univariate model, we let

$$\mathbf{Q}=\begin{pmatrix}\mathbf{R}_1\\ \mathbf{C}\end{pmatrix},\ \mathbf{P}=(\mathbf{R}_2,\mathbf{A})\text{ and }\eta=\begin{pmatrix}\mathbf{\Theta} & \mathbf{K}\\ \mathbf{\Pi} & \mathbf{\Gamma}\end{pmatrix}$$

so that

$$\mathbf{QBP}=\begin{pmatrix}\mathbf{R}_1\mathbf{BR}_2 & \mathbf{R}_1\mathbf{BA}\\ \mathbf{CBR}_2 & \mathbf{CBA}\end{pmatrix}=\begin{pmatrix}\mathbf{\Theta} & \mathbf{K}\\ \mathbf{\Pi} & \mathbf{\Gamma}\end{pmatrix}$$

Then, the restricted LS estimator of $\mathbf{B}$ under the hypothesis $H(\mathbf{CBA}=\mathbf{\Gamma})$ for the restriction $\mathbf{R}_1\mathbf{BA}=\mathbf{\Theta}$ is

$$\hat{\mathbf{B}}_{r,\omega}=\hat{\mathbf{B}}-(\mathbf{X'X})^{-1}\mathbf{Q}'(\mathbf{Q}(\mathbf{X'X})^{-1}\mathbf{Q}')^{-1}(\mathbf{Q}\hat{\mathbf{B}}\mathbf{P}-\eta)(\mathbf{P'P})^{-1}\mathbf{P}' \tag{7.8}$$

and the error SSP matrix is

$$\begin{aligned}\mathbf{E}_r^* &= (\mathbf{Y}-\hat{\mathbf{B}}_{r,\omega})'(\mathbf{Y}-\hat{\mathbf{B}}_{r,\omega})\\ &= (\mathbf{Y}-\mathbf{X}\hat{\mathbf{B}})'(\mathbf{Y}-\mathbf{X}\hat{\mathbf{B}})+(\hat{\mathbf{B}}_r-\hat{\mathbf{B}}_{r,\omega})'(\mathbf{X'X})(\hat{\mathbf{B}}_r-\hat{\mathbf{B}}_{r,\omega})\\ &= \mathbf{E}+\mathbf{A}(\mathbf{A'A})^{-1}(\mathbf{C}\hat{\mathbf{B}}\mathbf{A}-\mathbf{\Gamma})'(\mathbf{C}(\mathbf{X'X})^{-1}\mathbf{C}')^{-1}(\mathbf{C}\hat{\mathbf{B}}\mathbf{A}-\mathbf{\Gamma})'(\mathbf{A'A})^{-1}\mathbf{A}'\end{aligned} \tag{7.9}$$

where $\mathbf{E}=\mathbf{Y}'(\mathbf{I}-\mathbf{X}(\mathbf{X'X})^{-1}\mathbf{X}')\mathbf{Y}$, $\hat{\mathbf{B}}=(\mathbf{X'X})^{-1}\mathbf{X'Y}$ and $\hat{\mathbf{B}}_r$ is the restricted LS estimator defined in (7.4) (Timm, 1993a). In addition to these matrices, to test the hypothesis

$$H\colon\ \mathbf{CBA}=\mathbf{\Gamma} \tag{7.10}$$

where $g=\text{rank}(\mathbf{C})=v_h$ and $u=\text{rank}(\mathbf{A})$, we use the following matrices

$$\begin{aligned}\mathbf{H}_r &= (\mathbf{C}\hat{\mathbf{B}}_r\mathbf{A}-\mathbf{\Gamma})'(\mathbf{CFGC}')^{-1}(\mathbf{C}\hat{\mathbf{B}}_r\mathbf{A}-\mathbf{\Gamma})\\ \mathbf{F} &= \mathbf{I}-(\mathbf{X'X})^{-1}\mathbf{R}_1'(\mathbf{R}_1(\mathbf{X'X})^{-1}\mathbf{R}_1')^{-1}\mathbf{R}_1\\ \mathbf{E}_r &= \mathbf{A}'\mathbf{E}_r^*\mathbf{A}=\mathbf{A'EA}+(\mathbf{C}\hat{\mathbf{B}}\mathbf{A}-\mathbf{\Theta})'(\mathbf{C}(\mathbf{X'X})^{-1}\mathbf{C}')^{-1}(\mathbf{C}\hat{\mathbf{B}}\mathbf{A}-\mathbf{\Theta})\end{aligned} \tag{7.11}$$

where $\mathbf{F}=\mathbf{F}^2$ is idempotent, $\mathbf{G}=(\mathbf{X'X})^{-1}$ and $\mathbf{H}_r$ and $\mathbf{E}_r$ have independent Wishart distributions

$$\begin{aligned}\mathbf{E}_r &\sim W_u(v_e=n-k+r_1,\ \mathbf{A}'\mathbf{\Sigma}\mathbf{A})\\ \mathbf{H}_r &\sim W_u(v_h=g,\ \mathbf{A}'\mathbf{\Sigma}\mathbf{A},\ (\mathbf{A}'\mathbf{\Sigma}\mathbf{A})^{-1}\mathbf{\Delta})\end{aligned} \tag{7.12}$$

with noncentrality matrix

$$\mathbf{\Delta}=(\mathbf{\Gamma}-\mathbf{\Gamma}_o)'(\mathbf{CFG}^{-1}\mathbf{C}')^{-1}(\mathbf{\Gamma}-\mathbf{\Gamma}_o).$$

The test statistic assuming $\mathbf{A}=\mathbf{R}_2$ by (7.6) is

$$\lambda_*=\left\{\frac{|(\mathbf{Y}-\mathbf{X}\hat{\mathbf{B}}_r)'(\mathbf{Y}-\mathbf{X}\hat{\mathbf{B}}_r)|}{|(\mathbf{Y}-\mathbf{X}\hat{\mathbf{B}}_{r,\omega})'(\mathbf{Y}-\mathbf{X}\hat{\mathbf{B}}_{r,\omega})|}\right\}^{n/2} \tag{7.13}$$

On simplifying (7.13) using (3.8) for partitioned matrices, the U-statistic becomes

$$\Lambda=\lambda_*^{2/n}=\frac{|\mathbf{E}_r|}{|\mathbf{E}_r+\mathbf{H}_r|} \tag{7.14}$$

where $v_h=g=\text{rank}(\mathbf{C})$ and $v_e=n-k+r_1$. The other multivariate criteria follow by using $\mathbf{E}_r$ and $\mathbf{H}_r$, in place of $\mathbf{H}$ and $\mathbf{E}$ in (5.29).

To obtain $1-\alpha$ simultaneous confidence intervals for bilinear forms $\psi = \mathbf{c}'\mathbf{B}\mathbf{a}$, one finds that the best estimator for ψ is $\hat{\psi} = \mathbf{c}'\hat{\mathbf{B}}_r\mathbf{a}$ and $\hat{\sigma}_\psi^2 = (\mathbf{a}'\mathbf{S}\mathbf{a})\mathbf{c}'(\mathbf{F}\mathbf{G})\mathbf{c}$ where

$$\mathbf{S} = \mathbf{E}_r / (n-k+r_1) \tag{7.15}$$

so that $E(\mathbf{S}) = \mathbf{\Sigma}$. Extended linear hypotheses may also be tested using the restricted model.

Assuming (7.1), we have discussed how to test the hypothesis $H(\mathbf{CBA} = \mathbf{\Gamma})$ in the situation where the matrix $\mathbf{R}_2$ in the restriction $\mathbf{R}_1\mathbf{B}\mathbf{R}_2 = \mathbf{\Theta}$ is chosen equal to $\mathbf{A}$ in the hypothesis. The situation where one would like to test $\mathbf{CBA} = \mathbf{\Gamma}$ assuming $\mathbf{R}_1\mathbf{B}\mathbf{R}_2 = \mathbf{\Theta}$ for arbitrary $\mathbf{R}_2$ and $\mathbf{A}$ is more difficult since $\mathbf{A}$ and the structure of $\mathbf{\Sigma}$ are associated (Timm, 1996). This association occurs, for example, in the SUR or multiple design multivariate (MDM) model and the growth curve model. To see this, suppose we associate with $\mathbf{R}_1\mathbf{B}\mathbf{R}_2$ the more general structure $\mathbf{R}_*\mathbf{B} = \mathbf{\Theta}$ and assume $\mathbf{\Theta} = \mathbf{0}$. Then if we incorporate the restriction $\mathbf{R}_*\mathbf{B} = \mathbf{0}$ into (7.2) by setting appropriate parameters to zero and modifying the columns of $\mathbf{D} = \mathbf{I}_p \otimes \mathbf{X}$, (7.2) becomes

$$\underset{\mathbf{y}}{\begin{pmatrix} \mathbf{y}_1 \\ \mathbf{y}_2 \\ \vdots \\ \vdots \\ \vdots \\ \mathbf{y}_p \end{pmatrix}} = \underset{\mathbf{D}}{\begin{pmatrix} \mathbf{X}_1 & \mathbf{0} & \cdots & \cdots & \cdots & \mathbf{0} \\ \mathbf{0} & \mathbf{X}_2 & \cdots & \cdots & \cdots & \mathbf{0} \\ \vdots & \cdots & \ddots & \cdots & \cdots & \vdots \\ \vdots & \cdots & \cdots & \ddots & \cdots & \vdots \\ \vdots & \cdots & \cdots & \cdots & \ddots & \vdots \\ \mathbf{0} & \mathbf{0} & \cdots & \cdots & \cdots & \mathbf{X}_p \end{pmatrix}} \underset{\beta}{\begin{pmatrix} \beta_{11} \\ \beta_{22} \\ \vdots \\ \vdots \\ \vdots \\ \beta_{pp} \end{pmatrix}} + \underset{\mathbf{e}}{\begin{pmatrix} \mathbf{e}_1 \\ \mathbf{e}_2 \\ \vdots \\ \vdots \\ \vdots \\ \mathbf{e}_p \end{pmatrix}} \tag{7.16}$$

where $\mathbf{e} \sim \mathbf{N}_{np}(\mathbf{0}, \mathbf{\Omega} = \mathbf{\Sigma} \otimes \mathbf{I}_n)$. Then, the BLUE of β is

$$\hat{\beta} = (\mathbf{D}'\mathbf{\Omega}^{-1}\mathbf{D})^{-1}\mathbf{D}'\mathbf{\Omega}^{-1}\mathbf{y} \tag{7.17}$$

so that $\mathbf{\Sigma}$ does not drop out of equation (7.17). Model (7.16) is called the SUR model, Zellner's model, or the MDM model. Using matrix notation, we may write (7.16) as

$$\begin{aligned} \mathbf{Y}_{n \times p} &= [\mathbf{X}_1\beta_{11}, \mathbf{X}_2\beta_{22}, \ldots, \mathbf{X}_p\beta_{pp}] + \mathbf{U} \\ &= [\mathbf{X}_1, \mathbf{X}_2, \ldots, \mathbf{X}_p]\tilde{\mathbf{B}} + \mathbf{U} \\ \text{cov}(\mathbf{Y}) &= \mathbf{I}_n \otimes \mathbf{\Sigma} \end{aligned} \tag{7.18}$$

where

$$\tilde{\mathbf{B}} = \begin{pmatrix} \beta_{11} & \cdots & \cdots & \cdots & \cdots & \mathbf{0} \\ \vdots & \beta_{22} & \cdots & \cdots & \cdots & \vdots \\ \vdots & \cdots & \ddots & \cdots & \cdots & \vdots \\ \vdots & \cdots & \cdots & \ddots & \cdots & \vdots \\ \vdots & \cdots & \cdots & \cdots & \ddots & \vdots \\ \mathbf{0} & \cdots & \cdots & \cdots & \cdots & \beta_{pp} \end{pmatrix} \tag{7.19}$$

$\beta_{ii}(k_i \times 1), i = 1, 2, \ldots, p$ and $k = \sum_i k_i$. When $\mathbf{X}_1 = \mathbf{X}_2 = \ldots = \mathbf{X}_p = \mathbf{X}$, (7.18) reduces to the MANOVA or univariate regression model discussed in Chapter 5. If the design matrices are not equal, but $\mathbf{\Omega} = \sigma^2\mathbf{I}$ we have the independent regressions model; when the variables are independent, the problem is greatly simplified (Timm, 1975, pp. 331-347). The SUR model is discussed in Chapter 8.

7.3 The GMANOVA Model

Testing the restriction $\mathbf{R}_1\mathbf{B}\mathbf{R}_2 = \mathbf{\Theta}$ in (7.1) when $\mathbf{R}_2$ and $\mathbf{\Sigma}$ are associated is a special case of the Potthoff and Roy (1964) growth curve model also called the generalized MANOVA (GMANOVA) model. To see this, let the reduced form of the MANOVA model be defined as

$$\mathbf{Y}_{n\times p} = \mathbf{K}_{n\times q}\, \xi_{q\times p} + \mathbf{U}_{n\times p} \tag{7.20}$$

subject to the joint linear restrictions

$$\mathbf{R}_{*(q-k)\times q}\, \xi_{q\times p} = \mathbf{0} \;\text{ and }\; \xi_{q\times p}\mathbf{Q}_{*\,p\times(p-q)} = \mathbf{0} \tag{7.21}$$

where the $\text{rank}(\mathbf{R}_*) = q-k$ and the $\text{rank}(\mathbf{Q}_*) = p-q < p$. Then we may find a matrix $\mathbf{X}_2'$ $(p\times q)$ orthogonal to $\mathbf{Q}_*$ and a matrix $\mathbf{D}'(k\times q)$ orthogonal to $\mathbf{R}_*$ so that ξ may be represented

$$\xi = \mathbf{D}_{q\times k}\, \mathbf{B}_{k\times q}\, \mathbf{X}_{2(q\times p)} \tag{7.22}$$

where $\mathbf{B}_{k\times q}$ is a reparameterized matrix that is unrestricted. Then, (7.20) becomes

$$\begin{aligned} \mathbf{Y} &= \mathbf{KDBX}_2 + \mathbf{U} \\ \underset{n\times p}{\mathbf{Y}} &= \underset{n\times k}{\mathbf{X}_1}\,\underset{k\times q}{\mathbf{B}}\,\underset{q\times p}{\mathbf{X}_2} + \underset{n\times p}{\mathbf{U}} \\ \text{cov}(\mathbf{Y}) &= \mathbf{I}_n \otimes \mathbf{\Sigma} \end{aligned} \tag{7.23}$$

where $\mathbf{KD} = \mathbf{X}_1(\text{rank }\mathbf{X}_1 = k)$ and $\mathbf{X}_2(\text{rank }\mathbf{X}_2 = q)$ are known and $\mathbf{B}$ is the matrix of parameters. This is called the (unrestricted) GMANOVA model. Assuming (7.23), we can test the linear hypothesis

$$\mathbf{R}_1\mathbf{B}\mathbf{R}_2 = \mathbf{\Theta} \tag{7.24}$$

which is usually written as $\mathbf{CBA} = \mathbf{\Gamma}$. Letting $\mathbf{X}_2 = \mathbf{I}$, the GMANOVA model reduces to the MANOVA model. To show that (7.23) is again a special case of the GLM defined by (1.1), we define

$$\mathbf{X} = \mathbf{X}_2' \otimes \mathbf{X}_1,\; \mathbf{\Omega} = \mathbf{\Sigma}\otimes\mathbf{I}_n \;\text{ and }\; \mathbf{y}' = (\mathbf{y}_1', \mathbf{y}_2', \ldots, \mathbf{y}_p') \tag{7.25}$$

where $\mathbf{y}_i$ $(n\times 1)$, $\mathbf{y} = \text{vec}(\mathbf{Y})$, $\beta = \text{vec}(\mathbf{B})$, and $\mathbf{e} = \text{vec}(\mathbf{U})$ so that $\mathbf{y} = \mathbf{X}\beta + \mathbf{e}$

The GMANOVA model is a special case of the extended GMANOVA model. The extended or restricted GMANOVA model has the general form

$$\mathbf{Y} = \mathbf{X}_1\mathbf{B}\mathbf{X}_2 + \mathbf{U} \;\text{ with }\; \mathbf{X}_3\mathbf{B}\mathbf{X}_4 = \mathbf{\Theta} \tag{7.26}$$

where the restrictions are part of the model. Given (7.26), one may test hypotheses of the form

$$H\text{: } \mathbf{X}_5\mathbf{B}\mathbf{X}_6 = \mathbf{\Gamma} \text{ or } \mathbf{CBA} = \mathbf{\Gamma} \tag{7.27}$$

Setting $\mathbf{X}_2 = \mathbf{I}$ in (7.26), the MANOVA model with general double linear restrictions is a special case of the extended GMANOVA model where $\mathbf{X}_6 \neq \mathbf{X}_4$. Thus, we have that the MANOVA model is a special case of the GMANOVA model and that the general double linear restricted MANOVA model is a special case of the extended (restricted) GMANOVA model. Before discussing hypothesis testing using the MANOVA model with double linear restrictions which is complicated by the association of $\mathbf{R}_2$, $\mathbf{A}$ and $\mathbf{\Sigma}$, we will review the GMANOVA model with no restrictions.

Given the GMANOVA model in (7.23) without restrictions, the BLUE or Gauss-Markov (GM) estimate of $\mathbf{B}$ is obtained by using representation (7.25) and (4.6) so that

$$\begin{aligned}
\text{vec}(\hat{\mathbf{B}}) = \hat{\beta} &= (\mathbf{X}'\mathbf{\Omega}^{-1}\mathbf{X})^{-1}\mathbf{X}'\mathbf{\Omega}^{-1}\mathbf{y} \\
&= [(\mathbf{X}_2' \otimes \mathbf{X}_1)'(\mathbf{\Sigma} \otimes \mathbf{I}_n)^{-1}(\mathbf{X}_2' \otimes \mathbf{X}_1)]^{-1}(\mathbf{X}_2' \otimes \mathbf{X}_1)'(\mathbf{\Sigma} \otimes \mathbf{I}_n)^{-1}\mathbf{y} \\
&= [(\mathbf{X}_2\mathbf{\Sigma}^{-1}\mathbf{X}_2')^{-1}\mathbf{X}_2\mathbf{\Sigma}^{-1} \otimes (\mathbf{X}_1'\mathbf{X}_1)^{-1}\mathbf{X}_1']\mathbf{y} \qquad (7.28) \\
&= \text{vec}[(\mathbf{X}_1'\mathbf{X}_1)^{-1}\mathbf{X}_1'\mathbf{Y}\mathbf{\Sigma}^{-1}\mathbf{X}_2'(\mathbf{X}_2\mathbf{\Sigma}^{-1}\mathbf{X}_2')^{-1}] \\
\hat{\mathbf{B}}_{GM} &= (\mathbf{X}_1'\mathbf{X}_1)^{-1}\mathbf{X}_1'\mathbf{Y}\mathbf{\Sigma}^{-1}\mathbf{X}_2'(\mathbf{X}_2\mathbf{\Sigma}^{-1}\mathbf{X}_2')^{-1}
\end{aligned}$$

Adding the restriction $\mathbf{R}\beta = (\mathbf{R}_2' \otimes \mathbf{R}_1)\beta = \theta$ to the growth curve model and using (3.1) with β defined in (7.28), the BLUE or Gauss-Markov (GM) estimator of β, using matrix notation, is

$$\hat{\mathbf{B}}_{GM_r} = \hat{\mathbf{B}}_{GM} - \mathbf{P}_1(\mathbf{R}_1\hat{\mathbf{B}}\mathbf{R}_2 - \mathbf{\Theta})\mathbf{P}_2 \qquad (7.29)$$

where

$$\begin{aligned}
\mathbf{P}_1 &= (\mathbf{X}_1'\mathbf{X}_1)^{-1}\mathbf{R}_1'[\mathbf{R}_1(\mathbf{X}_1'\mathbf{X}_1)^{-1}\mathbf{R}_1']^{-1} \\
\mathbf{P}_2 &= [\mathbf{R}_2'(\mathbf{X}_2\mathbf{\Sigma}^{-1}\mathbf{X}_2')^{-1}\mathbf{R}_2]^{-1}\mathbf{R}_2'(\mathbf{X}_2\mathbf{\Sigma}^{-1}\mathbf{X}_2')^{-1}
\end{aligned}$$

From (7.28) we see that if $\mathbf{X}_2 = \mathbf{I}$ the GMANOVA estimator for $\mathbf{B}$ reduces to the OLSE for $\mathbf{B}$ under the MANOVA model, $\hat{\mathbf{B}} = (\mathbf{X}'\mathbf{X})^{-1}\mathbf{X}'\mathbf{Y}$. Furthermore, the BLUE estimator for $\mathbf{B}$, $\hat{\mathbf{B}}_{GM_r}$ in (7.29) reduces to the ML estimate for $\mathbf{B}$ in (7.5) obtained in the MANOVA model with double linear restrictions and known $\mathbf{\Sigma}$.

Potthoff and Roy (1964) recognized (7.28) and recommended reducing the GMANOVA model to the MANOVA model through the transformation

$$\mathbf{Y}_o = \mathbf{Y}\mathbf{G}^{-1}\mathbf{X}_2'(\mathbf{X}_2\mathbf{G}^{-1}\mathbf{X}_2')^{-1} \qquad (7.30)$$

where $\mathbf{G}_{p\times p}$ is any symmetric nonsingular matrix. If we use (7.30), the GMANOVA model is reduced to the MANOVA model where $E(\mathbf{Y}_o) = \mathbf{X}_1\mathbf{B}$, $\text{cov}(\mathbf{Y}_o) = \mathbf{I}_n \otimes \mathbf{\Sigma}_o$ and $\mathbf{\Sigma}_o$ is

$$\mathbf{\Sigma}_{o(q\times q)} = (\mathbf{X}_2\mathbf{G}^{-1}\mathbf{X}_2')^{-1}\mathbf{X}_2\mathbf{G}^{-1}\mathbf{\Sigma}\mathbf{G}^{-1}\mathbf{X}_2'(\mathbf{X}_2\mathbf{G}^{-1}\mathbf{X}_2')^{-1} \qquad (7.31)$$

Hence, the MANOVA theory discussed in Chapter 5 may be used to test

$$H\colon\ \mathbf{CBA} = \mathbf{\Gamma}$$

by using the transformed data matrix $\mathbf{Y}_o$ given the unrestricted GMANOVA model (7.23). From (7.30), the optimal choice of $\mathbf{G}$ is $\mathbf{\Sigma}$, but in practice $\mathbf{\Sigma}$ is unknown. When $p = q$, $\hat{\mathbf{B}}$ does not depend on $\mathbf{\Sigma}$ for either the restricted GMANOVA model or the (unrestricted) GMANOVA model and the transformation becomes

$$\mathbf{Y}_o = \mathbf{Y}\mathbf{X}_2^{-1} \qquad (7.32)$$

so that there is no need to choose $\mathbf{G}$. If $q < p$, however, the choice of $\mathbf{G}$ is important since it affects the power of the MANOVA tests, the variance of the parameters, and the widths of confidence intervals. Potthoff and Roy recommended selecting $\mathbf{G} = \hat{\mathbf{\Sigma}}$ where $\hat{\mathbf{\Sigma}}$ was a consistent estimator of $\mathbf{\Sigma}$ based on a different, independent set of data.

To avoid the arbitrary choice of $\mathbf{G}$, Rao (1965, 1966) and Khatri (1966) derived an LR test for the GMANOVA model. Rao transformed the model to a MANCOVA model and developed a conditional LR test for the GMANOVA model. Khatri used the ML method and a conditional argument to obtain the LR test criterion. Gleser and Olkin (1970) reduced the model to canonical form and using invariance arguments derived a LR test; they also showed that the Rao-Khatri test was the unconditional LR test.

To develop an LR ratio test of $H(\mathbf{CBA} = \mathbf{\Gamma})$ under the growth curve model defined in (7.23), Khatri (1966) derived the ML estimate of $\mathbf{B}$ and $\mathbf{\Sigma}$ under multivariate normality

$$\begin{aligned} \hat{\mathbf{B}}_{ML} &= (\mathbf{X}_1'\mathbf{X}_1)^{-1}\mathbf{X}_1\mathbf{Y}\mathbf{E}\mathbf{X}_2'(\mathbf{X}_2\mathbf{E}^{-1}\mathbf{X}_2')^{-1} \\ n\hat{\mathbf{\Sigma}} &= (\mathbf{Y} - \mathbf{X}_1\hat{\mathbf{B}}_{ML}\mathbf{X}_2)'(\mathbf{Y} - \mathbf{X}_1\hat{\mathbf{B}}_{ML}\mathbf{X}_2) \end{aligned} \tag{7.33}$$

where $\mathbf{E} = \mathbf{Y}'(\mathbf{I} - \mathbf{X}_1(\mathbf{X}_1'\mathbf{X}_1)^{-1}\mathbf{X}_1')\mathbf{Y}$. Comparing (7.33) with (7.28), we see that the BLUE or Gauss-Markov (GM) estimate of $\mathbf{B}$ and the ML estimate have the same form, but that they are not equal since $\mathbf{\Sigma}$ is usually unknown and not equal to its estimate. Rao (1967) derived the covariance matrix of $\mathbf{C}\hat{\mathbf{B}}_{ML}\mathbf{A}$:

$$\text{cov}(\mathbf{C}\hat{\mathbf{B}}_{ML}\mathbf{A}) = \mathbf{C}(\mathbf{X}_1'\mathbf{X}_1)\mathbf{C}' \otimes \frac{n-k-1}{n-k-(p-q)-1}\mathbf{A}'(\mathbf{X}_2\mathbf{\Sigma}^{-1}\mathbf{X}_2')^{-1}\mathbf{A} \tag{7.34}$$

The corresponding covariance matrix for $\mathbf{C}\hat{\mathbf{B}}_{GM}$ $\mathbf{A}$ is

$$\text{cov}(\mathbf{C}\hat{\mathbf{B}}_{GM}\mathbf{A}) = \mathbf{C}(\mathbf{X}_1'\mathbf{X})^{-1}\mathbf{C}' \otimes \mathbf{A}'(\mathbf{X}_2\mathbf{\Sigma}^{-1}\mathbf{X}_2')^{-1}\mathbf{A} \tag{7.35}$$

which agrees with the ML estimate except for the multiplication factor. The two are equal if $p = q$; otherwise, the GM estimator is best. Because $\mathbf{\Sigma}$ is unknown, an estimate of the covariance matrix is obtained by substituting an estimate $\hat{\mathbf{\Sigma}}$ for $\mathbf{\Sigma}$. One can also easily establish the class of covariance matrices that ensure equality of the GM estimator $\hat{\mathbf{B}}_{GM}$ and the OLSE $\hat{\mathbf{B}}$ for the GMANOVA model by applying Theorem 4.1 to the GMANOVA case. Then, $\mathbf{\Sigma} = \mathbf{X}_2'\mathbf{\Delta}\mathbf{X}_2 + \mathbf{Z}_2'\mathbf{\Gamma}\mathbf{Z}_2 + \sigma^2\mathbf{I}_n$, (Kariya, 1985, pp. 19-31).

To test the hypothesis $H(\mathbf{CBA} = \mathbf{\Gamma})$ under the GMANOVA model in (7.23), Khatri (1966), using maximum likelihood methods, and Rao (1965, 1966), using covariance adjustment, reduced the GMANOVA model to the conditional MANCOVA model

$$E(\mathbf{Y}_o|\mathbf{Z}) = \mathbf{X}_1\mathbf{B} + \mathbf{Z}\mathbf{\Gamma}_o \tag{7.36}$$

where $\mathbf{Y}_{o(n\times q)}$ is the transformed data matrix, $\mathbf{X}_1$ $(n \times k)$ is the known design matrix, $\mathbf{B}_{k\times q}$ is a matrix of parameters, $\mathbf{Z}_{n\times h}$ is a matrix of covariates, and $\mathbf{\Gamma}_{o(h\times q)}$ is a matrix of unknown regression coefficients.

To reduce (7.23) to (7.36), a $(p \times p)$ nonsingular matrix $\mathbf{H} = (\mathbf{H}_1, \mathbf{H}_2)$ is constructed such that $\mathbf{X}_2\mathbf{H}_1 = \mathbf{I}$ and $\mathbf{X}_2\mathbf{H}_2 = \mathbf{0}$. If the rank of $\mathbf{X}_2$ is q, $\mathbf{H}_1$ and $\mathbf{H}_2$ can be selected

$$\mathbf{H}_1 = \mathbf{G}^{-1}\mathbf{X}_2'(\mathbf{X}_2\mathbf{G}^{-1}\mathbf{X}_2')^{-1}, \text{ and } \mathbf{H}_2 = \mathbf{I} - \mathbf{H}_1\mathbf{X}_2 \tag{7.37}$$

where $\mathbf{G}$ is an arbitrary nonsingular matrix. Such a matrix $\mathbf{H}$ is not unique; however, estimates and tests are invariant for all choices of $\mathbf{H}$ satisfying the specified conditions (Khatri, 1966). Hence, $\mathbf{G}$ in the expressions for $\mathbf{H}_1$ does not affect estimates or tests under (7.36). If we equate

$$\begin{aligned} \mathbf{Y}_o &= \mathbf{Y}\mathbf{H}_1 = \mathbf{Y}\mathbf{G}^{-1}\mathbf{X}_2'(\mathbf{X}_2\mathbf{G}^{-1}\mathbf{X}_2')^{-1} \\ \mathbf{Z} &= \mathbf{Y}\mathbf{H}_2 = \mathbf{Y}(\mathbf{I} - \mathbf{H}_1\mathbf{X}_2) = \mathbf{Y}[\mathbf{I} - \mathbf{G}^{-1}\mathbf{X}_2'(\mathbf{X}_2\mathbf{G}^{-1}\mathbf{X}_2')^{-1}\mathbf{X}_2] \end{aligned} \tag{7.38}$$

$\mathbf{E}(\mathbf{Y}_o) = \mathbf{X}_1\mathbf{B}$ and $\mathbf{E}(\mathbf{Z}) = \mathbf{0}$; thus, the expected value of $\mathbf{Y}$ given $\mathbf{Z}$ is seen to be of the form specified in (7.36) (Grizzle and Allen, 1969, p. 362). Using this Rao-Khatri conditional model, the information in the covariates **Z,** which is ignored in the Potthoff-Roy reduction, is utilized. To find the BLUE of $\mathbf{B}$ and $\mathbf{\Gamma}_o$ under (7.36), we minimize the $\text{Tr}\{[\mathbf{Y} - \mathbf{E}(\mathbf{Y}_o|\mathbf{Z})]'[\mathbf{Y} - \mathbf{E}(\mathbf{Y}_o|\mathbf{Z})]\}$ to obtain the normal equations. The normal equations are

$$\begin{pmatrix} \mathbf{X}_1'\mathbf{X}_1 & \mathbf{X}_1'\mathbf{Z} \\ \mathbf{Z}'\mathbf{X}_1 & \mathbf{Z}'\mathbf{Z} \end{pmatrix} \begin{pmatrix} \mathbf{B} \\ \mathbf{\Gamma}_o \end{pmatrix} = \begin{pmatrix} \mathbf{X}_1'\mathbf{Y}_o \\ \mathbf{Z}'\mathbf{Y}_o \end{pmatrix} \tag{7.39}$$

Using (3.8), we have that

$$\begin{pmatrix} \mathbf{X}_1'\mathbf{X}_1 & \mathbf{X}_1'\mathbf{Z} \\ \mathbf{Z}'\mathbf{X}_1 & \mathbf{Z}'\mathbf{Z} \end{pmatrix}^{-1} = \begin{bmatrix} (\mathbf{X}_1'\mathbf{X}_1)^{-1} & \mathbf{0} \\ \mathbf{0} & \mathbf{0} \end{bmatrix} + \begin{bmatrix} -(\mathbf{X}_1'\mathbf{X}_1)^{-1}\mathbf{X}_1'\mathbf{Z} \\ \mathbf{I} \end{bmatrix} \left(\mathbf{Z}'\overline{\mathbf{X}}_1\mathbf{Z}\right)^{-1} \left[-\mathbf{Z}'\mathbf{X}_1(\mathbf{X}_1'\mathbf{X}_1)^{-1}, \mathbf{I}\right] \tag{7.40}$$

where $\overline{\mathbf{X}}_1 = \mathbf{I} - \mathbf{X}_1(\mathbf{X}_1'\mathbf{X}_1)^{-1}\mathbf{X}_1'$ so that the BLUE of $\mathbf{B}$ and $\mathbf{\Gamma}$ for the MANCOVA model are

$$\begin{aligned} \hat{\mathbf{B}} &= (\mathbf{X}_1'\mathbf{X}_1)^{-1}\mathbf{X}_1'\mathbf{Y}_o - (\mathbf{X}_1'\mathbf{X}_1)^{-1}\mathbf{X}_1'\mathbf{Z}\hat{\mathbf{\Gamma}}_o \\ \hat{\mathbf{\Gamma}}_o &= (\mathbf{Z}'\overline{\mathbf{X}}_1\mathbf{Z})^{-1}\mathbf{Z}'\overline{\mathbf{X}}_1\mathbf{Y}_o \end{aligned} \tag{7.41}$$

Using the reductions in (7.38) and the following theorem regarding projection matrices, we show that the BLUE of **B** under the MANCOVA model (7.36) is equal to the ML estimate for **B** using the GMANOVA model.

Theorem 7.1 Let $\mathbf{E}_{p \times p}$ be a nonsingular matrix, $\mathbf{X}_2$ $(q \times p)$ be of rank $q \leq p$, $\mathbf{H}_2'$ $(h \times p)$ be of rank h where $h = p - q$, and $\mathbf{X}_2\mathbf{H}_2 = \mathbf{0}$. Then,

$$\mathbf{H}_2(\mathbf{H}_2'\mathbf{E}\mathbf{H}_2)^{-1}\mathbf{H}_2' = \mathbf{E}^{-1} - \mathbf{E}^{-1}\mathbf{X}_2'(\mathbf{X}_2\mathbf{E}^{-1}\mathbf{X}_2')^{-1}\mathbf{X}_2\mathbf{E}^{-1}$$

Proceeding, from (7.41) we have:

$$\begin{aligned} \hat{\mathbf{B}} &= (\mathbf{X}_1'\mathbf{X}_1)^{-1}\mathbf{X}_1'\mathbf{Y}_o - (\mathbf{X}_1'\mathbf{X}_1)^{-1}\mathbf{X}_1'\mathbf{Z}[(\mathbf{Z}'\overline{\mathbf{X}}_1\mathbf{Z})^{-1}\mathbf{Z}'\overline{\mathbf{X}}_1\mathbf{Y}_o] \\ &= (\mathbf{X}_1'\mathbf{X}_1)^{-1}\mathbf{X}_1'\mathbf{Y}\mathbf{H}_1 - (\mathbf{X}_1'\mathbf{X}_1)^{-1}\mathbf{X}_1'\mathbf{Y}[\mathbf{H}_2(\mathbf{H}_2'\mathbf{E}\mathbf{H}_2)^{-1}\mathbf{H}_2']\mathbf{E}\mathbf{H}_1 \\ &= (\mathbf{X}_1'\mathbf{X}_1)^{-1}\mathbf{X}_1'\mathbf{Y}\mathbf{H}_1 - (\mathbf{X}_1'\mathbf{X}_1)^{-1}\mathbf{X}_1'\mathbf{Y}[\mathbf{E}^{-1} - \mathbf{E}^{-1}\mathbf{X}_2'(\mathbf{X}_2\mathbf{E}^{-1}\mathbf{X}_2')^{-1}\mathbf{X}_2\mathbf{E}^{-1}]\mathbf{E}\mathbf{H}_1 \\ &= (\mathbf{X}_1'\mathbf{X}_1)^{-1}\mathbf{X}'\mathbf{Y}\mathbf{E}^{-1}\mathbf{X}_2'(\mathbf{X}_2\mathbf{E}^{-1}\mathbf{X}_2')^{-1}\mathbf{X}_2\mathbf{H}_1 \\ &= (\mathbf{X}_1'\mathbf{X}_1)^{-1}\mathbf{X}_1'\mathbf{Y}\mathbf{S}^{-1}\mathbf{X}_2'(\mathbf{X}_2\mathbf{S}^{-1}\mathbf{X}_2')^{-1} \end{aligned} \tag{7.42}$$

the ML estimate of **B** under the GMANOVA model. Thus, if $q < p$, Rao's procedure using $p - q$ covariates, Khatri's ML estimate of **B,** and Potthoff-Roy's estimate weighting by $\mathbf{G}^{-1} = \mathbf{S}^{-1}$ all produce identical estimates. Setting $\mathbf{G} = \mathbf{I}$

in the Potthoff-Roy estimate is equivalent to not including any covariates in the Rao-Khatri reduction. When $p = q$, $\mathbf{H}_2$ does not exist and there is no need to choose $\mathbf{G}$ so that $\mathbf{Y}_o = \mathbf{Y}\mathbf{X}_2^{-1}$.

To obtain the likelihood ratio test of $H(\mathbf{CBA} = \mathbf{\Gamma})$ under the GMANOVA model, we use the standard theory in Chapter 5, the MANCOVA model where the design matrix is $\mathbf{X} = \binom{\mathbf{X}_1}{\mathbf{Z}}$, the rank$(\mathbf{X}_1) = n - k - h$, $h = p - q$, and the rank$(\mathbf{C}) = g = v_h$. To find the error sums of squares and products (SSP) matrix, observe that

$$
\begin{aligned}
\tilde{\mathbf{E}} &= \mathbf{A}'\mathbf{Y}_o(\mathbf{I} - \mathbf{X}(\mathbf{X}'\mathbf{X})^{-1}\mathbf{X}')\mathbf{Y}_o\mathbf{A} \\
&= \mathbf{A}'[\mathbf{Y}_o'\overline{\mathbf{X}}_1\mathbf{Y}_o - \hat{\mathbf{\Gamma}}_o'(\mathbf{Z}'\overline{\mathbf{X}}_1\mathbf{Z})\hat{\mathbf{\Gamma}}_o]\mathbf{A} \\
&= \mathbf{A}'[\mathbf{H}_1'\mathbf{Y}'\overline{\mathbf{X}}_1\mathbf{Y}\mathbf{H}_1 - \hat{\mathbf{\Gamma}}_o'\mathbf{Z}'\overline{\mathbf{X}}\mathbf{Y}_o]\mathbf{A} \\
&= \mathbf{A}'[\mathbf{H}_1'\mathbf{Y}'\overline{\mathbf{X}}_1\mathbf{Y}\mathbf{H}_1 - (\mathbf{H}_1'\mathbf{E}\mathbf{H}_2)(\mathbf{H}_2'\mathbf{E}\mathbf{H}_2)^{-1}\mathbf{H}_2'\mathbf{Y}'\overline{\mathbf{X}}_1\mathbf{Y}\mathbf{H}_1]\mathbf{A} \\
&= \mathbf{A}'[(\mathbf{H}_1'\mathbf{E}\mathbf{H}_1) - \mathbf{H}_1'\mathbf{E}\mathbf{H}_2(\mathbf{H}_2'\mathbf{E}\mathbf{H}_2)^{-1}(\mathbf{H}_2'\mathbf{E}\mathbf{H}_1)]\mathbf{A} \\
&= \mathbf{A}'[\mathbf{H}_1'\mathbf{E}\mathbf{H}_1 - \mathbf{H}_1'\mathbf{E}[\mathbf{H}_2(\mathbf{H}_2'\mathbf{H}_2)^{-1}\mathbf{H}_2']\mathbf{E}\mathbf{H}_1]\mathbf{A} \\
&= \mathbf{A}'(\mathbf{H}_1'\mathbf{X}_2(\mathbf{X}_2\mathbf{E}^{-1}\mathbf{X}_2')^{-1}\mathbf{X}_2'\mathbf{H}_1)\mathbf{A} \\
&= \mathbf{A}'(\mathbf{X}_2\mathbf{E}^{-1}\mathbf{X}_2')^{-1}\mathbf{A}
\end{aligned}
\tag{7.43}
$$

where $v_e = n - k - h = n - k - p + q$. To find the hypothesis test matrix, let $\tilde{\mathbf{C}} = (\mathbf{C}, \mathbf{I})$ to test the hypothesis regarding $\mathbf{B}$ in the MANCOVA model. The hypothesis test matrix is

$$
\begin{aligned}
\tilde{\mathbf{H}} &= (\mathbf{C}\hat{\mathbf{B}}\mathbf{A} - \mathbf{\Gamma})'[\mathbf{C}(\mathbf{X}_1'\mathbf{X}_1)^{-1}\mathbf{C} + \mathbf{C}(\mathbf{X}_1'\mathbf{X}_1)^{-1}\mathbf{X}_1'\mathbf{Z}(\mathbf{Z}'\overline{\mathbf{X}}_1\mathbf{Z})^{-1}\mathbf{Z}'\mathbf{X}_1(\mathbf{X}_1'\mathbf{X}_1)^{-1}\mathbf{C}]^{-1}(\mathbf{C}\hat{\mathbf{B}}\mathbf{A} - \mathbf{\Gamma}) \\
&= (\mathbf{C}\hat{\mathbf{B}}\mathbf{A} - \mathbf{\Gamma})'[\mathbf{C}\{(\mathbf{X}_1'\mathbf{X}_1)^{-1} + (\mathbf{X}_1'\mathbf{X}_1)^{-1}\mathbf{X}_1'\mathbf{Z}(\mathbf{Z}'\overline{\mathbf{X}}_1\mathbf{Z})^{-1}\mathbf{Z}'\mathbf{X}_1(\mathbf{X}_1'\mathbf{X}_1)^{-1}\}\mathbf{C}]^{-1}(\mathbf{C}\hat{\mathbf{B}}\mathbf{A} - \mathbf{\Gamma}) \\
&= (\mathbf{C}\hat{\mathbf{B}}\mathbf{A} - \mathbf{\Gamma})'[\mathbf{C}\{(\mathbf{X}_1'\mathbf{X}_1)^{-1} + (\mathbf{X}_1'\mathbf{X}_1)^{-1}\mathbf{X}_1'\mathbf{Y}\mathbf{H}_2(\mathbf{H}_2'\mathbf{E}\mathbf{H}_2)^{-1}\mathbf{H}_2'\mathbf{Y}\mathbf{X}_1(\mathbf{X}_1'\mathbf{X}_1)^{-1}\}\mathbf{C}]^{-1}(\mathbf{C}\hat{\mathbf{B}}\mathbf{A} - \mathbf{\Gamma}) \\
&= (\mathbf{C}\hat{\mathbf{B}}\mathbf{A} - \mathbf{\Gamma})'(\mathbf{C}\mathbf{R}^{-1}\mathbf{C}')^{-1}(\mathbf{C}\hat{\mathbf{B}}\mathbf{A} - \mathbf{\Gamma})
\end{aligned}
\tag{7.44}
$$

where

$$
\begin{aligned}
\mathbf{R}^{-1} &= (\mathbf{X}_1'\mathbf{X}_1)^{-1} + (\mathbf{X}_1'\mathbf{X}_1)^{-1}\mathbf{X}_1'\mathbf{Y}[\mathbf{E}^{-1} - \mathbf{E}^{-1}\mathbf{X}_2'(\mathbf{X}_2\mathbf{E}^{-1}\mathbf{X}_2')^{-1}\mathbf{X}_2\mathbf{E}^{-1}]\mathbf{Y}'\mathbf{X}_1(\mathbf{X}_1'\mathbf{X}_1)^{-1} \\
\hat{\mathbf{B}} &= (\mathbf{X}_1'\mathbf{X}_1)^{-1}\mathbf{X}_1'\mathbf{Y}\mathbf{S}^{-1}\mathbf{X}_2'(\mathbf{X}_2\mathbf{S}^{-1}\mathbf{X}_2')^{-1} \\
\mathbf{E} &= \mathbf{Y}'(\mathbf{I} - \mathbf{X}_1(\mathbf{X}_1'\mathbf{X}_1)^{-1}\mathbf{X}_1')\mathbf{Y}
\end{aligned}
\tag{7.45}
$$

Thus, with $\tilde{\mathbf{H}}$ defined in (7.44) and $\tilde{\mathbf{E}}$ defined in (7.43), the likelihood ratio criterion for testing the hypothesis $H(\mathbf{CBA} = \mathbf{\Gamma})$ for the GMANOVA model is

$$
\Lambda = \frac{|\tilde{\mathbf{E}}|}{|\tilde{\mathbf{E}} + \tilde{\mathbf{H}}|} \sim U_{u, v_h, v_e} \tag{7.46}
$$

The $1 - \alpha$ simultaneous confidence intervals for parametric functions $\psi = \mathbf{c}'\hat{\mathbf{B}}\mathbf{a}$ are given by

$$
\hat{\psi} - c_o\hat{\sigma}_{\hat{\psi}} < \psi < \hat{\psi} + c_o\hat{\sigma}_{\hat{\psi}} \tag{7.47}
$$

where

$$\hat{\sigma}_{\hat{\psi}}^2 = (\mathbf{a}'\tilde{\mathbf{E}}\mathbf{a})(\mathbf{c}'\mathbf{R}^{-1}\mathbf{c}) \tag{7.48}$$

and c_o is the appropriate constant chosen to agree with the test criteria

$$\begin{aligned}
c_o^2 &= \frac{1-U^a}{U^a} && \text{Wilks} \\
c_o^2 &= \frac{\theta^\alpha}{1-\theta^\alpha} && \text{Roy} \\
c_o^2 &= U_o^\alpha = T_{o,\alpha}^2 && \text{Bartlett - Lawley - Hotelling} \\
c_o^2 &= \frac{V^\alpha}{1-V^\alpha} && \text{Bartlett - Nanda - Pillai}
\end{aligned}$$

In testing $H(\mathbf{CBA} = \mathbf{\Gamma})$ under the GMANOVA model, we discussed the conditional likelihood ratio test that depends on selecting the appropriate number of $p-q$ covariates. If $p=q$, $\mathbf{H}_2$ does not exist, and we may use the Potthoff-Roy procedure which no longer depends on the arbitrary matrix $\mathbf{G}$. To test the adequacy of the model fit to the data, we must choose the most appropriate set of covariates. Recall that the GMANOVA model has the form

$$\begin{aligned}
E(\mathbf{Y}) &= \mathbf{X}_1\mathbf{B}\mathbf{X}_2 = \mathbf{X}_1\mathbf{\Theta} \\
E(\mathbf{Z}) &= \mathbf{X}_1\mathbf{B}\mathbf{X}_2\mathbf{H}_2 = \mathbf{X}_1\mathbf{\Theta}\mathbf{H}_2 = \mathbf{0}
\end{aligned} \tag{7.49}$$

where the $\text{rank}(\mathbf{X}_1) = k$ and $\mathbf{X}_1\mathbf{\Theta}\mathbf{H}_2 = \mathbf{0}$ if and only if $\mathbf{\Theta}\mathbf{H}_2 = \mathbf{0}$. Hence, we may evaluate the fit of the model by testing whether $E(\mathbf{Z}) = \mathbf{0}$ or equivalently that H: $\mathbf{\Theta}\mathbf{H}_2 = \mathbf{0}$ for the model $E(\mathbf{Y}) = \mathbf{X}_1\mathbf{\Theta}$. This is the MANOVA model discussed in Chapter 5 where now the hypothesis and error SSP matrices are

$$\begin{aligned}
\mathbf{H} &= (\hat{\mathbf{\Theta}}\mathbf{H}_2)'(\mathbf{X}_1'\mathbf{X}_1)(\hat{\mathbf{\Theta}}\mathbf{H}_2) \\
\mathbf{E} &= \mathbf{H}_2'\mathbf{Y}'(\mathbf{I} - \mathbf{X}_1(\mathbf{X}_1'\mathbf{X}_1)^{-1}\mathbf{X}_1')\mathbf{Y}\mathbf{H}_2
\end{aligned} \tag{7.50}$$

with $u = \text{rank}(\mathbf{H}_2)$, $v_h = \text{rank}(\mathbf{C}) = g$, $v_e = n - \text{rank}(\mathbf{X}_1) = n-k$, and where $\hat{\mathbf{\Theta}} = (\mathbf{X}_1'\mathbf{X}_1)^{-1}\mathbf{X}_1'\mathbf{Y}$, is given in (7.42). For additional detail see Grizzle and Allen (1969) and Seber (1984, Chapter 9.7). A discussion of growth curves may also be found in Srivastava and Carter (1983, Chapter 6) and Kshirsager and Smith (1995).

Gleser and Olkin (1970) develop a canonical form for the GMANOVA model and show that the Rao-Khatri conditional LR test is also the unconditional LR test. Tubbs et al. (1975) claim that their proposed test is an unconditional test of $H(\mathbf{CBA} = \mathbf{\Gamma})$ for the GMANOVA model; however, as noted by von Rosen (1989), the estimator of $\mathbf{B}$ estimated under the hypothesis $\mathbf{CBA} = \mathbf{\Gamma}$ with $\hat{\mathbf{\Sigma}}$ replacing $\mathbf{\Sigma}$ is not the constrained maximum likelihood estimator of $\mathbf{B}$. The expression for $\mathbf{B}$ derived by Tubbs et al. (1975, equation 2.15) may yield a nonoptimal solution. Their test procedure is not an LR test, but a test created using the substitution method as defined by Arnold (1981, p. 363) which is equivalent to setting $\mathbf{G} = \mathbf{S}$ in the Potthoff-Roy model. From (5.34), their statistic may be written as

$$X^2 = \text{Tr}[(\mathbf{C}\hat{\mathbf{B}}\mathbf{A} - \mathbf{\Gamma})'(\mathbf{C}(\mathbf{X}_1'\mathbf{X}_1)^{-1}\mathbf{C}')^{-1}(\mathbf{C}\hat{\mathbf{B}}\mathbf{A} - \mathbf{\Gamma})(\mathbf{A}'(\mathbf{X}_1\mathbf{E}\mathbf{X}_1')^{-1}\mathbf{A})^{-1}] \tag{7.51}$$

where $\hat{\mathbf{B}}$ and $\mathbf{E}$ are given in (7.45). X^2 converges in distribution to a chi-square distribution with $v =$ rank($\mathbf{C}$) rank($\mathbf{A}$) $= gu$ degrees of freedom. If $\mathbf{E}$ is replaced by an estimator $\hat{\mathbf{\Sigma}}$ of $\mathbf{\Sigma}$ so that $E(\hat{\mathbf{\Sigma}}) = K\mathbf{\Sigma}$, then $K^{-1}X^2 \sim \chi^2_{gu}$ as $n \to \infty$. The statistic in (7.51) has the same form as T_o^2, but it is unknown how the statistic behaves in small samples.

7.4 Canonical Form of the GMANOVA Model

Gleser and Olkin (1970), Srivastava and Khatri (1979), and Kariya (1985) discuss the growth curve model using an equivalent canonical form for the model, a common practice in the study of multivariate linear hypotheses. See Lehmann (1994, Chapter 8) or Anderson (1984, p. 296). To reduce the GMANOVA model to canonical form, we recall a few results from matrix algebra and vector spaces (Timm, 1975, Chapter 1; Stapleton, 1995).

Theorem 7.2 (Singular-value decomposition) Let $\mathbf{A}_{m\times n}$ be a matrix of rank r, with $r \le m \le n$. Then there exist orthogonal matrices $\mathbf{P}_{m\times m}$ and $\mathbf{Q}_{n\times n}$ such that $\mathbf{P}'\mathbf{A}\mathbf{Q} = \begin{pmatrix} \Lambda & \mathbf{0} \\ \mathbf{0} & \mathbf{0} \end{pmatrix}$ or $\mathbf{A} = \mathbf{P}\begin{pmatrix} \Lambda & \mathbf{0} \\ \mathbf{0} & \mathbf{0} \end{pmatrix}\mathbf{Q}'$ where $\Lambda = \text{Diag}(\lambda_1, \ldots, \lambda_r)$ is a diagonal matrix of nonnegative elements and $\mathbf{0}$ is a matrix of order $m \times (n-m)$. The diagonal elements of Λ^2 are the eigenvalues of $\mathbf{A}\mathbf{A}'$ and $\mathbf{A}'\mathbf{A}$.

Lemma 7.2 For the matrix $\mathbf{A}_{m\times m}$ in Theorem 7.2, there exists a nonsingular matrix $\mathbf{M}_{m\times m}$ such that $\mathbf{A} = \mathbf{M}\begin{pmatrix} \mathbf{I}_r & \mathbf{0} \\ \mathbf{0} & \mathbf{0} \end{pmatrix}\mathbf{Q}'$ where $\mathbf{Q}$ is an orthogonal matrix.

Using Lemma 7.2 we may write $\mathbf{X}_1$ of rank k and $\mathbf{X}_2$ of rank q for the GMANOVA model following Kariya (1985)

$$\mathbf{X}_1 = \mathbf{Q}_1\begin{pmatrix} \mathbf{I}_k \\ \mathbf{0} \end{pmatrix}\mathbf{P}_1 \text{ and } \mathbf{X}_2 = \mathbf{P}_2(\mathbf{I}_q, \mathbf{0})\mathbf{Q}_2$$

so that $\mathbf{Y} = \mathbf{Q}_1\begin{pmatrix} \mathbf{I}_k \\ \mathbf{0} \end{pmatrix}\mathbf{P}_1\mathbf{B}\mathbf{P}_2(\mathbf{I}_q, \mathbf{0})\mathbf{Q}_2$. Since $\mathbf{Q}_1$ and $\mathbf{Q}_2$ are known, the matrix

$$\mathbf{Y}^* = \mathbf{Q}_1'\mathbf{Y}\mathbf{Q}_2 = \begin{pmatrix} \mathbf{I}_k \\ \mathbf{0} \end{pmatrix}\mathbf{P}_1\mathbf{B}\mathbf{P}_2(\mathbf{I}_q, \mathbf{0})\mathbf{Q}_2$$

The transformed matrix $\mathbf{Y}^*$ is a random matrix with a matrix normal distribution with covariance matrix $\boldsymbol{\Sigma}^* = \mathbf{Q}_2'\boldsymbol{\Sigma}\mathbf{Q}_2$ and mean

$$E(\mathbf{Y}^*) = \begin{pmatrix} \mathbf{I}_k \\ \mathbf{0} \end{pmatrix}\mathbf{P}_1\mathbf{B}\mathbf{P}_2(\mathbf{I}_q, \mathbf{0}) \equiv \begin{pmatrix} \mathbf{I}_k \\ \mathbf{0} \end{pmatrix}\mathbf{B}^*(\mathbf{I}_p, \mathbf{0}) = \begin{pmatrix} \mathbf{B}^* & \mathbf{0} \\ \mathbf{0} & \mathbf{0} \end{pmatrix}$$

so that

$$\mathbf{Y}^* \sim \mathbf{N}_{n,p}\left[\begin{pmatrix} \mathbf{B}^* & \mathbf{0} \\ \mathbf{0} & \mathbf{0} \end{pmatrix}, \boldsymbol{\Sigma}^*, \mathbf{I}_p\right]$$

where $\mathbf{B}^*_{(k_1+k_2)\times(q_1+q_2)} = \mathbf{P}_1\mathbf{B}\mathbf{P}_2$, $k = k_1 + k_2$, $q = q_1 + q_2$, $p = q_1 + q_2 + q_3$ and $n = k_1 + k_2 + k_3$.

The GMANOVA hypothesis

$$H\colon\ \mathbf{C}\mathbf{B}\mathbf{A} = \boldsymbol{\Gamma} \equiv \mathbf{X}_3\mathbf{B}\mathbf{X}_4 \tag{7.52}$$

may be similarly reduced

$$\mathbf{X}_3\mathbf{B}\mathbf{X}_4 = \mathbf{X}_3\mathbf{P}_1^{-1}\mathbf{P}_1\mathbf{B}\mathbf{P}_2\mathbf{P}_2^{-1}\mathbf{X}_4 = \mathbf{X}_3\mathbf{P}_1^{-1}\mathbf{B}^*\mathbf{P}_2^{-1}\mathbf{X}_4.$$

Applying Lemma 7.2 to $\mathbf{X}_3\mathbf{P}_1^{-1}$ and $\mathbf{P}_2^{-1}\mathbf{X}_4$ yields

$$\mathbf{X}_3\mathbf{P}_1^{-1} = \mathbf{P}_3(\mathbf{I}_{k_1},\mathbf{0})\mathbf{Q}_3 \text{ and } \mathbf{P}_2^{-1}\mathbf{X}_4 = \mathbf{Q}_4\begin{pmatrix}\mathbf{0}\\ \mathbf{I}_{q_2}\end{pmatrix}\mathbf{P}_4$$

so that the hypothesis in (7.52) becomes

$$H\colon\ (\mathbf{I}_{k_1},\mathbf{0})\boldsymbol{\Theta}\begin{pmatrix}\mathbf{0}\\ \mathbf{I}_{q_2}\end{pmatrix} = \mathbf{P}_3^{-1}\mathbf{X}_o\mathbf{P}_4^{-1} \tag{7.53}$$

where $\boldsymbol{\Theta}_{k\times q} = \mathbf{Q}_3\mathbf{B}_*\mathbf{Q}_4$.

Because $\mathbf{Q}_3$ and $\mathbf{Q}_4$ are nonsingular and known, we may use them to transform $\mathbf{Y}^*$ to a new random matrix $\mathbf{Z}$

$$\mathbf{Z}_{n\times p} = \begin{pmatrix}\mathbf{Q}_3 & \mathbf{0}\\ \mathbf{0} & \mathbf{I}\end{pmatrix}\mathbf{Y}^*\begin{pmatrix}\mathbf{Q}_4 & \mathbf{0}\\ \mathbf{0} & \mathbf{I}\end{pmatrix} = \begin{pmatrix}\mathbf{Z}_{11} & \mathbf{Z}_{12} & \mathbf{Z}_{13}\\ \mathbf{Z}_{21} & \mathbf{Z}_{23} & \mathbf{Z}_{23}\\ \mathbf{Z}_{31} & \mathbf{Z}_{32} & \mathbf{Z}_{33}\end{pmatrix}\begin{matrix}k_1\\ k_2\\ k_3\end{matrix} \quad (\text{columns } q_1,\ q_2,\ q_3) \tag{7.54}$$

where each row of $\mathbf{Z}$ is normally distributed with covariance matrix

$$\boldsymbol{\Sigma}_{p\times p} = \left[\begin{pmatrix}\mathbf{Q}_4' & \mathbf{0}\\ \mathbf{0} & \mathbf{I}\end{pmatrix}\boldsymbol{\Sigma}^*\begin{pmatrix}\mathbf{Q}_4 & \mathbf{0}\\ \mathbf{0} & \mathbf{I}\end{pmatrix}\right] = \begin{pmatrix}\boldsymbol{\Sigma}_{11} & \boldsymbol{\Sigma}_{12} & \boldsymbol{\Sigma}_{13}\\ \boldsymbol{\Sigma}_{21} & \boldsymbol{\Sigma}_{22} & \boldsymbol{\Sigma}_{23}\\ \boldsymbol{\Sigma}_{31} & \boldsymbol{\Sigma}_{32} & \boldsymbol{\Sigma}_{33}\end{pmatrix}\begin{matrix}q_1\\ q_2\\ q_3\end{matrix} \quad (\text{columns } q_1,\ q_2,\ q_3) \tag{7.55}$$

$\boldsymbol{\Sigma}_i q_i = p$, $\boldsymbol{\Sigma}_{ij}(q_i \times q_j)$, $k_3 = n-(k_1+k_2)$, and $q_3 = p-q = p-q_1-q_2$. The matrix $\mathbf{Z} \sim N_{n,p}(\tilde{\Theta},\Sigma,\mathbf{I}_p)$ with Σ defined in (7.55) has mean matrix

$$\tilde{\boldsymbol{\Theta}} = \begin{pmatrix}\boldsymbol{\Theta}_{11} & \boldsymbol{\Theta}_{12} & \boldsymbol{\Theta}_{13}\\ \boldsymbol{\Theta}_{21} & \boldsymbol{\Theta}_{22} & \boldsymbol{\Theta}_{23}\\ \mathbf{0} & \mathbf{0} & \mathbf{0}\end{pmatrix} = \begin{pmatrix}\boldsymbol{\Theta}_{11} & \boldsymbol{\Theta}_{12} & \mathbf{0}\\ \boldsymbol{\Theta}_{21} & \boldsymbol{\Theta}_{23} & \mathbf{0}\\ \mathbf{0} & \mathbf{0} & \mathbf{0}\end{pmatrix}\begin{matrix}k_1\\ k_2\\ k_3\end{matrix} \quad (\text{columns } q_1,\ q_2,\ q_3) \tag{7.56}$$

where $\Theta = \begin{pmatrix}\Theta_{11} & \Theta_{12}\\ \Theta_{21} & \Theta_{22}\end{pmatrix} = \mathbf{Q}_3\mathbf{B}^*\mathbf{Q}_4$ and $\Sigma_i k_i = n$. Substituting $\boldsymbol{\Theta}$ into (7.53), the canonical form of the growth curve hypothesis is equivalent to testing

$$H\colon\ \Theta_{12} = \mathbf{P}_2^{-1}\mathbf{X}_o\mathbf{P}_4^{-1}$$

or letting $\mathbf{X}_o = \mathbf{0}$, the following equivalent hypothesis is

$$H\colon\ \mathbf{X}_3\mathbf{B}\mathbf{X}_4 \equiv \Theta_{12} = \mathbf{0} \tag{7.57}$$

This is the canonical form of the null hypothesis where the transformed data matrix is the canonically reduced matrix **Z**. For the canonical form of the GMANOVA model, we are interested in testing (7.57) given **Z**, the canonical data matrix. To test $H(\boldsymbol{\Theta}_{12} = \mathbf{0})$, we further observe from the canonical form of the model that

$$\overline{\mathbf{Z}} = \begin{pmatrix} \mathbf{Z}_{11} & \mathbf{Z}_{12} & \mathbf{Z}_{13} \\ \mathbf{Z}_{21} & \mathbf{Z}_{22} & \mathbf{Z}_{23} \end{pmatrix} \sim N_{(k_1+k_2),p}\left(\tilde{\boldsymbol{\Theta}} = \begin{pmatrix} \boldsymbol{\Theta}_{11} & \boldsymbol{\Theta}_{12} & \mathbf{0} \\ \boldsymbol{\Theta}_{21} & \boldsymbol{\Theta}_{22} & \mathbf{0} \end{pmatrix}, \boldsymbol{\Sigma}, \mathbf{I}_p\right)$$

$$\mathbf{Z}_3 = \begin{pmatrix} \mathbf{Z}_{31} & \mathbf{Z}_{32} & \mathbf{Z}_{33} \end{pmatrix} \sim N_{k_3,p}\left(\mathbf{0}, \ \boldsymbol{\Sigma}, \ \mathbf{I}_p\right) \tag{7.58}$$

$$\mathbf{Z}_3'\mathbf{Z}_3 = \mathbf{V} = \begin{pmatrix} \mathbf{V}_{11} & \mathbf{V}_{12} & \mathbf{V}_{13} \\ \mathbf{V}_{21} & \mathbf{V}_{22} & \mathbf{V}_{23} \\ \mathbf{V}_{31} & \mathbf{V}_{32} & \mathbf{V}_{33} \end{pmatrix} \sim W_p(k_3, \boldsymbol{\Sigma})$$

and that $\overline{\mathbf{Z}}$ and $\mathbf{V}$ are independent. Hence $(\overline{\mathbf{Z}}, \mathbf{V})$ are sufficient statistics for $(\boldsymbol{\Theta}, \boldsymbol{\Sigma})$.

Using the canonical form of the GMANOVA model, Gleser and Olkin (1970, p. 280), Srivastava and Khatri (1979, p. 194), and Kariya (1985, p. 104) show that the LR test statistic for testing $H(\boldsymbol{\Theta}_{12} = \mathbf{0})$ is

$$\lambda^{2/n} = \Lambda = \frac{|\mathbf{I} + \mathbf{Z}_{13}\mathbf{V}_{33}^{-1}\mathbf{Z}_{13}'|}{|\mathbf{I} + (\mathbf{Z}_{12}, \mathbf{Z}_{13})\begin{pmatrix} \mathbf{V}_{22} & \mathbf{V}_{23} \\ \mathbf{V}_{32} & \mathbf{V}_{33} \end{pmatrix}^{-1}(\mathbf{Z}_{12}, \mathbf{Z}_{13})'|}$$

$$= \frac{\left|\begin{pmatrix} \mathbf{V}_{22} & \mathbf{V}_{23} \\ \mathbf{V}_{32} & \mathbf{V}_{33} \end{pmatrix}\right| |\mathbf{V}_{33} + \mathbf{Z}_{13}'\mathbf{Z}_{13}|}{|\mathbf{V}_{33}| \left|\begin{pmatrix} \mathbf{V}_{22} & \mathbf{V}_{23} \\ \mathbf{V}_{32} & \mathbf{V}_{33} \end{pmatrix} + (\mathbf{Z}_{12}, \mathbf{Z}_{13})'(\mathbf{Z}_{12}, \mathbf{Z}_{13})\right|}$$

$$= \frac{|\mathbf{V}_{33}||\mathbf{V}_{22} - \mathbf{V}_{23}\mathbf{V}_{33}^{-1}\mathbf{V}_{32}||\mathbf{V}_{33} + \mathbf{Z}_{13}'\mathbf{Z}_{13}|}{|\mathbf{V}_{33}| \begin{vmatrix} \mathbf{V}_{22} + \mathbf{Z}_{12}'\mathbf{Z}_{12} & \mathbf{V}_{23} + \mathbf{Z}_{12}'\mathbf{Z}_{13} \\ \mathbf{V}_{32} + \mathbf{Z}_{13}'\mathbf{Z}_{12} & \mathbf{V}_{33} + \mathbf{Z}_{13}'\mathbf{Z}_{13} \end{vmatrix}} \tag{7.59}$$

$$= \frac{|\mathbf{V}_{22} - \mathbf{V}_{23}\mathbf{V}_{33}^{-1}\mathbf{V}_{32}|}{|(\mathbf{V}_{22} + \mathbf{Z}_{12}'\mathbf{Z}_{12}) - (\mathbf{V}_{23} + \mathbf{Z}_{12}'\mathbf{Z}_{13})(\mathbf{V}_{33} + \mathbf{Z}_{13}'\mathbf{Z}_{13})^{-1}(\mathbf{V}_{23} + \mathbf{Z}_{12}'\mathbf{Z}_{13})'|}$$

$$= \frac{|\mathbf{V}_{22.3}|}{|\mathbf{V}_{22.3} + \mathbf{X}'\mathbf{X}|} = \frac{1}{|\mathbf{I} + \mathbf{X}\mathbf{V}_{22.3}^{-1}\mathbf{X}'|} = \frac{1}{|\mathbf{I} + \mathbf{T}_1|}$$

by using the following:

$$|\mathbf{V}| = \left|\begin{pmatrix} \mathbf{V}_{11} & \mathbf{V}_{12} \\ \mathbf{V}_{21} & \mathbf{V}_{22} \end{pmatrix}\right| = |\mathbf{V}_{22}||\mathbf{V}_{11} - \mathbf{V}_{12}\mathbf{V}_{22}^{-1}\mathbf{V}_{21}|$$

if

$$\mathbf{P} \equiv (\mathbf{Y} - \mathbf{B}\xi\mathbf{A})(\mathbf{Y} - \mathbf{B}\xi\mathbf{A})' = \mathbf{S} + (\mathbf{Y} - \mathbf{B}\xi)\mathbf{A}\mathbf{A}'(\mathbf{Y} - \mathbf{B}\xi)'$$

for

$$\mathbf{Y}_{p\times n}, \ \mathbf{B}_{q\times p}, \ \xi_{p\times k}, \text{ and } \mathbf{A}_{k\times n} \tag{7.60}$$

then $\quad |\mathbf{P}| = |\mathbf{S}||\mathbf{I}_p + \mathbf{S}^{-1}(\mathbf{Y}-\mathbf{B}\xi)\mathbf{A}\mathbf{A}'(\mathbf{Y}-\mathbf{B}\xi)|$

$$= |\mathbf{S}||\mathbf{I}_k + (\mathbf{A}\mathbf{A}')(\mathbf{Y}-\mathbf{B}\xi)\mathbf{S}^{-1}(\mathbf{Y}-\mathbf{B}\xi)|;$$

and, defining

$$\mathbf{X} = (\mathbf{I}+\mathbf{Z}_{13}\mathbf{V}_{33}^{-1}\mathbf{Z}'_{13})^{-1/2}(\mathbf{Z}_{12}-\mathbf{Z}_{13}\mathbf{V}_{33}^{-1}\mathbf{V}_{32})$$

$$\mathbf{T}_1 = \mathbf{X}\mathbf{V}_{22.3}^{-1}\mathbf{X}', \text{ and } \mathbf{T}_2 = \mathbf{Z}_{13}\mathbf{V}_{33}^{-1}\mathbf{Z}'_{13}$$

The matrices $\mathbf{X}$ and $\mathbf{V}_{22.3}$ given $\mathbf{T}_2$ are independent and

$$\begin{aligned}
\mathbf{X}|\mathbf{T}_2 &\sim N_{k_1,p}[(\mathbf{I}+\mathbf{T}_2)^{-1/2}\Theta_{12},\mathbf{\Sigma}_{22.3},\mathbf{I}_p] \\
\mathbf{V}_{22.3} &\sim W_{q_2}(k_3-q_3,\mathbf{V}_{22.3}) \\
\mathbf{X}_{13} &\sim N_{k_1,p}[\mathbf{0},\mathbf{\Sigma}_{33},\mathbf{I}_p] \\
\mathbf{V}_{33} &\sim W_{q_3}(q_2,\mathbf{\Sigma}_{22})
\end{aligned} \tag{7.61}$$

where $v_e = n-k-p+q = k_3-q_3$.

To transform the canonical form of the test back into the original variables, we again recall some results from vector algebra.

Theorem 7.3 (The spectral decomposition theorem). Let $\mathbf{A}$ be a symmetric matrix of order n. Then there exists an orthogonal matrix $\mathbf{P}$ ($\mathbf{PP}' = \mathbf{I}$) such that $\mathbf{P}'\mathbf{AP} = \Lambda$ or $\mathbf{A} = \mathbf{P}\Lambda\mathbf{P}'$ where $\Lambda = \text{Diag}(\lambda_1, \lambda_2, \dots, \lambda_n)$ is a diagonal matrix whose diagonal elements are the eigenvalues of $\mathbf{A}$ and the columns of $\mathbf{P}$ correspond to the orthonormal eigenvectors of $\mathbf{A}$.

Using Theorem 7.3, let $\mathbf{X}_1$ $(n\times k)$ be a design matrix of rank k, then $\mathbf{X}_1'\mathbf{X}_1$ is symmetric and of rank k. Furthermore, there exists an orthogonal matrix $\mathbf{P}$ such that $\mathbf{P}'(\mathbf{X}_1'\mathbf{X}_1)\mathbf{P} = \Lambda$ or $(\mathbf{X}_1'\mathbf{X}_1) = \mathbf{P}\Lambda\mathbf{P}' = \mathbf{P}\Lambda^{1/2}\mathbf{P}'\mathbf{P}\Lambda^{1/2}\mathbf{P}'$. The positive square root matrix of $(\mathbf{X}_1'\mathbf{X}_1)$ is defined

$$(\mathbf{X}_1'\mathbf{X}_1)^{1/2} = \mathbf{P}\Lambda^{1/2}\mathbf{P}' \tag{7.62}$$

Theorem 7.4 Let $\mathbf{A}_{n\times k}$ be a matrix of rank k. Then the matrix $\mathbf{A}(\mathbf{A}'\mathbf{A})\mathbf{A} = \mathbf{P}$ is an orthogonal projection matrix if and only if $\mathbf{P}$ is symmetric and idempotent, $\mathbf{P}^2 = \mathbf{P}$. Furthermore, if $\mathbf{P}$ is a projection matrix, so is $\mathbf{I}-\mathbf{P}$, and $(\mathbf{I}-\mathbf{P})\mathbf{P} = \mathbf{0}$ where $\mathbf{P}$ and $\mathbf{I}-\mathbf{P}$ are disjoint.

Theorem 7.5 (Gram-Schmidt). Let $\mathbf{a}_1, \mathbf{a}_2, \dots, \mathbf{a}_n$ be a basis for a vector subspace $\mathbf{A}$ of R^n. Then there exists an orthonormal basis, $\mathbf{e}_1, \mathbf{e}_2, \dots, \mathbf{e}_k, \mathbf{e}_{k+1}, \mathbf{e}_{k+2}, \dots, \mathbf{e}_n$ such that $R^n = \mathbf{A}\oplus\mathbf{A}^\perp$ where $\mathbf{A}^\perp$ is the orthogonal complement of $\mathbf{A}$, $\mathbf{e}_i \,\varepsilon\, \mathbf{A}$ and $\mathbf{e}_j \,\varepsilon\, \mathbf{A}^\perp$ such that $\mathbf{e}_i'\mathbf{e}_j = \mathbf{0}$, and $\mathbf{e}_{k+1}, \dots, \mathbf{e}_n$ is associated with $\mathbf{A}^\perp$.

Associating the design matrix $\mathbf{X}_1$ with $\mathbf{A}$ and $\tilde{\mathbf{X}}_1$ with $\mathbf{A}^{\perp}$ in Theorem 7.5, we see for the GMANOVA model that

$$\begin{aligned} \operatorname{rank}(\mathbf{X}_1, \tilde{\mathbf{X}}_1) &= n \text{ and } \tilde{\mathbf{X}}_1'\mathbf{X}_1 = \mathbf{0} \\ \operatorname{rank}\begin{pmatrix}\mathbf{X}_2 \\ \tilde{\mathbf{X}}_2\end{pmatrix} &= q \text{ and } \mathbf{X}_2'\tilde{\mathbf{X}}_2 = \mathbf{0} \end{aligned}$$

so that we may form projection matrices

$$\begin{aligned} \overline{\mathbf{X}}_1 &= \tilde{\mathbf{X}}_1'(\tilde{\mathbf{X}}_1'\tilde{\mathbf{X}}_1)^{-1}\tilde{\mathbf{X}}_1 = \mathbf{I}_n - \mathbf{X}_1(\mathbf{X}_1'\mathbf{X}_1)^{-1}\mathbf{X}_1' \\ \overline{\mathbf{X}}_2 &= \tilde{\mathbf{X}}_1'(\tilde{\mathbf{X}}_2'\tilde{\mathbf{X}}_2)^{-1}\tilde{\mathbf{X}}_2 = \mathbf{I}_n - \mathbf{X}_2(\mathbf{X}_2'\mathbf{X}_2)^{-1}\mathbf{X}_2' \end{aligned}$$

Similarly, for the hypothesis $H(\mathbf{X}_3\mathbf{B}\mathbf{X}_4 = \mathbf{0})$, we form matrices $\tilde{\mathbf{X}}_3$ and $\tilde{\mathbf{X}}_4$ such that

$$\begin{aligned} \operatorname{rank}\begin{pmatrix}\mathbf{X}_3 \\ \tilde{\mathbf{X}}_3\end{pmatrix} &= k_1 + k_2 \text{ and } \mathbf{X}_3'\tilde{\mathbf{X}}_3 = \mathbf{0} \\ \operatorname{rank}(\mathbf{X}_4, \tilde{\mathbf{X}}_4) &= q_1 + q_2 \text{ and } \tilde{\mathbf{X}}_4'\mathbf{X}_4 = \mathbf{0} \end{aligned}$$

Following Kariya (1985, p. 81), we associate with the $\mathbf{P}_i$'s and $\mathbf{Q}_i$'s with the following matrices

$$\begin{aligned} \mathbf{P}_1 &= (\mathbf{X}_1'\mathbf{X}_1)^{1/2} \qquad \mathbf{P}_2 = (\mathbf{X}_2\mathbf{X}_2')^{1/2} \\ \mathbf{P}_3 &= (\mathbf{X}_3\mathbf{P}_1^{-2}\mathbf{X}_3)^{1/2} \qquad \mathbf{P}_4 = (\mathbf{X}_4'\mathbf{P}_2^{-2}\mathbf{X}_4)^{1/2} \\ \mathbf{Q}_1 &= [\mathbf{X}_1(\mathbf{X}_1'\mathbf{X}_1)^{-1/2}, \tilde{\mathbf{X}}_1(\tilde{\mathbf{X}}_1'\tilde{\mathbf{X}}_1)^{-1/2}] \\ \mathbf{Q}_2 &= \begin{pmatrix}(\mathbf{X}_2\mathbf{X}_2')^{-1/2}\mathbf{X}_2 \\ (\tilde{\mathbf{X}}_2\tilde{\mathbf{X}}_2')^{-1/2}\tilde{\mathbf{X}}_2\end{pmatrix} \\ \mathbf{Q}_3 &= \begin{pmatrix}(\mathbf{X}_3\mathbf{P}_1^{-2}\mathbf{X}_3')^{-1/2}\mathbf{X}_3\mathbf{P}_1^{-1} \\ (\tilde{\mathbf{X}}_3\mathbf{P}_1^{2}\tilde{\mathbf{X}}_3')^{-1/2}\tilde{\mathbf{X}}_3\mathbf{P}_1\end{pmatrix} \\ \mathbf{Q}_4 &= [\mathbf{P}_2\tilde{\mathbf{X}}_4(\tilde{\mathbf{X}}_4'\mathbf{P}_2^{2}\tilde{\mathbf{X}}_4)^{-1/2}, \ \mathbf{P}_2^{-1}\mathbf{X}_4(\mathbf{X}_4'\mathbf{P}_2^{-2}\mathbf{X}_4)^{-1/2}] \end{aligned}$$

where $\mathbf{Y}^* = \mathbf{Q}_1'\mathbf{Y}\mathbf{Q}_2$, $\mathbf{B}^* = \mathbf{P}_1\mathbf{B}\mathbf{P}_2$ and $\boldsymbol{\Sigma}^* = \mathbf{Q}_2'\boldsymbol{\Sigma}\mathbf{Q}_2$. Then, the $\mathbf{Z}_{ij}$ in (7.54) become

$$\begin{aligned} \mathbf{Z}_{12} &= (\mathbf{X}_3(\mathbf{X}_1'\mathbf{X}_1)^{-1}\mathbf{X}_3')^{-1/2}\mathbf{X}_3\hat{\mathbf{B}}\mathbf{X}_4(\mathbf{X}_4'(\mathbf{X}_2\mathbf{X}_2')^{-1}\mathbf{X}_4)^{-1/2} \\ \hat{\mathbf{B}} &= (\mathbf{X}_1'\mathbf{X}_1)^{-1}\mathbf{X}_1'\mathbf{Y}\mathbf{X}_2'(\mathbf{X}_2\mathbf{X}_2')^{-1} \\ \mathbf{Z}_{13} &= (\mathbf{X}_3(\mathbf{X}_1'\mathbf{X}_1)^{-1}\mathbf{X}_3')^{-1/2}\mathbf{X}_3(\mathbf{X}_1'\mathbf{X}_1)^{-1}\mathbf{X}_1'\mathbf{Y}\tilde{\mathbf{X}}_2'(\tilde{\mathbf{X}}_2\tilde{\mathbf{X}}_2')^{-1/2} \\ \mathbf{Z}_{32} &= (\tilde{\mathbf{X}}_1'\tilde{\mathbf{X}}_1)^{-1/2}\mathbf{X}_1'\mathbf{Y}\mathbf{X}_2'(\mathbf{X}_2\mathbf{X}_2')^{-1}\mathbf{X}_4(\mathbf{X}_4'(\mathbf{X}_2\mathbf{X}_2')\mathbf{X})^{-1/2} \\ \mathbf{Z}_{33} &= (\tilde{\mathbf{X}}_1'\mathbf{X}_1)^{-1/2}\tilde{\mathbf{X}}_1\mathbf{Y}\tilde{\mathbf{X}}_2'(\tilde{\mathbf{X}}_2\tilde{\mathbf{X}}_2')^{-1/2} \end{aligned} \tag{7.63}$$

in terms of the original variables. To obtain the test statistic, let

$$\begin{aligned} \overline{\mathbf{X}}_1 &= \mathbf{I}_n - \mathbf{X}_1(\mathbf{X}_1'\mathbf{X}_1)^{-1}\mathbf{X}_1' \\ \mathbf{E} &= \mathbf{Y}'\overline{\mathbf{X}}_1\mathbf{Y} \\ \mathbf{A}_1 &= (\mathbf{X}_3(\mathbf{X}_1'\mathbf{X}_1)^{-1}\mathbf{X}_3')^{-1/2}\mathbf{X}_3(\mathbf{X}_1'\mathbf{X}_1)^{-1}\mathbf{X}_1' \\ \mathbf{A}_2 &= \mathbf{X}_2'(\mathbf{X}_2\mathbf{X}_2')^{-1}\mathbf{X}_4(\mathbf{X}_4'(\mathbf{X}_2\mathbf{X}_2')\mathbf{X}_4)^{-1/2} \end{aligned} \tag{7.64}$$

then by (7.63),

$$
\begin{aligned}
\mathbf{V}_{22.3} &= (\mathbf{X}_4'(\mathbf{X}_2\mathbf{X}_2')^{-1}\mathbf{X}_4)^{-1/2}\mathbf{X}_4'(\mathbf{X}_2\mathbf{E}^{-1}\mathbf{X}_2')^{-1}\mathbf{X}_4(\mathbf{X}_4'(\mathbf{X}_2\mathbf{X}_2')^{-1}\mathbf{X}_4)^{-1/2} \\
\mathbf{Z}_{12} - \mathbf{Z}_{13}\mathbf{V}_{33}^{-1}\mathbf{V}_{33} &= \mathbf{A}_1\mathbf{Y}\mathbf{E}^{-1}\mathbf{X}_2'(\mathbf{X}_2\mathbf{E}^{-1}\mathbf{X}_2')^{-1}\mathbf{X}_2\mathbf{A}_2 \\
\mathbf{T}_2 &= \mathbf{A}_1\mathbf{Y}[\mathbf{E}^{-1} - \mathbf{E}^{-1}\mathbf{X}_2'(\mathbf{X}_2\mathbf{E}^{-1}\mathbf{X}_2')^{-1}\mathbf{X}_2\mathbf{E}^{-1}]\mathbf{Y}'\mathbf{A}_1 \\
\mathbf{T}_1 &= (\mathbf{I}+\mathbf{T}_2)^{-1/2}\mathbf{A}_1\mathbf{Y}\mathbf{E}^{-1}\mathbf{X}_2'(\mathbf{X}_2\mathbf{E}^{-1}\mathbf{X}_2')^{-1}\mathbf{X}_4[\mathbf{X}_4'(\mathbf{X}_2\mathbf{E}^{-1}\mathbf{X}_2')^{-1}\mathbf{X}_4]^{-1}\mathbf{X}_4'(\mathbf{X}_2\mathbf{E}^{-1}\mathbf{X}_2')^{-1}\mathbf{X}_2\mathbf{E}^{-1}\mathbf{Y}'\mathbf{A}_1'(\mathbf{I}+\mathbf{T}_2)^{-1/2}
\end{aligned}
\tag{7.65}
$$

For the MANOVA model, $\mathbf{X}_2 = \mathbf{I}$ and $\mathbf{T}_2 = \mathbf{0}$ so that

$$
\mathbf{T}_1 = \mathbf{A}_1\mathbf{Y}\mathbf{X}_4(\mathbf{X}_4'\mathbf{E}\mathbf{X}_4)^{-1}\mathbf{X}_4'\mathbf{Y}'\mathbf{A}_1' \tag{7.66}
$$

To verify the hypothesis and error SSP matrices for the MANOVA model, we use the Bartlett-Lawley-Hotelling trace criterion. Hence,

$$
\begin{aligned}
\mathrm{Tr}(\mathbf{T}_1) &= \mathrm{Tr}[(\mathbf{X}_3(\mathbf{X}_1'\mathbf{X}_1)^{-1}\mathbf{X}_3')^{-1/2}\mathbf{X}_3(\mathbf{X}_1'\mathbf{X}_1)^{-1}\mathbf{X}_1'\mathbf{Y}\mathbf{X}_4(\mathbf{X}_4'\mathbf{E}\mathbf{X}_4)^{-1}\mathbf{X}_4'\mathbf{Y}\mathbf{X}_1(\mathbf{X}_1'\mathbf{X}_1)^{-1}\mathbf{X}_3'(\mathbf{X}_3(\mathbf{X}_1'\mathbf{X}_1)^{-1}\mathbf{X}_3')^{-1/2}] \\
&= \mathrm{Tr}[(\mathbf{X}_3(\mathbf{X}_1'\mathbf{X}_1)^{-1}\mathbf{X}_3')^{-1/2}(\mathbf{X}_3\hat{\mathbf{B}}\mathbf{X}_4)(\mathbf{X}_4'\mathbf{E}\mathbf{X}_4)^{-1}(\mathbf{X}_3\hat{\mathbf{B}}\mathbf{X}_4)'(\mathbf{X}_3(\mathbf{X}_1'\mathbf{X}_1)^{-1}\mathbf{X}_3')^{-1/2}] \\
&= \mathrm{Tr}[(\mathbf{X}_4'\mathbf{E}\mathbf{X}_4)^{-1}(\mathbf{X}_3\hat{\mathbf{B}}\mathbf{X}_4)'(\mathbf{X}_3(\mathbf{X}_1'\mathbf{X}_1)^{-1}\mathbf{X}_3')^{-1}(\mathbf{X}_3\hat{\mathbf{B}}\mathbf{X}_4)] \\
&= \mathrm{Tr}(\mathbf{E}^{-1}\mathbf{H})
\end{aligned}
\tag{7.67}
$$

where

$$
\begin{aligned}
\mathbf{H} &= (\mathbf{X}_3\hat{\mathbf{B}}\mathbf{X}_4)'(\mathbf{X}_3(\mathbf{X}_1'\mathbf{X}_1)^{-1}\mathbf{X}_3')^{-1}(\mathbf{X}_3\hat{\mathbf{B}}\mathbf{X}_4) \\
\mathbf{E} &= \mathbf{X}_4'\mathbf{E}\mathbf{X}_4 \\
\hat{\mathbf{B}} &= (\mathbf{X}_1'\mathbf{X}_1)^{-1}\mathbf{X}_1'\mathbf{Y}
\end{aligned}
\tag{7.68}
$$

H and **E** in (7.68) are equal to the expressions given in (5.27) for testing $H(\mathbf{CBA} = \mathbf{0})$ in the MANOVA model with $\mathbf{X}_4 = \mathbf{A}$ and $\mathbf{X}_3 = \mathbf{C}$. In a similar manner, substituting $\mathbf{T}_1$ in (7.65) into the test criterion for the GMANOVA model, the canonical form of the test statistic in (7.59) reduces to Λ given in (7.46) using the MANCOVA model after some matrix algebra (Fujikoshi, 1974; Khatri, 1966).

7.5 Restricted Nonorthogonal Three-Factor Factorial MANOVA

The standard application of the RMGLM is to formulate a full-rank model for the MANOVA and, using restrictions, to constrain the interactions to equal zero to obtain an additive model. We impose restrictions on the linear model $\mathbf{Y} = \mathbf{XB} + \mathbf{U}$ of the form $\mathbf{R}_1\mathbf{B}\mathbf{R}_2 = \mathbf{0}$ where $\mathbf{R}_2 = \mathbf{A}$ or $\mathbf{R}_2 = \mathbf{I}$ when testing $H{:}\,\mathbf{CBA} = \mathbf{0}$. When $\mathbf{R}_2 \neq \mathbf{A}$ or **I**, the RMGLM becomes a mixture of the GMANOVA and the MANOVA models. Applications of the more general restricted model are discussed in Chapter 8.

In Chapter 5, we illustrated how to analyze a complete three-factor factorial design which included all factors A, B, C, AB, AC, BC, and ABC in the design. This design may also be analyzed using a multivariate full-rank model. The full-rank model for the three-factor factorial design with no empty cells is

$$
\begin{aligned}
&\mathbf{y}_{ijkm} = \mu_{ijk} + \mathbf{e}_{ijm} \\
&\qquad i = 1, 2, \dots, I; \quad j = 1, 2, \dots, J; \quad k = 1, 2, \dots, K; \quad m = 1, 2, \dots, M_{ijk} \\
&\mathbf{e}_{ijkm} \sim N_p(\mathbf{0}, \ \Sigma)
\end{aligned}
\tag{7.69}
$$

Given (7.69), either weighted or unweighted hypotheses about the main effects may be tested. For a 2 by 3 design, for example,

$$\bar{\mu}_{1.} = \frac{w_{11}\mu_{11} + w_{12}\mu_{12} + w_{13}\mu_{13}}{w_{11} + w_{12} + w_{13}}$$

and

$$\bar{\mu}_{2.} = \frac{w_{21}\mu_{21} + w_{22}\mu_{22} + w_{23}\mu_{23}}{w_{21} + w_{22} + w_{23}}$$

are the means for populations A_1 and A_2 collapsing across, or ignoring, factor B. Forming a weighted mean is essentially collapsing across or ignoring the factor or factors over which the mean is being found. However, other methods of weighting are available in the three-factor design. To test a hypothesis about factor A in a three-factor design, one can consider three different ways of collapsing across factors B and C: both factors B and C are collapsed or just factor B or just factor C. For an I by J by K design, the mean

$$\bar{\bar{\mu}}_{i..} = \sum_j \sum_k \frac{M_{ijk}\mu_{ijk}}{M_{i++}} \qquad i = 1, 2, \ldots, I \tag{7.70}$$

where M_{i++} is the total number of observations at the ith level of factor A, may be constructed to test

$$H_{A \text{ ignoring } B\&C} : \text{all } \bar{\bar{\mu}}_{i..} \text{ are equal} \tag{7.71}$$

Alternatively, means

$$\bar{\mu}_{i..} = \frac{1}{J}\sum_j \left(\frac{\sum_k M_{ijk}\mu_{ijk}}{M_{ij+}} \right) \qquad i = 1, 2, \ldots, I \tag{7.72}$$

where M_{ij+} is the total number of observations at the ith level of A and the jth level of B, are formed to test

$$H_{A \text{ given } B, \text{ ignoring } C} : \text{all } \bar{\mu}_{i..} \text{ are equal} \tag{7.73}$$

Finally, the mean

$$\tilde{\mu}_{i..} = \frac{1}{K}\sum_k \left(\frac{\sum_j \mu_{ijk}\mu_{ijk}}{M_{i+k}} \right) \qquad i = 1, 2, \ldots, I \tag{7.74}$$

where M_{i+k} is the total number of observations at the ith level of A and the kth level of C, is used to test

$$H_{A \text{ given } C, \text{ ignoring } B} : \text{ all } \tilde{\mu}_{i..} \text{ are equal} \tag{7.75}$$

In a similar fashion three different hypotheses can be tested by considering the main effects of factor B or by considering the main effects of factor C. Unweighted tests of hypotheses, as well as weighted tests, may be developed for each factor A, B, and C. To perform tests of either weighted or unweighted hypotheses about main effects in a three-factor design, one can construct hypothesis test matrices using the same procedures discussed in the analysis of univariate designs.

To test for two-factor interactions, we average cell means to produce a two-factor table of means and then express the hypotheses as we did for two-factor designs. For example, with an I by J by K design we form the hypothesis of no AB interaction by finding the average at each AB treatment combination as shown in Table 7.1.

Table 7.1 *AB Treatment Combination Means: Three-Factor Design*

	B_1	B_2	...	B_J
A_1	$\mu_{11.} = \sum_k \mu_{11k} / K$	$\mu_{12.} = \sum_k \mu_{12k} / K$	...	$\mu_{1J.} = \sum_k \mu_{1Jk} / K$
A_2	$\mu_{21.} = \sum_k \mu_{21k} / K$	$\mu_{22.} = \sum_k \mu_{22k} / K$	...	$\mu_{2J.} = \sum_k \mu_{2Jk} / K$
⋮	⋮	⋮	...	⋮
A_I	$\mu_{I1.} = \sum_k \mu_{I1k} / K$	$\mu_{I2.} = \sum_k \mu_{I2k} / K$	...	$\mu_{IJ.} = \sum_k \mu_{IJk} / K$

A general form for the hypothesis of no AB interaction is

$$H_{AB}: \quad \mu_{ij.} - \mu_{i'j.} - \mu_{ij'.} + \mu_{i'j'.} = 0 \qquad \text{for all } i, i', j, j' \tag{7.76}$$

Similarly we may form AC and BC interaction hypotheses by defining means for each AC and BC treatment combination and express the hypotheses of no AC and BC interaction as

$$\begin{aligned} H_{AC}: &\quad \mu_{i.k} - \mu_{i'.k} - \mu_{i.k'} + \mu_{i'.k'} = 0 \qquad \text{for all i, } i', k, k' \\ H_{BC}: &\quad \mu_{.jk} - \mu_{.j'k} - \mu_{.jk'} + \mu_{.j'k'} = 0 \qquad \text{for all j, } j', k, k' \end{aligned} \tag{7.77}$$

When the cell sizes are unequal we may also use weighted averages of the cell means as the two-factor treatment combination means and test weighted hypotheses of no two-factor interactions. For the AB interaction, for example, we define

$$\bar{\mu}_{ij.} = \sum_k M_{ijk}\mu_{ijk} / M_{ij+} \qquad i = 1, 2, \ldots, I; \quad j = 1, 2, \ldots, J$$

and then test the hypothesis

$$H_{AB^*}: \quad \bar{\mu}_{ij.} - \bar{\mu}_{i'j.} - \bar{\mu}_{ij'.} + \bar{\mu}_{i'j'.} = 0 \qquad \text{for all } i, i', j, j' \tag{7.78}$$

Similar hypotheses can be formed for the AC^* and BC^* interactions. Weighted hypotheses should not be tested in a three-factor design (or any other design, for that matter) unless the weights have meaning in the populations to which inferences are to be made.

In the three-factor design, the three-way interaction is zero if all the linear combinations of cell means used in the two-factor interaction are identical at all levels of the third factor. Thus, the hypothesis of no three-factor interaction can be stated as an equality of, for example, combinations of AB means at each level of C. For an I by J by K design, we would have

$$H_{ABC}:\ (\mu_{ijk} - \mu_{i'jk} - \mu_{ij'k} + \mu_{i'j'k}) - (\mu_{ijk'} - \mu_{i'jk'} - \mu_{ij'k'} + \mu_{i'j'k'}) = \mathbf{0} \quad \text{for all } i, i', j, j', k, k' \tag{7.79}$$

The hypothesis of no three-factor interaction can also be stated as an equality of the AC interaction combinations at different levels of B or the equality of the BC interaction combinations at different levels of A. All three forms of the hypothesis are equivalent.

To illustrate the analysis of data from a three-factor design with unequal cell frequencies, we use the data shown in Table 7.2, and analyzed in Section 5.7.

Table 7.2 *Data for 3 by 3 by 2 MANOVA Design*

	C_1			C_2		
	B_1	B_2	B_3	B_1	B_2	B_3
A_1	$Y1$: 1, 2, 3 $Y2$: 22, 21, 14	$Y1$: 3, 4 $Y2$: 31, 25	$Y1$: 1, 2 $Y2$: 31, 41	$Y1$: 3, 4 $Y2$: 66, 55	$Y1$: 4, 5, 6 $Y2$: 61, 11, 14	$Y1$: 1, 2 $Y2$: 41, 21
A_2	$Y1$: 3, 5, $Y2$: 31, 66,	$Y1$: 2 $Y2$: 45	$Y1$: 4, 5, 6 $Y2$: 21, 21, 31	$Y1$: 2 $Y2$: 41	$Y1$: 2, 3 $Y2$: 47, 61	$Y1$: 4, 5 $Y2$: 41, 55
A_3	$Y1$: 2, 3 $Y2$: 21, 31	$Y1$: 4, 5, 6 $Y2$: 66, 41, 51	$Y1$: 1, 2, 3 $Y2$: 61, 47, 35	$Y1$: 1 $Y2$: 41	$Y1$: 2, 3, 4 $Y2$: 18, 21, 31	$Y1$: 5, 6, 7 $Y2$: 57, 64, 77

If the first column in the design matrix is associated with cell $A_1B_1C_1$, the second with $A_1B_2C_1$, etc., then the parameter vector is μ in (7.80). Thus, the first six means are from level 1 of A, the second six from level 2 of A, and the last six are from level 3 of A, and we can test

$$H_A:\ \text{all } \mu_{i..} \text{ equal}$$
$$H_B:\ \text{all } \mu_{.j.} \text{ equal}$$
$$H_C:\ \text{all } \mu_{..k} \text{ equal}$$

using $\mathbf{C}_A$, $\mathbf{C}_B$ and $\mathbf{C}_C$, respectively, in (7.80). The matrices to test H_A, H_B and H_C are not unique.

If we wish to test weighted rather than unweighted hypotheses, we can choose the appropriate hypothesis test matrices from among those shown in (7.81), (7.82), and (7.83). We introduce some new notation here in order to differentiate among the matrices. The ordinary subscripts, A, B, and C, indicate an unweighted hypothesis. An asterisk on a subscript, as in A^*, B^*, and C^*, indicates a weighted hypothesis where all other factors are ignored in the hypothesis. A letter in square brackets indicates a factor that is not ignored in the weighted hypothesis and implies that all other factors are ignored. Thus, $\mathbf{C}_{A[B]}$ means the weighted hypothesis test matrix for testing A given B but ignoring C, $\mathbf{C}_{A^*}$ means the weighted hypothesis test matrix for testing A ignoring both B and C, and $\mathbf{C}_A$ means the hypothesis test matrix for testing A given B and C (unweighted test).

$$\mu = \begin{bmatrix} \mu_{111} \\ \mu_{121} \\ \mu_{131} \\ \mu_{112} \\ \mu_{122} \\ \mu_{132} \\ \mu_{211} \\ \mu_{221} \\ \mu_{231} \\ \mu_{212} \\ \mu_{222} \\ \mu_{232} \\ \mu_{311} \\ \mu_{321} \\ \mu_{331} \\ \mu_{312} \\ \mu_{322} \\ \mu_{332} \end{bmatrix} \quad \mathbf{C}'_A = \begin{bmatrix} 1/6 & 0 \\ 1/6 & 0 \\ 1/6 & 0 \\ 1/6 & 0 \\ 1/6 & 0 \\ 1/6 & 0 \\ 0 & 1/6 \\ 0 & 1/6 \\ 0 & 1/6 \\ 0 & 1/6 \\ 0 & 1/6 \\ 0 & 1/6 \\ -1/6 & -1/6 \\ -1/6 & -1/6 \\ -1/6 & -1/6 \\ -1/6 & -1/6 \\ -1/6 & -1/6 \\ -1/6 & -1/6 \end{bmatrix} \quad \mathbf{C}'_B = \begin{bmatrix} 1/6 & 0 \\ 0 & 1/6 \\ -1/6 & -1/6 \\ 1/6 & 0 \\ 0 & 1/6 \\ -1/6 & -1/6 \\ 1/6 & 0 \\ 0 & 1/6 \\ -1/6 & -1/6 \\ 1/6 & 0 \\ 0 & 1/6 \\ -1/6 & -1/6 \\ 1/6 & 0 \\ 0 & 1/6 \\ -1/6 & -1/6 \\ 1/6 & 0 \\ 0 & 1/6 \\ -1/6 & -1/6 \end{bmatrix} \quad \mathbf{C}'_C = \begin{bmatrix} 1/9 \\ 1/9 \\ 1/9 \\ -1/9 \\ -1/9 \\ -1/9 \\ 1/9 \\ 1/9 \\ 1/9 \\ -1/9 \\ -1/9 \\ -1/9 \\ 1/9 \\ 1/9 \\ 1/9 \\ -1/9 \\ -1/9 \\ -1/9 \end{bmatrix} \tag{7.80}$$

$$\mathbf{C}'_{A[B]} = \begin{bmatrix} 3/15 & 0 \\ 2/15 & 0 \\ 2/12 & 0 \\ 2/15 & 0 \\ 3/15 & 0 \\ 2/12 & 0 \\ 0 & 2/9 \\ 0 & 1/9 \\ 0 & 3/15 \\ 0 & 1/9 \\ 0 & 2/9 \\ 0 & 2/15 \\ -2/9 & -2/9 \\ -3/18 & -3/18 \\ -3/18 & -3/18 \\ -1/9 & -1/9 \\ -3/18 & -3/18 \\ -3/18 & -3/18 \end{bmatrix} \quad \mathbf{C}'_{A[C]} = \begin{bmatrix} 3/14 & 0 \\ 2/14 & 0 \\ 2/14 & 0 \\ 2/14 & 0 \\ 3/14 & 0 \\ 2/14 & 0 \\ 0 & 2/12 \\ 0 & 1/12 \\ 0 & 3/12 \\ 0 & 1/10 \\ 0 & 2/10 \\ 0 & 2/10 \\ -2/16 & -2/16 \\ -3/16 & -3/16 \\ -3/16 & -3/16 \\ -1/14 & -1/14 \\ -3/14 & -3/14 \\ -3/14 & -3/14 \end{bmatrix} \quad \mathbf{C}'_{A*} = \begin{bmatrix} 3/14 & 0 \\ 2/14 & 0 \\ 2/14 & 0 \\ 2/14 & 0 \\ 3/14 & 0 \\ 2/14 & 0 \\ 0 & 2/11 \\ 0 & 1/11 \\ 0 & 3/11 \\ 0 & 1/11 \\ 0 & 2/11 \\ 0 & 2/11 \\ -2/15 & -2/15 \\ -3/15 & -3/15 \\ -3/15 & -3/15 \\ -1/15 & -1/15 \\ -3/15 & -3/15 \\ -3/15 & -3/15 \end{bmatrix} \tag{7.81}$$

$$
\mathbf{C}'_{B[A]} = \begin{bmatrix} 3/15 & 0 \\ 0 & 2/15 \\ -2/12 & -2/12 \\ 2/15 & 0 \\ 0 & 3/15 \\ -2/12 & -2/12 \\ 2/9 & 0 \\ 0 & 1/9 \\ -3/15 & -3/15 \\ 1/9 & 0 \\ 0 & 3/18 \\ -2/15 & -3/18 \\ 2/9 & 0 \\ 0 & 3/18 \\ -3/18 & -3/18 \\ 1/9 & 0 \\ 0 & 3/81 \\ -3/18 & -3/18 \end{bmatrix} \quad
\mathbf{C}'_{B[C]} = \begin{bmatrix} 3/14 & 0 \\ 0 & 2/12 \\ -2/16 & -2/16 \\ 2/8 & 0 \\ 0 & 3/16 \\ -2/14 & -2/14 \\ 2/14 & 0 \\ 0 & 1/12 \\ -3/16 & -3/16 \\ 1/8 & 0 \\ 0 & 2/16 \\ -2/14 & -2/14 \\ 2/14 & 0 \\ 0 & 3/12 \\ -3/16 & -3/16 \\ 1/8 & 0 \\ 0 & 3/16 \\ -3/14 & -3/14 \end{bmatrix} \quad
\mathbf{C}'_{B^{*}} = \begin{bmatrix} 3/11 & 0 \\ 0 & 2/14 \\ -2/15 & -2/15 \\ 2/11 & 0 \\ 0 & 3/14 \\ -2/15 & -2/15 \\ 2/11 & 0 \\ 0 & 1/14 \\ -3/15 & -3/15 \\ 1/11 & 0 \\ 0 & 2/14 \\ -2/15 & -2/15 \\ 2/11 & 0 \\ 0 & 3/14 \\ -3/15 & -3/15 \\ 1/11 & 0 \\ 0 & 3/14 \\ -3/15 & -3/15 \end{bmatrix} \tag{7.82}
$$

$$
\mathbf{C}'_{C[A]} = \begin{bmatrix} 3/21 \\ 2/21 \\ 2/21 \\ -2/21 \\ -3/21 \\ -2/21 \\ 2/18 \\ 1/18 \\ 3/18 \\ -1/15 \\ -2/15 \\ -2/15 \\ 2/24 \\ 3/24 \\ 3/24 \\ -1/21 \\ -3/21 \\ -3/21 \end{bmatrix} \quad
\mathbf{C}'_{C[B]} = \begin{bmatrix} 3/21 \\ 2/18 \\ 2/24 \\ -2/12 \\ -3/24 \\ -2/21 \\ 2/21 \\ 1/18 \\ 3/24 \\ -1/12 \\ -2/24 \\ -2/21 \\ 2/21 \\ 3/18 \\ 3/24 \\ -1/12 \\ -3/24 \\ -3/21 \end{bmatrix} \quad
\mathbf{C}'_{C*} = \begin{bmatrix} 3/21 \\ 2/21 \\ 2/21 \\ -2/19 \\ -3/19 \\ -2/19 \\ 2/21 \\ 1/21 \\ 3/21 \\ -1/19 \\ -2/19 \\ -1/19 \\ 2/21 \\ 3/21 \\ 3/21 \\ -1/19 \\ -3/19 \\ -3/19 \end{bmatrix} \tag{7.83}
$$

If one wants to see how the two-factor interactions are tested, it is convenient to form 2 by 2 tables of means. A table of means for the AB treatment combinations is shown in Table 7.3. Each mean in Table 7.3 is accompanied by two numbers, the frequency (M_{ijk}) of the cell used in estimating the mean, and a number in parentheses indicating the position of the mean in the parameter vector. The latter number helps to place coefficients in the correct positions in hypothesis test matrices.

Table 7.3 *Means of AB Treatment Combinations*

	B_1	B_2	B_3
A_1	$\mu_{111}, M_{111} = 3, (\mu 1)$ $\mu_{112}, M_{112} = 2, (\mu 4)$	$\mu_{121}, M_{121} = 2, (\mu 2)$ $\mu_{122}, M_{122} = 3, (\mu 5)$	$\mu_{131}, M_{131} = 2, (\mu 3)$ $\mu_{132}, M_{132} = 2, (\mu 6)$
A_2	$\mu_{211}, M_{211} = 3, (\mu 7)$ $\mu_{212}, M_{212} = 2, (\mu 10)$	$\mu_{221}, M_{221} = 1, (\mu 8)$ $\mu_{222}, M_{222} = 2, (\mu 11)$	$\mu_{231}, \mu_{231} = 3, (\mu 9)$ $\mu_{232}, \mu_{232} = 2, (\mu 12)$
A_3	$\mu_{311}, M_{311} = 2, (\mu 13)$ $\mu_{312}, M_{312} = 1, (\mu 16)$	$\mu_{321}, M_{321} = 3, (\mu 14)$ $\mu_{322}, M_{322} = 3, (\mu 17)$	$\mu_{331}, M_{331} = 3, (\mu 15)$ $\mu_{332}, M_{332} = 3, (\mu 18)$

Using Table 7.3, we can form hypothesis test matrices for either weighted or unweighted tests of no AB interaction. One example of each is shown in (7.84). Note that each coefficient in $\mathbf{C}'_{AB}$ is 1/2 since each unweighted AB mean is the simple average of two cell means. The coefficients in $\mathbf{C}'_{AB^*}$ are those required to form weighted AB means. In a manner similar to the interpretation of the weighted main effect tests, the weighted interaction tests may be interpreted as tests of two factor interactions ignoring the factor not entering into the interaction.

$$\mathbf{C}'_{AB} = \begin{bmatrix} 1/2 & 0 & 0 & 0 \\ -1/2 & 1/2 & 0 & 0 \\ 0 & -1/2 & 0 & 0 \\ 1/2 & 0 & 1/2 & 0 \\ -1/2 & 1/2 & -1/2 & 0 \\ 0 & -1/2 & 0 & 0 \\ -1/2 & 0 & 1/2 & 0 \\ 1/2 & -1/2 & -1/2 & 1/2 \\ 0 & 1/2 & 0 & -1/2 \\ -1/2 & 0 & -1/2 & 0 \\ 1/2 & -1/2 & 1/2 & 1/2 \\ 0 & 1/2 & 0 & -1/2 \\ 0 & 0 & -1/2 & 0 \\ 0 & 0 & 1/2 & -1/2 \\ 0 & 0 & 0 & 1/2 \\ 0 & 0 & -1/2 & 0 \\ 0 & 0 & 1/2 & -1/2 \\ 0 & 0 & 0 & 1/2 \end{bmatrix} \quad \mathbf{C}'_{AB^*} = \begin{bmatrix} 3/5 & 0 & 0 & 0 \\ -2/5 & 2/5 & 0 & 0 \\ 0 & -2/4 & 0 & 0 \\ 2/5 & 0 & 0 & 0 \\ -3/5 & 3/5 & 0 & 0 \\ 0 & -2/4 & 0 & 0 \\ -2/3 & 0 & 2/3 & 0 \\ 1/3 & -1/3 & -1/3 & 1/3 \\ 0 & 3/5 & 0 & -3/5 \\ -1/3 & 0 & 1/3 & 0 \\ 2/3 & -2/3 & -2/3 & 2/3 \\ 0 & 2/5 & 0 & -2/5 \\ 0 & 0 & -2/3 & 0 \\ 0 & 0 & 3/6 & -3/6 \\ 0 & 0 & 0 & 3/6 \\ 0 & 0 & -1/3 & 0 \\ 0 & 0 & 3/6 & -3/6 \\ 0 & 0 & 0 & 3/6 \end{bmatrix} \tag{7.84}$$

Using a similar procedure, we form hypothesis test matrices for weighted or unweighted tests of the AC and BC interactions, examples of which are shown in (7.85).

$$
\mathbf{C}'_{AC} = \begin{bmatrix} 1/3 & 0 \\ 1/3 & 0 \\ 1/3 & 0 \\ -1/3 & 0 \\ -1/3 & 0 \\ -1/3 & 0 \\ -1/3 & 1/3 \\ -1/3 & 1/3 \\ -1/3 & 1/3 \\ 1/3 & -1/3 \\ 1/3 & -1/3 \\ 1/3 & -1/3 \\ 0 & -1/3 \\ 0 & -1/3 \\ 0 & -1/3 \\ 0 & 1/3 \\ 0 & 1/3 \\ 0 & 1/3 \end{bmatrix}
\quad
\mathbf{C}'_{AC^{\bullet}} = \begin{bmatrix} 3/7 & 0 \\ 2/7 & 0 \\ 2/7 & 0 \\ -2/7 & 0 \\ -3/7 & 0 \\ -2/7 & 0 \\ -2/6 & 2/6 \\ -1.6 & 1/6 \\ -3/6 & 3/6 \\ 1/5 & -1/5 \\ 2/5 & -2/5 \\ 2/5 & -2/5 \\ 0 & -2/8 \\ 0 & -3/8 \\ 0 & -3/8 \\ 0 & 1/7 \\ 0 & 3/7 \\ 0 & 3/7 \end{bmatrix}
\quad
\mathbf{C}'_{BC} = \begin{bmatrix} 1/3 & 0 \\ -1/3 & 1/3 \\ 0 & -1/3 \\ -1/3 & 0 \\ 1/3 & -1/3 \\ 0 & 1/3 \\ 1/3 & 0 \\ -1/3 & 1/3 \\ 0 & -1/3 \\ -1/3 & 0 \\ 1/3 & -1/3 \\ 0 & 1/3 \\ 1/3 & 0 \\ -1/3 & 1/3 \\ 0 & -1/3 \\ -1/3 & 0 \\ 1/3 & -1/3 \\ 0 & 1/3 \end{bmatrix}
\quad
\mathbf{C}'_{BC^{\bullet}} = \begin{bmatrix} 3/7 & 0 \\ -2/6 & 2/6 \\ 0 & -2/8 \\ -2/4 & 0 \\ 3/8 & -3/8 \\ 0 & 2/7 \\ 2/7 & 0 \\ -1/6 & 1/6 \\ 0 & -3/8 \\ -1/4 & 0 \\ 2/8 & -2/8 \\ 0 & 2/7 \\ 2/7 & 0 \\ -3/6 & 3/6 \\ 0 & -3/8 \\ -1/4 & 0 \\ 3/8 & -3/8 \\ 0 & 3/7 \end{bmatrix}
\qquad (7.85)
$$

To illustrate the formation of hypotheses of no three-factor interaction, we write out linear combinations of cell means like those for contrasts for any one of the two-factor interactions, and state that those combinations are identical at the different levels of the third factor. Choosing the AC interaction, for example, we write out linear combinations for each level of B and state that they are equal, as follows.

$$
\begin{array}{ccc}
\underline{\text{Level } B_1} & \underline{\text{Level } B_2} & \underline{\text{Level } B_3} \\
(\mu_{111}-\mu_{112}-\mu_{211}+\mu_{212}) = & (\mu_{121}-\mu_{122}-\mu_{221}+\mu_{222}) = & (\mu_{131}-\mu_{132}-\mu_{231}+\mu_{232}) \\
(\mu_{211}-\mu_{212}-\mu_{311}+\mu_{312}) = & (\mu_{221}-\mu_{222}-\mu_{321}+\mu_{322}) = & (\mu_{231}-\mu_{232}-\mu_{331}+\mu_{332})
\end{array}
$$

From each of these two sets of equalities we then form two contrasts, for example:

$$
\begin{aligned}
(\mu_{111}-\mu_{112}-\mu_{211}+\mu_{212})-(\mu_{121}-\mu_{122}-\mu_{221}+\mu_{212}) &= 0 \\
(\mu_{121}-\mu_{122}-\mu_{221}+\mu_{222})-(\mu_{131}-\mu_{132}-\mu_{231}+\mu_{232}) &= 0 \\
(\mu_{211}-\mu_{212}-\mu_{311}+\mu_{312})-(\mu_{221}-\mu_{222}-\mu_{321}+\mu_{322}) &= 0 \\
(\mu_{221}-\mu_{222}-\mu_{321}+\mu_{322})-(\mu_{231}-\mu_{232}-\mu_{331}+\mu_{332}) &= 0
\end{aligned}
$$

These contrasts are then used to form the hypothesis test matrix shown in (7.86).

$$\mathbf{C}'_{ABC} = \begin{bmatrix} 1 & 0 & 0 & 0 \\ -1 & 1 & 0 & 0 \\ 0 & -1 & 0 & 0 \\ -1 & 0 & 0 & 0 \\ 1 & -1 & 0 & 0 \\ 0 & 1 & 0 & 0 \\ -1 & 0 & 1 & 0 \\ 1 & -1 & -1 & 1 \\ 0 & 1 & 0 & -1 \\ 1 & 0 & -1 & 0 \\ -1 & 1 & 1 & -1 \\ 0 & -1 & 0 & 1 \\ 0 & 0 & -1 & 0 \\ 0 & 0 & 1 & -1 \\ 0 & 0 & 0 & 1 \\ 0 & 0 & 1 & 0 \\ 0 & 0 & -1 & 1 \\ 0 & 0 & 0 & -1 \end{bmatrix} \tag{7.86}$$

The data in Table 7.2 were analyzed in Chapter 5 using PROC GLM which is equivalent to using an unrestricted full rank cell means model. The RMGLM is appropriate whenever some terms are to be excluded from the cell means model. The terms are excluded by incorporating restrictions into the full-rank means model. When restrictions are used, the estimates of model parameters are no longer the cell means.

To illustrate the use of restrictions, we will assume that all interactions involving factor C are zero. Hence, the matrix of restrictions $\mathbf{R}$ must restrict the AC, BC, and ABC interactions to zero. To accomplish this, the matrix $\mathbf{R}$ would be constructed from $\mathbf{C}_{AC}$, $\mathbf{C}_{BC}$ and $\mathbf{C}_{ABC}$:

$$\mathbf{R} = \begin{pmatrix} \mathbf{C}_{AC} \\ \mathbf{C}_{BC} \\ \mathbf{C}_{ABC} \end{pmatrix} \tag{7.87}$$

and the matrix $\mathbf{A} = \mathbf{I}$. Given $\mathbf{R}$, we may test A, B, C, and AB using the matrices $\mathbf{C}_A, \mathbf{C}_B, \mathbf{C}_C$ and $\mathbf{C}_{AB}$, respectively, as defined for the full-rank means model.

PROC GLM does not allow one to impose general restrictions on the model parameters, except to exclude the intercept from the model–the intercept equal to zero restriction. To demonstrate the use of general restrictions on model parameters, we must use PROC REG with the MTEST statement. To test H_A, H_B, H_C and H_{AB} using PROC REG, we use the RESTRICT statement to formulate the matrix $\mathbf{R}$ which operates on the matrix of means $\mathbf{B}$. The hypotheses are tested using the MTEST statement and the hypothesis test matrices $\mathbf{C}_A, \mathbf{C}_B, \mathbf{C}_C$ and $\mathbf{C}_{AB}$. The PROC REG step and the PROC GLM step are included in Program 7_5.sas to complete the comparison with the less-than-full-rank model.

Program 7_5.sas

```
/* Program 7_5.sas */
options ls=80 ps=60 nodate nonumber;
title1 'Output 7.5: Restricted Nonorthogonal 3 Factor MANOVA Design';

data three;
   infile 'c:\7_5.dat';
   input a b c y1 y2 u1 - u18;
proc print data=three;
run;

/* Using PROC REG */
proc reg;
   title2 'Analysis using PROC REG';
   model y1 y2 = u1 - u18/noint;
   restrict .3333*u1+.3333*u2+.3333*u3-.3333*u4-
            .3333*u5-.3333*u6-.3333*u7-.3333*u8-
            .3333*u9+.3333*u10+.3333*u11+.3333*u12=0,
            .3333*u7+.3333*u8+.3333*u9-.3333*u10-
            .3333*u11-.3333*u12-.3333*u13-.3333*u14-
            .3333*u15+.3333*u16+.3333*u17+.3333*u18=0,
            .3333*u1-.3333*u2-.3333*u4+.3333*u5+
            .3333*u7-.3333*u8-.3333*u10+.3333*u11+
            .3333*u13-.3333*u14-.3333*u16+.3333*u17=0,
            .3333*u2-.3333*u3-.3333*u5+.3333*u6+
            .3333*u8-.3333*u9-.3333*u11+.3333*u12+
            .3333*u14-.3333*u15-.3333*u17+.3333*u18,
            u1-u2-u4+u5-u7+u8+u10-u11=0,
            u2-u3-u5+u6-u8+u9+u11-u12=0,
            u7-u8-u10+u11-u13+u14+u16-u17=0,
            u8-u9-u11+u12-u14+u15+u17-u18=0;

   A: mtest u1+u2+u3+u4+u5+u6-u13-u14-u15-u16-u17-u18=0,
            u7+u8+u9+u10+u11+u12-u13-u14-u15-u16-u17-u18=0/print details;

   B: mtest u1-u3+u4-u6+u7-u9+u10-u12+u13-u15+u16-u18=0,
            u2-u3+u5-u6+u8-u9+u11-u12+u14-u15+u17-u18=0/print details;

   C: mtest u1+u2+u3-u4-u5-u6+u7+u8+u9-u10-u11-u12+u13+u14+u15-u16-u17-u18=0
      /print details;

   AB: mtest u1-u2+u4-u5-u7+u8-u10+u11=0,
             u2-u3+u5-u6-u8+u9-u11+u12=0,
             u7-u8+u10-u11-u13+u14-u16+u17=0,
             u8-u9+u11-u12-u14+u15-u17+u18=0/print details;
run;

/* Using PROC GLM */
proc glm;
title2 'Analysis using PROC GLM';
   class a b c;
   model y1 y2 = a b c a*b/ss3;
   lsmeans a b c a*b;
   manova h=a b c a*b/printe printh;
run;
```

To analyze the three-factor design using PROC REG, the data and design matrix must be input. The data are contained in the data file 7_5.dat where the variables U1 to U28 are the design matrix. The MTEST statements are used to test the unweighted tests H_A, H_B, H_C, and H_{AB}.

Result and Interpretation 7_5.sas

PROC REG outputs the matrix $\hat{\mathbf{B}}_r$ given in (7.4) with $\mathbf{R}_2 = \mathbf{I}$ and $\mathbf{R}_1 = \mathbf{R}$ defined in (7.87) and $\boldsymbol{\Theta} = \mathbf{0}$, a variable at a time. An estimate for each restriction is also calculated. The hypothesis and error test matrices $\mathbf{H}_r$ and $\mathbf{E}_r$ as defined in (7.11) are output for each MTEST statement. Output 7.5 includes the data for the analysis and only the unweighted test of A using both PROC REG and PROC GLM.

Output 7.5: *Restricted Nonorthogonal 3 Factor MANOVA Design*

Output 7.5: Restricted Nonorthogonal 3 Factor MANOVA Design

OBS	A	B	C	Y1	Y2	U1	U2	U3	U4	U5	U6	U7	U8	U9	U10	U11	U12	U13	U14	U15	U16	U17	U18
1	1	1	1	1	22	1	0	0	0	0	0	0	0	0	0	0	0	0	0	0	0	0	0
2	1	1	1	2	21	1	0	0	0	0	0	0	0	0	0	0	0	0	0	0	0	0	0
3	1	1	1	3	14	1	0	0	0	0	0	0	0	0	0	0	0	0	0	0	0	0	0
4	1	2	1	3	31	0	1	0	0	0	0	0	0	0	0	0	0	0	0	0	0	0	0
5	1	2	1	4	25	0	1	0	0	0	0	0	0	0	0	0	0	0	0	0	0	0	0
6	1	3	1	1	31	0	0	1	0	0	0	0	0	0	0	0	0	0	0	0	0	0	0
7	1	3	1	2	41	0	0	1	0	0	0	0	0	0	0	0	0	0	0	0	0	0	0
8	1	1	2	3	66	0	0	0	1	0	0	0	0	0	0	0	0	0	0	0	0	0	0
9	1	1	2	4	55	0	0	0	1	0	0	0	0	0	0	0	0	0	0	0	0	0	0
10	1	2	2	4	61	0	0	0	0	1	0	0	0	0	0	0	0	0	0	0	0	0	0
11	1	2	2	5	11	0	0	0	0	1	0	0	0	0	0	0	0	0	0	0	0	0	0
12	1	2	2	6	21	0	0	0	0	1	0	0	0	0	0	0	0	0	0	0	0	0	0
13	1	3	2	1	41	0	0	0	0	0	1	0	0	0	0	0	0	0	0	0	0	0	0
14	1	3	2	2	21	0	0	0	0	0	1	0	0	0	0	0	0	0	0	0	0	0	0
15	2	1	1	3	31	0	0	0	0	0	0	1	0	0	0	0	0	0	0	0	0	0	0
16	2	1	1	5	66	0	0	0	0	0	0	1	0	0	0	0	0	0	0	0	0	0	0
17	2	2	1	2	45	0	0	0	0	0	0	0	1	0	0	0	0	0	0	0	0	0	0
18	2	3	1	4	21	0	0	0	0	0	0	0	0	1	0	0	0	0	0	0	0	0	0
19	2	3	1	5	21	0	0	0	0	0	0	0	0	1	0	0	0	0	0	0	0	0	0
20	2	3	1	6	31	0	0	0	0	0	0	0	0	1	0	0	0	0	0	0	0	0	0
21	2	1	2	2	41	0	0	0	0	0	0	0	0	0	1	0	0	0	0	0	0	0	0
22	2	2	2	2	47	0	0	0	0	0	0	0	0	0	0	1	0	0	0	0	0	0	0
23	2	2	2	3	61	0	0	0	0	0	0	0	0	0	0	1	0	0	0	0	0	0	0
24	2	3	2	4	41	0	0	0	0	0	0	0	0	0	0	0	1	0	0	0	0	0	0
25	2	3	2	5	55	0	0	0	0	0	0	0	0	0	0	0	1	0	0	0	0	0	0
26	3	1	1	2	21	0	0	0	0	0	0	0	0	0	0	0	0	1	0	0	0	0	0
27	3	1	1	3	31	0	0	0	0	0	0	0	0	0	0	0	0	1	0	0	0	0	0
28	3	2	1	4	66	0	0	0	0	0	0	0	0	0	0	0	0	0	1	0	0	0	0
29	3	2	1	5	41	0	0	0	0	0	0	0	0	0	0	0	0	0	1	0	0	0	0
30	3	2	1	6	51	0	0	0	0	0	0	0	0	0	0	0	0	0	1	0	0	0	0
31	3	3	1	1	61	0	0	0	0	0	0	0	0	0	0	0	0	0	0	1	0	0	0
32	3	3	1	2	47	0	0	0	0	0	0	0	0	0	0	0	0	0	0	1	0	0	0
33	3	3	1	3	35	0	0	0	0	0	0	0	0	0	0	0	0	0	0	1	0	0	0
34	3	1	2	1	41	0	0	0	0	0	0	0	0	0	0	0	0	0	0	0	1	0	0
35	3	2	2	2	18	0	0	0	0	0	0	0	0	0	0	0	0	0	0	0	0	1	0
36	3	2	2	3	21	0	0	0	0	0	0	0	0	0	0	0	0	0	0	0	0	1	0
37	3	2	2	4	31	0	0	0	0	0	0	0	0	0	0	0	0	0	0	0	0	1	0
38	3	3	2	5	57	0	0	0	0	0	0	0	0	0	0	0	0	0	0	0	0	0	1
39	3	3	2	6	64	0	0	0	0	0	0	0	0	0	0	0	0	0	0	0	0	0	1
40	3	3	2	7	77	0	0	0	0	0	0	0	0	0	0	0	0	0	0	0	0	0	1

Output 7.5 (*continued*)

```
                          Analysis using PROC REG

Multivariate Test: A

                           L Ginv(X'X) L'   LB-cj

         5.2782407407      2.6712962963      -3.138888889      -56.34236111
         2.6712962963      6.1351851852      0.8777777778      8.9430555556

                     Inv(L Ginv(X'X) L')     Inv()(LB-cj)

         0.2430049584      -0.105805811      -0.855639554      -14.63770036
         -0.105805811      0.2090627476      0.5156233191      7.8310089877

                            E, the Error Matrix

                          57.866666667      226.69166667
                          226.69166667      7927.0822917

                          H, the Hypothesis Matrix

                           3.138360179      52.820000701
                          52.820000701      894.75574822

                            T, the H + E Matrix

                          61.005026846      279.51166737
                          279.51166737      8821.8380399

                                Eigenvectors

                          0.0465491876      0.0085524051
                          0.1304185966      -0.007711139

                                Eigenvalues

                                  0.1143020468
                                  0.0003444607
```

Output 7.5 (*continued*)

```
          Output 7.5: Restricted Nonorthogonal 3 Factor MANOVA Design
                            Analysis using PROC REG

                 Multivariate Statistics and F Approximations

                            S=2    M=-0.5    N=13.5

Statistic                      Value           F      Num DF    Den DF  Pr > F

Wilks' Lambda               0.88539287    0.9099          4        58  0.4643
Pillai's Trace              0.11464651    0.9121          4        60  0.4628
Hotelling-Lawley Trace      0.12939766    0.9058          4        56  0.4669
Roy's Greatest Root         0.12905308    1.9358          2        30  0.1619

        NOTE: F Statistic for Roy's Greatest Root is an upper bound.
               NOTE: F Statistic for Wilks' Lambda is exact.

                            Analysis using PROC GLM

                        General Linear Models Procedure
                       Multivariate Analysis of Variance

                        H = Type III SS&CP Matrix for A

                                   Y1                Y2

                 Y1       3.138360179      52.820000701
                 Y2      52.820000701      894.75574822


         Characteristic Roots and Vectors of: E Inverse * H, where
          H = Type III SS&CP Matrix for A    E = Error SS&CP Matrix

        Characteristic    Percent        Characteristic Vector  V'EV=1
             Root
                                                  Y1              Y2

          0.12905308      99.73            0.04946173      0.00908752
          0.00034458       0.27            0.13044106     -0.00771247

               Manova Test Criteria and F Approximations for
                   the Hypothesis of no Overall A Effect
          H = Type III SS&CP Matrix for A    E = Error SS&CP Matrix

                            S=2    M=-0.5    N=13.5

Statistic                      Value           F      Num DF    Den DF  Pr > F

Wilks' Lambda               0.88539287    0.9099          4        58  0.4643
Pillai's Trace              0.11464651    0.9121          4        60  0.4628
Hotelling-Lawley Trace      0.12939766    0.9058          4        56  0.4669
Roy's Greatest Root         0.12905308    1.9358          2        30  0.1619

        NOTE: F Statistic for Roy's Greatest Root is an upper bound.
               NOTE: F Statistic for Wilks' Lambda is exact.
```

As expected, the multivariate test criteria for PROC REG and PROC GLM agree. The purpose for using PROC REG in this example was to demonstrate the equivalence between the two approaches. Often using the RMGLM with PROC REG to test hypotheses of the form given in (7.70) is easier than trying to determine estimable functions using contrasts in PROC GLM when standard TYPE I, II, III, and IV sum of squares do not yield the hypothesis test desired.

7.6 Restricted Intraclass Covariance Design

Other important applications of the restricted model are to formulate an intraclass regression model, which permits different regression equations for each class, and to add restrictions for each class to ensure parallelism of slopes. This results in the standard MANCOVA model. The MANCOVA design, discussed in Chapter 5, requires the slopes of the regression lines within each population to be identical; that

$$\boldsymbol{\Gamma}_1 = \boldsymbol{\Gamma}_2 = \ldots = \boldsymbol{\Gamma}_I = \boldsymbol{\Gamma} \tag{7.88}$$

for I populations. More generally letting $\mathbf{B}_i$ and $\boldsymbol{\Gamma}_i$ represent the slope and intercept for the regression of Y on Z for the ith population, $i = 1, 2, \ldots, I$, we have the intraclass covariance model

$$\underset{J_i \times p}{E(\mathbf{Y}_i)} = \underset{J_i \times k}{\mathbf{X}_i} \; \underset{k \times p}{\mathbf{B}_i} + \underset{J_i \times h}{\mathbf{Z}_i} \; \underset{h \times p}{\boldsymbol{\Gamma}_i} \tag{7.89}$$

where the $\sum_i J_i = N$, also called the heterogeneous data model. Given (7.89), three natural tests are evident. The test of coincidence

$$\begin{aligned} H_c\colon \; & \mathbf{B}_1 = \mathbf{B}_2 = \ldots = \mathbf{B}_I \\ & \boldsymbol{\Gamma}_1 = \boldsymbol{\Gamma}_2 = \ldots = \boldsymbol{\Gamma}_I \end{aligned} \tag{7.90}$$

the test of parallelism,

$$H_p\colon \; \boldsymbol{\Gamma}_1 = \boldsymbol{\Gamma}_2 = \ldots = \boldsymbol{\Gamma}_I = \boldsymbol{\Gamma} \tag{7.91}$$

and the test of intercepts given that $\boldsymbol{\Gamma}_1 = \boldsymbol{\Gamma}_2 = \ldots = \boldsymbol{\Gamma}_I$, the MANCOVA tests for group differences

$$H_g\colon \; \mathbf{B}_1 = \mathbf{B}_2 = \ldots = \mathbf{B}_I \tag{7.92}$$

In Chapter 3, a single-factor restricted intraclass covariance model was discussed. In this chapter we illustrate the analysis of a multivariate two-factor nonorthogonal intraclass covariance design and show how restrictions are used to produce a nonorthogonal MANCOVA design, a restricted intraclass covariance design.

The data given in Table 7.4 are utilized for the analysis

Table 7.4 *Two-Factor Intraclass design*

	B_1	B_2	B_3
A_1	Z: 1, 2, 4, 4, 5	Z: 3, 5, 6, 5	Z: 6, 5, 6
	Y1: 2, 2, 3, 3, 4	Y1: 3, 4, 5, 5	Y1: 5, 6, 8
	Y2: 22, 21, 14, 25, 31	Y2: 31, 25, 30, 31	Y2: 31, 41, 50
A_2	Z: 2, 4, 3	Z: 5, 8, 4, 6	Z: 3, 3, 4
	Y1: 3, 5, 5	Y1: 5, 6, 6, 7	Y1: 3, 4, 4
	Y2: 31, 66, 45	Y2: 45, 34, 31, 30	Y2: 21, 21, 31

For this example the combined design matrix has the general form

$$\mathbf{X} = \begin{pmatrix} \mathbf{X}_1 & \mathbf{0} & \cdots & \cdots & \cdots & \mathbf{0} & \mathbf{Z}_1 & \mathbf{0} & \cdots & \cdots & \cdots & \mathbf{0} \\ \mathbf{0} & \mathbf{X}_2 & \cdots & \cdots & \cdots & \mathbf{0} & \mathbf{0} & \mathbf{Z}_2 & \cdots & \cdots & \cdots & \mathbf{0} \\ \vdots & \vdots & \ddots & \cdots & \cdots & \vdots & \vdots & \vdots & \ddots & \cdots & \cdots & \vdots \\ \vdots & \vdots & \cdots & \ddots & \cdots & \vdots & \vdots & \vdots & \cdots & \ddots & \cdots & \vdots \\ \vdots & \vdots & \cdots & \cdots & \ddots & \vdots & \vdots & \vdots & \cdots & \cdots & \ddots & \vdots \\ \mathbf{0} & \mathbf{0} & \cdots & \cdots & \cdots & \mathbf{X}_{IJ} & \mathbf{0} & \mathbf{0} & \cdots & \cdots & \cdots & \mathbf{Z}_{IJ} \end{pmatrix} \tag{7.93}$$

The design matrix, $\mathbf{X}$, is

$$\mathbf{X} = \begin{bmatrix} 1 & 0 & 0 & 0 & 0 & 0 & 1 & 0 & 0 & 0 & 0 & 0 \\ 1 & 0 & 0 & 0 & 0 & 0 & 2 & 0 & 0 & 0 & 0 & 0 \\ 1 & 0 & 0 & 0 & 0 & 0 & 4 & 0 & 0 & 0 & 0 & 0 \\ 1 & 0 & 0 & 0 & 0 & 0 & 4 & 0 & 0 & 0 & 0 & 0 \\ 1 & 0 & 0 & 0 & 0 & 0 & 5 & 0 & 0 & 0 & 0 & 0 \\ 0 & 1 & 0 & 0 & 0 & 0 & 0 & 3 & 0 & 0 & 0 & 0 \\ 0 & 1 & 0 & 0 & 0 & 0 & 0 & 5 & 0 & 0 & 0 & 0 \\ 0 & 1 & 0 & 0 & 0 & 0 & 0 & 6 & 0 & 0 & 0 & 0 \\ 0 & 1 & 0 & 0 & 0 & 0 & 0 & 5 & 0 & 0 & 0 & 0 \\ 0 & 0 & 1 & 0 & 0 & 0 & 0 & 0 & 6 & 0 & 0 & 0 \\ 0 & 0 & 1 & 0 & 0 & 0 & 0 & 0 & 5 & 0 & 0 & 0 \\ 0 & 0 & 1 & 0 & 0 & 0 & 0 & 0 & 6 & 0 & 0 & 0 \\ 0 & 0 & 0 & 1 & 0 & 0 & 0 & 0 & 0 & 2 & 0 & 0 \\ 0 & 0 & 0 & 1 & 0 & 0 & 0 & 0 & 0 & 4 & 0 & 0 \\ 0 & 0 & 0 & 1 & 0 & 0 & 0 & 0 & 0 & 3 & 0 & 0 \\ 0 & 0 & 0 & 0 & 1 & 0 & 0 & 0 & 0 & 0 & 5 & 0 \\ 0 & 0 & 0 & 0 & 1 & 0 & 0 & 0 & 0 & 0 & 8 & 0 \\ 0 & 0 & 0 & 0 & 1 & 0 & 0 & 0 & 0 & 0 & 4 & 0 \\ 0 & 0 & 0 & 0 & 1 & 0 & 0 & 0 & 0 & 0 & 6 & 0 \\ 0 & 0 & 0 & 0 & 0 & 1 & 0 & 0 & 0 & 0 & 0 & 3 \\ 0 & 0 & 0 & 0 & 0 & 1 & 0 & 0 & 0 & 0 & 0 & 3 \\ 0 & 0 & 0 & 0 & 0 & 1 & 0 & 0 & 0 & 0 & 0 & 4 \end{bmatrix} \tag{7.94}$$

and the parameter matrix **B** is

$$\mathbf{B}=\begin{pmatrix}\mathbf{B}_1\\ \mathbf{B}_2\\ \mathbf{B}_3\\ \mathbf{B}_4\\ \mathbf{B}_5\\ \mathbf{B}_6\\ \mathbf{\Gamma}_1\\ \mathbf{\Gamma}_2\\ \mathbf{\Gamma}_3\\ \mathbf{\Gamma}_4\\ \mathbf{\Gamma}_5\\ \mathbf{\Gamma}_6\end{pmatrix}=\begin{bmatrix}\mu_{111} & \mu_{112}\\ \mu_{121} & \mu_{122}\\ \mu_{131} & \mu_{132}\\ \mu_{211} & \mu_{212}\\ \mu_{221} & \mu_{222}\\ \mu_{231} & \mu_{232}\\ \gamma_{111} & \gamma_{112}\\ \gamma_{121} & \gamma_{122}\\ \gamma_{131} & \gamma_{132}\\ \gamma_{211} & \gamma_{212}\\ \gamma_{221} & \gamma_{222}\\ \gamma_{231} & \gamma_{232}\end{bmatrix} \tag{7.95}$$

The parameters in (7.95) are associated with the cells in the design, as given in Table 7.5.

Table 7.5 *Intraclass Covariance Parameters*

	B_1	B_2	B_3
A_1	$\mathbf{B}_1,\mathbf{\Gamma}_1$	$\mathbf{B}_2,\mathbf{\Gamma}_2$	$\mathbf{B}_3,\mathbf{\Gamma}_3$
A_2	$\mathbf{B}_4,\mathbf{\Gamma}_4$	$\mathbf{B}_5,\mathbf{\Gamma}_5$	$\mathbf{B}_6,\mathbf{\Gamma}_6$

Given (7.89), the first hypothesis of interest is the test of parallelism (7.91), also written

$$H_{\text{Parallel}}:\ \text{all } \mathbf{\Gamma}_i \text{ are equal} \tag{7.96}$$

If (7.96) is true, (7.89) reduces to the full-rank MANCOVA model, a restricted intraclass covariance model. The test of parallelism for these data thus becomes

$$H_{\text{Parallel}}:\ \mathbf{\Gamma}_1=\mathbf{\Gamma}_2=\ldots=\mathbf{\Gamma}_6=\mathbf{\Gamma} \tag{7.97}$$

To test (7.97), the hypothesis test matrix is

$$\mathbf{C}=\begin{pmatrix} & 1 & 0 & 0 & 0 & 0 & -1\\ & 0 & 1 & 0 & 0 & 0 & -1\\ \mathbf{0}_{5\text{x}6} & 0 & 0 & 1 & 0 & 0 & -1\\ & 0 & 0 & 0 & 1 & 0 & -1\\ & 0 & 0 & 0 & 0 & 1 & -1\end{pmatrix} \tag{7.98}$$

The SAS code to perform the analyses is in Program 7_6.sas.

Program 7_6.sas

```
/*Program 7_6.sas */

options  ls=80 ps=60 nodate nonumber;
title1 'Output 7.6: Restricted Intra-class Covariance Design';
```

```
data intra;
   infile 'c:\7_6.dat';
   input a b y1 y2 u1-u6 z1-z6;
proc print data=intra;
run;

/* Testing Equality of intra-class regression coefficients  */
proc reg;
   title2 'Testing Equality of Inta-class Regression Coefficients';
   model y1 y2 = u1 - u6 z1 - z6/noint;
   Parallel: mtest z1-z6=0,
              z2-z6=0,
              z3-z6=0,
              z4-z6=0,
              z5-z6=0;
   Coin: mtest z1-z6=0,z2-z6=0,z3-z6=0,z4-z6=0,z5-z6=0,
                u1-u6=0,u2-u6=0,u3-u6=0,u4-z6=0,u5-u6=0;
run;

/* MANCOVA model with unweighted and weighted tests*/
proc reg;
   title2 'MANCOVA Analyis using PROC REG';
   model y1 y2= u1 - u6 z1 - z6/noint;
   restrict z1-z6=0,
            z2-z6=0,
            z3-z6=0,
            z4-z6=0,
            z5-z6=0;
   A: mtest u1+u2+u3-u4-u5-u6=0;
   B: mtest u1-u3+u4-u6=0,
            u2-u3+u5-u6=0;
   AB: mtest u1-u3-u4+u6=0,
             u2-u3-u5+u6=0;
   Reg: mtest z6=0;
   Awt: mtest .4167*u1+.3333*u2+.25*u3-.3*u4-.4*u5-.3*u6=0;
   Bwt: mtest .625*u1-.5*u3+.375*u4-.5*u6=0,
              .500*u2-.5*u3+.500*u5-.5*u6=0;
run;

/* MANOVA model with unweighted tests */
proc reg;
   title2 ' Restricted Unweighted MANOVA tests using PROC REG';
   model y1 y2= u1 - u6 z1 -z6/noint;
   restrict z1=z2=z3=z4=z5=z6=0;
   A: mtest u1+u2+u3-u4-u5-u6=0;
   B: mtest u1-u3+u4-u6=0,
            u2-u3+u5-u6=0;
   AB: mtest u1-u3-u4+u6=0,
            u2-u3-u5+u6=0;
run;

data mancova;
   infile 'c:\7_6a.dat';
   input a b y1 y2 z;
proc print data=mancova;
run;
```

```
/* Unweighted MANOVA Analysis using GLM */
proc glm;
   title2 'Unweighted MANOVA tests using PROC GLM';
   class a b;
   model y1 y2=a b a*b/ss3;
   lsmeans a b a*b;
   manova h= a b a*b/htype=3 etype=3;
run;

/* Unweighted MANCOVA Analysis using GLM */
proc glm;
   title2 'Unweighted MANCOVA tests using PROC GLM';
   class a b;
   model y1 y2=a b a*b z/ss3;
   manova h=a b a*b z/htype=3 etype=3;
run;
```

Result and Interpretation 7_6.sas

The SAS code to perform the tests of parallelism and coincidence is given in the first part of Program 7_6.sas. PROC REG is used with MTEST statements to perform the tests. The results of these tests are found in Output 7.6.

Output 7.6: *Restricted Intra-class Covariance Design Testing Equality of Intra-class Regression Coefficients*

```
              Output 7.6: Restricted Intra-class Covariance Design
             Testing Equality of Inta-class Regression Coefficients

Multivariate Test: PARALLEL

                  Multivariate Statistics and F Approximations

                               S=2     M=1     N=3.5

  Statistic                     Value          F       Num DF     Den DF   Pr > F

  Wilks' Lambda               0.38546201     1.0992        10         18   0.4126
  Pillai's Trace              0.68602300     1.0442        10         20   0.4446
  Hotelling-Lawley Trace      1.40883655     1.1271        10         16   0.4009
  Roy's Greatest Root         1.26186988     2.5237         5         10   0.0998

        NOTE: F Statistic for Roy's Greatest Root is an upper bound.
               NOTE: F Statistic for Wilks' Lambda is exact.

Multivariate Test: COIN

                  Multivariate Statistics and F Approximations

                               S=2    M=3.5     N=3.5

  Statistic                     Value          F       Num DF     Den DF   Pr > F

  Wilks' Lambda               0.07952611     2.2914        20         18   0.0412
  Pillai's Trace              1.33467229     2.0060        20         20   0.0640
  Hotelling-Lawley Trace      6.36615446     2.5465        20         16   0.0313
  Roy's Greatest Root         5.40200678     5.4020        10         10   0.0067

        NOTE: F Statistic for Roy's Greatest Root is an upper bound.
               NOTE: F Statistic for Wilks' Lambda is exact.
```

Because we do not reject the null hypothesis of parallelism, we would not perform the test of coincidence. Instead, the matrix used to test for parallelism would be included in the analysis as a restriction for the intraclass covariance design to create a nonorthogonal MANCOVA design. The SAS code to perform the MANCOVA, or RMGLM, analysis performing weighted and unweighted tests of main effects and interaction is found in Program 7_6.sas. Also performed is the test of $\mathbf{\Gamma} = \mathbf{0}$, labeled REG in the output. If $\mathbf{\Gamma} = \mathbf{0}$, the restricted MANCOVA model becomes a MANOVA model. The output for the restricted MANCOVA analysis follows.

Output 7.6: *(continued)*

```
              Output 7.6: Restricted Intra-class Covariance Design
                         MANCOVA Analyis using PROC REG

Multivariate Test: A
                     Multivariate Statistics and Exact F Statistics

                              S=1     M=0     N=6

  Statistic                     Value           F       Num DF    Den DF  Pr > F

  Wilks' Lambda              0.85235220      1.2126          2        14  0.3268
  Pillai's Trace             0.14764780      1.2126          2        14  0.3268
  Hotelling-Lawley Trace     0.17322393      1.2126          2        14  0.3268
  Roy's Greatest Root        0.17322393      1.2126          2        14  0.3268

Multivariate Test: B

                     Multivariate Statistics and F Approximations

                             S=2     M=-0.5     N=6

  Statistic                     Value           F       Num DF    Den DF  Pr > F

  Wilks' Lambda              0.70266524      1.3507          4        28  0.2761
  Pillai's Trace             0.30143190      1.3310          4        30  0.2813
  Hotelling-Lawley Trace     0.41732193      1.3563          4        26  0.2762
  Roy's Greatest Root        0.40284784      3.0214          2        15  0.0790


Multivariate Test: AB

                     Multivariate Statistics and F Approximations

                             S=2     M=-0.5     N=6

  Statistic                     Value           F       Num DF    Den DF  Pr > F

  Wilks' Lambda              0.36903633      4.5229          4        28  0.0061
  Pillai's Trace             0.72734922      4.2864          4        30  0.0073
  Hotelling-Lawley Trace     1.44857857      4.7079          4        26  0.0054
  Roy's Greatest Root        1.23752726      9.2815          2        15  0.0024
```

Output 7.6 *(continued)*

```
Multivariate Test: REG

              Multivariate Statistics and Exact F Statistics

                            S=1    M=0    N=6

 Statistic                    Value          F      Num DF    Den DF  Pr > F

 Wilks' Lambda           0.65831474     3.6332          2        14  0.0536
 Pillai's Trace          0.34168526     3.6332          2        14  0.0536
 Hotelling-Lawley Trace  0.51903024     3.6332          2        14  0.0536
 Roy's Greatest Root     0.51903024     3.6332          2        14  0.0536


Multivariate Test: AWT

              Multivariate Statistics and Exact F Statistics

                            S=1    M=0    N=6

 Statistic                    Value          F      Num DF    Den DF  Pr > F

 Wilks' Lambda           0.75008682     2.3323          2        14  0.1336
 Pillai's Trace          0.24991318     2.3323          2        14  0.1336
 Hotelling-Lawley Trace  0.33317901     2.3323          2        14  0.1336
 Roy's Greatest Root     0.33317901     2.3323          2        14  0.1336


Multivariate Test: BWT

               Multivariate Statistics and F Approximations

                           S=2    M=-0.5    N=6

 Statistic                    Value          F      Num DF    Den DF  Pr > F

 Wilks' Lambda           0.70044260     1.3640          4        28  0.2716
 Pillai's Trace          0.30279252     1.3380          4        30  0.2788
 Hotelling-Lawley Trace  0.42305005     1.3749          4        26  0.2699
 Roy's Greatest Root     0.41183518     3.0888          2        15  0.0753
```

It is also possible to specify a model with pooled estimates of the regression matrices for each level of a factor, called intrarow or intracolumn covariance models (Searle, 1987, p. 451). For example, if $\mathbf{\Gamma}_1 = \mathbf{\Gamma}_2 = \mathbf{\Gamma}_3 = \mathbf{\Gamma}_{1A}$ and $\mathbf{\Gamma}_4 = \mathbf{\Gamma}_5 = \mathbf{\Gamma}_6 = \mathbf{\Gamma}_{2A}$ we would have an intrarow covariance model. Of course, if $\mathbf{\Gamma}_{1A} = \mathbf{\Gamma}_{2A} = \mathbf{\Gamma}$, we again have the standard MANCOVA model with restrictions. Similar models may be hypothesized for columns, or we may formulate an intrarow plus intracolumn model (Searle, 1987, p. 451).

Because the test of interaction is significant, both the weighted and unweighted tests of A and B are confounded. Some may consider the test REG ($\mathbf{\Gamma} = \mathbf{0}$) to be nonsignificant ($p$ = .0536). If $\mathbf{\Gamma} = \mathbf{0}$, then a MANOVA model is appropriate for the analysis. This may be accomplished by using the intraclass covariance model and imposing the restriction that all $\mathbf{\Gamma}_i = \mathbf{0}$. The SAS code to perform this restricted MANOVA analysis is included in

Program 7_6.sas. Unweighted tests of *A*, *B*, and *AB* are performed using PROC REG with MTEST statements. The results are found in the continuation of Output 7.6 that follows.

Output 7.6 (*continued*)

```
        Restricted Unweighted MANOVA tests using PROC REG

Multivariate Test: A

                 Multivariate Statistics and Exact F Statistics

                           S=1      M=0      N=6.5

  Statistic                     Value          F       Num DF    Den DF   Pr > F

  Wilks' Lambda               0.90935857     0.7476         2        15   0.4904
  Pillai's Trace              0.09064143     0.7476         2        15   0.4904
  Hotelling-Lawley Trace      0.09967622     0.7476         2        15   0.4904
  Roy's Greatest Root         0.09967622     0.7476         2        15   0.4904

Multivariate Test: B

                 Multivariate Statistics and F Approximations

                           S=2      M=-0.5      N=6.5

  Statistic                     Value          F       Num DF    Den DF   Pr > F

  Wilks' Lambda               0.47779057     3.3503         4        30   0.0221
  Pillai's Trace              0.52223783     2.8272         4        32   0.0409
  Hotelling-Lawley Trace      1.09290775     3.8252         4        28   0.0133
  Roy's Greatest Root         1.09285335     8.7428         2        16   0.0027

Multivariate Test: AB

                 Multivariate Statistics and F Approximations

                           S=2      M=-0.5      N=6.5

  Statistic                     Value          F       Num DF    Den DF   Pr > F

  Wilks' Lambda               0.28662870     6.5088         4        30   0.0007
  Pillai's Trace              0.87267678     6.1929         4        32   0.0008
  Hotelling-Lawley Trace      1.93304375     6.7657         4        28   0.0006
  Roy's Greatest Root         1.58164312    12.6531         2        16   0.0005
```

Program 7_6.sas also contains statements for an unweighted analysis of the data using PROC GLM for the MANOVA design and using PROC GLM for the MANCOVA design. Again, the MANOVA output from the GLM and REG procedures is identical. However, with the RGMLM you do not have the complicated estimable function problem since the regression model is full rank. To save space, only the output for the MANOVA model using the REG procedure is provided.

7.7 Growth Curve Analysis

While the theory associated with the growth curve model is complicated, application of the model in practice is straightforward. A convenient form of the growth curve model given in (7.23) is

$$\begin{aligned} E(\mathbf{Y}_{n\times p}) &= \mathbf{X}_{n\times k}\mathbf{B}_{k\times q}\mathbf{P}_{q\times p} \\ \operatorname{cov}(\mathbf{Y}) &= \mathbf{I}_n \otimes \mathbf{\Sigma}. \end{aligned} \tag{7.99}$$

More specifically, suppose we have k groups with n_i subjects per group and suppose that the growth curve for this i^{th} group is

$$\beta_{io} + \beta_{i1}t + \beta_{i2}t^2 + \ldots + \beta_{i,q-1}t^{q-1} \tag{7.100}$$

Letting $\mathbf{Y}_{n\times p}$ be the data matrix, we define $n = \sum_i n_i, \mathbf{X}, \mathbf{B}$ and $\mathbf{P}$

$$\mathbf{X}_{n\times k} = \begin{pmatrix} \mathbf{1}_{n_1} & \mathbf{0} & \cdots & \cdots & \cdots & \mathbf{0} \\ \mathbf{0} & \mathbf{1}_{n_2} & \cdots & \cdots & \cdots & \mathbf{0} \\ \mathbf{0} & \mathbf{0} & \ddots & \cdots & \cdots & \mathbf{0} \\ \vdots & \vdots & \cdots & \ddots & \cdots & \vdots \\ \vdots & \vdots & \cdots & \cdots & \ddots & \vdots \\ \mathbf{0} & \mathbf{0} & \cdots & \cdots & \cdots & \mathbf{1}_{n_k} \end{pmatrix} \tag{7.101}$$

$$\mathbf{B}_{k\times q} = \begin{pmatrix} \beta_{10} & \beta_{11} & \cdots & \cdots & \cdots & \beta_{1,q-1} \\ \beta_{20} & \beta_{21} & \cdots & \cdots & \cdots & \beta_{2,q-1} \\ \vdots & \vdots & \ddots & \cdots & \cdots & \vdots \\ \vdots & \vdots & \cdots & \ddots & \cdots & \vdots \\ \vdots & \vdots & \cdots & \cdots & \ddots & \vdots \\ \beta_{k0} & \beta_{k1} & \cdots & \cdots & \cdots & \beta_{k,q-1} \end{pmatrix} \tag{7.102}$$

$$\mathbf{P}_{q\times p} = \begin{pmatrix} 1 & 1 & \cdots & \cdots & \cdots & 1 \\ t_1 & t_2 & \cdots & \cdots & \cdots & t_p \\ t_1^2 & t_2^2 & \ddots & \cdots & \cdots & t_p^2 \\ \vdots & \vdots & \cdots & \ddots & \cdots & \vdots \\ \vdots & \vdots & \cdots & \cdots & \ddots & \vdots \\ t_1^{q-1} & t_2^{q-1} & \cdots & \cdots & \cdots & t_p^{q-1} \end{pmatrix} \tag{7.103}$$

where $\mathbf{1}_{n_i}$ is a vector of n_i 1's, a unit vector.

Here we assumed the repeated observations follow a polynomial growth curve and that we have only one dependent variable observed over time. To expand the model to the double multivariate design where we have vector-valued observations at each time point is straightforward. For simplicity, suppose we have two measurements at each time point where for each variable we have the growth curves

$$\begin{aligned} &\beta_{io} + \beta_{i1}t + \ldots + \beta_{i,q_1-1}t_{q_1-1} \\ &\Theta_{io} + \Theta_{i1}t + \ldots + \Theta_{i,q_2-1}t_{q_2-1} \end{aligned} \tag{7.104}$$

Then,

$$\mathbf{B}_{k\times(q_1+q_2)} = \begin{pmatrix} \beta_{10} & \beta_{11} & \cdots & \cdots & \cdots & \beta_{1,q_1-1} & \Theta_{10} & \Theta_{11} & \cdots & \cdots & \cdots & \Theta_{1,q_2-1} \\ \beta_{20} & \beta_{21} & \cdots & \cdots & \cdots & \beta_{2,q_1-1} & \Theta_{20} & \Theta_{21} & \cdots & \cdots & \cdots & \Theta_{2,q_2-1} \\ \vdots & \vdots & \ddots & \cdots & \cdots & \vdots & \vdots & \vdots & \ddots & \cdots & \cdots & \vdots \\ \vdots & \vdots & \cdots & \ddots & \cdots & \vdots & \vdots & \vdots & \cdots & \ddots & \cdots & \vdots \\ \vdots & \vdots & \cdots & \cdots & \ddots & \vdots & \vdots & \vdots & \cdots & \cdots & \ddots & \vdots \\ \beta_{k0} & \beta_{k1} & \cdots & \cdots & \cdots & \beta_{k,q_1-1} & \Theta_{k0} & \Theta_{k1} & \cdots & \cdots & \cdots & \Theta_{k,q_2-1} \end{pmatrix} \tag{7.105}$$

$$\mathbf{P}_{(q_1+q_2)\times 2p} = \begin{pmatrix} 1 & 1 & \cdots & \cdots & \cdots & 1 & 0 & 0 & \cdots & \cdots & \cdots & 0 \\ t_1 & t_2 & \cdots & \cdots & \cdots & t_p & \vdots & \vdots & \cdots & \cdots & \cdots & \vdots \\ \vdots & \vdots & \cdots & \cdots & \cdots & \vdots & \vdots & \vdots & \cdots & \cdots & \cdots & \vdots \\ \vdots & \vdots & \cdots & \cdots & \cdots & \vdots & \vdots & \vdots & \cdots & \cdots & \cdots & \vdots \\ \vdots & \vdots & \cdots & \cdots & \cdots & \vdots & \vdots & \vdots & \cdots & \cdots & \cdots & \vdots \\ t_1^{q_1-1} & t_2^{q_1-1} & \cdots & \cdots & \cdots & t_p^{q_1-1} & 0 & 0 & \cdots & \cdots & \cdots & 0 \\ 0 & 0 & \cdots & \cdots & \cdots & 0 & 1 & 1 & \cdots & \cdots & \cdots & 1 \\ \vdots & \vdots & \cdots & \cdots & \cdots & \vdots & t_1 & t_2 & \cdots & \cdots & \cdots & t_p \\ \vdots & \vdots & \cdots & \cdots & \cdots & \vdots & \vdots & \vdots & \cdots & \cdots & \cdots & \vdots \\ \vdots & \vdots & \cdots & \cdots & \cdots & \vdots & \vdots & \vdots & \cdots & \cdots & \cdots & \vdots \\ \vdots & \vdots & \cdots & \cdots & \cdots & \vdots & \vdots & \vdots & \cdots & \cdots & \cdots & \vdots \\ 0 & 0 & \cdots & \cdots & \cdots & 0 & t_1^{q_2-1} & t_2^{q_2-1} & \cdots & \cdots & \cdots & t_p^{q_2-1} \end{pmatrix}$$

where $\mathbf{X}$ is as in (7.101) and $\mathbf{Y}_{n\times 2p}$ is the data matrix with the first p columns associated with variable 1 and the second p columns associated with variable 2.

Given (7.99) with only one dependent variable observed over time, we are interested in testing hypotheses of the form $H\colon \mathbf{CBA} = \mathbf{0}$. Some hypotheses of interest may be

(1) the regression equations are coincident (equal) across the k groups

(2) the regression equations are parallel.

For test (1), the matrices $\mathbf{C}$ and $\mathbf{A}$ are

$$\mathbf{C}_{(k-1)\times k} = \begin{pmatrix} 1 & 0 & \cdots & \cdots & \cdots & -1 \\ 0 & 1 & \cdots & \cdots & \cdots & -1 \\ \vdots & \vdots & \ddots & \cdots & \cdots & \vdots \\ \vdots & \vdots & \cdots & \ddots & \cdots & \vdots \\ \vdots & \vdots & \cdots & \cdots & \ddots & \vdots \\ 0 & 0 & \cdots & \cdots & \cdots & -1 \end{pmatrix} \text{ and } \mathbf{A} = \mathbf{I}_q \tag{7.106}$$

For test (2), **C** is defined as in (7.106) and **A** becomes

$$\mathbf{A}_{q\times(q-1)} = \begin{pmatrix} 0 & 0 & \cdots & \cdots & \cdots & 0 \\ 1 & 0 & \cdots & \cdots & \cdots & 0 \\ 0 & 1 & \ddots & \cdots & \cdots & 0 \\ \vdots & \vdots & \cdots & \ddots & \cdots & \vdots \\ \vdots & \vdots & \cdots & \cdots & \ddots & \vdots \\ 0 & 0 & \cdots & \cdots & \cdots & 1 \end{pmatrix} \tag{7.107}$$

We may also evaluate the degree of the polynomial. For example, to test that the growth curves are of degree $q-2$ or less, we set

$$\mathbf{C} = \mathbf{I}_k,\ \mathbf{A}_{q\times 1} = \begin{pmatrix} 0 \\ \vdots \\ 0 \\ 1 \end{pmatrix} \tag{7.108}$$

To test hypotheses $\mathbf{CBA} = \mathbf{0}$ given (7.99), Potthoff and Roy (1964) suggested the transformation

$$\mathbf{Y}_o = \mathbf{Y}\mathbf{G}^{-1}\mathbf{P}'(\mathbf{P}\mathbf{G}^{-1}\mathbf{P}')^{-1} \tag{7.109}$$

where $\mathbf{G}_{p\times p}$ is any symmetric, positive definite matrix, that is independent of **Y**. Under the transformation, the GMANOVA model reduces to the MANOVA model. However, the analysis depends on the unknown matrix **G** which is critical when $q < p$ which occurs frequently in practice.

To avoid the choice of **G**, we may use the Rao-Khatri reduction which reduces the analysis of (7.99) to the analysis of a MANCOVA model. To accomplish the reduction, a matrix $\mathbf{H}_{p\times p} = (\mathbf{H}_1, \mathbf{H}_2)$ where **H** is nonsingular and $\mathbf{Y}_o = \mathbf{YH}_1$ and $\mathbf{Z} = \mathbf{YH}_2$ is created. The matrix $\mathbf{H}_1$ $(p\times q)$ is any matrix that forms a basis for the vector space generated by the rows of **P**, rank($\mathbf{H}_1$) = q and $\mathbf{P}\,\mathbf{H}_1 = \mathbf{I}_q$. $\mathbf{H}_2$ $[p\times(p-q)]$ is created such that the rank($\mathbf{H}_2$) = p – q and $\mathbf{P}\ \mathbf{P}\ \mathbf{H}_2 = \mathbf{0}_{p-q}$. The GMANOVA model is thus reduced to the MANCOVA model

$$E(\mathbf{Y}_o|\mathbf{Z}) = \mathbf{XB} + \mathbf{Z\Gamma} \tag{7.110}$$

To form $\mathbf{H}_1$ and $\mathbf{H}_2$, we can set $\mathbf{G} = \mathbf{I}$ in (7.109) so that $\mathbf{H}_1 = \mathbf{P}'(\mathbf{PP}')^{-1}$ and let $\mathbf{H}_2$ be any $p-q$ linearly independent columns of $\mathbf{I} - \mathbf{P}'(\mathbf{PP}')^{-1}\mathbf{P}$. More conveniently, suppose **P** is a matrix of orthogonal polynomials; then $\mathbf{PP}' = \mathbf{I}$ so that $\mathbf{H}_1 = \mathbf{P}'$ and $\mathbf{H}_2$ can be formed such that $\mathbf{H} = (\mathbf{H}_1, \mathbf{H}_2)$ is an orthogonal matrix. That is, select $\mathbf{H}_1$ to be the matrix of orthogonalized polynomials of degree *0* to $q-1$ and $\mathbf{H}_2$ to be the similar matrix of the higher polynomials of order q to $p-1$.

A limitation of the application of the Rao-Khatri model lies in the determination of the number of (not necessarily all) higher order terms to include in the model as covariates. To test that the higher order polynomials are zero, the test of model adequacy using (7.50) is usually employed using several combinations of the higher order terms.

For our first example of the growth curve model, we analyze the dog data of Grizzle and Allen (1969). The data consist of four groups of dogs (control and three treatments) with responses of coronary sinus potassium (MIL equivalents per liter) at times 1 (2) 13 minutes after coronary occlusion. The SAS code to perform the analysis is provided in Program 7_7.sas.

Program 7_7.sas

```
/* Program 7_7.sas */
/* Growth Curve Analysis (Grizzle and Allen, 1969) */

options ls=80 ps=60 nodate nonumber;
title1 'Output 7.7: Growth Curve Analysis - Grizzle and Allen Data';

data grizzle;
   infile'c:\7_7.dat';
   input group y1 y2 y3 y4 y5 y6 y7;
proc print data=grizzle;
run;

proc summary data=grizzle;
   class group;
   var y1 -y7;
   output out=new mean=mr1-mr7;
proc print data=new;
run;
data plot;
   set new;
   array mr(7) mr1-mr7;
   do time = 1 to 7;
      response = mr(time);
      output;
   end;
drop mr1-mr7;
run;
proc plot;
   title2 'Growth Curve Plots of Group Means';
   plot response*time=group;
run;

proc glm data=grizzle;
   title2 'Transformed Data Polynomials';
    class group;
    model y1 - y7 = group/nouni;
    repeated time polynomial/summary nom nou;
run;

proc iml;
   use grizzle;
   read all var {y1 y2 y3 y4 y5 y6 y7} into y;
   read all var {Group} into gr;
   /* Create Orthogonal Polynomials of degree  q-1=6 */
   x={1 2 3 4 5 6 7};
   P_prime=orpol(x,6);
   Y0=Y*P_prime;
   print 'P prime matrix',P_prime;
   t=y0||gr;
   /* Create New Transformed data set */
   varnames={yt1 yt2 yt3 yt4 yt5 yt6 yt7 group};
   create trans from t (|colname=varnames|);
   append from t;
quit;
run;
proc print data=trans; run;
```

```
/* Test of model fit  */
proc glm data=trans;
   title2 'Test of Cubic Fit';
   model yt5 - yt7=/nouni;
   manova h=intercept;
run;
proc glm data=trans;
   title2 'Test of Quadratic Fit';
   model yt4 - yt7=/nouni;
   manova h=intercept;
run;

/* Using a Rao-Khatri MANCOVA model          */
/* Test for group differences (coincidence) */
proc glm data=trans;
   title2 'Test of Group Differences';
   class group;
   model yt1 - yt4 = group yt5 - yt7/nouni;
   manova h=group/printh printe;
run;
/*Test for Parallelism of Profiles */
proc glm data=Trans;
   title2 ' Test of Parallelism';
   class group;
   model yt1 - yt4 = group yt5 - yt7/noint nouni;
   contrast 'parallel' group 1 -1 0 0,
                       group 1 0 -1 0,
                       group 1 0 0 -1;
   manova m=(0 1 0 0,
             0 0 1 0,
             0 0 0 1) prefix = parll/printe printh;
run;

/* Estimate of matrix B and contrasts */
proc glm data=trans;
   title2 ' Estimate of Matrix B';
   class group;
   model yt1 - yt4 = group yt5 - yt7;
   estimate 'beta1' intercept 1 group 1 0 0 0;
   estimate 'beta2' intercept 1 group 0 1 0 0;
   estimate 'beta3' intercept 1 group 0 0 1 0;
   estimate 'beta4' intercept 1 group 0 0 0 1;
   estimate '1vs2'  group 1 -1 0 0;
   estimate '1vs3'  group 1 0 -1 0;
   estimate '1vs4'  group 1 0 0 -1;
run;
```

Result and Interpretation 7_7.sas

The data are listed in Output 7.7 and include corrections of errors in the published data set. For dog 2, the value 3.7 has been changed to 4.7. For dog 14, the value 2.9 has been changed to 3.9 and for dog 19, the value of 5.3 has been changed to 4.3. For dog 20, the value 3.7 has been changed to 4.0.

Output 7.7: *Growth Curve Analysis - Grizzle and Allen Data*

Output 7.7: Growth Curve Analysis - Grizzle and Allen Data

OBS	GROUP	Y1	Y2	Y3	Y4	Y5	Y6	Y7
1	1	4.0	4.0	4.1	3.6	3.6	3.8	3.1
2	1	4.2	4.3	4.7	4.7	4.8	5.0	5.2
3	1	4.3	4.2	4.3	4.3	4.5	5.8	5.4
4	1	4.2	4.4	4.6	4.9	5.3	5.6	4.9
5	1	4.6	4.4	5.3	5.6	5.9	5.9	5.3
6	1	3.1	3.6	4.9	5.2	5.3	4.2	4.1
7	1	3.7	3.9	3.9	4.8	5.2	5.4	4.2
8	1	4.3	4.2	4.4	5.2	5.6	5.4	4.7
9	1	4.6	4.6	4.4	4.6	5.4	5.9	5.6
10	2	3.4	3.4	3.5	3.1	3.1	3.7	3.3
11	2	3.0	3.2	3.0	3.0	3.1	3.2	3.1
12	2	3.0	3.1	3.2	3.0	3.3	3.0	3.0
13	2	3.1	3.2	3.2	3.2	3.3	3.1	3.1
14	2	3.8	3.9	4.0	3.9	3.5	3.5	3.4
15	2	3.0	3.6	3.2	3.1	3.0	3.0	3.0
16	2	3.3	3.3	3.3	3.4	3.6	3.1	3.1
17	2	4.2	4.0	4.2	4.1	4.2	4.0	4.0
18	2	4.1	4.2	4.3	4.3	4.2	4.0	4.2
19	2	4.5	4.4	4.3	4.5	4.3	4.4	4.4
20	3	3.2	3.3	3.8	3.8	4.4	4.2	4.0
21	3	3.3	3.4	3.4	3.7	3.7	3.6	3.7
22	3	3.1	3.3	3.2	3.1	3.2	3.1	3.1
23	3	3.6	3.4	3.5	4.6	4.9	5.2	4.4
24	3	4.5	4.5	5.4	5.7	4.9	4.0	4.0
25	3	3.7	4.0	4.4	4.2	4.6	4.8	5.4
26	3	3.5	3.9	5.8	5.4	4.9	5.3	5.6
27	3	3.9	4.0	4.1	5.0	5.4	4.4	3.9
28	4	3.1	3.5	3.5	3.2	3.0	3.0	3.2
29	4	3.3	3.2	3.6	3.7	3.7	4.2	4.4
30	4	3.5	3.9	4.7	4.3	3.9	3.4	3.5
31	4	3.4	3.4	3.5	3.3	3.4	3.2	3.4
32	4	3.7	3.8	4.2	4.3	3.6	3.8	3.7
33	4	4.0	4.6	4.8	4.9	5.4	5.6	4.8
34	4	4.2	3.9	4.5	4.7	3.9	3.8	3.7
35	4	4.1	4.1	3.7	4.0	4.1	4.6	4.7
36	4	3.5	3.6	3.6	4.2	4.8	4.9	5.0

When performing a growth curve analysis, one must first determine the degree of polynomial to be fit to the data. To do this one can plot the data and perform tests of fit. Using the SAS code suggested by Freund et al. (1991, p. 281) included in Program 7_7.sas, we summarize the data for this example using PROC SUMMARY and PROC PLOT. The results are found in the continuation of Output 7.7 that follows.

Output 7.7 *(continued)*

Output 7.7: Growth Curve Analysis - Grizzle and Allen Data
Growth Curve Plots of Group Means

OBS	GROUP	_TYPE_	_FREQ_	MR1	MR2	MR3	MR4	MR5	MR6	MR7
1	.	0	36	3.72222	3.82500	4.06944	4.18333	4.25000	4.25278	4.10000
2	1	1	9	4.11111	4.17778	4.51111	4.76667	5.06667	5.22222	4.72222
3	2	1	10	3.54000	3.63000	3.62000	3.56000	3.56000	3.50000	3.46000
4	3	1	8	3.60000	3.72500	4.20000	4.43750	4.50000	4.32500	4.26250
5	4	1	9	3.64444	3.77778	4.01111	4.06667	3.97778	4.05556	4.04444

Output 7.7 (*continued*)

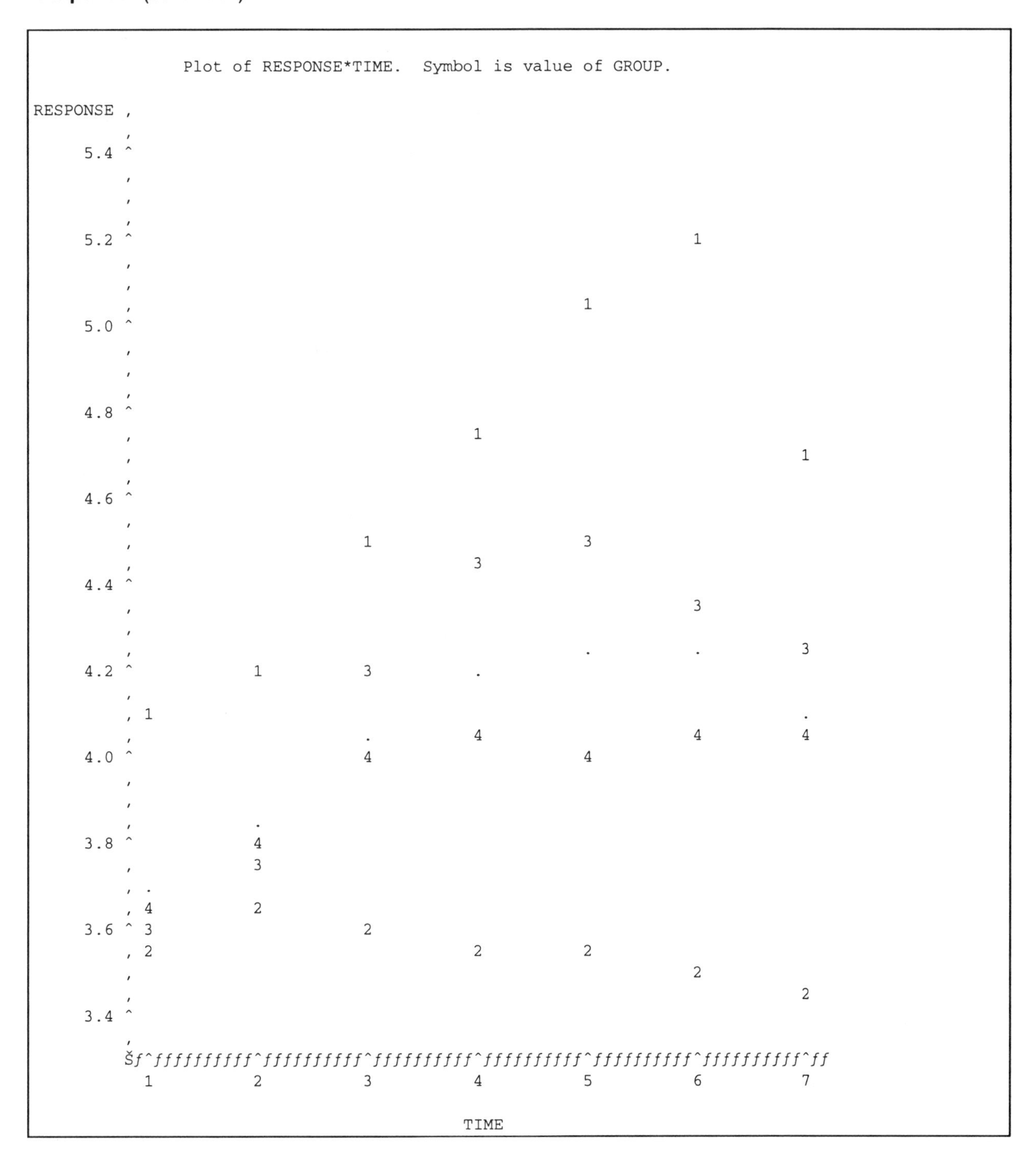

The plot suggests that a third degree polynomial should fit the data. To again try to determine the order of polynomial to fit to the data, we evaluate polynomial contrasts fit to the seven time points. This is done in Program 7_7.sas using PROC GLM with the POLYNOMIAL option in the REPEATED statement. The results are found in the continuation of Output 7.7 that follows.

Output 7.7 (*continued*)

```
           Output 7.7: Growth Curve Analysis - Grizzle and Allen Data
                          Transformed Data Polynomials

                         General Linear Models Procedure
                      Repeated Measures Analysis of Variance
                    Analysis of Variance of Contrast Variables

         TIME.N represents the nth degree polynomial contrast for TIME

Contrast Variable: TIME.1

Source                  DF       Type III SS      Mean Square   F Value      Pr > F

MEAN                     1        6.65307281       6.65307281     18.95      0.0001
GROUP                    3        4.93401885       1.64467295      4.69      0.0080

Error                   32       11.23299504       0.35103109

Contrast Variable: TIME.2

Source                  DF       Type III SS      Mean Square   F Value      Pr > F

MEAN                     1        3.07031855       3.07031855     13.97      0.0007
GROUP                    3        0.93658036       0.31219345      1.42      0.2549

Error                   32        7.03361475       0.21980046

Contrast Variable: TIME.3

Source                  DF       Type III SS      Mean Square   F Value      Pr > F

MEAN                     1        0.34404995       0.34404995      2.73      0.1080
GROUP                    3        1.27958796       0.42652932      3.39      0.0298

Error                   32        4.02647685       0.12582740

Contrast Variable: TIME.4

Source                  DF       Type III SS      Mean Square   F Value      Pr > F

MEAN                     1        0.04159233       0.04159233      0.47      0.4982
GROUP                    3        0.41893921       0.13964640      1.58      0.2144

Error                   32        2.83604455       0.08862639

Contrast Variable: TIME.5

Source                  DF       Type III SS      Mean Square   F Value      Pr > F

MEAN                     1        0.08284696       0.08284696      1.30      0.2622
GROUP                    3        0.07749041       0.02583014      0.41      0.7496

Error                   32        2.03484755       0.06358899
```

Output 7.7 *(continued)*

Contrast Variable: TIME.6

Source	DF	Type III SS	Mean Square	F Value	Pr > F
MEAN	1	0.00956449	0.00956449	0.36	0.5531
GROUP	3	0.02366098	0.00788699	0.30	0.8277
Error	32	0.85161649	0.02661302		

The significance of the TIME.1, TIME.2, and TIME.3 polynomial contrasts for the mean effect suggests a third degree polynomial will fit the data. Assuming a third degree polynomial fits the data, the GMANOVA model can be written

$$E(\mathbf{Y}_{36\times 7}) = \mathbf{X}_{36\times 4}\,\mathbf{B}_{4\times 4}\mathbf{P}_{4\times 7}$$

where

$$\mathbf{X} = \begin{pmatrix} \mathbf{1}_9 & \mathbf{0} & \mathbf{0} & \mathbf{0} \\ \mathbf{0} & \mathbf{1}_{10} & \mathbf{0} & \mathbf{0} \\ \mathbf{0} & \mathbf{0} & \mathbf{1}_8 & \mathbf{0} \\ \mathbf{0} & \mathbf{0} & \mathbf{0} & \mathbf{1}_7 \end{pmatrix}$$

and $\mathbf{P}' \equiv \mathbf{H}_1$ consists of the first four columns of the 7 x 7 matrix of normalized orthogonal polynomials (constant, linear, quadratic, cubic, 4th, 5th, 6th). In Program 7_7.sas we generate $\mathbf{H} = (\mathbf{H}_1, \mathbf{H}_2)$ using PROC IML with the ORPOL function. The transformed data set $\mathbf{Y}_o = \mathbf{YH}$ is created and named TRANS. It is this data set that is used throughout the rest of Program 7_7.sas to perform the Rao-Khatri analysis.

Using the Rao-Khatri model to perform the overall test of fit for the third degree polynomial model we have $\mathbf{Y}_o = \mathbf{YH} = (\mathbf{YH}_1 \vdots \mathbf{YH}_2) = (\mathbf{y1}, \mathbf{y2}, \mathbf{y3}, \mathbf{y4} \vdots \mathbf{y5}, \mathbf{y6}, \mathbf{y7})$; the Rao-Khatri reduction

$$E(\mathbf{Y}_o | \mathbf{Z} = \mathbf{YH}_2) = \mathbf{XB} + \mathbf{Z\Gamma}.$$

Using (7.50) we test $H{:}\mathbf{BH}_2 = \mathbf{0}$ where $\mathbf{H}_2$ (7×3) corresponds to the normalized polynomials in $\mathbf{H}$ of degree 4, 5, and 6. This analysis is performed in Program 7_7.sas using PROC GLM with the transformed data and the MODEL statement `model yt5-yt7=/nouni`. The adequacy of a second degree polynomial is also tested. The results of these tests are included in the following output.

Output 7.7 *(continued)*

```
          Output 7.7: Growth Curve Analysis - Grizzle and Allen Data
                              Test of Cubic Fit

              Manova Test Criteria and Exact F Statistics for
               the Hypothesis of no Overall INTERCEPT Effect
       H = Type III SS&CP Matrix for INTERCEPT   E = Error SS&CP Matrix

                          S=1    M=0.5    N=15.5

Statistic                    Value         F      Num DF    Den DF  Pr > F

Wilks' Lambda            0.95296544    0.5429         3        33  0.6563
Pillai's Trace           0.04703456    0.5429         3        33  0.6563
Hotelling-Lawley Trace   0.04935599    0.5429         3        33  0.6563
Roy's Greatest Root      0.04935599    0.5429         3        33  0.6563

                          Test of Quadratic Fit
              Manova Test Criteria and Exact F Statistics for
               the Hypothesis of no Overall INTERCEPT Effect
       H = Type III SS&CP Matrix for INTERCEPT   E = Error SS&CP Matrix

                            S=1    M=1    N=15

Statistic                    Value         F      Num DF    Den DF  Pr > F

Wilks' Lambda            0.78985294    2.1285         4        32  0.1001
Pillai's Trace           0.21014706    2.1285         4        32  0.1001
Hotelling-Lawley Trace   0.26605847    2.1285         4        32  0.1001
Roy's Greatest Root      0.26605847    2.1285         4        32  0.1001
```

The Wilks' Λ criterion for the test of cubic fit is $\Lambda = 0.95296544$ with p–value 0.6563. Hence, we do not reject the null hypothesis that the third degree polynomial adequately fits the data. Testing the adequacy of a second degree polynomial, the p–value is 0.1001.

Having fit a polynomial growth curve to the groups, one may next be interested in testing the hypothesis of coincidence, that the growth curves are the same for the groups with respect to all polynomial components: constant, linear, quadratic, and cubic. The hypothesis test matrix from (7.104) to test this hypothesis is

$$\mathbf{C} = \begin{pmatrix} 1 & -1 & 0 & 0 \\ 1 & 0 & -1 & 0 \\ 1 & 0 & 0 & -1 \end{pmatrix}$$

and $\mathbf{A} = \mathbf{I}_4$. Another hypothesis is that the growth curves for the groups are parallel, that all polynomial components are coincident except for the intercept term. The hypothesis test matrices for the test of parallelism from (7.105) are

$$\mathbf{C} = \begin{pmatrix} 1 & -1 & 0 & 0 \\ 1 & 0 & -1 & 0 \\ 1 & 0 & 0 & -1 \end{pmatrix} \text{ and } \mathbf{A} = \begin{pmatrix} 0 & 0 & 0 \\ 1 & 0 & 0 \\ 0 & 1 & 0 \\ 0 & 0 & 1 \end{pmatrix}$$

The results of these tests are found in the following output.

Output 7.7 *(continued)*

```
          Output 7.7: Growth Curve Analysis - Grizzle and Allen Data
                           Test of Group Differences

                 Manova Test Criteria and F Approximations for
                   the Hypothesis of no Overall GROUP Effect
          H = Type III SS&CP Matrix for GROUP    E = Error SS&CP Matrix

                               S=3     M=0     N=12

Statistic                     Value          F       Num DF    Den DF   Pr > F

Wilks' Lambda              0.37304366     2.6000         12  69.08104   0.0065
Pillai's Trace             0.73453561     2.2696         12        84   0.0151
Hotelling-Lawley Trace     1.39788294     2.8734         12        74   0.0027
Roy's Greatest Root        1.16390208     8.1473          4        28   0.0002

        NOTE: F Statistic for Roy's Greatest Root is an upper bound.

                             Test of Parallelism
                  Manova Test Criteria and F Approximations for
                   the Hypothesis of no Overall parallel Effect
             on the variables defined by the M Matrix Transformation
        H = Contrast SS&CP Matrix for parallel    E = Error SS&CP Matrix

                            S=3     M=-0.5     N=12.5

Statistic                     Value          F       Num DF    Den DF   Pr > F

Wilks' Lambda              0.44098587     2.9266          9  65.86151   0.0056
Pillai's Trace             0.63307895     2.5855          9        87   0.0109
Hotelling-Lawley Trace     1.10030957     3.1379          9        77   0.0029
Roy's Greatest Root        0.91977535     8.8912          3        29   0.0002

        NOTE: F Statistic for Roy's Greatest Root is an upper bound.
```

From the output we see that the test of coincidence or group differences in growth curves results in a Wilks' Λ of $\Lambda = 0.373$ with a p-value of 0.0065; thus, we reject the null hypothesis of coincidence. The result of the test of parallelism is likewise significant with a Wilks' Λ of $\Lambda = .441$ and p-value of 0.0056.

Finally, the code in Program 7_7.sas also outputs the columns of the parameter matrix **B** and differences in trends comparing group 1 with 2, 3, and 4 for which Grizzle and Allen (1969) establish Bonferroni type confidence intervals.

Output 7.7 (*continued*)

```
          Output 7.7: Growth Curve Analysis - Grizzle and Allen Data
                              Estimate of Matrix B

                        General Linear Models Procedure

Dependent Variable: YT1
                                    Sum of              Mean
Source                  DF         Squares            Square   F Value      Pr > F

Model                    6     47.28980503        7.88163417      4.34      0.0030

Error                   29     52.64015528        1.81517777

Corrected Total         35     99.92996032

                  R-Square            C.V.          Root MSE              YT1 Mean

                  0.473229        12.55012         1.3472853             10.735241

Source                  DF       Type I SS       Mean Square   F Value      Pr > F

GROUP                    3     41.59394444       13.86464815      7.64      0.0007
YT5                      1      2.70233149        2.70233149      1.49      0.2322
YT6                      1      1.51537133        1.51537133      0.83      0.3684
YT7                      1      1.47815777        1.47815777      0.81      0.3743

Source                  DF     Type III SS       Mean Square   F Value      Pr > F

GROUP                    3     40.58112688       13.52704229      7.45      0.0008
YT5                      1      2.80584489        2.80584489      1.55      0.2237
YT6                      1      1.68117916        1.68117916      0.93      0.3438
YT7                      1      1.47815777        1.47815777      0.81      0.3743

                                          T for H0:    Pr > |T|    Std Error of
Parameter                   Estimate     Parameter=0                 Estimate

beta1                     12.3600560           26.52     0.0001      0.46597989
beta2                      9.4760269           22.10     0.0001      0.42876012
beta3                     10.8041790           21.38     0.0001      0.50535322
beta4                     10.2489211           22.18     0.0001      0.46198282
1vs2                       2.8840290            4.58     0.0001      0.62927959
1vs3                       1.5558770            2.22     0.0346      0.70167366
1vs4                       2.1111349            3.24     0.0030      0.65227900

Dependent Variable: YT2
                                    Sum of              Mean
Source                  DF         Squares            Square   F Value      Pr > F

Model                    6      5.97929186        0.99654864      2.84      0.0269

Error                   29     10.18772202        0.35130076

Corrected Total         35     16.16701389

                  R-Square            C.V.          Root MSE              YT2 Mean

                  0.369845        144.5673         0.5927063             0.4099865
```

Output 7.7 *(continued)*

```
Source                    DF        Type I SS      Mean Square    F Value       Pr > F

GROUP                      3       4.93401885       1.64467295       4.68       0.0087
YT5                        1       0.99025257       0.99025257       2.82       0.1039
YT6                        1       0.00269973       0.00269973       0.01       0.9307
YT7                        1       0.05232072       0.05232072       0.15       0.7024

Source                    DF      Type III SS      Mean Square    F Value       Pr > F

GROUP                      3       5.12314533       1.70771511       4.86       0.0074
YT5                        1       1.03802325       1.03802325       2.95       0.0963
YT6                        1       0.00153507       0.00153507       0.00       0.9477
YT7                        1       0.05232072       0.05232072       0.15       0.7024

                                         T for H0:    Pr > |T|    Std Error of
Parameter                  Estimate     Parameter=0                  Estimate

beta1                    0.78158437            3.81     0.0007      0.20499682
beta2                   -0.12980854           -0.69     0.4968      0.18862287
beta3                    0.77508977            3.49     0.0016      0.22231819
beta4                    0.36053612            1.77     0.0866      0.20323841
1vs2                     0.91139291            3.29     0.0026      0.27683667
1vs3                     0.00649460            0.02     0.9834      0.30868472
1vs4                     0.42104825            1.47     0.1531      0.28695471

Dependent Variable: YT3
                                           Sum of             Mean
Source                    DF              Squares           Square    F Value       Pr > F

Model                      6       4.56724448       0.76120741       6.49       0.0002

Error                     29       3.40295062       0.11734312

Corrected Total           35       7.97019511

                    R-Square                C.V.         Root MSE             YT3 Mean

                    0.573040           -121.6621        0.3425538           -0.2815617

Source                    DF        Type I SS      Mean Square    F Value       Pr > F

GROUP                      3       0.93658036       0.31219345       2.66       0.0667
YT5                        1       3.47601447       3.47601447      29.62       0.0001
YT6                        1       0.10423858       0.10423858       0.89       0.3537
YT7                        1       0.05041107       0.05041107       0.43       0.5173

Source                    DF      Type III SS      Mean Square    F Value       Pr > F

GROUP                      3       0.84143745       0.28047915       2.39       0.0891
YT5                        1       3.32599423       3.32599423      28.34       0.0001
YT6                        1       0.11210910       0.11210910       0.96       0.3364
YT7                        1       0.05041107       0.05041107       0.43       0.5173
```

Output 7.7 (*continued*)

Parameter	Estimate	T for H0: Parameter=0	Pr > |T|	Std Error of Estimate
beta1	-0.51060254	-4.31	0.0002	0.11847765
beta2	-0.10861098	-1.00	0.3273	0.10901434
beta3	-0.26207919	-2.04	0.0506	0.12848851
beta4	-0.16898628	-1.44	0.1610	0.11746137
1vs2	-0.40199156	-2.51	0.0178	0.15999739
1vs3	-0.24852335	-1.39	0.1742	0.17840393
1vs4	-0.34161626	-2.06	0.0485	0.16584510

Dependent Variable: YT4

Source	DF	Sum of Squares	Mean Square	F Value	Pr > F
Model	6	3.33912123	0.55652020	8.21	0.0001
Error	29	1.96694358	0.06782564		
Corrected Total	35	5.30606481			

R-Square	C.V.	Root MSE	YT4 Mean
0.629303	-276.6922	0.2604336	-0.0941239

Source	DF	Type I SS	Mean Square	F Value	Pr > F
GROUP	3	1.27958796	0.42652932	6.29	0.0020
YT5	1	0.39681664	0.39681664	5.85	0.0221
YT6	1	1.17923881	1.17923881	17.39	0.0003
YT7	1	0.48347782	0.48347782	7.13	0.0123

Source	DF	Type III SS	Mean Square	F Value	Pr > F
GROUP	3	1.46775111	0.48925037	7.21	0.0009
YT5	1	0.14043009	0.14043009	2.07	0.1609
YT6	1	1.09338492	1.09338492	16.12	0.0004
YT7	1	0.48347782	0.48347782	7.13	0.0123

Parameter	Estimate	T for H0: Parameter=0	Pr > |T|	Std Error of Estimate
beta1	-0.46611264	-5.17	0.0001	0.09007506
beta2	0.04481510	0.54	0.5928	0.08288038
beta3	-0.19615826	-2.01	0.0540	0.09768602
beta4	0.00177679	0.02	0.9843	0.08930241
1vs2	-0.51092774	-4.20	0.0002	0.12164129
1vs3	-0.26995438	-1.99	0.0561	0.13563524
1vs4	-0.46788944	-3.71	0.0009	0.12608713

From (7.42), and summarizing from Output 7.7

$$\hat{\mathbf{B}} = (\mathbf{X}'\mathbf{X})^{-1}\mathbf{X}'\mathbf{Y}\mathbf{S}^{-1}\mathbf{P}'(\mathbf{P}\mathbf{S}^{-1}\mathbf{P}')^{-1}$$

$$= \begin{pmatrix} 12.36 & 0.78 & -0.51 & -0.47 \\ 9.48 & -0.13 & -0.11 & 0.04 \\ 10.80 & 0.78 & -0.26 & -0.20 \\ 10.25 & 0.36 & -0.17 & 0.00 \end{pmatrix} \tag{7.111}$$

This matrix does not agree with the matrix reported by Grizzle and Allen (1969). Their estimate of regression is

$$\beta = \begin{pmatrix} 4.671 & 0.148 & -0.056 & -0.190 \\ 3.582 & -0.025 & -0.012 & 0.018 \\ 4.084 & 0.146 & -0.029 & -0.080 \\ 3.874 & 0.068 & -0.018 & 0.001 \end{pmatrix} \quad \text{or} \quad \Xi = \begin{pmatrix} 32.701 & 4.135 & -4.680 & -1.142 \\ 25.071 & -0.687 & -0.995 & 0.110 \\ 28.585 & 4.101 & -2.402 & -0.480 \\ 27.116 & 1.908 & -1.549 & 0.004 \end{pmatrix} = \beta(\mathbf{Q}_1'\mathbf{Q}_1) \tag{7.112}$$

This difference is due to the fact that the matrix $\mathbf{P} = \mathbf{Q}_1$ of orthogonal polynomials was not normalized to unity in the Grizzle and Allen analysis where

$$\mathbf{Q}' = \left(\begin{array}{cccc|ccc} 1 & -3 & 5 & -1 & 3 & -1 & 1 \\ 1 & -2 & 0 & 1 & -7 & 4 & -6 \\ 1 & -1 & -3 & 1 & 1 & -5 & 15 \\ 1 & 0 & -4 & 0 & 6 & 0 & -20 \\ 1 & 1 & -3 & -1 & 1 & 5 & 15 \\ 1 & 2 & 0 & -1 & -7 & -4 & 6 \\ 1 & 3 & 5 & 1 & 3 & 1 & 1 \end{array}\right) = (\mathbf{Q}_1 \vdots \mathbf{Q}_2) = \mathbf{P}'\mathbf{D}^{-1} \tag{7.113}$$

In our SAS analysis each column of $\mathbf{Q}'$ was normalized to unity. For additional detail see Seber (1984, p. 491). Seber does not use the modified data set, but rather the published data which contain printing errors.

7.8 Multiple Response Growth Curves

For our second application of the growth curve model, the dental data of Dr. Thomas Zullo, analyzed in Chapter 6, are again analyzed. The data were partially discussed by Timm (1980a). The design consists of three dependent variables obtained over three time points. Our goal is to analyze the trend for the two groups over the three dependent variables simultaneously and to determine the best model for the study. The SAS code to perform the multiple response growth curve analysis is provided in Program 7_8.sas.

Program 7_8.sas

```
/* Program 7_8.sas                          */
/* Growth Curve Analysis (Timm, 1980a) */

options ls=80 ps=60 nodate nonumber;
title1 'Output 7.8: Multiple Response Growth Curve Analysis- Zullo data';

data zullo;
   infile 'c:\mmm.dat';
   input group y1 - y9;
proc print data=zullo;
run;

data test;
   set zullo;
   array z{9} y1 - y9;
   k=1;
   do rvar = 1,2,3;
      do time = 1,2,3;
         grmeans = z{k};
         k=k+1;
         output;
      end;
   end;
   keep rvar group time grmeans;
run;
proc summary nway;
   class group rvar time;
   var grmeans;
   output out=new mean=grmeans;
proc print data = new;
run;
proc plot;
   title2 'Plot of Means';
   plot grmeans*time=group;
run;

proc iml;
   use zullo;
   read all var {y1 y2 y3 y4 y5 y6 y7 y8 y9} into y;
   read all var {Group} into gr;
   /* Create Orthogonal Polynomials of degree  q-1=2 */
   x={1 2 3};
   P=orpol(x,2);
   print P;
   P_prime=block(P,P,P);
   Y0=Y*P_prime;
   t=y0||gr;
   /* Create New Transformed data set */
   varnames={yt1 yt2 yt3 yt4 yt5 yt6 yt7 yt8 yt9 group};
   create trans from t (|colname=varnames|);
   append from t;
quit;
proc print data=trans;
run;

/* Test model adequacy or fit */
proc glm data=trans;
   title2 'Test of Fit - Linear';
   model yt3 yt6 yt9=;
   manova h=intercept;
run;
```

```
proc glm data=trans;
   title2 'Test of Fit - Quadratic(1)/Linear(2)';
   model yt6 yt9=/nouni;
   manova h=intercept;
run;

/* Test of significance of Beta weights */
proc glm data=trans;
   title2 'Test of Quadratic Beta Weights';
   class group;
   model yt1 - yt9 = group/noint nouni;
   contrast 'order' group 1 0,
                    group 0 1;
   manova m=(0 0 1 0 0 0 0 0 0,
             0 0 0 0 0 1 0 0 0,
             0 0 0 0 0 0 0 0 1);
run;

/* Using a Rao-Khatri MANCOVA model                     */
/* Test for group differences (coincidence) p=q and q<p */
proc glm data=trans;
   title2 'Coincidence p=q';
   class group;
   model yt1 - yt9 = group/nouni;
   manova h=group/printh printe;
run;

proc glm data=trans;
   title2 'Coincidence q<p';
   class group;
   model yt1 yt2 yt3 yt4 yt5 yt7 yt8 = group yt6 yt9/nouni;
   manova h=group/printh printe;
run;

/*Test for Parallelism of Profiles p=q and q<p  */
proc glm data=trans;
   title2 'Parallelism p=q';
   class group;
   model yt1 -yt9 = group/noint nouni;
   contrast 'parl(p=q)' group 1 -1;
   manova m=(0 1 0 0 0 0 0 0 0,
             0 0 1 0 0 0 0 0 0,
             0 0 0 0 0 0 0 0 0,
             0 0 0 0 1 0 0 0 0,
             0 0 0 0 0 1 0 0 0,
             0 0 0 0 0 0 0 1 0,
             0 0 0 0 0 0 0 0 1) prefix = parll/printe printh;
run;

proc glm data = trans;
   title2 'Parallelism q<p';
   class group;
   model yt1 yt2 yt3 yt4 yt5 yt7 yt8 = group yt6 yt9/noint nouni;
   contrast 'parl(q<p)' group 1 -1;
   manova m=(0 1 0 0 0 0 0,
             0 0 1 0 0 0 0,
             0 0 0 0 1 0 0,
             0 0 0 0 0 0 1) prefix = parl/printe printh;
run;
```

```
/* Estimate of matrix B and contrasts p=q and q<p */
proc glm data=trans;
   title2 'Beta Matrix p=q';
   class group;
   model yt1-yt9 =group;
   estimate 'beta1' intercept 1 group 1 0;
   estimate 'beta2' intercept 1 group 0 1;
   estimate '1vs2'  group 1 -1;
run;

proc glm data=trans;
   title2 'Beta Matrix q<p';
   class group;
   model yt1 yt2 yt3 yt4 yt5 yt7 yt8 = group yt6 yt9;
   estimate 'beta1' intercept 1 group 1 0;
   estimate 'beta2' intercept 1 group 0 1;
   estimate '1vs2'  group 1 -1;
run;
```

Result and Interpretation 7_8.sas

The data file, mmm.dat, which was used for the mixed model analysis, must first be reorganized to nest the time dimension within each variable. This was accomplished in Program 7_8.sas using a DO loop in the DATA step. Next, using PROC SUMMARY and PLOT, the means are plotted for each variable against time. The resulting plot is found in Output 7.8.

Output 7.8: *Multiple Response Growth Curve Analysis- Zullo data Plot of Means*

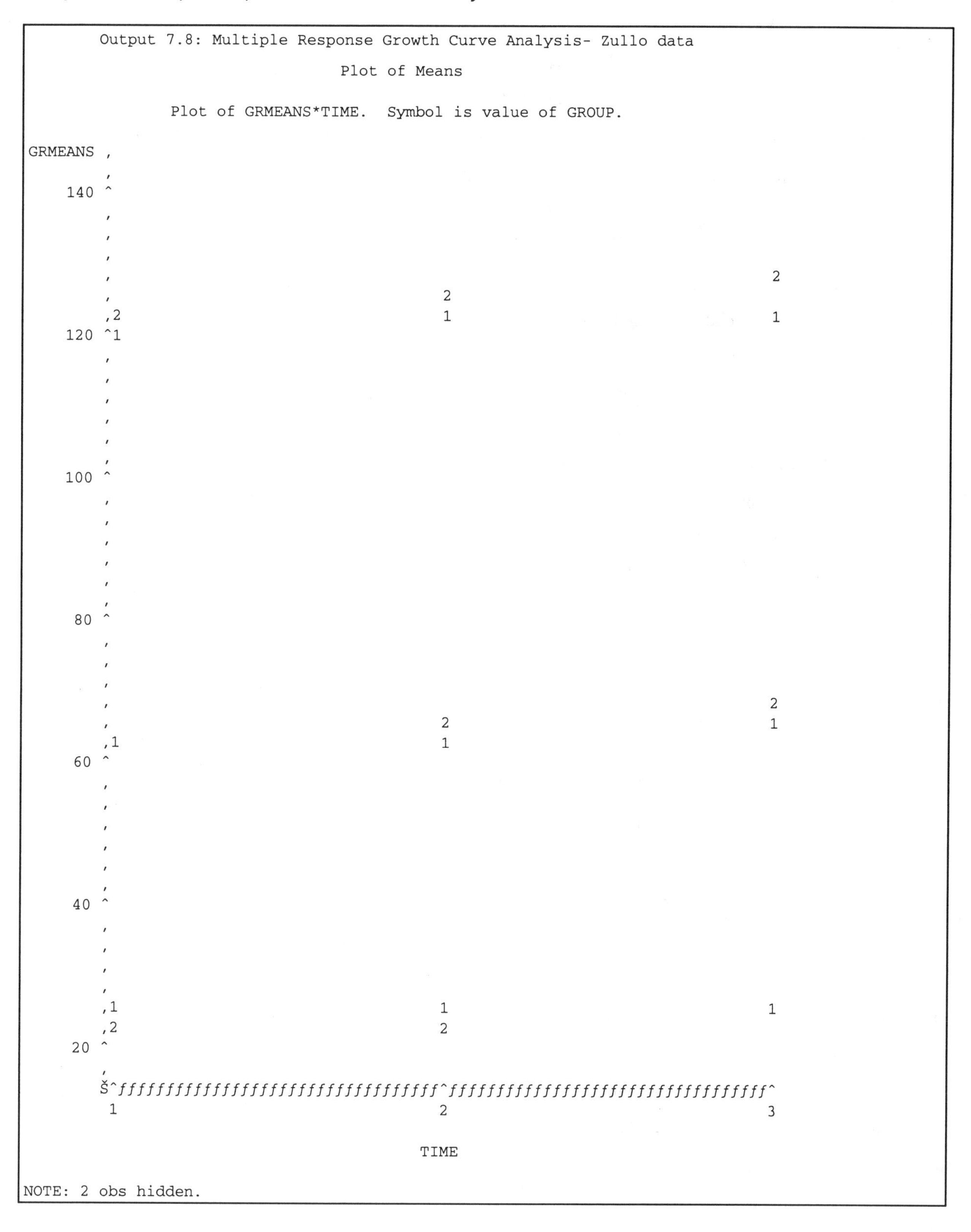

The plot suggests that a first degree (linear) polynomial should be fit to the data.

Assuming that $p = q$, the matrix of parameters **B** would have the form

$$\mathbf{B} = \begin{pmatrix} \beta_{10} & \beta_{11} & \beta_{12} & \theta_{10} & \theta_{11} & \theta_{12} & \xi_{10} & \xi_{11} & \xi_{12} \\ \beta_{20} & \beta_{21} & \beta_{22} & \theta_{20} & \theta_{21} & \theta_{22} & \xi_{21} & \xi_{21} & \xi_{22} \end{pmatrix} \tag{7.114}$$

where **B** is partitioned for variables 1, 2, and 3 represented by β, θ, and ξ, respectively. If $q < p$ for all variables, and if we assume a linear model as we have in this example, then

$$\mathbf{B} = \begin{pmatrix} \beta_{10} & \beta_{11} & \theta_{10} & \theta_{11} & \xi_{10} & \xi_{11} \\ \beta_{20} & \beta_{21} & \theta_{20} & \theta_{21} & \xi_{20} & \xi_{21} \end{pmatrix} \tag{7.115}$$

For the GMANOVA model we may also fit different curves to each variable. For example, variable one may be fit best by a quadratic polynomial and variables two and three by a linear trend; then, the matrix of parameters would be

$$\mathbf{B} = \begin{pmatrix} \beta_{10} & \beta_{11} & \beta_{12} & \theta_{10} & \theta_{11} & \xi_{10} & \xi_{11} \\ \beta_{20} & \beta_{21} & \beta_{22} & \theta_{20} & \theta_{21} & \xi_{20} & \xi_{21} \end{pmatrix} \tag{7.116}$$

If we fit the full model, $p = q$, quadratic trend, for each variable, with **B** defined in (7.114), the normalized matrix of orthogonal polynomials has the form

$$\mathbf{P}' = \begin{pmatrix} \mathbf{P}_1' & \mathbf{0} & \mathbf{0} \\ \mathbf{0} & \mathbf{P}_2' & \mathbf{0} \\ \mathbf{0} & \mathbf{0} & \mathbf{P}_3' \end{pmatrix} \text{ where } \mathbf{P}_1' = \mathbf{P}_2' = \mathbf{P}_3' = \begin{pmatrix} .577 & -0.707 & 0.408 \\ .577 & 0 & -0.816 \\ .577 & 0.707 & 0.408 \end{pmatrix} \tag{7.117}$$

The matrix $\mathbf{P}_i'$ is obtained from the orthogonal polynomial matrix

$$\mathbf{Q}' = \begin{pmatrix} 1 & -1 & 1 \\ 1 & 0 & -2 \\ 1 & 1 & 1 \end{pmatrix}$$

by normalizing the columns of $\mathbf{Q}'$. The matrix $\mathbf{Q}'$ is generated by linear and quadratic polynomials (x – 2) and $3x^2 - 12x + 10$. For $\mathbf{Q}'$, the polynomial has the form

$$y = \alpha_o + \alpha_1(x-2) + \alpha_3(3x^2 - 12x + 10)$$

for each variable. Normalizing the polynomial, we have that

$$\begin{aligned} y &= \alpha_o / \sqrt{3} + \alpha_1(x-2)/\sqrt{2} + \alpha_3(3x^2 - 12x + 10)/\sqrt{6} \\ &= \alpha_o' + \alpha_1'(x-2) + \alpha_3'(3x^2 - 12x + 10) \end{aligned} \tag{7.118}$$

In Program 7_8.sas, PROC IML is used to create a new transformed data set using these orthogonal polynomials. The transformed data are $\mathbf{Y}_o = \mathbf{YP}'$. It is this transformed data set that is used for the analysis in the remainder of Program 7_8.sas. The test of model adequacy or fit is next performed. PROC GLM is used to test that a linear trend is appropriate for each of the three variables by testing that *yt*3, *yt*6 and *yt*9 are equal to zero. The result of this test is found in Output 7.8.

Output 7.8 *(continued)*

```
       Output 7.8: Multiple Response Growth Curve Analysis- Zullo data
                              Test of Fit - Linear

                          General Linear Models Procedure

Dependent Variable: YT3
                                     Sum of            Mean
Source                  DF          Squares          Square   F Value      Pr > F

Model                    1       4.89814815      4.89814815     15.32      0.0011

Error                   17       5.43518519      0.31971678

Uncorrected Total       18      10.33333333

                R-Square             C.V.          Root MSE              YT3 Mean

                0.000000        -108.3934         0.5654350            -0.5216506


Source                  DF        Type I SS     Mean Square   F Value      Pr > F

INTERCEPT                1       4.89814815      4.89814815     15.32      0.0011

Source                  DF      Type III SS     Mean Square   F Value      Pr > F

INTERCEPT                1       4.89814815      4.89814815     15.32      0.0011

                                          T for H0:    Pr > |T|   Std Error of
Parameter                   Estimate     Parameter=0               Estimate

INTERCEPT               -.5216505934           -3.91     0.0011     0.13327432


Dependent Variable: YT6
                                     Sum of            Mean
Source                  DF          Squares          Square   F Value      Pr > F

Model                    1       0.05787037      0.05787037      0.05      0.8263

Error                   17      19.81712963      1.16571351

Uncorrected Total       18      19.87500000

                R-Square             C.V.          Root MSE              YT6 Mean

                0.000000         1904.163         1.0796821             0.0567012

Source                  DF        Type I SS     Mean Square   F Value      Pr > F

INTERCEPT                1       0.05787037      0.05787037      0.05      0.8263

Source                  DF      Type III SS     Mean Square   F Value      Pr > F

INTERCEPT                1       0.05787037      0.05787037      0.05      0.8263

                                          T for H0:    Pr > |T|   Std Error of
Parameter                   Estimate     Parameter=0               Estimate

INTERCEPT               0.0567011515            0.22     0.8263     0.25448352
```

Output 7.8 (*continued*)

```
Dependent Variable: YT9
                                        Sum of                Mean
Source                     DF          Squares              Square    F Value      Pr > F

Model                       1       0.37925926          0.37925926       1.03      0.3239

Error                      17       6.24740741          0.36749455

Uncorrected Total          18       6.62666667

                     R-Square             C.V.            Root MSE                YT9 Mean

                     0.000000        -417.6318           0.6062133              -0.1451549

Source                     DF        Type I SS         Mean Square    F Value      Pr > F

INTERCEPT                   1       0.37925926          0.37925926       1.03      0.3239

Source                     DF      Type III SS         Mean Square    F Value      Pr > F

INTERCEPT                   1       0.37925926          0.37925926       1.03      0.3239

                                           T for H0:     Pr > |T|     Std Error of
Parameter                     Estimate    Parameter=0                    Estimate

INTERCEPT                -.1451549477          -1.02       0.3239      0.14288584


                    Manova Test Criteria and Exact F Statistics for
                     the Hypothesis of no Overall INTERCEPT Effect
          H = Type III SS&CP Matrix for INTERCEPT    E = Error SS&CP Matrix

                               S=1     M=0.5     N=6.5

  Statistic                       Value          F       Num DF     Den DF   Pr > F

  Wilks' Lambda              0.43811339     6.4126            3         15   0.0052
  Pillai's Trace             0.56188661     6.4126            3         15   0.0052
  Hotelling-Lawley Trace     1.28251413     6.4126            3         15   0.0052
  Roy's Greatest Root        1.28251413     6.4126            3         15   0.0052
```

We see from the output labeled `no Overall INTERCEPT Effect` that we reject the null hypothesis that the third degree polynomial equals zero ($p = 0.0052$). Thus the linear trend is not an adequate fit to the data. Upon investigating the results of the univariate tests of fit of the linear trend for each variable found in the output, we see that the linear trend is adequate for variables two ($p = 0.8263$) and three ($p = 0.3239$) but not for variable one ($p = 0.0011$). Next PROC GLM is used to test the overall fit of a model assuming a linear trend for variables two and three and a quadratic trend for variable one. This is done with PROC GLM by testing that $yt6$ and $yt9$ are equal to zero. The results of the overall test of fit is found in the following output.

Output 7.8 *(continued)*

```
              Manova Test Criteria and Exact F Statistics for
                the Hypothesis of no Overall INTERCEPT Effect
       H = Type III SS&CP Matrix for INTERCEPT    E = Error SS&CP Matrix

                              S=1     M=0     N=7

 Statistic                     Value          F       Num DF    Den DF  Pr > F
 Wilks' Lambda              0.94208276    0.4918         2        16  0.6205
 Pillai's Trace             0.05791724    0.4918         2        16  0.6205
 Hotelling-Lawley Trace     0.06147786    0.4918         2        16  0.6205
 Roy's Greatest Root        0.06147786    0.4918         2        16  0.6205
```

Since the *p*-value is 0.6205, this model seems to be an adequate fit to the data. However, Program 7_8.sas also contains the SAS code to perform tests of quadratic beta weights. The results are found in the following output.

Output 7.8 *(continued)*

```
         Output 7.8: Multiple Response Growth Curve Analysis- Zullo data
                        Test of Quadratic Beta Weights

                       General Linear Models Procedure
                      Multivariate Analysis of Variance

           Characteristic Roots and Vectors of: E Inverse * H, where
         H = Contrast SS&CP Matrix for order    E = Error SS&CP Matrix

                Variables have been transformed by the M Matrix

Characteristic    Percent                Characteristic Vector  V'EV=1
     Root
                                          MVAR1           MVAR2           MVAR3

    1.28518252      93.33             0.44640405     -0.03876216      0.22044188
    0.09184121       6.67             0.06301927      0.20410811     -0.13065167
    0.00000000       0.00            -0.08289664      0.11161858      0.34151082


                 Manova Test Criteria and F Approximations for
                   the Hypothesis of no Overall order Effect
            on the variables defined by the M Matrix Transformation
         H = Contrast SS&CP Matrix for order    E = Error SS&CP Matrix

                              S=2     M=0     N=6

 Statistic                     Value          F       Num DF    Den DF  Pr > F

 Wilks' Lambda              0.40079253    2.7047         6        28  0.0337
 Pillai's Trace             0.64651410    2.3883         6        30  0.0526
 Hotelling-Lawley Trace     1.37702373    2.9836         6        26  0.0236
 Roy's Greatest Root        1.28518252    6.4259         3        15  0.0052

         NOTE: F Statistic for Roy's Greatest Root is an upper bound.
               NOTE: F Statistic for Wilks' Lambda is exact.
```

Given **B** in (7.144) and using (7.106), from the previous output we see that the beta weights for quadratic trend across all variables simultaneously are significant for all criteria except perhaps Pillai's Trace (p-value = 0.0526). However, the univariate tests seem to confirm the overall test of fit that the quadratic terms for variables two and three are not significant; hence, the null hypothesis that they are zero is not rejected.

Because our analysis is inconclusive on the degree of polynomial to use, we analyze the data assuming $p = q$ for all variables and also assuming $q < p$ with the parameter matrix given in (7.116). Program 7_8.sas contains the SAS code to perform the test of coincidence, the test of parallelism, and to estimate the model parameters, **B**, for both the model with $p = q$ and the model with $q < p$. Only the output continuing the test results for the model with $q < p$ follows.

Output 7.8 (*continued*)

```
        Output 7.8: Multiple Respnose Growth Curve Analysis-Zullo Data
                              Coincidence q<p

                        General Linear Models Procedure
                       Multivariate Analysis of Variance

                 Manova Test Criteria and Exact F Statistics for
                    the Hypothesis of no Overall GROUP Effect
         H = Type III SS&CP Matrix for GROUP    E = Error SS&CP Matrix

                              S=1     M=2.5     N=3

Statistic                      Value           F      Num DF    Den DF  Pr > F

Wilks' Lambda             0.46153844      1.3333           7         8  0.3456
Pillai's Trace            0.53846156      1.3333           7         8  0.3456
Hotelling-Lawley Trace    1.16666678      1.3333           7         8  0.3456
Roy's Greatest Root       1.16666678      1.3333           7         8  0.3456


                              Parallelism q<p

                        General Linear Models Procedure
                       Multivariate Analysis of Variance

                 Manova Test Criteria and Exact F Statistics for
                   the Hypothesis of no Overall parl(q<p) Effect
               on the variables defined by the M Matrix Transformation
        H = Contrast SS&CP Matrix for parl(q<p)    E = Error SS&CP Matrix

                              S=1     M=1     N=4.5

Statistic                      Value           F      Num DF    Den DF  Pr > F

Wilks' Lambda             0.63725850      1.5654           4        11  0.2513
Pillai's Trace            0.36274150      1.5654           4        11  0.2513
Hotelling-Lawley Trace    0.56922190      1.5654           4        11  0.2513
Roy's Greatest Root       0.56922190      1.5654           4        11  0.2513
```

Output 7.8 (*continued*)

```
                              Beta Matrix q<p

Dependent Variable: YT1
                                     Sum of              Mean
Source                  DF          Squares            Square   F Value       Pr > F

Model                    3     405.39749688      135.13249896      0.88       0.4761

Error                   14    2155.02842904      153.93060207

Corrected Total         17    2560.42592593

                  R-Square             C.V.          Root MSE              YT1 Mean

                  0.158332         5.809675         12.406877             213.55545

Source                  DF        Type I SS       Mean Square   F Value       Pr > F

GROUP                    1     208.07407407      208.07407407      1.35       0.2644
YT6                      1      73.72788092       73.72788092      0.48       0.5002
YT9                      1     123.59554189      123.59554189      0.80       0.3854

Source                  DF      Type III SS       Mean Square   F Value       Pr > F

GROUP                    1     219.88968975      219.88968975      1.43       0.2519
YT6                      1      87.93710591       87.93710591      0.57       0.4623
YT9                      1     123.59554189      123.59554189      0.80       0.3854

                                        T for H0:     Pr > |T|    Std Error of
Parameter                   Estimate   Parameter=0                  Estimate

beta1                     210.430524         48.01      0.0001      4.38324739
beta2                     217.738945         51.80      0.0001      4.20349669
1vs2                       -7.308421         -1.20      0.2519      6.11481882

Dependent Variable: YT2
                                     Sum of              Mean
Source                  DF          Squares            Square   F Value       Pr > F

Model                    3       4.04513609        1.34837870      2.54       0.0983

Error                   14       7.42708613        0.53050615

Corrected Total         17      11.47222222

                  R-Square             C.V.          Root MSE              YT2 Mean

                  0.352603         25.39860         0.7283585             2.8677108

Source                  DF        Type I SS       Mean Square   F Value       Pr > F

GROUP                    1       1.77777778        1.77777778      3.35       0.0885
YT6                      1       0.02469758        0.02469758      0.05       0.8323
YT9                      1       2.24266074        2.24266074      4.23       0.0589

Source                  DF      Type III SS       Mean Square   F Value       Pr > F

GROUP                    1       1.12707175        1.12707175      2.12       0.1670
YT6                      1       0.00221646        0.00221646      0.00       0.9494
YT9                      1       2.24266074        2.24266074      4.23       0.0589
```

Output 7.8 (*continued*)

Parameter	Estimate	T for H0: Parameter=0	Pr > |T|	Std Error of Estimate
beta1	2.69477045	10.47	0.0001	0.25732306
beta2	3.21800558	13.04	0.0001	0.24677061
1vs2	-0.52323513	-1.46	0.1670	0.35897675

Dependent Variable: YT3

Source	DF	Sum of Squares	Mean Square	F Value	Pr > F
Model	3	0.67599665	0.22533222	0.66	0.5885
Error	14	4.75918853	0.33994204		
Corrected Total	17	5.43518519			

R-Square	C.V.	Root MSE	YT3 Mean
0.124374	-111.7694	0.5830455	-0.5216506

Source	DF	Type I SS	Mean Square	F Value	Pr > F
GROUP	1	0.03703704	0.03703704	0.11	0.7462
YT6	1	0.11068175	0.11068175	0.33	0.5773
YT9	1	0.52827786	0.52827786	1.55	0.2330

Source	DF	Type III SS	Mean Square	F Value	Pr > F
GROUP	1	0.00013336	0.00013336	0.00	0.9845
YT6	1	0.07760457	0.07760457	0.23	0.6402
YT9	1	0.52827786	0.52827786	1.55	0.2330

Parameter	Estimate	T for H0: Parameter=0	Pr > |T|	Std Error of Estimate
beta1	-0.56523589	-2.74	0.0158	0.20598516
beta2	-0.57092743	-2.89	0.0119	0.19753800
1vs2	0.00569154	0.02	0.9845	0.28735817

Dependent Variable: YT4

Source	DF	Sum of Squares	Mean Square	F Value	Pr > F
Model	3	170.17364396	56.72454799	0.85	0.4908
Error	14	937.46061530	66.96147252		
Corrected Total	17	1107.63425926			

R-Square	C.V.	Root MSE	YT4 Mean
0.153637	7.263222	8.1829990	112.66349

Source	DF	Type I SS	Mean Square	F Value	Pr > F
GROUP	1	31.89351852	31.89351852	0.48	0.5014
YT6	1	0.04190001	0.04190001	0.00	0.9804
YT9	1	138.23822543	138.23822543	2.06	0.1727

Output 7.8 (*continued*)

```
Source                  DF       Type III SS      Mean Square   F Value      Pr > F

GROUP                    1       16.89430656      16.89430656      0.25      0.6233
YT6                      1        1.13433129       1.13433129      0.02      0.8983
YT9                      1      138.23822543     138.23822543      2.06      0.1727

                                     T for H0:     Pr > |T|    Std Error of
Parameter                 Estimate   Parameter=0                  Estimate

beta1                   112.327788         38.85      0.0001      2.89098605
beta2                   114.353563         41.25      0.0001      2.77243085
1vs2                     -2.025775         -0.50      0.6233      4.03305001

Dependent Variable: YT5
                                        Sum of             Mean
Source                  DF             Squares           Square   F Value      Pr > F

Model                    3          2.41087614       0.80362538      1.60      0.2345

Error                   14          7.04051275       0.50289377

Corrected Total         17          9.45138889

                  R-Square                C.V.         Root MSE              YT5 Mean

                  0.255082            38.00424        0.7091500             1.8659762

Source                  DF         Type I SS      Mean Square   F Value      Pr > F

GROUP                    1        0.56250000       0.56250000      1.12      0.3081
YT6                      1        0.57129256       0.57129256      1.14      0.3045
YT9                      1        1.27708358       1.27708358      2.54      0.1334

Source                  DF       Type III SS      Mean Square   F Value      Pr > F

GROUP                    1        0.14828157       0.14828157      0.29      0.5957
YT6                      1        0.45032608       0.45032608      0.90      0.3600
YT9                      1        1.27708358       1.27708358      2.54      0.1334

                                     T for H0:     Pr > |T|    Std Error of
Parameter                 Estimate   Parameter=0                  Estimate

beta1                   1.84643088          7.37      0.0001      0.25053686
beta2                   2.03621711          8.47      0.0001      0.24026270
1vs2                   -0.18978623         -0.54      0.5957      0.34950970


Dependent Variable: YT6


Dependent Variable: YT7
                                        Sum of             Mean
Source                  DF             Squares           Square   F Value      Pr > F

Model                    3        423.12572558     141.04190853      1.66      0.2214

Error                   14       1190.96631146      85.06902225

Corrected Total         17       1614.09203704
```

Output 7.8 (*continued*)

```
              R-Square            C.V.          Root MSE              YT7 Mean

              0.262145        21.66782         9.2232870             42.566752


Source                  DF         Type I SS      Mean Square    F Value      Pr > F

GROUP                    1        8.88166667       8.88166667       0.10      0.7514
YT6                      1       60.61774343      60.61774343       0.71      0.4128
YT9                      1      353.62631548     353.62631548       4.16      0.0608

Source                  DF       Type III SS      Mean Square    F Value      Pr > F

GROUP                    1       50.31905717      50.31905717       0.59      0.4546
YT6                      1       40.80534544      40.80534544       0.48      0.4999
YT9                      1      353.62631548     353.62631548       4.16      0.0608

                                         T for H0:    Pr > |T|   Std Error of
Parameter                 Estimate     Parameter=0                 Estimate

beta1                   45.5052400           13.97      0.0001     3.25851121
beta2                   42.0091117           13.44      0.0001     3.12488433
1vs2                     3.4961283            0.77      0.4546     4.54576343

Dependent Variable: YT8
                                      Sum of             Mean
Source                  DF           Squares           Square    F Value      Pr > F

Model                    3       16.83145197       5.61048399      42.72      0.0001

Error                   14        1.83854803       0.13132486

Corrected Total         17       18.67000000

              R-Square            C.V.          Root MSE              YT8 Mean

              0.901524        139.7710         0.3623877             0.2592725

Source                  DF         Type I SS      Mean Square    F Value      Pr > F

GROUP                    1        0.05444444       0.05444444       0.41      0.5301
YT6                      1        0.04250686       0.04250686       0.32      0.5784
YT9                      1       16.73450067      16.73450067     127.43      0.0001

Source                  DF       Type III SS      Mean Square    F Value      Pr > F

GROUP                    1        0.09034805       0.09034805       0.69      0.4208
YT6                      1        0.00882027       0.00882027       0.07      0.7993
YT9                      1       16.73450067      16.73450067     127.43      0.0001

                                         T for H0:    Pr > |T|   Std Error of
Parameter                 Estimate     Parameter=0                 Estimate

beta1                  -0.05408199           -0.42      0.6791     0.12802859
beta2                   0.09406080            0.77      0.4563     0.12277832
1vs2                   -0.14814279           -0.83      0.4208     0.17860540
```

With $p = q = 3$, the matrix $\hat{\mathbf{B}}$ is

$$\hat{\mathbf{B}} = \begin{pmatrix} 210.155 & 2.553 & -0.476 & 111.332 & 1.689 & 0.340 & 43.269 & 0.314 & -0.227 \\ 216.955 & 3.182 & -0.567 & 113.994 & 2.043 & -0.227 & 41.864 & 0.204 & -0.064 \end{pmatrix}$$

using normalized orthogonal polynomials. To test for parallelism with $p = q = 3$

$$H_p{:}(\beta_{11} \quad \beta_{12} \quad \theta_{11} \quad \theta_{12} \quad \xi_{11} \quad \xi_{12}) = (\beta_{21} \quad \beta_{22} \quad \theta_{21} \quad \theta_{22} \quad \xi_{21} \quad \xi_{22})$$

the matrices used are

$$\mathbf{C} = (1 - 1) \text{ and } \mathbf{A} = \mathbf{I}_3 \otimes \mathbf{A}_1 \text{ where } \mathbf{A}_1 = \begin{pmatrix} 0 & 0 \\ 1 & 0 \\ 0 & 1 \end{pmatrix} \tag{7.119}$$

Using Wilks' Λ – criterion, $\Lambda = 0.583$ with p-value of 0.3292. Given parallelism, we next test for coincidence or for group differences again assuming $p = q = 3$. For this test, $\mathbf{C} = (1, -1)$ and $\mathbf{A} = \mathbf{I}_3$. The Wilks' Λ– criterion is $\Lambda =$ 0.422 with p–value 0.3965.

For $q < p$ and with $\mathbf{B}$ defined in (7.116), the estimate of $\mathbf{B}$ is from the output labeled `Beta Matrix q < p`

$$\hat{\mathbf{B}} = \begin{pmatrix} 210.430 & 2.695 & -0.565 & 112.328 & 1.846 & 45.505 & -0.054 \\ 217.738 & 3.218 & -0.571 & 114.353 & 2.036 & 42.009 & 0.094 \end{pmatrix} \tag{7.120}$$

Using Wilks' Λ criterion, we do not reject the null hypothesis of coincidence since from the output we see that $\Lambda = 0.461$ with p-value 0.3456. Likewise, we do not reject the null hypothesis of parallelism, $\Lambda = 0.637$ with p-value of 0.2513. These results are consistent with the $p = q$ model analysis. However, we have not conclusively determined whether the full model, $p = q$, or the model $p < q$ best fits the data.

7.9 Single Growth Curve

In the previous example, we had difficulty establishing the best covariates for the analysis. For this last application, we consider the Zullo dental data as if it were obtained from a single population and we show how the REG procedure may be used to help determine the best set of covariates. Program 7_9.sas contains the SAS code for this analysis.

Program 7_9.sas

```
/* Program 7_9.sas                         */
/* Growth Curve Analysis (Timm, 1980a) */

options ls=80 ps=60 nodate nonumber;
title1 'Output 7.9: Single Growth Curve Analysis';

data zullo;
   infile 'c:\mmm.dat';
   input group y1 - y9;
run;
```

```
data test;
   set zullo;
   array z{9} y1 - y9;
   k=1;
   do rvar = 1,2,3;
      do time = 1,2,3;
         grmeans = z{k};
         k=k+1;
         output;
      end;
   end;
   keep rvar time grmeans;
run;
proc summary nway;
   class rvar time;
   var grmeans;
   output out=new mean=grmeans;
run;
proc print data = new;
proc plot;
   title2 'Plot of Variable Means';
   plot grmeans*time;
run;

proc iml;
   use zullo;
   read all var {y1 y2 y3 y4 y5 y6 y7 y8 y9} into y;
   /* Create Orthogonal Polynomials of degree  q-1=2 */
   x={1 2 3};
   P=orpol(x,2);
   print P;
   P_prime=block(P,P,P);
   y0=y*P_prime;
   /* Create New Transformed data set */
   varnames={yt1 yt2 yt3 yt4 yt5 yt6 yt7 yt8 yt9 group};
   create trans from y0 (|colname=varnames|);
   append from y0;
   proc print data=trans;
quit;
run;

/* Test model adequacy or fit */
proc glm data=trans;
   title2 'Test of Model Adequacy';
   model yt6 yt9=/nouni;
   manova h=intercept;
run;

/* Using a Rao-Khatri MANCOVA model                */
/* Estimate of the matrix B using reg p=q and q<p */
proc reg data=trans;
   title2 'Estimate of B using PROC REG';
   model yt1 - yt5 yt7 yt8=/P;
   model yt1 - yt5 yt7 yt8=yt6/P;
   model yt1 - yt5 yt7 yt8=yt9/P;
   model yt1 - yt5 yt7 yt8=yt6 yt9/P;
run;

/* Using a Rao-Khatri MANCOVA model                */
/* Estimate of the matrix B using glm p=q and q<p */
proc glm data=trans;
   title2 'Estimate of B p=q using PROC GLM';
   model yt1 - yt5 yt7 yt8 = ;
   estimate 'beta' intercept 1;
run;
```

```
proc glm data=trans;
   title2 'Estimate of B q<p with var ty9 using PROC GLM';
   model yt1 - yt5 yt7 yt8 = yt9;
   estimate 'beta' intercept 1;
run;
```

In Program 7_9.sas, the first several sections of code are the same as in Program 7_8.sas. Recall from the results in Output 7.8 that both *yt*6 and *yt*9 are candidates for covariates. The question that needs to be addressed is this: which fits the data better – a model with both of these covariates or a model with only one covariate? The best model should have the smallest residuals and standard errors for the estimates of the coefficients of the parameter matrix **B**. To obtain these estimates we use PROC REG with the P option which causes the observed and fitted values to be output.

Result and Interpretation 7_9.sas

To save space, the long output for this example is not included but the results are summarized as follows. For the full model, the matrix $\hat{\mathbf{B}}$ and standard errors are

$$\hat{\mathbf{B}}(s.e.) = \begin{pmatrix} 213.555 & 2.868 & -0.522 & \vdots & 112.663 & 1.866 & 0.057 & \vdots & 42.567 & 0.259 & -0.145 \\ (2.892) & (0.194) & (0.133) & \vdots & (1.903) & (0.176) & (0.254) & \vdots & (2.297) & (0.247) & (0.142) \end{pmatrix}$$

Using *yt*6 as a covariate, we have that

$$\hat{\mathbf{B}}(s.e.) = \begin{pmatrix} 213.500 & 2.874 & -0.526 & \vdots & 112.680 & 1.877 & \vdots & 42.651 & 0.256 \\ (2.975) & (0.197) & (0.083) & \vdots & (1.962) & (0.173) & \vdots & (2.338) & (0.255) \end{pmatrix}$$

Using *yt*9 as a covariate, we have that

$$\hat{\mathbf{B}}(s.e.) = \begin{pmatrix} 214.273 & 2.965 & -0.566 & \vdots & 113.382 & 1.940 & \vdots & 43.654 & 0.219 \\ (2.977) & (0.179) & (0.134) & \vdots & (1.874) & (0.170) & \vdots & (2.157) & (0.085) \end{pmatrix}$$

Finally using both *yt*6 and *yt*9 as covariates,

$$\hat{\mathbf{B}}(s.e.) = \begin{pmatrix} 214.236 & 2.967 & -0.568 & \vdots & 113.382 & 1.945 & \vdots & 43.685 & 0.023 \\ (3.056) & (0.183) & (0.137) & \vdots & (1.937) & (0.168) & \vdots & (2.209) & (0.087) \end{pmatrix}$$

Reviewing the standard errors, we see that the model with only one covariate (*yt*9) appears to be better than the model using two covariates. Rao (1987) suggests other procedures for selecting covariates in the application of the GMANOVA model.

While the Rao-Khatri approach to fitting polynomial growth curves is straightforward, there are serious limitations. These include

(1) The covariates are not independent of treatment.

(2) Different covariates may not be applied to different treatments.

(3) The set of covariates may vary from sample to sample.

(4) The observation vectors do not contain missing data.

More important, this approach does not model individual effects, but rather group effects, does not work well with long series, and does not provide a mechanism for modeling the structure of $\mathbf{\Sigma}$, Jöreskog (1973). The mixed linear model discussed in Chapter 9 provides a mechanism for modeling $\mathbf{\Sigma}$, using PROC MIXED. Furthermore, Galecki

(1994) discusses a general model that permits one to specify structured covariance matrices in the analysis of double multivariate linear models. A comprehensive review of growth curve models is presented by von Rosen (1991).

8. The SUR Model and the Restricted GMANOVA Model

Theory

Applications

8.1 Introduction

In this chapter, we show that the restricted multivariate general linear model is really a mixture of the MANOVA and GMANOVA models. For this model, approximate likelihood ratio tests are discussed. More generally, the MANOVA-GMANOVA model is shown to be a sum-profiles model which may be analyzed using the completely general MANOVA (CGMANOVA) model. Because likelihood ratio tests for these models are difficult to construct, we show how the SUR model may be used to estimate model parameters and to test hypotheses. We also show how the SUR model may be used to analyze the restricted GMANOVA model.

8.2 The MANOVA-GMANOVA Model

Our discussion in Chapter 7 began with the analysis of the restricted multivariate GLM (RMGLM) model where we were interested in testing the general hypothesis $H(\mathbf{CBA} = \mathbf{\Gamma})$ using the model

$$\mathbf{Y}_{n\times p} = \mathbf{X}_{n\times k}\mathbf{B}_{k\times p} + \mathbf{U}_{n\times p} \text{ with } \mathbf{R}_{1_{r_1\times k}}\mathbf{B}_{k\times p}\mathbf{R}_{2_{p\times r_2}} = \mathbf{\Theta}_o \tag{8.1}$$

where $\mathbf{R}_1, \mathbf{R}_2, \mathbf{C}, \mathbf{A}, \mathbf{\Gamma}$, and $\mathbf{\Theta}_o$ are known. The rank$(\mathbf{R}_1) = r_1 \le k$, the rank$(\mathbf{R}_2) = r_2 \le p$, $\mathbf{C}$ is $(g\times k)$ with rank$(\mathbf{C}) = g \le k$, and $\mathbf{A}$ is $(p\times u)$ with rank$(\mathbf{A}) = u \le p$. To test $H(\mathbf{CBA} = \mathbf{\Gamma})$, we assumed in Chapter 7 that $\mathbf{A} = \mathbf{R}_2$ since this eliminated the association of $\mathbf{R}_2$ with $\mathbf{\Sigma}$. Testing the restriction when $\mathbf{\Sigma}$ and $\mathbf{R}_2$ are associated led to the GMANOVA model. Using (8.1) with $\mathbf{\Theta}_o = \mathbf{0}$, we may incorporate the restrictions into the model. To do this, we expand the matrices $\mathbf{R}_1$ and $\mathbf{R}_2$ to full rank using Theorem 7.5 and construct new matrices $\mathbf{R}$ and $\mathbf{P}$ such that

$$\begin{aligned} \mathbf{R}_{k\times k} &= \begin{pmatrix} \mathbf{R}_1 \\ \tilde{\mathbf{R}}_1 \end{pmatrix}; \qquad \text{rank}(\mathbf{R}) = k \\ \mathbf{P}_{p\times p} &= (\mathbf{R}_2, \tilde{\mathbf{R}}_2); \quad \text{rank}(\mathbf{P}) = p \end{aligned} \tag{8.2}$$

Since $\mathbf{R}^{-1}$ and $\mathbf{P}^{-1}$ exist, $\mathbf{XB}$ may be written as

$$\begin{aligned} \mathbf{XB} &= \mathbf{XR}^{-1}\mathbf{RBPP}^{-1} \\ &= \mathbf{XR}^{-1}\begin{pmatrix} \mathbf{R}_1\mathbf{BR}_2 & \mathbf{R}_1\mathbf{B}\tilde{\mathbf{R}}_2 \\ \tilde{\mathbf{R}}_1\mathbf{BR}_2 & \tilde{\mathbf{R}}_1\mathbf{B}\tilde{\mathbf{R}}_2 \end{pmatrix}\mathbf{P}^{-1} \end{aligned} \tag{8.3}$$

Letting $\mathbf{X}^* = \mathbf{XR}^{-1} = (\mathbf{X}_1^*, \mathbf{X}_2^*)$ and $\mathbf{P}^{-1} = \begin{pmatrix} \mathbf{P}^{11} \\ \mathbf{P}^{21} \end{pmatrix}$, (8.3) becomes

$$\begin{aligned} \mathbf{XB} &= (\mathbf{X}_1^*, \mathbf{X}_2^*)\begin{pmatrix} \mathbf{0} & \beta_{12} \\ \beta_{21} & \beta_{22} \end{pmatrix}\begin{pmatrix} \mathbf{P}^{11} \\ \mathbf{P}^{21} \end{pmatrix} \\ &= \mathbf{X}_1^*\beta_{12}\mathbf{P}^{21} + \mathbf{X}_2^*(\beta_{21}, \beta_{22})\mathbf{P}^{-1} \end{aligned} \tag{8.4}$$

Finally, if we let $\mathbf{X}_1 = \mathbf{X}_1^*, \mathbf{\Theta} = \beta_{12} = \mathbf{R}_1\mathbf{B}\tilde{\mathbf{R}}_2, \mathbf{X}_2 = \mathbf{P}^{21}, \mathbf{Z} = \mathbf{X}_2^*$, and $\mathbf{\Gamma}_o = (\beta_{21}, \beta_{22})\mathbf{P}^{-1} = (\tilde{\mathbf{R}}_1\mathbf{BR}_2, \tilde{\mathbf{R}}_1\mathbf{BR}_2)\mathbf{P}^{-1}$, the restricted multivariate GLM has the form

$$E(\mathbf{Y}|\mathbf{Z}) = \mathbf{X}_1\mathbf{\Theta}\mathbf{X}_2 + \mathbf{Z}\mathbf{\Gamma}_o \tag{8.5}$$

This is a mixture of the GMANOVA and the MANOVA models (Chinchilli and Elswick, 1985). The matrix $\mathbf{X}_1(n\times m_1)$ is a known matrix of rank $m_1 \le n$, $\mathbf{X}_2\ (q\times p)$ is a matrix of rank $q \le p$, $\mathbf{\Theta}_{m_1\times q}$ is a matrix of unknown parameters, $\mathbf{Z}_{n\times m_2}$ is a known matrix of rank $m_2 \le n$ and $\mathbf{\Gamma}_{0(m_2\times p)}$ is a matrix of unknown parameters, where $q = p - m_2$ and $n > m_1 + m_2 + p$.

In Chapter 7, we saw how to test hypotheses of the form $H(\mathbf{CBA} = \mathbf{\Gamma} = \mathbf{0})$ for the GMANOVA model. Recall that the ML estimates of $\mathbf{B}$ and $\mathbf{\Sigma}$ are

$$\begin{aligned} \hat{\mathbf{B}} &= (\mathbf{X}_1'\mathbf{X}_1)^{-1}\mathbf{X}_1'\mathbf{YE}^{-1}\mathbf{X}_2'(\mathbf{X}_2\mathbf{E}^{-1}\mathbf{X}_2')^{-1} \\ n\hat{\mathbf{\Sigma}} &= (\mathbf{Y} - \mathbf{X}_1\hat{\mathbf{B}}\mathbf{X}_2)'(\mathbf{Y} - \mathbf{X}_1\hat{\mathbf{B}}\mathbf{X}_2) \\ &= (\mathbf{Y} - \mathbf{X}_1\hat{\mathbf{B}}\mathbf{X}_2)'(\mathbf{I}_n - \mathbf{X}_1(\mathbf{X}_1'\mathbf{X}_1)^{-1}\mathbf{X}_1')(\mathbf{Y} - \mathbf{X}_1\hat{\mathbf{B}}\mathbf{X}_2) + (\mathbf{Y} - \mathbf{X}_1\hat{\mathbf{B}}\mathbf{X}_2)'(\mathbf{X}_1(\mathbf{X}_1'\mathbf{X}_1)^{-1}\mathbf{X}_1')(\mathbf{Y} - \mathbf{X}_1\hat{\mathbf{B}}\mathbf{X}_2) \\ &= \mathbf{E} + \mathbf{W}'\mathbf{Y}'\mathbf{X}_1(\mathbf{X}_1'\mathbf{X}_1)^{-1}\mathbf{X}_1'\mathbf{YW} \end{aligned} \tag{8.6}$$

where $\mathbf{E} = \mathbf{Y}'(\mathbf{I}_n - \mathbf{X}_1(\mathbf{X}_1'\mathbf{X}_1)^{-1}\mathbf{X}_1')\mathbf{Y}$ and $\mathbf{W} = \mathbf{I}_p - \mathbf{E}^{-1}\mathbf{X}_2'(\mathbf{X}_2\mathbf{E}^{-1}\mathbf{X}_2')^{-1}\mathbf{X}_2$. To obtain the LR test, we employed Rao's conditional model which reduced the GMANOVA model $E(\mathbf{Y}) = \mathbf{X}_1\mathbf{BX}_2$ to a MANCOVA model using the transformation $\mathbf{Y}_o = \mathbf{YH}_1 = \mathbf{YG}^{-1}\mathbf{X}_2'(\mathbf{X}_2\mathbf{G}^{-1}\mathbf{X}_2')^{-1}$ and incorporating a set of covariates $\mathbf{Z} = \mathbf{YH}_2 = \mathbf{Y}(\mathbf{I} - \mathbf{H}_1\mathbf{X}_2) = \mathbf{Y}(\mathbf{I} - \mathbf{G}^{-1}\mathbf{X}_2'(\mathbf{X}_2\mathbf{G}^{-1}\mathbf{X}_2')^{-1}\mathbf{X}_2)$ into the model. Khatri, using the ML estimates in (8.6) and a conditional argument also derived an LR test of $H(\mathbf{CBA} = \mathbf{\Gamma})$, (Srivastava and Khatri, 1979, p. 193). For both, the LR test is to reject H if

$$\Lambda = \frac{|\tilde{\mathbf{E}}|}{|\tilde{\mathbf{E}} + \tilde{\mathbf{H}}|} < U^{\alpha}_{u, v_h, v_e} \tag{8.7}$$

where

$$\begin{aligned} \tilde{\mathbf{E}} &= \mathbf{A}'(\mathbf{X}_2 \mathbf{E}^{-1} \mathbf{X}_2')\mathbf{A} \\ \tilde{\mathbf{H}} &= (\mathbf{C}\hat{\mathbf{B}}\mathbf{A})'(\mathbf{C}\mathbf{R}^{-1}\mathbf{C}')^{-1}(\mathbf{C}\hat{\mathbf{B}}\mathbf{A}) \end{aligned} \tag{8.8}$$

with $\mathbf{E}$ and $\hat{\mathbf{B}}$ defined in (8.6), and $\mathbf{R}^{-1}$ defined as

$$\mathbf{R}^{-1} = (\mathbf{X}_1'\mathbf{X}_1)^{-1} + (\mathbf{X}_1'\mathbf{X}_1)^{-1}\mathbf{X}_1'\mathbf{Y}[\mathbf{E}^{-1} - \mathbf{E}^{-1}\mathbf{X}_2'(\mathbf{X}_2\mathbf{E}^{-1}\mathbf{X}_2')^{-1}]\mathbf{Y}'\mathbf{X}_1(\mathbf{X}_1'\mathbf{X}_1)^{-1} \tag{8.9}$$

For this LR test $u = \text{rank}(\mathbf{A}), g = \text{rank}(\mathbf{C}) \equiv v_h$, and $v_e = n - \text{rank}(\mathbf{X}_1) - p + \text{rank}(\mathbf{X}_2)$.

To obtain LR tests of the parameters in the MANOVA-GMANOVA model (8.5), Chinchilli and Elswick (1985) derive the ML estimate of $\mathbf{\Theta}$, $\mathbf{\Gamma}_o$, and $\mathbf{\Sigma}$. Given model (8.5), they form the augmented matrix $\mathbf{X} = (\mathbf{X}_1, \mathbf{Z})$ and associate

$$(\mathbf{X}'\mathbf{X}) = \begin{pmatrix} \mathbf{X}_1'\mathbf{X}_1 & \mathbf{X}_1'\mathbf{Z} \\ \mathbf{Z}'\mathbf{X}_1 & \mathbf{Z}'\mathbf{Z} \end{pmatrix} = \begin{pmatrix} \mathbf{X}_{11} & \mathbf{X}_{12} \\ \mathbf{X}_{21} & \mathbf{X}_{22} \end{pmatrix}$$

with

$$(\mathbf{X}'\mathbf{X})^{-1} = \begin{pmatrix} \mathbf{X}^{11} & \mathbf{X}^{12} \\ \mathbf{X}^{21} & \mathbf{X}^{22} \end{pmatrix}$$

where the partitioned matrices $\mathbf{X}^{ij}$ are defined in (3.8). Then, the ML estimates of the model parameters are

$$\begin{aligned} \hat{\mathbf{\Theta}} &= (\mathbf{X}^{11}\mathbf{X}_1' + \mathbf{X}^{12}\mathbf{Z}')\mathbf{Y}\mathbf{E}^{-1}\mathbf{X}_2'(\mathbf{X}_2\mathbf{E}^{-1}\mathbf{X}_2')^{-1} \\ \hat{\mathbf{\Gamma}}_o &= (\mathbf{X}^{21}\mathbf{X}_1' + \mathbf{X}^{22}\mathbf{Z}')\mathbf{Y}\mathbf{E}^{-1}\mathbf{X}_2'(\mathbf{X}_2\mathbf{E}^{-1}\mathbf{X}_2')^{-1}\mathbf{X}_2 + (\mathbf{Z}'\mathbf{Z})^{-1}\mathbf{Z}'\mathbf{Y}\mathbf{W} \\ n\hat{\mathbf{\Sigma}} &= \mathbf{E} + \mathbf{W}'\mathbf{Y}'(\mathbf{X}(\mathbf{X}'\mathbf{X})^{-1}\mathbf{X}' - \mathbf{Z}(\mathbf{Z}'\mathbf{Z})^{-1}\mathbf{Z}')\mathbf{Y}\mathbf{W} \end{aligned} \tag{8.10}$$

where $\mathbf{E} = \mathbf{Y}'(\mathbf{I}_n - \mathbf{X}(\mathbf{X}'\mathbf{X})^{-1}\mathbf{X}')\mathbf{Y}$ and $\mathbf{W} = \mathbf{I}_p - \mathbf{E}^{-1}\mathbf{X}_2'(\mathbf{X}_2\mathbf{E}^{-1}\mathbf{X}_2')^{-1}\mathbf{X}_2$

For the MANOVA-GMANOVA model (8.5), three hypotheses are of interest

$$\begin{aligned} H_1&: \ \mathbf{C}_1\mathbf{\Theta}\mathbf{A}_1 &= \mathbf{0} \\ H_2&: \ \mathbf{C}_2\mathbf{\Gamma}_o &= \mathbf{0} \\ H_3&: \ \mathbf{H}_1 \cap \mathbf{H}_2 &= \mathbf{0} \end{aligned} \tag{8.11}$$

where $\mathbf{C}_1(g_1 \times m_1)$ of rank $g_1 \le m_1$, $\mathbf{\Theta}_{m_1 \times g_1}$, $\mathbf{A}_1(g_1 \times u)$ of rank $u \le g_1$, $\mathbf{C}_2(g_2 \times m_2)$ of rank $g_2 \le m_2$ and $\mathbf{\Gamma}_{o(m_2 \times p)}$.

To obtain the LR test for H_1, one may employ the direct approach used by Khatri or the conditional approach used by Rao. If one uses the conditional approach, the data matrix $\mathbf{Y}$ is transformed to the matrix

$$\mathbf{Y}_o = (\mathbf{Y}_1, \mathbf{Y}_2) = (\mathbf{Y}\mathbf{H}_1', \mathbf{Y}\mathbf{H}_2') \tag{8.12}$$

(Chinchilli and Elswick, 1985) where

$$\underset{p \times q}{\mathbf{H}_1'} = \mathbf{X}_2'(\mathbf{X}_2\mathbf{X}_2')^{-1} \qquad \underset{p \times (p-q)}{\mathbf{H}_2'} = \mathbf{I}_p - \mathbf{X}_2'(\mathbf{X}_2\mathbf{X}_2')^{-1}\mathbf{X}_2 \tag{8.13}$$

so that $\mathbf{H}_1\mathbf{X}_2' = \mathbf{I}, \mathbf{H}_2\mathbf{X}_2' = \mathbf{0}$, and $\mathbf{H}_2'$ is any $(p-q)$ columns of $\mathbf{I}_p - \mathbf{H}_1'\mathbf{X}_2$. Assuming $\mathbf{Y}$ has a matrix multivariate normal distribution, by Definition 5.1 we write the density for the MANOVA-GMANOVA model as

$$(2\pi)^{np/2}|\mathbf{\Sigma}|^{-n/2} \operatorname{etr}\left[-\frac{1}{2}(\mathbf{Y}-\mathbf{X}_1\mathbf{\Theta}\mathbf{X}_2-\mathbf{Z}\mathbf{\Gamma}_o)\mathbf{\Sigma}^{-1}(\mathbf{Y}-\mathbf{X}_1\mathbf{\Theta}\mathbf{X}_2-\mathbf{Z}\mathbf{\Gamma}_o)'\right] \tag{8.14}$$

Using Theorem 6.1 with $\mathbf{H}_2 \equiv \mathbf{H}_2'$, observe that

$$\mathbf{\Sigma} = \mathbf{X}_2'(\mathbf{X}_2\mathbf{\Sigma}^{-1}\mathbf{X}_2')^{-1}\mathbf{X}_2 + \mathbf{\Sigma}\mathbf{H}_2'(\mathbf{H}_2\mathbf{\Sigma}\mathbf{H}_2')^{-1}\mathbf{H}_2\mathbf{\Sigma} \tag{8.15}$$

On postmultiplying (8.15) by $\mathbf{X}_2'(\mathbf{X}_2\mathbf{X}_2')^{-1}$ and premultiplying by $\mathbf{\Sigma}^{-1}$, we have that

$$\mathbf{\Sigma}^{-1}\mathbf{X}_2'(\mathbf{X}_2\mathbf{\Sigma}^{-1}\mathbf{X}_2')^{-1} = \mathbf{X}_2'(\mathbf{X}_2\mathbf{X}_2')^{-1} - \mathbf{H}_2'(\mathbf{H}_2\mathbf{\Sigma}\mathbf{H}_2')^{-1}\mathbf{H}_2\mathbf{\Sigma}\mathbf{X}_2'(\mathbf{X}_2\mathbf{X}_2')^{-1} \tag{8.16}$$

If one uses the identities (8.15) and (8.16), and results of conditional normal distributions, the joint density of $\mathbf{Y}_o = (\mathbf{Y}_1, \mathbf{Y}_2)$ in (8.12) may be expressed as a product of $\mathbf{Y}_2$ and the conditional density of $\mathbf{Y}_1$ given $\mathbf{Y}_2$ where

$$\begin{aligned} \mathbf{Y}_1|\mathbf{Y}_2 &\sim N_{n,p}(\mathbf{X}_1\mathbf{\Theta}+\mathbf{Z}\xi+\mathbf{Y}_2\gamma,(\mathbf{X}_2\mathbf{\Sigma}^{-1}\mathbf{X}_2)^{-1},\mathbf{I}_n) \\ \mathbf{Y}_2 &\sim N_{n,p-q}(\mathbf{Z}\mathbf{\Gamma}_o\mathbf{H}_2',\mathbf{H}_2\mathbf{\Sigma}\mathbf{H}_2',\mathbf{I}_n) \end{aligned} \tag{8.17}$$

and

$$\begin{aligned} \xi &= \mathbf{\Gamma}_o\mathbf{\Sigma}^{-1}\mathbf{X}_2'(\mathbf{X}_2\mathbf{\Sigma}^{-1}\mathbf{X}_2')^{-1} \\ \gamma &= (\mathbf{H}_2\mathbf{\Sigma}\mathbf{H}_2')^{-1}\mathbf{H}_2\mathbf{\Sigma}\mathbf{X}_2' \end{aligned} \tag{8.18}$$

The density of $\mathbf{Y}_2$ does not involve $\mathbf{\Theta}$; thus, to test $H_1(\mathbf{C}_1\mathbf{\Theta}\mathbf{A}_1 = \mathbf{0})$, we use the MANOVA model and the conditional distribution of $\mathbf{Y}_1|\mathbf{Y}_2$. The LR test of H_1 under the MANOVA-GMANOVA model is to reject $H_1(\mathbf{C}_1\mathbf{\Theta}\mathbf{A}_1 = \mathbf{0})$ if

$$\Lambda = \frac{|\tilde{\mathbf{E}}|}{|\tilde{\mathbf{E}}+\tilde{\mathbf{H}}|} < U^{\alpha}_{u,v_h,v_e} \tag{8.19}$$

where the maximum likelihood estimate of $\hat{\mathbf{\Theta}}$ is given in (8.10). Using identities (8.15) and (8.16), $\hat{\mathbf{\Theta}}$ becomes

$$\hat{\mathbf{\Theta}} = (\mathbf{X}^{11}\mathbf{X}_1' + \mathbf{X}^{12}\mathbf{Z}')[\mathbf{I}_n - \mathbf{Y}_2(\mathbf{Y}_2'\mathbf{Y}_2 - \mathbf{Y}_2'\mathbf{X}(\mathbf{X}'\mathbf{X})^{-1}\mathbf{X}'\mathbf{Y}_2)^{-1}\mathbf{Y}_2'\tilde{\mathbf{X}}]\mathbf{Y}_1 \tag{8.20}$$

where $\mathbf{X} = (\mathbf{X}_1, \mathbf{Z})$, $\mathbf{Y}_2 = \mathbf{Y}\mathbf{H}_2'$, $\mathbf{Y}_1 = \mathbf{Y}\mathbf{H}_1' = \mathbf{Y}\mathbf{X}_2'(\mathbf{X}_2\mathbf{X}_2')^{-1}$, and $\tilde{\mathbf{X}} = \mathbf{I}_n = \mathbf{X}(\mathbf{X}'\mathbf{X})^{-1}\mathbf{X}'$. The hypothesis test matrices are

$$\begin{aligned} \tilde{\mathbf{E}} &= \mathbf{A}_1'(\mathbf{X}_2\mathbf{E}^{-1}\mathbf{X}_2')^{-1}\mathbf{A}_1 \\ \tilde{\mathbf{H}} &= (\mathbf{C}_1\hat{\mathbf{\Theta}}\mathbf{A}_1)'(\mathbf{C}_1\mathbf{R}_1^{-1}\mathbf{C}_1')^{-1}(\mathbf{C}_1\hat{\mathbf{\Theta}}\mathbf{A}_1) \end{aligned} \tag{8.21}$$

where $\mathbf{R}_1^{-1}$ is

$$\mathbf{R}_1^{-1} = \mathbf{X}^{11} + (\mathbf{X}^{11}\mathbf{X}_1' + \mathbf{X}^{12}\mathbf{Z}')\mathbf{Y}_2(\mathbf{Y}_2'\mathbf{Y}_2 - \mathbf{Y}_2'\mathbf{X}(\mathbf{X}'\mathbf{X})^{-1}\mathbf{X}'\mathbf{Y}_2)^{-1}\mathbf{Y}_2'(\mathbf{X}_1\mathbf{X}^{11} + \mathbf{Z}\mathbf{X}^{21}) \tag{8.22}$$

and $\mathbf{E} = \mathbf{Y}(\mathbf{I} - \mathbf{X}(\mathbf{X}'\mathbf{X})^{-1}\mathbf{X}')\mathbf{Y}$. The matrices $\tilde{\mathbf{E}}$ and $\tilde{\mathbf{H}}$ have independent Wishart distributions

$$\begin{aligned} \tilde{\mathbf{E}} &\sim W_u(v_e = n - m_1 - m_2 - p + q, \mathbf{A}_1'(\mathbf{X}_2\mathbf{\Sigma}^{1}\mathbf{X}_2')^{-1}\mathbf{A}_1, \mathbf{\Delta} = \mathbf{0}) \\ \tilde{\mathbf{H}} &\sim W_u(v_h = g_1, \mathbf{A}_1'(\mathbf{X}_2\mathbf{\Sigma}^{-1}\mathbf{X}_2)^{-1}\mathbf{A}_1, [\mathbf{A}_1'(\mathbf{X}_2\mathbf{\Sigma}^{-1}\mathbf{X}_2')^{-1}\mathbf{A}_1]^{-1}\mathbf{\Delta}) \end{aligned} \tag{8.23}$$

where the noncentrality matrix is

$$\mathbf{\Delta} = (\mathbf{C}_1\mathbf{\Theta}\mathbf{A}_1)'(\mathbf{C}_1\mathbf{R}_1^{-1}\mathbf{C}_1')^{-1}(\mathbf{C}_1\mathbf{\Theta}\mathbf{A}_1) \tag{8.24}$$

For $\mathbf{Z} = \mathbf{0}$, the LR test $H(\mathbf{C}_1\mathbf{\Theta}\mathbf{A}_1 = \mathbf{0})$ reduces to the GMANOVA model.

If one wants to test the hypothesis H_2 under the MANOVA-GMANOVA model, the LR test of $H_2(\mathbf{C}_2\boldsymbol{\Gamma}_o = \mathbf{0})$ is to reject the null hypothesis for small values of

$$\Lambda = \frac{|\widetilde{\mathbf{E}}_1|}{|\widetilde{\mathbf{E}}_1 + \widetilde{\mathbf{H}}_1|} \frac{|\widetilde{\mathbf{E}}_2|}{|\widetilde{\mathbf{E}}_2 + \widetilde{\mathbf{H}}_2|} = \lambda_1\lambda_2 \tag{8.25}$$

where

$$\begin{aligned}
\mathbf{X} &= (\mathbf{X}_1, \mathbf{Z}) \\
\mathbf{E} &= \mathbf{Y}'(\mathbf{I}_n - \mathbf{X}(\mathbf{X}'\mathbf{X})^{-1}\mathbf{X}')\mathbf{Y} \\
\widetilde{\mathbf{E}}_1 &= (\mathbf{X}_2\mathbf{E}^{-1}\mathbf{X}_2')^{-1} \\
\widetilde{\mathbf{H}}_1 &= (\mathbf{C}_2\hat{\xi})'(\mathbf{C}_2\mathbf{R}_2\mathbf{C}_2')^{-1}(\mathbf{C}_2\hat{\xi}) \\
\hat{\xi} &= \hat{\boldsymbol{\Gamma}}_o\mathbf{E}^{-1}\mathbf{X}_2'(\mathbf{X}_2\mathbf{E}^{-1}\mathbf{X}_2')^{-1} \\
\widetilde{\mathbf{E}}_2 &= \mathbf{H}_2\mathbf{E}_2\mathbf{H}_2' \\
\widetilde{\mathbf{H}}_2 &= (\mathbf{C}_2\hat{\boldsymbol{\Gamma}}_o\mathbf{H}_2')'(\mathbf{C}_2(\mathbf{Z}'\mathbf{Z})^{-1}\mathbf{C}_2')^{-1}(\mathbf{C}_2\hat{\boldsymbol{\Gamma}}_o\mathbf{H}_2') \\
\mathbf{R}_2 &= (\mathbf{X}^{22} + \mathbf{X}^{21}\mathbf{X}_1' + \mathbf{X}^{22}\mathbf{Z}')\mathbf{Y}(\mathbf{E}^{-1} - \mathbf{E}^{-1}\mathbf{X}_2'(\mathbf{X}_2\mathbf{E}^{-1}\mathbf{X}_2')^{-1}\mathbf{X}_2\mathbf{E}^{-1})\mathbf{Y}'(\mathbf{X}_1\mathbf{X}^{12} + \mathbf{Z}\mathbf{X}^{22}) \\
\mathbf{E}_2 &= \mathbf{Y}'(\mathbf{I}_n - \mathbf{Z}(\mathbf{Z}'\mathbf{Z})^{-1}\mathbf{Z}')\mathbf{Y} \\
\hat{\boldsymbol{\Gamma}}_o\mathbf{H}_2' &= (\mathbf{Z}'\mathbf{Z})^{-1}\mathbf{Z}'\mathbf{Y}'\mathbf{H}_2' (\text{since } \mathbf{X}_2\mathbf{H}_2' = \mathbf{0})
\end{aligned} \tag{8.26}$$

and $\hat{\boldsymbol{\Gamma}}_o$ is defined in (8.10). The matrices $\widetilde{\mathbf{E}}_1$, $\widetilde{\mathbf{H}}_1$, $\widetilde{\mathbf{E}}_2$, and $\widetilde{\mathbf{H}}_2$ have independent Wishart distributions

$$\begin{aligned}
\widetilde{\mathbf{E}}_1 &\sim W_q(v_e = n - m_1 - m_2 - p + q, (\mathbf{X}_2\boldsymbol{\Sigma}^{1}\mathbf{X}_2')^{-1}, \boldsymbol{\Delta} = \mathbf{0}) \\
\widetilde{\mathbf{H}}_1 &\sim W_q(v_h = g_1, \mathbf{X}_2\boldsymbol{\Sigma}^{-1}\mathbf{X}_2', (\mathbf{X}_2\boldsymbol{\Sigma}^{-1}\mathbf{X}_2')^{-1}\boldsymbol{\Delta}) \\
\widetilde{\mathbf{E}}_2 &\sim W_{p-q}(v_e = n - m_2, \mathbf{H}_2\boldsymbol{\Sigma}\mathbf{H}_2', \boldsymbol{\Delta} = \mathbf{0}) \\
\widetilde{\mathbf{H}}_2 &\sim W_{p-q}(v_h = g_2, \mathbf{H}_2\boldsymbol{\Sigma}\mathbf{H}_2', (\mathbf{H}_2\boldsymbol{\Sigma}\mathbf{H}_2')^{-1}\boldsymbol{\Delta})
\end{aligned} \tag{8.27}$$

The distribution of Λ in (8.25) follows a matrix inverted Dirichlet distribution (a generalization of the multivariate beta II density, also called the inverted multivariate beta density) which arises when one tests the equality of means and covariances of several multivariate normal distributions (Anderson, 1984, p. 409; Johnson and Kotz, 1972, p. 237).

Since λ_1 and λ_2 are independent likelihood ratio statistics, we may use Theorem 5.3 with $v_e \equiv v_1 = n - m_1 - m_2 - p + q$ and $v_e \equiv v_2 = n - m_2$ to obtain an approximate critical value for the test of H_2 in that

$$\begin{aligned}
X^2 &= [v_1 - (q - g_1 + 1)/2] \ln \lambda_1 + [v_2 - (p - q - g_2 + 1)/2] \ln \lambda_2 \\
&\mathrel{\dot\sim} \chi^{\alpha}_{(qg_1)} + \chi^{\alpha}_{[(p-q_1)g_2]}
\end{aligned} \tag{8.28}$$

Alternatively to obtain an approximate p-value, we employ Theorem 5.4 using the sum of two independent chi-square distributions where for λ_1, $k = m_1 + m_2 + p - q$, $g \equiv g_1$ and $u \equiv q$ and for λ_2, $k = m_2$, $g \equiv g_2$ and $u \equiv p - q$ for the MANOVA-GMANOVA model.

The LR test of $H_3(\mathbf{C}_1\mathbf{\Theta}\mathbf{A}_1 = \mathbf{0},\ \mathbf{C}_2\mathbf{\Gamma}_o = \mathbf{0})$ rejects H_3 for small values of $\mathbf{\Lambda}$ with

$$\mathbf{\Lambda} = \frac{|\mathbf{E}_1^*|}{|\mathbf{E}_1^* + \mathbf{H}_1^*|}\frac{|\widetilde{\mathbf{E}}_2|}{|\widetilde{\mathbf{E}}_2 + \widetilde{\mathbf{H}}_2|} = \lambda_1\lambda_2 \tag{8.29}$$

where

$$\begin{aligned}\mathbf{E}_1^* &= (\mathbf{X}_2\mathbf{E}^{-1}\mathbf{X}_2')^{-1}\\ \mathbf{H}_1^* &= (\mathbf{C}^*\hat{\mathbf{\Theta}}^*)'(\mathbf{C}^*\mathbf{R}^{-1}\mathbf{C}^{*\prime})^{-1}(\mathbf{C}^*\hat{\mathbf{\Theta}}^*)\end{aligned} \tag{8.30}$$

and $\widetilde{\mathbf{E}}_2$ and $\widetilde{\mathbf{H}}_2$ are given in (8.26), $\mathbf{R}^{-1}$ is defined in (8.9) with $\mathbf{X}_1$ replaced by $\mathbf{X} = (\mathbf{X}_1, \mathbf{Z})$

$$\mathbf{C}^* = \begin{pmatrix} \mathbf{C}_1 & \mathbf{0} \\ \mathbf{0} & \mathbf{C}_2 \end{pmatrix} \text{ and } \hat{\mathbf{\Theta}}^* = \begin{pmatrix} \hat{\mathbf{\Theta}} \\ \hat{\xi} \end{pmatrix} \tag{8.31}$$

with $\hat{\mathbf{\Theta}}$ defined in (8.10), $\hat{\xi}$ defined in (8.26), and $\hat{\mathbf{\Gamma}}_o$ given in (8.10), and $\mathbf{E} = \mathbf{Y}'(\mathbf{I}_n - \mathbf{X}(\mathbf{X}'\mathbf{X})^{-1}\mathbf{X}')\mathbf{Y}$. The distributions of the test matrices are identical to those given in (8.27) with $\widetilde{\mathbf{E}}_1$ and $\widetilde{\mathbf{H}}_1$ replaced by $\mathbf{E}_1^*$ and $\mathbf{H}_1^*$. The degrees of freedom for $\mathbf{H}_1^*$ are $g_1 + g_2$, not g_2. To evaluate the test statistic in (8.29), we may alternatively use a chi-square distribution to approximate the test of $\mathbf{H}_3(\mathbf{H}_1 \cap \mathbf{H}_2)$ as given in (8.29) with g_1 replaced by $g_1 + g_2$ in (8.28). Theorem 5.4 is used to obtain an approximate p-value for the test. This is illustrated later in this chapter when applications of the MANOVA-GMANOVA model are discussed.

8.3 Tests of Fit

In our discussion of the GMANOVA model in Chapter 7, we presented a test (7.49) for the inclusion of the higher order polynomials as covariates in the MANCOVA model. The test evaluates whether the best model for the data is a MANOVA or GMANOVA model. To see this, recall that a MANOVA model $E(\mathbf{Y}) = \mathbf{X}_1\mathbf{B}$ may be represented as a GMANOVA model if $\mathbf{B}$ can be rewritten as $\mathbf{B} = \mathbf{\Theta}\mathbf{X}_2$. If $\mathbf{H}_2[(p-q)\times p]$ is a matrix of full rank such that $\mathbf{B}\mathbf{H}_2' = \mathbf{0}$, the condition $\mathbf{B} = \mathbf{\Theta}\mathbf{X}_2$ is equivalent to the requirement that $\mathbf{B}\mathbf{H}_2' = \mathbf{\Theta}\mathbf{X}_2\mathbf{H}_2' = \mathbf{0}$. Thus, to evaluate the adequacy of the GMANOVA model versus the MANOVA model, we tested the hypothesis that $\mathbf{B}\mathbf{H}_2' = \mathbf{0}$ under the GMANOVA model. Failing to reject the null hypothesis supports the adequacy of the GMANOVA model. For example, one may evaluate whether a quadratic model is adequate for a set of data which would suggest that a subset of higher order polynomials should possibly be included as covariates in the GMANOVA model. If no covariates are included in the model, then one has a MANOVA model. Thus, one needs to test the null hypothesis that the model is GMANOVA versus the alternative that the model is MANOVA. For MANOVA-GMANOVA models we may evaluate the adequacy of the GMANOVA model versus the MANOVA-GMANOVA model and the adequacy of the MANOVA-GMANOVA model versus the MANOVA model. These goodness-of-fit tests were derived by Chinchilli and Elswick (1985).

An alternative approach for evaluating the adequacy of the MANOVA model versus the GMANOVA model is to test

$$\begin{aligned} \mathbf{\Omega}_o&: E(\mathbf{Y}) = \mathbf{X}_1\mathbf{\Theta}\mathbf{X}_2 \\ \omega&: E(\mathbf{Y}) = \mathbf{X}_1\mathbf{\Theta}^* \end{aligned} \tag{8.32}$$

using the likelihood ratio test directly. We compare the ML estimate of $\mathbf{\Sigma}$ under $\mathbf{\Omega}_o$, $\hat{\mathbf{\Sigma}}_{\mathbf{\Omega}_o}$, with the corresponding estimate under ω, $\hat{\mathbf{\Sigma}}_\omega$. From (8.6),

$$n\hat{\mathbf{\Sigma}}_{\mathbf{\Omega}_o} = \mathbf{E} + \mathbf{W}'\mathbf{Y}'\mathbf{X}_1(\mathbf{X}_1'\mathbf{X}_1)^{-1}\mathbf{X}_1'\mathbf{Y}\mathbf{W} \tag{8.33}$$

Under ω, $\mathbf{\Theta}^* = \mathbf{\Theta}\mathbf{X}_2$. Hence,

$$n\hat{\mathbf{\Sigma}}_\omega = \mathbf{E} \tag{8.34}$$

so that

$$\Lambda = \lambda^{2/n} \frac{|\mathbf{E}|}{|\mathbf{E} + \mathbf{W}'\mathbf{Y}'\mathbf{X}_1(\mathbf{X}_1'\mathbf{X}_1)^{-1}\mathbf{X}_1'\mathbf{Y}\mathbf{W}|} \tag{8.35}$$

If we let $\mathbf{M}_{p\times p} = (\mathbf{H}_2', \ \mathbf{E}^{-1}\mathbf{X}_2'(\mathbf{X}_2\mathbf{X}_2')^{-1})$ be a matrix of full rank, and $\mathbf{H}_2$ be any set of $p-q$ independent rows of $\mathbf{I}_p - \mathbf{X}_2'(\mathbf{X}_2\mathbf{X}_2')^{-1}\mathbf{X}_2$ so that $\mathbf{X}_2\mathbf{H}_2' = \mathbf{0}$, we may reduce Λ in (8.35) to Λ in (7.50). That is,

$$\begin{aligned}
\Lambda &= \frac{|\mathbf{E}|}{|\mathbf{E} + \mathbf{W}'\mathbf{Y}'\mathbf{X}_1(\mathbf{X}_1'\mathbf{X}_1)^{-1}\mathbf{X}_1'\mathbf{Y}\mathbf{W}|} \\
&= \frac{|\mathbf{M}'||\mathbf{E}||\mathbf{M}|}{|\mathbf{M}'||\mathbf{E} + \mathbf{W}'\mathbf{Y}'(\mathbf{X}_1(\mathbf{X}_1'\mathbf{X}_1)^{-1}\mathbf{X}_1'\mathbf{Y})\mathbf{W}||\mathbf{M}|} \\
&= \frac{|\mathbf{M}'\mathbf{E}\mathbf{M}|}{|\mathbf{M}'\mathbf{E}\mathbf{M} + \mathbf{M}'\mathbf{W}'\mathbf{Y}'(\mathbf{X}_1(\mathbf{X}_1'\mathbf{X}_1)^{-1}\mathbf{X}_1'\mathbf{Y})\mathbf{W}\mathbf{M}|} \\
&= \frac{|\mathrm{Diag}(\mathbf{H}_2\mathbf{E}\mathbf{H}_2', (\mathbf{X}_2\mathbf{X}_2')^{-1}\mathbf{X}_2\mathbf{E}^{-1}\mathbf{X}_2'(\mathbf{X}_2\mathbf{X}_2')^{-1}|}{|\mathrm{Diag}(\mathbf{H}_2\mathbf{E}\mathbf{H}_2' + \mathbf{H}_2\mathbf{Y}'\mathbf{X}_1(\mathbf{X}_1'\mathbf{X}_1)^{-1}\mathbf{X}_1'\mathbf{Y}\mathbf{H}_2', (\mathbf{X}_2\mathbf{X}_2')^{-1}\mathbf{X}_2\mathbf{E}^{-1}\mathbf{X}_2'(\mathbf{X}_2\mathbf{X}_2')^{-1})|} \\
&= \frac{|\mathbf{H}_2\mathbf{E}\mathbf{H}_2'|}{|\mathbf{H}_2\mathbf{E}\mathbf{H}_2' + \mathbf{H}_2\mathbf{Y}'\mathbf{X}_1(\mathbf{X}_1'\mathbf{X}1)^{-1}\mathbf{X}_1\mathbf{Y}\mathbf{H}_2'|} \\
&= \frac{|\mathbf{H}_2\mathbf{Y}'\mathbf{X}_1(\mathbf{X}_1'\mathbf{X}_1)^{-1}\mathbf{X}_1\mathbf{Y}\mathbf{H}_2'|}{|\mathbf{H}_2\mathbf{Y}'(\mathbf{I} - \mathbf{X}_1(\mathbf{X}_1'\mathbf{X}_1')^{-1}\mathbf{X}_1')\mathbf{Y}\mathbf{H}_2'|}
\end{aligned} \tag{8.36}$$

which is the same statistic given in (7.50) since $\mathbf{H}_2$ in (8.36) is $\mathbf{H}_2'$ in (7.50). This is the approach employed by Chinchilli and Elswick (1985) to test model adequacy for the MANOVA-GMANOVA model.

To test

$$\begin{aligned} \mathbf{\Omega}_o&: E(\mathbf{Y}) = \mathbf{X}_1\mathbf{\Theta}\mathbf{X}_2 + \mathbf{Z}\mathbf{\Gamma}_o \\ \omega&: E(\mathbf{Y}) = \mathbf{X}_1\mathbf{\Theta}\mathbf{X}_2 + \mathbf{Z}\mathbf{\Gamma}_o^*\mathbf{X}_2 \end{aligned} \tag{8.37}$$

the null hypothesis that the model is GMANOVA versus the alternative that the model is MANOVA-GMANOVA, we derive the LR test by using $\hat{\boldsymbol{\Sigma}}_{\Omega_1}$ in (8.10) and $\hat{\boldsymbol{\Sigma}}_{\omega}$ in (8.6) with $\mathbf{X} = (\mathbf{X}_1, \mathbf{Z})$. The statistic is

$$\begin{aligned}\Lambda &= \frac{|\mathbf{E} + \mathbf{W}'\mathbf{Y}'(\mathbf{X}(\mathbf{X}'\mathbf{X})^{-1}\mathbf{X}' - \mathbf{Z}(\mathbf{Z}'\mathbf{Z})^{-1}\mathbf{Z}')\mathbf{Y}\mathbf{W}|}{|\mathbf{E} + \mathbf{W}'\mathbf{Y}'\mathbf{X}(\mathbf{X}'\mathbf{X})^{-1}\mathbf{X}'\mathbf{Y}\mathbf{W}|} \\ &= |\tilde{\mathbf{E}}|/|\tilde{\mathbf{E}} + \tilde{\mathbf{H}}|\end{aligned} \tag{8.38}$$

where

$$\begin{aligned}\tilde{\mathbf{H}} &= \mathbf{H}_2\mathbf{Y}'\mathbf{Z}(\mathbf{Z}'\mathbf{Z})^{-1}\mathbf{Z}'\mathbf{Y}\mathbf{H}_2' \\ \tilde{\mathbf{E}} &= \mathbf{H}_2\mathbf{Y}'(\mathbf{I} - \mathbf{Z}(\mathbf{Z}'\mathbf{Z})^{-1}\mathbf{Z}')\mathbf{Y}\mathbf{H}_2'\end{aligned} \tag{8.39}$$

have independent Wishart distributions with m_2 and $n - m_2$ degrees of freedom, respectively. The parameter matrix $\mathbf{H}_2\boldsymbol{\Sigma}\mathbf{H}_2'$ is of rank $p - q$. Hence $\boldsymbol{\Lambda}$ may be compared to a U-distribution with $u = p - q$, $v_h = m_2$, and $v_e = n - m_2$ degrees of freedom.

To test

$$\begin{aligned}\Omega_o&: E(\mathbf{Y}) = \mathbf{X}_1\boldsymbol{\Theta}^* + \mathbf{Z}\boldsymbol{\Gamma}_o \\ \omega&: E(\mathbf{Y}) = \mathbf{X}_1\boldsymbol{\Theta}\mathbf{X}_2 + \mathbf{Z}\boldsymbol{\Gamma}_o\end{aligned} \tag{8.40}$$

that the model is mixed MANOVA-GMANOVA versus the alternative that the model is MANCOVA, we use the LR test

$$\Lambda = |\tilde{\mathbf{H}}|/|\tilde{\mathbf{H}} + \tilde{\mathbf{E}}| \tag{8.41}$$

where

$$\begin{aligned}\tilde{\mathbf{H}} &= \mathbf{H}_2\mathbf{Y}'(\mathbf{X}(\mathbf{X}'\mathbf{X})^{-1}\mathbf{X}' - \mathbf{Z}(\mathbf{Z}'\mathbf{Z})^{-1}\mathbf{Z}')\mathbf{Y}\mathbf{H}_2' \\ \tilde{\mathbf{E}} &= \mathbf{H}_2\mathbf{E}\mathbf{H}_2'\end{aligned} \tag{8.42}$$

have independent Wishart distribution with degrees of freedom m_1 and $n - m_1 - m_2$, respectively, with parameter matrix $\mathbf{H}_2\boldsymbol{\Sigma}\mathbf{H}_2'$ of rank $p - q$. For this test $u = p - q$, $v_h = m_1$ and $v_e = n - m_1 - m_2$ for the U-distribution.

The RMGLM may be used to test hypotheses on the between-groups design of a repeated measurement experiment with covariates, subject to the restriction of equality of slopes over groups. One application of the MANOVA-GMANOVA model is in experiments where data are collected on a response made before the application of repeated treatments and this initial response is then used as a covariate. Letting the data matrix $\mathbf{Y}$ contain differences between consecutive time points and $\mathbf{Z}_{n \times 1}$ be the a vector of initial response data, we have a simple application of the model.

8.4 Sum of Profiles and CGMANOVA Models

An extension of the MANOVA-GMANOVA model which allows for both between-design and within-design covariates was proposed by Patel (1986). The model is written

$$E(\mathbf{Y}_{n \times p}) = \mathbf{X}\mathbf{B} + \sum_{j=1}^{r} \mathbf{Z}_j\boldsymbol{\Gamma}_j \tag{8.43}$$

where $\mathbf{Y}_{n \times p}$ is a data matrix on p occasions for n subjects, $\mathbf{X}_{n \times k}$ is a known between-groups design matrix, $\mathbf{B}_{k \times p}$ is a matrix of parameters, $\mathbf{Z}_j(n \times p)$ is a within-design matrix of covariates of full rank p that change with time, and

$\mathbf{\Gamma}_j(p \times p)$ are diagonal matrices with diagonal elements γ_{jk} for $k = 1, 2, \ldots, p$ and $j = 1, 2, \ldots, r$. Letting $\mathbf{XB} = \mathbf{X}_1\mathbf{B}_1 + \mathbf{X}_2\mathbf{B}_2$, we see that we have both between-design and within-design covariates in that $\mathbf{X}_1$ is the between-MANOVA design matrix with covariates $\mathbf{X}_2$. To estimate the parameters in (8.43) is complicated using ML methods. Verbyla (1988) relates (8.43) to a SUR model. To establish the relationship, Verbyla lets

$$\mathbf{B} = (\beta_1,\ \beta_2, \ldots, \beta_p) \text{ and } \mathbf{Z}_j = (\mathbf{Z}_{j1},\ \mathbf{Z}_{j2}, \ldots, \mathbf{Z}_{jp})$$

Then, (8.43) is written as

$$\begin{aligned} E(\mathbf{Y}) &= \mathbf{X}(\beta_1,\ \beta_2, \ldots, \beta_p) + \sum_{j=1}^{r} (\mathbf{Z}_{j1}\gamma_{j2}, \ldots, \mathbf{Z}_{jp}\gamma_{jp}) \\ &= (\mathbf{X}\beta_1 + \mathbf{Z}_{.1}\gamma_{.1}, \ldots, \mathbf{X}\beta_p + \mathbf{Z}_{.p}\gamma_{.p}) \\ &= (\mathbf{X}_1\theta_{11}, \mathbf{X}_2\theta_{22}, \ldots, \mathbf{X}_p\theta_{pp}) \\ &= (\mathbf{X}_1, \mathbf{X}_2, \ldots, \mathbf{X}_p)\tilde{\mathbf{\Theta}} \end{aligned} \tag{8.44}$$

where

$$\tilde{\mathbf{\Theta}} = \begin{pmatrix} \theta_{11} & \cdots & \cdots & \cdots & \cdots & \mathbf{0} \\ \vdots & \theta_{22} & \cdots & \cdots & \cdots & \vdots \\ \vdots & \cdots & \ddots & \cdots & \cdots & \vdots \\ \vdots & \cdots & \cdots & \ddots & \cdots & \vdots \\ \vdots & \cdots & \cdots & \cdots & \ddots & \vdots \\ \mathbf{0} & \cdots & \cdots & \cdots & \cdots & \theta_{pp} \end{pmatrix}$$

$$\begin{aligned} \mathbf{Z}_{.k} &= (\mathbf{Z}_{1k}, \mathbf{Z}_{2k}, \ldots, \mathbf{Z}_{rk}) \\ \gamma'_{.k} &= (\gamma_{1k}, \gamma_{2k}, \ldots, \gamma_{rk}) \\ \mathbf{X}_k &= (\mathbf{X}, \mathbf{Z}_{.k}) \text{ and } \theta'_{kk} = (\beta'_k, \gamma'_{.k}) \end{aligned}$$

which is a SUR model.

Verbyla and Venables (1988) extended the MANOVA-GMANOVA model further to a sum of GMANOVA profiles given by

$$E(\mathbf{Y}_{n\times p}) = \sum_{i=1}^{r} \mathbf{X}_i\mathbf{B}_i\mathbf{M}_i \tag{8.45}$$

where $\mathbf{X}_i(n \times k_i)$ and $\mathbf{M}_i(q_i \times p)$ are known matrices, and $\mathbf{B}_i(k_i \times q_i)$ is a matrix of unknown parameters. Reducing (8.45) to a canonical form, Verbyla and Venables show that (8.45) is a multivariate version of the SUR model calling their generalization the extended GMANOVA model or sum of profiles model. More recently, von Rosen (1989) has found closed form maximum likelihood estimates for the parameters $\mathbf{B}_i$ in (8.45) under the restrictions (1) $\text{rank}(\mathbf{X}_1) + p \le n$ or (2) $V(\mathbf{X}_r) \subseteq, \ldots, \subseteq V(\mathbf{X}_1)$ where $V(\mathbf{X}_i)$ represents the column space of $\mathbf{X}_i$. Using von Rosen's expressions for the maximum likelihood estimate $\hat{\mathbf{B}}_i$ and the unique maximum likelihood estimate of $\mathbf{\Sigma}$, one may obtain likelihood ratio tests of hypotheses of the form $H{:}\, \mathbf{A}_i\mathbf{B}_i\mathbf{C}_i = \mathbf{0}$ $i = 1, 2, \ldots, r$ when the nested condition $V(\mathbf{A}'_r) \subseteq V(\mathbf{A}'_{r-1}) \subseteq, \ldots, \subseteq V(\mathbf{A}'_1)$ is satisfied. However, the distribution of the likelihood statistic was not studied.

A further generalization of (8.45) was considered by Hecker (1987) as was briefly discussed in Chapter 6. To see this generalization, we roll out the matrices $\mathbf{M}_i$, $\mathbf{X}_i$ and $\mathbf{B}_i$ as vectors, columnwise. Then (8.45) becomes

$$
\begin{aligned}
E[\text{vec}(\mathbf{Y}')] &= \left[\sum_{i=1}^{r} (\mathbf{X}_i \otimes \mathbf{M}_i')\right]\text{vec}(\mathbf{B}_i') \\
&= [(\mathbf{X}_1 \otimes \mathbf{M}_1')(\mathbf{X}_2 \otimes \mathbf{M}_2'), \ldots, (\mathbf{X}_r \otimes \mathbf{M}_r')] \begin{pmatrix} \text{vec}\mathbf{B}_1' \\ \text{vec}\mathbf{B}_2' \\ \vdots \\ \vdots \\ \vdots \\ \text{vec}\mathbf{B}_r' \end{pmatrix}
\end{aligned} \tag{8.46}
$$

so that Hecker's model generalizes (8.45). Hecker calls his model the completely general MANOVA model (CGMANOVA).

Maximum likelihood estimators of model parameters for general sum of profile models of the form proposed by Verbyla and Venables (1988) involve solving likelihood equations of the form

$$
\begin{aligned}
n\mathbf{\Sigma} &= (\mathbf{Y} - \mathbf{XBM})'(\mathbf{Y} - \mathbf{XBM}) \\
\mathbf{M\Sigma}^{-1} & (\mathbf{Y} - \mathbf{XBM})\mathbf{X}' = \mathbf{0}
\end{aligned} \tag{8.47}
$$

These were solved by Chinchilli and Elswick (1985) for the MANOVA-GMANOVA model. von Rosen (1989) showed that closed form maximum likelihood solutions to the likelihood equations for models of the form

$$
E(\mathbf{Y}) = \sum_{i=1}^{r} \mathbf{X}_i\mathbf{B}_i\mathbf{M}_i + \mathbf{X}_{r+1}\mathbf{B}_{r+1} \tag{8.48}
$$

have closed form if and only if the $\mathbf{X}_i$, $i = 1, 2, \ldots, r$ are nested. An alternative approach to the estimation problem is to utilize the SUR model, which only involves regression calculations.

8.5 The SUR Model

Before using the SUR model for general sum of profile models, we first more completely discuss the SUR model which was introduced in Chapter 7. For convenience, we repeat the model here. The SUR or MDM model is

$$
\begin{aligned}
E(\mathbf{Y}_{n\times p}) &= (\mathbf{X}_1\theta_{11}, \mathbf{X}_2\theta_{22}, \ldots, \mathbf{X}_p\theta_{pp}) \\
\text{cov}(\mathbf{Y}) &= \mathbf{I}_n \otimes \mathbf{\Sigma} \\
E(\mathbf{y}_j) &= \mathbf{X}_j\theta_{jj}, \ \text{cov}(\mathbf{y}_i, \mathbf{y}_j) = \sigma_{ij}\mathbf{I}_n
\end{aligned} \tag{8.49}
$$

Letting $\mathbf{y} = \text{vec}(\mathbf{Y})$ and $\theta' = (\theta_{11}', \theta_{22}', \ldots, \theta_{pp}')$, (8.49) becomes

$$
\begin{aligned}
E(\mathbf{y}) &= \mathbf{D}\theta \\
\text{cov}(\mathbf{y}) &= \mathbf{\Sigma} \otimes \mathbf{I}_n \\
\underset{np\times\sum_{j=1}^{p}k_j}{\mathbf{D}} &= \begin{pmatrix} \mathbf{X}_1 & \mathbf{0} & \cdots & \cdots & \cdots & \mathbf{0} \\ \mathbf{0} & \mathbf{X}_2 & \mathbf{0} & \cdots & \cdots & \mathbf{0} \\ \vdots & \cdots & \ddots & \cdots & \cdots & \vdots \\ \vdots & \cdots & \cdots & \ddots & \cdots & \vdots \\ \vdots & \cdots & \cdots & \cdots & \ddots & \vdots \\ \mathbf{0} & \cdots & \cdots & \cdots & \mathbf{0} & \mathbf{X}_p \end{pmatrix}
\end{aligned} \tag{8.50}
$$

where each $\mathbf{X}_j (n \times k_j)$ has full rank k_j, and $\theta_{jj}(k_j \times 1)$ are fixed parameters to be estimated.

Several BAN estimators for θ have been suggested in the literature. Kmenta and Gilbert (1968) compare the ML estimate with Telser's iterative estimator (TIE), Telser (1964), Zellner's two-stage Aitken estimator (ZEF), also called the FGLS estimator, and Zellner's iterative Aitken estimator (IZEF), Zellner (1962, 1963).

From the vector version of the GLM, the BLUE of θ by (4.6) is

$$\hat{\theta} = [\mathbf{D}'(\mathbf{\Sigma} \otimes \mathbf{I}_n)^{-1}\mathbf{D}]^{-1}\mathbf{D}'(\mathbf{\Sigma} \otimes \mathbf{I}_n)^{-1}\mathbf{y} \tag{8.51}$$

if $\mathbf{\Sigma}$ is known. Zellner proposed replacing $\mathbf{\Sigma}$ by a consistent estimator $\hat{\mathbf{\Sigma}} = (\hat{\sigma}_{ij})$ where

$$\hat{\sigma}_{ij} = \frac{1}{n-m}\mathbf{y}_i'(\mathbf{I}_n - \mathbf{X}_i(\mathbf{X}_i'\mathbf{X}_i)^{-1}\mathbf{X}_i')(\mathbf{I}_n - \mathbf{X}_j(\mathbf{X}_j'\mathbf{X}_j)^{-1}\mathbf{X}_j')\mathbf{y}_j \tag{8.52}$$

where $m =$ rank$(\mathbf{X}_i)$ assuming that the number of variables in each regression is the same. Relaxing this assumption we may let $m = 0$ or let

$$m_{ij} = \text{Tr}[(\mathbf{I}_n - \mathbf{X}_i(\mathbf{X}_i'\mathbf{X}_i)^{-1}\mathbf{X}_i')(\mathbf{I}_n - \mathbf{X}_j(\mathbf{X}_j'\mathbf{X}_j)^{-1}\mathbf{X}_j')] \tag{8.53}$$

Substituting $\hat{\mathbf{\Sigma}}$ for $\mathbf{\Sigma}$ in (8.51), the ZEF or FGLS estimator of θ is $\hat{\hat{\theta}}$ defined as

$$\hat{\hat{\theta}} = \begin{pmatrix} \hat{\hat{\theta}}_{11} \\ \hat{\hat{\theta}}_{22} \\ \vdots \\ \vdots \\ \vdots \\ \hat{\hat{\theta}}_{pp} \end{pmatrix} = \begin{pmatrix} \hat{\sigma}^{11}\mathbf{X}_1'\mathbf{X}_1 & \hat{\sigma}^{12}\mathbf{X}_1'\mathbf{X}_2 & \cdots & \cdots & \cdots & \hat{\sigma}^{1p}\mathbf{X}_1'\mathbf{X}_p \\ \hat{\sigma}^{12}\mathbf{X}_2'\mathbf{X}_1 & \hat{\sigma}^{22}\mathbf{X}_2'\mathbf{X}_2 & \cdots & \cdots & \cdots & \hat{\sigma}^{2p}\mathbf{X}_2'\mathbf{X}_p \\ \vdots & \vdots & \cdots & \cdots & \cdots & \vdots \\ \vdots & \vdots & \cdots & \cdots & \cdots & \vdots \\ \vdots & \vdots & \cdots & \cdots & \cdots & \vdots \\ \hat{\sigma}^{1p}\mathbf{X}_p'\mathbf{X}_1 & \hat{\sigma}^{2p}\mathbf{X}_p'\mathbf{X}_2 & \cdots & \cdots & \cdots & \hat{\sigma}^{pp}\mathbf{X}_p'\mathbf{X}_p \end{pmatrix}^{-1} \begin{pmatrix} \mathbf{X}_1' \sum_{j=1}^{p} \hat{\sigma}^{1j}\mathbf{y}_j \\ \mathbf{X}_2' \sum_{j=1}^{p} \hat{\sigma}^{2j}\mathbf{y}_j \\ \vdots \\ \vdots \\ \vdots \\ \mathbf{X}_p' \sum_{j=1}^{p} \hat{\sigma}^{pj}\mathbf{y}_j \end{pmatrix} \tag{8.54}$$

where $\hat{\mathbf{\Sigma}}^{-1} = (\hat{\sigma}^{ij})$. The asymptotic covariance matrix of $\hat{\hat{\theta}}$ is

$$(\mathbf{D}'(\mathbf{\Sigma}^{-1} \otimes \mathbf{I}_n)\mathbf{D})^{-1} = \begin{pmatrix} \sigma^{11}\mathbf{X}_1'\mathbf{X}_1 & \cdots & \cdots & \cdots & \cdots & \sigma^{1p}\mathbf{X}_1'\mathbf{X}_p \\ \vdots & \ddots & \cdots & \cdots & \cdots & \vdots \\ \vdots & \cdots & \ddots & \cdots & \cdots & \vdots \\ \vdots & \cdots & \cdots & \ddots & \cdots & \vdots \\ \vdots & \cdots & \cdots & \cdots & \ddots & \vdots \\ \sigma^{1p}\mathbf{X}_p'\mathbf{X}_1 & \cdots & \cdots & \cdots & \cdots & \sigma^{pp}\mathbf{X}_p'\mathbf{X}_p \end{pmatrix}^{-1} \tag{8.55}$$

To estimate θ we completed the step of estimating $\mathbf{\Sigma}$ with $\hat{\mathbf{\Sigma}}$ and then estimated $\hat{\hat{\theta}}$, a two-stage process. The ZEF estimator may also be used to calculate an estimate of $\mathbf{\Omega} = \mathbf{\Sigma} \otimes \mathbf{I}_n$:

$$\hat{\mathbf{\Omega}}_2 = (\mathbf{y} - \mathbf{D}\hat{\hat{\theta}}_1)(\mathbf{y} - \mathbf{D}\hat{\hat{\theta}}_1)' \tag{8.56}$$

where $\hat{\hat{\theta}}_1$ is the ZEF estimator in (8.54). Then, a new estimate of $\hat{\hat{\theta}}$ is provided

$$\hat{\hat{\theta}}_2 = (\mathbf{D}'\hat{\mathbf{\Omega}}_2^{-1}\mathbf{D})^{-1}\mathbf{D}'\hat{\mathbf{\Omega}}_2^{-1}\mathbf{y} \tag{8.57}$$

Continuing in this manner, the ith iteration is

$$\begin{aligned}\hat{\Omega}_i &= (\mathbf{y} - \mathbf{D}\hat{\hat{\theta}}_{i-1})(\mathbf{y} - \mathbf{D}\hat{\hat{\theta}}_{i-1})' \\ \hat{\hat{\theta}}_i &= (\mathbf{D}'\hat{\Omega}_i^{-1}\mathbf{D})^{-1}\mathbf{D}'\hat{\Omega}_i^{-1}\mathbf{y}\end{aligned} \tag{8.58}$$

The process continues until $\hat{\hat{\theta}}_i$ converges in that $\|\hat{\hat{\theta}}_i - \hat{\hat{\theta}}_{i-1}\|^2 < \varepsilon$. This process results in the IZEF estimator of θ, also called the iterative feasible generalized least squares (IFGLS) estimate.

Telser (1964) proposed an alternative method for estimating θ_{jj} in (8.49). He introduced into the system a new set of errors $\mathbf{v}_j$, disturbances of all the other equations in the system. Letting $\mathbf{y}_j = \mathbf{X}_j\theta_{jj} + \mathbf{e}_j$ for $j = 1, 2, \ldots, p$, Telser defines

$$\begin{array}{lllllll} \mathbf{e}_1 = & a_{12}\mathbf{e}_2 + a_{13}\mathbf{e}_3 & + \ldots + & a_{1p}\mathbf{e}_p & + & \mathbf{v}_1 \\ \mathbf{e}_2 = & a_{21}\mathbf{e}_1 + a_{23}\mathbf{e}_3 & + \ldots + & a_{2p}\mathbf{e}_p & + & \mathbf{v}_2 \\ \vdots & \vdots \quad \vdots & \cdots & \vdots & + & \vdots \\ \mathbf{e}_p = & a_{p1}\mathbf{e}_1 + a_{p2}\mathbf{e}_2 & + \ldots + & a_{p-1,p-1}\mathbf{e}_{p-1} & + & \mathbf{v}_p \end{array} \tag{8.59}$$

where the a_{ij} are scalars and $\mathbf{v}_j$ are random errors such that $\text{cov}(\mathbf{e}_i, \mathbf{v}_j) = \mathbf{0}$ and Telser shows that $a_{jp} = -\sigma^{jp} / \sigma^{jj}$. Telser substitutes $\mathbf{e}_j$ into the equation $\mathbf{y}_j = \mathbf{X}_j\mathbf{e}_{jj} + \mathbf{e}_j$ for $j = 1, 2, \ldots, p$. The iterative process yields estimates of a_{ij} and θ_{jj}, and continues until convergence in the θ_{jj}. Telser showed that his estimator has the same asymptotic properties as the IZEF estimator.

To obtain ML estimators of θ and $\mathbf{\Sigma}$, Kmenta and Gilbert (1968) employed an iterative procedure to solve the likelihood equations using the likelihood

$$L(\theta, \mathbf{\Sigma} | \mathbf{y}) = (2\pi)^{-np/2} |\mathbf{\Sigma}|^{-n/2} \text{ etr}[-\frac{1}{2}(\mathbf{\Sigma}^{-1} \otimes \mathbf{I})(\mathbf{y} - \mathbf{D}\theta)(\mathbf{y} - \mathbf{D}\theta)'] \tag{8.60}$$

In their study, they compared these estimators and found that they appeared to give similar results, with the ZEF estimator favored in small samples. Because the ZEF estimator is always unbiased and requires the least calculation, they recommended the ZEF estimator for the SUR model. Park (1993) showed that the ML and IZEF estimators are mathematically equivalent.

To test hypotheses of the form $H(\mathbf{C}_*\theta = \xi)$ given the SUR model in (8.49), we utilize Wald's statistic defined in (4.15). We use $\hat{\mathbf{\Sigma}} = (\hat{\sigma}_{ij})$ defined in (8.52), $\hat{\hat{\theta}}$ given in (8.54) and the design matrix $\mathbf{D}$ specified in (8.49). Then

$$W = (\mathbf{C}_*\hat{\hat{\theta}} - \xi)'[\mathbf{C}_*(\mathbf{D}'(\hat{\mathbf{\Sigma}} \otimes \mathbf{I}_n)^{-1}\mathbf{D})^{-1}\mathbf{C}_*']^{-1}(\mathbf{C}_*\hat{\hat{\theta}} - \xi) \tag{8.61}$$

Under the null hypothesis, the statistic W has an asymptotic chi-square distribution with degrees of freedom $v_h = \text{rank}(\mathbf{C}_*)$. Alternatively, we may write W as T_o^2 so that

$$T_o^2 = \text{Tr}\{[\mathbf{C}_*(\mathbf{D}'(\hat{\mathbf{\Sigma}} \otimes \mathbf{I})^{-1}\mathbf{D})^{-1}\mathbf{C}_*']^{-1}(\mathbf{C}_*\hat{\hat{\theta}} - \xi)(\mathbf{C}_*\hat{\hat{\theta}} - \xi)'\} \tag{8.62}$$

For small sample sizes it is more conservative to use W / v_h which is approximately distributed as an F-distribution with $v_h = \text{rank}(\mathbf{C}_*)$ and $v_e = n - \sum_j k_j / p$ where $v_e = n - m$ if the rank $(\mathbf{X}_i) = m$ for each of the p equations (Theil, 1971, pp. 402-403).

Comparing (8.49) with (8.44), we may use the SUR model to estimate the parameters in the model proposed by Patel (1986) for the analysis of repeated measures designs with changing covariates. The estimators using the SUR model are always unbiased, generally have smaller variances than the ML estimator, and are easy to compute since they involve only regression calculations (Kmenta and Gilbert, 1968). Patel (1986) derived LR tests of

$$\begin{aligned} &H_{01}: \mathbf{CBA} = \mathbf{0} \\ &H_{02}: \mathbf{\Gamma}_j = \mathbf{0} \qquad \text{for } j > r_2 < r \end{aligned} \tag{8.63}$$

and other tests for two-period and three-period crossover designs. For higher order designs he constructed a Wald statistic using the asymptotic covariance matrix of the ML estimators. While both approaches are asymptotically equivalent, it is unknown which performs best in small samples.

The GMANOVA model and the MANOVA-GMANOVA models may also be represented as SUR models (Stanek and Koch, 1985; Verbyla, 1988). To establish the equivalence, recall that the GMANOVA model is

$$\mathbf{Y}_{n\times p} = \mathbf{X}_{n\times k}\mathbf{B}_{k\times q}\mathbf{P}_{q\times p} + \mathbf{U}_{n\times p} \tag{8.64}$$

Next, consider the SUR model

$$\begin{aligned} \underset{n\times 1}{\mathbf{y}_j} &= \underset{n\times k}{\mathbf{X}}\ \underset{k\times 1}{\beta_j} + \underset{n\times 1}{\mathbf{u}_j} \qquad &&\text{for } j = 1,\ldots,q \\ \underset{n\times 1}{\mathbf{y}_j} &= \underset{n\times 1}{\mathbf{u}_j} &&\text{for } j = q+1,\ldots,p \end{aligned} \tag{8.65}$$

where the $\text{rank}(\mathbf{X}) = k$, $E(\mathbf{u}_j) = \mathbf{0}$, $\text{cov}(\mathbf{u}_j, \mathbf{u}_{j'}) = \sigma_{jj'}\mathbf{I}_n$. Expressing (8.65) as a univariate model by stacking the $\mathbf{y}_j$ vectors, we have that

$$\mathbf{y}_{np\times 1} = (\mathbf{P}'\otimes\mathbf{X})_{np\times qk}\,\beta_{qk\times 1} + \mathbf{u}_{np\times 1} \tag{8.66}$$

where

$$\mathbf{P}_{q\times p} = (\mathbf{I}_q, \mathbf{0}_{q\times(p\times q)}) \tag{8.67}$$

and the $\text{cov}(\mathbf{y}) = \mathbf{\Sigma}\otimes\mathbf{I}_n$. Applying (4.6) to (8.66),

$$\begin{aligned} \hat{\beta} &= [(\mathbf{P}'\otimes\mathbf{X})'(\mathbf{\Sigma}\otimes\mathbf{I}_n)^{-1}(\mathbf{P}'\otimes\mathbf{X})]^{-1}(\mathbf{P}'\otimes\mathbf{X})'(\mathbf{\Sigma}\otimes\mathbf{I})^{-1}\mathbf{y} \\ &= [(\mathbf{P\Sigma}^{-1}\mathbf{P}')^{-1}\otimes(\mathbf{X}'\mathbf{X})^{-1}](\mathbf{P\Sigma}^{-1}\otimes\mathbf{X}')\mathbf{y} \\ &= [(\mathbf{P\Sigma}^{-1}\mathbf{P}')^{-1}\mathbf{P\Sigma}^{-1}\otimes(\mathbf{X}'\mathbf{X})^{-1}\mathbf{X}']\mathbf{y} \end{aligned} \tag{8.68}$$

Substituting $\mathbf{S}$ for $\mathbf{\Sigma}$ in (8.68) with $\mathbf{S} = \mathbf{Y}'(\mathbf{I} - \mathbf{X}(\mathbf{X}'\mathbf{X})^{-1}\mathbf{X}')\mathbf{Y}/(n-k)$, the two-stage ZEF estimator is equal to the ML estimate for $\mathbf{B}$ in (8.68) since

$$\hat{\beta} = \text{vec}(\hat{\mathbf{B}}) = \text{vec}[(\mathbf{X}'\mathbf{X})^{-1}\mathbf{X}'\mathbf{Y}\mathbf{S}^{-1}\mathbf{P}'(\mathbf{P}\mathbf{S}^{-1}\mathbf{P}')^{-1}] \tag{8.69}$$

for the GMANOVA model (8.64) as shown in (7.42).

To establish the equivalence of the GMANOVA and SUR models for arbitrary $\mathbf{P}$, let

$$\mathbf{H} = (\mathbf{H}_1, \mathbf{H}_2) = (\mathbf{S}^{-1}\mathbf{P}'(\mathbf{P}\mathbf{S}^{-1}\mathbf{P}')^{-1}, \mathbf{I}_p - \mathbf{H}_1\mathbf{P}) \tag{8.70}$$

and $\mathbf{Y}_o = \mathbf{YH}$, $\mathbf{X}_2 = \mathbf{PH}$ and $\mathbf{U}_o = \mathbf{UH}$ in (8.64):

$$\mathbf{Y}_{o_{n\times p}} = \mathbf{X}_{n\times k}\mathbf{B}_{k\times p}\mathbf{X}_{2_{q\times p}} + \mathbf{U}_{o_{n\times p}} \tag{8.71}$$

with **P** defined in (8.67). Then, the ML estimate of **B** is

$$\begin{aligned}\hat{\mathbf{B}} &= (\mathbf{X}'\mathbf{X})^{-1}\mathbf{X}'\mathbf{Y}_o(\mathbf{H}'\mathbf{S}\mathbf{H})^{-1}\mathbf{X}_2'(\mathbf{X}_2(\mathbf{H}'\mathbf{S}\mathbf{H})^{-1}\mathbf{X}_2')^{-1} \\ &= (\mathbf{X}'\mathbf{X})^{-1}\mathbf{X}'\mathbf{Y}\mathbf{S}^{-1}\mathbf{P}'(\mathbf{P}\mathbf{S}^{-1}\mathbf{P}')^{-1}\end{aligned} \tag{8.72}$$

Hence, model (8.71) is equivalent to the SUR model given in (8.66) using the transformed data matrices $\mathbf{Y}_o = \mathbf{YH}$ and $\mathbf{X}_2 = \mathbf{PH}$.

The MANOVA-GMANOVA model, or more generally the sum of profiles model, may be represented as a multivariate SUR (MSUR) model (Verbyla and Venables,1988). With the MSUR model, as with the SUR model, we may estimate parameters using regression equations rather than solving likelihood equations. The MSUR model has the following general form

$$\begin{aligned}E(\mathbf{Y}_1, \mathbf{Y}_2, \ldots, \mathbf{Y}_q) &= (\mathbf{X}_1\mathbf{\Theta}_{11}, \mathbf{X}_2\mathbf{\Theta}_{22}, \ldots, \mathbf{X}_q\mathbf{\Theta}_{qq}) \\ \mathbf{Y}_{i_{n\times p_i}} &= \mathbf{X}_{i_{n\times k_i}}\mathbf{\Theta}_{i_{k_i\times p_i}} + \mathbf{U}_{i_{n\times p_i}} \qquad i = 1, 2, \ldots, q\end{aligned} \tag{8.73}$$

where the $\text{cov}(\mathbf{Y}_i, \mathbf{Y}_j) = \mathbf{\Sigma}_{ij}$ is of order $(p_i \times p_j)$, $\mathbf{\Sigma} = (\mathbf{\Sigma}_{ij})$ of order $(p \times p)$ and $p = \sum_{i=1}^{q} p_i$. If $p_1 = p_2 = \ldots = p_q = p_o$, then (8.73) becomes

$$\begin{pmatrix}\mathbf{Y}_1 \\ \mathbf{Y}_2 \\ \vdots \\ \vdots \\ \vdots \\ \mathbf{Y}_q\end{pmatrix} = \begin{pmatrix}\mathbf{X}_1 & \mathbf{0} & \cdots & \cdots & \mathbf{0} & \mathbf{0} \\ \mathbf{0} & \mathbf{X}_2 & \mathbf{0} & \cdots & \cdots & \mathbf{0} \\ \vdots & \cdots & \ddots & \cdots & \cdots & \vdots \\ \vdots & \cdots & \cdots & \ddots & \cdots & \vdots \\ \vdots & \cdots & \cdots & \cdots & \ddots & \vdots \\ \mathbf{0} & \mathbf{0} & \cdots & \cdots & \mathbf{0} & \mathbf{X}_q\end{pmatrix}\begin{pmatrix}\mathbf{\Theta}_{11} \\ \mathbf{\Theta}_{22} \\ \vdots \\ \vdots \\ \vdots \\ \mathbf{\Theta}_{qq}\end{pmatrix} + \begin{pmatrix}\mathbf{U}_1 \\ \mathbf{U}_2 \\ \vdots \\ \vdots \\ \vdots \\ \mathbf{U}_q\end{pmatrix}$$

a matrix generalization of (7.16).

Representing (8.73) as a GLM, let

$$\begin{aligned}&\mathbf{U} = (\mathbf{E}_1, \mathbf{E}_2, \ldots, \mathbf{E}_q) && \mathbf{U}_i = (\mathbf{e}_{i1}, \mathbf{e}_{i2}, \ldots, \mathbf{e}_{ip_i}) \\ &\mathbf{Y} = (\mathbf{Y}_1, \mathbf{Y}_2, \ldots, \mathbf{Y}_q) && \mathbf{Y}_i = (\mathbf{y}_{i1}, \mathbf{y}_{i2}, \ldots, \mathbf{y}_{ip_i}) \\ &\mathbf{\Theta} = (\mathbf{\Theta}_{11}, \mathbf{\Theta}_{22}, \ldots, \mathbf{\Theta}_q) && \mathbf{\Theta}_{jj} = (\mathbf{\Theta}_{jj1}, \mathbf{\Theta}_{jj2}, \ldots, \mathbf{\Theta}_{jjp_j})\end{aligned} \tag{8.74}$$

where

$$\begin{aligned}\mathbf{e} &= \text{vec}(\mathbf{U}) = [\text{vec}(\mathbf{U}_1)\ , \ldots, \text{vec}(\mathbf{U}_q)] \\ \mathbf{y} &= \text{vec}(\mathbf{Y}) = [\text{vec}(\mathbf{Y}_1)\ , \ldots, \text{vec}(\mathbf{Y}_q)] \\ \mathbf{\Theta} &= \text{vec}(\mathbf{\Theta}) = [\text{vec}(\mathbf{\Theta}_{11}), \ldots, \text{vec}(\mathbf{\Theta}_{qq})]\end{aligned} \tag{8.75}$$

so that $\mathbf{y}_{ij}$ is the jth column of $\mathbf{Y}_i$ and θ_{jjt} is the tth column of $\mathbf{\Theta}_{jj}$, then (8.73) becomes

$$\begin{aligned}\mathbf{y} &= \mathbf{D}\theta + \mathbf{e} \\ \mathbf{D} &= \text{diag}(\mathbf{I}_{p_1} \otimes \mathbf{X}_1\ , \ldots, \mathbf{I}_{p_q} \otimes \mathbf{X}_q)\end{aligned} \tag{8.76}$$

and diag $[(\mathbf{A}_1, \mathbf{A}_2\ , \ldots, \mathbf{A}_q)]$ is a block diagonal matrix with $\mathbf{A}_i = \mathbf{I}_{p_i} \otimes \mathbf{X}_i$. Applying (4.6) to the GLM, we find that the BLUE of θ is

$$\hat{\theta} = (\mathbf{D}'(\mathbf{\Sigma} \otimes \mathbf{I}_n)^{-1}\mathbf{D})^{-1}\mathbf{D}'(\mathbf{\Sigma} \otimes \mathbf{I}_n)^{-1}\mathbf{y} \tag{8.77}$$

so that the multivariate FGLS estimator for θ is

$$\hat{\hat{\theta}} = (\mathbf{D}'(\hat{\mathbf{\Sigma}}^{-1} \otimes \mathbf{I}_n)\mathbf{D})^{-1}\mathbf{D}'(\hat{\mathbf{\Sigma}}^{-1} \otimes \mathbf{I}_n)\mathbf{y} \tag{8.78}$$

where $\hat{\mathbf{\Sigma}}$ is any consistent estimator of $\mathbf{\Sigma}$. Following (8.52), we may set

$$\hat{\Sigma}_{ij} = \frac{1}{n-m_{ij}}[\mathbf{Y}_i'(\mathbf{I}_n - \mathbf{X}_i(\mathbf{X}_i'\mathbf{X}_i)^{-1}\mathbf{X}_i')(\mathbf{I}_n - \mathbf{X}_j(\mathbf{X}_j'\mathbf{X}_j)^{-1}\mathbf{X}_j')\mathbf{Y}_j] \tag{8.79}$$

where $m_{ij} = 0$ or

$$m_{ij} = \text{Tr}[(\mathbf{I}_n - \mathbf{X}_i(\mathbf{X}_i'\mathbf{X}_i)^{-1}\mathbf{X}_i')(\mathbf{I}_n - \mathbf{X}_j(\mathbf{X}_j'\mathbf{X}_j)^{-1}\mathbf{X}_j')] \tag{8.80}$$

Since (8.76) is a GLM, the Wald statistic

$$X^2 = (\mathbf{C}_*\hat{\hat{\theta}} - \xi)'\{\mathbf{C}_*[\mathbf{D}'(\hat{\Sigma} \otimes \mathbf{I}_n)^{-1}\mathbf{D}]^{-1}\mathbf{C}_*\}^{-1}(\mathbf{C}_*\hat{\hat{\theta}} - \xi)' \tag{8.81}$$

may be used to test hypotheses $H\colon \mathbf{C}_*\theta = \xi$ where the $\text{rank}(\mathbf{C}_*) = v_h$ with $\mathbf{D}$ is defined in (8.76).

That the parameters in (8.45) may be estimated using the MSUR model was shown by Verbyla and Venables (1988) employing a canonical reduction of (8.45). Using a conditional argument to partition the data matrix, they relate the MANOVA-GMANOVA model to an MSUR analysis as suggested by Stanek and Koch (1985). For the sum of profiles model,

$$E(\mathbf{Y}_{n \times p}) = \sum_{i=1}^{r} \mathbf{X}_{i_{n \times k_i}} \mathbf{B}_{i_{k_i \times q_i}} \mathbf{M}_{i_{q_i \times p}} \tag{8.82}$$

where the $\text{rank}(\mathbf{X}_i) = \text{k}_i$, $\text{rank}(\mathbf{M}_i) = \text{q}_i$ where $q_i \leq p$, so that $\mathbf{X}_i$ and $\mathbf{M}_i'$ have full-column rank. Letting $r = 2$, we may consider the vector space $\mathbf{M}_1' + \mathbf{M}_2'$. An orthogonal decomposition of $\mathbf{M}_1' + \mathbf{M}_2'$ where $\mathbf{A}|\mathbf{B}$ represents the orthocomplement of $\mathbf{A}$ relative to $\mathbf{B}$ is

$$\begin{aligned} \mathbf{M}_1' + \mathbf{M}_2' &= \mathbf{M}_1' \oplus (\mathbf{M}_1' + \mathbf{M}_2')|\mathbf{M}_1' \\ \mathbf{M}_1' + \mathbf{M}_2' &= \mathbf{M}_2' \oplus (\mathbf{M}_1' + \mathbf{M}_2')|\mathbf{M}_2' \end{aligned}$$

which is not unique since $\mathbf{M}_1' \cap \mathbf{M}_2' \neq \{\mathbf{0}\}$. Using this fact, we can obtain a basis for $\mathbf{M}_1' \cap \mathbf{M}_2' = \mathbf{N}_{12}$. Similarly, we can construct a basis of $\mathbf{M}_1'$ and $\mathbf{M}_2'$, and $\mathbf{N}_1$ and $\mathbf{N}_2$, respectively. Hence, there exist nonsingular matrices $\mathbf{A}_1'$ and $\mathbf{A}_2'$ such that $\mathbf{M}_1' = (\mathbf{N}_1', \mathbf{N}_{12}')\mathbf{A}_1'$ and $\mathbf{M}_2' = (\mathbf{N}_{12}', \mathbf{N}_1')\mathbf{A}_2'$. The jth column of $\mathbf{M}_i'$ is a linear combination of the rows of $\mathbf{A}_i'$ by construction. Then (8.82) becomes

$$\begin{aligned} E(\mathbf{Y}) = \mathbf{X}_1\mathbf{B}_1\mathbf{M}_1 + \mathbf{X}_2\mathbf{B}_2\mathbf{M}_2 &= \mathbf{X}_1\mathbf{B}_1\mathbf{A}_1\begin{pmatrix} \mathbf{N}_1 \\ \mathbf{N}_{12} \end{pmatrix} + \mathbf{X}_2\mathbf{B}_2\mathbf{A}_2\begin{pmatrix} \mathbf{N}_{12} \\ \mathbf{N}_2 \end{pmatrix} \\ &= (\mathbf{X}_1\mathbf{X}_2)\begin{bmatrix} \mathbf{B}_{11}^* & \mathbf{B}_{12}^* & \mathbf{0} \\ \mathbf{0} & \mathbf{B}_{21}^* & \mathbf{B}_{22}^* \end{bmatrix}\begin{pmatrix} \mathbf{N}_1 \\ \mathbf{N}_{12} \\ \mathbf{N}_2 \end{pmatrix} \\ &= \mathbf{X}\mathbf{B}^*\mathbf{Q} \end{aligned} \tag{8.83}$$

where $\begin{pmatrix} \mathbf{B}_{11}^* \\ \mathbf{B}_{12}^* \end{pmatrix} = \mathbf{B}_1\mathbf{A}_1$ and $\begin{pmatrix} \mathbf{B}_{21}^* \\ \mathbf{B}_{22}^* \end{pmatrix} = \mathbf{B}_2\mathbf{A}_2$. The design matrix $\mathbf{M}' = \mathbf{M}_1' + \mathbf{M}_2'$ is of full rank since $\mathbf{N}_1'$, $\mathbf{N}_{12}'$, and $\mathbf{N}_{12}'$ form a basis for the sum. If $\mathbf{Q}$ is less than full rank, $q < p$, by the GMANOVA model, there exists a matrix $\mathbf{H} = (\mathbf{H}_1, \mathbf{H}_2)$ such that

$$\mathbf{Y}_o = (\mathbf{Y}_1, \mathbf{Y}_2) = [\mathbf{YH}_1, \mathbf{Y}(\mathbf{I} - \mathbf{H}_1\mathbf{Q})] = (\mathbf{YH}_1, \mathbf{YH}_2) \tag{8.84}$$

so that (8.83) becomes

$$\begin{aligned} E(\mathbf{Y}_o) = \mathbf{E}(\mathbf{Y}_1, \mathbf{Y}_2) &= \left[\mathbf{X}_1\mathbf{B}_{11}^*, (\mathbf{X}_1, \mathbf{X}_2)\begin{pmatrix} \mathbf{B}_{12}^* \\ \mathbf{B}_{21}^* \end{pmatrix}, \mathbf{X}_2\mathbf{B}_{22}^*, \mathbf{0} \right] \\ &= (\mathbf{X}_1^*\Theta_{11}, \mathbf{X}_2^*\Theta_{22}, \mathbf{X}_3^*\Theta_{33}, \mathbf{0}) \end{aligned} \tag{8.85}$$

which has the form of (8.73), an MSUR model. To estimate Θ_{11}, Θ_{22}, and Θ_{33} we would use (8.78) to determine the multivariate FGLS estimate (Park and Woolson, 1992).

8.6 The Restricted GMANOVA Model

We saw in Chapter 7 that the MANOVA model is a special case of the GMANOVA model and that estimation for the GMANOVA model may be accomplished using the SUR model. We also showed that the restricted GLM is a special case of the MANOVA-GMANOVA model which is a special case of the sum of profiles model whose parameters may be estimated using the MSUR model. However, we also noted in Chapter 7 that the restricted multivariate GLM is a special case of the GMANOVA model with double restrictions termed the extended GMANOVA model by Kariya (1985). To develop the restricted GMANOVA model, recall that the hypothesis for the extended GMANOVA model was written as

$$H\colon \mathbf{X}_5\mathbf{B}\mathbf{X}_6 = \mathbf{0} \tag{8.86}$$

where the $\text{rank}(\mathbf{X}_5) = r_5 \leq k$ and the $\text{rank}(\mathbf{X}_6) = r_6 \leq q$. The restricted GMANOVA model is defined

$$\begin{aligned} \mathbf{Y}_{n \times p} &= \mathbf{X}_{1_{n\times k}}\mathbf{B}_{k\times q}\mathbf{X}_{2_{q\times p}} + \mathbf{U}_{n\times p} \\ \mathbf{X}_{3_{r_3\times k}}\mathbf{B}_{k\times q}\mathbf{X}_{4_{q\times r_4}} &= \mathbf{0} \end{aligned} \tag{8.87}$$

where the $\text{rank}(\mathbf{X}_1) = k$, $\text{rank}(\mathbf{X}_2) = q$, and the $\text{rank}(\mathbf{X}_i) = r_i$ $(i = 3, 4, 5, 6)$. When $\mathbf{X}_2 = \mathbf{I}$, (8.87) reduces to (8.1). To test (8.86), we allow $\mathbf{X}_2$ to be arbitrary and $\mathbf{X}_4 \neq \mathbf{X}_6$. We may not allow the matrices $\mathbf{X}_3$, $\mathbf{X}_4$, and $\mathbf{X}_5$ to be entirely arbitrary but have to impose restrictions on the matrices to ensure that the parameters are identifiable and testable. This was also the case for the RMGLM discussed in Chapter 7 with $\mathbf{X}_4 = \mathbf{X}_6$ and $\mathbf{X}_2 = \mathbf{I}$. Recall that we required the matrices $\mathbf{X}_3$ and $\mathbf{X}_5$ to be disjoint (linearly independent or that the intersection of the column space of $\mathbf{X}_3'$ and $\mathbf{X}_5'$ must be null, $V(\mathbf{X}_3') \cap V(\mathbf{X}_5') = \mathbf{0}$). Furthermore, the parameter space $\boldsymbol{\Omega}_o$ was restricted by $\mathbf{X}_3$, the restricted parameter space $\tilde{\boldsymbol{\Omega}}$ is the intersection of the column space of $\mathbf{X}_1$ and the null space of $\mathbf{X}_3(\mathbf{X}_1'\mathbf{X}_1)^{-1}\mathbf{X}_1'$, $[N(\mathbf{X}_3\mathbf{X}_1'\mathbf{X}_1)^{-1}\mathbf{X}_1']$, and the hypothesis test space $\omega = N(\mathbf{X}_5(\mathbf{X}_1'\mathbf{X}_1)^{-1}\mathbf{X}_1') \cap \boldsymbol{\Omega}_o$. Thus, for the RMGLM, $\mathbf{X}_5$ is a subspace of $\mathbf{X}_3$ relative to $(\mathbf{X}_1'\mathbf{X}_1)^{-1}$ so that the column space of $(\mathbf{X}_1'\mathbf{X}_1)^{-1}\mathbf{X}_5'$ is included in the column space of $(\mathbf{X}_1'\mathbf{X}_1)^{-1}\mathbf{X}_3'$.

Testing (8.86) given (8.87) is complicated. The first solution appears to be that of Gleser and Olkin (1970, p. 288) which was developed using the canonical form of the model as an "aside". Kabe (1975, p. 36 eq. 7) claims to have solved the problem using the original variables, but his results do not appear to agree with those of Gleser and Olkin. The test developed by Kabe is not an LR test. Kabe's (1981, p. 2549) result for the restricted MANOVA model also seems to be incorrect. A complete analysis of the model in canonical form is provided by Kariya (1985, Chapter 4). Testing hypotheses for the extended GMANOVA model is complicated since it is really a sum of profiles model. Following (8.2), observe that by incorporating the restrictions into the model that

$$\begin{aligned}
\mathbf{X}_1\mathbf{B}\mathbf{X}_2 &= (\mathbf{X}_{11}^*, \mathbf{X}_{12}^*)\begin{pmatrix} \mathbf{0} & \beta_{12} \\ \beta_{21} & \beta_{22} \end{pmatrix}\begin{pmatrix} \mathbf{Q}^{11} \\ \mathbf{Q}^{21} \end{pmatrix}\begin{pmatrix} \mathbf{X}_{21}^* \\ \mathbf{X}_{22}^* \end{pmatrix} \\
&= \mathbf{X}_{11}^*\beta_{12}\mathbf{Q}^{21}\mathbf{X}_{22}^* + \mathbf{X}_{12}^*(\beta_{12}, \beta_{22})\mathbf{Q}^{-1}\mathbf{X}_2 \\
&= \mathbf{A}_1\mathbf{B}_1\mathbf{C}_1 + \mathbf{A}_2\mathbf{B}_2\mathbf{C}_2
\end{aligned} \tag{8.88}$$

a sum of profiles model which can be expressed as an MSUR model, Timm (1996).

To ensure identifiability and hence testability of (8.86) given (8.87), Kariya (1985) uses the following result from matrix algebra to reduce the restricted GMANOVA model to canonical form.

Theorem 8.1 Given two symmetric matrices **A** and **B**, the matrices **A** and **B** may be simultaneously diagonalized by an orthogonal matrix **P** such that $\mathbf{P}'\mathbf{A}\mathbf{P} = \mathbf{D}_1$ and $\mathbf{P}'\mathbf{A}\mathbf{P} = \mathbf{D}_2$ if and only if **A** and **B** commute, $\mathbf{AB} = \mathbf{BA}$.

Following Kariya (1985, p. 143), we have the following result.

Theorem 8.2 The restricted GMANOVA model may be reduced to canonical form if the matrices $\mathbf{X}_3, \mathbf{X}_4, \mathbf{X}_5$, and $\mathbf{X}_6$ and satisfy the condition that

$$\mathbf{M}_3\mathbf{M}_5 = \mathbf{M}_5\mathbf{M}_3 \text{ and } \mathbf{M}_4\mathbf{M}_6 = \mathbf{M}_6\mathbf{M}_4$$

where

$$\begin{aligned}
\mathbf{M}_i &= (\mathbf{X}_1'\mathbf{X}_1)^{-1/2}\mathbf{X}_i'(\mathbf{X}_i(\mathbf{X}_1'\mathbf{X}_1)^{-1}\mathbf{X}_i)^{-1}\mathbf{X}_i(\mathbf{X}_1'\mathbf{X}_1)^{-1/2} & (i = 3, 5) \\
\mathbf{M}_j &= (\mathbf{X}_2\mathbf{X}_2')^{-1/2}\mathbf{X}_j(\mathbf{X}_j'(\mathbf{X}_2\mathbf{X}_2')^{-1}\mathbf{X}_j')^{-1}\mathbf{X}_j'(\mathbf{X}_2\mathbf{X}_2')^{-1/2} & (j = 4, 6).
\end{aligned}$$

Using Theorem 8.2, a canonical form for the restricted GMANOVA problem, leads one to consider four cases: (1) $\mathbf{M}_3\mathbf{M}_5 = \mathbf{M}_5$ and $\mathbf{M}_4\mathbf{M}_6 = \mathbf{M}_4$, (2) $\mathbf{M}_3\mathbf{M}_5 = \mathbf{M}_5$ and $\mathbf{M}_4\mathbf{M}_6 = \mathbf{0}$, (3) $\mathbf{M}_3\mathbf{M}_5 = \mathbf{M}_3$ and $\mathbf{M}_4\mathbf{M}_6 = \mathbf{M}_6$, and (4) $\mathbf{M}_3\mathbf{M}_5 = \mathbf{0}$ and $\mathbf{M}_4\mathbf{M}_6 = \mathbf{M}_6$. When $\mathbf{X}_2 = \mathbf{I}$, we have as a special case the MANOVA model with double linear restrictions. For this problem,

$$\begin{aligned}
&\underset{q\times q}{\mathbf{M}_4} = \mathbf{X}_4(\mathbf{X}_4'\mathbf{X}_4)^{-1}\mathbf{X}_4' \text{ and } \underset{q\times q}{\mathbf{M}_6} = \mathbf{X}_6(\mathbf{X}_6'\mathbf{X}_6)^{-1}\mathbf{X}_6' \\
&\underset{q\times q}{\mathbf{M}_3} = (\mathbf{X}_1'\mathbf{X}_1)^{-1/2}\mathbf{X}_3'(\mathbf{X}_3'(\mathbf{X}_1'\mathbf{X}_1)^{-1}\mathbf{X}_3')^{-1}\mathbf{X}_3(\mathbf{X}_1'\mathbf{X}_1)^{-1/2} \\
&\underset{k\times k}{\mathbf{M}_5} = (\mathbf{X}_1'\mathbf{X}_1)^{-1/2}\mathbf{X}_5'(\mathbf{X}_5'(\mathbf{X}_1'\mathbf{X}_1)^{-1}\mathbf{X}_5')^{-1}\mathbf{X}_5(\mathbf{X}_1'\mathbf{X}_1)^{-1/2}
\end{aligned}$$

and by Theorem 8.2, we must have $\mathbf{M}_3\mathbf{M}_5 = \mathbf{M}_5\mathbf{M}_3$ and $\mathbf{M}_4\mathbf{M}_6 = \mathbf{M}_6\mathbf{M}_4$. If we further restrict the matrices to the situation discussed in Chapter 7 where $\mathbf{X}_4 = \mathbf{X}_6 = \mathbf{I}$ or $\mathbf{X}_4 = \mathbf{X}_6 = \mathbf{A}$ (say), we observe that $\mathbf{M}_4 = \mathbf{M}_6 = \mathbf{I}$ so that $\mathbf{M}_4\mathbf{M}_6 = \mathbf{M}_6\mathbf{M}_4$. In addition, we require $\mathbf{M}_3\mathbf{M}_5 = \mathbf{M}_5\mathbf{M}_3$. This condition implies that the row space of $\mathbf{X}_5(\mathbf{X}_1'\mathbf{X}_1)^{-1/2}$ is a subspace of the row space of $\mathbf{X}_3(\mathbf{X}_1'\mathbf{X}_1)^{-1/2}$ or that $\mathbf{X}_5$ is a subspace of $\mathbf{X}_3$ relative to $(\mathbf{X}_1'\mathbf{X}_1)^{-1}$ as required in the development of this special case.

While Kariya (1985) has been able to reduce the restricted GMANOVA model to canonical form, using Lemma 7.2 and Theorem 8.2 and develop LR tests, the distribution of the LR criterion is unknown. An alternative analysis is to utilize the SUR model to estimate parameters and use the large sample Wald statistic to test hypotheses. The more general approach is to use the CGMANOVA model (8.46) proposed by Hecker (1987). Since the model is linear, we may estimate parameters of the sum of profiles model and test hypotheses using the Wald statistic. Timm (1996) developed goodness-of-fit tests.

Returning to the SUR model (8.49), recall that the FGLS estimator for θ,

$$\hat{\hat{\theta}} = [\mathbf{D}'(\hat{\boldsymbol{\Sigma}}^{-1} \otimes \mathbf{I}_n)\mathbf{D}]^{-1}\mathbf{D}'(\hat{\boldsymbol{\Sigma}}^{-1} \otimes \mathbf{I}_n)\mathbf{y} \tag{8.89}$$

was obtained with no restrictions on θ. We can also test hypotheses of the form $H{:}\,\mathbf{C}_*\theta = \theta_o$ with the restriction

$$\mathbf{R}\theta = \xi \tag{8.90}$$

added to the SUR model. Again, the rank of $\mathbf{R}$ is s where the rows of $\mathbf{R}$ are linearly independent, the intersection of $\mathbf{R}$ and $\mathbf{C}_*$. Following the arguments for the RGLM, if $\hat{\boldsymbol{\Sigma}}$ is a consistent estimator of $\boldsymbol{\Sigma}$, the restricted FGLS estimate of θ is

$$\hat{\hat{\theta}}_r = \hat{\hat{\theta}} - \hat{\mathbf{W}}^{-1}\mathbf{R}(\mathbf{R}\hat{\mathbf{W}}^{-1}\mathbf{R}')^{-1}(\mathbf{R}\hat{\hat{\theta}} - \xi) \tag{8.91}$$

where

$$\hat{\mathbf{W}}^{-1} = [\mathbf{D}'(\hat{\boldsymbol{\Sigma}}^{-1} \otimes \mathbf{I}_n)\mathbf{D}]^{-1} \tag{8.92}$$

Furthermore, the covariance matrix for $\hat{\hat{\theta}}_r$ is

$$\begin{aligned} \operatorname{cov}(\hat{\hat{\theta}}_r) &= \boldsymbol{\Omega}^{-1}\boldsymbol{\Omega}^{-1}\mathbf{R}'(\mathbf{R}\boldsymbol{\Omega}^{-1}\mathbf{R}')^{-1}\mathbf{R}\boldsymbol{\Omega}^{-1} \\ &= [\mathbf{I} - \boldsymbol{\Omega}^{-1}\mathbf{R}'(\mathbf{R}\boldsymbol{\Omega}^{-1}\mathbf{R}')^{-1}\mathbf{R}]\boldsymbol{\Omega}^{-1} \\ &= \mathbf{F}\boldsymbol{\Omega}^{-1} \end{aligned} \tag{8.93}$$

where $\boldsymbol{\Omega}^{-1} = [\mathbf{D}'(\boldsymbol{\Sigma}^{-1} \otimes \mathbf{I}_n)\mathbf{D}]^{-1}$.

To test the hypotheses $H(\mathbf{C}_*\theta = \theta_o)$, assuming multivariate normality, we again may use the Wald statistic

$$X^2 = (\mathbf{C}_*\hat{\hat{\theta}}_r - \theta_o)'[\mathbf{C}_*(\hat{\mathbf{F}}\hat{\mathbf{W}}^{-1}\hat{\mathbf{F}}')\mathbf{C}_*']^{-1}(\mathbf{C}_*\hat{\hat{\theta}}_r - \theta_o) \tag{8.94}$$

where $\hat{\mathbf{F}} = \mathbf{I} - \hat{\mathbf{W}}^{-1}\mathbf{R}'(\mathbf{R}\hat{\mathbf{W}}^{-1}\mathbf{R}')\mathbf{R}$ and $\hat{\mathbf{W}}^{-1} = [\mathbf{D}'(\hat{\boldsymbol{\Sigma}}^{-1} \otimes \mathbf{I}_n)\mathbf{D}]^{-1}$. The asymptotic null distribution of X^2 is chi-square with $v_h = \operatorname{rank}(\mathbf{C}_*)$ degrees of freedom. If $H(\mathbf{C}_*\theta = \theta_o)$ is rejected, approximate $1-\alpha$ simultaneous intervals for parametric functions $\psi = \mathbf{c}'\theta$ are obtained using the formula:

$$\hat{\psi} - c_o\hat{\sigma}_{\hat{\psi}} \le \psi \le \hat{\psi} + c_o\hat{\sigma}_{\hat{\psi}} \tag{8.95}$$

where

$$\hat{\psi} = \mathbf{c}'\hat{\hat{\theta}}_r \text{ and } \hat{\sigma}^2_{\hat{\psi}} = \mathbf{c}'\hat{\mathbf{F}}\hat{\mathbf{W}}^{-1}\mathbf{c} \tag{8.96}$$

and c_o^2 is the $\chi^2_\alpha(v_h)$ critical value for level α.

Restrictions may also be added to the MSUR model (8.73). Because of the complications involved with double restrictions for the MANOVA and GMANOVA model, it is convenient to utilize the vector form (8.76) of the model and modify the Wald statistic in (8.81) as shown in (8.94) to test hypotheses of the form $H(\mathbf{C}_*\theta = \theta_o)$ where $\hat{\hat{\theta}}$ is defined in (8.78) and the restricted estimator has the form given in (8.91).

8.7 GMANOVA-SUR: One Population

In Section 7.7, we illustrated how to estimate the parameter matrix **B** of the GMANOVA model using the Rao-Khatri reduction which utilized the MANCOVA model, following a transformation of the data matrix **Y**. Since the GMANOVA model may be represented as a SUR model as in (8.66), we have shown in (8.71) that the estimate of **B** may be obtained from the SUR model for general **P**. To illustrate the equivalence between the two models we use the Elston and Grizzle (1962) ramus height data. These data were also examined by Lindsey (1993, p. 90).

The data consist of ramus heights, measured in mm, of a single cohort of boys aged 8, 8.5, 9, and 9.5 years of age. The study was conducted to establish a "normal" growth curve for boys to be used by orthodontists. The data are listed in Output 8.7. The SAS code to perform the analysis of the data utilizing the two models is found in Program 8_7.sas.

Program 8_7.sas

```
/* Program 8_7.sas                              */
/* GMANOVA - SUR (Elston and Grizzle, 1962) */

options ls=80 ps=60 nodate nonumber;
title1 'Output 8.7: GMANOVA-SUR One Population Ramus Height Data';

data ramus;
    infile 'c:\8_7.dat';
    input group y1 y2 y3 y4;
proc print data=ramus;
run;
proc summary data=ramus;
   class group;
   var y1 -y4;
   output out=new mean=mr1-mr4;
proc print data=new;
run;
data plot;
   set new;
   array mr(4) mr1-mr4;
   do time = 1 to 4;
      response = mr(time);
      output;
   end;
   drop mr1-mr4;
run;
```

```
proc plot;
   title2 'Plot of Ramus Heights';
    plot response*time=group;
run;

proc iml;
   use ramus;
   read all var {y1 y2 y3 y4} into y;
   read all var {Group} into x;
   y1=y[1:20,1:1]; y2=y[1:20,2:2]; y3=y[1:20,3:3];y4=y[1:20,4:4];
   vecy=y1//y2//y3//y4;
   n=nrow(x);
   k=ncol(x);
   xpx=x`*x;
   xpxi=inv(xpx);
   s=(y`*y-y`*x*xpxi*x`*y)/(n-k);
   si=inv(s);
   print s;
   /*Input Polynomials of degree  q-1=3 */
   P_prime={1 -3 1 -1, 1 -1 -1 3, 1 1 -1 -3,1 3 1 1};
   p=p_prime[1:4,1:2];
   /* Estimate of B using SUR model with 2sls  */
   beta=((inv(p`*si*p)*p`*si)@(xpxi*x`))*vecy;
   print 'Estimate of B using SUR estimate and PROC IML', beta;
   Y0=Y*P_prime*inv(p_prime`*p_prime);
   print 'P prime matrix',P_prime;
   t=y0||x;
   varnames={yt1 yt2 yt3 yt4 x};
   create trans from t (|colname=varnames|);
   append from t;
quit;
run;

/* Test linear model fit or adequacy  */
proc glm data=trans;
   title2 'Test of Linear Fit';
   model yt3 yt4=/nouni;
   manova h=intercept;
run;

/* Estimate of B using a Rao-Khatri MANCOVA model  */
proc glm data=trans;
   title2 'Estimate of B using PROC GLM';
   class x;
   model yt1 yt2 = x yt3 yt4;
   estimate 'beta' intercept 1 x 1;
run;

/* Estimate of B using syslin procedure */
proc syslin sur data=trans;
   title2 'Estimate of B using PROC SYSLIN';
   model yt1 = x yt3 yt4/noint;
   model yt2 = x yt3 yt4/noint;
run;
```

Result and Interpretation 8_7.sas

Output 8.7: *GMANOVA-SUR One Population Ramus Height Data*

```
        Output 8.7: GMANOVA-SUR One Population Ramus Height Data

             OBS     GROUP       Y1        Y2        Y3        Y4

               1       1        47.8      48.8      49.0      49.7
               2       1        46.4      47.3      47.7      48.4
               3       1        46.3      46.8      47.8      48.5
               4       1        45.1      45.3      46.1      47.2
               5       1        47.6      48.5      48.9      49.3
               6       1        52.5      53.2      53.3      53.7
               7       1        51.2      53.0      54.3      54.5
               8       1        49.8      50.0      50.3      52.7
               9       1        48.1      50.8      52.3      54.4
              10       1        45.0      47.0      47.3      48.3
              11       1        51.2      51.4      51.6      51.9
              12       1        48.5      49.2      53.0      55.5
              13       1        52.1      52.8      53.7      55.0
              14       1        48.2      48.9      49.3      49.8
              15       1        49.6      50.4      51.2      51.8
              16       1        50.7      51.7      52.7      53.3
              17       1        47.2      47.7      48.4      49.5
              18       1        53.3      54.6      55.1      55.3
              19       1        46.2      47.5      48.1      48.4
              20       1        46.3      47.6      51.3      51.8
```

To estimate **B** using the SUR model, we calculate the FGLS estimator of vec(**B**) = β. We use PROC IML to transform the matrix **Y** to a vector, vec(**Y**), and to calculate **S**. A matrix of orthogonal polynomials **P** is created and **X** is a $\mathbf{1}_{20\times1}$ design matrix. An estimate of β is then computed as

$$\hat{\beta} = [(\mathbf{PS}^{-1}\mathbf{P}')^{-1}\mathbf{PS}^{-1} \otimes (\mathbf{X}'\mathbf{X})^{-1}\mathbf{X}']\text{vec}(\mathbf{Y}) = \begin{pmatrix} 50.0496 \\ 0.4654 \end{pmatrix} \tag{8.97}$$

The results are given in the continuation of Output 8.7 that follows.

Output 8.7 (*continued*)

```
        Output 8.7  GMANOVA-SUR One Population Ramus Height Data

             Estimate of B using SUR estimate and PROC IML

                                     BETA
                                50.049582
                                0.4654022

                              P prime matrix

                  P_PRIME
                         1         -3          1         -1
                         1         -1         -1          3
                         1          1         -1         -3
                         1          3          1          1
```

The SAS code to perform the GMANOVA analysis is similar to that used in Section 7.7. From the results of the test of linear fit, we see that a linear model fits the data. The estimate of **B** using the GMANOVA model and PROC GLM are found in the continuation of Output 8.7 that follows.

Output 8.7 (*continued*)

```
              Output 8.7  GMANOVA-SUR One Population Ramus Height Data

                               Test of Linear Fit

                           General Linear Models Procedure
                          Multivariate Analysis of Variance

                  Manova Test Criteria and Exact F Statistics for
                    the Hypothesis of no Overall INTERCEPT Effect
        H = Type III SS&CP Matrix for INTERCEPT    E = Error SS&CP Matrix

                             S=1     M=0     N=8

  Statistic                      Value          F       Num DF    Den DF  Pr > F

  Wilks' Lambda              0.98952499     0.0953          2         18  0.9096
  Pillai's Trace             0.01047501     0.0953          2         18  0.9096
  Hotelling-Lawley Trace     0.01058590     0.0953          2         18  0.9096
  Roy's Greatest Root        0.01058590     0.0953          2         18  0.9096


                         Estimate of B using PROC GLM

Dependent Variable: YT1
                                     Sum of            Mean
Source                  DF          Squares          Square    F Value       Pr > F

Model                    2       1.43697269      0.71848635       0.10       0.9019

Error                   17     117.59427731      6.91731043

Corrected Total         19     119.03125000

                  R-Square             C.V.        Root MSE                YT1 Mean

                  0.012072         5.252278       2.6300780               50.075000

Source                  DF        Type I SS     Mean Square    F Value       Pr > F

X                        0       0.00000000           .           .            .
YT3                      1       0.81507834      0.81507834       0.12       0.7356
YT4                      1       0.62189436      0.62189436       0.09       0.7679

Source                  DF      Type III SS     Mean Square    F Value       Pr > F

X                        0       0.00000000           .           .            .
YT3                      1       0.96629413      0.96629413       0.14       0.7132
YT4                      1       0.62189436      0.62189436       0.09       0.7679

                                       T for H0:      Pr > |T|     Std Error of
Parameter                 Estimate    Parameter=0                   Estimate

beta                    50.0495824          84.66       0.0001     0.59120793
```

Output 8.7 *(continued)*

```
Dependent Variable: YT2
                                      Sum of                Mean
Source                      DF       Squares              Square    F Value        Pr > F

Model                        2    0.57620272          0.28810136       4.72        0.0234

Error                       17    1.03670228          0.06098249

Corrected Total             19    1.61290500

                      R-Square          C.V.            Root MSE                 YT2 Mean

                      0.357245      52.93598           0.2469463                0.4665000

Source                      DF      Type I SS        Mean Square    F Value        Pr > F

X                            0    0.00000000                .           .             .
YT3                          1    0.03710447          0.03710447       0.61        0.4461
YT4                          1    0.53909825          0.53909825       8.84        0.0085

Source                      DF    Type III SS        Mean Square    F Value        Pr > F

X                            0    0.00000000                .           .             .
YT3                          1    0.01251627          0.01251627       0.21        0.6563
YT4                          1    0.53909825          0.53909825       8.84        0.0085

                                              T for H0:     Pr > |T|    Std Error of
Parameter                       Estimate     Parameter=0                  Estimate

beta                          0.46540223            8.38      0.0001      0.05551038
```

As expected, the estimate of **B** from the GMANOVA model and the estimate of β from the SUR model agree. We next estimate **B** using PROC SYSLIN which estimates parameters for systems of equations using OLS, 2SLS, 3SLS, SUR and many other estimation procedures. To estimate **B** for the GMANOVA model using the SUR model, we use the transformed data matrix $\mathbf{Y}_o = \mathbf{YH}$ where **H** is defined in (8.70). For PROC SYSLIN, the parameter estimates are associated with the independent (regression) variable x. As expected the PROC SYSLIN output agrees with the IML and GLM procedures. For an alternative approach, see Verbyla (1986).

Output 8.7 *(continued)*

```
          Output 8.7 GMANOVA-SUR One Population Ramus Heights Data
                             SYSLIN Procedure
                      Ordinary Least Squares Estimation

Model: YT1
Dependent variable: YT1

                             Analysis of Variance

                               Sum of        Mean
        Source          DF    Squares       Square       F Value       Prob>F

        Model            3  50151.54947  16717.18316     2416.717       0.0001
        Error           17    117.59428      6.91731
        U Total         20  50269.14375

                  Root MSE      2.63008     R-Square       0.9977
                  Dep Mean     50.07500     Adj R-SQ       0.9972
                  C.V.          5.25228

NOTE: The NOINT option changes the definition of the R-Square statistic to:
       1 - (Residual Sum of Squares/Uncorrected Total Sum of Squares).


                             Parameter Estimates

                    Parameter      Standard    T for H0:
      Variable  DF   Estimate         Error   Parameter=0    Prob > |T|

      X          1   50.049582     0.591208        84.656        0.0001
      YT3        1   -0.959930     2.568346        -0.374        0.7132
      YT4        1   -1.909606     6.368749        -0.300        0.7679

                               SYSLIN Procedure
                      Ordinary Least Squares Estimation

Model: YT2
Dependent variable: YT2

                             Analysis of Variance

                               Sum of        Mean
        Source          DF    Squares       Square       F Value       Prob>F

        Model            3      4.92865      1.64288       26.940       0.0001
        Error           17      1.03670      0.06098
        U Total         20      5.96535
Root MSE      0.24695      R-Square       0.8262
                  Dep Mean      0.46650     Adj R-SQ       0.7955
                  C.V.         52.93598

NOTE: The NOINT option changes the definition of the R-Square statistic to:
       1 - (Residual Sum of Squares/Uncorrected Total Sum of Squares).
```

Output 8.7 *(continued)*

Parameter Estimates

Variable	DF	Parameter Estimate	Standard Error	T for H0: Parameter=0	Prob > |T|
X	1	0.465402	0.055510	8.384	0.0001
YT3	1	0.109250	0.241150	0.453	0.6563
YT4	1	-1.777950	0.597982	-2.973	0.0085

8.8 GMANOVA-SUR: Several Populations

As a second illustration of the equivalence of the GMANOVA and SUR models, we reanalyze the Grizzle and Allen (1969) dog data analyzed in Section 7.7 using the GMANOVA model. PROC SYSLIN is now used to analyze the same data using the SUR model. The SAS code for the analysis is given in Program 8_8.sas.

Program 8_8.sas

```
/* Program 8_8.sas                                          */
/* GMANOVA/SUR Several Groups (Grizzle and Allen, 1969) */

options ls = 80 ps = 60 nodate nonumber;
title1 'Output 8.8: GMANOVA-SUR Several Population Dog Data';

data grizzle;
   infile 'c:\8_8.dat';
   input group y1 y2 y3 y4 y5 y6 y7 x1 x2 x3 x4;
proc print data=grizzle;
run;
proc summary data=grizzle;
   class group;
   var y1 -y7;
   output out=new mean=mr1-mr7;
proc print data=new;
run;
proc glm data=grizzle;
   class group;
   model y1 - y7 = group/nouni;
run;

proc iml;
   use grizzle;
   read all var {y1 y2 y3 y4 y5 y6 y7} into y;
   read all var {Group} into gr;
   read all var {x1 x2 x3 x4} into x;
   /* Create Orthogonal Polynomials of degree  q-1=6 */
   z={1 2 3 4 5 6 7};
   P_prime=orpol(z,6);
   Y0=Y*P_prime;
   print 'P prime matrix',P_prime;
   t=y0||x||gr;
   /* Create New Transformed data set */
   varnames={yt1 yt2 yt3 yt4 yt5 yt6 yt7 x1 x2 x3 x4 group};
   create trans from t (|colname=varnames|);
   append from t;
quit;
proc print data=trans;
run;
```

```
/* Test model fit  */
proc glm data=trans;
   title2 ' Test of Model Fit';
   model yt5 - yt7=/nouni;
   manova h=intercept;
run;

/* Using a Rao-Khatri MANCOVA model           */
/* Test for group differences (coincidence) */
proc glm data=trans;
   class group;
   model yt1 - yt4 = group yt5 - yt7/nouni;
   manova h=group/printh printe;
run;

/* Estimate of the matrix B */
proc glm data=trans;
   title2 'Estimate of B using PROC GLM';
   class group;
   model yt1 - yt4 = group yt5 - yt7;
   estimate 'beta1' intercept 1 group 1 0 0 0;
   estimate 'beta2' intercept 1 group 0 1 0 0;
   estimate 'beta3' intercept 1 group 0 0 1 0;
   estimate 'beta4' intercept 1 group 0 0 0 1;

/* Estimate of B using syslin procedure */
proc syslin sur data=trans;
   title2 'Estimate of B using PROC SYSLIN';
   model yt1 = x1 x2 x3 x4 yt5 yt6 yt7/noint;
   model yt2 = x1 x2 x3 x4 yt5 yt6 yt7/noint;
   model yt3 = x1 x2 x3 x4 yt5 yt6 yt7/noint;
   model yt4 = x1 x2 x3 x4 yt5 yt6 yt7/noint;
   coin: test x1=x2=x3=x4/print;
run;
```

Result and Interpretation 8_8.sas

Recall from Section 7.7 that this application involves fitting a third-degree polynomial to the four groups, a control group and three treatments. For convenience, we have included the code to perform the GMANOVA analysis in Program 8_8; however, only the SUR analysis using PROC SYSLIN results are given in Output 8.8.

Output 8.8: *GMANOVA-SUR Several Population Dog Data, Estimate of B using PROC SYSLIN*

```
              Output 8.8: GMANOVA-SUR Several Population Dog Data
                        Estimate of B using PROC SYSLIN

                                SYSLIN Procedure
                   Seemingly Unrelated Regression Estimation

                             Cross Model Covariance

 Sigma              YT1              YT2              YT3              YT4

 YT1       1.9650232293     0.4472613197     -0.028890743     -0.174835642
 YT2       0.4472613197     0.3628622408     0.0124628405     -0.069577226
 YT3       -0.028890743     0.0124628405      0.129361123     0.0293035358
 YT4       -0.174835642     -0.069577226     0.0293035358     0.0751637245

                            Cross Model Correlation

  Corr              YT1              YT2              YT3              YT4

  YT1                 1     0.5296715024     -0.057302421     -0.454927282
  YT2      0.5296715024                1     0.0575233619     -0.421300597
  YT3      -0.057302421     0.0575233619                1     0.2971761767
  YT4      -0.454927282     -0.421300597     0.2971761767                1

                        Cross Model Inverse Correlation

Inv Corr              YT1              YT2              YT3              YT4

YT1          1.5287571169     -0.627153903     -0.004914667     0.4327135282
YT2          -0.627153903     1.5297353658     -0.253014029     0.4343587442
YT3          -0.004914667     -0.253014029       1.14799848     -0.449988577
YT4          0.4327135282     0.4343587442     -0.449988577     1.5135746723

                         Cross Model Inverse Covariance

 Inv Sigma              YT1              YT2              YT3              YT4

 YT1          0.7779842468     -0.742710213     -0.009747839     1.1259328306
 YT2          -0.742710213      4.215746897     -1.167809018     2.6301077064
 YT3          -0.009747839     -1.167809018     8.8743700832     -4.563472676
 YT4          1.1259328306     2.6301077064     -4.563472676     20.137036622

          System Weighted MSE:  1 with 116 degrees of freedom.
          System Weighted R-Square:  0.9634

Model: YT1
Dependent variable: YT1

                              Parameter Estimates

                      Parameter      Standard    T for H0:
     Variable  DF      Estimate         Error  Parameter=0     Prob > |T|

     X1         1     12.367336      0.488032       25.341         0.0001
     X2         1      9.448517      0.488482       19.343         0.0001
     X3         1     10.748460      0.522161       20.585         0.0001
     X4         1     10.359835      0.477636       21.690         0.0001
     YT5        1      1.012136      0.795901        1.272         0.2136
     YT6        1     -0.208192      0.894399       -0.233         0.8176
     YT7        1      0.266191      1.338977        0.199         0.8438
```

Output 8.8 (*continued*)

Model: YT2
Dependent variable: YT2

Parameter Estimates

Variable	DF	Parameter Estimate	Standard Error	T for H0: Parameter=0	Prob > |T|
X1	1	0.800823	0.209718	3.819	0.0007
X2	1	-0.111754	0.209911	-0.532	0.5985
X3	1	0.736464	0.224384	3.282	0.0027
X4	1	0.352830	0.205250	1.719	0.0963
YT5	1	-0.470402	0.342016	-1.375	0.1795
YT6	1	0.064871	0.384343	0.169	0.8671
YT7	1	-0.091378	0.575387	-0.159	0.8749

Model: YT3
Dependent variable: YT3

Parameter Estimates

Variable	DF	Parameter Estimate	Standard Error	T for H0: Parameter=0	Prob > |T|
X1	1	-0.489032	0.125218	-3.905	0.0005
X2	1	-0.139863	0.125333	-1.116	0.2736
X3	1	-0.277891	0.133975	-2.074	0.0470
X4	1	-0.141670	0.122550	-1.156	0.2571
YT5	1	-1.185126	0.204210	-5.803	0.0001
YT6	1	-0.031095	0.229482	-0.136	0.8932
YT7	1	-0.109814	0.343551	-0.320	0.7515

Model: YT4
Dependent variable: YT4

Parameter Estimates

Variable	DF	Parameter Estimate	Standard Error	T for H0: Parameter=0	Prob > |T|
X1	1	-0.466621	0.095448	-4.889	0.0001
X2	1	-0.024373	0.095536	-0.255	0.8004
X3	1	-0.197087	0.102123	-1.930	0.0635
X4	1	0.004771	0.093415	0.051	0.9596
YT5	1	0.185747	0.155661	1.193	0.2424
YT6	1	-0.696930	0.174925	-3.984	0.0004
YT7	1	0.649324	0.261875	2.480	0.0192

L Inv(X'X) L'			LB-C
0.0169534913	-0.00790216	-0.000953421	-0.442248
-0.00790216	0.0195702868	-0.009686663	0.1727133893
-0.000953421	-0.009686663	0.018741844	-0.201857467

Inv(L Inv(X'X) L')			Lagrangians
82.877973197	47.773017218	28.907439425	-34.23676231
47.773017218	96.201168727	52.151543654	-15.03947088
28.907439425	52.151543654	81.781461196	-20.28518662

Output 8.8 *(continued)*

```
Test: COIN
     Numerator:    5.546113   DF:     3  F Value:   5.5461
     Denominator:         1   DF:   116   Prob>F:   0.0014
```

From Output 8.8, we see that PROC SYSLIN yields the same estimate for the matrix **B** as the GMANOVA model, again fitting a third-degree polynomial.

While the estimate of **B** is identical for the two models, this is not the case for tests of hypotheses. For the SUR model, we usually employ the Wald statistic (8.61) which has an asymptotic chi-square distribution. For the GMANOVA model, we have an exact test. PROC SYSLIN employs an F-statistic

$$F = (\mathbf{L}\hat{\hat{\beta}} - \mathbf{c})'[\mathbf{L}(s^2(\mathbf{X}'\mathbf{X})^{-1})\mathbf{L}'](\mathbf{L}\hat{\hat{\beta}} - \mathbf{c}) / \operatorname{rank}(\mathbf{L}) \tag{8.98}$$

which is valid only for OLS. For our example, the test of coincidence is tested using PROC SYSLIN with the statement

```
TEST x1 = x2 = x3 = x4/print;
```

The p-value for the F-test is $p = 0.0014$ which compares favorably with the GMANOVA exact test of coincidence of profiles which (from Section 7.7) was $p = .0065$.

8.9 SUR Model

Applications 8.7 and 8.8 illustrated the equivalence of the estimates of **B** from the GMANOVA model and the SUR model in the analysis of growth curve data. The SUR model may also be used to estimate model parameters for multiple design multivariate (MDM) regression models. PROC SYSLIN performs analysis of a system of equations with one dependent (endogenous) variable and several regressors (instrument) variables. The procedure provides numerous methods for estimating model parameters for a system of multiple regression equations. For a detailed discussion of several of the estimation methods, see Greene (1993). PROC SYSLIN provides the SUR estimator of β which substitutes for $\mathbf{\Sigma}$ the unbiased covariance estimate **S**. Replacing **S** with the ML estimate $\hat{\mathbf{\Sigma}}$, the Feasible Generalized Least Squares (FGLS) estimator iterates to the ML estimate of β, the FIML option in PROC SYSLIN. For the SUR model, the 2SLS option in PROC SYSLIN is equal to the OLS estimator, and the 3SLS estimator is equal to the SUR estimator. The IT3SLS estimator does not converge to the ML estimator since **S**, and not the ML estimate of $\mathbf{\Sigma}$, is used at each step of the iteration process.

To illustrate the analysis of the SUR model using PROC SYSLIN, we use the Grunfeld investment data given in Greene (1993, p. 445). The data consist of 20 yearly observations of gross investment, firm value, and stock value of plant and equipment for five firms. The model to be estimated is

$$I_{it} = \beta_o + \beta_1 F_{it} + \beta_2 C_{it} + e_{it} \tag{8.99}$$

and $i = 1, 2, \ldots, 5$ companies. The companies are General Motors (GM), Chrysler (Ch), General Electric (GE), Westinghouse Electric (WE), and U.S. Steel (US). Program 8_9.sas contains the SAS code to obtain the estimated regression equations using two estimation methods SUR and FIML.

Program 8_9.sas

```
/* Program 8_9.sas           */
/* Seemingly Unrelated Regressions (Greene, 1993, p.445) */

options ls=80 ps=60 nodate nonumber;
title1 'Output 8.9: SUR - Greene-Grunfeld Investment Data';

data greene;
   infile'c:\8_9.dat';
   input year gm_i gm_f gm_c ch_i ch_f ch_c
              ge_i ge_f ge_c we_i we_f we_c
              us_i us_f us_c;
   label gm_i='Gross Investment GM'
         gm_f='Market Value GM prior yr'
         gm_c='Stock Value GM prior yr'
         ch_i='Gross Investment CH'
         ch_f='Market Value CH prior yr'
         ch_c='Stock Value CH prior yr'
         ge_i='Gross Investment GE'
         ge_f='Market Value GE prior yr'
         ge_c='Stock Value GE prior yr'
         we_i='Gross Investment WE'
         we_f='Market Value WE prior yr'
         we_c='Stock Value WE prior yr'
         us_i='Gross Investment US'
         us_f='Market Value US prior yr'
         us_c='Stock Value US prior yr';
proc print data=Greene;
run;

/* GM=General Moters, Ch=Chrysler, GE=General Electric,  */
/* WE=Westinghouse, and US=U.S. Steel                     */

/* SYSLIN procedure with SUR option */
proc syslin sur;
            title2 'Using PROC SYSLIN with SUR Option';
   gm: model gm_i=gm_f gm_c;
   ch: model ch_i=ch_f ch_c;
   ge: model ge_i=ge_f ge_c;
   we: model we_i=we_f we_c;
   us: model us_i=us_f us_c;
   coin: stest gm.intercept-us.intercept,ch.intercept-us.intercept,
         ge.intercept-us.intercept,we.intercept-us.intercept,
         gm.gm_f-us.us_f,ch.ch_f-us.us_f,ge.ge_f-us.us_f,we.we_f-us.us_f,
         gm.gm_c-us.us_c,ch.ch_c-us.us_c,ge.ge_c-us.us_c,we.we_c-us.us_c;
run;
```

```
/* SYSLIN procedure using FIML option */
proc syslin fiml;
   title2 'Using PROC SYSLIN with FIML Option';
   endogenous gm_i ch_i ge_i we_i us_i;
   instruments gm_f gm_c ch_f ch_c ge_f ge_c we_f we_c us_f us_c;
   gm: model gm_i=gm_f gm_c;
   ch: model ch_i=ch_f ch_c;
   ge: model ge_i=ge_f ge_c;
   we: model we_i=we_f we_c;
   us: model us_i=us_f us_c;
run;
```

Result and Interpretation 8_9.sas

The results are found in Output 8.9.

Output 8.9: *SUR - Greene-Grunfeld Investment Data*

Output 8.9: SUR - Greene-Grunfeld Investment Data

OBS	YEAR	GM_I	GM_F	GM_C	CH_I	CH_F	CH_C	GE_I
1	1935	317.6	3078.5	2.8	40.29	417.5	10.5	33.1
2	1936	391.8	4661.7	52.6	72.76	837.8	10.2	45.0
3	1937	410.6	5387.1	156.9	66.26	883.9	34.7	77.2
4	1938	257.7	2792.2	209.2	51.60	437.9	51.8	44.6
5	1939	330.8	4313.2	203.4	52.41	679.7	64.3	48.1
6	1940	461.2	4643.9	207.2	69.41	727.8	67.1	74.4
7	1941	512.0	4551.2	255.2	68.35	643.6	75.2	113.0
8	1942	448.0	3244.1	303.7	46.80	410.9	71.4	91.9
9	1943	499.6	4053.7	264.1	47.40	588.4	67.1	61.3
10	1944	547.5	4379.3	201.6	59.57	698.4	60.5	56.8
11	1945	561.2	4840.9	265.0	88.78	846.4	54.6	93.6
12	1946	688.1	4900.9	402.2	74.12	893.8	84.8	159.9
13	1947	568.9	3526.5	761.5	62.68	579.0	96.8	147.2
14	1948	529.2	3254.7	922.4	89.36	694.6	110.2	146.3
15	1949	555.1	3700.2	1020.1	78.98	590.3	147.4	98.3
16	1950	642.9	3755.6	1099.0	100.66	693.5	163.2	93.5
17	1951	755.9	4833.0	1207.7	160.62	809.0	203.5	135.2
18	1952	891.2	4924.9	1430.5	145.00	727.0	290.6	157.3
19	1953	1304.4	6241.7	1777.3	174.93	1001.5	346.1	179.5
20	1954	1486.7	5593.6	2226.3	172.49	703.2	414.9	189.6

Output 8.9: *(continued)*

```
OBS     GE_F      GE_C     WE_I      WE_F     WE_C     US_I     US_F     US_C

  1    1170.6     97.8    12.93     191.5      1.8    209.9   1362.4     53.8
  2    2015.8    104.4    25.90     516.0      0.8    355.3   1807.1     50.5
  3    2803.3    118.0    35.05     729.0      7.4    469.9   2676.3    118.1
  4    2039.7    156.2    22.89     560.4     18.1    262.3   1801.9    260.2
  5    2256.2    172.6    18.84     519.9     23.5    230.4   1957.3    312.7
  6    2132.2    186.6    28.57     628.5     26.5    261.6   2202.9    254.2
  7    1834.1    220.9    48.51     537.1     36.2    472.8   2380.5    261.4
  8    1588.0    287.8    43.34     561.2     60.8    445.6   2168.6    298.7
  9    1749.4    319.9    37.02     617.2     84.4    361.6   1985.1    301.8
 10    1687.2    321.3    37.81     626.7     91.2    288.2   1813.9    279.1
 11    2007.7    319.6    39.27     737.2     92.4    258.7   1850.2    213.8
 12    2208.3    346.0    53.46     760.5     86.0    420.3   2067.7    232.6
 13    1656.7    456.4    55.56     581.4    111.1    420.5   1796.7    264.8
 14    1604.4    543.4    49.56     662.3    130.6    494.5   1625.8    306.9
 15    1431.8    618.3    32.04     583.8    141.8    405.1   1667.0    351.1
 16    1610.5    647.4    32.24     635.2    136.7    418.8   1677.4    357.8
 17    1819.4    671.3    54.38     723.8    129.7    588.2   2289.5    342.1
 18    2079.7    726.1    71.78     864.1    145.5    645.2   2159.4    444.2
 19    2371.6    800.3    90.08    1193.5    174.8    641.0   2031.3    623.6
 20    2759.9    888.9    68.60    1188.9    213.5    459.3   2115.5    669.7

                         Using PROC SYSLIN with SUR Option
                                SYSLIN Procedure
                         Ordinary Least Squares Estimation

Model: GM
Dependent variable: GM_I Gross Investment GM

                              Analysis of Variance

                                 Sum of        Mean
        Source          DF      Squares       Square        F Value       Prob>F

        Model            2 1677686.6746 838843.33729         99.579       0.0001
        Error           17 143205.87741   8423.87514
        C Total         19 1820892.5520

                  Root MSE     91.78167      R-Square        0.9214
                  Dep Mean    608.02000      Adj R-SQ        0.9121
                  C.V.         15.09517

                              Parameter Estimates

                          Parameter       Standard    T for H0:
      Variable   DF        Estimate          Error   Parameter=0    Prob > |T|

      INTERCEP    1     -149.782453     105.842125        -1.415        0.1751
      GM_F        1        0.119281       0.025834         4.617        0.0002
      GM_C        1        0.371445       0.037073        10.019        0.0001

                      Variable
      Variable   DF      Label

      INTERCEP    1   Intercept
      GM_F        1   Market Value GM prior yr
      GM_C        1   Stock Value GM prior yr
```

Output 8.9 *(continued)*

```
Model: CH
Dependent variable: CH_I Gross Investment CH

                              Analysis of Variance

                                 Sum of         Mean
        Source          DF      Squares       Square       F Value        Prob>F

        Model            2  31686.54369  15843.27185        89.855        0.0001
        Error           17   2997.44436    176.32026
        C Total         19  34683.98805

                  Root MSE      13.27856      R-Square       0.9136
                  Dep Mean      86.12350      Adj R-SQ       0.9034
                  C.V.          15.41805

                              Parameter Estimates

                         Parameter      Standard    T for H0:
      Variable  DF        Estimate         Error   Parameter=0     Prob > |T|

      INTERCEP   1       -6.189961     13.506478        -0.458         0.6525
      CH_F       1        0.077948      0.019973         3.903         0.0011
      CH_C       1        0.315718      0.028813        10.957         0.0001

Variable
      Variable  DF     Label

      INTERCEP   1  Intercept
      CH_F       1  Market Value CH prior yr
      CH_C       1  Stock Value CH prior yr

Model: GE
Dependent variable: GE_I Gross Investment GE

                              Analysis of Variance

                                 Sum of         Mean
        Source          DF      Squares       Square       F Value        Prob>F

        Model            2  31632.03023  15816.01511        20.344        0.0001
        Error           17  13216.58777    777.44634
        C Total         19  44848.61800

                  Root MSE      27.88272      R-Square       0.7053
                  Dep Mean     102.29000      Adj R-SQ       0.6706
                  C.V.          27.25850

                              Parameter Estimates

                         Parameter      Standard    T for H0:
      Variable  DF        Estimate         Error   Parameter=0     Prob > |T|

      INTERCEP   1       -9.956306     31.374249        -0.317         0.7548
      GE_F       1        0.026551      0.015566         1.706         0.1063
      GE_C       1        0.151694      0.025704         5.902         0.0001
```

Output 8.9 (*continued*)

```
                    Variable
      Variable  DF     Label

      INTERCEP   1  Intercept
      GE_F       1  Market Value GE prior yr
      GE_C       1  Stock Value GE prior yr

Model: WE
Dependent variable: WE_I Gross Investment WE

                              Analysis of Variance

                                 Sum of        Mean
        Source            DF    Squares       Square        F Value        Prob>F

        Model              2  5165.55292  2582.77646         24.761        0.0001
        Error             17  1773.23393   104.30788
        C Total           19  6938.78686

                  Root MSE     10.21312     R-Square        0.7444
                  Dep Mean     42.89150     Adj R-SQ        0.7144
                  C.V.         23.81153

                              Parameter Estimates

                         Parameter      Standard    T for H0:
      Variable  DF        Estimate         Error   Parameter=0    Prob > |T|

      INTERCEP   1       -0.509390      8.015289        -0.064        0.9501
      WE_F       1        0.052894      0.015707         3.368        0.0037
      WE_C       1        0.092406      0.056099         1.647        0.1179

Variable
      Variable  DF     Label

      INTERCEP   1  Intercept
      WE_F       1  Market Value WE prior yr
      WE_C       1  Stock Value WE prior yr

Model: US
Dependent variable: US_I Gross Investment US

                              Analysis of Variance

                                 Sum of        Mean
        Source            DF    Squares       Square        F Value        Prob>F

        Model              2 139978.07436  69989.03718        6.687        0.0072
        Error             17 177928.31364  10466.37139
        C Total           19 317906.38800

                  Root MSE    102.30529     R-Square        0.4403
                  Dep Mean    405.46000     Adj R-SQ        0.3745
                  C.V.         25.23191
```

Output 8.9 *(continued)*

```
                              Parameter Estimates

                       Parameter       Standard     T for H0:
     Variable  DF       Estimate          Error   Parameter=0     Prob > |T|

     INTERCEP   1     -30.368532     157.047695        -0.193         0.8490
     US_F       1       0.156571       0.078886         1.985         0.0635
     US_C       1       0.423866       0.155216         2.731         0.0142

                    Variable
     Variable  DF     Label

     INTERCEP   1  Intercept
     US_F       1  Market Value US prior yr
     US_C       1  Stock Value US prior yr


                                SYSLIN Procedure
                      Seemingly Unrelated Regression Estimation

                              Cross Model Covariance

Sigma              GM             CH             GE             WE             US

GM       8423.8751418   -332.6546159   714.74486532    148.4425554   -2614.188281
CH       -332.6546159   176.32025657   -25.14782439   15.655238013   491.85723205
GE       714.74486532   -25.14782439   777.44633943   207.58713102   1064.6491135
WE        148.4425554   15.655238013   207.58713102   104.30787826   642.57124214
US       -2614.188281   491.85723205   1064.6491135   642.57124214    10466.37139

                             Cross Model Correlation

Corr              GM             CH             GE             WE             US

GM                 1   -0.272952131   0.2792928849   0.1583594212   -0.278408719
CH      -0.272952131              1   -0.067922569   0.1154383244   0.3620677859
GE      0.2792928849   -0.067922569              1   0.7289649707   0.3732271381
WE      0.1583594212   0.1154383244   0.7289649707              1   0.6149851921
US      -0.278408719   0.3620677859   0.3732271381   0.6149851921              1

                         Cross Model Inverse Correlation

 Inv Corr            GM             CH             GE             WE             US

 GM        1.4116044913   0.1464942472   -0.326666835   -0.460562271   0.7451220552
 CH        0.1464942472   1.2337288685   0.2761456044   -0.086697321   -0.455655669
 GE        -0.326666835   0.2761456044   2.3305486284   -1.651171334   -0.045308397
 WE        -0.460562271   -0.086697321   -1.651171334   3.1636695944   -1.426182247
 US        0.7451220552   -0.455655669   -0.045308397   -1.426182247   2.2664180025
```

Output 8.9 (*continued*)

```
                       Cross Model Inverse Covariance

Inv Sigma            GM             CH             GE             WE             US

GM         0.0001675719  0.0001202025  -0.000127648  -0.000491331  0.0000793548
CH         0.0001202025  0.0069970909  0.0007458506  -0.000639287  -0.000335419
GE         -0.000127648  0.0007458506  0.0029976971  -0.005798269  -0.000015883
WE         -0.000491331  -0.000639287  -0.005798269  0.0303301117  -0.001364955
US         0.0000793548  -0.000335419  -0.000015883  -0.001364955  0.0002165429


           System Weighted MSE:  0.94013 with 85 degrees of freedom.
           System Weighted R-Square:  0.8707


Model: GM
Dependent variable: GM_I Gross Investment GM

                              Parameter Estimates

                          Parameter      Standard    T for H0:
      Variable  DF         Estimate         Error   Parameter=0    Prob > |T|

      INTERCEP   1      -162.364105     97.032161        -1.673        0.1126
      GM_F       1         0.120493      0.023460         5.136        0.0001
      GM_C       1         0.382746      0.035542        10.769        0.0001

                     Variable
      Variable  DF     Label

      INTERCEP   1   Intercept
      GM_F       1   Market Value GM prior yr
      GM_C       1   Stock Value GM prior yr


Model: CH
Dependent variable: CH_I Gross Investment CH

                              Parameter Estimates

                          Parameter      Standard    T for H0:
      Variable  DF         Estimate         Error   Parameter=0    Prob > |T|

      INTERCEP   1         0.504304     12.487416         0.040        0.9683
      CH_F       1         0.069546      0.018328         3.795        0.0014
      CH_C       1         0.308545      0.028053        10.999        0.0001

                     Variable
      Variable  DF     Label

      INTERCEP   1   Intercept
      CH_F       1   Market Value CH prior yr
      CH_C       1   Stock Value CH prior yr
```

Output 8.9 (*continued*)

```
Model: GE
Dependent variable: GE_I Gross Investment GE

                              Parameter Estimates

                     Parameter      Standard    T for H0:
      Variable  DF    Estimate         Error   Parameter=0    Prob > |T|

      INTERCEP   1  -22.438913     27.678793        -0.811        0.4287
      GE_F       1    0.037291      0.013301         2.804        0.0122
      GE_C       1    0.130783      0.023916         5.468        0.0001

                  Variable
      Variable  DF   Label

      INTERCEP   1  Intercept
      GE_F       1  Market Value GE prior yr
      GE_C       1  Stock Value GE prior yr

Model: WE
Dependent variable: WE_I Gross Investment WE

                              Parameter Estimates

                     Parameter      Standard    T for H0:
      Variable  DF    Estimate         Error   Parameter=0    Prob > |T|

      INTERCEP   1    1.088877      6.788627         0.160        0.8745
      WE_F       1    0.057009      0.012324         4.626        0.0002
      WE_C       1    0.041506      0.044689         0.929        0.3660

                  Variable
      Variable  DF   Label

      INTERCEP   1  Intercept
      WE_F       1  Market Value WE prior yr
      WE_C       1  Stock Value WE prior yr

Model: US
Dependent variable: US_I Gross Investment US

                              Parameter Estimates

                     Parameter      Standard    T for H0:
      Variable  DF    Estimate         Error   Parameter=0    Prob > |T|

      INTERCEP   1   85.423255    121.348101         0.704        0.4910
      US_F       1    0.101478      0.059421         1.708        0.1059
      US_C       1    0.399991      0.138613         2.886        0.0103

                  Variable
      Variable  DF   Label

      INTERCEP   1  Intercept
      US_F       1  Market Value US prior yr
      US_C       1  Stock Value US prior yr

Test: COIN
     Numerator:     109.794   DF:    12   F Value: 116.7857
     Denominator: 0.940132    DF:    85    Prob>F:   0.0001
```

Output 8.9 (*continued*)

```
                      Using PROC SYSLIN With FIML Option
                               SYSLIN Procedure
                 Full-Information Maximum Likelihood Estimation

Model: GM
Dependent variable: GM_I Gross Investment GM

                                  Parameter Estimates

                           Parameter      Standard    T for H0:
      Variable  DF          Estimate         Error   Parameter=0    Prob > |T|

      INTERCEP   1       -173.037991     84.279578        -2.053        0.0558
      GM_F       1          0.121953      0.020243         6.024        0.0001
      GM_C       1          0.389452      0.031852        12.227        0.0001

                        Variable
      Variable  DF         Label

      INTERCEP   1      Intercept
      GM_F       1      Market Value GM prior yr
      GM_C       1      Stock Value GM prior yr


Model: CH
Dependent variable: CH_I Gross Investment CH

                                  Parameter Estimates

                           Parameter      Standard    T for H0:
      Variable  DF          Estimate         Error   Parameter=0    Prob > |T|

      INTERCEP   1          2.378386     11.631365         0.204        0.8404
      CH_F       1          0.067451      0.017102         3.944        0.0010
      CH_C       1          0.305066      0.026067        11.703        0.0001

                        Variable
      Variable  DF         Label

      INTERCEP   1      Intercept
      CH_F       1      Market Value CH prior yr
      CH_C       1      Stock Value CH prior yr

Model: GE
Dependent variable: GE_I Gross Investment GE

                                  Parameter Estimates

                           Parameter      Standard    T for H0:
      Variable  DF          Estimate         Error   Parameter=0    Prob > |T|

      INTERCEP   1        -16.375793     24.960830        -0.656        0.5206
      GE_F       1          0.037019      0.011770         3.145        0.0059
      GE_C       1          0.116953      0.021731         5.382        0.0001
```

Output 8.9 (*continued*)

```
                      Variable
     Variable  DF       Label

     INTERCEP   1   Intercept
     GE_F       1   Market Value GE prior yr
     GE_C       1   Stock Value GE prior yr

Model: WE
Dependent variable: WE_I Gross Investment WE

                              Parameter Estimates

                       Parameter      Standard    T for H0:
     Variable  DF       Estimate         Error   Parameter=0     Prob > |T|

     INTERCEP   1       4.489292      6.022069         0.745         0.4662
     WE_F       1       0.053860      0.010294         5.232         0.0001
     WE_C       1       0.026469      0.037038         0.715         0.4845

                      Variable
     Variable  DF       Label

     INTERCEP   1   Intercept
     WE_F       1   Market Value WE prior yr
     WE_C       1   Stock Value WE prior yr

Model: US
Dependent variable: US_I Gross Investment US

                              Parameter Estimates

                       Parameter      Standard    T for H0:
     Variable  DF       Estimate         Error   Parameter=0     Prob > |T|

     INTERCEP   1     138.013543     94.607500         1.459         0.1628
     US_F       1       0.088600      0.045278         1.957         0.0670
     US_C       1       0.309293      0.117830         2.625         0.0177

                      Variable
     Variable  DF       Label

     INTERCEP   1   Intercept
     US_F       1   Market Value US prior yr
     US_C       1   Stock Value US prior yr
```

When using PROC SYLIN, one is always provided with the OLS estimates which ignore the dependency across models. The parameter estimates from Output 8.9 are summarized in Table 8.1.

Table 8.1 *OLS Estimates (Standard Errors)*

	β_0		β_1		β_2	
GM	-149.78200	(105.84)	0.11928	(0.0258)	0.37145	(0.0371)
CH	-6.18996	(13.51)	0.77950	(0.01997)	0.31572	(0.02881)
GE	-9.95630	(31.37)	0.02655	(0.01557)	0.15169	(0.02570)
WE	-0.50939	(8.015)	0.05289	(0.01571)	0.09241	(0.05610)
US	-30.36850	(157.05)	0.15657	(0.07889)	0.42387	(0.15522)

Taking into account the covariance matrix across equations, we summarize the SUR estimates obtained from Ouput 8.9 in Table 8.2.

Table 8.2 *SUR Estimates (Standard Errors)*

	β_0		β_1		β_2	
GM	-162.3641	(97.03)	0.12049	(0.0235)	0.38275	(0.0355)
CH	0.5043	(12.49)	0.06955	(0.01833)	0.30850	(0.02805)
GE	- 22.4390	(27.68)	0.03729	(0.01330)	0.13078	(0.02392)
WE	1.0889	(6.788)	0.05701	(0.01232)	0.04150	(0.04469)
US	85.4230	(121.35)	0.10145	(0.05942)	0.39990	(0.13861)

The FIML estimates can be found likewise in Output 8.9, but are not summarized in a table here.

Comparing the entries in the two tables, we see that there is a small gain in efficiency by using the SUR model. The gain is usually measured by investigating the diagonal elements of the Cross Model Inverse Correlation matrix, the model inflation factor (MIF). For our example, the values are between 1.23 and 3.16. The ML estimate of β is found in Output 8.9 under the FIML results. The differences between the SUR and FIML estimates are small. Because the asymptotic properties of the SUR and ML estimates are identical, and because no test of the model parameters is provided using PROC SYSLIN, the SUR estimator is preferred. For very small samples, SUR is less efficient than OLS.

For the investment data, we may also test the equivalence of the regression equations

$$H\colon \beta_1 = \beta_2 = \beta_3 = \beta_4 = \beta_5 \tag{8.100}$$

for the five firms. This test was performed by using PROC SYSLIN with the STEST statement as shown in Program 8_9.sas. The result is found in Output 8.9. The approximate F-statistic is 116.7857 with p-value of 0.0001. Hence, the null hypothesis of coincidence is rejected.

In our analysis of the investment data, we fit to the data a model that allowed for the variability of the β_i across firms by estimating the fixed parameters β_i for each firm. Another approach to modeling parameter variation across firms as examined by Greene (1993, p. 459) represents parameter heterogeneity as stochastic or random variation. Thus, the parameters β_i are represented as a fixed effect plus a vector for random variation

$$\beta_i = \beta + \mathbf{v}_i \qquad i = 1, 2, \ldots, n \tag{8.101}$$

Then for each firm, we have the linear model

$$\begin{aligned} \mathbf{y}_i &= \mathbf{X}_i\beta_i + \mathbf{e}_i \\ &= \mathbf{X}_i\beta + (\mathbf{X}_i\mathbf{v}_i + \mathbf{e}_i) \\ &= \mathbf{X}_i\beta + \mathbf{u}_i \end{aligned} \tag{8.102}$$

More generally, (8.102) is a special case of the random coefficient regression model

$$\mathbf{y}_i = \mathbf{X}_i\beta + \mathbf{Z}_i\mathbf{b}_i + \mathbf{e}_i \tag{8.103}$$

(Swamy, 1971). In (8.103), $\mathbf{y}_i$ $(n_i \times 1)$ is a vector of repeated measurements at n_i time points for the ith firm (or subject), so the number of points may be different for each firm;, $\mathbf{X}_i$ $(n_i \times p)$ and $\mathbf{Z}_i$ $(n_i \times k)$ are known design matrices; $\beta_{p\times 1}$ is a fixed vector of parameters common for all firms; $\mathbf{b}_i$ $(k \times 1)$ is a random component representing stochastic variation across firms; and $\mathbf{e}_i$ $(n_i \times 1)$ is a vector of random errors. For the random coefficient model, we further assume that the random vectors $\mathbf{b}_i$ have mean zero, common covariance structure $\mathbf{D}$ and that the $\mathbf{b}_i$ are uncorrelated, $\text{cov}(\mathbf{b}_i, \mathbf{b}_j) = \mathbf{0}$. However, we assume that the errors $\mathbf{e}_i$ have zero mean and covariance structure $\boldsymbol{\Psi}_i$ which may be different across firms, but that the errors are independent of each other and of the random components so that the $\text{cov}(\mathbf{b}_i, \mathbf{b}_{i'}) = \mathbf{0}$, $\text{cov}(\mathbf{e}_i, \mathbf{e}_{i'}) = \mathbf{0}$, and $\text{cov}(\mathbf{b}_i, \mathbf{e}_{i'}) = \mathbf{0}$ for all $i \neq i'$. Thus, we have that the covariance structure of $\mathbf{y}_i$ is

$$\text{cov}(\mathbf{y}_i) = \Omega_i = \mathbf{Z}_i \mathbf{D} \mathbf{Z}_i' + \Psi_i. \tag{8.104}$$

Stacking the vectors in (8.102) into a single vector, the general linear model results

$$\begin{pmatrix} \mathbf{y}_1 \\ \mathbf{y}_2 \\ \vdots \\ \vdots \\ \vdots \\ \mathbf{y}_n \end{pmatrix} = \begin{pmatrix} \mathbf{X}_1 \\ \mathbf{X}_2 \\ \vdots \\ \vdots \\ \vdots \\ \mathbf{X}_n \end{pmatrix} \beta + \begin{pmatrix} \mathbf{Z}_1 & \mathbf{0} & \mathbf{0} & \cdots & \cdots & \mathbf{0} \\ \mathbf{0} & \mathbf{Z}_2 & \mathbf{0} & \cdots & \cdots & \mathbf{0} \\ \vdots & \cdots & \ddots & \cdots & \cdots & \vdots \\ \vdots & \cdots & \cdots & \ddots & \cdots & \vdots \\ \vdots & \cdots & \cdots & \cdots & \ddots & \vdots \\ \mathbf{0} & \mathbf{0} & \mathbf{0} & \cdots & \cdots & \mathbf{Z}_n \end{pmatrix} \begin{pmatrix} \mathbf{b}_1 \\ \mathbf{b}_2 \\ \vdots \\ \vdots \\ \vdots \\ \mathbf{b}_n \end{pmatrix} + \begin{pmatrix} \mathbf{e}_1 \\ \mathbf{e}_2 \\ \vdots \\ \vdots \\ \vdots \\ \mathbf{e}_n \end{pmatrix} \tag{8.105}$$

$$\underset{N\times 1}{\mathbf{y}} = \underset{N\times p}{\mathbf{X}} \underset{p\times 1}{\beta} + \underset{N\times k}{\mathbf{Z}} \underset{k\times 1}{\mathbf{b}} + \underset{N\times 1}{\mathbf{e}}$$

where $N = \sum_i n_i$. Employing the vector form of the model, we have that the

$$\text{cov}(\mathbf{b}) = \begin{pmatrix} \mathbf{D} & \cdots & \cdots & \cdots & \cdots & \mathbf{0} \\ \vdots & \mathbf{D} & \cdots & \cdots & \cdots & \vdots \\ \vdots & \cdots & \ddots & \cdots & \cdots & \vdots \\ \vdots & \cdots & \cdots & \ddots & \cdots & \vdots \\ \vdots & \cdots & \cdots & \cdots & \ddots & \vdots \\ \mathbf{0} & \cdots & \cdots & \cdots & \cdots & \mathbf{D} \end{pmatrix} = \mathbf{I}_n \otimes \mathbf{D} = \mathbf{V}$$

$$\text{cov}(\mathbf{e}) = \begin{pmatrix} \Psi_1 & \cdots & \cdots & \cdots & \cdots & \mathbf{0} \\ \vdots & \Psi_2 & \cdots & \cdots & \cdots & \vdots \\ \vdots & \cdots & \ddots & \cdots & \cdots & \vdots \\ \vdots & \cdots & \cdots & \ddots & \cdots & \vdots \\ \vdots & \cdots & \cdots & \cdots & \ddots & \vdots \\ \mathbf{0} & \cdots & \cdots & \cdots & \cdots & \Psi_n \end{pmatrix} = \Psi \tag{8.106}$$

so that the

$$\text{cov}(\mathbf{y}) = \mathbf{ZVZ}' + \mathbf{\Psi} = \begin{pmatrix} \mathbf{\Omega}_1 & \cdots & \cdots & \cdots & \cdots & \mathbf{0} \\ \vdots & \mathbf{\Omega}_2 & \cdots & \cdots & \cdots & \vdots \\ \vdots & \cdots & \ddots & \cdots & \cdots & \vdots \\ \vdots & \cdots & \cdots & \ddots & \cdots & \vdots \\ \vdots & \cdots & \cdots & \cdots & \ddots & \vdots \\ \mathbf{0} & \cdots & \cdots & \cdots & \cdots & \mathbf{\Omega}_n \end{pmatrix} = \mathbf{\Omega} \tag{8.107}$$

Under multivariate normality of $\mathbf{y}$,

$$\mathbf{y} \sim N(\mathbf{X}\beta, \mathbf{ZVZ}' + \mathbf{\Psi})$$

The ML estimate of β, the mean of the random process $\beta_i = \beta + \mathbf{v}_i$ is

$$\begin{aligned} \hat{\beta} &= [\mathbf{X}'(\mathbf{ZVZ}' + \mathbf{\Psi})^{-1}\mathbf{X}'(\mathbf{ZVZ}' + \mathbf{\Psi})^{-1}]\mathbf{y} \\ &= \left(\sum_{i=1}^{n} \mathbf{X}_i'\mathbf{\Omega}_i^{-1}\mathbf{X}_i\right)^{-1}\left[\sum_{i=1}^{n} \mathbf{X}_i'\mathbf{\Omega}_i^{-1}\mathbf{y}_i\right] \end{aligned} \tag{8.108}$$

where $\mathbf{\Omega}_i = \mathbf{Z}_i\mathbf{D}\mathbf{Z}_i' + \mathbf{\Psi}_i$ for $i = 1, 2, \ldots, n$. Because $\mathbf{D}$ and $\mathbf{\Psi}_i$ are usually unknown in practice, we cannot apply the model directly, but must estimate $\mathbf{D}$ and $\mathbf{\Psi}_i$ in order to obtain an estimate of β. More will be said about this model in Chapter 9 since the random coefficient model is equivalent to the two-stage hierarchical linear model. In SAS, we use PROC MIXED to analyze these designs.

8.10 Two-Period Crossover Design with Changing Covariates

We have illustrated how the SUR model may be used to estimate the parameters of the growth curve model and how to analyze data containing both cross-sectional and time series data. We now show how the SUR model given in (8.49) and (7.18) may be used to analyze repeated measures data with changing covariates. To illustrate the theory, we consider a simple two-group, cross-sectional example with $p = 3$ repeated measures, and $k = 2$ covariates.

From (8.43), the model for our example has the form

$$E(\underset{n\times 3}{\mathbf{Y}}) = \underset{n\times 2}{\mathbf{X}}\ \underset{2\times 3}{\mathbf{B}} + \underset{n\times 3}{\mathbf{Z}_1}\ \underset{3\times 3}{\mathbf{\Gamma}_1} + \underset{n\times 3}{\mathbf{Z}_2}\ \underset{3\times 3}{\mathbf{\Gamma}_2} \tag{8.109}$$

$$E\begin{pmatrix} \mathbf{Y}_1 \\ \mathbf{Y}_2 \end{pmatrix} = \begin{pmatrix} \mathbf{1} & \mathbf{0} \\ \mathbf{0} & \mathbf{1} \end{pmatrix}\begin{pmatrix} \mu_{11} & \mu_{12} & \mu_{13} \\ \mu_{21} & \mu_{22} & \mu_{23} \end{pmatrix} + (\mathbf{z}_{11}, \ \mathbf{z}_{12}, \ \mathbf{z}_{13})\begin{pmatrix} \gamma_{11} & 0 & 0 \\ 0 & \gamma_{12} & 0 \\ 0 & 0 & \gamma_{13} \end{pmatrix} + (\mathbf{z}_{21}, \ \mathbf{z}_{22}, \ \mathbf{z}_{23})\begin{pmatrix} \gamma_{21} & 0 & 0 \\ 0 & \gamma_{22} & 0 \\ 0 & 0 & \gamma_{23} \end{pmatrix}$$

$$\begin{aligned} &= \mathbf{X}[\beta_1, \beta_2, \beta_3] + \sum_{j=1}^{2}[\mathbf{z}_{j1}\gamma_{j1}, \ \mathbf{z}_{j2}\gamma_{j2}, \ \mathbf{z}_{j3}\gamma_{j3}] \\ &= [\mathbf{X}\beta_1 + \mathbf{Z}_{.1}\gamma_{.1}, \ \ \mathbf{X}\beta_2 + \mathbf{Z}_{.2}\gamma_{.2}, \ \ \mathbf{X}\beta_3 + \mathbf{Z}_{.3}\gamma_{.3}] \\ &= [\mathbf{X}_1\theta_{11}, \ \ \mathbf{X}_2\theta_{22}, \ \ \mathbf{X}_3\theta_{33}] \end{aligned}$$

where

$$\begin{aligned} \mathbf{Z}_{.1} &= (\mathbf{z}_{11}, \mathbf{z}_{21}) & \gamma'_{.1} &= (\gamma_{11}, \gamma_{21}) \\ \mathbf{Z}_{.2} &= (\mathbf{z}_{12}, \mathbf{z}_{22}) & \gamma'_{.2} &= (\gamma_{12}, \gamma_{22}) \\ \mathbf{Z}_{.3} &= (\mathbf{z}_{13}, \mathbf{z}_{23}) & \gamma'_{.3} &= (\gamma_{13}, \gamma_{23}) \end{aligned} \tag{8.110}$$

and

$$\begin{aligned} \mathbf{X}_1 &= (\mathbf{X}, \mathbf{Z}_{.1}) = (\mathbf{X}, \mathbf{z}_{11}, \mathbf{z}_{21}) & \theta'_{11} &= (\beta'_1, \gamma'_{.1}) = (\mu_{11}, \mu_{21}, \gamma_{11}\gamma_{21}) \\ \mathbf{X}_2 &= (\mathbf{X}, \mathbf{Z}_{.2}) = (\mathbf{X}, \mathbf{z}_{12}, \mathbf{z}_{22}) & \theta'_{22} &= (\beta'_2, \gamma'_{.2}) = (\mu_{12}, \mu_{22}, \gamma_{12}\gamma_{22}) \\ \mathbf{X}_3 &= (\mathbf{X}, \mathbf{Z}_{.3}) = (\mathbf{X}, \mathbf{z}_{13}, \mathbf{z}_{23}) & \theta'_{33} &= (\beta'_3, \gamma'_{.3}) = (\mu_{13}, \mu_{23}, \gamma_{13}\gamma_{23}) \end{aligned} \tag{8.111}$$

Thus, letting $\mathbf{y}_i$ be the ith column of $\mathbf{Y}$, we have the SUR model in (8.49) with

$$\mathbf{y}_i = \mathbf{X}_i \theta_{ii} + \mathbf{e}_{ii} \qquad i = 1, 2, 3 \tag{8.112}$$

To illustrate the application of the SUR model for a design with changing covariates, we consider the two-period crossover design analyzed by Patel (1986) using likelihood ratio methods. Table 8.3 contains data on the FEV_1 (forced expired volume in one second) obtained from a crossover trial involving 17 patients with mild-to-moderate bronchial asthma. The data are provided in Patel (1983). Patients were randomly assigned to two treatment sequences in which acute effects of single oral doses of two active drugs are compared. The variable z_i contains baseline values for Period i, and the variable y_i contains the average of the responses obtained two and three hours after the initiation of a dose for Period i (i = 1, 2). Thus we have a repeated measure design with $p = 2$ variables and one covariate. The model is a special case of (8.109):

$$\underset{n\times 2}{E(\mathbf{Y})} = \begin{pmatrix} \mathbf{1} & \mathbf{0} \\ \mathbf{0} & \mathbf{1} \end{pmatrix} \begin{pmatrix} \mu_{11} & \mu_{12} \\ \mu_{21} & \mu_{22} \end{pmatrix} + (\mathbf{z}_{11}, \ \mathbf{z}_{12}) \begin{pmatrix} \gamma_{11} & 0 \\ 0 & \gamma_{12} \end{pmatrix} \tag{8.113}$$

Table 8.3 *FEV_1 Measurements in a Crossover Trial of Two Active Drugs in Asthmatic Patients*

Treatment Sequence 1					Treatment Sequence 2				
Patient	Period 1 Drug A		Period 2 Drug B		Patient	Period 1 Drug B		Period 2 Drug A	
	z_1^*	y_1^*	z_2	y_2		z_1	y_1	z_1	y_1
1	1.09	1.28	1.24	1.33	9	1.74	3.06	1.54	1.38
2	1.38	1.60	1.90	2.21	10	2.41	2.68	2.13	2.10
3	2.27	2.46	2.19	2.43	11	3.05	2.60	2.18	2.32
4	1.34	1.41	1.47	1.81	12	1.20	1.48	1.41	1.30
5	1.31	1.40	0.85	0.85	13	1.70	2.08	2.21	2.34
6	0.96	1.12	1.12	1.20	14	1.89	2.72	2.05	2.48
7	0.66	0.90	0.78	0.90	15	0.89	1.94	0.72	1.11
8	1.69	2.41	1.90	2.79	16	2.41	3.35	2.83	3.23
					17	0.96	1.16	1.01	1.25
Mean	1.34	1.57	1.43	1.69		1.81	2.34	1.79	1.95
S.D.	0.49	0.57	0.52	0.73		0.73	0.73	0.67	0.72

[*] **z** = baseline measurement
y = average of measurements made two and three hours after administration of the drug.

These data are analyzed using the SAS code in Program 8_10.sas.

Program 8_10.sas

```
/* Program 8_10.sas                          */
/* SUR Changing Covariates (Patel, 1986) */

options ls=80 ps=60 nodate nonumber;
title 'Output 8.10: SUR - 2-Period Crossover Design with Changing Covariate';

data patel;
   infile'c:\8_10.dat';
   input group x y1 z1 y2 z2 x1 x2;
proc print data=patel;
run;

proc summary data=patel;
   class x;
   var y1 z1 y2 z2;
   output out=new mean=y1 z1 y2 z2;
proc print data=new;
run;

proc syslin sur data=patel;
   title2 'Esimates using PROC SYSLIN';
   s1: model y1=x z1/covb;
   s2: model y2=x z2/covb;
run;

proc iml;
   title2 'Estimates using PROC IML';
   use patel;
```

```
   read all var {y1 z1 y2 z2} into y;
   read all var {x} into m;
   read all var {x1 x2} into d;
   n=nrow(y);
   y1=y[1:17,1:1];y2=y[1:17,3:3];
   yt=y1||y2;
   z1=y[1:17,2:2];z2=y[1:17,4:4];
   z=z1||z2;
   x1=d||z1;x2=d||z2;
   y11=y[1:8,1:1];  z11=y[1:8,2:2];  y12=y[1:8,3:3];  z12=y[1:8,4:4];
   y21=y[9:17,1:1]; z21=y[9:17,2:2]; y22=y[9:17,3:3]; z22=y[9:17,4:4];
   m1=m[1:8,1:1]; m2=m[9:17,1:1];
   p1=m1*inv(m1`*m1)*m1`; p2=m2*inv(m2`*m2)*m2`;
   s=i(4);
   s[1,1]=y11`*y11-y11`*p1*y11+y21`*y21-y21`*p2*y21;
   s[2,2]=z11`*z11-z11`*p1*z11+z21`*z21-z21`*p2*z21;
   s[3,3]=y12`*y12-y12`*p1*y12+y22`*y22-y22`*p2*y22;
   s[4,4]=z12`*z12-z12`*p1*z12+z22`*z22-z22`*p2*z22;
   s[1,2]=y11`*z11-y11`*p1*z11+y21`*z21-y21`*p2*z21;
   s[1,3]=y11`*y12-y11`*p1*y12+y21`*y22-y21`*p2*y22;
   s[1,4]=y11`*z12-y11`*p1*z12+y21`*z22-y21`*p2*z22;
   s[2,3]=z11`*y12-z11`*p1*y12+z21`*y22-z21`*p2*y22;
   s[2,4]=z11`*z12-z11`*p1*z12+z21`*z22-z21`*p2*z22;
   s[3,4]=y12`*z12-y12`*p1*z12+y22`*z22-y22`*p2*z22;
   do i=1 to 4;
      do j=1 to 4;
         s[j,i]=s[i,j];
      end;
   end;
   print 'SS and SP matrix of (y1,z1,y2,z2)',s;
    /* Input SUR estimates from SYLIN output */
   bhat={0.442258 .097582,.815344 -.052338};
   ghat={0.84504 0,0 1.115726};
   vhat=(yt-d*bhat-z*ghat)`*(yt-d*bhat-z*ghat);
   sigma=vhat/n;
   print 'Residual SS matrix',vhat;
   print 'Estimate of SUR Covariance Matrix using vhat/n',sigma;
   covehat=inv(inv(sigma)@(x1`*x2));
   print 'Asymptotic Covariance Matrix of Parameters',covehat;
   /* Hypothesis test matrices */
   ct={1 -1 0 0 0 0,0 0 0 -1 1 0}; dft=nrow(ct);
   cgamma={0 0 1 0 0 1}; dfgamma=nrow(cgamma);
   ctxp={1 -1 0 1 1 0}; dftp=nrow(ctxp);
   theta={0.442258 .815344 .84504 .097582 -.052338 1.115726};
   /*  Wald test of treatment */
   chi_t=(ct*theta`)`*inv(ct*covehat*ct`)*(ct*theta`);
   p_t=1-probchi(chi_t,dft);
   print chi_t dft p_t;
   /* Wald test of no covariates */
   chi_g=(cgamma*theta`)`*inv(cgamma*covehat*cgamma`)*(cgamma*theta`);
   p_g=1-probchi(chi_g,dfgamma);
   print chi_g dfgamma p_g;
   /* Wald test of no treatment x period interaction */
   chi_tp=(ctxp*theta`)`*inv(ctxp*covehat*ctxp`)*(ctxp*theta`);
   p_tp=1-probchi(chi_tp,dftp);
   print chi_tp dftp p_tp;
quit;
run;
```

Result and Interpretation 8_10.sas

To analyze the data in Table 8.3 using PROC SYSLIN we rearrange the data as shown in Output 8.10.

Output 8.10: *SUR - 2-Period Crossover Design with Changing Covariate*

```
Output 8.10: SUR - 2-Period Crossover Design with Changing Covariate

 OBS    GROUP    X     Y1     Z1     Y2     Z2    X1    X2

   1      1      1   1.28   1.09   1.33   1.24    1     0
   2      1      1   1.60   1.38   2.21   1.90    1     0
   3      1      1   2.46   2.27   2.43   2.19    1     0
   4      1      1   1.41   1.34   1.81   1.47    1     0
   5      1      1   1.40   1.31   0.85   0.85    1     0
   6      1      1   1.12   0.96   1.20   1.12    1     0
   7      1      1   0.90   0.66   0.90   0.78    1     0
   8      1      1   2.41   1.69   2.79   1.90    1     0
   9      1     -1   3.06   1.74   1.38   1.54    0     1
  10      1     -1   2.68   2.41   2.10   2.13    0     1
  11      1     -1   2.60   3.05   2.32   2.18    0     1
  12      1     -1   1.48   1.20   1.30   1.41    0     1
  13      1     -1   2.08   1.70   2.34   2.21    0     1
  14      1     -1   2.72   1.89   2.48   2.05    0     1
  15      1     -1   1.94   0.89   1.11   0.72    0     1
  16      1     -1   3.35   2.41   3.23   2.83    0     1
  17      1     -1   1.16   0.96   1.25   1.01    0     1

OBS    X    _TYPE_   _FREQ_      Y1         Z1         Y2         Z2

 1     .      0        17     1.97941    1.58529    1.82529    1.61941
 2    -1      1         9     2.34111    1.80556    1.94556    1.78667
 3     1      1         8     1.57250    1.33750    1.69000    1.43125
```

Because PROC SYLIN fits an intercept, the design matrix must be reparameterized to a constant term plus the treatment effect. Thus, the parameter matrix is defined:

$$\begin{pmatrix} \mu_{11} & \mu_{12} \\ \mu_{21} & \mu_{22} \end{pmatrix} \equiv \begin{pmatrix} \mu_1 & \alpha_{11} \\ \mu_2 & \alpha_{21} \end{pmatrix}$$

where $\delta = \alpha_{11} - \alpha_{21}$ is the difference in the treatment effects.

Output 8.10 (*continued*)

```
Output 8.10:  SUR-2-Period Crossover Design with Changing Covariates

                  Estimates Using PROC SYSLIN
                       SYSLIN Procedure
             Seemingly Unrelated Regression Estimation

                      Cross Model Covariance

             Sigma               S1                S2

             S1             0.17797788      0.0343445072
             S2           0.0343445072      0.0610892275
```

Output 8.10 *(continued)*

```
                         Cross Model Correlation

                 Corr                  S1                  S2

                 S1                     1        0.3293760055
                 S2          0.3293760055                   1

                     Cross Model Inverse Correlation

               Inv Corr                S1                  S2

               S1            1.1216905889        -0.369457966
               S2            -0.369457966        1.1216905889


                      Cross Model Inverse Covariance

              Inv Sigma                S1                  S2

              S1             6.3024157216        -3.543232927
              S2             -3.543232927         18.36151209

         System Weighted MSE:  0.99757 with 28 degrees of freedom.
         System Weighted R-Square:  0.8285

Model: S1
Dependent variable: Y1

                           Parameter Estimates

                     Parameter      Standard    T for H0:
      Variable  DF    Estimate         Error   Parameter=0    Prob > |T|

      INTERCEP   1    0.628801      0.287230         2.189        0.0460
      X          1   -0.186543      0.110010        -1.696        0.1121
      Z1         1    0.845040      0.170738         4.949        0.0002

Model: S2
Dependent variable: Y2

                           Parameter Estimates

                     Parameter      Standard    T for H0:
      Variable  DF    Estimate         Error   Parameter=0    Prob > |T|

      INTERCEP   1    0.022622      0.177420         0.128        0.9004
      X          1    0.070496      0.062817         1.122        0.2806
      Z2         1    1.115726      0.103762        10.753        0.0001

Covariance of Estimates

         COVB            INTERCEP                  X                   Z1

         INTERCEP    0.0825008297       -0.010103355         -0.045812229
         X           -0.010103355       0.0121022316         0.0068222365
         Z1          -0.045812229       0.0068222365         0.0291513965
         INTERCEP    0.0142527015       -0.001701326         -0.007779323
         X           -0.001231038       0.0022283615         0.0008592208
         Z2          -0.007598346       0.0011315256         0.0048350056
```

Output 8.10 *(continued)*

COVB	INTERCEP	X	Z2
INTERCEP	0.0142527015	-0.001231038	-0.007598346
X	-0.001701326	0.0022283615	0.0011315256
Z1	-0.007779323	0.0008592208	0.0048350056
INTERCEP	0.031477939	-0.002866325	-0.017322995
X	-0.002866325	0.003945973	0.0019133128
Z2	-0.017322995	0.0019133128	0.0107665904

From the continuation of Output 8.10 we see that the unconstrained design parameters for the model are

$$\begin{pmatrix} \hat{\mu}_1 & \hat{\alpha}_{11} \\ \hat{\mu}_2 & \hat{\alpha}_{21} \end{pmatrix} = \begin{pmatrix} 0.628801 & -0.186543 \\ 0.022622 & 0.07496 \end{pmatrix}$$

and the covariate parameters are estimated

$$\hat{\gamma}_{11} = 0.84504 \quad \hat{\gamma}_{12} = 1.115726$$

To estimate the difference between the two drugs, we have that

$$\hat{\delta} = \hat{\alpha}_{11} - \hat{\alpha}_{21} = -0.257039$$

Since $\hat{\alpha}_{11}$ is the average difference of drug A – drug B and $\hat{\alpha}_{21}$ is the average difference of drug B – drug A, the difference of the differences is the treatment effect, $\hat{\delta}$.

Patel (1986) proposed an iterative scheme to obtain ML estimates for the model parameters where the initial estimates are obtained from the sum of squares and products (SSP) matrix of y_1, z_1, y_2 and z_2. With his procedure, he obtained estimates for μ_{ij} where i = sequence and j = period of

$$\begin{pmatrix} \hat{\mu}_{11} & \hat{\mu}_{12} \\ \hat{\mu}_{21} & \hat{\mu}_{22} \end{pmatrix} = \begin{pmatrix} 0.422 & 0.093 \\ 0.815 & -0.048 \end{pmatrix} \quad \text{and} \quad \begin{matrix} \hat{\gamma}_{11} = 0.845 \\ \hat{\gamma}_{12} = 1.150 \end{matrix}$$

so that the treatment effect $\hat{\delta} = \left(\hat{\mu}_{11} + \hat{\mu}_{22} - \hat{\mu}_{12} - \hat{\mu}_{21}\right)/2 = -0.257$. The corresponding estimates for the μ_{ij} parameters from the SUR output are

$$\begin{aligned} \hat{\mu}_{11} &= 0.628801 - 0.186543 = 0.442258 \\ \hat{\mu}_{12} &= 0.628801 + 0.186543 = 0.815343 \\ \hat{\mu}_{21} &= 0.022622 + 0.07496 = 0.097582 \\ \hat{\mu}_{22} &= 0.022622 - 0.07496 = -0.052338 \end{aligned}$$

Hence, for this example, there is close agreement between Patel's iterative ML estimates and the SUR estimates, (these data are reanalyzed in Chapter 9 using PROC MIXED).

To obtain an estimate of the covariance matrix $\mathbf{\Sigma}$, Patel (1986) suggests dividing the residual matrix

$$\mathbf{V} = (\mathbf{Y} - \mathbf{X}\hat{\mathbf{B}} - \mathbf{Z}\hat{\mathbf{\Gamma}})'(\mathbf{Y} - \mathbf{X}\hat{\mathbf{B}} - \mathbf{Z}\hat{\mathbf{\Gamma}})$$

by n so that $\hat{\mathbf{\Sigma}} = \hat{\mathbf{V}} / n$. The degrees of freedom associated with this estimator are $n - m = n -$ rank$(\mathbf{X}) - r$. Program 8_10.sas contains the PROC IML code to perform this analysis. The results are found in the continuation of Output 8.10 that follows.

Output 8.10 *(continued)*

```
Output 8.10: SUR-2-Period Crossover Design with Changing Covariates
                  Estimates using PROC IML


              SS and SP matrix of (y1,z1,y2,z2)

                        S
             6.5142389 4.8637944 5.4266444 4.5598083
             4.8637944 5.8809722 5.0529222 4.6975917
             5.4266444 5.0529222 7.9154222 6.2118667
             4.5598083 4.6975917 6.2118667 5.4654875

                        Residual SS matrix

                          VHAT
                       2.4935959 0.4982712
                       0.4982712 0.8579608

         Estimate of SUR Covariance Matrix using vhat/n

                          SIGMA
                       0.1466821 0.0293101
                       0.0293101 0.0504683

          Asymptotic Covariance Matrix of Parameters

  COVEHAT
 0.0781091 0.0806916 -0.044691 0.0156078 0.0161238  -0.00893
 0.0746172 0.1170274 -0.055789   0.01491 0.0233845 -0.011148
 -0.041763 -0.056378  0.031225 -0.008345 -0.011266 0.0062394
 0.0156078 0.0161238  -0.00893 0.0268747 0.0277632 -0.015377
   0.01491 0.0233845 -0.011148 0.0256732 0.0402651 -0.019195
 -0.008345 -0.011266 0.0062394 -0.014369 -0.019398 0.0107434

                   CHI_T       DFT        P_T
               7.6545847         2 0.0217685

                   CHI_G   DFGAMMA        P_G
               70.611637         1          0

                  CHI_TP      DFTP       P_TP
               0.7420086         1 0.3890182
```

From the PROC IML output, we see that the estimates of $\mathbf{V}$ and $\boldsymbol{\Sigma}$ using the SUR estimates for $\hat{\mathbf{B}}$ and $\hat{\boldsymbol{\Gamma}}$ are

$$\hat{\mathbf{V}} = \begin{pmatrix} 2.494 & 0.498 \\ 0.498 & 0.858 \end{pmatrix} \quad \text{and} \quad \hat{\boldsymbol{\Sigma}} = \begin{pmatrix} 0.1467 & 0.0293 \\ 0.0293 & 0.0505 \end{pmatrix}$$

which are in agreement with Patel's estimates to three significant figures. The estimate of $\boldsymbol{\Sigma}$ computed by PROC SYLIN is defined in (8.52), and is labeled the `Cross Model Covariance` matrix in the output. It is approximately equal to the unbiased estimator

$$\mathbf{S} = \hat{\mathbf{V}} / (n - m)$$

$\hat{\boldsymbol{\Sigma}}$ and $\mathbf{S}$ are asymptotically equivalent.

To obtain the asymptotic covariance matrix of $\theta' = (\theta'_{11}, \theta'_{12}, \ldots, \theta'_{pp})$ using the SUR model, we use expression (8.55). The covariance matrix is

$$(\hat{\mathbf{\Sigma}}^{-1} \otimes \mathbf{X}_i'\mathbf{X}_j)^{-1}$$

where $\hat{\mathbf{\Sigma}}$ is any consistent estimator of $\mathbf{\Sigma}$, $\mathbf{X}_1 = (\mathbf{X}, \mathbf{z}_{11})$ and $\mathbf{X}_2 = (\mathbf{X}, \mathbf{z}_{12})$ for our example. The asymptotic covariance matrix of

$$\theta' = (\hat{\mu}_{11}, \hat{\mu}_{21}, \hat{\gamma}_{11}, \hat{\mu}_{12}, \hat{\mu}_{22}, \hat{\gamma}_{12})$$

is given in the output using $\hat{\mathbf{\Sigma}} = \hat{\mathbf{V}} / n$. The asymptotic variance of the treatment effect $\hat{\delta} = (\hat{\mu}_{11} + \hat{\mu}_{22} - \hat{\mu}_{12} - \hat{\mu}_{21}) / 2$ is 0.00736 which compares favorably with the ML estimate of 0.0094 obtained by Patel. The ease of the estimation procedure for the SUR model supports its use over the iterative ML procedure.

To test hypotheses given in (8.43), Patel (1986) derived likelihood ratio tests of

$$H_{01}: \quad \mathbf{CBA} = \mathbf{0}$$

$$H_{02}: \quad \mathbf{\Gamma}_i = \mathbf{0} \qquad \text{for all } i > r_1 < r$$

and H_{03} the equality of equal carryover effects as measured in terms of the baseline values. Because the exact distribution of the likelihood ratio statistic is unknown, he used the asymptotic chi-square distribution to obtain approximate critical values for the statistics. Patel also developed approximate tests based on Wilks' Λ-criterion. Using the latter approach, Verbyla (1988) suggests obtaining SUR estimates with constraints imposed by the hypothesis to obtain approximate tests based on Wilks' Λ-criterion. In particular, for H_{01},

$$\mathbf{E} = \mathbf{A}'(\mathbf{Y} - \mathbf{X}\hat{\mathbf{B}}_{\Omega} - \mathbf{Z}\hat{\mathbf{\Gamma}}_{\Omega})'(\mathbf{Y} - \mathbf{X}\hat{\mathbf{B}}_{\Omega} - \mathbf{Z}\hat{\mathbf{\Gamma}}_{\Omega})\mathbf{A}$$

$$\mathbf{H} = \mathbf{A}'(\mathbf{Y} - \mathbf{X}\hat{\mathbf{B}}_{\omega} - \mathbf{Z}\hat{\mathbf{\Gamma}}_{\omega})'(\mathbf{Y} - \mathbf{X}\hat{\mathbf{B}}_{\omega} - \mathbf{Z}\hat{\mathbf{\Gamma}}_{\omega})\mathbf{A}$$

so that

$$U = \frac{|\mathbf{E}|}{|\mathbf{H}|} \mathrel{\dot\sim} U(u, v_h, v_e)$$

where $u = \text{rank}(\mathbf{A})$, $v_h = \text{rank}(\mathbf{C})$, $k = \text{rank}(\mathbf{X})$, and $v_e = n - k - r$, and $\hat{\mathbf{B}}_{\omega}$ and $\hat{\mathbf{\Gamma}}_{\omega}$ are SUR estimates subject to the restriction of the hypothesis $\mathbf{CB} = \mathbf{0}$. However, this process is also complicated since it involves the sum of profiles model.

Instead of either of these approaches, we propose using the Wald statistic given in (8.61). In particular,

$$W = (\mathbf{C}_* \hat{\hat{\theta}} - \xi)'[\mathbf{C}_*(\hat{\mathbf{\Sigma}}^{-1} \otimes \mathbf{X}_i'\mathbf{X}_j)^{-1}\mathbf{C}_*]^{-1}(\mathbf{C}_* \hat{\hat{\theta}} - \xi)$$

has an asymptotic chi-square distribution with degrees of freedom $v = \text{rank}(\mathbf{C}_*)$. To test for differences in treatments, we write no covariates or $\mathbf{\Gamma}_i = \mathbf{0}$, and for no treatment by period interaction, the hypothesis test matrices are

$$H_{ot}: \ \mathbf{C}_* = \begin{pmatrix} 1 & -1 & 0 & 0 & 0 & 0 \\ 0 & 0 & 0 & -1 & 1 & 0 \end{pmatrix}$$

$$H_{\Gamma}: \ \mathbf{C}_* = \begin{pmatrix} 0 & 0 & 1 & 0 & 0 & 1 \end{pmatrix}$$

$$H_{t\times p}: \ \mathbf{C}_* = \begin{pmatrix} 1 & -1 & 0 & 1 & -1 & 0 \end{pmatrix}$$

where the Wald statistics have asymptotic chi-square distributions with 2, 1, and 1 degrees of freedom, respectively. The PROC IML code to perform these tests is included in Program 8_10.sas. From the PROC IML output, the p-value for the test of interaction is 0.389. The corresponding p-values obtained by Patel were 0.351 using his chi-square LR approximation and 0.404 using his approximate Wilks' Λ-criterion. For the test of treatment the p-value for the Wald statistic is 0.022. Patel obtained p-values of 0.001 and 0.020 for his two methods. While all three procedures are asymptotically equivalent, it is unknown how each performs for small samples.

8.11 Repeated Measurements with Changing Covariates

As our next example, we analyze repeated measures data with changing covariates. The SUR procedure and the Wald statistic provide researchers with a tool for the analysis of complete multivariate data with time varying covariates. The structure of the covariance matrix can be arbitrary, and we do not require equality of the covariates across variables. To illustrate, we use the data from Winer (1971, p. 806). The design is a pretest, posttest design with varying covariates. An exact solution exists assuming a common regression equation. The circularity condition is always met since the covariance of the difference is a constant; hence, all we require is homogeneity of variances across groups. We use the data to show how one may test for equal slopes and intercepts. The SAS code to perform the analysis is given in Program 8_11.sas.

Program 8_11.sas

```
/* Program 8_11.sas             */
/* SUR Winer(1971, page 806) */

options ls=80 ps=60 nodate nonumber;
title1 'Output 8.11:SUR - Repeated Measures with Changing Covariates';

data winer;
   infile'c:\8_11.dat';
   input group x1 x2 z1 y1 z2 y2 gr1 gr2 gr3;
proc print data=winer;
run;
proc summary data=winer;
   class x1;
   var z1 y1 z2 y2;
   output out=new mean= z1 y1 z2 y2;
```

```
proc print data=new;
run;

proc syslin sur data=winer;
   title2 'Estimate of B using SUR';
   s1: model y1=x1 x2 z1/covb;
   s2: model y2=x1 x2 z2/covb;
proc iml;
   title2 'Using Proc IML';
   use winer;
   read all var {y1 z1 y2 z2} into y;
   read all var {gr1 gr2 gr3} into d;
   n=nrow(y);
   y1=y[1:9,1:1];y2=y[1:9,3:3];
   yt=y1||y2;
   z1=y[1:9,2:2];z2=y[1:9,4:4];
   z=z1||z2;
   x1=d||z1;x2=d||z2;
   /* Input SUR estimates from SYLIN output */
   bhat={6.324944 11.102363, 3.361562 5.913461, 5.287068 9.017636};
   ghat={0.890147 0,0 0.804161};
   vhat=(yt-d*bhat-z*ghat)`*(yt-d*bhat-z*ghat);
   sigma=vhat/n;
   print 'Residual SS matrix',vhat;
   print 'Estimate of SUR Covariance Matrix using vhat/n',sigma;
   covehat=inv(inv(sigma)@(x1`*x2));
   print 'Asymptotic Covariance Matrix of Parameters',covehat;
   /* Hypothesis test matrices */
   mt={1 0 -1 0 0  0  0 0,
       0 1 -1 0 0  0  0 0,
       0 0  0 0 1 -1  0 0,
       0 0  0 0 0  1 -1 0}; dfmt=nrow(mt);
   cgamma={0 0 0 1 0 0 0 -1}; dfgamma=nrow(cgamma);
   theta={6.324944 3.361562 5.287068 0.890147
          11.102363 5.913461 9.017636 0.804161};
   print theta;
   /*  Wald test of multivariate treatment */
   chi_mt=(mt*theta`)`*inv(mt*covehat*mt`)*(mt*theta`);
   p_mt=1-probchi(chi_mt,dfmt);
   print 'Multivariate test of intercepts',mt;
   print chi_mt dfmt p_mt;
   /* Wald test of equal covariates */
   print 'Equality of Covariates', cgamma;
   chi_g=(cgamma*theta`)`*inv(cgamma*covehat*cgamma`)*(cgamma*theta`);
   p_g=1-probchi(chi_g,dfgamma);
   print chi_g dfgamma p_g;
quit;
run;
```

Result and Interpretation 8_11.sas

The results are found in Output 8.11.

Output 8.11: *SUR - Repeated Measures with Changing Covariates*

Output 8.11:SUR - Repeated Measures with Changing Covariates

OBS	GROUP	X1	X2	Z1	Y1	Z2	Y2	GR1	GR2	GR3
1	1	1	0	3	8	4	14	1	0	0
2	1	1	0	5	11	9	18	1	0	0
3	1	1	0	11	16	14	22	1	0	0
4	1	0	1	2	6	1	8	0	1	0
5	1	0	1	8	12	9	14	0	1	0

Output 8.11 *(continued)*

```
    6       1       0     1    10     9     9    10      0      1      0
    7       1      -1    -1     7    10     4    10      0      0      1
    8       1      -1    -1     8    14    10    18      0      0      1
    9       1      -1    -1     9    15    12    22      0      0      1

  OBS     X1    _TYPE_    _FREQ_       Z1          Y1          Z2          Y2

   1       .      0         9       7.00000     11.2222     8.00000     15.1111
   2      -1      1         3       8.00000     13.0000     8.66667     16.6667
   3       0      1         3       6.66667      9.0000     6.33333     10.6667
   4       1      1         3       6.33333     11.6667     9.00000     18.0000

                         Estimate of B using SUR

                              SYSLIN Procedure
                  Seemingly Unrelated Regression Estimation

                           Cross Model Covariance

                  Sigma                 S1                  S2

                  S1          4.0358255452        3.8600512143
                  S2          3.8600512143        5.3640488656

                          Cross Model Correlation

                   Corr                 S1                  S2

                   S1                    1        0.8296224802
                   S2         0.8296224802                   1

                      Cross Model Inverse Correlation

                 Inv Corr                 S1                  S2

                 S1             3.2079398794        -2.661379039
                 S2             -2.661379039        3.2079398794

                       Cross Model Inverse Covariance

                Inv Sigma                 S1                  S2

                S1              0.7948658443        -0.571997561
                S2              -0.571997561        0.5980444921

        System Weighted MSE:  0.82556 with 10 degrees of freedom.
        System Weighted R-Square:  0.8608

                             SYSLIN Procedure
                  Seemingly Unrelated Regression Estimation

Model: S1
Dependent variable: Y1

                            Parameter Estimates

                     Parameter       Standard    T for H0:
      Variable  DF    Estimate          Error  Parameter=0    Prob > |T|

      INTERCEP   1    4.991191       1.541579        3.238        0.0230
      X1         1    1.037876       0.956210        1.085        0.3273
      X2         1   -1.925506       0.949327       -2.028        0.0983
      Z1         1    0.890147       0.198363        4.487        0.0065
```

Output 8.11 *(continued)*

Model: S2
Dependent variable: Y2

Parameter Estimates

Variable	DF	Parameter Estimate	Standard Error	T for H0: Parameter=0	Prob > |T|
INTERCEP	1	8.677820	1.571956	5.520	0.0027
X1	1	2.084727	1.105128	1.886	0.1179
X2	1	-3.104175	1.128447	-2.751	0.0403
Z2	1	0.804161	0.171165	4.698	0.0053

Covariance of Estimates

COVB	INTERCEP	X1	X2	Z1
INTERCEP	2.3764642781	-0.183622783	-0.091811391	-0.275434174
X1	-0.183622783	0.9143380052	-0.439681119	0.0262318261
X2	-0.091811391	-0.439681119	0.9012220922	0.013115913
Z1	-0.275434174	0.0262318261	0.013115913	0.0393477391
INTERCEP	1.8412535567	-0.134510379	-0.067255189	-0.201765568
X1	0.1765448722	0.8409753614	-0.437301478	-0.025220696
X2	-0.294241454	-0.400871584	0.8718006565	0.0420344934
Z2	-0.176544872	0.0168137973	0.0084068987	0.025220696

COVB	INTERCEP	X1	X2	Z2
INTERCEP	1.8412535567	0.1765448722	-0.294241454	-0.176544872
X1	-0.134510379	0.8409753614	-0.400871584	0.0168137973
X2	-0.067255189	-0.437301478	0.8718006565	0.0084068987
Z1	-0.201765568	-0.025220696	0.0420344934	0.025220696
INTERCEP	2.4710453069	0.2343799847	-0.390633308	-0.234379985
X1	0.2343799847	1.2213083571	-0.644834593	-0.029297498
X2	-0.390633308	-0.644834593	1.2733927981	0.0488291635
Z2	-0.234379985	-0.029297498	0.0488291635	0.0292974981

The data are listed in Output 8.11. Using PROC SYLIN, we present the estimates for the model parameters in Output 8.11 and summarized them as follows:

$$\begin{pmatrix} \mu_1 & \mu_2 \\ \hat{\alpha}_{11} & \hat{\alpha}_{21} \\ \hat{\alpha}_{12} & \hat{\alpha}_{22} \end{pmatrix} = \begin{pmatrix} 4.991191 & 8.677820 \\ 1.037876 & 2.084727 \\ -1.925506 & -3.104175 \end{pmatrix}$$

Transforming these parameters to cell means, we have that

$$\hat{\mathbf{B}} = \begin{pmatrix} \hat{\mu}_{11} & \hat{\mu}_{12} \\ \hat{\mu}_{21} & \hat{\mu}_{22} \\ \hat{\mu}_{31} & \hat{\mu}_{32} \end{pmatrix} = \begin{pmatrix} 6.325 & 11.102 \\ 3.362 & 5.913 \\ 5.287 & 9.018 \end{pmatrix} \text{ and } \hat{\boldsymbol{\Gamma}} = \begin{pmatrix} \hat{\gamma}_{11} & 0 \\ 0 & \hat{\gamma}_{12} \end{pmatrix} = \begin{pmatrix} 0.890147 & 0 \\ 0 & 0.804161 \end{pmatrix}$$

The estimates of $\mathbf{B}$ and $\boldsymbol{\Gamma}$ are FGLS estimates and not ML estimates. It is a more complicated process to obtain ML estimates. This will be discussed in Chapter 9. To test the equality of intercepts and slopes, we use the Wald statistic. The hypothesis test matrices in Program 8_11.sas to test H: $\mathbf{CBA} = \mathbf{0}$ are obtained by recalling that the test of H is equivalent to the test of $\mathbf{C}_* \text{vec}(\mathbf{B}) = \mathbf{0}$ where $\mathbf{C}_* = \mathbf{A}' \otimes \mathbf{C}$. The PROC IML output containing the results follows.

Output 8.11 *(continued)*

```
                              Using Proc IML
                            Residual SS matrix

                                  VHAT
                           22.474349  23.51342
                            23.51342 29.623092

              Estimate of SUR Covariance Matrix using vhat/n

                                 SIGMA
                           2.4971499 2.6126022
                           2.6126022 3.2914546

                Asymptotic Covariance Matrix of Parameters

 COVEHAT
2.5004014 1.7558086 2.1069703 -0.263371  2.616004 1.8369859 2.2043831 -0.275548
1.1737905 2.0679523 1.4826828 -0.185335 1.2280591 2.1635612 1.5512325 -0.193904
1.6062397 1.6907786 2.8613176 -0.253617 1.6805019 1.7689494 2.9936067 -0.265342
-0.185335  -0.19509 -0.234108 0.0292635 -0.193904  -0.20411 -0.244931 0.0306164
 2.616004 1.8369859 2.2043831 -0.275548 3.2957404  2.314304 2.7771648 -0.347146
1.2280591 2.1635612 1.5512325 -0.193904 1.5471551 2.7257359 1.9543012 -0.244288
1.6805019 1.7689494 2.9936067 -0.265342 2.1171596 2.2285891 3.7714584 -0.334288
-0.193904  -0.20411 -0.244931 0.0306164 -0.244288 -0.257145 -0.308574 0.0385717

   THETA
 6.324944  3.361562  5.287068  0.890147 11.102363  5.913461  9.017636  0.804161

                     Multivariate test of intercepts

      MT
       1         0        -1         0         0         0         0         0
       0         1        -1         0         0         0         0         0
       0         0         0         0         1        -1         0         0
       0         0         0         0         0         1        -1         0

                         CHI_MT       DFMT        P_MT
                      17.331075          4 0.0016666

                          Equality of Covariates

     CGAMMA
       0         0         0         1         0         0         0        -1

                          CHI_G    DFGAMMA         P_G
                      1.1198432          1 0.2899522
```

The Wald statistic for the equality of intercepts is $X^2 = 17.33$ with p-value 0.0017. The test of equal slopes is not rejected (p-value = 0.2900); hence, a common slope model is appropriate for the design. Using the matrix $\hat{\Gamma}$, the between and within regression slopes given by Winer (1971, p. 806) are easily obtained

$$\hat{\gamma}_b = (\hat{\gamma}_{11} + \hat{\gamma}_{12})/2 = 0.847$$

$$\hat{\gamma}_\omega = (8\hat{\gamma}_{11} + 9\hat{\gamma}_{12})/17 = 0.845$$

A reanalysis of the data with equal slopes is illustrated in Chapter 9 using a mixed model, PROC MIXED, and PROC GLM. An analysis of these data using the ANCOVA model is provided in BMDP Statistical Software, Inc. (1992, p. 1289). The SUR methodology provides a procedure for the analysis of complex multivariate designs that do not require iterative methods. How well the procedure performs for small samples is, in general, unknown. However, for very small samples, its use should be avoided.

8.12 MANOVA-GMANOVA Model

Given Patel's model with no covariates so that $E(\mathbf{Y}) = \mathbf{XB}$ subject to the restriction $\mathbf{R}_1\mathbf{B}\mathbf{R}_2 = \mathbf{0}$, Verbyla (1988) following (8.4) showed that the restricted MANOVA model reduces to a mixture of the GMANOVA and MANOVA models as given in (8.5). In particular

$$\begin{aligned}\mathbf{XB} &= \mathbf{X}_1^*\beta_{12}\mathbf{P}^{21} + \mathbf{X}_2^*(\beta_{21}, \beta_{22})\mathbf{P}^{-1} \\ &= \mathbf{X}_1^*\beta_{12}\mathbf{P}^{21} + \mathbf{X}_2^*\beta_{2.}\end{aligned} \tag{8.114}$$

where $\beta_{2.} = (\beta_{21}, \beta_{22})\mathbf{P}^{-1}$, $\mathbf{X}^* = \mathbf{X}\mathbf{R}^{-1} = (\mathbf{X}_1^*, \mathbf{X}_2^*)$, $\mathbf{P}^{-1} = \begin{pmatrix}\mathbf{P}^{11} \\ \mathbf{P}^{21}\end{pmatrix}$ and $\mathbf{P}$ and $\mathbf{R}$ are defined in (8.2). Adding covariates to the model, $\mathbf{XB}$ in Patel's model can be replaced by (8.114) with $\mathbf{X}_2^*\beta$ and the covariates $\sum_j \mathbf{Z}_j\mathbf{\Gamma}_j$ incorporated into a SUR model as in (8.44). Thus, Patel's model with restrictions has the general form under the hypothesis

$$E(\mathbf{Y}) = \mathbf{X}_1^*\beta_{12}\mathbf{P}^{21} + (\mathbf{Z}_1^*\gamma_1^*, \ldots, \mathbf{Z}_p^*\gamma_p^*) \tag{8.115}$$

where $\mathbf{Z}_j^* = (\mathbf{X}_j, \mathbf{X}_2^*)$, $\gamma_j^{*\prime} = (\gamma_j', \beta_{2.j}')$ and $\beta_{2.j}$ is the jth column of $\beta_{2.}$, a mixture of the GMANOVA and the MANOVA models.

Verbyla (1988) recommends writing (8.114) as a sum of profiles model. In particular, $\mathbf{P}^{21}$ is partitioned as $(\mathbf{P}^{211}, \mathbf{P}^{212}, \ldots, \mathbf{P}^{21p})$ so that (8.114) becomes

$$E(\mathbf{Y}) = \sum_{j=1}^{p} \mathbf{W}_j\theta_{jj}\mathbf{e}_j' \tag{8.116}$$

where $\mathbf{W}_j = (\mathbf{X}_1^*\mathbf{Z}_i^*)$, $\theta_{jj} = [(\beta_{12}\mathbf{P}^{21k})', \gamma_k^{*\prime}]$ and $\mathbf{e}_j$ is the jth vector of the normalized basis of a p-dimensional Euclidean space. Writing (8.115) as an MSUR model, Verbyla recommends estimating the parameters using an iterative least squares procedures following Verbyla and Venables (1988). We will not pursue this approach here since an alternative approach which seems to work reasonably well is to use the CGMANOVA model developed by Hecker (1987). Before discussing this approach, we turn to the analysis of the MANOVA-GMANOVA model considered by Chinchilli and Elswick (1985), a special case of the Patel model with closed form ML estimates and LR tests.

A simple application of the GMANOVA-MANOVA model is a design with m_1 groups, m_2 covariates, and p repeated measurements. The MANOVA model contains the covariates and GMANOVA model the growth curves. This design was considered by Chinchilli and Elswick (1985) using data from Danford et al. (1960). The Danford et al. (1960) data set consists of 45 patients with cancerous lesions who were subjected to whole-body x-radiation. The radiation dosage was at four levels, a control and three treatment levels. A baseline measurement was taken at day 0 (pretreatment), and then measurements were taken daily for ten consecutive days (posttreatment). For our example, only the first five measurements are used. Program 8_12.sas contains the SAS code to analyze both the GMANOVA model

$$\underset{45\times5}{E(\mathbf{Y})} = \underset{45\times4}{\mathbf{X}_1}\ \underset{4\times2}{\boldsymbol{\Theta}}\ \underset{2\times5}{\mathbf{X}_2} \tag{8.117}$$

and the MANOVA-GMANOVA model

$$\underset{45\times5}{E(\mathbf{Y})} = \underset{45\times4}{\mathbf{X}_1}\ \underset{4\times2}{\boldsymbol{\Theta}}\ \underset{2\times5}{\mathbf{X}_2} + \underset{45\times1}{\mathbf{Z}}\ \underset{1\times5}{\boldsymbol{\Gamma}_o} \tag{8.118}$$

where $\mathbf{X}_2$ is a matrix of orthogonal polynomials, and $\mathbf{Z}$ contains the baseline, covariate measurements.

Program 8_12.sas

```
/* Program 8_12.sas                                      */
/* MANOVA-GMANOVA, Chinchilli and Elswick (1985) */

options ls=80 ps=60 nodate nonumber;
title1 'Output 8.12: MANOVA-GMANOVA Model';

data chinels;
   infile'c:\8_12.dat';
   input group d1 d2 d3 d4 z1 y1 - y5;
proc print data=chinels;
run;
proc summary data=chinels;
   class group;
   var y1 - y5 z1;
   output out=new1 mean= y1 - y5 z1;
proc print data=new1;
run;
data plot;
   set new1;
   array mr(5) y1 - y5;
   do time = 1 to 5;
      response = mr(time);
      output;
   end;
   drop mr1 - mr5;
proc plot;
   plot response*time=group;
run;

proc glm data=chinels;
   model y1 - y5 = group/nouni;
   repeated time polynomial/summary nom nou;
run;

proc iml;
   use chinels;
   read all var {y1 y2 y3 y4 y5} into y;
   read all var {d1 d2 d3 d4} into x1;
   read all var {group} into gr;
   read all var {z1} into z;
   /* Create Orthogonal Polynomials of degree 9  */
   t={1 2 3 4 5};
   p_prime=orpol(t,4);
   y0=y*p_prime;
   new=y0||x1||gr||z;
   /*Create Transformed data set */
   varnames={yt1 yt2 yt3 yt4 yt5 d1 d2 d3 d4 group z};
   create trans from new (|colname=varnames|);
   append from new;
quit;
```

```
proc print data=trans;
run;

/*Test of fit, ignoring covariates */
proc glm data=trans;
   title2 'Test of Fit, Ignoring Covariate';
   model yt5=/nouni;
   manova h=intercept;
run;

/*Estimate of growth matrix B ignoring covariate */
proc glm data=trans;
   title2 'Estimate of B ignoring Covariate';
   class group;
   model yt1 yt2 = group yt3-yt5;
   estimate 'beta1' intercept 1 group 1 0 0 0;
   estimate 'beta2' intercept 1 group 0 1 0 0;
   estimate 'beta3' intercept 1 group 0 0 1 0;
   estimate 'beta4' intercept 1 group 0 0 0 1;
run;

/* Goodness-of-Fit Test of GMANOVA Model vs mixed GMANOVA-MANOVA */
/* using PROC IML */
proc iml;
   title2 'Test of Fit of GMANOVA vs GMANOVA-MANOVA, Using PROC IML';
   use trans;
   read all var {yt3 yt4 yt5} into y;
   read all var {z} into z;
   gam_str=inv(z`*z)*z`*y;print 'Gamma Star',gam_str;
   h=y`*z*inv(z`*z)*z`*y; print 'Hypothesis Matrix',h;
   e=y`*y-h;print 'Error Matrix',e;
   den=h+e;print 'H + E Matrix',den;
   lamda=det(e)/det(e+h);
   c=root(e);
   g=(inv(c`)*h*inv(c));
   values=eigval(round(g,.00001));
   print 'Eigenvalues',values;
   print 'Wilks Lamda',lamda;

/* Using PROC GLM */
proc glm data=trans;
   title2 'Test of GMANOVA vs GMANOVA-MANOVA, Using PROC GLM';
   model yt3 yt4 yt5 =z/noint;
   manova h=z/printe printh;
run;

/*Goodness-of-Fit test of mixed GMANOVA-MANOVA model vs MANOVA */
/* Using PROC IML */
proc iml;
   title2 'Test of GMANOVA-MANOVA vs MANOVA, Using PROC IML';
   use trans;
   read all var {yt3 yt4 yt5} into y;
   read all var {d1 d2 d3 d4 z} into x;
   read all var {z} into z;
   h=y`*x*inv(x`*x)*x`*y-y`*z*inv(z`*z)*z`*y;
   print 'Hypothesis Matrix ', h;
   e=y`*y-y`*x*inv(x`*x)*x`*y;
   print 'Error Matrix',e;
   den=h+e; print 'H + E Matrix',den;
   lamda=det(e)/det(e+h);
   c=root(e);
   g=(inv(c`)*h*inv(c));
   values=eigval(round(g,.0001));
   print 'Eigenvalues',values;
```

```
   print 'Wilks Lamda',lamda;
quit;
run;

/*Using PROC GLM */
proc glm data=trans;
   title2 'Test of GMANOVA-MANOVA vs MANOVA, Using PROC GLM';
   class group;
   model yt3 yt4 yt5 = group z/noint nouni;
   manova h=group/printe printh;
run;

/*Obtain mixed GMANOVA-MANOVA parameter estimates and Tests*/
proc iml;
   title2 'GMANOVA-MANOVA Parameter Estimates and Tests';
   use chinels;
   read all var {y1 y2 y3 y4 y5} into y;
   read all var {d1 d2 d3 d4}  into x1;
   read all var {d1 d2 d3 d4 z1} into x;
   read all var {z1} into z;
   use trans;
   read all var {yt1 yt2} into y0;
   read all var {yt3 yt4 yt5} into y0c;
   t={1 2 3 4 5};
   p_prime=orpol(t,4);
   p=p_prime[1:5,1:2];
   pc=p_prime[1:5,3:5];
   h1=p`; x2=p`;
   h2=pc`;
   n=nrow(y); m1=ncol(x1); m2=ncol(z); p=ncol(y); q=nrow(h1);
   print n m1 m2 p q;
   e1=y`*y-y`*x1*inv(x1`*x1)*x1`*y;
   w=i(5)-inv(e1)*x2`*inv(x2*inv(e1)*x2`)*x2;
   sigma1=(e1+w`*y`*x1*inv(x1`*x1)*x1`*y*w)/n;
   print 'ML estimate of sigma Gmanova Model',sigma1;
   sigma=sigma1-(w`*y`*z*inv(z`*z)*z`*y*w)/n;
   print 'ML estimate of sigma Gmanova-Manova Model', sigma;
   c=inv(x`*x);
   x11=c[1:4,1:4]; x12=c[1:4,5:5];
   x21=c[5:5,1:4]; x22=c[5:5,5:5];
   theta=(x11*x1`+x12*z`)*y*inv(e1)*x2`*inv(x2*inv(e1)*x2`);
   print 'Theta Hat Matrix',theta;
   gamma0=(x21*x1`+x22*z`)*y*inv(e1)*x2`*
          inv(x2*inv(e1)*x2`)*x2+inv(z`*z)*z`*y*w;
   print 'Gamma Zero Hat Matrix',gamma0;
   /*Begin test gamma0 equal to zero  */
   e=y`*y-y`*x*inv(x`*x)*x`*y;
   e1t=inv(x2*inv(e)*x2`);print e1t;
   e2=y`*y-y`*z*inv(z`*z)*z`*y;
   e2t=h2*e2*h2`;print e2t;
   eta=gamma0*inv(e)*x2`*inv(x2*inv(e)*x2`);
   r2=(x22+x21*x1`+x22*z`)*y*(inv(e)-inv(e)*x2`*inv(x2*inv(e)*x2`)*x2*inv
      (e))*y`*(x1*x12+z*x22);
   h1t=eta`*inv(r2)*eta;print h1t;
   h2t=(gamma0*h2`)`*(z`*z)*(gamma0*h2`);print h2t;
   lamda1=det(e1t)/(det(e1t+h1t)); lamda2=det(e2t)/(det(e2t+h2t));
   new1=n-m1-m2-p+q; g1=3; u1=q; v1=(u1-g1+1)/2;
   chi1=-(new1-v1)#log(lamda1); df1=u1#g1;
   new2=n-m2; g2=1; u2=p-q; v2=(u2-g2+1)/2;
   chi2=-(new2-v2)#log(lamda2); df2=u2#g2;
   print 'For test of gamma0', lamda1 chi1 df1;
   print                       lamda2 chi2 df2;
   pval1=1-probchi(chi1,df1);pval2=1-probchi(chi2,df2);k=m1+m2+p-q;
   rho1=1-(k-g1+v1)/n; gamma1=new1#(u1**2+g1**2-5)/(48#(rho1#n)**2);
```

```
      rho2=1-(m2-g2+v2)/n;gamma2=new2#(u2**2+g2**2-5)/(48#(rho2#n)**2);
      print rho1 rho2 gamma1 gamma2;
      pval1=(1-gamma1)#(1-probchi(chi1,df1))+gamma1#(1-probchi(chi1,df1+4));
      pval2=(1-gamma2)#(1-probchi(chi2,df2))+gamma2#(1-probchi(chi2,df2+4));
      pvalue=pval1+pval2;
      print 'p-value of test H2: gamma0=zero' pval1 pval2 pvalue;
      /*Begin test of Coincidence of profiles */
      c1={1 -1 0 0, 1 0 -1 0, 1 0 0 -1};
      a1=i(2);
      e=a1`*e1t*a1;
      r1=x11+(x11*x1`+x12*z`)*y0c*inv(y0c`*y0c-y0c`*x*inv(x`*x)*x`*y0c)*y0c`*
         (x1*x11+z*x21);
      h=(c1*theta*a1)`*inv(c1*r1*c1`)*(c1*theta*a1);
      lamda=det(e)/det(e+h);
      dfe=n-m1-m2-p+q; dfh=3; u=2;
      chi_2=-(dfe-(u-dfh+1)/2)#log(lamda); df=u#dfh; v1=(u+dfh+1)/2;
      print 'Test of Coincidence Wilks and Chi-square' lamda dfe dfh chi_2 df;
      new=u*dfh; rho=1-(k-dfh+v1)/n; omega=new#(u**2+dfh**2-5)/(48#(rho#n)**2);
      print rho omega;
      pval=(1-omega)#(1-probchi(chi_2,df))+omega#(1-probchi(chi_2,df+4));
      print 'p-value of test H1: Coincidence' pval;
   quit;
   run;
```

Result and Interpretation 8_12.sas

Plotting the data for model (8.117), we see that the data appear linear. This is confirmed by the sequential polynomial tests and the test of model adequacy, ignoring the baseline data (p-value = 0.9162). The matrix of parameters $\mathbf{\Theta}$ represents our assumption that the model is linear, ignoring the covariates. This result is consistent with the test of model adequacy, again ignoring the covariates (p-value = 0.9162). Finally, the unadjusted matrix of parameters is

$$\hat{\mathbf{\Theta}}' = \begin{pmatrix} 399.67 & 312.96 & 418.73 & 453.73 \\ 49.06 & 31.35 & 34.01 & 26.90 \end{pmatrix} \tag{8.119}$$

The question that now arises is whether we can improve upon these estimates by reducing their variability by including a baseline covariate in the model. This would also improve the power of tests regarding model parameters. To continue, we must evaluate whether the model (8.118) fits the data. To evaluate the adequacy of the model, we perform the goodness-of-fit tests given in (8.37) and (8.40). In (8.37), ω states that the model associated with the covariates is the same as the model for the repeated measures, linear for our example. Letting $\mathbf{Y}_2 = \mathbf{Y}\mathbf{H}_2'$, ignoring the growth curve model, we see that $E(\mathbf{Y}_2) = \mathbf{Z}\mathbf{\Gamma}^*$ where $\mathbf{\Gamma}^* = \mathbf{\Gamma}_o\mathbf{H}_2'$ is a $m_2 \times p$ matrix of parameters. To test $H{:}\mathbf{\Gamma}^* = \mathbf{0}$, from (8.38) we have

$$\tilde{\mathbf{H}} = \mathbf{Y}_2'\mathbf{Z}(\mathbf{Z}'\mathbf{Z})^{-1}\mathbf{Z}'\mathbf{Y}_2$$
$$\mathbf{E} = \mathbf{Y}_2'\mathbf{Y}_2 - \tilde{\mathbf{H}}$$

with $\mathbf{\Lambda} = |\tilde{\mathbf{E}}|/|\tilde{\mathbf{H}} + \tilde{\mathbf{E}}| \sim U_{p-q}(v_n = m_2,\ v_e = n - m_2)$. In Program 8_12.sas, we perform these calculations using PROC IML and PROC GLM using the NOINT option in the MODEL statement and the MANOVA statement. The results are given in Output 8.12.

Output 8.12 *Test of Fit of GMANOVA vs GMANOVA-MANOVA, Using PROC IML*

```
          Test of Fit of GMANOVA vs GMANOVA-MANOVA, Using PROC IML

                              Gamma Star

                         GAM_STR
                        -0.030038 0.0412179 0.0086033

                          Hypothesis Matrix

                               H
                        792.41047  -1087.35 -226.9609
                         -1087.35 1492.0675 311.43704
                        -226.9609 311.43704 65.005793

                             Error Matrix

                               E
                        20686.875 4854.0334 1398.7245
                        4854.0334 10126.233 -144.6413
                        1398.7245 -144.6413 5541.2085

                             H + E Matrix

                             DEN
                        21479.286 3766.6835 1171.7635
                        3766.6835   11618.3 166.79572
                        1171.7635 166.79572 5606.2143

                             Eigenvalues

                                 VALUES
                              0.2940672
                              9.5286E-7
                              -8.187E-6

                             Wilks Lamda

                                  LAMDA
                               0.772756

               Test of GMANOVA vs GMANOVA-MANOVA, Using PROC GLM

                     General Linear Models Procedure
                    Multivariate Analysis of Variance

               Manova Test Criteria and Exact F Statistics for
                    the Hypothesis of no Overall Z Effect
           H = Type III SS&CP Matrix for Z   E = Error SS&CP Matrix

                          S=1    M=0.5    N=20

Statistic                    Value          F       Num DF    Den DF  Pr > F

Wilks' Lambda              0.77275595     4.1170         3        42  0.0120
Pillai's Trace             0.22724405     4.1170         3        42  0.0120
Hotelling-Lawley Trace     0.29406962     4.1170         3        42  0.0120
Roy's Greatest Root        0.29406962     4.1170         3        42  0.0120
```

The null hypothesis that the covariate for the model is also linear is rejected ($p = 0.0120$).

We next want to evaluate whether the growth curve portion of the model is a linear function of time with the covariate term in the model. This is not the same as the test of model adequacy ignoring the covariate, but is the test given in (8.40). The null hypothesis ω is that we have a MANOVA-GMANOVA model versus the alternative that the model is MANOVA. Again letting $\mathbf{Y}_2 = \mathbf{Y}\mathbf{H}_2'$, we have that

$$\tilde{\mathbf{H}} = \mathbf{Y}_2'(\mathbf{X}(\mathbf{X}'\mathbf{X})^{-1}\mathbf{X}' - \mathbf{Z}(\mathbf{Z}'\mathbf{Z})^{-1}\mathbf{Z}')\mathbf{Y}_2$$
$$\tilde{\mathbf{E}} = \mathbf{H}_2\mathbf{E}\mathbf{H}_2 = \mathbf{Y}_2'(\mathbf{I} - \mathbf{X}(\mathbf{X}'\mathbf{X})^{-1}\mathbf{Z}')\mathbf{Y}_2$$

where $\mathbf{X} = (\mathbf{X}_1, \mathbf{Z})$ and $\mathbf{E} = \mathbf{Y}'(\mathbf{I} - \mathbf{X}(\mathbf{X}'\mathbf{X})^{-1}\mathbf{X}')\mathbf{Y}$. Alternatively, we may employ PROC GLM with the NOINT option in the MODEL statement. For this test, $\Lambda = |\tilde{\mathbf{E}}|/|\tilde{\mathbf{H}} + \tilde{\mathbf{E}}| \sim U_{p-q}(v_n = m_1,\ v_e = n - m_1 - m_2)$. The results are given in the output that follows.

Output 8.12 (*continued*)

```
                  Output 8.12:  MANOVA-GMANOVA Model

           Test of GMANOVA-MANOVA vs MANOVA, Using PROC IML

                          Hypothesis Matrix

                            H
                  788.50214 392.37063 -22.72803
                  392.37063 2522.6501 -577.5972
                  -22.72803 -577.5972 731.82554

                            Error Matrix

                            E
                  19898.373 4461.6628 1421.4525
                  4461.6628 7603.5825 432.95587
                  1421.4525 432.95587  4809.383

                            H + E Matrix

                          DEN
                  20686.875 4854.0334 1398.7245
                  4854.0334 10126.233 -144.6413
                  1398.7245 -144.6413 5541.2085

                            Eigenvalues

                               VALUES
                            0.4133794
                            0.1106642
                            0.0384564

                            Wilks Lamda

                               LAMDA
                             0.613405

           Test of GMANOVA-MANOVA vs MANOVA, Using PROC GLM

                  General Linear Models Procedure
                  Multivariate Analysis of Variance
```

Output 8.12 (*continued*)

```
               Manova Test Criteria and F Approximations for
                  the Hypothesis of no Overall GROUP Effect
         H = Type III SS&CP Matrix for GROUP    E = Error SS&CP Matrix

                              S=3     M=0     N=18

 Statistic                     Value          F       Num DF     Den DF   Pr > F

 Wilks' Lambda              0.61340500     1.7047         12   100.8301   0.0766
 Pillai's Trace             0.42919846     1.6695         12        120   0.0819
 Hotelling-Lawley Trace     0.56255143     1.7189         12        110   0.0721
 Roy's Greatest Root        0.41335519     4.1336          4         40   0.0068

NOTE: F Statistic for Roy's Greatest Root is an upper bound.
```

Because the p-value for the test is nonsignificant ($p = 0.0766$), the MANOVA-GMANOVA model appears reasonable for the experiment.

To obtain the ML estimates of the parameter matrix $\mathbf{\Theta}$, $\mathbf{\Gamma}_o$ and $\mathbf{\Sigma}$, we use PROC IML and the expressions given in (8.10).

Output 8.12 (*continued*)

```
                        Output 8.12:  MANOVA-GMANOVA Model
                     GMANOVA-MANOVA Parameter Estimates and Tests

                        N          M1          M2          P          Q
                       45           4           1          5          2

                         ML estimate of sigma Gmanova Model

                 SIGMA1
                4485.756  4171.786 4767.3581  4927.917 4439.5311
                4171.786 4374.0923 4866.1494 4977.0937  4503.038
               4767.3581 4866.1494  5704.589 5753.2162 5222.0549
                4927.917 4977.0937 5753.2162 6112.5601 5548.1936
               4439.5311  4503.038 5222.0549 5548.1936  6001.389

                  ML estimate of sigma Gmanova-Manova Model

                  SIGMA
               4411.6058 4170.1745 4759.1262 4869.0095 4407.5881
               4170.1745 4374.0573 4865.9705 4975.8135 4502.3439
               4759.1262 4865.9705 5703.6751 5746.6766 5218.5087
               4869.0095 4975.8135 5746.6766 6065.7619 5522.8169
               4407.5881 4502.3439 5218.5087 5522.8169 5987.6283

                                Theta Hat Matrix

                                  THETA
                             150.83889 32.387865
                             102.79981 17.265428
                             139.86553 15.326229
                             132.18866 5.3569949

                              Gamma Zero Hat Matrix

                                   GAMMA0
               0.7825576 0.8870342 0.9257084 0.9277608 0.9943531
```

The estimates from the output are

$$\hat{\Theta}' = \begin{pmatrix} 150.84 & 102.80 & 139.86 & 132.19 \\ 32.39 & 17.27 & 15.33 & 5.36 \end{pmatrix}$$

$$\hat{\Gamma}_o' = (0.783,\quad 0.887,\quad 0.926,\quad 0.928,\quad 0.994). \tag{8.120}$$

Comparing (8.119) with (8.120), we see that the estimator for $\mathbf{\Theta}$ differs for the two models, GMANOVA versus the MANOVA-GMANOVA model. More important, inclusion of the covariate reduces the standard errors of the estimates which should also increase the precision of the parameter estimates.

We next test hypotheses for the MANOVA-GMANOVA model. LR tests for $\mathbf{\Theta}$ and $\mathbf{\Gamma}_o$ are given in (8.19) and (8.25), respectively. A joint LR test of $\mathbf{\Theta}$ and $\mathbf{\Gamma}_o$ is provided in (8.29). To perform these tests in SAS, we utilize PROC IML, since standard routines are not yet available. For our example, the primary tests of interest are

$$H_1:\quad \mathbf{C}_1\mathbf{\Theta}\mathbf{A}_1 = \mathbf{0}$$
$$H_2:\quad \mathbf{C}_2\mathbf{\Gamma}_o = \mathbf{0}$$

If we select $\mathbf{C}_2 = \mathbf{I}$, the test of H_2 evaluates the significance of the covariates in the mixed GMANOVA-MANOVA model. To perform the test of $H_2: \mathbf{\Gamma}_o = \mathbf{0}$, we use the expressions given in (8.26) to obtain λ_1 and λ_2, as shown in Program 8_12.sas. The results are given in the output.

Output 8.12 (*continued*)

```
                         E1T
                190246.04 37086.825
                37086.825 27303.515

                         E2T
           20686.875 4854.0334 1398.7245
           4854.0334 10126.233 -144.6413
           1398.7245 -144.6413 5541.2085

                          H1T
                2973202.7 199190.21
                199190.21 13344.781


                          H2T
           792.41047   -1087.35 -226.9609
            -1087.35 1492.0675 311.43704
           -226.9609 311.43704 65.005793

                 For test of gamma0

             LAMDA1        CHI1         DF1
           0.0524856 109.04703           6

              LAMDA2        CHI2         DF2
            0.772756   10.95616           3

           RHO1       RHO2     GAMMA1     GAMMA2
      0.8888889 0.9666667 0.0038542 0.0024222

 p-value of test H2: gamma0=zero PVAL1       PVAL2       PVALUE
                                     0   0.0122767   0.0122767
```

From the output we see that $\lambda_1 = 0.0524$ and $\lambda_2 = 0.7728$. Using Theorem 5.3, we see that the chi-square statistics are

$$\begin{aligned} X_1^2 &= -[v_e - (u - v_h + 1)/2]\ln \lambda_1 \\ &= -[(n - m_1 - m_2 - p + q) - (u_1 - g_1 + 1)/2]\ln \lambda_1 \\ &= -37 \ln \lambda_1 \\ &= 109.05 \end{aligned}$$

$$\begin{aligned} X_2^2 &= -[v_e - (u - v_h + 1)/2]\ln \lambda_2 \\ &= -[(n - m_2) - (u_2 - g_2 + 1)/2]\ln \lambda_2 \\ &= -[44 - 1.5]\ln \lambda_2 \\ &= -42.5 \ln \lambda_2 \\ &= 10.96 \end{aligned}$$

since for our example, $n = 45$, $m_1 = 4$, $m_2 = 1$, $p = 5$ and $q = 2$. Thus, for λ_1, $v_e = 37 = v_1$, $u = \text{rank}(\mathbf{X}_2) = 2 = q = u_1$ and $v_h = \text{rank}(\mathbf{C}_1) = 3 = g_1$. For $\lambda_2, v_e = 44 = v_2$, $u = p - q = 3 = u_2$ and $v_h = \text{rank}(\mathbf{C}_2) = 1 = g_2$. If we invoke formula (8.28),

$$\begin{aligned} X_1^2 &= 109.05 > \chi^{.05}_{(qg_1)} = \chi^{.05}_{(6)} = 14.4494 \\ X_2^2 &= \ \ 10.96 > \chi^{.05}_{[(p-q)g_2]} = \chi^{.05}_{(3)} = 9.34840 \end{aligned}$$

so that H_2 is rejected, the covariates should remain in the model. The p-value for the test of H_2 is approximated using Theorem 5.4. That is, the sum of the probabilities

$$P(\rho_1 n \ln \lambda_1 > 109.05) \doteq (1 - \gamma_1) P(X^2_{v_1} > 109.05) + \gamma_1 P(X^2_{v_1+4} > 109.05) \doteq 0$$

where $v_1 = u_1 g_1 = 6, \gamma_1 = v_1(u_1^2 + g_1^2 - 5)/48(\rho_1 n)^2 = 0.0039$ since $k = m_1 + m_2 + p - q = 8$, $g_1 = 3$, $u_1 = 2$, $\rho_1 = 1 - n^{-1}[k - g_1 + (u_1 + g_1 + 1)/2] = 0.8889$ and the

$$P(\rho_2 n \ln \lambda_2 > 10.96) \doteq (1 - \gamma_2) P(X^2_{v_2} > 10.96) + \gamma_2 P(X^2_{v_2+4} > 10.96) \doteq 0.0123$$

where $v_2 = u_2 g_2 = (p - q) g_2 = 3$, $\gamma_2 = v_2(u_2^2 + g_2^2 - 5)/48(\rho_2 n)^2 = 0.0024$ since $k = m_2 = 1$, $g_2 = 1$ and $\rho_2 = 0.9667$ so that the estimated p-value is 0.0123 as seen in the output.

We next test the hypothesis H_1. Since we have fit linear polynomials to each group, the test of group differences is equivalent to testing that the polynomials are coincident. For this hypothesis,

$$\mathbf{C}_1 = \begin{pmatrix} 1 & -1 & 0 & 0 \\ 1 & 0 & -1 & 0 \\ 1 & 0 & 0 & -1 \end{pmatrix} \text{ and } \mathbf{A}_1 = \mathbf{I}_2$$

which compares the control to each treatment. To perform the test of coincidence, $H_1\colon \mathbf{C}_1 \boldsymbol{\Theta} \mathbf{A}_1 = \mathbf{0}$, we evaluate the expressions given in (8.21) to obtain Λ given in (8.19) as shown in Program 8_12.sas. The output follows.

Output 8.12 (*continued*)

```
                Test of Coincidence Wilks and Chi-square

                   LAMDA        DFE          DFH       CHI_2
               0.8321483         37            3  6.7985487

                                     DF
                                      6

                               RHO      OMEGA
                         0.8222222  0.0007305

                                                       PVAL
               p-value of test H1: Coincidence  0.3401753
```

Using Theorem 5.3, we see that the p-value for the test is 0.3402 so that the test is not significant.

To test for parallelism of profiles, one would select the matrices $\mathbf{C}_1$ and $\mathbf{A}_1$

$$\mathbf{C}_1 = \begin{pmatrix} 1 & -1 & 0 & 0 \\ 1 & 0 & -1 & 0 \\ 1 & 0 & 0 & -1 \end{pmatrix} \quad \mathbf{A}_1 = \begin{pmatrix} 0 \\ 1 \end{pmatrix}$$

with the test following that outlined for the test of coincidence.

Because the GMANOVA-MANOVA model allows one to incorporate a covariate into the model, the mixed GMANOVA-MANOVA model is usually more resolute than the GMANOVA model when the covariate is significant. For our example, this was not the case because of the large variability within patients.

8.13 CGMANOVA Model

The SUR model may be used to represent numerous multivariate linear models as special cases. We have already demonstrated how the model may be used to analyze the GMANOVA model and Patel's model which allows for changing covariates. The SUR model may be used to analyze the MANOVA model with double restrictions, the extended GMANOVA model

$$E(\mathbf{Y})_{n\times p} = \sum_{i=1}^{r} (\mathbf{X}_i)_{n\times k_i} (\mathbf{B}_i)_{k_i \times q_i} (\mathbf{M}_i)_{q_i \times p} \tag{8.121}$$

proposed by Verbyla and Venables (1988), and the mixed GMANOVA-MANOVA model. To see this, we let the SUR model in (8.49) be written in its most general form

$$\underset{p\times 1}{\mathbf{y}_i} = \underset{p\times m}{\mathbf{D}_i} \; \underset{m\times 1}{\theta} + \underset{p\times 1}{\mathbf{e}_i} \tag{8.122}$$

where $i = 1, 2, \ldots, n$ and the $\text{var}(\mathbf{e}_i) = \mathbf{\Sigma}$, the structure of the p-variates. Letting

$$\mathbf{D}_i = \begin{pmatrix} \mathbf{x}'_{i1} & 0 & \cdots & \cdots & \cdots & 0 \\ 0 & \mathbf{x}'_{i2} & \cdots & \cdots & \cdots & 0 \\ \vdots & \cdots & \ddots & \cdots & \cdots & \vdots \\ \vdots & \cdots & \cdots & \ddots & \cdots & \vdots \\ \vdots & \cdots & \cdots & \cdots & \ddots & \vdots \\ 0 & 0 & \cdots & \cdots & \cdots & \mathbf{x}'_{ip} \end{pmatrix}$$

where $\mathbf{x}_{ij}$ is a $k_j \times 1$ vector of variables for the jth variable $(j = 1, 2, \ldots, p)$ and the ith subject $(i = 1, 2, \ldots, n)$, and $m = \sum_{j=1}^{p} k_j$, we have the MDM model in another form. When $\mathbf{x}_{ij} = \mathbf{x}_i$ so that $\mathbf{D}_i = \mathbf{I}_p \otimes \mathbf{x}_i'$, the MDM model reduces to the standard multivariate GLM. The most general SUR model may be represented as

$$\underset{np\times 1}{\mathbf{y}} = \underset{np\times m}{\mathbf{D}} \underset{m\times 1}{\theta} + \underset{np\times 1}{\mathbf{e}} \tag{8.123}$$

where $\mathbf{y} = (\mathbf{y}_1', \mathbf{y}_2', \ldots, \mathbf{y}_n')'$, $\mathbf{D} = (\mathbf{D}_1', \mathbf{D}_2', \ldots, \mathbf{D}_n')'$ and $\mathbf{e} = (\mathbf{e}_1', \mathbf{e}_2', \ldots, \mathbf{e}_n')'$ and the $\text{var}(\mathbf{y}) = \mathbf{I}_n \otimes \mathbf{\Sigma}$. This is the exact form of the CGMANOVA (8.49) proposed by Hecker (1987). Hence, the CGMANOVA model is also a SUR model. Using (8.123) we may represent (8.121) as a CGMANOVA/SUR model. To see this, we let the ith row of $\mathbf{Y}_{n\times p}$ be $\mathbf{y}_i'$ in $\mathbf{y}$, the design matrix $\mathbf{D} = [(\mathbf{X}_1 \otimes \mathbf{M}_1'), \ldots, (\mathbf{X}_r \otimes \mathbf{M}_r')]$ and $\theta = (\theta_1', \theta_2', \ldots, \theta_r')'$ where $\theta_i = \text{vec}(\mathbf{B}_i)$. Hence, to estimate θ we may use the most general SUR model instead of the MSUR model. By (8.52), an estimate of θ is

$$\hat{\hat{\theta}} = (\mathbf{D}'(\mathbf{I}_n \otimes \hat{\mathbf{\Sigma}}^{-1})\mathbf{D})^{-1}\mathbf{D}'(\mathbf{I}_n \otimes \hat{\mathbf{\Sigma}})^{-1}\mathbf{y} \tag{8.124}$$

where $\hat{\mathbf{\Sigma}}$ is any consistent estimator of $\mathbf{\Sigma}$. A convenient candidate is

$$\hat{\mathbf{\Sigma}} = \mathbf{Y}'(\mathbf{I} - \mathbf{1}(\mathbf{1}'\mathbf{1})^{-1}\mathbf{1}')\mathbf{Y}/n \tag{8.125}$$

based on the within-subject variation.

To test hypotheses, we use the Wald statistic given in (8.61), instead of LR tests. Thus,

$$W = (\mathbf{C}\hat{\hat{\theta}} - \xi)'[\mathbf{C}(\mathbf{D}'(\mathbf{I}_n \otimes \hat{\mathbf{\Sigma}}^{-1})\mathbf{D})^{-1}\mathbf{C}']^{-1}(\mathbf{C}\hat{\hat{\theta}} - \xi) \tag{8.126}$$

We may analyze the data previously analyzed in Section 8.12 now using Hecker's CGMANOVA model. The SAS code to do the analysis is found in Program 8_13.sas. To represent the model in (8.117) as a SUR model, we have from (8.46) that

$$\underset{225\times 1}{\begin{pmatrix} \mathbf{y}_1 \\ \mathbf{y}_2 \\ \vdots \\ \vdots \\ \vdots \\ \mathbf{y}_{45} \end{pmatrix}} = \underset{225\times 13}{[\mathbf{X}_1 \otimes \mathbf{X}_2', \mathbf{Z} \otimes \mathbf{I}_5]} \begin{pmatrix} \theta_{11} \\ \theta_{12} \\ \theta_{21} \\ \theta_{22} \\ \theta_{31} \\ \theta_{32} \\ \theta_{41} \\ \theta_{42} \\ \gamma_1 \\ \gamma_2 \\ \gamma_3 \\ \gamma_4 \end{pmatrix} + \underset{225\times 1}{\begin{pmatrix} \mathbf{e}_1 \\ \mathbf{e}_2 \\ \vdots \\ \vdots \\ \vdots \\ \mathbf{e}_{45} \end{pmatrix}} \tag{8.127}$$

Alternatively, we have that

$$\underset{5\times 1}{\mathbf{y}_i} = \underset{5\times 13}{\mathbf{D}_i} \underset{13\times 1}{\theta} + \underset{5\times 1}{\mathbf{e}_i} \qquad i = 1, 2, \ldots, n \tag{8.128}$$

where $\mathbf{D}_i = (\mathbf{I}_p \otimes \mathbf{x}'_{ip})$ are sets of five rows of the design matrix $\mathbf{D}$, where $\mathbf{x}_{ij}$ is a vector of independent variables for response j on subject i, for the most general SUR model (Hecker, 1987).

Program 8_13.sas

```
/* Program 8_13.sas                 */
/* GMANOVA Model,  Hecker(1987)     */

options ls=80 ps=60 nodate nonumber;
title1 'Output 8.13: CGMANOVA Model';

data gman;
   infile'c:\8_13.dat';
   input group d1 d2 d3 d4 z1 y1 - y5;
run;

proc iml;
   use gman;
   read all var {y1 y2 y3 y4 y5} into y;
   read all var {d1 d2 d3 d4}  into x1;
   read all var {d1 d2 d3 d4 z1} into x;
   read all var {z1} into z;
   t={1 2 3 4 5};
   p_prime=orpol(t,4);
   p=p_prime[1:5,1:2];
   x2=p`;
   n=nrow(y);
   k=ncol(y);
   one=j(n,1,1);
   sigma=(y`*y-y`*one*inv(one`*one)*one`*y)/n;
   print 'Consistent Estimator of Covariance Matrix', sigma;
   d1=x1@x2`; d2=z@i(5);
   d=d1||d2;
   yc=shape(y,225,1,0);
   v=i(n)@inv(sigma);
   theta=inv(d`*v*d)*d`*inv((i(n)@sigma))*yc;
   print 'GLSE of Parameter Matrix', theta;
   /* Test of Coincidence  */
   c={1 0 0 0 0 0 -1  0 0 0 0 0 0,
      0 1 0 0 0 0  0 -1 0 0 0 0 0,
      0 0 1 0 0 0 -1  0 0 0 0 0 0,
      0 0 0 1 0 0  0 -1 0 0 0 0 0,
      0 0 0 0 1 0 -1  0 0 0 0 0 0,
      0 0 0 0 0 1  0 -1 0 0 0 0 0};
   w=(c*theta)`*inv(c*inv(d`*v*d)*c`)*(c*theta);
   df=6;
   pvalue=1-probchi(w,df);
   print 'Wald Statistic for Coincidence',w df pvalue;
   /* Test Gamma0 equals zero   */
   c={0 0 0 0 0 0 0 0 1 0 0 0 0,
      0 0 0 0 0 0 0 0 0 1 0 0 0,
      0 0 0 0 0 0 0 0 0 0 1 0 0,
      0 0 0 0 0 0 0 0 0 0 0 1 0,
      0 0 0 0 0 0 0 0 0 0 0 0 1};
   w=(c*theta)`*inv(c*inv(d`*v*d)*c`)*(c*theta);
   df=5;
   pvalue =1-probchi(w,df);
   print 'Wald Statistic for Gamma0=0',w df pvalue;
quit;
run;
```

Result and Interpretation 8_13.sas

The results of the analysis are found in Output 8_13.sas.

Output 8.13: *CGMANOVA Model*

```
              Output 8.13: CGMANOVA Model

        Consistent Estimator of Covariance Matrix

        SIGMA
    4877.4588 4691.8854  5197.364 5306.9077 4903.0178
    4691.8854 4897.6277  5337.644 5469.7165   5013.28
     5197.364  5337.644  6109.998 6166.5551 5671.2533
    5306.9077 5469.7165 6166.5551 6516.3388 6028.1244
    4903.0178   5013.28 5671.2533 6028.1244 6520.4711

              GLSE of Parameter Matrix

                          THETA
                      147.99995
                       32.04349
                      107.71277
                      18.078296
                      141.46387
                      15.817764
                      132.79786
                      5.5771335
                      0.7795546
                      0.8832298
                      0.9211025
                      0.9223535
                      0.9881443

             Wald Statistic for Coincidence

                     W         DF     PVALUE
             4.7744378          6 0.5730516

              Wald Statistic for Gamma0=0

                     W         DF     PVALUE
             42.329283          5 5.0524E-8
```

In Program 8_13.sas, we estimate θ using formula (8.124). The estimators of $\boldsymbol{\Theta}$ and $\boldsymbol{\Gamma}_o$, are from the output

$$\hat{\boldsymbol{\Theta}}' = \begin{pmatrix} 148.00 & 107.71 & 141.46 & 132.80 \\ 32.04 & 18.07 & 15.82 & 5.57 \end{pmatrix} \tag{8.129}$$

$$\hat{\boldsymbol{\Gamma}}_o' = (0.780, \quad 0.883, \quad 0.921, \quad 0.922, \quad 0.988)$$

and $\hat{\boldsymbol{\Sigma}}$ is estimated using (8.125). Comparing (8.129) with the ML estimates given in (8.118), we have close agreement between the FGLS and the ML estimate for this example, even using a naive estimate of $\boldsymbol{\Sigma}$.

To test hypotheses, we use formula (8.134). To illustrate, we perform the test of coincidence. Using the CGMANOVA model, the hypothesis test matrix is

$$\mathbf{C}=\begin{pmatrix} 1 & 0 & 0 & 0 & 0 & 0 & -1 & 0 & 0 & 0 & 0 & 0 & 0 \\ 0 & 1 & 0 & 0 & 0 & 0 & 0 & -1 & 0 & 0 & 0 & 0 & 0 \\ 0 & 0 & 1 & 0 & 0 & 0 & -1 & 0 & 0 & 0 & 0 & 0 & 0 \\ 0 & 0 & 0 & 1 & 0 & 0 & 0 & -1 & 0 & 0 & 0 & 0 & 0 \\ 0 & 0 & 0 & 0 & 1 & 0 & -1 & 0 & 0 & 0 & 0 & 0 & 0 \\ 0 & 0 & 0 & 0 & 0 & 1 & 0 & -1 & 0 & 0 & 0 & 0 & 0 \end{pmatrix}$$

where the rank($\mathbf{C}$) = 6. From Output 8.13, the Wald statistic is 4.774. Comparing W to the chi-square distribution with $df = 6$, the p-value for the test is 0.5731. Because the Wald statistic is a large sample result, the p-value of 0.57 is larger than the corresponding LR test p-value of 0.34 obtained using the MANOVA-GMANOVA model. It is currently unknown under what conditions the CGMANOVA model is optimal. Hecker (1987) obtained bounds for an LR test as discussed in Section 5.5.

Finally, we illustrate the test that $\mathbf{\Gamma}_o = \mathbf{0}$ using the CGMANOVA model and the Wald statistic. For this hypothesis, the hypothesis test matrix is

$$\mathbf{C} = (\underset{5\times 8}{\mathbf{O}}, \mathbf{I}_5)$$

where the rank($\mathbf{C}$) = 5. From Output 8.13, the Wald statistic is W = 42.33 with p-value < 0.0001 for the test. This result is again consistent with the LR test result.

In conclusion, we see that the SUR model is a very flexible model for the analysis of multivariate data; however, the precision of the result is unknown for small samples when compared to corresponding LR tests. Finally, to incorporate restrictions into the SUR model for known $\mathbf{\Sigma}$, one may transform $\mathbf{y}_i$ by $\mathbf{\Sigma}^{-1/2}$. Then $\mathbf{z} = (\mathbf{I}_n \otimes \mathbf{\Sigma}^{-1/2})\,\mathbf{y}$ has a multivariate normal distribution with mean $\mu = (\mathbf{I}_n \otimes \mathbf{\Sigma}^{1/2})\theta$ and covariance matrix $\mathbf{\Omega} = \mathbf{I}_n \otimes \mathbf{\Sigma}^{-1/2}\mathbf{\Sigma}\mathbf{\Sigma}^{-1/2} = \mathbf{I}_n \otimes \mathbf{I}_p$ so that the restricted linear model theory in Chapter 3 may be applied to the transformed model. Thus, depending on $\mathbf{\Sigma}$,

$$Q_h(\mathbf{\Sigma}) = (\mathbf{C}\hat{\hat{\theta}})'(\mathbf{C}[\mathbf{D}'(\mathbf{\Sigma})\mathbf{D}(\mathbf{\Sigma})]^{-1}\mathbf{C}')^{-1}(\mathbf{C}\hat{\hat{\theta}})$$

with $\hat{\hat{\theta}}$ defined in (8.124) and under H_o, $\mathbf{C}\theta = \mathbf{0}$, $Q_h(\mathbf{\Sigma})$ has a chi-square distribution with df = rank($\mathbf{C}$). The theory in Chapter 3 may be followed exactly to obtain approximate tests for a restricted SUR model.

9. Two-Stage Hierarchical Linear Models

Theory

Applications

9.1 Introduction

In our formulation of a growth curve model for a single group, the same model is fit to each subject and the variation within subjects has a structured covariance matrix. The growth curve model does not model random variation among subjects in terms of model parameters. Because subjects are really not the same within a group, we may allow the model for each subject to have two components: fixed effects that are common to subjects and random effects that are unique to subjects. In this way, we may formulate a general linear mixed model (GLMM) for each subject.

More generally suppose we have n groups indexed by j where subjects are nested within groups. Then for an observation y_{ij}, the subjects are indexed by i and groups are indexed by j. Letting β_j represent parameters at the group level, even though the β_j are not observed, one may model the β_j hierarchically, using observed data to structure models conditionally on fixed parameters. Such models are called hierarchical linear models or multilevel models. In this chapter, we discuss estimation and hypothesis testing for two-stage hierarchical linear models (TSHLM). Analysis of these models is illustrated using PROC MIXED.

9.2 Two-Stage Hierarchical Linear Models

The classical approach to the analysis of repeated measures data is to use growth curve models or repeated measures models for groups of subjects over a fixed number of trials. A more general approach is to formulate growth as a two-stage hierarchical linear model (TSHLM) where population parameters, individual effects, and within-subject variation are defined at the first stage of model development, and between subject variation is modeled at the second stage (Laird and Ware, 1982; Bock, 1989; Bryk and Raudenbush, 1992; Longford, 1993; Diggle et al., 1994; Ware, 1985; and Vonesh and Chinchilli, 1997).

For a simple curvilinear regression model, one may postulate a quadratic model of the form

$$y_i = \beta_o + \beta_1 x + \beta_2 x^2 + e_i \qquad i = 1, 2, \ldots, n \tag{9.1}$$

where $e_i \sim N(0,\sigma^2)$ and $\beta_i' = (\beta_{oi}, \beta_{1i}, \beta_{2i})$ is the vector of regression parameters for the ith subject. The OLSE of β_i is

$$\hat{\beta}_i = (\mathbf{X}_i'\mathbf{X}_i)^{-1}\mathbf{X}_i'\mathbf{y}_i \tag{9.2}$$

where $\mathbf{y}_i(n_i \times 1)$ is a data vector of measurements and $\mathbf{X}_i(n_i \times 3)$ is a matrix with rows $(1, \mathrm{x}, \mathrm{x}^2)$ for $\mathrm{x} = 1, 2, \ldots, \mathrm{n}_i$. The parameters β_i contain information about the variation between individuals that is not specified in (9.2). Each β_i contains information regarding an overall population mean $\beta' = (\beta_o, \beta_1, \beta_2)$ and $\mathbf{b}_i$, the random departures of the individual variation from the population mean

$$\beta_i = \beta + \mathbf{b}_i \tag{9.3}$$

where $\mathbf{b}_i' = (b_{oi}, b_{1i}, b_{2i})$. The model for $\mathbf{y}_i$ given $\mathbf{b}_i$ is for $i = 1, 2, \ldots, n$:

$$y_i|\mathbf{b}_i = (\beta_o + b_{oi}) + (\beta_1 + b_{1i})\mathrm{x} + (\beta_2 + b_{2i})\mathrm{x}^2 + e_i \tag{9.4}$$

Further, we may assume a distribution for $\mathbf{b}_i$, for example

$$\mathbf{b}_i \sim \mathbf{N}_3(\mathbf{0}, \mathbf{D}) \tag{9.5}$$

This regression model may be formulated as a two-stage hierarchical linear model (TSHLM). The population parameters, individual effects, and within-subject variation are specified in Stage 1 (9.4), and the between-subject variation in Stage 2, (9.5).

Stage 1. Conditionally on β and $\mathbf{b}_1, \mathbf{b}_2, \ldots, \mathbf{b}_n$, the vectors $\mathbf{y}_i$ are independent with structure

$$\mathbf{y}_i|(\beta, \mathbf{b}_i) = \mathbf{X}_i(\beta + \mathbf{b}_i) + \mathbf{e}_i \tag{9.6}$$

where $\mathbf{e}_i \sim N_{n_i}(\mathbf{0}, \sigma^2\mathbf{I}_{n_i})$.

Stage 2. The vectors $\mathbf{b}_i \sim N_k(\mathbf{0}, \mathbf{D})$ and are independent of $\mathbf{e}_i$.

For this illustration, the design matrix $\mathbf{X}_i$ was common to β and $\mathbf{b}_i$. More generally, we may let $\beta_{p\times 1}$ be a vector of unknown population parameters and $\mathbf{X}_i(n_i \times p)$ the design matrix linking β to $\mathbf{y}_i$. Let $\mathbf{b}_i(k \times 1)$ be a vector of unknown individual effects and $\mathbf{Z}_i(n_i \times k)$ a design matrix linking $\mathbf{b}_i$ and $\mathbf{y}_i$. For this general case we have

Stage 1. for each subject i,

$$\mathbf{y}_i = \mathbf{X}_i\beta + \mathbf{Z}_i\mathbf{b}_i + \mathbf{e}_i \tag{9.7}$$

where $\mathbf{e}_i \sim N_{n_i}(\mathbf{0}, \mathbf{\Psi}_i)$. At this stage β and $\mathbf{b}_i$ are considered fixed and $\mathbf{\Psi}_i$ depends on i through n_i. When $\mathbf{\Psi}_i = \sigma^2\mathbf{I}_{n_i}$, the TSHLM is said to be *conditional-independent* since the responses for each individual are independent, conditional on β and $\mathbf{b}_i$ and

Stage 2. the $\mathbf{b}_i \sim N_k(\mathbf{0}, \mathbf{D})$ are independent of each other and of $\mathbf{e}_i$. The population parameters β are treated as fixed.

Thus it follows from Stage 1 and Stage 2 that for $i = 1, 2, \ldots, n$

$$\begin{aligned} E(\mathbf{y}_i) &= \mathbf{X}_i\beta \\ \mathrm{var}(\mathbf{y}_i) &= \mathbf{Z}_i\mathbf{D}\mathbf{Z}_i' + \mathbf{\Psi}_i = \mathbf{\Omega}_i \\ \mathrm{cov}(\mathbf{y}_i, \mathbf{y}_j) &= \mathbf{0} \qquad i \neq j \end{aligned} \tag{9.8}$$

Assuming all covariances in (9.8) are known and β is fixed, the BLUE of β by (4.6) is

$$\hat{\beta} = \left(\sum_{i=1}^{n} \mathbf{X}_i' \boldsymbol{\Omega}_i^{-1} \mathbf{X}_i \right)^{-1} \sum_{i=1}^{n} \mathbf{X}_i' \boldsymbol{\Omega}_i^{-1} \mathbf{y}_i \tag{9.9}$$

To estimate $\mathbf{b}_i$ is more complicated. By an extension of the Gauss-Markov theorem for general linear mixed models, Harville (1976) showed that the BLUE of $\mathbf{b}_i$ is

$$\hat{\mathbf{b}}_i = \mathbf{D}\mathbf{Z}_i' \boldsymbol{\Omega}_i^{-1} (\mathbf{y}_i - \mathbf{X}_i \hat{\beta}) \tag{9.10}$$

and that the

$$\begin{aligned} \text{var}(\hat{\beta}) &= \left(\sum_{i=1}^{n} \mathbf{X}_i' \boldsymbol{\Omega}_i^{-1} \mathbf{X}_i \right)^{-1} \\ \text{var}(\hat{\mathbf{b}}_i) &= \mathbf{D} - \mathbf{D}\mathbf{Z}_i' \mathbf{P}_i \mathbf{Z}_i \mathbf{D} \\ \mathbf{P}_i &= \boldsymbol{\Omega}_i^{-1} - \boldsymbol{\Omega}_i^{-1} \mathbf{X}_i \left(\sum_{i=1}^{n} \mathbf{X}_i' \boldsymbol{\Omega}_i^{-1} \mathbf{X}_i \right)^{-1} \mathbf{X}_i' \boldsymbol{\Omega}_i^{-1} \end{aligned} \tag{9.11}$$

Motivation for (9.10) follows from the fact that if

$$\begin{aligned} \mathbf{y} &= \mathbf{X}\beta + \mathbf{Z}\mathbf{b} + \mathbf{e} \\ \text{var}(\mathbf{y}) &= \mathbf{Z}\mathbf{V}\mathbf{Z}' + \boldsymbol{\Psi} = \boldsymbol{\Omega} \\ \text{cov}(\mathbf{y}, \mathbf{b}) &= \mathbf{Z}\mathbf{V} \text{ and } \mathbf{V} = \mathbf{I}_n \otimes \mathbf{D} \end{aligned} \tag{9.12}$$

and

$$\begin{pmatrix} \mathbf{b} \\ \mathbf{y} \end{pmatrix} \sim \mathbf{N}\left\{ \begin{pmatrix} \mathbf{0} \\ \mathbf{X}\beta \end{pmatrix}, \begin{pmatrix} \mathbf{V} & \mathbf{V}\mathbf{Z}' \\ \mathbf{Z}\mathbf{V} & \boldsymbol{\Omega} \end{pmatrix} \right\}$$

then the conditional mean of the distribution of $\mathbf{b}|\mathbf{y}$ is

$$E(\mathbf{b}|\mathbf{y}) = \mathbf{V}\mathbf{Z}'\boldsymbol{\Omega}^{-1}(\mathbf{y} - \mathbf{X}\beta) \tag{9.13}$$

the best linear unbiased predictor (BLUP) of $\mathbf{b}$. Furthermore, the variance of $\mathbf{b}|\mathbf{y}$ is

$$\text{var}(\mathbf{b}|\mathbf{y}) = \mathbf{V} - \mathbf{V}\mathbf{Z}'(\mathbf{Z}\mathbf{V}\mathbf{Z}' + \boldsymbol{\Psi})^{-1}\mathbf{Z}\mathbf{V} \tag{9.14}$$

Using the matrix identity

$$(\mathbf{Z}\mathbf{V}\mathbf{Z}' + \boldsymbol{\Psi})^{-1} = \boldsymbol{\Psi}^{-1} - \boldsymbol{\Psi}^{-1}\mathbf{Z}(\mathbf{Z}'\boldsymbol{\Psi}^{-1}\mathbf{Z} + \mathbf{V}^{-1})^{-1}\mathbf{Z}'\boldsymbol{\Psi}^{-1} \tag{9.15}$$

in (9.14), and letting $\boldsymbol{\Phi} = \mathbf{Z}'\boldsymbol{\Psi}^{-1}\mathbf{Z}$, the

$$\begin{aligned} \text{var}(\mathbf{b}|\mathbf{y}) &= \mathbf{V} - \mathbf{V}\boldsymbol{\Phi}[\mathbf{I} - (\boldsymbol{\Phi} + \mathbf{V}^{-1})^{-1}\boldsymbol{\Phi}]\mathbf{V} \\ &= \mathbf{V} - \mathbf{V}\boldsymbol{\Phi}[(\boldsymbol{\Phi} + \mathbf{V}^{-1})(\boldsymbol{\Phi} + \mathbf{V}^{-1})^{-1} - \boldsymbol{\Phi}]\mathbf{V} \\ &= \mathbf{V} - \mathbf{V}\boldsymbol{\Phi}[(\boldsymbol{\Phi} + \mathbf{V}^{-1})^{-1}(\boldsymbol{\Phi} + \mathbf{V}^{-1} - \boldsymbol{\Phi})\mathbf{V}] \\ &= \mathbf{V} - \mathbf{V}\boldsymbol{\Phi}(\boldsymbol{\Phi} + \mathbf{V}^{-1})^{-1} \\ &= \mathbf{V}[\mathbf{I} - \boldsymbol{\Phi}[(\boldsymbol{\Phi} + \mathbf{V}^{-1})^{-1}] \\ &= \mathbf{V}[(\boldsymbol{\Phi} + \mathbf{V}^{-1})(\boldsymbol{\Phi} + \mathbf{V}^{-1})^{-1} - \boldsymbol{\Phi}(\boldsymbol{\Phi} + \mathbf{V}^{-1})^{-1}] \\ &= [\mathbf{V}(\boldsymbol{\Phi} + \mathbf{V}^{-1} - \boldsymbol{\Phi})](\boldsymbol{\Phi} + \mathbf{V}^{-1})^{-1} \\ &= (\boldsymbol{\Phi} + \mathbf{V}^{-1})^{-1} \\ &= (\mathbf{Z}'\boldsymbol{\Psi}^{-1}\mathbf{Z} + \mathbf{V}^{-1})^{-1} \end{aligned} \tag{9.16}$$

has a simple form using $\mathbf{\Psi}^{-1}$ and $\mathbf{V}^{-1}$. If $\mathbf{\Psi} = \sigma^2\mathbf{I}$, then the $\text{var}(\mathbf{b}|\mathbf{y}) = \sigma^2(\mathbf{Z}'\mathbf{Z}+\sigma^2\mathbf{V}^{-1})^{-1}$.

Because the population covariance matrices $\mathbf{\Psi}_i$ and $\mathbf{D}$ are unknown, we may not directly use (9.9) and (9.10) to estimate β and $\mathbf{b}_i$. There are two strategies for the estimation of model parameters for the TSHLM, either classical inference using least squares and maximum likelihood or the empirical Bayes method using as prior information the distribution of the parameters defined in Stage 2.

To obtain parameter estimates it is convenient to let $\theta_{q\times1}$ be a vector of the nonredundant variances and covariances $\mathbf{\Psi}_i$, $i = 1, \ldots, n$ and $\mathbf{D}$. With a consistent estimator for θ and hence $\mathbf{\Psi}_i$ and $\mathbf{D}$, we may let

$$\hat{\mathbf{\Omega}}_i^{-1} = (\hat{\mathbf{\Psi}}_i + \mathbf{Z}_i\hat{\mathbf{D}}\mathbf{Z}_i')^{-1} \tag{9.17}$$

and estimate β and $\mathbf{b}_i$ using the least squares equations (9.9) and (9.10) by replacing , $\mathbf{\Omega}_i^{-1}$ with $\hat{\mathbf{\Omega}}_i^{-1}$, (Harville, 1977). Let these estimates be represented by $\hat{\beta}(\hat{\theta})$ and $\hat{\mathbf{b}}_i(\hat{\theta})$.

To estimate θ and β using ML, the marginal normal distribution of $\mathbf{y}' = (\mathbf{y}_1', \mathbf{y}_2', \ldots, \mathbf{y}_n')$ is used

$$\mathbf{y}_{N\times1} \sim N_N(\mu, \mathbf{\Omega}) \tag{9.18}$$

where $\mathbf{\Omega} = \text{diag}(\mathbf{\Omega}_1, \mathbf{\Omega}_2, \ldots, \mathbf{\Omega}_n)$, $\mu' = (\mu_1, \mu_2, \ldots, \mu_n)$, $\mu_i = \mathbf{X}_i\beta$ and $N = \sum_i n_i$. By maximizing the likelihood function given in (10.14) or equivalently

$$L = L(\beta, \mathbf{\Omega}|\mathbf{y}) = (2\pi)^{N/2}|\mathbf{\Omega}|^{-1/2}\exp\left[-\frac{1}{2}(\mathbf{y}-\mu)'\mathbf{\Omega}^{-1}(\mathbf{y}-\mu)\right] \tag{9.19}$$

The ML estimates of β and θ , $\hat{\beta}_{ML}$ and $\hat{\theta}_{ML}$, satisfy the relationship $\hat{\beta}_{ML} = \hat{\beta}(\hat{\theta}_{ML})$ for the parameter β. See Searle et al. Chapter 6 (1992).

The ML approach provides an estimate of β and θ, however, in balanced ANOVA designs, ML estimates of variance components fail to take into account the degrees of freedom lost in estimating β, and are hence biased downward (Searle et al., 1992, p. 250). This is overcome using classical methods by employing a restricted (residual) maximum likelihood estimator (REML) for θ, $\hat{\theta}_R$. Using (9.9) and (9.10) estimates $\hat{\beta}_{ML}$ and $\hat{\mathbf{b}}_{ML} = \hat{\mathbf{b}}_i(\hat{\theta}_{ML})$ are realized. The REML estimate is obtained by maximizing the likelihood (10.17) of θ based on independent residual contrasts of $\mathbf{y}$, $\psi = \mathbf{C}\mathbf{y}$ where the rows of $\mathbf{C}$ are in the orthocomplement of the design space (Searle et al., 1992, pp. 252 and 323). These estimators $\hat{\beta}(\hat{\theta}_R)$, and $\hat{\mathbf{b}}_i(\hat{\theta}_R)$ provide alternative estimates of β and θ . Letting $\mathbf{b}' = (\mathbf{b}_1', \mathbf{b}_2', \ldots, \mathbf{b}_n')$ and equating $\hat{\mathbf{b}}_{ML} = E(\mathbf{b}|\tilde{\mathbf{y}}, \hat{\beta}_{ML}, \hat{\theta}_{ML})$ gives $\hat{\mathbf{b}}_{ML} = \hat{\mathbf{b}}(\hat{\theta}_{ML})$, yields the empirical Bayes estimate of $\mathbf{b}$ when $\hat{\theta}_{ML}$ is ML. Hence, for the TSHLM one may estimate β using the ML method and $\mathbf{b}$ by the empirical Bayes procedure which is equivalent to the REML estimate, $\hat{\mathbf{b}}(\hat{\theta}_R)$, (Harville, 1976).

An alternative approach is to use the full Bayes method of estimation. Then we have to introduce a prior distribution on the location parameter in Stage 2 (Gelman et al., 1996). Recall that if θ and Y are two random variables, then the density of θ given Y is

$$\begin{aligned} f(\theta \mid Y) &= \frac{f(Y,\theta)}{f(Y)} = \frac{f(Y|\theta)f(\theta)}{\int f(Y,\theta)d\theta} \\ &= \frac{f(Y|\theta)f(\theta)}{\int f(Y|\theta)f(\theta)d\theta} \end{aligned} \tag{9.20}$$

However, $f(\theta \mid Y)$ may be regarded as a function of θ called the likelihood where if $Y_i \sim f(Y \mid \theta)$, $L(\theta|Y) = \prod_i f(Y_i|\theta)$.

Hence, $f(\theta|Y)$ in (9.20) is proportional to a likelihood times a prior:

$$f(\theta \mid Y) \propto L(\theta|Y)f(\theta) \tag{9.21}$$

Knowing the posterior density $f(\theta \mid Y)$, we define the Bayes estimator for θ as

$$E(\theta \mid Y) = \int \theta f(\theta|Y)d\theta \tag{9.22}$$

Suppose $\theta' = (\theta_1, \theta_2)$, then to find the joint density $f(\theta_1, \theta_2|Y)$ we have that

$$f(\theta_1, \theta_2 \mid Y) = \frac{f(Y, \theta_1, \theta_2)}{f(Y)} = \frac{f(Y|\theta_1, \theta_2)f(\theta_1|\theta_2)f(\theta_2)}{\int f(Y|\theta_1, \theta_2)f(\theta_1|\theta_2)f(\theta_2)d\theta_1 d\theta_2} \tag{9.23}$$

Hence to estimate θ using the exact Bayes procedure requires specification of a prior for θ_2 and the density of $f(\theta_1|\theta_2)$ where θ_2 represents parameters of the distribution of θ_1 called hyperparameters; to calculate the joint posterior of θ_1 and θ_2 given Y may be problematic. However, suppose we can specify the distribution $f(\theta_1, \theta_2)$, then the posterior density for θ_1 is

$$f(\theta_1 \mid Y, \theta_2) = \frac{f(Y|\theta_1, \theta_2)f(\theta_1|\theta_2)}{\int f(Y|\theta_1, \theta_2)f(\theta_1|\theta_2)d\theta_1} \tag{9.24}$$

and the Bayes estimator of θ_1 is

$$\theta_1 = E(\theta_1 \mid Y_1, \theta_1) \tag{9.25}$$

Because θ_2 is unknown, we may estimate θ_2 from the marginal distribution $f(Y|\theta_2)$ where

$$f(Y|\theta_2) = \int f(Y|\theta_1, \theta_2)f(\theta_1|\theta_2)d\theta_1 \tag{9.26}$$

and the empirical Bayes estimate of θ_1 is

$$\hat{\theta}_1 = E(\theta_1 \mid Y, \hat{\theta}_2) \tag{9.27}$$

For a Bayesian formulation of the TSHLM, Stage 1 remains unchanged. For Stage 2, we let β and $\mathbf{b}_i$ be independent and normally distributed with means $\mathbf{0}$ and the $\text{var}(\beta) = \mathbf{\Gamma}$, $\text{var}(\mathbf{b}_i) = \mathbf{D}$ and the $\text{cov}(\beta, \mathbf{b}_i) = \mathbf{0}$. Then the marginal distribution of $\mathbf{y}_i$ is

$$\mathbf{y}_i \sim N_{n_i}(\mathbf{0}, \mathbf{X}_i\mathbf{\Gamma}\mathbf{X}_i' + \mathbf{Z}_i\mathbf{D}\mathbf{Z}_i' + \mathbf{\Psi}_i) \tag{9.28}$$

Again, θ is equal to the unknown nonredundant parameters in $\mathbf{D}$ and $\mathbf{\Psi}_i$, $i = 1, 2, \ldots, n$.

If θ and $\boldsymbol{\Gamma}$ were known, Bayes estimates of β and $\mathbf{b}$ could be obtained as the posterior expectations given $\mathbf{y}, \theta$ and $\boldsymbol{\Gamma}^{-1}$.

$$\begin{aligned} E(\beta | \mathbf{y}, \boldsymbol{\Gamma}^{-1}, \theta) &= \left(\sum_{i=1}^{n} \mathbf{X}_i' \boldsymbol{\Omega}_i^{-1} \mathbf{X}_i + \boldsymbol{\Gamma}^{-1} \right)^{-1} \sum_{i=1}^{n} \mathbf{X}_i' \boldsymbol{\Omega}_i^{1} \mathbf{y}_i = \hat{\beta} \\ E(\mathbf{b}_i | \mathbf{y}_i, \boldsymbol{\Gamma}^{-1}, \theta) &= \mathbf{D} \mathbf{Z}_i' \boldsymbol{\Omega}_i^{-1} (\mathbf{y}_i - \mathbf{X}_i \hat{\beta}) = \hat{\mathbf{b}}_i \end{aligned} \tag{9.29}$$

where

$$\begin{aligned} \boldsymbol{\Omega}_i^{-1} &= (\mathbf{Z}_i \mathbf{D} \mathbf{Z}_i' + \boldsymbol{\Psi}_i + \boldsymbol{\Gamma}^{-1})^{-1} \\ \text{var}(\beta | \mathbf{y}, \theta) &= \left(\sum_{i=1}^{n} \mathbf{X}_i' \boldsymbol{\Omega}_i^{-1} \mathbf{X}_i + \boldsymbol{\Gamma}^{-1} \right)^{-1} \\ \text{var}(\mathbf{b}_i | \mathbf{y}_i, \theta) &= \mathbf{D} - \mathbf{D} \mathbf{Z}_i' \mathbf{P}_i \mathbf{Z}_i \mathbf{D} \\ \mathbf{P}_i &= \mathbf{X} \boldsymbol{\Omega}_i^{-1} - \boldsymbol{\Omega}_i^{-1} \mathbf{X}_i \left(\boldsymbol{\Gamma}^{1} + \sum_{i=1}^{n} \mathbf{X}_i' \boldsymbol{\Omega}_i^{-1} \mathbf{X}_i \right)^{-1} \mathbf{X}_i' \boldsymbol{\Omega}_i^{-1} \end{aligned} \tag{9.30}$$

Because $\boldsymbol{\Gamma}$ and θ are unknown, the empirical Bayes approach is to replace $\boldsymbol{\Gamma}$ and θ with maximum likelihood estimates obtained by maximizing their marginal normal likelihoods based on y, integrating over β and $\mathbf{b}$.

For the Bayes formulation of the TSHLM we have no information about the variation in $\boldsymbol{\Gamma}$, we only have information regarding the between and within variation. Hence, Harville (1976), Laird and Ware (1982), and Dempster et al. (1981) propose setting $\boldsymbol{\Gamma}^{-1} = \mathbf{0}$, indicating vague prior information about β. Then, from the limiting marginal distribution of θ given $\mathbf{y}$, the estimate of θ is REML, $\hat{\theta}_R$, (Harville, 1976). Thus, we can replace θ for $\hat{\theta}_R$ in (9.29) and (9.30) to obtain $\beta(\hat{\theta}_R)$ and $\hat{\mathbf{b}}_i(\hat{\theta}_R)$, the empirical Bayes estimates:

$$\begin{aligned} E(\beta | \mathbf{y}, \boldsymbol{\Gamma}^{-1} = \mathbf{0}, \hat{\theta}_R) &= \hat{\beta}(\hat{\theta}_R) \\ E(\mathbf{b}_i | \mathbf{y}_i, \boldsymbol{\Gamma}^{-1} = \mathbf{0}, \hat{\theta}_R) &= E\{\mathbf{b}_i | \mathbf{y}_i, \hat{\beta}(\hat{\theta}_R), \hat{\theta}_R\} = \hat{\mathbf{b}}_i(\hat{\theta}_R) \end{aligned} \tag{9.31}$$

Substituting $\hat{\theta}_R$ in (9.30) causes the variances to be underestimated. This problem was studied by Kackar and Harville (1984) and Kass and Steffey (1989). The revised estimate involves the estimated variance as the first term plus a quadratic form $\delta' \mathbf{F}_{(\theta)}^{-1} \delta$ where δ are first-order partial derivatives of a real valued function $G(\mathbf{y}, \hat{\theta})$ with regard to θ evaluated at $\theta = \hat{\theta}$ and $\mathbf{F}^{-1}(\theta)$ is the inverse of the second-order partial derivatives of the ln-likelihood of the REML estimate:

$$\text{var}\, G(\mathbf{y}, \hat{\theta}) \doteq \text{var}[G(\mathbf{y}, \theta) | \hat{\theta}] + \delta' \mathbf{F}^{-1}(\theta) \delta \tag{9.32}$$

(Searle et al., 1992, p. 341-343).

To compute the ML or REML estimates of θ, one may use the EM algorithm developed by Dempster et al. (1977) to obtain ML estimates of μ and $\boldsymbol{\Sigma}$ when data are missing at random. The EM algorithm consists of two steps, the expectation or E-step and the maximization or M-step of sufficient statistics $\mathbf{t}$ of the parameters:

$$\begin{aligned} E\text{-step}&: \hat{\mathbf{t}} \equiv E\{\mathbf{t} | \mathbf{y}, \hat{\theta}\} \\ M\text{-step}&: \hat{\theta} \equiv M(\mathbf{t}) \end{aligned} \tag{9.33}$$

The algorithm treats the actual data as the *incomplete* set and the "actual + missing" data as the *complete* (augmented) set. The algorithm alternates between calculating conditional expected values (predictions) and maximizing likelihoods. If we use the procedure for mixed model estimation, the observed $\mathbf{y}$ are the *incomplete* data, and the complete data are $\mathbf{y}$ and the unobserved parameters β, $\mathbf{b}_i$ and θ, (Dempster et al., 1981; Laird et al., 1987).

To utilize the EM algorithm for the TSHLM, we let

$$\mathbf{\Psi}_i = \sigma^2\mathbf{I}_{n_i} \text{ and } \mathbf{\Omega}_i = \sigma^2\mathbf{I}_n + \mathbf{Z}_i\mathbf{D}\mathbf{Z}_i' \tag{9.34}$$

Then by (9.15) and (9.16),

$$\begin{aligned}\mathbf{\Omega}_i^{-1} &= [\mathbf{I} - \mathbf{Z}_i(\sigma^2\mathbf{D}^{-1}\mathbf{Z}_i'\mathbf{Z}_i)^{-1}\mathbf{Z}_i']/\sigma^2 \\ \mathbf{b}_i|\mathbf{y}_i &\sim N[\mathbf{D}\mathbf{Z}_i'\mathbf{\Omega}_i^{-1}(\mathbf{y}_i - \mathbf{X}_i\hat{\beta}_i), \sigma^2\mathbf{Z}_i(\sigma^2\mathbf{D}^{-1} + \mathbf{Z}_i'\mathbf{Z}_i)^{-1}\mathbf{Z}_i']\end{aligned} \tag{9.35}$$

and

$$E(\mathbf{b}_i'\mathbf{b}_i|\mathbf{y}_i) = [E(\mathbf{b}_i|\mathbf{y}_i)]'[E(\mathbf{b}_i|\mathbf{y}_i)] + \sigma^2\text{Tr}[\mathbf{Z}_i(\mathbf{Z}_i' + \sigma^2\mathbf{D}^{-1})^{-1}\mathbf{Z}_i'] \tag{9.36}$$

using the fact that if $\mathbf{y} \sim N(\mu, \mathbf{\Sigma})$, then $E(\mathbf{y}'\mathbf{A}\mathbf{y}) = \mu'\mathbf{A}\mu + \text{Tr}(\mathbf{A\Sigma})$, Timm (1975, p. 112). Furthermore, observe that by (9.35)

$$\sigma^2\text{Tr}[\mathbf{Z}_i(\mathbf{Z}_i'\mathbf{Z}_i + \sigma^2\mathbf{D}^{-1})^{-1}\mathbf{Z}_i'] = \sigma^2\text{Tr}(\mathbf{I} - \sigma^2\mathbf{\Omega}_i^{-1}) \tag{9.37}$$

and that sufficient statistics for θ are t_1 and the $k(k+1)/2$ nonredundant elements of $\mathbf{T}_2$:

$$\begin{aligned}\hat{\sigma}^2 &= \sum_{i=1}^{n}\mathbf{e}_i'\mathbf{e}_i \Big/ \sum_{i=1}^{n} n_i = t_1/N \\ \hat{\mathbf{D}} &= n^{-1}\sum_{i=1}^{n}\mathbf{b}_i\mathbf{b}_i' = \mathbf{T}_2/n\end{aligned} \tag{9.38}$$

If an estimate of θ is available, we may calculate estimates of the missing sufficient statistics by equating them to their expectations conditioned on the incomplete, observed data $\mathbf{y}$. Letting $\hat{\theta}$ be an estimate of θ, $\hat{\beta}(\theta)$, and $\hat{\mathbf{b}}_i(\hat{\theta})$ are the estimates of β and $\mathbf{b}_i$, and the estimated sufficient statistics are

$$\begin{aligned}\hat{t}_1 &= E\left\{\sum_{i=1}^{n}\mathbf{e}_i'\mathbf{e}_i|\mathbf{y}_i, \hat{\beta}(\hat{\theta}), \hat{\theta}\right\} \\ &= \sum_{i=1}^{n}\left[\hat{\mathbf{e}}_i'(\hat{\theta})\hat{\mathbf{e}}_i(\hat{\theta}) + \text{Tr var}\{\mathbf{e}_i|\mathbf{y}_i, \hat{\beta}(\hat{\theta}), \hat{\theta}\}\right] \\ \hat{\mathbf{T}}_2 &= E\left\{\sum_{i=1}^{n}\mathbf{b}_i\mathbf{b}_i'|\mathbf{y}_i, \hat{\beta}(\hat{\theta}), \hat{\theta}\right\} \\ &= \sum_{i=1}^{n}\left[\hat{\mathbf{b}}_i(\hat{\theta})\mathbf{b}_i'(\hat{\theta}) + \text{var}(\mathbf{b}_i|\mathbf{y}_i, \hat{\beta}(\hat{\theta}), \hat{\theta})\right]\end{aligned} \tag{9.39}$$

where $\hat{\mathbf{e}}_i(\hat{\theta}) = E(\mathbf{e}_i|\mathbf{y}_i, \hat{\beta}(\hat{\theta}), \hat{\theta}) = \mathbf{y}_i - \mathbf{X}_i\hat{\beta}(\hat{\theta}) - \mathbf{Z}_i\hat{\mathbf{b}}_i(\hat{\theta})$.

To convert these expressions into useful computational form following Laird et al. (1987), let ω ($\omega = 0, 1, 2, \ldots, \infty$) index the iterations where $\omega = 0$ is a starting value. For

$$
\begin{aligned}
\beta^{(\omega)} &= \left(\sum_{i=1}^{n} \mathbf{X}_i' \mathbf{W}_i^{(\omega)} \mathbf{X}_i \right)^{-1} \sum_{i=1}^{n} \mathbf{X}_i' \mathbf{W}_i^{(\omega)} \mathbf{y}_i \\
\mathbf{W}_i^{(\omega)} &= \left[\mathbf{\Omega}_i^{(\omega)} \right]^{-1} \\
\mathbf{\Omega}_i^{(\omega)} &= \sigma^{(\omega)2} \mathbf{I}_{n_i} + \mathbf{Z}_i \mathbf{D}^{(\omega)} \mathbf{Z}_i' \\
\mathbf{b}_i^{\omega} &= \mathbf{D}^{(\omega)} \mathbf{Z}_i' \mathbf{W}_i^{(\omega)} \mathbf{r}_i^{(\omega)} \\
\mathbf{r}_i^{(\omega)} &= \mathbf{y}_i - \mathbf{X}_i \beta^{(\omega)}
\end{aligned}
\tag{9.40}
$$

by (9.37) and (9.36), we have, substituting into (9.39), that the ML estimates of θ are

$$
\begin{aligned}
\sigma^{(\omega+1)^2} &= \left\{ \sum_{i=1}^{n} (\mathbf{r}_i^{(\omega)} - \mathbf{Z}_i^{(\omega)} \mathbf{b}_i^{(\omega)})(\mathbf{r}_i^{(\omega)} - \mathbf{Z}_i \mathbf{b}_i^{(\omega)}) + \sigma^{(\omega)^2} \mathrm{Tr}(\mathbf{I} - \sigma^{(\omega)^2} \mathbf{W}_i^{(\omega)})] \right\} / N \\
\mathbf{D}^{(\omega+1)} &= \sum_{i=1}^{n} [\mathbf{b}_i^{(\omega)} \mathbf{b}_i'^{(\omega)} + \mathbf{D}^{(\omega)} (\mathbf{I} - \mathbf{Z}_i' \mathbf{W}_i^{(\omega)} \mathbf{Z}_i \mathbf{D}^{(\omega)})] / n
\end{aligned}
\tag{9.41}
$$

To obtain the REML estimates of θ, $\hat{\theta}_R$, we replace the matrix $\mathbf{W}_i^{(\omega)}$ in (9.41) with $\mathbf{P}_i^{(\omega)}$ where

$$
\mathbf{P}_i^{(\omega)} = \mathbf{W}_i^{(\omega)} \left[\mathbf{I} - \mathbf{X}_i \left(\sum_{i=1}^{n} \mathbf{X}_i' \mathbf{W}_i^{(\omega)} \mathbf{X}_i \right)^{-1} \mathbf{X}_i' \mathbf{W}_i^{(\omega)} \right]
\tag{9.42}
$$

For starting values, one uses the OLSE of β, $\mathbf{b}_i$, σ^2, and $\mathbf{D}$:

$$
\begin{aligned}
\hat{\beta}_o &= \left(\sum_{i=1}^{n} \mathbf{X}_i' \mathbf{X}_i \right)^{-1} \sum_{i=1}^{n} \mathbf{X}_i' \mathbf{y}_i \\
\hat{\mathbf{b}}_i &= (\mathbf{Z}_i' \mathbf{Z}_i)^{-1} \mathbf{Z}_i' (\mathbf{y}_i - \mathbf{X}_i \hat{\beta}_o) \\
\hat{\sigma}_o^2 &= \left[\sum_{i=1}^{n} \mathbf{y}_i' \mathbf{y}_i - \hat{\beta}_o' \sum_{i=1}^{n} \mathbf{X}_i' \mathbf{y}_i - \sum_{i=1}^{n} \mathbf{b}_i' \mathbf{Z}_i' (\mathbf{y}_i - \mathbf{X}_i \hat{\beta}_o)^2 \right] / [N - (n-1)k - p] \\
\hat{\mathbf{D}}_o &= \sum_{i=1}^{n} \mathbf{b}_i \mathbf{b}_i' / n - \hat{\sigma}_o^2 \sum_{i=1}^{n} (\mathbf{Z}_i' \mathbf{Z}_i)^{-1} / n
\end{aligned}
\tag{9.43}
$$

The EM algorithm allows one to obtain parametric empirical Bayes (PEB) estimates of θ which results in an REML estimate of θ, $\hat{\theta}_R$, or an ML estimate of θ, $\hat{\theta}_{ML}$, for the variance and covariance components of the mixed model. With $\hat{\theta}_{ML}$ substituted into (9.9), $\hat{\beta}_{ML} = \hat{\beta}(\theta_{ML})$ is the maximum likelihood estimate of the fixed effect β; $\hat{\mathbf{b}}_{ML}^{(i)} = \hat{\mathbf{b}}_i(\hat{\theta}_{ML})$ is the empirical Bayes estimate of $\mathbf{b}_i$ by (9.10). While the ML procedure yields estimates of β and $\mathbf{b}_i$, they are not unbiased for the components of variance; however, the REML estimate $\hat{\theta}_R$ is unbiased for the components, but does not directly include a procedure for estimating β and $\mathbf{b}_i$. Harville (1976) showed that an estimate of θ obtained by maximizing the limiting marginal likelihood of θ given $\mathbf{y}$ is equivalent to the REML likelihood in the classical case. Thus, the effects $\hat{\beta} = \hat{\beta}(\theta_R)$ and $\hat{\mathbf{b}}_i = \hat{\mathbf{b}}_i(\theta_R)$ are called REML estimates or parametric empirical Bayes(PEB) estimates. More important Harville (1976) showed that if we use the $\mathrm{var}(\beta|\mathbf{y}, \hat{\theta}_R)$

and the $\text{var}(\mathbf{b}|\mathbf{y},\ \hat{\theta}_R)$ to estimate the variances of $\hat{\beta}(\hat{\theta}_R)$ and $\hat{\mathbf{b}}(\hat{\theta}_R)$ that these are the same as the sample-theory variances of $\text{var}(\hat{\beta})$ and the $\text{var}(\hat{\mathbf{b}}-\mathbf{b})$ where from (9.11) the

$$\begin{aligned}\text{var}(\hat{\beta}) &= \left(\sum_{i=1}^{n} \mathbf{X}_i'\boldsymbol{\Omega}_i^{-1}\mathbf{X}_i\right)^{-1} \\ var(\hat{\mathbf{b}}_i-\mathbf{b}_i) &= \mathbf{D}-\mathbf{D}\mathbf{Z}_i'\boldsymbol{\Omega}_i^{-1}\mathbf{Z}_i\mathbf{D}+\mathbf{D}\mathbf{Z}_i'\boldsymbol{\Omega}_i^{-1}\mathbf{X}_i\left(\sum_{i=1}^{n} \mathbf{X}_i'\boldsymbol{\Omega}_i^{-1}\mathbf{X}_i\right)^{-1}\mathbf{X}_i'\boldsymbol{\Omega}_i\mathbf{Z}_i\mathbf{D}\end{aligned} \tag{9.44}$$

If the $\text{var}(\hat{\mathbf{b}}_i)$ is used, the variation in $\hat{\mathbf{b}}_i-\mathbf{b}_i$ would be underestimated since it ignores the variation in $\mathbf{b}_i$. However, as with the weighted linear model with fixed effects and unknown covariance matrix, substituting $\hat{\boldsymbol{\Omega}}_i$ based on $\hat{\theta}_R$ for $\boldsymbol{\Omega}_i^{-1}$ in (9.44) underestimates the variance of $\hat{\beta}$ and the variance of $\hat{\mathbf{b}}_i-\mathbf{b}$ since for example, if $\boldsymbol{\Gamma}^{-1}=\mathbf{0}$ the

$$\text{var}(\beta|\mathbf{y}) = E[\text{var}(\beta|\mathbf{y},\theta)] + \text{var}[E(\beta|\mathbf{y},\theta)] \tag{9.45}$$

and only the first term in the expression is being estimated by the "substitution principle." Because the second term is ignored, variances are underestimated (Searle et al., 1992, p. 339-343).

Estimation of model parameters using the EM algorithm have been improved upon by Longford (1987) who incorporated Fisher scoring. An iterative generalized least squares procedure has been developed by Goldstein (1986). Park (1993) shows that the EM algorithm and the iterative two-stage IZEF algorithm are equivalent. Kreft et al. (1994) compare several computer packages using various methods of estimation. Lindstrom and Bates (1988) develop a Newton-Raphson (NR) algorithm to estimate the model parameters. Their algorithm is efficient, effective, and consistently converges, and it shows that NR is better than EM. For a comprehensive discussion of the EM method, see Liu and Rubin (1994), and McLachlan and Krishnan (1997). Wolfinger et al. (1994) developed an NR sweep algorithm that is employed in PROC MIXED.

For the TSHLM we have three groups of parameters that are estimated: the fixed effects β, the random effects $\mathbf{b}_i,\ i=1,2,\dots,n,$ and the variance components θ. To test hypotheses we write the model as a general linear mixed model:

$$\mathbf{y}_{N\times 1} = \mathbf{X}_{N\times p}\beta_{p\times 1} + \mathbf{Z}_{N\times k}\mathbf{b}_{k\times 1} + \mathbf{e}_{N\times 1} \tag{9.46}$$

where

$$\mathbf{y}'_{1\times N} = (\mathbf{y}'_1, \mathbf{y}'_2, \ldots, \mathbf{y}'_n); \;\; \underset{n_i\times 1}{\mathbf{y}_i}\,; \;\; \sum_{i=1}^{n} n_i = N$$

$$\mathbf{b}'_{1\times N} = (\mathbf{b}'_1, \mathbf{b}'_2, \ldots, \mathbf{b}'_n); \;\; \underset{k_i\times 1}{\mathbf{b}_i}\,; \;\; \sum_{i=1}^{n} k_i = k$$

$$\underset{N\times p}{\mathbf{X}} = \begin{pmatrix} \mathbf{X}_1 \\ \mathbf{X}_2 \\ \vdots \\ \vdots \\ \vdots \\ \mathbf{X}_n \end{pmatrix}; \;\; \underset{n_i\times p}{\mathbf{X}_i}$$

For random coefficient models, $\mathbf{Z}_{N\times k}$ takes the general form

$$\mathbf{Z} = \begin{bmatrix} \mathbf{Z}_1 & \mathbf{0} & \cdots & \cdots & \cdots & \mathbf{0} \\ \mathbf{0} & \mathbf{Z}_2 & \cdots & \cdots & \cdots & \mathbf{0} \\ \vdots & \vdots & \cdots & \cdots & \cdots & \vdots \\ \vdots & \vdots & \cdots & \cdots & \cdots & \vdots \\ \vdots & \vdots & \cdots & \cdots & \cdots & \vdots \\ \mathbf{0} & \cdots & \cdots & \cdots & \cdots & \mathbf{Z}_n \end{bmatrix} \tag{9.47}$$

where $\mathbf{Z}_i(n_i \times k_i)$. Alternatively, for variance component models, $\mathbf{Z}$ is partitioned such that

$$\mathbf{Z} = (\mathbf{Z}_1, \mathbf{Z}_2, \ldots, \mathbf{Z}_n); \;\; \mathbf{Z}_i(n_i \times k_i). \tag{9.48}$$

Then

$$\mathbf{y} = \mathbf{X}\beta + \sum_{i=1}^{n} \mathbf{Z}_i \mathbf{b}_i + \mathbf{e}$$

$$\text{var}(\mathbf{y}) = \mathbf{ZVZ}' + \sigma^2\mathbf{I} = \sum_{i=1}^{n} \sigma_i^2 \mathbf{Z}_i\mathbf{Z}'_i + \sigma_e^2 \mathbf{I}_N$$

More generally,

$$\begin{gathered} \mathbf{y} = \mathbf{X}\beta + \mathbf{Zb} + \mathbf{e} \\ \text{cov}(\mathbf{y}) = \mathbf{ZVZ}' + \sigma^2\mathbf{I}_N \text{ and } \mathbf{V} = \mathbf{I}_n \otimes \mathbf{D} \\ \text{cov}(\mathbf{b}_i, \mathbf{b}_j) = \mathbf{0} \quad \text{cov}(\mathbf{b}, \mathbf{e}) = \mathbf{0} \quad \text{cov}(\mathbf{b}_i) = \mathbf{D} \end{gathered} \tag{9.49}$$

For developing tests of hypotheses for the mixed model, it is convenient to use the hierarchical normal form of the model

$$\begin{gathered} (\mathbf{y}|\beta, \mathbf{b}) \sim N(\mathbf{X}\beta + \mathbf{Zb}, \mathbf{\Omega} = \mathbf{ZVZ}' + \sigma^2\mathbf{I}_N) \\ \beta \sim N(\beta_o, \mathbf{\Gamma}), \;\; \mathbf{b} \sim \mathbf{N}(\mathbf{0}, \mathbf{V} = \mathbf{I}_n \otimes \mathbf{D}) \end{gathered} \tag{9.50}$$

and assume $\mathbf{\Gamma}^{-1} = \mathbf{0}$, then we have that

$$\begin{aligned} E(\beta|\mathbf{y}) &= (\mathbf{X}'\mathbf{\Omega}^{-1}\mathbf{X} + \mathbf{\Gamma}^{-1})^{-1}(\mathbf{X}'\mathbf{\Omega}^{-1}\mathbf{y} + \mathbf{\Gamma}^{-1}\beta_o) \\ &= (\mathbf{X}'\mathbf{\Omega}^{-1}\mathbf{X})^{-1}\mathbf{X}'\mathbf{\Omega}^{-1}\mathbf{y} \\ \text{var}(\beta|\mathbf{y}) &= (\mathbf{X}'\mathbf{\Omega}^{-1}\mathbf{X} + \mathbf{\Gamma}^{-1})^{-1} = (\mathbf{X}'\mathbf{\Omega}^{-1}\mathbf{X})^{-1} \end{aligned}$$

$$\begin{aligned} E(\mathbf{b}|\mathbf{y}) &= \mathbf{VZ}'(\mathbf{\Omega} + \mathbf{X\Gamma X}')^{-1}(\mathbf{y} - \mathbf{X}\beta_o) \\ &= \mathbf{VZ}'\mathbf{\Omega}^{-1}(\mathbf{y} - \mathbf{X}\hat{\beta}) \\ &= \hat{\mathbf{b}}(\text{BLUP}) \\ \text{var}(\mathbf{b}|\mathbf{y}) &= \mathbf{V} - \mathbf{VZ}'(\mathbf{X\Gamma X}' + \mathbf{ZVZ}' + \sigma^2\mathbf{I}_N)^{-1}\mathbf{ZV} \\ &= [\mathbf{Z}'(\mathbf{X\Gamma X}' + \sigma^2\mathbf{I}_N)^{-1}\mathbf{Z} + \mathbf{V}^{-1}]^{-1} = \sigma^2(\mathbf{Z}'\mathbf{Z} + \sigma^2\mathbf{V}^{-1})^{-1} \end{aligned} \tag{9.51}$$

Testing hypotheses using the mixed model or the parametric empirical Bayes method is complicated by the estimation of the variance components θ. Tests depend on the large sample covariance matrix of the maximum likelihood or empirical Bayes estimates.

If one tests the hypothesis regarding the fixed effects β, $H\colon \mathbf{C}\beta = \mathbf{0}$, the statistic

$$X^2 = (\mathbf{C}\hat{\beta})'(\mathbf{C}(\mathbf{X}'\hat{\mathbf{\Omega}}^{-1}\mathbf{C})^{-1}\mathbf{C}\hat{\beta} \tag{9.52}$$

has a chi-square distribution with $v = \text{rank}(\mathbf{C})$ when H is true. Alternatively, one may also use an F-approximation, X^2 / v.

A very approximate test of $H\colon \mathbf{Cb} = \mathbf{0}$ is to use the empirical Bayes estimate for **b**. Then to test H, one again forms a chi-square statistic

$$\begin{aligned} X^2 &= (\mathbf{C}\hat{\mathbf{b}})'(\mathbf{C}\hat{\mathbf{\Omega}}^*\mathbf{C}')^{-1}\mathbf{C}\hat{\mathbf{b}} \\ \hat{\mathbf{\Omega}}^* &= \hat{\mathbf{V}} - \hat{\mathbf{V}}\mathbf{Z}'[\hat{\mathbf{\Omega}}^{-1} - \hat{\mathbf{\Omega}}^{-1}\mathbf{X}(\mathbf{X}'\hat{\mathbf{\Omega}}^{-1}\mathbf{X})^{-1}\mathbf{X}'\hat{\mathbf{\Omega}}^{-1}]\mathbf{Z}\hat{\mathbf{V}} \end{aligned} \tag{9.53}$$

where X^2 has a chi-square distribution under the null hypothesis with degrees of freedom $v = \text{rank}(\mathbf{C})$.

The last hypothesis of interest is whether some variances or covariance components of θ are perhaps zero. Clearly whether the variances are zero is of interest since it indicates whether there is random variation in $\mathbf{b}_i$ of $\mathbf{b}$. However, there is no adequate test of $H_\theta\colon \mathbf{C}\theta = \mathbf{0}$, even for large samples since the normal approximation is very poor. Bryk and Raudenbush (1992, p. 55) propose a goodness-of-fit test.

9.3 Random Coefficient Model: One Population

Given a group of n_j subjects, the classical form of the general linear model

$$\begin{aligned} &\mathbf{y} = \mathbf{X}\beta + \mathbf{e} \\ &E(\mathbf{e}) = \mathbf{0} \qquad \text{var}(\mathbf{e}) = \mathbf{\Omega} = \sigma^2\mathbf{I}_{n_j} \end{aligned} \tag{9.54}$$

assumes a known design matrix $\mathbf{X}_{n_j \times p}$ that is under the control of the experimenter, where the elements are measured without error, homogeneity of variance, and a fixed parameter vector β. Until now, most of our discussion regarding the GLM has allowed the structure Ω to change, but β has remained fixed, the same for all subjects.

Because subjects are really different within a group, we may allow β to be random and different among subjects. Then,

$$y_i = \mathbf{x}_i'\beta_i + e_i$$
$$\beta_i = \beta + \mathbf{b}_i$$

where the vectors $\mathbf{b}_i$ are assumed to be independent and identically distributed with zero mean and covariance structure $\mathbf{D}$. Then, we have a fixed mean

$$E(\mathbf{y}) = \mathbf{X}\beta$$

but heteroscedasticity among the y_i:

$$\text{var}(y_i) = \mathbf{x}_i'\mathbf{D}\mathbf{x}_i + \sigma^2 = \sigma_i^2$$

More generally, suppose we have n groups indexed by j. Then (9.54) becomes

$$\mathbf{y}_j = \mathbf{X}_j\beta_j + \mathbf{e}_j \tag{9.55}$$

where $\mathbf{y}_j$ and $\mathbf{e}_j$ are vectors of length n_j, the number of subjects in the jth group. Furthermore, assume

$$E(\mathbf{e}_j) = \mathbf{0}$$
$$\text{var}(\mathbf{e}_j) = \sigma_j^2\mathbf{I}$$

where the $\mathbf{e}_j$ are again independent, but not identically distributed since the variances are not homogenous across groups. Letting index i represent classes and index j subjects within classes, (9.55) and (9.56) ignore the fact that classes are part of the same school, a level of commonality for the model equations.

To model group differences, one may assume an ANCOVA model where the β_j in (9.55) have equal slopes with different intercepts and homogenous variances. That is

$$\beta_j = \mathbf{Z}_j\gamma = \begin{pmatrix} \mathbf{1}_j' & \mathbf{0} \\ \mathbf{0} & \mathbf{I} \end{pmatrix}\begin{pmatrix} \alpha \\ \beta \end{pmatrix} \tag{9.56}$$

where α contains n intercepts, and $\mathbf{1}_j$ is the jth unit vectors, and α is a vector of n fixed intercepts. Combining (9.56) with (9.55), we have

$$\mathbf{y}_j = \mathbf{X}_j\mathbf{Z}_j\gamma + \mathbf{e}_j \tag{9.57}$$

where the $\text{var}(\mathbf{y}_j) = \sigma_j^2\mathbf{I}$, the classical linear model, results if $\sigma_1^2 = \sigma_2^2 = \ldots = \sigma_n^2 = \sigma^2$.

Associating a random coefficient model to (9.55), we may postulate a different model for each classroom where the structure of the coefficients is not fixed but random, varying over replications:

$$\begin{aligned} \mathbf{y}_j &= \mathbf{X}_j\beta_j + \mathbf{e}_j \\ \beta_j &= \beta + \mathbf{b}_j \end{aligned} \tag{9.58}$$

so that

$$\begin{aligned} \mathbf{y}_j &= \mathbf{X}_j\beta + \mathbf{X}_j\mathbf{b}_j + \mathbf{e}_j \\ \text{cov}(\mathbf{y}_j) &= \mathbf{X}_j'\mathbf{D}\mathbf{X}_j + \sigma_j^2 I. \end{aligned} \tag{9.59}$$

In (9.59), we assume the $\mathbf{b}_j$ are independent of one another, independent of the errors $\mathbf{e}_j$, have zero mean, and have common covariance matrix $\mathbf{D}$. Model (9.59) has the form of the mixed model. If the random errors at the second level are all zero, (9.59) reduces to (9.55).

Model (9.59) is a hierarchical linear model, sometimes called a multilevel model or a nested mixed linear model (Goldstein, 1986, and Longford, 1987). In particular, we have that

$$\begin{aligned} \mathbf{y}_j &= \mathbf{X}_j\beta_j + \mathbf{e}_j \\ \beta_j &= \mathbf{Z}_j\gamma + \mathbf{b}_j \end{aligned}$$

so that

$$\begin{aligned} \mathbf{y}_j &= (\mathbf{X}_j\mathbf{Z}_j)\gamma + \mathbf{X}_j\mathbf{b}_j + \mathbf{e}_j \\ E(\mathbf{y}_j) &= (\mathbf{X}_j\mathbf{Z}_j)\gamma \\ \text{cov}(\mathbf{y}_j) &= \mathbf{X}_j\mathbf{D}\mathbf{X}'_j + \sigma_j^2\mathbf{I} \end{aligned} \tag{9.60}$$

a mixed linear model.

Given (9.60), an estimate of β_j is

$$\hat{\beta}_j = (\mathbf{X}'_j\mathbf{X}_j)^{-1}\mathbf{X}'_j\mathbf{y}_j \tag{9.61}$$

so that

$$\hat{\gamma} = \left(\sum_{j=1}^{n} \mathbf{Z}'_j\mathbf{Z}_j\right)^{-1} \sum_{j=1}^{n} \mathbf{Z}'_j\beta_j$$

Alternatively, using (9.57),

$$\begin{aligned} \hat{\gamma} &= \left(\sum_{j=1}^{n} \mathbf{Z}'_j\mathbf{X}'_j\mathbf{X}_j\mathbf{Z}_j\right)^{-1} \sum_{j=1}^{n} \mathbf{Z}_j\mathbf{X}_j\mathbf{y}_j \\ &= \left(\sum_{j=1}^{n} \mathbf{Z}'_j\mathbf{X}'_j\mathbf{X}_j\mathbf{Z}_j\right)^{-1} \sum_{j=1}^{n} \mathbf{Z}_j(\mathbf{X}'_j\mathbf{X}_j)\hat{\beta}_j \end{aligned} \tag{9.62}$$

by (9.61). However, this estimator of γ is neither a BLUE nor a best linear unbiased prediction (BLUP) estimator (de Leeuw and Kreft, 1986; Vonesh and Carter, 1987). The BLUE of γ, assuming σ_j^2 and $\mathbf{D}$ are known, is

$$\begin{aligned} \hat{\gamma} &= \left(\sum_{j=1}^{n} \mathbf{Z}'_j\mathbf{W}_j^{-1}\mathbf{Z}_j\right)^{-1} \sum_{j=1}^{n} \mathbf{Z}'_j\mathbf{W}_j\hat{\beta}_j \\ \mathbf{W}_j &= \mathbf{D} + \sigma_j^2(\mathbf{X}'_j\mathbf{X}_j)^{-1} \end{aligned} \tag{9.63}$$

where $\mathbf{W}_j$ is the covariance matrix for the OLSE $\hat{\beta}_j$. Because $\mathbf{D}$ and σ_j^2 are unknown, estimation of model parameters for hierarchical linear models is complicated with the analysis dependent upon the algorithm and the computer program (Kreft et al., 1994; Gelfand et al., 1995).

In 9.2, the two-stage hierarchical linear model was introduced as a repeated measurement experiment in which n_i repeated observations are obtained on $i = 1, 2, \ldots, n$ subjects which permitted one to estimate individual models by modeling within-subject and between-subject variation as a mixed model. From (9.60), we see that the presentation of the HLM of Bryk and Raudenbush (1992) is also a mixed linear model.

There are many issues associated with the analysis of hierarchical linear models in the social sciences. Articles in Volume 20, Number 2 of the *Journal of Educational and Behavioral Statistics* review several problems and concerns. For example, the article by Draper (1995) reviews inference issues. A discussion of nonlinear models is examined by Davidian and Giltinan (1995). Longford (1993) discusses numerous algorithms. Everson and Morris (in press) develop a procedure for making inferences about the parameters of the two-stage normal HLM which outperforms REML and Gibbs sampling procedures. Their procedure is based on independent draws from an exact posterior distribution, a full Bayesian model. Thum (1997) develops an HLM for multivariate observations.

We now turn to the analysis of several data sets using the MIXED procedure, documented in SAS Institute Inc. (1992) and PROC GENMOD, discussed in SAS Institute Inc. (1993) to analyze the random coefficient model, with repeated measures and hierarchical data models.

In Chapter 8, we used the Elston and Grizzle (1962) ramus heights data to illustrate the SUR model in the analysis of cross-sectional time series data. Recall that the data are the ramus heights (in mm) of 20 boys measured at four time points (8, 8.5, 9, and 9.5 years of age). The SUR model fit a model to the entire group with an unknown covariance structure $\Sigma_{p \times p}$. With a consistent estimator for Σ, we used the generalized least squares procedure to estimate the model parameters for the group.

Plotting the data for each boy (Output 9.3), observe that two boys show rapid growth during the period 8.5 to 9 years, while the growth of the others is more gradual. Hence, the variation among the boys suggests fitting a model that considers stochastic variation of the n slopes, a random coefficient or mixed model. For this analysis, we have $i = 1, 2, \ldots, 20$ boys and $j = 1, 2, 3, 4$ time points. For each boy, we fit the linear model

$$y_i = \beta_o + \beta_1 x_i + e_i \qquad i = 1, 2, \ldots, n \tag{9.64}$$

where $e_i \sim IN(0, \sigma^2)$, β_o is the group mean common to all boys and β_i is specific to a boy. That is, β_i is random with linear structure $\beta_i = \beta_1 + b_i$ where $E(\beta_i) = \beta_1$, $E(b_i) = 0$ and the variance of b_i is σ_b^2. Using the random coefficient model with orthogonal polynomials for a linear model, the model in (9.64) becomes

$$\underset{4\times 1}{\mathbf{y}_i} = \underset{4\times 2}{\mathbf{X}_i}\ \underset{2\times 1}{\beta} + \underset{4\times 1}{\mathbf{Z}_i}\ \underset{1\times 1}{\mathbf{b}_i} + \underset{4\times 1}{\mathbf{e}_i}$$

$$\mathbf{y}_i = \begin{pmatrix} 1 & -3 \\ 1 & -1 \\ 1 & -1 \\ 1 & 3 \end{pmatrix} \begin{pmatrix} \beta_0 \\ \beta_1 \end{pmatrix} + \begin{pmatrix} 1 \\ 1 \\ 1 \\ 1 \end{pmatrix} \mathbf{b}_i + \mathbf{e}_i \tag{9.65}$$

$$\text{cov}(\mathbf{y}_i) = \mathbf{Z}_i \mathbf{D} \mathbf{Z}_i' + \Psi_i = \sigma_b^2 J_4 + \sigma^2 \mathbf{I}_4.$$

Stacking the vectors $\mathbf{y}_i$ into a vector $\mathbf{y}$, the GLM for the problems is

$$\begin{pmatrix} \mathbf{y}_1 \\ \mathbf{y}_2 \\ \vdots \\ \vdots \\ \vdots \\ \mathbf{y}_n \end{pmatrix} = \begin{pmatrix} \mathbf{X}_1 \\ \mathbf{X}_2 \\ \vdots \\ \vdots \\ \vdots \\ \mathbf{X}_n \end{pmatrix} \beta + \begin{pmatrix} \mathbf{Z}_1 & \mathbf{0} & \cdots & \cdots & \cdots & \mathbf{0} \\ \mathbf{0} & \mathbf{Z}_2 & \cdots & \cdots & \cdots & \mathbf{0} \\ \vdots & \cdots & \ddots & \cdots & \cdots & \vdots \\ \vdots & \cdots & \cdots & \ddots & \cdots & \vdots \\ \vdots & \cdots & \cdots & \cdots & \ddots & \vdots \\ \mathbf{0} & \cdots & \cdots & \cdots & \mathbf{0} & \mathbf{Z}_n \end{pmatrix} \begin{pmatrix} \mathbf{b}_1 \\ \mathbf{b}_2 \\ \vdots \\ \vdots \\ \vdots \\ \mathbf{b}_n \end{pmatrix} + \begin{pmatrix} \mathbf{e}_1 \\ \mathbf{e}_2 \\ \vdots \\ \vdots \\ \vdots \\ \mathbf{e}_n \end{pmatrix} \tag{9.66}$$

$$\mathbf{y} = \mathbf{X} \quad \beta + \quad \mathbf{Z} \quad \mathbf{b} + \mathbf{e}$$

$$\mathrm{cov}(\mathbf{b}) = \begin{pmatrix} \mathbf{D} & \cdots & \cdots & \cdots & \cdots & \mathbf{0} \\ \vdots & \mathbf{D} & \cdots & \cdots & \cdots & \vdots \\ \vdots & \cdots & \ddots & \cdots & \cdots & \vdots \\ \vdots & \cdots & \cdots & \ddots & \cdots & \vdots \\ \vdots & \cdots & \cdots & \cdots & \ddots & \vdots \\ \mathbf{0} & \cdots & \cdots & \cdots & \cdots & \mathbf{D} \end{pmatrix} = \mathbf{I}_n \otimes \sigma_b^2 = \sigma_b^2\, \mathbf{I}_{20} = \mathbf{V}$$

$$\mathrm{cov}(\mathbf{e}) = (\mathbf{I}_n \otimes \Psi_i) = \begin{pmatrix} \Psi_1 & \cdots & \cdots & \cdots & \cdots & \mathbf{0} \\ \vdots & \Psi_2 & \cdots & \cdots & \cdots & \vdots \\ \vdots & \cdots & \ddots & \cdots & \cdots & \vdots \\ \vdots & \cdots & \cdots & \ddots & \cdots & \vdots \\ \vdots & \cdots & \cdots & \cdots & \ddots & \vdots \\ \mathbf{0} & \cdots & \cdots & \cdots & \cdots & \Psi_n \end{pmatrix} = \Psi = \sigma^2 \mathbf{I}_{80}$$

so that the $\mathrm{cov}(\mathbf{y}) = \mathbf{ZVZ}' + \Psi = \Omega_{80\times 80} = \mathbf{I}_n \otimes \Omega_i$ where $\mathbf{\Omega}_i = \mathbf{Z}_i\mathbf{D}\mathbf{Z}_i' + \Psi_i = \sigma_b^2\mathbf{J}_4 + \sigma^2\mathbf{I}_4$ and σ_b^2 and σ^2 are the variance components.

To analyze the data using PROC MIXED, we reorganize the data for the SUR model so that each boy has four repeated measurements at four ages, and the design matrix of orthogonal polynomials for a linear model is input. PROC MIXED has three essential statements: MODEL, RANDOM, and REPEATED. PROC MIXED provides both REML and ML estimates of model parameters, as well as minimum variance quadratic unbiased estimates (MINQUE). The latter procedure is useful for large data sets when REML and ML methods fail to converge. The SAS code to performs the analysis is provided in Program 9_3.sas.

Program 9_3.sas

```
/* Program 9_3.sas                                        */
/* Random Coefficient Model - Elston and Grizzle (1962) */

options ls=80 ps=60 nodate nonumber;
title1 'Output 9.3: Random Coefficient Model One Population';

data ramus;
   infile 'c:\ramus.dat';
   input group y1 y2 y3 y4 boys;
   y=y1; age=8;   x=-3; output;
   y=y2; age=8.5; x=-1; output;
```

```
      y=y3; age=9;   x=1;  output;
      y=y4; age=9.5; x=3;  output;
   proc print data=ramus;
      title2 'Original Ramus Height Data for Growth Curve Analysis';
   run;
   proc summary data=ramus;
      class boys;
      var y1-y4;
      output out=new mean=mr1-mr4;
   proc print data=new;
      title2 'Reformated Ramus Height Data for PROC MIXED';
   run;
   data plot;
      set new;
      array mr(4) mr1-mr4;
      do time = 1 to 4;
         response= mr(time);
         output;
      end;
      drop mr1-mr4;
   proc plot;
      title2 'Individual Profiles';
      plot response*time=boys;
   run;
   proc mixed data=ramus method=reml covtest ratio;
      title2 'Linear Growth with Random Slope and Variance Components';
      class boys;
      model y=x/s p;
      random boys/type=vc subject=boys;
      repeated /type=simple subject=boys r;
   run;
   proc mixed data=ramus method=reml covtest ratio;
      title2 'Linear Growth with Random Slope and Unknown Structure';
      class boys;
      model y=x/s p;
      repeated /type=un subject=boys r;
   run;
   proc mixed data=ramus method=reml covtest ratio;
      title2 'Linear Growth with Random Slope and Compound Symmetry Structure';
      class boys;
      model y=x/s p;
      repeated /type=cs subject=boys r;
   run;
   proc genmod data=ramus;
      title2 'Linear Growth and Homogeniety of Variance';
      class boys;
      model y=x/dist=normal link=identity lrci waldci obstats;
   run;
```

The MODEL statement of PROC MIXED is used to specify the design matrix **X** of the mixed model, where the intercept in the model is always included by default. The S option requests the fixed effect estimate $\hat{\beta}$ to be printed and the P option prints the predicted values. The RANDOM statement defines the random effects in the vector **b** and is used to define the **Z** matrix in the mixed model. The structure of **V** is defined by using the TYPE = option. In our example, $\mathbf{V} = \sigma_b^2 \mathbf{I}_{20}$ so that TYPE = VC, for variance component. The REPEATED statement is used to establish the structure of Ψ and the R option requests the first block of Ψ, Ψ_i to be printed. In PROC MIXED, the $\text{cov}(\mathbf{e}) = \mathbf{I} \otimes \Psi_i \equiv \mathbf{R}$ and the $\text{cov}(\mathbf{b}) = \mathbf{I} \otimes \mathbf{D} \equiv \mathbf{G}$.

Result and Interpretation 9_3.sas

The results are found in Output 9.3.

Output 9.3: *Random Coefficient Model One Population*

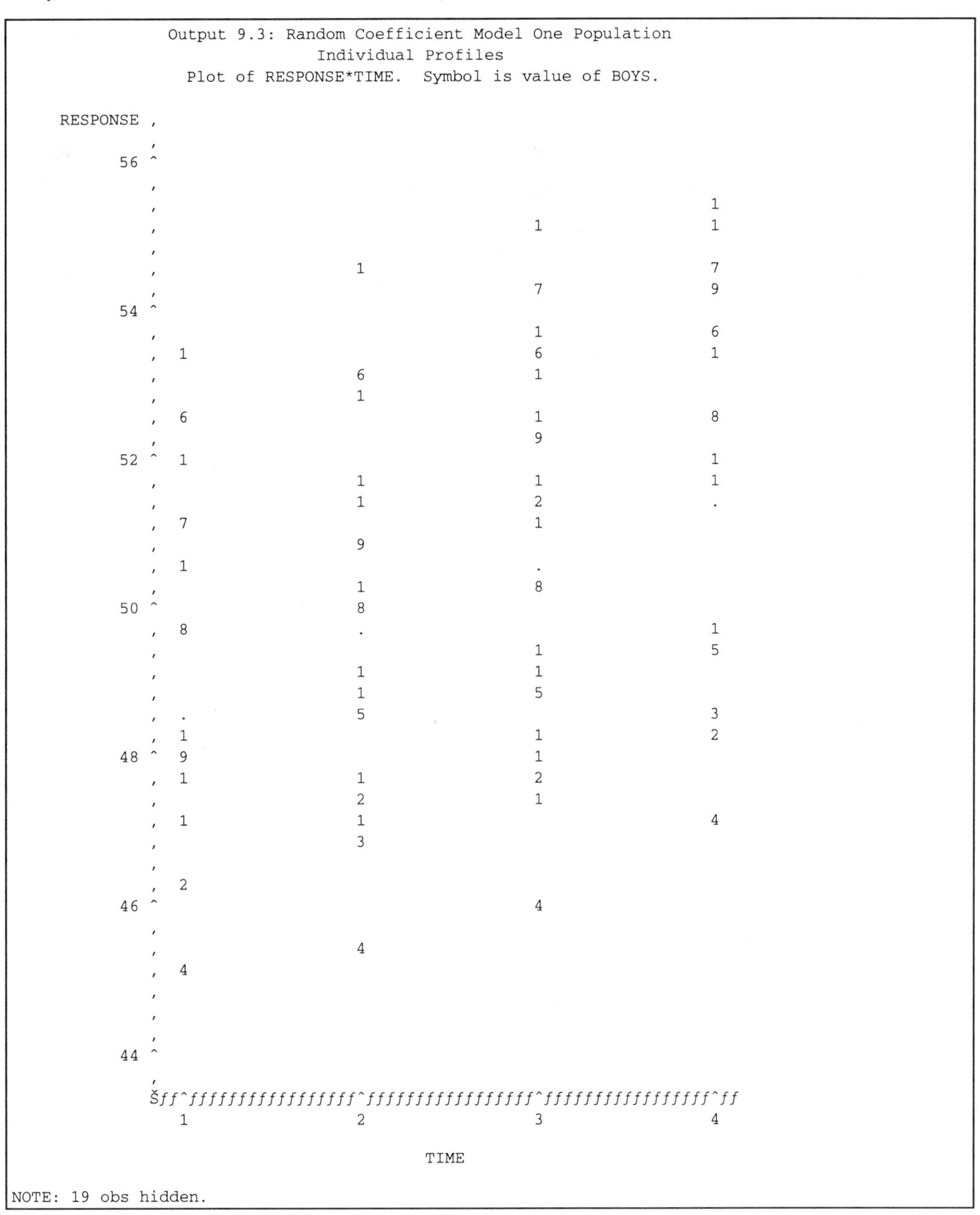

```
                Output 9.3: Random Coefficient Model One Population
                               Individual Profiles
                  Plot of RESPONSE*TIME.  Symbol is value of BOYS.

     RESPONSE ,
              ,
           56 ^
              ,
              ,                                                        1
              ,                                     1                  1
              ,
              ,                   1                                    7
              ,                                     7                  9
           54 ^
              ,                                     1                  6
              ,  1                                  6                  1
              ,                   6                 1
              ,                   1
              ,  6                                  1                  8
              ,                                     9
           52 ^  1                                                     1
              ,                   1                 1                  1
              ,                   1                 2                  .
              ,  7                                  1
              ,                   9
              ,  1                                  .
              ,                   1                 8
           50 ^                   8
              ,  8                .                                    1
              ,                                     1                  5
              ,                   1                 1
              ,                   1                 5
              ,  .                5                                    3
              ,  1                                  1                  2
           48 ^  9                                  1
              ,  1                1                 2
              ,                   2                 1
              ,  1                1                                    4
              ,                   3
              ,
              ,  2
           46 ^                                     4
              ,
              ,                   4
              ,  4
              ,
              ,
              ,
           44 ^
              ,
              Šff^fffffffffffffffff^fffffffffffffffff^fffffffffffffffff^ff
                 1                 2                 3                 4

                                         TIME

NOTE: 19 obs hidden.
```

Output 9.3 (*continued*)

```
         Linear Growth with Random Slope and Variance Components
                           The MIXED Procedure
                         Class Level Information

             Class      Levels  Values

             BOYS           20  1 2 3 4 5 6 7 8 9 10 11 12 13
                                14 15 16 17 18 19 20

                    REML Estimation Iteration History

            Iteration  Evaluations     Objective     Criterion

                    0            1  235.76483407
                    1            1  126.64050447    0.00000000

                         Convergence criteria met.
                            R Matrix for BOYS 1

         Row          COL1           COL2           COL3           COL4

           1    0.67790000
           2                   0.67790000
           3                                  0.67790000
           4                                                 0.67790000

                  Covariance Parameter Estimates (REML)

      Cov Parm  Subject      Estimate      Std Error        Z  Pr > |Z|

      BOYS      BOYS       6.09532763     2.03280986     3.00    0.0027
      DIAG      BOYS       0.67790000     0.12481151     5.43    0.0001

                     Model Fitting Information for Y

                  Description                          Value
                  Observations                       80.0000
                  Res Log Likelihood                -134.997
                  Akaike's Information Criterion    -136.997
                  Schwarz's Bayesian Criterion      -139.354
                  -2 Res Log Likelihood             269.9949

                       Solution for Fixed Effects

       Effect         Estimate      Std Error     DF        t  Pr > |t|

       INTERCEPT   50.07500000     0.55967860     19    89.47    0.0001
       X            0.46650000     0.04116734     59    11.33    0.0001

         Linear Growth with Random Slope and Variance Components
                          Tests of Fixed Effects

                 Source      NDF   DDF  Type III F  Pr > F

                 X             1    59      128.41  0.0001
```

Output 9.3 (*continued*)

```
            Linear Growth with Random Slope and Unknown Structure
                           The MIXED Procedure
                         Class Level Information

               Class     Levels  Values

               BOYS          20  1 2 3 4 5 6 7 8 9 10 11 12 13
                                 14 15 16 17 18 19 20

REML Estimation Iteration History

              Iteration  Evaluations      Objective     Criterion

                      0            1  235.76483407
                      1            2   84.57219879    0.00000794
                      2            1   84.57186064    0.00000000

                         Convergence criteria met.
                            R Matrix for BOYS 1

          Row          COL1          COL2          COL3          COL4

            1    6.32689828    6.18361405    5.78265481    5.55567213
            2    6.18361405    6.43386452    6.15472695    5.93555543
            3    5.78265481    6.15472695    6.89417252    6.93059298
            4    5.55567213    5.93555543    6.93059298    7.44609958

                  Covariance Parameter Estimates (REML)

       Cov Parm  Subject      Estimate     Std Error        Z  Pr > |Z|

       UN(1,1)   BOYS       6.32689828    2.05177287     3.08    0.0020
       UN(2,1)   BOYS       6.18361405    2.03610051     3.04    0.0024
       UN(2,2)   BOYS       6.43386452    2.08273266     3.09    0.0020
       UN(3,1)   BOYS       5.78265481    2.01256175     2.87    0.0041
       UN(3,2)   BOYS       6.15472695    2.07756587     2.96    0.0031
       UN(3,3)   BOYS       6.89417252    2.22960982     3.09    0.0020
       UN(4,1)   BOYS       5.55567213    2.02515482     2.74    0.0061
       UN(4,2)   BOYS       5.93555543    2.09081641     2.84    0.0045
       UN(4,3)   BOYS       6.93059298    2.28129267     3.04    0.0024
       UN(4,4)   BOYS       7.44609958    2.41021860     3.09    0.0020

                     Model Fitting Information for Y

                  Description                         Value
                  Observations                      80.0000
                  Res Log Likelihood               -113.963
                  Akaike's Information Criterion   -123.963
                  Schwarz's Bayesian Criterion     -135.747
                  -2 Res Log Likelihood            227.9263

            Linear Growth with Random Slope and Unknown Structure
                     Model Fitting Information for Y

                  Description                         Value

                  Null Model LRT Chi-Square        151.1930
                  Null Model LRT DF                  9.0000
                  Null Model LRT P-Value             0.0000
```

Output 9.3 (*continued*)

```
                         Solution for Fixed Effects

         Effect          Estimate      Std Error     DF         t  Pr > |t|

         INTERCEPT    50.04958234     0.55629000     19     89.97     0.0001
         X             0.46540220     0.05223065     19      8.91     0.0001

                           Tests of Fixed Effects

                   Source       NDF    DDF  Type III F  Pr > F

                   X              1     19       79.40  0.0001

Linear Growth with Random Slope and Compound Symmetry Structure
                            The MIXED Procedure
                          Class Level Information

                Class      Levels  Values
                BOYS           20  1 2 3 4 5 6 7 8 9 10 11 12 13
                                   14 15 16 17 18 19 20

                      REML Estimation Iteration History

               Iteration  Evaluations     Objective     Criterion

                       0            1  235.76483407
                       1            1  126.64050447    0.00000000

                         Convergence criteria met.
                            R Matrix for BOYS 1

           Row            COL1           COL2           COL3           COL4

             1      6.77322763     6.09532763     6.09532763     6.09532763
             2      6.09532763     6.77322763     6.09532763     6.09532763
             3      6.09532763     6.09532763     6.77322763     6.09532763
             4      6.09532763     6.09532763     6.09532763     6.77322763

                    Covariance Parameter Estimates (REML)

 Cov Parm   Subject          Ratio       Estimate      Std Error        Z  Pr > |Z|

 CS         BOYS        8.99148493     6.09532763     2.03280986     3.00    0.0027
 Residual               1.00000000     0.67790000     0.12481151     5.43    0.0001

                       Model Fitting Information for Y

                    Description                          Value
                    Observations                       80.0000
                    Res Log Likelihood                -134.997
                    Akaike's Information Criterion    -136.997
                    Schwarz's Bayesian Criterion      -139.354
                    -2 Res Log Likelihood             269.9949
                    Null Model LRT Chi-Square         109.1243
                    Null Model LRT DF                   1.0000
                    Null Model LRT P-Value              0.0000
```

Output 9.3 *(continued)*

```
        Linear Growth with Random Slope and Compound Symmetry Structure
                         Solution for Fixed Effects

         Effect             Estimate      Std Error      DF          t   Pr > |t|

         INTERCEPT    50.07500000    0.55967860      19      89.47      0.0001
         X             0.46650000    0.04116734      59      11.33      0.0001

                           Tests of Fixed Effects

                  Source         NDF    DDF   Type III F   Pr > F

                  X                1     59       128.41   0.0001
```

From Output 9.3, the REML estimate of β is

$$\hat{\beta}_{REML} = \left(\sum_{i=1}^{n} \mathbf{X}_i'\hat{\Omega}_i^{-1}\mathbf{X}_i\right)^{-1}\left(\sum_{i=1}^{n} \mathbf{X}_i'\hat{\Omega}_i^{-1}\mathbf{y}_i\right)$$

$$= \begin{pmatrix} 50.0750 \\ 0.4665 \end{pmatrix}$$

where $\hat{\Omega}_i = \hat{\sigma}_b \mathbf{J} + \sigma^2 \mathbf{I}$, and $\hat{\sigma}_b^2 = 6.0953$ and $\hat{\sigma}^2 = 0.6779$ are REML estimates of variance components. The corresponding ML estimates are 5.7849 and 0.6666, respectively, which are biased downward. Comparing $\hat{\beta}_{\text{REML}}$ with the estimate of $\hat{\beta}_{\text{FGLS}}$ using the SUR model, we have close agreement between the two sets of parameter estimates. To obtain the SUR model result using PROC MIXED, we remove the RANDOM statement which models $\mathbf{D}$ and use the TYPE = UN option in the REPEATED statement. Because $\mathbf{S} \doteq \sum_{i=1}^{n} \hat{\Psi}_i / n$, $\hat{\beta}'_{\text{REML}} = \hat{\beta}'_{\text{FGLS}} =$ (50.0496, 0.4654) as shown in Output 9.3. To invoke compound symmetry structure for the diagonal block Ψ_i of Ψ, we use the TYPE = CS option in the RANDOM statement. The common covariance matrix is output with the R option. In the output, `CS` contains the common covariance REML estimate; `Residual` contains the estimate of common variance. Even though the random coefficient and compound symmetry models are not the same, in general, since $\mathbf{\Sigma} = \mathbf{ZDZ}' + \mathbf{\Psi} \neq \mathbf{\Psi}$ we have that $\mathbf{Z}_i\mathbf{DZ}_i' + \sigma^2\mathbf{I} = \mathbf{\Psi}_i$ and (diag $[\mathbf{\Psi}_i]) = \mathbf{\Psi}$ so that the random coefficient model and the model fit assuming compound symmetry of Σ give identical results in this example.

To evaluate the model fit to the data using PROC MIXED, numerous model fitting statistics are provided. The likelihood ratio test (LRT) compares the null model $\Omega = \sigma^2\mathbf{I}$ with the structured model input. For the ramus example, the model with compound symmetry structure appears to be the best model. This follows from the observation that the two model-fit criteria provided by PROC MIXED, Akaike's Information Criterion (AIC) and Schwarz's Bayesian Criterion (SBC), are closer to zero for the compound symmetry structure. For a discussion regarding "fit" statistics, see Littell et al. (1996) and SAS Institute Inc. (1992).

Alternatively, one may fit a model to the ramus data using the generalized linear model and the GENMOD procedure. The generalized linear model extends the GLM to a wider class of models. The components of the model are

$$(1)\quad E(Y_i) = \mu_i$$
$$(2)\quad g(\mu_i) = \mathbf{x}_i'\beta = \eta_i$$
$$(3)\quad \text{var}(Y_i) = \phi V(\mu_i) / w_i$$

where ϕ is a known or must be estimated, w_i are known weights for each observation, and g is a monotonic differentiable link function (McCullagh and Nelder, 1989). For additional detail, see SAS Institute Inc. (1993).

9.4 Random Coefficient Model: Several Populations

In Chapter 7, we analyzed the Grizzle and Allen (1969) dog data by fitting a third-degree polynomial to each group using a full multivariate growth curve model of the form $E(\mathbf{Y}) = \mathbf{XBP}$ with large unknown covariance matrix Σ. For such a model, the matrix Σ may be poorly estimated given its size. In addition, the growth curve model does not permit the specification and estimation of random individual characteristics or permit one to specify a known structure for Σ. We now analyze the same data using PROC MIXED which permits the specification of fixed and random effects and the characterization of $\mathbf{\Sigma}$ using a hierarchical linear model. The SAS code to perform the analysis is given in Program 9_4.sas.

Program 9_4.sas

```
/* Program 9_4.sas                                          */
/* Random Coefficient Growth Curve (Grizzle and Allen,1969) */

options ls=80 ps=60 nodate nonumber;
title 'Output 9.4: Random Coefficient Model Several Populations';

data grizzle;
   infile 'c:\7_7.dat';
   input treat y1 y2 y3 y4 y5 y6 y7 dog;
   y=y1; time= 1; x1=-3; x2= 5; x3=-1; output;
   y=y2; time= 3; x1=-2; x2= 0; x3= 1; output;
   y=y3; time= 5; x1=-1; x2=-3; x3= 1; output;
   y=y4; time= 7; x1= 0; x2=-4; x3= 0; output;
   y=y5; time= 9; x1= 1; x2=-3; x3=-1; output;
   y=y6; time=11; x1= 2; x2= 0; x3=-1; output;
   y=y7; time=13; x1= 3; x2= 5; x3= 1; output;
   drop y1- y7;
proc print data=grizzle;
   title2 'Grizzle and Allan Dog Data';
run;
proc mixed data=Grizzle method=reml covtest ratio;
   title2 'Hierarchical Random Coefficient Growth Curve Model';
   class treat dog;
   model y=treat x1 x2 x3/s p;
   random intercept x1 x2 x3/type=un subject=dog g s;
   lsmeans treat;
run;
proc mixed data=grizzle method=reml covtest ratio;
   title2 'Growth Curve Model Unknown Structure';
   class treat dog;
```

```
   model y=treat x1 x2 x3/s p;
   repeated/type=un subject=dog r;
   lsmeans treat;
run;
proc mixed data=grizzle method=reml covtest ratio;
   title2 'Growth Curve Model Compound Symmetry';
   class treat dog;
   model y=treat x1 x2 x3/s p;
   repeated/type=cs subject=dog r;
   lsmeans treat;
run;
proc mixed data=grizzle method=reml covtest ratio;
   title2 'Growth Curve Model with Heterogeneous Compound Symmetry';
   class treat dog;
   model y=treat x1 x2 x3/noint s;
   repeated intercept diag/subject=dog group=treat;
   contrast 'group diff'
            treat 1 0 0 -1,
            treat 0 1 0 -1,
            treat 0 0 1 -1;
run;
```

For Stage 1 of the TSHLM, we assume the 7x1 vector of repeated measures for each dog has the general form

$$\underset{7\times 1}{\mathbf{y}_i} = \underset{7\times 7}{\mathbf{X}_i}\ \underset{7\times 1}{\beta} + \underset{7\times 4}{\mathbf{Z}_i}\ \underset{4\times 1}{\mathbf{b}_i} + \underset{7\times 1}{\mathbf{e}_i} \tag{9.67}$$

where $\mathbf{X}_i$ is the design matrix of the fixed effects, β contains the intercept, independent treatment effects and common slopes represented by linear (L), quadratic (Q), and cubic (C) polynomials; $\mathbf{b}_i$ is a random vector of trends, intercept (I), L, Q, C; $\mathbf{e}_i$ is a random vector of errors with structure $\Psi_i = \sigma^2\mathbf{I}$ and $\mathbf{Z}_i$ are covariates. For Stage 2 of the model, we assume that $\mathbf{b}_i$ has mean zero and covariance matrix $\mathbf{D}_{4\times 4}$ so that the $\text{cov}(\mathbf{b}) = \mathbf{I}_n \otimes \mathbf{D} = \mathbf{V} \equiv \mathbf{G}$, $\mathbf{b}' = (\mathbf{b}_1', \ldots, \mathbf{b}_n')$ and $\Psi \equiv \mathbf{R}$. Model (9.67) is a random coefficient growth curve model. To analyze this model using PROC MIXED, we create the matrices

$$\underset{N\times p}{\mathbf{X}} = \begin{pmatrix} \mathbf{X}_1 \\ \mathbf{X}_2 \\ \vdots \\ \vdots \\ \vdots \\ \mathbf{X}_n \end{pmatrix}, \quad \underset{N\times k}{\mathbf{Z}} = \begin{pmatrix} \mathbf{Z}_1 & \cdots & \cdots & \cdots & \cdots & \mathbf{0} \\ \vdots & \mathbf{Z}_2 & \cdots & \cdots & \cdots & \vdots \\ \vdots & \cdots & \ddots & \cdots & \cdots & \vdots \\ \vdots & \cdots & \cdots & \ddots & \cdots & \vdots \\ \vdots & \cdots & \cdots & \cdots & \ddots & \vdots \\ \mathbf{0} & \cdots & \cdots & \cdots & \cdots & \mathbf{Z}_n \end{pmatrix} \text{ and } \mathbf{ZGZ}' + \mathbf{R} \tag{9.68}$$

The MODEL statement of PROC MIXED defines **X**, where an intercept is included in the model by default. To remove the intercept, we use the NOINT option. The S option causes the fixed effects β to be printed. The RANDOM statement defines the random effects in **b** by creating the matrix **Z** and the structure of **G**. The S option on the RANDOM statement prints the EBLUPs of the individual dogs expressed as deviations about the population estimate. The structure of **G** is defined using the TYPE = option where G requests that the estimate of **G** be printed. PROC MIXED does not include an intercept in the RANDOM statement. The structure of $\mathbf{G} = \mathbf{I}_n \otimes \mathbf{D}$ and the $\text{cov}(\mathbf{e}) = \mathbf{I}_n \otimes \sigma^2\mathbf{I}$, so that $\Omega_i = \mathbf{Z}_i\mathbf{D}\mathbf{Z}_i' + \sigma^2\mathbf{I}$ and $\mathbf{R} = \sigma^2\mathbf{I}$.

Result and Interpretation 9_4.sas

The results of the analysis are given in Output 9.4.

Output 9.4: *Random Coefficient Model Several Populations*

```
Output 9.4: Random Coefficient Model Several Populations

          Hierarchical Random Coefficient Growth Curve Model
                        The MIXED Procedure

                      Class Level Information

         Class       Levels   Values

         TREAT            4   1 2 3 4
         DOG             36   1 2 3 4 5 6 7 8 9 10 11 12 13
                              14 15 16 17 18 19 20 21 22 23
                              24 25 26 27 28 29 30 31 32 33
                              34 35 36
```

Output 9.4: *(continued)*

```
                         REML Estimation Iteration History

                 Iteration  Evaluations      Objective      Criterion

                         0            1    47.52285001
                         1            4  -168.8922386     0.00171031
                         2            1  -169.0519552     0.00005843
                         3            1  -169.0570159     0.00000009
                         4            1  -169.0570231     0.00000000

                            Convergence criteria met.

                                     G Matrix

  Effect      DOG   Row          COL1          COL2          COL3          COL4

  INTERCEPT   1       1    0.28863879    0.03629268   -0.00752803   -0.02382261
  X1          1       2    0.03629268    0.01439433    0.00013304   -0.01008394
  X2          1       3   -0.00752803    0.00013304    0.00201008    0.00055450
  X3          1       4   -0.02382261   -0.01008394    0.00055450    0.01545473

                    Covariance Parameter Estimates (REML)

 Cov Parm    Subject         Ratio      Estimate     Std Error       Z  Pr > |Z|

 UN(1,1)     DOG        4.90269718    0.28863879    0.08299251    3.48    0.0005
 UN(2,1)     DOG        0.61645227    0.03629268    0.01504828    2.41    0.0159
 UN(2,2)     DOG        0.24449596    0.01439433    0.00395389    3.64    0.0003
 UN(3,1)     DOG       -0.12786797   -0.00752803    0.00525077   -1.43    0.1517
 UN(3,2)     DOG        0.00225969    0.00013304    0.00113061    0.12    0.9063
 UN(3,3)     DOG        0.03414231    0.00201008    0.00065502    3.07    0.0021
 UN(4,1)     DOG       -0.40464086   -0.02382261    0.01686017   -1.41    0.1577
 UN(4,2)     DOG       -0.17128163   -0.01008394    0.00384898   -2.62    0.0088
 UN(4,3)     DOG        0.00941849    0.00055450    0.00140209    0.40    0.6925
 UN(4,4)     DOG        0.26250752    0.01545473    0.00618580    2.50    0.0125
 Residual               1.00000000    0.05887347    0.00801166    7.35    0.0001

               Hierarchical Random Coefficient Growth Curve Model
                        Model Fitting Information for Y

                   Description                              Value

                   Observations                          252.0000
                   Res Log Likelihood                    -140.611
                   Akaike's Information Criterion        -151.611
                   Schwarz's Bayesian Criterion          -170.868
                   -2 Res Log Likelihood                 281.2229
                   Null Model LRT Chi-Square             216.5799
                   Null Model LRT DF                      10.0000
                   Null Model LRT P-Value                  0.0000
```

From Output 9.4, we see that the 4×4 block-diagonal matrix **D** is printed, the first block of **G.** From the output, the estimate of the residual variance $\hat{\sigma}^2$ is found in the `Residual` row and the `Estimate` column; the value is $\hat{\sigma}^2 = 0.0589$. The asymptotic *Z*-tests show that the variance components in **D** are significantly different from zero so that the random coefficient growth curve model is better than the null model, $\Psi \equiv \mathbf{R} = \sigma^2\mathbf{I}$. This is supported by the LRT ($p < 0.0001$).

The fixed effects for the model are the intercept (μ), treatment effects $\hat{\alpha}_i = \hat{\mu}_i - \hat{\mu}$ for $i = 1, 2, 3$, and slopes (linear, quadratic, and cubic). PROC MIXED follows McLean et al. (1991) to test hypotheses regarding the fixed effects using approximate F-tests. In particular, under the null hypotheses $H\colon \mathbf{L}\mu = \mathbf{0}$ is tested using

$$F = (\mathbf{L}\hat{\mu})'(\mathbf{L}\hat{\mathbf{C}}\mathbf{L}')^{-}(\mathbf{L}\hat{\mu}) / v_h$$
$$\hat{\mathbf{C}} = \begin{pmatrix} \mathbf{X}'\hat{\mathbf{R}}^{-1}\mathbf{X} & \mathbf{X}'\hat{\mathbf{R}}^{-1}\mathbf{Z} \\ \mathbf{Z}'\hat{\mathbf{R}}^{-1}\mathbf{X} & \mathbf{Z}'\hat{\mathbf{R}}^{-1}\mathbf{Z} + \hat{\mathbf{G}}^{-1} \end{pmatrix}^{-} \tag{9.69}$$

with $v_h = \text{rank}\,(\mathbf{L})$, and $\hat{\mu}' = (\hat{\beta}', \hat{\mathbf{b}}')$. The approximate t-test p-values show that the control differs from the treatments and that the slopes are significantly different from zero.

The matrix $\hat{\mathbf{C}}$ is an approximate estimate of the covariance matrix of $\hat{\mu}$ which tends to underestimate the true sampling variability (Kass and Stefffey, 1989). The test statistic in (9.69) is a Wald-like statistic. In large samples, results using PROC MIXED and the SUR model are equivalent.

For the random coefficient growth curve model, we modeled variation in the dogs by varying the designs $\mathbf{Z}_i$ across dogs. An alternative specification is to assume $\mathbf{Z}_1 = \mathbf{Z}_2 = \ldots = \mathbf{Z}_n = \mathbf{0}$ and to model $\mathbf{R}$. In particular, assume that we have a repeated measured design where $\mathbf{R}$ is unknown or satisfies the structure of compound symmetry. As part of the model, we must estimate $\mathbf{R}$. Thus we have a restricted growth curve model with restrictions on Σ. To analyze this design in PROC MIXED, we remove the RANDOM statement and use the REPEATED statement with the TYPE = UN or TYPE = CS options. Program 9_4.sas contains the SAS code to do these analyses.

Output 9.4 (*continued*)

```
              Growth Curve Model Unknown Structure
                      The MIXED Procedure
                    Class Level Information

          Class      Levels   Values

          TREAT           4   1 2 3 4
          DOG            36   1 2 3 4 5 6 7 8 9 10 11 12 13
                              14 15 16 17 18 19 20 21 22 23
                              24 25 26 27 28 29 30 31 32 33
                              34 35 36

                REML Estimation Iteration History

         Iteration  Evaluations      Objective      Criterion

                 0            1    47.52285001
                 1            4   -246.4322048     0.00256840
                 2            1   -246.7881583     0.00014461
                 3            1   -246.8067882     0.00000071
                 4            1   -246.8068765     0.00000000
```

Output 9.4 (*continued*)

Convergence criteria met.

R Matrix for DOG 1

Row	COL1	COL2	COL3	COL4	COL5
1	0.21981908	0.17078592	0.17263552	0.20365675	0.17454016
2	0.17078592	0.16926627	0.18543546	0.19325868	0.16563409
3	0.17263552	0.18543546	0.41304150	0.39328176	0.30060339
4	0.20365675	0.19325868	0.39328176	0.51776582	0.47680504
5	0.17454016	0.16563409	0.30060339	0.47680504	0.58109539
6	0.19903768	0.17237010	0.25665004	0.40604340	0.53698634
7	0.18363637	0.16903574	0.27869339	0.35155947	0.42876758

R Matrix for DOG 1

COL6	COL7
0.19903768	0.18363637
0.17237010	0.16903574
0.25665004	0.27869339
0.40604340	0.35155947
0.53698634	0.42876758
0.69611421	0.56620613
0.56620613	0.58328244

Growth Curve Model Unknown Structure

Covariance Parameter Estimates (REML)

Cov Parm	Subject	Estimate	Std Error	Z	Pr > |Z|
UN(1,1)	DOG	0.21981908	0.05372978	4.09	0.0001
UN(2,1)	DOG	0.17078592	0.04507062	3.79	0.0002
UN(2,2)	DOG	0.16926627	0.04226141	4.01	0.0001
UN(3,1)	DOG	0.17263552	0.06249636	2.76	0.0057
UN(3,2)	DOG	0.18543546	0.05884701	3.15	0.0016
UN(3,3)	DOG	0.41304150	0.10501262	3.93	0.0001
UN(4,1)	DOG	0.20365675	0.07364235	2.77	0.0057
UN(4,2)	DOG	0.19325868	0.06763833	2.86	0.0043
UN(4,3)	DOG	0.39328176	0.11266636	3.49	0.0005
UN(4,4)	DOG	0.51776582	0.13632544	3.80	0.0001
UN(5,1)	DOG	0.17454016	0.07879292	2.22	0.0267
UN(5,2)	DOG	0.16563409	0.07243302	2.29	0.0222
UN(5,3)	DOG	0.30060339	0.11190649	2.69	0.0072
UN(5,4)	DOG	0.47680504	0.13981581	3.41	0.0006
UN(5,5)	DOG	0.58109539	0.15895327	3.66	0.0003
UN(6,1)	DOG	0.19903768	0.08676428	2.29	0.0218
UN(6,2)	DOG	0.17237010	0.07865722	2.19	0.0284
UN(6,3)	DOG	0.25665004	0.11646491	2.20	0.0275
UN(6,4)	DOG	0.40604340	0.14120769	2.88	0.0040
UN(6,5)	DOG	0.53698634	0.16241543	3.31	0.0009
UN(6,6)	DOG	0.69611421	0.18678589	3.73	0.0002
UN(7,1)	DOG	0.18363637	0.07500938	2.45	0.0144
UN(7,2)	DOG	0.16903574	0.06815608	2.48	0.0131
UN(7,3)	DOG	0.27869339	0.10642072	2.62	0.0088
UN(7,4)	DOG	0.35155947	0.12454505	2.82	0.0048
UN(7,5)	DOG	0.42876758	0.13946635	3.07	0.0021
UN(7,6)	DOG	0.56620613	0.16036647	3.53	0.0004
UN(7,7)	DOG	0.58328244	0.15103118	3.86	0.0001

Output 9.4 (*continued*)

Model Fitting Information for Y

Description	Value
Observations	252.0000
Res Log Likelihood	-101.737
Akaike's Information Criterion	-129.737
Schwarz's Bayesian Criterion	-178.754
-2 Res Log Likelihood	203.4730
Null Model LRT Chi-Square	294.3297
Null Model LRT DF	27.0000
Null Model LRT P-Value	0.0000

Solution for Fixed Effects

Effect	TREAT	Estimate	Std Error	DF	t	Pr > |t|
INTERCEPT		3.94745169	0.14495142	32	27.23	0.0001
TREAT	1	0.44932994	0.18808302	32	2.39	0.0230
TREAT	2	-0.07516963	0.18332065	32	-0.41	0.6845
TREAT	3	-0.02810960	0.19387154	32	-0.14	0.8856

Growth Curve Model Unknown Structure
Solution for Fixed Effects

Effect	TREAT	Estimate	Std Error	DF	t	Pr > |t|
TREAT	4	0.00000000	.	.	.	.
X1		0.07748159	0.02083219	32	3.72	0.0008
X2		-0.02790417	0.00633260	32	-4.41	0.0001
X3		-0.05870152	0.02131486	32	-2.75	0.0096

Tests of Fixed Effects

Source	NDF	DDF	Type III F	Pr > F
TREAT	3	32	3.39	0.0298
X1	1	32	13.83	0.0008
X2	1	32	19.42	0.0001
X3	1	32	7.58	0.0096

Least Squares Means

Effect	TREAT	LSMEAN	Std Error	DF	t	Pr > |t|
TREAT	1	4.39678164	0.14495142	32	30.33	0.0001
TREAT	2	3.87228207	0.13871609	32	27.92	0.0001
TREAT	3	3.91934209	0.15238722	32	25.72	0.0001
TREAT	4	3.94745169	0.14495142	32	27.23	0.0001

Growth Curve Model Compound Symmetry
The MIXED Procedure
Class Level Information

Class	Levels	Values
TREAT	4	1 2 3 4
DOG	36	1 2 3 4 5 6 7 8 9 10 11 12 13 14 15 16 17 18 19 20 21 22 23 24 25 26 27 28 29 30 31 32 33 34 35 36

Output 9.4 (*continued*)

```
                        REML Estimation Iteration History

                Iteration  Evaluations     Objective     Criterion

                        0            1   47.52285001
                        1            1  -78.76183362    0.00000000

                            Convergence criteria met.
                                R Matrix for DOG 1

    Row          COL1           COL2           COL3           COL4           COL5

      1    0.40449950     0.23641683     0.23641683     0.23641683     0.23641683
      2    0.23641683     0.40449950     0.23641683     0.23641683     0.23641683
      3    0.23641683     0.23641683     0.40449950     0.23641683     0.23641683
      4    0.23641683     0.23641683     0.23641683     0.40449950     0.23641683
      5    0.23641683     0.23641683     0.23641683     0.23641683     0.40449950
      6    0.23641683     0.23641683     0.23641683     0.23641683     0.23641683
      7    0.23641683     0.23641683     0.23641683     0.23641683     0.23641683

                                R Matrix for DOG 1

                                    COL6           COL7

                              0.23641683     0.23641683
                              0.23641683     0.23641683
                              0.23641683     0.23641683
                              0.23641683     0.23641683
                              0.23641683     0.23641683
                              0.40449950     0.23641683
                              0.23641683     0.40449950

                      Covariance Parameter Estimates (REML)

 Cov Parm   Subject          Ratio       Estimate      Std Error        Z  Pr > |Z|

 CS         DOG         1.40655098     0.23641683     0.06514872     3.63    0.0003
 Residual               1.00000000     0.16808266     0.01628726    10.32    0.0001

                       Growth Curve Model Compound Symmetry
                         Model Fitting Information for Y

                     Description                          Value

                     Observations                      252.0000
                     Res Log Likelihood                -185.759
                     Akaike's Information Criterion    -187.759
                     Schwarz's Bayesian Criterion      -191.260
                     -2 Res Log Likelihood             371.5180
                     Null Model LRT Chi-Square         126.2847
                     Null Model LRT DF                   1.0000
                     Null Model LRT P-Value              0.0000
```

Comparing Akaike's Information Criterion (AIC) for the three models, the constrained model with unknown Σ is best where AIC = -129.737.

Finally by using the GROUP option in PROC MIXED, we may fit a model that permits heterogeneity of compound symmetry across groups. To accomplish this, we alter the form of the REPEATED statement to allow for different diagonal elements across treatments. The SAS code is given in Program 9_4.sas, and the output follows.

Output 9.4 (*continued*)

```
             Growth Curve Model with Heterogeneous Compound Symmetry
                            The MIXED Procedure
                          Class Level Information

              Class     Levels  Values

              TREAT          4  1 2 3 4
              DOG           36  1 2 3 4 5 6 7 8 9 10 11 12 13
                                14 15 16 17 18 19 20 21 22 23
                                24 25 26 27 28 29 30 31 32 33
                                34 35 36

                     REML Estimation Iteration History

             Iteration  Evaluations     Objective     Criterion

                     0            1   47.52285001
                     1            2  -123.1835414    0.10881828
                     2            1  -125.0099170    0.04231734
                     3            1  -128.1288074    0.00544475
                     4            1  -128.5162367    0.00022210
                     5            1  -128.5309819    0.00000060
                     6            1  -128.5310208    0.00000000

                         Convergence criteria met.
                   Covariance Parameter Estimates (REML)

 Cov Parm     Subject  Group        Estimate     Std Error       Z  Pr > |Z|

 INTERCEPT    DOG      TREAT 1    0.15016452    0.09853647    1.52    0.1275
 INTERCEPT    DOG      TREAT 2    0.25278112    0.12087989    2.09    0.0365
 INTERCEPT    DOG      TREAT 3    0.31618457    0.19242330    1.64    0.1003
 INTERCEPT    DOG      TREAT 4    0.21923435    0.12127759    1.81    0.0707
 DIAG         DOG      TREAT 1    0.32234043    0.06436038    5.01    0.0001
 DIAG         DOG      TREAT 2    0.02546869    0.00552806    4.61    0.0001
 DIAG         DOG      TREAT 3    0.30385085    0.06341532    4.79    0.0001
 DIAG         DOG      TREAT 4    0.16206587    0.03164439    5.12    0.0001

                       Model Fitting Information for Y

                  Description                          Value

                  Observations                      252.0000
                  Res Log Likelihood                -160.874
                  Akaike's Information Criterion    -168.874
                  Schwarz's Bayesian Criterion      -182.879
                  -2 Res Log Likelihood             321.7489
                  Null Model LRT Chi-Square         176.0539
                  Null Model LRT DF                   7.0000
                  Null Model LRT P-Value              0.0000

          Growth Curve Model with Heterogeneous Compound Symmetry
                         Solution for Fixed Effects

     Effect  TREAT        Estimate      Std Error     DF         t  Pr > |t|

     TREAT   1          4.65396825     0.14765318     32     31.52    0.0001
     TREAT   2          3.55285714     0.16013104     32     22.19    0.0001
     TREAT   3          4.15000000     0.21201174     32     19.57    0.0001
     TREAT   4          3.93968254     0.16410925     32     24.01    0.0001
     X1                 0.00658289     0.00843115    213      0.78    0.4358
     X2                -0.01484508     0.00486773    213     -3.05    0.0026
     X3                 0.00595951     0.01821337    213      0.33    0.7438
```

Output 9.4 (*continued*)

```
                         Tests of Fixed Effects

               Source    NDF    DDF  Type III F   Pr > F

               TREAT       4     32      611.31   0.0001
               X1          1    213        0.61   0.4358
               X2          1    213        9.30   0.0026
               X3          1    213        0.11   0.7438

                       CONTRAST Statement Results

          Source                  NDF    DDF        F   Pr > F

          group diff                3     32     8.92   0.0002
```

From the output, we see that the AIC statistic is larger for the heterogeneous model. This is also supported by the approximate LRT using -2 REML likelihood. If we subtract the two values (371.52 - 321.75 = 49.77) and compare that result with a chi-square distribution with degree of freedom 8 - 2 = 6, the number of parameters estimated in Ω minus the number estimated under the hypothesis ω, this result favors the heterogeneous model over the model with compound symmetry. The model with unknown Σ appears optimal. The test of `group diff` is also significant for the heterogeneous model. This was tested using a contrast since the NOINT option was used in the MODEL statement. In the section of output labeled `Tests of Fixed Effects`, in the row `TREAT` the test is that all treatment means are equal and equal to zero.

9.5 Mixed Model Repeated Measures

In Chapter 3 we analyzed a split-plot repeated measures design with PROC REG and PROC GLM using an RGLM and the GLM. Both the full-rank model and the overparameterized less-than-full-rank model required the within-subjects covariance matrix Σ to be homogeneous and have circularity structure for exact F-tests. While some authors recommend using adjusted F-tests when the circularity condition does not hold, we recommend that a multivariate analysis be used as illustrated in Chapter 5.

The split-plot repeated measures design may also be analyzed using PROC MIXED. This approach allows one to jointly test hypotheses regarding fixed effects and to evaluate the structure of the within-subjects covariance matrix by estimating its components. To illustrate, we reanalyze the data in Table 3.13 in Section 3.5.2 using PROC MIXED. Recall that these data were used to evaluate memory capacity based on probe-word position. In Program 9_5.sas, we analyze the problem using two structures for the within-subjects covariance matrix: compound symmetry and unknown structure. In Section 3.5.3, we found the structure to satisfy the circularity condition.

Program 9_5.sas

```
/* Program 9_5.sas                  */
/* Data from Timm(1975, page 244) */

options ls=80 ps=60 nodate nonumber;
title1 'Output 9.5: Analysis of Repeated Measurements Using PROC MIXED';
```

```
data timm;
   infile 'c:\9_5.dat';
   input group subj probe y;
proc print data=timm;
proc mixed covtest ratio;
   title2 'Compound Symmetry';
   class group subj probe;
   model y = group probe group*probe;
   repeated/type=cs subject=subj(group) r;
run;
proc mixed covtest ratio;
   title2 'Unknown Structure';
   class group subj probe;
   model y= group probe group*probe/ddfm=satterth;
   repeated/type=un subject=subj(group) r;
run;
```

To analyze the data in Table 3.13 using PROC MIXED, we organize the data as a three-factor design with three class variables: GROUP, SUBJ, and PROBE. The MODEL statement specifies the linear model and the REPEATED statement models the within-subjects covariance matrix. For this design, subjects are nested within groups. The REPEATED statement defines the block diagonal matrix $\mathbf{\Psi} \equiv \mathbf{R}$. The two covariance structures are specified using the TYPE = option where CS and UN define compound symmetry and general unknown structure, respectively.

Result and Interpretation 9_5.sas

The results are in Output 9.5 below.

Output 9.5: *Analysis of Repeated Measurements Using PROC MIXED*

```
          Output 9.5: Analysis of Repeated Measurements Using PROC MIXED


                              Compound Symmetry
                             The MIXED Procedure
                           Class Level Information

                Class      Levels  Values

                GROUP           2  1 2
                SUBJ           20  1 2 3 4 5 6 7 8 9 10 11 12 13
                                   14 15 16 17 18 19 20
                PROBE           5  1 2 3 4 5

                       REML Estimation Iteration History

               Iteration  Evaluations     Objective     Criterion

                       0            1  514.32432448
                       1            1  500.93761018    0.00000000
```

Output 9.5 *(continued)*

```
                     Convergence criteria met.
                    R Matrix for SUBJ(GROUP) 1 1

Row          COL1          COL2          COL3          COL4          COL5

  1   86.39000000   28.16833333   28.16833333   28.16833333   28.16833333
  2   28.16833333   86.39000000   28.16833333   28.16833333   28.16833333
  3   28.16833333   28.16833333   86.39000000   28.16833333   28.16833333
  4   28.16833333   28.16833333   28.16833333   86.39000000   28.16833333
  5   28.16833333   28.16833333   28.16833333   28.16833333   86.39000000


                Covariance Parameter Estimates (REML)

Cov Parm   Subject              Ratio      Estimate     Std Error        Z

CS         SUBJ(GROUP)     0.48381187   28.16833333   13.41204290     2.10
Residual                   1.00000000   58.22166667    9.70361111     6.00

                Covariance Parameter Estimates (REML)

                Pr > |Z|

                  0.0357
                  0.0001

                   Model Fitting Information for Y

               Description                          Value

               Observations                      100.0000
               Res Log Likelihood                -333.173
               Akaike's Information Criterion    -335.173

                        Compound Symmetry
                   Model Fitting Information for Y

               Description                          Value

               Schwarz's Bayesian Criterion      -337.673
               -2 Res Log Likelihood             666.3465
               Null Model LRT Chi-Square          13.3867
               Null Model LRT DF                   1.0000
               Null Model LRT P-Value              0.0003

                       Tests of Fixed Effects

             Source          NDF   DDF  Type III F  Pr > F

             GROUP             1    18        8.90  0.0080
             PROBE             4    72       14.48  0.0001
             GROUP*PROBE       4    72        0.34  0.8479

                        Unknown Structure
                       The MIXED Procedure
                     Class Level Information

            Class     Levels  Values

            GROUP          2  1 2
            SUBJ          20  1 2 3 4 5 6 7 8 9 10 11 12 13
                              14 15 16 17 18 19 20
            PROBE          5  1 2 3 4 5
```

Output 9.5 (*continued*)

```
                     REML Estimation Iteration History

           Iteration  Evaluations     Objective     Criterion

                   0            1  514.32432448
                   1            1  485.68180840    0.00000000

                         Convergence criteria met.
                        R Matrix for SUBJ(GROUP) 1 1

 Row          COL1          COL2          COL3          COL4          COL5

   1  141.49444444   33.16666667   52.58888889   14.32777778   21.43888889
   2   33.16666667   53.35555556   31.62222222    8.62222222   16.63333333
   3   52.58888889   31.62222222  122.44444444   38.12222222   33.32222222
   4   14.32777778    8.62222222   38.12222222   64.69444444   31.83888889
   5   21.43888889   16.63333333   33.32222222   31.83888889   49.96111111

                   Covariance Parameter Estimates (REML)

   Cov Parm  Subject          Estimate     Std Error        Z  Pr > |Z|

   UN(1,1)   SUBJ(GROUP)  141.49444444   47.16481481     3.00    0.0027
   UN(2,1)   SUBJ(GROUP)   33.16666667   21.92099763     1.51    0.1303
   UN(2,2)   SUBJ(GROUP)   53.35555556   17.78518519     3.00    0.0027
   UN(3,1)   SUBJ(GROUP)   52.58888889   33.40891421     1.57    0.1155
   UN(3,2)   SUBJ(GROUP)   31.62222222   20.45734899     1.55    0.1222
   UN(3,3)   SUBJ(GROUP)  122.44444444   40.81481481     3.00    0.0027
   UN(4,1)   SUBJ(GROUP)   14.32777778   22.80252141     0.63    0.5298
   UN(4,2)   SUBJ(GROUP)    8.62222222   13.99633035     0.62    0.5379
   UN(4,3)   SUBJ(GROUP)   38.12222222   22.82150440     1.67    0.0948
   UN(4,4)   SUBJ(GROUP)   64.69444444   21.56481481     3.00    0.0027
   UN(5,1)   SUBJ(GROUP)   21.43888889   20.45163077     1.05    0.2945
   UN(5,2)   SUBJ(GROUP)   16.63333333   12.78534451     1.30    0.1933
   UN(5,3)   SUBJ(GROUP)   33.32222222   20.03861687     1.66    0.0963
   UN(5,4)   SUBJ(GROUP)   31.83888889   15.35853214     2.07    0.0382
   UN(5,5)   SUBJ(GROUP)   49.96111111   16.65370370     3.00    0.0027

                            Unknown Structure
                       Model Fitting Information for Y

                 Description                           Value

                 Observations                       100.0000
                 Res Log Likelihood                 -325.545
                 Akaike's Information Criterion     -340.545
                 Schwarz's Bayesian Criterion       -359.294
                 -2 Res Log Likelihood              651.0907
                 Null Model LRT Chi-Square           28.6425
                 Null Model LRT DF                   14.0000
                 Null Model LRT P-Value               0.0117

                         Tests of Fixed Effects

               Source         NDF    DDF  Type III F  Pr > F

               GROUP            1     18        8.90  0.0080
               PROBE            4     18       15.97  0.0001
               GROUP*PROBE      4     18        0.86  0.5047
```

Comparing the results for the two models, observe that the F-tests are identical for the between-subjects factor group, and that the result agrees with PROC GLM. Between tests do not depend on the structure of Σ; only the within-subjects tests depend on Σ. For this example, the within-subjects tests under compound symmetry for PROC GLM and PROC MIXED are identical. This is the case by Theorem 4.1 since we have an equal number of subjects for each group and compound symmetry (McLean and Sanders, 1988). When Σ is unknown, one may use Satterthwaite's formula to adjust the *p*-values by using the DDMF = SATTERTH option on the MODEL statement as illustrated in Program 9_5.sas. The adjusted *p*-value for the interaction test is 0.8479. Without Satterthwaite's adjustment, the *p*-value is 0.4903. For large data sets, Satterthwaite's formula significantly increases the execution time.

Finally, comparing the AIC values for each model, we see that the model with compound symmetry appears to fit the data better since $-333.173 > -340.545$. These results are consistent with those obtained in Chapter 3. The multivariate analysis in Chapter 5 is very conservative for these data. When analyzing repeated measures data one should consider several models.

9.6 Mixed Model Repeated Measures with Changing Covariates

Mixed model analysis of repeated measures data allows the researcher to build alternative models for the within-subjects covariance matrix. To evaluate model fit, we use the AIC and SBC criteria. The model with a value closest to zero (least negative) is usually the best model. To illustrate, we now analyze with PROC MIXED the Winer data with time-varying covariates, previously analyzed using the SUR model in Chapter 8. The SAS code to perform the analysis is given in Program 9_6.sas.

Program 9_6.sas

```
/* Program 9_6.sas                                      */
/* Mixed Model - Winer(1971, page 806) and Patel (1986) */

options ls=80 ps=60 nodate nonumber;
title1 'Output 9.6: Mixed Model Repeated Measures with Changing Covariates';

data winer;
   input person g b x y @@;
   cards;
   1 1 1  3  8 1 1 2  4 14
   2 1 1  5 11 2 1 2  9 18
   3 1 1 11 16 3 1 2 14 22
   4 2 1  2  6 4 2 2  1  8
   5 2 1  8 12 5 2 2  9 14
   6 2 1 10  9 6 2 2  9 10
   7 3 1  7 10 7 3 2  4 10
   8 3 1  8 14 8 3 2 10 18
   9 3 1  9 15 9 3 2 12 22
   ;
```

```
proc print data=winer;
proc mixed data=winer method=ml covtest ratio;
   title2 'Variance Components';
   class person g b;
   model y =  g b g*b x;
   random person/type=vc subject=person;
   repeated/type=simple subject=person r;
run;
proc mixed data=winer method=ml covtest ratio;
   title2 'Unknown Covariance Matrix';
   class person g b;
   model y = g b g*b x;
   repeated/type=un subject=person r;
run;
proc mixed data=winer method=ml covtest ratio;
   title2 'Compound Symmetry';
   class person g b;
   model y = g*b x/noint s;
   contrast 'G'  g*b 1 1 0 0 -1 -1,
                 g*b 0 0 1 1 -1 -1;
   contrast 'B'  g*b 1 -1 1 -1 1 -1;
   contrast 'BG' g*b 1 -1 -1 1 0 0,
                 g*b 0  0  1 -1 -1 1;
   repeated/type=cs subject=person r;
run;
proc glm;
   title2 'Mixed Model Compound Symmetry';
   class person g b;
   model y= g person(g) b g*b x;
   test  h=g e=person(g);
run;

data patel;
   infile'c:\patel.dat';
   input treat person period y z1 z2;
proc print data=patel;
   title2 'Patel data';
proc mixed data=patel method=ml covtest ratio;
   class treat person period;
   model y=treat*period z1 z2/noint s;
   contrast 'treat' treat*period 1 1 -1 -1;
   estimate 'treat(delta)' treat*period 1 1 -1 -1;
   contrast 'inter' treat*period 1 -1 -1 1;
   repeated /type=un subject=person r;
run;
```

The analysis is performed for three models: (1) variance components, (2) unknown covariance structure, and (3) compound symmetry structure. For the variance components model, the RANDOM and REPEATED statements generate spherical structure for each subject. The components of variance are $\sigma^2_{\text{subjects}}$ and σ^2. To generate compound symmetry structure, there are no random effects in the model; the REPEATED statement generates a covariance matrix, $\text{cov}(\mathbf{y}) = \mathbf{\Psi} \equiv \mathbf{R}$, with compound symmetry. This is not the same as the variance components model where the $\text{cov}(\mathbf{y}) = \mathbf{ZDZ}' + \mathbf{\Psi}$ and $\mathbf{\Psi} = \sigma^2\mathbf{I}$. Finally, we allow the $\text{cov}(\mathbf{y}) = \mathbf{\Omega} = \mathbf{ZDZ}' + \mathbf{\Psi}$ for arbitrary $\mathbf{\Psi}$, and unknown structure for $\mathbf{\Omega}$.

Result and Interpretation 9_6.sas

Output 9.6: *Mixed Model Repeated Measures with Changing Covariates*

```
Output 9.6: Mixed Model Repeated Measures with Changing Covariates

              OBS    PERSON    G    B     X     Y

                1       1      1    1     3     8
                2       1      1    2     4    14
                3       2      1    1     5    11
                4       2      1    2     9    18
                5       3      1    1    11    16
                6       3      1    2    14    22
                7       4      2    1     2     6
                8       4      2    2     1     8
                9       5      2    1     8    12
               10       5      2    2     9    14
               11       6      2    1    10     9
               12       6      2    2     9    10
               13       7      3    1     7    10
               14       7      3    2     4    10
               15       8      3    1     8    14
               16       8      3    2    10    18
               17       9      3    1     9    15
               18       9      3    2    12    22

                      Variance Components
                     The MIXED Procedure
                   Class Level Information

              Class      Levels  Values

              PERSON          9  1 2 3 4 5 6 7 8 9
              G               3  1 2 3
              B               2  1 2

                ML Estimation Iteration History

       Iteration  Evaluations      Objective      Criterion

               0            1    35.41656840
               1            2    22.46478372     0.00000000

                  Convergence criteria met.
                    R Matrix for PERSON 1

                 Row          COL1            COL2

                   1    0.33311460
                   2                    0.33311460
```

Output 9.6 (*continued*)

```
                Covariance Parameter Estimates (MLE)

Cov Parm  Subject      Estimate     Std Error        Z  Pr > |Z|

PERSON    PERSON     2.29848410    1.16468483     1.97    0.0484
DIAG      PERSON     0.33311460    0.15703216     2.12    0.0339

                   Model Fitting Information for Y

               Description                        Value

               Observations                     18.0000
               Log Likelihood                  -27.7733
               Akaike's Information Criterion  -29.7733
               Schwarz's Bayesian Criterion    -30.6637
               -2 Log Likelihood                55.5466

                     Tests of Fixed Effects

              Source      NDF   DDF  Type III F  Pr > F

              G             2     5        5.63  0.0525
              B             1     5      106.89  0.0001
              G*B           2     5        4.10  0.0883
              X             1     5       56.93  0.0006

                      Unknown Covariance Matrix
                         The MIXED Procedure
                       Class Level Information

                 Class     Levels  Values

                 PERSON         9  1 2 3 4 5 6 7 8 9
                 G              3  1 2 3
                 B              2  1 2

                    ML Estimation Iteration History

         Iteration  Evaluations     Objective     Criterion

                 0            1  35.41656840
                 1            2  21.64764523    0.00000143
                 2            1  21.64762970    0.00000000

                      Convergence criteria met.
                        R Matrix for PERSON 1

                    Row           COL1           COL2

                      1     2.24773306     2.30775884
                      2     2.30775884     3.03660666


                Covariance Parameter Estimates (MLE)

Cov Parm  Subject      Estimate     Std Error        Z  Pr > |Z|

UN(1,1)   PERSON     2.24773306    1.06085956     2.12    0.0341
UN(2,1)   PERSON     2.30775884    1.16424735     1.98    0.0475
UN(2,2)   PERSON     3.03660666    1.44831347     2.10    0.0360
```

Output 9.6 (*continued*)

```
                    Model Fitting Information for Y

               Description                          Value
               Observations                       18.0000
               Log Likelihood                    -27.3647
               Akaike's Information Criterion    -30.3647
               Schwarz's Bayesian Criterion      -31.7003
               -2 Log Likelihood                  54.7294
               Null Model LRT Chi-Square          13.7689
               Null Model LRT DF                   2.0000
               Null Model LRT P-Value              0.0010

                     Unknown Covariance Matrix
                       Tests of Fixed Effects

              Source        NDF   DDF  Type III F  Pr > F

              G               2     6        5.72  0.0407
              B               1     6      107.19  0.0001
              G*B             2     6        4.35  0.0680
              X               1     6       48.42  0.0004

                         Compound Symmetry
                        The MIXED Procedure
                      Class Level Information

                 Class      Levels  Values

                 PERSON          9  1 2 3 4 5 6 7 8 9
                 G               3  1 2 3
                 B               2  1 2

                   ML Estimation Iteration History

          Iteration  Evaluations      Objective      Criterion

                  0            1   35.41656840
                  1            2   22.46478372     0.00000000

                     Convergence criteria met.
                       R Matrix for PERSON 1

                    Row            COL1            COL2

                      1     2.63159870      2.29848410
                      2     2.29848410      2.63159870

                  Covariance Parameter Estimates (MLE)

Cov Parm   Subject        Ratio       Estimate      Std Error       Z  Pr > |Z|

CS         PERSON    6.89998006     2.29848410     1.16468483    1.97    0.0484
Residual             1.00000000     0.33311460     0.15703216    2.12    0.0339
```

Output 9.6 (*continued*)

Model Fitting Information for Y

Description	Value
Observations	18.0000
Log Likelihood	-27.7733
Akaike's Information Criterion	-29.7733
Schwarz's Bayesian Criterion	-30.6637
-2 Log Likelihood	55.5466
Null Model LRT Chi-Square	12.9518
Null Model LRT DF	1.0000
Null Model LRT P-Value	0.0003

Solution for Fixed Effects

Effect	G	B	Estimate	Std Error	DF	t	Pr > |t|
G*B	1	1	6.30682204	1.17551649	2	5.37	0.0330

Compound Symmetry
Solution for Fixed Effects

Effect	G	B	Estimate	Std Error	DF	t	Pr > |t|
G*B	1	2	10.38337869	1.37704892	2	7.54	0.0171
G*B	2	1	3.35805829	1.19848106	2	2.80	0.1073
G*B	2	2	5.30682204	1.17551649	2	4.51	0.0457
G*B	3	1	6.22966994	1.29706826	2	4.80	0.0407
G*B	3	2	9.33214244	1.34987977	2	6.91	0.0203
X			0.84629126	0.11216523	2	7.55	0.0171

Tests of Fixed Effects

Source	NDF	DDF	Type III F	Pr > F
G*B	6	2	21.29	0.0455
X	1	2	56.93	0.0171

CONTRAST Statement Results

Source	NDF	DDF	F	Pr > F
G	2	2	5.63	0.1509
B	1	2	106.89	0.0092
BG	2	2	4.10	0.1961

Comparing the AIC statistics, observe that the fit for the variance components model and compound symmetry model are equal, both better than the model with unknown structure. Also, the null hypothesis that $\mathbf{\Psi} = \sigma^2\mathbf{I}$ is rejected. It is more likely that $\mathbf{\Omega}$ has compound symmetry structure, since $\mathbf{ZDZ'} + \sigma^2\mathbf{I}$ may not have a simple structure, in general.

Selecting the ML method, $\hat{\sigma}^2_{\text{subjects}} = 2.2298$ and $\hat{\sigma}^2 = 0.3331$ (the corresponding REML estimates are 3.741 and 0.548, respectively). The ANCOVA table for the fixed effect approximate F-tests using both REML and ML methods is shown in Table 9.1. The results are not identical since $\hat{\mathbf{C}}$ in (9.69) are not equal for the two estimation procedures.

Table 9.1 *PROC MIXED*

	REML		ML	
	F	*p*-value	F	*p*-value
G	3.46	0.1142	5.63	.1509
B	65.05	0.0005	106.89	.0092
G*B	2.50	0.1771	4.10	.1961
X	34.78	0.0020	56.93	.0171

The REML output may be obtained by setting METHOD = REML on the PROC MIXED statement.

Using OLS and the ANCOVA model, we can obtain the results given in Winer (1971, p. 801) using PROC GLM as illustrated in the continuation of Output 9.6 that follows.

Output 9.6 (*continued*)

```
                    Mixed Model Compound Symmetry
                   General Linear Models Procedure
                      Class Level Information

                  Class     Levels     Values

                  PERSON         9     1 2 3 4 5 6 7 8 9

                  G              3     1 2 3

                  B              2     1 2

                Number of observations in data set = 18

                    Mixed Model Compound Symmetry
                   General Linear Models Procedure

Dependent Variable: Y
                                   Sum of           Mean
Source                  DF        Squares         Square   F Value      Pr > F

Model                   12   371.50198413    30.95849868     51.63      0.0002

Error                    5     2.99801587     0.59960317

Corrected Total         17   374.50000000

                R-Square           C.V.        Root MSE                 Y Mean

                0.991995       5.881067       0.7743405              13.166667


Source                  DF      Type I SS    Mean Square   F Value      Pr > F

G                        2   100.00000000    50.00000000     83.39      0.0001
PERSON(G)                6   177.00000000    29.50000000     49.20      0.0003
B                        1    68.05555556    68.05555556    113.50      0.0001
G*B                      2    16.44444444     8.22222222     13.71      0.0093
X                        1    10.00198413    10.00198413     16.68      0.0095
```

Output 9.6 (*continued*)

Source	DF	Type III SS	Mean Square	F Value	Pr > F
G	2	38.08303436	19.04151718	31.76	0.0014
PERSON(G)	6	44.37055236	7.39509206	12.33	0.0072
B	1	31.54702917	31.54702917	52.61	0.0008
G*B	2	2.33928571	1.16964286	1.95	0.2365
X	1	10.00198413	10.00198413	16.68	0.0095

Tests of Hypotheses using the Type III MS for PERSON(G) as an error term

Source	DF	Type III SS	Mean Square	F Value	Pr > F
G	2	38.08303436	19.04151718	2.57	0.1558

The ANCOVA Table 9.2 from PROC GLM follows from the output. Table 9.2 is similar to Table 3.10, except that we have a covariate in the model that is changing with time.

Table 9.2 *PROC GLM ANCOVA*

	F	*p*-value
G	2.57	0.1558
B	52.61	0.0072
G*B	1.95	0.2365
X	16.68	0.0095

Table 9.2 is in close agreement with Table 9.1. The results would be identical for an orthogonal design by Theorem 4.1.

In Section 8.11 we used the Winer data to evaluate the difference in intercepts using the SUR model to obtain tests. Because F in (9.69) is a Wald-like statistic, SUR and PROC MIXED give equivalent results in large samples. To compare the two procedures, we use PROC MIXED to analyze the Patel data given in Section 8.10, with unknown covariance structure. The SAS statements are included in Program 9_6.sas. The output for the analysis follows.

Output 9.6 (*continued*)

Patel data

OBS	TREAT	PERSON	PERIOD	Y	Z1	Z2
1	1	1	1	1.28	1.09	0.00
2	1	1	2	1.33	0.00	1.24
3	1	2	1	1.60	1.38	0.00
4	1	2	2	2.21	0.00	1.90
5	1	3	1	2.46	2.27	0.00
6	1	3	2	2.43	0.00	2.19
7	1	4	1	1.41	1.34	0.00
8	1	4	2	1.81	0.00	1.47
9	1	5	1	1.40	1.31	0.00
10	1	5	2	0.85	0.00	0.85
11	1	6	1	1.12	0.96	0.00
12	1	6	2	1.20	0.00	1.12
13	1	7	1	0.90	0.66	0.00
14	1	7	2	0.90	0.00	0.78
15	1	8	1	2.41	1.69	0.00
16	1	8	2	2.79	0.00	1.90
17	2	9	1	3.06	1.74	0.00
18	2	9	2	1.38	0.00	1.54
19	2	10	1	2.68	2.41	0.00
20	2	10	2	2.10	0.00	2.13
21	2	11	1	2.60	3.05	0.00
22	2	11	2	2.32	0.00	2.18
23	2	12	1	1.48	1.20	0.00
24	2	12	2	1.30	0.00	1.41
25	2	13	1	2.08	1.70	0.00
26	2	13	2	2.34	0.00	2.21
27	2	14	1	2.72	1.89	0.00
28	2	14	2	2.48	0.00	2.05
29	2	15	1	1.94	0.89	0.00
30	2	15	2	1.11	0.00	0.72
31	2	16	1	3.35	2.41	0.00
32	2	16	2	3.23	0.00	2.83
33	2	17	1	1.16	0.96	0.00
34	2	17	2	1.25	0.00	1.01

Patel data
The MIXED Procedure
Class Level Information

Class	Levels	Values
TREAT	2	1 2
PERSON	17	1 2 3 4 5 6 7 8 9 10 11 12 13 14 15 16 17
PERIOD	2	1 2

ML Estimation Iteration History

Iteration	Evaluations	Objective	Criterion
0	1	-44.82268202	
1	2	-51.50492225	0.00000010
2	1	-51.50492479	0.00000000

Output 9.6 (*continued*)

```
                    Convergence criteria met.
                     R Matrix for PERSON 1

                Row           COL1            COL2

                  1     0.14668567      0.02933787
                  2     0.02933787      0.05045803

           Covariance Parameter Estimates (MLE)

Cov Parm  Subject      Estimate     Std Error        Z  Pr > |Z|

UN(1,1)   PERSON     0.14668567    0.05035098     2.91    0.0036
UN(2,1)   PERSON     0.02933787    0.02228568     1.32    0.1880
UN(2,2)   PERSON     0.05045803    0.01735717     2.91    0.0036


              Model Fitting Information for Y

           Description                            Value
           Observations                         34.0000
           Log Likelihood                       -5.4914
           Akaike's Information Criterion       -8.4914
           Schwarz's Bayesian Criterion        -10.7810
           -2 Log Likelihood                    10.9829
           Null Model LRT Chi-Square             6.6822
           Null Model LRT DF                     2.0000
           Null Model LRT P-Value                0.0354

                         Patel data
                 Solution for Fixed Effects

Effect         TREAT  PERIOD      Estimate     Std Error    DF        t  Pr > |t|

TREAT*PERIOD   1      1         0.44187664    0.24736171    17     1.79    0.0919
TREAT*PERIOD   1      2         0.09413461    0.15642849    17     0.60    0.5553
TREAT*PERIOD   2      1         0.81482828    0.30723016    17     2.65    0.0168
TREAT*PERIOD   2      2        -0.04660479    0.18414554    17    -0.25    0.8032
Z1                              0.84532588    0.15477197    17     5.46    0.0001
Z2                              1.11501512    0.09416158    17    11.84    0.0001

                      Tests of Fixed Effects

          Source          NDF    DDF   Type III F   Pr > F

          TREAT*PERIOD      4     17         2.69   0.0666
          Z1                1     17        29.83   0.0001
          Z2                1     17       140.22   0.0001

                  ESTIMATE Statement Results

    Parameter               Estimate     Std Error     DF        t  Pr > |t|

    treat(delta)         -0.23221224    0.26103583     17    -0.89    0.3861

                  CONTRAST Statement Results

        Source                   NDF    DDF         F   Pr > F

        treat                      1     17      0.79   0.3861
        inter                      1     17      7.00   0.0170
```

The ML estimates of **B** and **Ω** using PROC MIXED are

$$\hat{\mathbf{B}}_{ML} = \begin{pmatrix} 0.442 & 0.094 \\ 0.815 & -0.047 \end{pmatrix}$$

$$\hat{\boldsymbol{\Omega}}_{ML} = \begin{pmatrix} 0.1467 & 0.0293 \\ 0.0293 & 0.0505 \end{pmatrix}$$

which agree with Patel's result using a different algorithm. The treatment effect $\hat{\delta} = -0.232$ is in close agreement with the value -0.257 obtained by Patel and the SUR model. The p-values for the approximate F-tests for treatment and interaction are 0.3927 and 0.0228, respectively. These p-values may be improved by using the Satterthwaite procedure to adjust the degrees of freedom for the approximate F-statistics. To make this correction, use the DDFM = SATTERTH option in the MODEL statement. Then, the p-values are 0.386 and 0.017, respectively. For this example, PROC MIXED and the SUR model yielded equivalent results. For very small samples, both procedures should be avoided.

The analysis of repeated measures data is complicated. If one does not impose any structure on the covariance matrix, a multivariate analysis is used. If one wants to model the covariance matrix and the mixed effects, a mixed model analysis is used. For an excellent discussion of the problems associated with the analysis of repeated measurement data using PROC MIXED and PROC GLM, see Littell et al. (1996).

9.7 Two-Level Hierarchical Linear Models

The hierarchical linear model (HLM) examples we have considered so far have been in the context of repeated measurement designs with subjects random and nested within fixed factors. We now turn our attention to the analysis of two-level nested experimental designs: classrooms nested within schools, schools nested within districts, departments nested within firms, etc., where the nested factor is always random. To illustrate these models, we use PROC MIXED to obtain results discussed by Raudenbush (1993). The SAS code to perform the analysis of several data sets is provided in Program 9_7.sas.

Program 9_7.sas

```
/* Program 9_7.sas                          */
/* ANOVA data HLM Model (Raudenbush,1993)   */

options ls=80 ps=60 nodate nonumber;
title1 'Output 9.7: HLM Examples';

data hlm1;
   infile 'c:\hlm.dat';
   input treat y;
proc print data=hlm1;
   title2 'HLM One-Way ANOVA';
run;
proc mixed data=hlm1 method=reml covtest ratio;
   class treat;
   model y=/s;
   random intercept/type=vc subject= treat g;
run;
```

```
data hlm2;
   infile 'c:\hlmnest.dat';
   input method classes y;
proc print data=hlm2;
   title2 'HLM Nested Two-Factor ANOVA';
proc mixed data=hlm2 method =reml covtest ratio;
   class method classes;
   model y = method/noint s;
   random classes/type=vc subject=method g;
   estimate 'overall mean' method .5 .5;
   estimate 'contrast diff method' method 1 -1;
run;

data hlm3;
   infile 'c:\hlmcross.dat';
   input practice tutor y;
proc print data=hlm3;
   title2 ' HLM Crossed Two-Factor Mixed MANOVA';
proc mixed data=hlm3 method=reml covtest ratio;
   class tutor;
   model y = practice/ s;
   random intercept practice/ type=un subject=tutor;
run;

data hlm4;
   infile 'c:\hlmsplit.dat';
   input person y1 y2 y3 y4;
   y=y1; dur=1; x1=-1.5; x2= .5; x3= -.5; output;
   y=y2; dur=2; x1= -.5; x2=-.5; x3= 1.5; output;
   y=y3; dur=3; x1=  .5; x2=-.5; x3=-1.5; output;
   y=y4; dur=4; x1= 1.5; x2= .5; x3=  .5; output;
proc print data=hlm4;
   title2 'HLM Repeated Measures';
proc mixed data=hlm4 method=reml covtest ratio;
   class person;
   model y= x1 x2 x3 / s;
   repeated /type=cs subject=person r;
   contrast 'duration' x1 1, x2 1 x3 1;
run;
```

Result and Interpretation 9_7.sas

To begin, we consider the HLM representation for the one-way random effects ANOVA model. For such a model, the observation of subject i in group j is represented

$$\begin{pmatrix} y_{1j} \\ y_{2j} \\ \vdots \\ \vdots \\ \vdots \\ y_{n_j j} \end{pmatrix} = \begin{pmatrix} 1 \\ 1 \\ \vdots \\ \vdots \\ \vdots \\ 1 \end{pmatrix} \beta_j + \begin{pmatrix} e_{1j} \\ e_{2j} \\ \vdots \\ \vdots \\ \vdots \\ e_{n_j j} \end{pmatrix} \tag{9.70}$$

and

$$\beta_j = \mu + b_j$$

where there are n_j subjects per group, $j = 1, \dots, J$ groups and $b_j \sim N(0, \sigma_b^2)$. The fixed j^{th} group mean β_j at level two is represented as an overall mean μ, plus the random effect for the jth group. Substituting β_j into the vector equation (9.70), we obtain

$$\mathbf{y}_j = \mathbf{1}_{n_j}\mu + \mathbf{1}_{n_j}\mathbf{b}_j + \mathbf{e}_j \qquad j = 1, 2, \dots, J$$

a mixed linear model with $\mathbf{X}_j = \mathbf{Z}_j = \mathbf{1}_{n_j}$. Each element y_{ij} of $\mathbf{y}_j$ may be represented as a random effects ANOVA model

$$y_{ij} = \mu + b_j + e_{ij}$$

where $e_{ij} \sim N(0, \sigma^2)$, $b_j \sim N(0, \sigma_b^2)$ and σ^2 and σ_b^2 are the variance components to be estimated.

To illustrate the analysis of this design using PROC MIXED, we use data for fifty children randomly assigned to five training methods developed to discriminate among blocks (Raudenbush, 1993, p. 466). The responses are the number of blocks correctly identified. Program 9_7.sas contains the SAS code to analyze this data set. The results are given in Output 9.7.

Output 9.7: *HLM Examples HLM One-Way ANOVA*

```
                         Output 9.7: HLM Examples
                            HLM One-Way ANOVA

                            OBS    TREAT    Y

                              1      1      0
                              2      1      1
                              3      1      3
                              4      1      1
                              5      1      1
                              6      1      2
                              7      1      2
                              8      1      1
                              9      1      1
                             10      1      2
                             11      2      2
                             12      2      3
                             13      2      4
                             14      2      2
                             15      2      1
                             16      2      1
                             17      2      2
                             18      2      2
                             19      2      3
                             20      2      4
                             21      3      2
                             22      3      3
                             23      3      4
                             24      3      4
                             25      3      2
                             26      3      1
                             27      3      2
                             28      3      3
                             29      3      2
                             30      3      2
                             31      4      2
                             32      4      4
```

Output 9.7 (*continued*)

```
                    33      4       5
                    34      4       3
                    35      4       2
                    36      4       1
                    37      4       3
                    38      4       3
                    39      4       2
                    40      4       4
                    41      5       1
                    42      5       0
                    43      5       2
                    44      5       1
                    45      5       1
                    46      5       2
                    47      5       1
                    48      5       0
                    49      5       1
                    50      5       3

                  HLM One-Way ANOVA
                 The MIXED Procedure
                Class Level Information

              Class     Levels  Values

             TREAT           5  1 2 3 4 5

           REML Estimation Iteration History

    Iteration  Evaluations     Objective     Criterion

            0            1   68.73757345
            1            1   60.50210953    0.00000000

             Convergence criteria met.

                      G Matrix

           Effect     TREAT   Row          COL1

           INTERCEPT  1          1    0.44522222
```

Output 9.7 *(continued)*

Covariance Parameter Estimates (REML)

Cov Parm	Subject	Ratio	Estimate	Std Error	Z	Pr > |Z|
INTERCEPT	TREAT	0.43744541	0.44522222	0.38738209	1.15	0.2504
Residual		1.00000000	1.01777778	0.21456640	4.74	0.0001

Model Fitting Information for Y

Description	Value
Observations	50.0000
Res Log Likelihood	-75.2790
Akaike's Information Criterion	-77.2790
Schwarz's Bayesian Criterion	-79.1709
-2 Res Log Likelihood	150.5581

Solution for Fixed Effects

Effect	Estimate	Std Error	DF	t	Pr > |t|
INTERCEPT	2.08000000	0.33075671	4	6.29	0.0033

The **G** matrix provides the REML estimate of σ_b^2, $\hat{\sigma}_b^2 = 0.4452$. The residual variance estimate is $\hat{\sigma}^2 = 1.018$. The ANOVA table for the fixed effects parameter for the null hypotheses $H{:}\,\mu = 0$ is also provided ($p = 0.0033$). One may also compute the intraclass correlation ρ which represents the proportion of the variance in the dependent variable Y (discrimination) accounted for by the second level (method) variable. Thus, $\hat{\rho} = 0.4452/(0.4452 + 1.0178) = 0.3043$. Hence, approximately 30% of block discrimination is due to method.

For our second example, the two-factor nested design analyzed by Raudenbush (1993, p. 475) is reanalyzed. Four classrooms are nested within two methods where the classroom factor is random and methods are fixed effects. As with the ANOVA model, the first level of the model is identical to (9.70) where y_{ij} is the observation of the ith subject ($i = 1, 2, \ldots, n_j$) for the jth classroom ($j = 1, 2, \ldots, J$). The fixed parameter β_j is the mean of the nested factor (class j). For level two, we let

$$\beta_j = \mu + \gamma_j + b_j \qquad b_j \sim N(0, \sigma_b^2)$$

so that the jth classroom mean equals an overall mean plus the effect due to the jth classroom and a random error term. Using (9.56), SAS forms a matrix $\mathbf{Z}_j$ orthogonal to $\mathbf{X}_j = \mathbf{1}_{n_j}$, a matrix of contrasts. The SAS code to analyze these data is given in Program 9_7.sas. The results are given in the continuation of Output 9.7 that follows.

Output 9.7 (*continued*)

```
              HLM Nested Two-Factor ANOVA

             OBS    METHOD    CLASSES       Y

               1       1          1         3
               2       1          1         6
               3       1          1         3
               4       1          1         3
               5       1          2         1
               6       1          2         2
               7       1          2         2
               8       1          2         2
               9       1          3         5
              10       1          3         6
              11       1          3         5
              12       1          3         6
              13       1          4         2
              14       1          4         3
              15       1          4         4
              16       1          4         3
              17       2          5         7
              18       2          5         8
              19       2          5         7
              20       2          5         6
              21       2          6         4
              22       2          6         5
              23       2          6         4
              24       2          6         3
              25       2          7         7
              26       2          7         8
              27       2          7         9
              28       2          7         8
              29       2          8        10
              30       2          8        10
              31       2          8         9
              32       2          8        11

              HLM Nested Two-Factor ANOVA
                 The MIXED Procedure
               Class Level Information

         Class      Levels  Values

         METHOD          2  1 2
         CLASSES         8  1 2 3 4 5 6 7 8
```

Output 9.7 (*continued*)

```
                     REML Estimation Iteration History

             Iteration  Evaluations     Objective     Criterion

                     0            1   77.87478666
                     1            1   46.44294870    0.00000000


                         Convergence criteria met.
                                 G Matrix

    Effect   METHOD  CLASSES   Row           COL1           COL2           COL3

    CLASSES  1       1           1    4.16145833
    CLASSES  1       2           2                   4.16145833
    CLASSES  1       3           3                                  4.16145833
    CLASSES  1       4           4
    CLASSES  1       5           5
    CLASSES  1       6           6
    CLASSES  1       7           7
    CLASSES  1       8           8

                                  G Matrix

             COL4           COL5           COL6           COL7           COL8

        4.16145833
                       4.16145833
                                      4.16145833
                                                     4.16145833
                                                                    4.16145833

                  Covariance Parameter Estimates (REML)

 Cov Parm    Subject          Ratio      Estimate     Std Error       Z  Pr > |Z|

 CLASSES     METHOD      5.39864865    4.16145833    2.51449475    1.65    0.0979
 Residual                1.00000000    0.77083333    0.22252042    3.46    0.0005

                       HLM Nested Two-Factor ANOVA
                      Model Fitting Information for Y

                   Description                           Value

                   Observations                        32.0000
                   Res Log Likelihood                 -50.7896
                   Akaike's Information Criterion     -52.7896
                   Schwarz's Bayesian Criterion       -54.1908
                   -2 Res Log Likelihood              101.5793

                         Solution for Fixed Effects

       Effect  METHOD       Estimate      Std Error     DF       t  Pr > |t|

       METHOD  1          3.50000000     1.04333200      6    3.35    0.0153
       METHOD  2          7.25000000     1.04333200      6    6.95    0.0004
```

Output 9.7 (*continued*)

```
                        Tests of Fixed Effects

              Source    NDF    DDF   Type III F   Pr > F

              METHOD      2      6        29.77   0.0008

                      ESTIMATE Statement Results

Parameter                   Estimate      Std Error     DF        t   Pr > |t|

overall mean              5.37500000     0.73774713      6     7.29     0.0003
contrast diff method     -3.75000000     1.47549427      6    -2.54     0.0440
```

The variance component REML estimates are $\hat{\sigma}_b^2 = 4.1615$ and $\hat{\sigma}_e^2 = 0.7708$. The approximate Z-test of $H_1\!:\sigma_b^2 = 0$ is not rejected ($p = 0.0979$). Furthermore, while the overall mean μ estimated by $\hat{\mu} = 5.374$ is significantly different from zero, the contrast $\psi = \gamma_1 - \gamma_2$ estimated by $\hat{\psi} = 3.75$ is not significant at the $\alpha = 0.01$ level, $p \doteq 0.0440$.

For our third example, we consider a two-factor crossed design (with replications within cells) (Raudenbush, 1993, p. 482). Forty subjects are assigned to 10 tutors with the subjects then randomized to two instructional tasks (practice and no practice). Writing this model as a HLM, we let

$$y_{ij} = \beta_j + \gamma_j X_j + e_{ij} \qquad i = 1, \ldots, n_j;\ j = 1, \ldots, J$$

where X_j is a contrast vector comparing subjects in tutor j at $K-1$ fixed effect levels. For the example, with $K = 2$ levels the linear model is

$$\begin{pmatrix} y_{1j} \\ y_{2j} \\ y_{3j} \\ y_{4j} \end{pmatrix} = \begin{pmatrix} 1 & 1 \\ 1 & 1 \\ 1 & -1 \\ 1 & -1 \end{pmatrix} \begin{pmatrix} \beta_{0j} \\ \beta_{1j} \end{pmatrix} + \begin{pmatrix} e_{1j} \\ e_{1j} \\ e_{2j} \\ e_{2j} \end{pmatrix}$$

$$\mathbf{y}_j \quad = \quad \mathbf{X}_j \quad \beta_j \quad + \quad \mathbf{e}_j$$

where β_{0j} is the mean for the jth tutor and β_{1j} is the practice contrast. For level two of the design, we assume a random coefficient model

$$\beta_{0j} = \mu + b_{0j}$$
$$\beta_{1j} = \mu_c + b_{1j}$$

where μ is the overall mean, μ_c is a treatment contrast (slope), and b_{0j} and b_{1j} are randomly correlated errors.

Thus, the mixed model for the design is

$$
\mathbf{y}_j = \begin{pmatrix} 1 & 1 \\ 1 & 1 \\ 1 & -1 \\ 1 & -1 \end{pmatrix} \begin{pmatrix} \mu \\ \mu_c \end{pmatrix} + \begin{pmatrix} 1 & 1 \\ 1 & 1 \\ 1 & -1 \\ 1 & -1 \end{pmatrix} \mathbf{b}_j + \mathbf{e}_j
$$

$$
\mathbf{y}_j = \mathbf{X}_j \quad \gamma + \mathbf{Z}_j \quad \mathbf{b}_j + \mathbf{e}_j \tag{9.71}
$$

where the $\text{cov}(\mathbf{y}_j) = \mathbf{Z}_j\mathbf{D}\mathbf{Z}_j' + \sigma^2\mathbf{I}$. The SAS code to perform this analysis is given in Program 9_ 7.sas. To analyze (9.71) using PROC MIXED, the MODEL statement again specifies the mean model effects. The intercept is included by default and the variable PRACTICE models μ_c. Because we exclude PRACTICE from the CLASS statement, it is considered as a continuous variable. The RANDOM statement models the covariance matrix and the matrix **Z**. The INTERCEPT option is required since RANDOM does not included the intercept by default. The TYPE = UN option specifies the covariance matrix. The results for this analysis follow.

Output 9.7 *(continued)*

HLM Crossed Two-Factor Mixed MANOVA

OBS	PRACTICE	TUTOR	Y
1	1	1	65
2	1	1	70
3	1	2	70
4	1	2	78
5	1	3	62
6	1	3	66
7	1	4	56
8	1	4	64
9	1	5	62
10	1	5	70
11	1	6	45
12	1	6	48
13	1	7	56
14	1	7	69
15	1	8	82
16	1	8	86
17	1	9	53
18	1	9	54
19	1	10	82
20	1	10	88
21	-1	1	140
22	-1	1	155
23	-1	2	159
24	-1	2	163
25	-1	3	163
26	-1	3	181
27	-1	4	139
28	-1	4	142
29	-1	5	127
30	-1	5	138
31	-1	6	141
32	-1	6	146
33	-1	7	130
34	-1	7	138

Output 9.7 (*continued*)

```
                         35       -1          8      139
                         36       -1          8      144
                         37       -1          9      128
                         38       -1          9      130
                         39       -1         10      156
                         40       -1         10      165

                       HLM Crossed Two-Factor Mixed MANOVA
                              The MIXED Procedure
                            Class Level Information

                    Class      Levels  Values

                    TUTOR          10  1 2 3 4 5 6 7 8 9 10

                       REML Estimation Iteration History

               Iteration  Evaluations     Objective     Criterion

                       0            1  243.19259881
                       1            1  219.51382614    0.00000000

                           Convergence criteria met.
                     Covariance Parameter Estimates (REML)

Cov Parm    Subject           Ratio       Estimate      Std Error       Z  Pr > |Z|

UN(1,1)     TUTOR        3.28870070   111.65138889   56.69757896    1.97    0.0489
UN(2,1)     TUTOR       -0.36491573   -12.38888889   27.02311444   -0.46    0.6466
UN(2,2)     TUTOR        1.32371952    44.94027778   25.32870301    1.77    0.0760
Residual                 1.00000000    33.95000000   10.73593266    3.16    0.0016

                          Model Fitting Information for Y

                    Description                          Value

                    Observations                       40.0000
                    Res Log Likelihood                -144.677
                    Akaike's Information Criterion    -148.677
                    Schwarz's Bayesian Criterion      -151.952
                    -2 Res Log Likelihood             289.3532
                    Null Model LRT Chi-Square          23.6788
                    Null Model LRT DF                   3.0000
                    Null Model LRT P-Value              0.0000

                           Solution for Fixed Effects

        Effect          Estimate      Std Error    DF        t  Pr > |t|

        INTERCEPT  106.25000000     3.46610572      9    30.65    0.0001
        PRACTICE   -39.95000000     2.31144495      9   -17.28    0.0001

                             Tests of Fixed Effects

                   Source       NDF    DDF  Type III F  Pr > F

                   PRACTICE       1      9      298.72  0.0001
```

SAS estimates the covariance term as −12.39, which is incorrectly set to zero by Raudenbush (1993, p. 482). For SAS to analyze this design correctly, the practice factor must have contrast values 1 and −1.

The last HLM example considered is a standard split-plot repeated measures design with constant, linear, quadratic, and cubic polynomials fit to the within-subject effect. The data are taken from Kirk (1982, p. 244) where eight subjects (blocks) are observed over four treatment conditions. The SAS code to analyze these data are found in Program 9_ 7.sas. The results using PROC MIXED, assuming compound symmetry follow.

Output 9.7 (*continued*)

HLM Repeated Measures

OBS	PERSON	Y1	Y2	Y3	Y4	Y	DUR	X1	X2	X3
1	1	3	4	7	7	3	1	-1.5	0.5	-0.5
2	1	3	4	7	7	4	2	-0.5	-0.5	1.5
3	1	3	4	7	7	7	3	0.5	-0.5	-1.5
4	1	3	4	7	7	7	4	1.5	0.5	0.5
5	2	6	5	8	8	6	1	-1.5	0.5	-0.5
6	2	6	5	8	8	5	2	-0.5	-0.5	1.5
7	2	6	5	8	8	8	3	0.5	-0.5	-1.5
8	2	6	5	8	8	8	4	1.5	0.5	0.5
9	3	3	4	7	9	3	1	-1.5	0.5	-0.5
10	3	3	4	7	9	4	2	-0.5	-0.5	1.5
11	3	3	4	7	9	7	3	0.5	-0.5	-1.5
12	3	3	4	7	9	9	4	1.5	0.5	0.5
13	4	3	3	6	8	3	1	-1.5	0.5	-0.5
14	4	3	3	6	8	3	2	-0.5	-0.5	1.5
15	4	3	3	6	8	6	3	0.5	-0.5	-1.5
16	4	3	3	6	8	8	4	1.5	0.5	0.5
17	5	1	2	5	10	1	1	-1.5	0.5	-0.5
18	5	1	2	5	10	2	2	-0.5	-0.5	1.5
19	5	1	2	5	10	5	3	0.5	-0.5	-1.5
20	5	1	2	5	10	10	4	1.5	0.5	0.5
21	6	2	3	6	10	2	1	-1.5	0.5	-0.5
22	6	2	3	6	10	3	2	-0.5	-0.5	1.5
23	6	2	3	6	10	6	3	0.5	-0.5	-1.5
24	6	2	3	6	10	10	4	1.5	0.5	0.5
25	7	2	4	5	9	2	1	-1.5	0.5	-0.5
26	7	2	4	5	9	4	2	-0.5	-0.5	1.5
27	7	2	4	5	9	5	3	0.5	-0.5	-1.5
28	7	2	4	5	9	9	4	1.5	0.5	0.5
29	8	2	3	6	11	2	1	-1.5	0.5	-0.5
30	8	2	3	6	11	3	2	-0.5	-0.5	1.5
31	8	2	3	6	11	6	3	0.5	-0.5	-1.5
32	8	2	3	6	11	11	4	1.5	0.5	0.5

Output 9.7 (*continued*)

```
                          HLM Repeated Measures
                          The MIXED Procedure
                        Class Level Information

                   Class     Levels  Values

                   PERSON         8  1 2 3 4 5 6 7 8

                   REML Estimation Iteration History

           Iteration  Evaluations     Objective     Criterion

                   0            1   51.60122794
                   1            1   51.39468046    0.00000000

                       Convergence criteria met.
                         R Matrix for PERSON 1

          Row          COL1          COL2          COL3          COL4

            1    1.46428571    0.10714286    0.10714286    0.10714286
            2    0.10714286    1.46428571    0.10714286    0.10714286
            3    0.10714286    0.10714286    1.46428571    0.10714286
            4    0.10714286    0.10714286    0.10714286    1.46428571

                 Covariance Parameter Estimates (REML)

Cov Parm   Subject        Ratio      Estimate     Std Error       Z  Pr > |Z|

CS         PERSON    0.07894737    0.10714286    0.26058729    0.41    0.6810
Residual             1.00000000    1.35714286    0.41882338    3.24    0.0012

                    Model Fitting Information for Y

                 Description                        Value

                 Observations                     32.0000
                 Res Log Likelihood              -51.4276
                 Akaike's Information Criterion  -53.4276
                 Schwarz's Bayesian Criterion    -54.7598
                 -2 Res Log Likelihood           102.8552
                 Null Model LRT Chi-Square         0.2065
                 Null Model LRT DF                 1.0000
                 Null Model LRT P-Value            0.6495

                      Solution for Fixed Effects

        Effect         Estimate      Std Error    DF        t  Pr > |t|

        INTERCEPT    5.37500000     0.23622780     7    22.75    0.0001

                         HLM Repeated Measures
                       Solution for Fixed Effects

        Effect         Estimate      Std Error    DF        t  Pr > |t|

        X1           2.15000000     0.18419710    21    11.67    0.0001
        X2           1.00000000     0.41187724    21     2.43    0.0243
        X3          -0.20000000     0.18419710    21    -1.09    0.2899
```

Output 9.7 *(continued)*

```
                       Tests of Fixed Effects

            Source       NDF    DDF  Type III F   Pr > F

            X1             1     21      136.24   0.0001
            X2             1     21        5.89   0.0243
            X3             1     21        1.18   0.2899

                      CONTRAST Statement Results

         Source                     NDF    DDF        F   Pr > F

         duration                     2     21    69.69   0.0001
```

The results indicate that the estimate of variance components for error and block are $\hat{\sigma}^2 = 1.36$ and $\hat{\sigma}_b^2 = 0.107$. From the test of fixed effects, there appears to be a significant linear effect in treatment duration, $t = 11.67$ with p-value < 0.0001. The quadratic and cubic trends are not significant.

9.8 HLM Rat Experiment

For the last example of this section, we consider the rat weight data from Byrk et al. (1988) and analyzed by Rogosa and Saner (1995). The data consist of ten rats (subjects) with repeated measurements taken five times (Weeks 0, 1, 2, 3, 4), and a covariate, the mother's weight (z). To analyze these data using PROC MIXED, we fit a linear model $E(y) = \alpha + \beta x + \delta z$ to the data using orthogonal polynomials and three different structures for the covariance matrix of the random intercept and slope parameters: TYPE = UN, TYPE = CS, and TYPE = VC. The SAS code to perform the data analysis is given in Program 9_8.sas.

Program 9_8.sas

```
/* Program 9_8.sas                  */
/* HLM Manual Byrk et al. (1988) */

options ls=80 ps=60 nodate nonumber;
title 'Output 9.8: HLM Manual Rat Data';

data hlm5;
   infile 'c:\hlmrat.dat';
   input rat y1 y2 y3 y4 y5 z;
         y=y1;week=0;x=-2;output;
         y=y2;week=1;x=-1;output;
         y=y3;week=2;x= 0;output;
         y=y4;week=3;x= 1;output;
         y=y5;week=4;x= 2;output;
proc print data=hlm5;
   title2 'Random Coefficient Model Unknown Covariance Structure';
run;
```

```
proc mixed data=hlm5 method=reml covtest ratio;
   class rat;
   model y=x z/s p;
   random intercept x/type=un subject=rat g s;
   repeated /type=simple subject=rat r;
   make 'solutionr' out=eblups;
run;

data extract;
   set eblups;
   retain constant;
   if _effect_='INTERCEPT' then constant=_est_;
      else do;
         slope=_est_; output;
      end;
   keep slope constant;
run;
data raw;
   infile 'c:\hlmrat.dat';
   input rat y1 y2 y3 y4 y5 z;
data new;
   merge raw extract;
proc print data=new;
   title2 'Rat Data and Random Model Estimates';
run;
proc plot data=new;
   plot slope*z / hpos=30 vpos=25;
proc reg data=new;
   title2 'Slope Regression Analysis';
   model slope=z/stb;
run;
proc mixed data=hlm5 method=reml covtest ratio;
   title2 'Covariance Structure Compound symmetry';
   class rat;
   model y = x z/s;
   random int x/type=cs subject=rat g;
   repeated /type=simple subject=rat;
run;
proc mixed data=hlm5 method=reml covtest ratio;
   title2 'Variance Components Model';
   class rat;
   model y = x z/s;
   random int x/type=vc subject=rat g;
   repeated /type=simple subject=rat;
run;
```

Result and Interpretation 9_8.sas

The results of the analysis aregiven in Output 9.8.

Output 9.8: *HLM Manual Rat Data*

Output 9.8: HLM Manual Rat Data
Random Coefficient Model Unknown Covariance Structure

OBS	RAT	Y1	Y2	Y3	Y4	Y5	Z	Y	WEEK	X
1	1	61	72	118	130	176	170	61	0	-2
2	1	61	72	118	130	176	170	72	1	-1
3	1	61	72	118	130	176	170	118	2	0
4	1	61	72	118	130	176	170	130	3	1
5	1	61	72	118	130	176	170	176	4	2
6	2	65	85	129	148	174	194	65	0	-2
7	2	65	85	129	148	174	194	85	1	-1
8	2	65	85	129	148	174	194	129	2	0
9	2	65	85	129	148	174	194	148	3	1
10	2	65	85	129	148	174	194	174	4	2
11	3	57	68	130	143	201	187	57	0	-2
12	3	57	68	130	143	201	187	68	1	-1
13	3	57	68	130	143	201	187	130	2	0
14	3	57	68	130	143	201	187	143	3	1
15	3	57	68	130	143	201	187	201	4	2
16	4	46	74	116	124	157	156	46	0	-2
17	4	46	74	116	124	157	156	74	1	-1
18	4	46	74	116	124	157	156	116	2	0
19	4	46	74	116	124	157	156	124	3	1
20	4	46	74	116	124	157	156	157	4	2
21	5	47	85	103	117	148	155	47	0	-2
22	5	47	85	103	117	148	155	85	1	-1
23	5	47	85	103	117	148	155	103	2	0
24	5	47	85	103	117	148	155	117	3	1
25	5	47	85	103	117	148	155	148	4	2
26	6	43	58	109	133	152	150	43	0	-2
27	6	43	58	109	133	152	150	58	1	-1
28	6	43	58	109	133	152	150	109	2	0
29	6	43	58	109	133	152	150	133	3	1
30	6	43	58	109	133	152	150	152	4	2
31	7	53	62	82	112	156	138	53	0	-2
32	7	53	62	82	112	156	138	62	1	-1
33	7	53	62	82	112	156	138	82	2	0
34	7	53	62	82	112	156	138	112	3	1
35	7	53	62	82	112	156	138	156	4	2
36	8	72	96	117	129	154	154	72	0	-2
37	8	72	96	117	129	154	154	96	1	-1
38	8	72	96	117	129	154	154	117	2	0
39	8	72	96	117	129	154	154	129	3	1
40	8	72	96	117	129	154	154	154	4	2
41	9	53	54	87	120	138	149	53	0	-2
42	9	53	54	87	120	138	149	54	1	-1
43	9	53	54	87	120	138	149	87	2	0
44	9	53	54	87	120	138	149	120	3	1
45	9	53	54	87	120	138	149	138	4	2
46	10	72	98	114	144	177	167	72	0	-2
47	10	72	98	114	144	177	167	98	1	-1
48	10	72	98	114	144	177	167	114	2	0
49	10	72	98	114	144	177	167	144	3	1
50	10	72	98	114	144	177	167	177	4	2

Output 9.8 (*continued*)

```
          Random Coefficient Model Unknown Covariance Structure
                          The MIXED Procedure
                        Class Level Information

                 Class     Levels  Values

                 RAT           10  1 2 3 4 5 6 7 8 9 10

                    REML Estimation Iteration History

            Iteration  Evaluations     Objective     Criterion

                    0            1  295.81311607
                    1            2  290.87319398    0.00002174
                    2            1  290.86993291    0.00000006
                    3            1  290.86992477    0.00000000

                       Convergence criteria met.
                           R Matrix for RAT 1

    Row          COL1          COL2          COL3          COL4          COL5

      1   91.74666667
      2                 91.74666667
      3                               91.74666667
      4                                             91.74666667
      5                                                           91.74666667

                                G Matrix

            Effect     RAT   Row          COL1          COL2

            INTERCEPT  1       1   29.40132798  -11.06651176
            X          1       2  -11.06651176   10.53908999

                  Covariance Parameter Estimates (REML)

        Cov Parm  Subject       Estimate     Std Error       Z  Pr > |Z|

        UN(1,1)   RAT        29.40132798   25.20891478    1.17    0.2435
        UN(2,1)   RAT       -11.06651176   12.82765059   -0.86    0.3883
        UN(2,2)   RAT        10.53908999    9.59031597    1.10    0.2718
        DIAG      RAT        91.74666667   23.68888747    3.87    0.0001

                     Model Fitting Information for Y

                 Description                        Value

                 Observations                     50.0000
                 Res Log Likelihood              -188.625

          Random Coefficient Model Unknown Covariance Structure
                     Model Fitting Information for Y

                 Description                        Value

                 Akaike's Information Criterion  -192.625
                 Schwarz's Bayesian Criterion    -196.325
                 -2 Res Log Likelihood           377.2501
                 Null Model LRT Chi-Square         4.9432
                 Null Model LRT DF                 3.0000
                 Null Model LRT P-Value            0.1760
```

Output 9.8 *(continued)*

Solution for Fixed Effects

Effect	Estimate	Std Error	DF	t	Pr > |t|
INTERCEPT	5.51762266	19.93675569	8	0.28	0.7890
X	26.76000000	1.40405686	9	19.06	0.0001
Z	0.62754554	0.12232493	30	5.13	0.0001

Solution for Random Effects

Effect	RAT	Estimate	SE Pred	DF	t	Pr > |t|
INTERCEPT	1	-0.95268206	3.52956728	30	-0.27	0.7891
X	1	1.04784344	2.21545903	30	0.47	0.6397
INTERCEPT	2	-4.27438909	4.11517247	30	-1.04	0.3072
X	2	1.49864597	2.26475000	30	0.66	0.5132
INTERCEPT	3	-4.07879030	3.88252170	30	-1.05	0.3019
X	3	4.81661064	2.24439491	30	2.15	0.0401
INTERCEPT	4	-0.11733139	3.51102149	30	-0.03	0.9736
X	4	0.20642752	2.21400490	30	0.09	0.9263
INTERCEPT	5	-0.72275567	3.51964419	30	-0.21	0.8387
X	5	-1.21683289	2.21468015	30	-0.55	0.5868
INTERCEPT	6	-0.99242398	3.58202621	30	-0.28	0.7836
X	6	1.26163398	2.21960842	30	0.57	0.5740
INTERCEPT	7	0.77952910	3.85298215	30	0.20	0.8410
X	7	-0.64864974	2.24188241	30	-0.29	0.7743
INTERCEPT	8	8.13737042	3.52956728	30	2.31	0.0282
X	8	-4.70140153	2.21545903	30	-2.12	0.0422
INTERCEPT	9	-4.02962106	3.59826279	30	-1.12	0.2717
X	9	-0.40039176	2.22090351	30	-0.18	0.8581
INTERCEPT	10	6.25109405	3.50370878	30	1.78	0.0845
X	10	-1.86388562	2.21343338	30	-0.84	0.4064

Tests of Fixed Effects

Source	NDF	DDF	Type III F	Pr > F
X	1	9	363.25	0.0001
Z	1	30	26.32	0.0001

Random Coefficient Model Unknown Covariance Structure
Predicted Values

Y	Predicted	SE Pred	L95	U95	Residual
61.0000	55.6320	6.1894	42.9915	68.2725	5.3680
72.0000	83.4398	4.4725	74.3058	92.5739	-11.4398
118.0000	111.2477	3.3528	104.4004	118.0950	6.7523
130.0000	139.0555	3.4674	131.9742	146.1368	-9.0555
176.0000	166.8634	4.7273	157.2089	176.5179	9.1366
65.0000	66.4698	6.2328	53.7406	79.1989	-1.4698
85.0000	94.7284	4.6318	85.2690	104.1878	-9.7284
129.0000	122.9871	3.7476	115.3335	130.6406	6.0129
148.0000	151.2457	4.0768	142.9198	159.5716	-3.2457
174.0000	179.5044	5.4020	168.4721	190.5367	-5.5044
57.0000	55.6366	6.2148	42.9443	68.3290	1.3634
68.0000	87.2132	4.5663	77.8877	96.5388	-19.2132
130.0000	118.7898	3.5888	111.4606	126.1191	11.2102
143.0000	150.3665	3.8353	142.5338	158.1991	-7.3665
201.0000	181.9431	5.1324	171.4614	192.4247	19.0569
46.0000	49.3645	6.1881	36.7267	62.0024	-3.3645
74.0000	76.3310	4.4678	67.2066	85.4554	-2.3310
116.0000	103.2974	3.3406	96.4751	110.1197	12.7026

Output 9.8 (*continued*)

```
124.0000   130.2638   3.4480  123.2221  137.3055   -6.2638
157.0000   157.2303   4.7062  147.6189  166.8416   -0.2303
 47.0000    50.9781   6.1887   38.3390   63.6172   -3.9781
 85.0000    76.5213   4.4700   67.3924   85.6501    8.4787
103.0000   102.0644   3.3462   95.2305  108.8983    0.9356
117.0000   127.6076   3.4570  120.5475  134.6677  -10.6076
148.0000   153.1508   4.7160  143.5193  162.7822   -5.1508
 43.0000    42.6138   6.1930   29.9659   55.2617    0.3862
 58.0000    70.6354   4.4860   61.4738   79.7970  -12.6354
109.0000    98.6570   3.3874   91.7390  105.5751   10.3430
133.0000   126.6787   3.5222  119.4854  133.8719    6.3213
152.0000   154.7003   4.7872  144.9235  164.4770   -2.7003
 53.0000    40.6757   6.2126   27.9879   53.3636   12.3243
 62.0000    66.7871   4.5581   57.4781   76.0961   -4.7871
 82.0000    92.8984   3.5688   85.6100  100.1868  -10.8984
112.0000   119.0098   3.8046  111.2399  126.7797   -7.0098
156.0000   145.1211   5.0983  134.7091  155.5332   10.8789
 72.0000    66.1798   6.1894   53.5393   78.8203    5.8202
 96.0000    88.2384   4.4725   79.1044   97.3725    7.7616
117.0000   110.2970   3.3528  103.4497  117.1443    6.7030
129.0000   132.3556   3.4674  125.2743  139.4369   -3.3556
154.0000   154.4142   4.7273  144.7597  164.0687   -0.4142
 53.0000    42.2731   6.1942   29.6229   54.9233   10.7269
 54.0000    68.6327   4.4902   59.4625   77.8029  -14.6327
 87.0000    94.9923   3.3982   88.0523  101.9323   -7.9923
120.0000   121.3519   3.5391  114.1240  128.5798   -1.3519
138.0000   147.7115   4.8057  137.8969  157.5261   -9.7115
 72.0000    66.7766   6.1877   54.1397   79.4135    5.2234
 98.0000    91.6727   4.4659   82.5521  100.7933    6.3273
114.0000   116.5688   3.3357  109.7563  123.3813   -2.5688
144.0000   141.4649   3.4403  134.4389  148.4910    2.5351
177.0000   166.3611   4.6979  156.7667  175.9554   10.6389
```

Rat Data and Random Model Estimates

OBS	RAT	Y1	Y2	Y3	Y4	Y5	Z	CONSTANT	SLOPE
1	1	61	72	118	130	176	170	-0.95268	1.04784
2	2	65	85	129	148	174	194	-4.27439	1.49865
3	3	57	68	130	143	201	187	-4.07879	4.81661
4	4	46	74	116	124	157	156	-0.11733	0.20643
5	5	47	85	103	117	148	155	-0.72276	-1.21683
6	6	43	58	109	133	152	150	-0.99242	1.26163
7	7	53	62	82	112	156	138	0.77953	-0.64865
8	8	72	96	117	129	154	154	8.13737	-4.70140
9	9	53	54	87	120	138	149	-4.02962	-0.40039
10	10	72	98	114	144	177	167	6.25109	-1.86389

Output 9.8 (*continued*)

```
                      Rat Data and Random Model Estimates
              Plot of SLOPE*Z.  Legend: A = 1 obs, B = 2 obs, etc.

                       5.0 ^                        A
                           ,
                           ,
                           ,
                           ,
                           ,
                       2.5 ^
                           ,
                           ,                          A
                     SLOPE ,              A     A
                           ,
                           ,
                       0.0 ^                A
                           ,              A
                           ,           A
                           ,               A
                           ,                   A
                           ,
                      -2.5 ^
                           ,
                           ,
                           ,
                           ,
                           ,               A
                      -5.0 ^
                           Š^ƒƒƒƒƒƒƒƒƒƒƒƒƒ^ƒƒƒƒƒƒƒƒƒƒƒƒƒ^ƒ
                           100           150           200

                                          Z
                           Slope Regression Analysis

Model: MODEL1
Dependent Variable: SLOPE

                              Analysis of Variance

                                 Sum of         Mean
        Source          DF      Squares       Square       F Value       Prob>F

        Model            1     17.11313     17.11313         3.537       0.0968
        Error            8     38.70385      4.83798
        C Total          9     55.81697

            Root MSE       2.19954     R-square       0.3066
            Dep Mean      -0.00000     Adj R-sq       0.2199
            C.V.      -1.193476E17

                              Parameter Estimates

                       Parameter       Standard    T for H0:
      Variable  DF      Estimate          Error   Parameter=0    Prob > |T|

      INTERCEP   1    -12.719498     6.79864241        -1.871        0.0983
      Z          1      0.078515     0.04174672         1.881        0.0968
                    Standardized
      Variable  DF      Estimate

      INTERCEP   1    0.00000000
      Z          1    0.55370897
```

Output 9.8 (*continued*)

```
                   Covariance Structure Compound symmetry
                        The MIXED Procedure
                      Class Level Information

               Class     Levels  Values

               RAT           10  1 2 3 4 5 6 7 8 9 10

                  REML Estimation Iteration History

          Iteration  Evaluations     Objective     Criterion

                  0            1  295.81311607
                  1            2  291.63265702    0.00001382
                  2            1  291.63061706    0.00000000

                      Convergence criteria met.
                              G Matrix

           Effect      RAT   Row           COL1           COL2

           INTERCEPT  1       1   15.63823073    -8.86649041
           X          1       2   -8.86649041    15.63823073

                 Covariance Parameter Estimates (REML)

   Cov Parm   Subject      Estimate      Std Error       Z  Pr > |Z|

   Variance   RAT       24.50472114   19.16159610    1.28    0.2010
   CS         RAT       -8.86649041   10.93403829   -0.81    0.4174
   DIAG       RAT       94.33622435   24.86487126    3.79    0.0001

                    Model Fitting Information for Y

               Description                          Value

               Observations                       50.0000
               Res Log Likelihood                -189.005
               Akaike's Information Criterion    -192.005
               Schwarz's Bayesian Criterion      -194.781
               -2 Res Log Likelihood             378.0108
               Null Model LRT Chi-Square           4.1825
               Null Model LRT DF                   2.0000
               Null Model LRT P-Value              0.1235


                      Solution for Fixed Effects

    Effect         Estimate     Std Error     DF        t  Pr > |t|

    INTERCEPT   10.45967797   17.32103235      8     0.60    0.5627
                 Covariance Structure Compound symmetry
                      Solution for Fixed Effects

    Effect         Estimate     Std Error     DF        t  Pr > |t|

    X           26.76000000    1.58340940      9    16.90    0.0001
    Z            0.59703902    0.10630332     30     5.62    0.0001
```

Output 9.8 (*continued*)

```
                 Tests of Fixed Effects

          Source       NDF    DDF  Type III F  Pr > F

          X              1      9      285.62  0.0001
          Z              1     30       31.54  0.0001

                 Variance Components Model
                    The MIXED Procedure
                  Class Level Information

           Class     Levels  Values

           RAT           10  1 2 3 4 5 6 7 8 9 10

              REML Estimation Iteration History

      Iteration  Evaluations     Objective     Criterion

              0            1  295.81311607
              1            1  291.71537286    0.00000000

                 Convergence criteria met.
                         G Matrix

       Effect      RAT   Row           COL1          COL2

       INTERCEPT   1       1   27.81933814
       X           1       2                  10.53911111

            Covariance Parameter Estimates (REML)

Cov Parm     Subject      Estimate     Std Error       Z  Pr > |Z|

INTERCEPT    RAT       27.81933814   23.56550640    1.18    0.2378
X            RAT       10.53911111    9.59033526    1.10    0.2718
DIAG         RAT       91.74666667   23.68888747    3.87    0.0001

               Model Fitting Information for Y

           Description                          Value

           Observations                       50.0000
           Res Log Likelihood                -189.048
           Akaike's Information Criterion    -192.048
           Schwarz's Bayesian Criterion      -194.823
           -2 Res Log Likelihood             378.0956
                 Solution for Fixed Effects

  Effect         Estimate      Std Error     DF        t  Pr > |t|

  INTERCEPT   18.87365994    21.00215302      8     0.90    0.3951
  X           26.76000000     1.40405761      9    19.06    0.0001
  Z            0.54510086     0.12896265     30     4.23    0.0002
```

Output 9.8 (*continued*)

```
                    Variance Components Model
                     Tests of Fixed Effects

            Source       NDF    DDF   Type III F   Pr > F

            X              1      9       363.25   0.0001
            Z              1     30        17.87   0.0002
```

Comparing the AIC statistics for the three covariance structures, we see that all AIC values are about the same. We first consider the model with unknown structure $\mathbf{\Omega}_i$. The random coefficient model fit to the rat data is a conditional HLM where

$$y_{ij} = a_i + b_i x_{ij} + \delta z_{ij} + e_{ij}$$

$$\begin{pmatrix} a_i \\ b_i \end{pmatrix} \sim N\left(\begin{pmatrix} \alpha \\ \beta \end{pmatrix}, \mathbf{\Omega}_i\right)$$

$$\mathbf{\Omega}_i = \begin{pmatrix} \sigma_a^2 & \sigma_{ab} \\ \sigma_{ab} & \sigma_b^2 \end{pmatrix}$$

$$e_{ij} \sim N(0, \sigma_e^2)$$

so that the fixed effects $\alpha = E(a_i)$, $\beta = E(b_i)$ and δ is a common slope parameter. When investigating change, we are interested in the variance of the slope parameter b, $\hat{\sigma}_b^2$. The REML estimates for a_i and b_i are provided in the matrix **G** by using the G option on the RANDOM statement. For our example, the average rate of change in growth for the ten rats is $\hat{\beta} = 26.76$ where the estimated variance in the random slopes is estimated by $\hat{\sigma}_b^2 = 10.539$ with standard error 9.590. This standard deviation is a bit larger than the bootstrap estimate of 7.707 obtained by Rogosa and Saner (1995). The p-value for testing H: $\sigma_b^2 = 0$ is nonsignificant, 0.2710; thus, there does not appear to be wide variation in growth. As pointed out by Rogosa and Saner (1995), the test of variance does not take into account the background variable z. More important, the estimate of γ in the equation $E(b|z) = \mu + \gamma z$ may be of primary interest. To investigate the relation of rates of change in growth to the weight of the mother, we use the EBLUP estimates of this slope b_i by using the S option on the MODEL statement. However, PROC MIXED calculates deviations $a_i - \alpha$ and $b_i - \beta$. These deviations may be output to a SAS data set by using the MAKE statement and read into PROC REG after some manipulation, as illustrated in Program 9_8.sas. This fits the model $E[(b - \beta)|z] = \alpha^* + \beta^* z$ to the random slope deviations. The estimates of α^* and β^* using OLS are found in Output 9.8. To obtain an estimate of γ, one must divide the estimate $\hat{\beta}^*$ by $\hat{\rho}(\hat{b})$, the reliability of the OLS estimate as defined by Rogosa and Saner (1995):

$$\hat{\rho}(\hat{b}) = \hat{\sigma}_b^2 / (\hat{\sigma}_b^2 + \hat{\sigma}_e^2 / n) = 0.534603$$

Hence,

$$\hat{\gamma} = \hat{\beta}^* / \hat{\rho}(\hat{b}) = 0.146866$$

which has estimated standard deviation $\hat{\sigma}_{\hat{\gamma}} = \hat{\sigma}_{\hat{\beta}^*} / \hat{\rho}(\hat{b}) = 0.0781$. The t-test for testing $H: \gamma = 0$ from Output 9.8 is 0.097; this indicates no significant linear relationship between growth rate in rats and the mother's weight (z).

10. Incomplete Repeated Measurement Data

Theory

Applications

10.1 Introduction

In our discussion of the GLM, we have assumed that the data matrix $\mathbf{Y}_{n\times p}$ is complete, i.e., that there are no missing data. However, in many studies involving the analysis of repeated measurement designs and longitudinal growth data, the data matrix is not complete. When data are incomplete, estimating model parameters and testing linear hypotheses becomes more difficult since bias may be introduced into the estimation procedure and test statistics with known exact small sample distributions are not easily obtained. A review of procedures for dealing with regression models where data are missing at random or by design is provided by Little (1992).

If the number of missing data points is very small, the simplest method of analysis is to discard the incomplete response vectors and include only the complete cases in the analysis; this is called the listwise deletion method. The primary concern with this procedure is whether bias is introduced into the sample estimates of population parameters (Timm, 1970 and Little, 1992, 1995). Furthermore, discarding data is a practice many would find undesirable. There are four general approaches to the missing data problem for growth or repeated-measures studies:

(1) imputation techniques which involve using estimates of the missing values to obtain parameter estimates, adjusting these estimates for bias, and adjusting the degrees of freedom of the test statistic sampling distribution;

(2) methods which estimate model parameters using maximum likelihood (ML) or restricted maximum likelihood (REML) from the incomplete data followed by LR tests, large sample chi-square tests, or goodness-of-fit tests;

(3) feasible generalized least squares (FGLS) approaches which involve finding a consistent estimate of the covariance matrix Σ from the incomplete data and estimating model parameters using GLSE with hypothesis testing performed by an asymptotic Wald-type statistic;

(4) Bayesian, parametric empirical Bayes (PEB) and complex simulation techniques such as the Gibbs sampling and multiple imputation methods.

In general, maximum likelihood procedures under multivariate normality use three general algorithms: Newton-Raphson, Fisher scoring, and the EM algorithm. FGLS procedures and Bayesian methods are superior to single-value imputation procedures (Little, 1992).

When missing data were encountered in the early years of data analysis, the approach was to estimate the missing data and, after inserting these estimates into the data matrix **Y**, to use this modified matrix for the analysis (Timm, 1970). Although this approach is not incorrect, iterative procedures–that estimate the model parameters from the complete data, then use the estimated parameters to estimate the missing data, estimate the model parameters again, and repeat the process until some convergence criterion is met–work much better. This iterative approach was first proposed by Orchard and Woodbury (1972); the process is called the missing information principle. Dempster et al. (1977) expanded upon the principle to include a larger class of problems and called their approach the EM algorithm. Under general regularity conditions, the approach leads to parameter estimates that converge to the ML estimates.

Kleinbaum (1973) proposed a large sample iterative method employing a generalized least squares procedure to solve missing data problems in growth curve studies. He showed that the resulting estimates were best asymptotically normal (BAN) and developed a Wald statistic to test hypotheses. The difference between this approach and the EM algorithm approach is that, under multivariate normality, the EM algorithm procedure is an exact method whereas Kleinbaum's (1973) is asymptotic. Liski (1985) developed an EM procedure for the GMANOVA model. In large samples, both are essentially equivalent. In this chapter, we show how to analyze repeated measures designs with missing data using PROC MIXED which uses a Newton-Raphson algorithm, (Wolfinger et al., 1994). We also illustrate Kleinbaum's procedure using PROC IML code.

10.2 An FGLS Procedure

One of the first solutions to the problem of analyzing growth curve models with missing data was proposed by Kleinbaum (1973). Kleinbaum's approach involved the use of an FGLS estimate for the model parameters. The FGLS estimate depends on obtaining a consistent estimate of Σ which is positive definite. With this approach one partitions the GMANOVA model (7.23) data matrix $\mathbf{Y}_{n\times p}$ into m mutually exclusive subsets according to the pattern of missing data where the jth subset is the data matrix $\mathbf{Y}_j(n_j \times p_j)$ with $p_j \leq p$. For each data pattern $\mathbf{Y}_j$, one formulates the model

$$\begin{aligned} E(\mathbf{Y}_j) &= \mathbf{X}_j\mathbf{B}\mathbf{P}\mathbf{A}_j \\ \operatorname{cov}(\mathbf{Y}_j) &= \mathbf{I}_{n_i} \otimes \mathbf{A}_j'\Sigma\mathbf{A}_j \qquad j = 1, 2, \ldots, m \end{aligned} \tag{10.1}$$

where $\mathbf{X}_j(n_j \times k)$, $\mathbf{B}(k \times q)$, $\mathbf{P} \equiv \mathbf{X}_2(q \times p)$ and Σ are as defined in the GMANOVA model (7.23) and $\mathbf{A}_j(p \times p_j)$ is an incidence matrix of 0' s and 1' s such that $\mathbf{A}_j = (a_{rs}^{(j)})$ where

$$a_{rs}^{(j)} = \begin{cases} 1 & \text{if the } r^{\text{th}} \text{ ordered observation over time in the} \\ & j^{\text{th}} \text{ data set corresponds to the } s^{\text{th}} \text{ ordered} \\ & \text{observation over time for the complete data set} \\ 0 & \text{otherwise} \end{cases}$$

Alternatively, using the vec(.) operator, we may let

$\mathbf{y}_{N\times 1} = (\mathbf{y}_1', \mathbf{y}_2', \ldots, \mathbf{y}_m')' = [\text{vec}'(\mathbf{Y}_1), \text{vec}'(\mathbf{Y}_2), \ldots, \text{vec}'(\mathbf{Y}_m)]'$ then

$$\begin{aligned} E(\mathbf{y}) &= \mathbf{D}\beta \quad \text{cov } \mathbf{y} = \Omega \\ \underset{N\times kq}{\mathbf{D}} &= \begin{pmatrix} \mathbf{A}_1'\mathbf{P}' \otimes \mathbf{X}_1 \\ \vdots \\ \mathbf{A}_m'\mathbf{P}' \otimes \mathbf{X}_m \end{pmatrix} \\ \underset{N\times N}{\Omega} &= \text{Diag}(\mathbf{A}_1'\Sigma\mathbf{A}_1 \otimes \mathbf{I}_{n_1}, \ldots, \mathbf{A}_m'\Sigma\mathbf{A}_m \otimes \mathbf{I}_{n_m}) \end{aligned} \tag{10.2}$$

where $N = \sum_{i=1}^{m} n_i p_i$ and $\beta = \text{vec}(\mathbf{B})$.

To obtain the incidence matrix $\mathbf{A}_j$, we take the identity matrix of order p_j, and insert a row vector of zeros between rows of the identity matrix whenever a missing value occurs. This results in a $p \times p_j$ matrix such that whenever the kth ordered observation is missing, the kth row of $\mathbf{A}_j$ will be all zeros.

Some examples are

(1) No missing data, $\mathbf{A}_j = \mathbf{I}_p$

(2) Only the last observation missing, $\mathbf{A}_j = \begin{pmatrix} \mathbf{I}_{p-1} \\ \mathbf{0}' \end{pmatrix}$

(3) Only the ith and j^{th} observation missing

$$\mathbf{A}_j = \begin{pmatrix} \mathbf{I}_{i-1} & \underset{(i-1)\times(p-i-1)}{\mathbf{0}} & \\ \underset{1\times(p-2)}{\mathbf{0}'} & & \\ \underset{(j-i-1)\times(i-1)}{\mathbf{0}} & \underset{(j-i-1)\times(j-i-1)}{\mathbf{I}} & \underset{(j-i-1)\times(p-j)}{\mathbf{0}} \\ \underset{1\times(p-2)}{\mathbf{0}'} & & \\ \underset{(p-j)\times(j-2)}{\mathbf{0}} & & \underset{(p-j)\times(p-j)}{\mathbf{I}} \end{pmatrix}$$

If $\hat{\Sigma}$ is a consistent estimate of Σ, then the FGLS estimate of β is

$$\begin{aligned} \hat{\hat{\beta}} &= (\mathbf{D}'\hat{\Omega}^{-1}\mathbf{D})^{-1}\mathbf{D}'\hat{\Omega}^{-1}\mathbf{y} \\ &= \left[\sum_{j=1}^{m} \mathbf{P}\mathbf{A}_j(\mathbf{A}_j'\hat{\Sigma}\mathbf{A}_j)^{-1}\mathbf{A}_j'\mathbf{P}' \otimes \mathbf{X}_j'\mathbf{X}_j\right]^{-} \sum_{j=1}^{m} \text{vec}\left[\mathbf{X}_j'\mathbf{Y}_j(\mathbf{A}_j'\hat{\Sigma}\mathbf{A}_j)^{-1}\mathbf{A}_j'\hat{\Sigma}\mathbf{A}_j)^{-1}\mathbf{A}_j\mathbf{P}\right] \\ &= \left[\sum_{j=1}^{m} \mathbf{P}\mathbf{A}_j(\mathbf{A}_j'\hat{\Sigma}\mathbf{A}_j)^{-1}\mathbf{A}_j'\mathbf{P}' \otimes \mathbf{X}_j'\mathbf{X}_j\right]^{-} \sum_{j=1}^{m} \left[\mathbf{P}\mathbf{A}_j(\mathbf{A}_j'\hat{\Sigma}\mathbf{A}_j)^{-1} \otimes \mathbf{X}_j'\right]\mathbf{y}_j \end{aligned} \tag{10.3}$$

Letting $\mathbf{C}_* \equiv \mathbf{A}' \otimes \mathbf{C}$ for parametric functions of the form $\mathbf{C}_*\beta$ where $\beta = \text{vec}(\mathbf{B})$, Kleinbaum showed that the asymptotic covariance matrix of $\mathbf{C}_*\beta$ is computed to be

$$\mathbf{C}_*(\mathbf{D}'\hat{\Omega}^{-1}\mathbf{D})^{-1}\mathbf{C}'_* = \mathbf{C}_*\left[\sum_{j=1}^{m} \mathbf{PA}_j(\mathbf{A}'_j\hat{\Sigma}\mathbf{A}_j)^{-1}\mathbf{A}_j\mathbf{P}' \otimes \mathbf{X}'_j\mathbf{X}_j\right]^{-} \mathbf{C}'_* \tag{10.4}$$

and that the Wald statistic for testing $H(\mathbf{C}_*\beta = \gamma)$ given by

$$X^2 = (\mathbf{C}_*\hat{\hat{\beta}} - \gamma)'(\mathbf{C}_*(\mathbf{D}'\hat{\Omega}^{-1}\mathbf{D})^{-1}\mathbf{C}'_*)^{-}\ (\mathbf{C}_*\hat{\hat{\beta}} - \gamma) \tag{10.5}$$

converges to a chi-square distribution with $v = \text{rank}(\mathbf{C}_*)$ degrees of freedom under the null hypothesis.

Kleinbaum (1973) suggested estimating Σ with the unbiased and consistent estimate $\hat{\Sigma} = (\hat{\sigma}_{rs})$ where $\hat{\sigma}_{rs}$ is estimated from the data matrix $\mathbf{Y}$ using variables r and s where both pairs are observed. Then

$$\hat{\sigma}_{rs} = \left(\frac{1}{N_{rs} - k}\right)\mathbf{y}'_{rs}(\mathbf{I}_{N_{rs}} - \mathbf{D}_{rs}(\mathbf{D}'_{rs}\mathbf{D}_{rs})^{-1}\mathbf{D}'_{rs})\mathbf{y}_{sr} \qquad r, s = 1, 2, \ldots, p \tag{10.6}$$

where

N_{rs} is the number of observations for which both variables r and s are observed,

$\mathbf{y}_{rs}$ $(N_{rs} \times 1)$ is the observation vector for variable r corresponding to those in which both variables r and s are observed, and

$\mathbf{D}_{rs}$ $(N_{rs} \times k)$ is the across-variables design matrix consisting of a row of $\mathbf{X}_j$ matrices corresponding to $\mathbf{y}_{rs}$.

Unfortunately, the sample MANOVA estimator for Σ is not necessarily nonsingular (positive definite) in small samples. While one may replace ordinary inverses with generalized inverses ($\mathbf{AA}^{-}\mathbf{A} = \mathbf{A}$) to obtain estimates of model parameters, the Wald statistic may lead to spurious results or even be negative. To address this problem, Woolson and Clarke (1987) recommend writing the incomplete MANOVA model as a SUR model and using OLS to estimate β. With β estimated, an estimate of Σ is obtained from the vector of residuals, $\mathbf{r}_j = \mathbf{y}_j - \mathbf{X}_j\hat{\beta}_j$, which is adjusted to ensure a positive definite estimate of Σ. More recently, Mensah et al. (1993) conducted a Monte Carlo simulation experiment to evaluate estimators of Σ using the GMANOVA model instead of the MANOVA model.

Using only complete data vectors $\mathbf{Y}_j (j = 1, 2, \ldots, p)$ and corresponding design matrices $\mathbf{Z}_j(n_j \times k)$ Mensah et al. (1993) constructed an unweighted and fully weighted GMANOVA estimator of $\mathbf{B}$:

$$\begin{aligned}\hat{\mathbf{B}}_{UW} &= [(\mathbf{Z}'_1\mathbf{Z}_1)^{-1}\mathbf{Z}_1\mathbf{Y}_1, \ldots, (\mathbf{Z}'_p\mathbf{Z}_p)^{-1}\mathbf{Z}'_p\mathbf{Y}_p]\mathbf{P}'(\mathbf{PP}')^{-1} \\ \hat{\mathbf{B}}_{FW} &= [(\mathbf{Z}'_1\mathbf{Z}_1)^{-1}\mathbf{Z}_1\mathbf{Y}_1, \ldots, (\mathbf{Z}'_p\mathbf{Z}_p)^{-1}\mathbf{Z}'_p\mathbf{Y}_p]\hat{\Sigma}_{KL}\mathbf{P}'(\mathbf{P}\hat{\Sigma}^{-1}_{KL}\mathbf{P}')^{-1}\end{aligned} \tag{10.7}$$

where $\hat{\Sigma}_{KL} = (\hat{\sigma}_{rs})$ is Kleinbaum's positive definite estimator of Σ. Following Woolson and Clarke (1987), unweighted and fully weighted estimators of Σ were constructed utilizing residual matrices for observation vectors of complete pairs:

$$\begin{aligned}\hat{\sigma}_{UW,rs} &= \left(\frac{1}{N_{rs} - k}\right)(\mathbf{y}_{rs} - \mathbf{D}_{rs}\hat{\hat{\beta}}_{UW})'(\mathbf{y}_{sr} - \mathbf{D}_{rs}\hat{\hat{\beta}}_{UW}) \\ \hat{\sigma}_{FW,rs} &= \left(\frac{1}{N_{rs} - k}\right)(\mathbf{y}_{rs} - \mathbf{D}_{rs}\hat{\hat{\beta}}_{FW})'(\mathbf{y}_{sr} - \mathbf{D}_{rs}\hat{\hat{\beta}}_{FW})\end{aligned} \tag{10.8}$$

where $\hat{\Sigma}_{UW} = (\hat{\sigma}_{UW,rs})$, $\hat{\Sigma}_{FW} = (\hat{\sigma}_{FW,rs})$ and $r, s = 1, 2, \ldots, p$. Recognizing that the ML estimate of Σ for the GMANOVA model has the form

$$\hat{\Sigma} = \hat{\Sigma}_{KL} + \mathbf{W}'\left(\frac{\mathbf{Y}'\mathbf{Y}}{n} - \hat{\Sigma}_{KL}\right)\mathbf{W} \tag{10.9}$$

where $\mathbf{W} = \mathbf{I}_p - \hat{\Sigma}_{KW}^{-1}\mathbf{P}'(\mathbf{P}\hat{\Sigma}_{KL}^{-1}\mathbf{P}')^{-1}\mathbf{P}$ as shown in (8.6) for the complete data matrix $\mathbf{Y}_{n\times p}$, they suggested two other consistent missing data estimators of Σ:

$$\begin{aligned}\hat{\Sigma}^*_{FW} &= \hat{\Sigma}_{KL} + \mathbf{W}'_{FW}(\mathbf{T}_{KL} - \hat{\Sigma}_{KL})\mathbf{W}_{FW}\\ \hat{\Sigma}^*_{UW} &= \hat{\Sigma}_{KL} + \mathbf{W}'_{UW}(\mathbf{T}_{KL} - \hat{\Sigma}_{KL})\mathbf{W}_{UW}\end{aligned} \tag{10.10}$$

where

$$\begin{aligned}\mathbf{W}_{FW} &= \mathbf{I}_p - \hat{\Sigma}_{KL}\mathbf{P}'(\mathbf{P}\hat{\Sigma}_{KL}^{-1}\mathbf{P}')^{-1}\mathbf{P}\\ \mathbf{W}_{UW} &= \mathbf{I}_p - \mathbf{P}'(\mathbf{P}\mathbf{P}')\mathbf{P}\\ \mathbf{T}_{KL} &= (\mathbf{y}'_{rs}\mathbf{y}_{rs} / N_{rs}) \qquad r, s = 1, 2, \ldots, p.\end{aligned} \tag{10.11}$$

While Mensah et al. (1993) investigated only one covariance matrix and two missing data patterns (10% and 40%), they found that the estimators $\hat{\Sigma}_{KL}$ and $\hat{\Sigma}^*_{\mathrm{UW}}$ were better than $\hat{\Sigma}_{FW}$, $\hat{\Sigma}^*_{\mathrm{FW}}$, and $\hat{\Sigma}_{UW}$, although both tended to underestimate Σ. The poorest estimator was $\hat{\Sigma}_{UW}$, with $\hat{\Sigma}_{FW}$, and $\hat{\Sigma}^*_{FW}$ overestimating Σ. Tests of hypotheses were also compared, with $\hat{\Sigma}_{KL}$ and $\hat{\Sigma}^*_{UW}$ yielding consistent results with some gains for the unweighted estimator over Kleinbaum's estimate. With 40% or more data missing, asymptotic tests using Wald statistics were found to be very unreliable and should not be used.

10.3 An ML Procedure

The FGLS methods fit a model using the incomplete data and between-subject variation for j patterns of missing data for the data matrix $\mathbf{Y}_{n\times p}$. An alternative approach is to develop a model that represents within-subject variation and to use maximum likelihood methods to estimate Σ_i, a submatrix of Σ; this is the approach taken by Jennrich and Schluchter (1986) and Wolfinger et al. (1994). They considered the n_i responses of each subject $\mathbf{y}_{n_i\times 1}$ where the model for the vector of responses is

$$\mathbf{y}_i = \mathbf{X}_i\beta + \mathbf{e}_i \qquad i = 1, 2, \ldots, n \tag{10.12}$$

where $\mathbf{X}_i(n_i \times p)$ is a known matrix, $\beta_{p\times 1}$ is a vector of unknown parameters, and $\mathbf{e}_i \sim IN(\mathbf{0}, \Sigma_i)$. In addition, the structure of Σ_i is assumed to be a known function of q unknown covariance parameters of the general form

$$\Sigma_i(\theta) = \mathbf{Z}_i\mathbf{D}\mathbf{Z}'_i + \Psi_i \tag{10.13}$$

contained in a vector θ.

To estimate the elements θ_i of θ and β using the method of maximum likelihood, recall that the normal density for $\mathbf{y}_i$ is

$$f(\mathbf{y}_i) = (2\pi)^{-n_i/2}|\Sigma_i|^{-1/2}\exp\left[-\frac{1}{2}(\mathbf{y}_i - \mathbf{X}_i\beta)'\Sigma_i^{-1}(\mathbf{y}_i - \mathbf{X}_i\beta)\right] \tag{10.14}$$

so that the ln-likelihood function is

$$L^*_{ML} = -\frac{1}{2}\sum_{i=1}^{n} n_i \ln(2\pi) - \frac{1}{2}\sum_{i=1}^{n} |\Sigma_i| + \sum_{i=1}^{n} (\mathbf{y}_i - \mathbf{X}_i\beta)'\Sigma_i^{-1}(\mathbf{y}_i - \mathbf{X}_i\beta) \tag{10.15}$$

for obtaining ML estimates of $\Sigma_i(\theta)$ and β. To obtain restricted maximum likelihood (REML) estimates, the OLSE of β is given by

$$\hat{\beta} = \left(\sum_{i=1}^{n} \mathbf{X}_i'\mathbf{X}_i\right)^{-1}\left(\sum_{i=1}^{n} \mathbf{X}_i\mathbf{y}_i\right) \tag{10.16}$$

so that the least squares residuals are $\mathbf{r}_i = \mathbf{y}_i - \mathbf{X}_i\hat{\beta}$, $i = 1, 2, \ldots, n$. The ln-likelihood of the residuals is

$$L^*_{REML} = \frac{-(N-p)}{2}\ln(2\pi) - \frac{1}{2}\sum_{i=1}^{n} \ln|\Sigma_i| - \frac{1}{2}\ln\left|\sum_{i=1}^{n} \mathbf{X}_i'\Sigma_i^{-1}\mathbf{X}_i\right| - \frac{1}{2}\sum_{i=1}^{n} \mathbf{r}_i'\Sigma_i^{-1}\mathbf{r}_i \tag{10.17}$$

where $N = \sum_{i=1}^{n} n_i$, Searle et al. (1992, p. 325). If $\hat{\theta}$ is the ML or REML estimate of θ and $\hat{\Sigma}_i = \Sigma_i(\hat{\theta})$, then the FGLS estimate of β is

$$\hat{\hat{\beta}} = \left(\sum_{i=1}^{n} \mathbf{X}_i'\hat{\Sigma}_i^{-1}\mathbf{X}_i\right)^{-1}\left(\sum_{i=1}^{n} \mathbf{X}'\hat{\Sigma}_i^{-1}\mathbf{y}_i\right) \tag{10.18}$$

The algorithms for computing ML estimates (Newton-Raphson, Fisher scoring, and hybrid scoring/generalized EM) are described in Jennich and Schluchter (1986). The ML procedure used in PROC MIXED is described in Wolfinger et al. (1994). In calculating the estimator of β given in (10.18), standard errors of the elements of β are obtained using the inverse of Fisher's expected information matrix

$$\text{var}\,\hat{\hat{\beta}} = \left(\sum_{i=1}^{n} \mathbf{X}_i'\hat{\Sigma}_i^{-1}\mathbf{X}_i\right)^{-1} \tag{10.19}$$

To test hypotheses regarding the elements of β, we again may use the Wald statistic. If we want to test $H(\mathbf{C}\beta = \xi_o)$:

$$X^2 = (\mathbf{C}\hat{\hat{\beta}} - \xi_o)'\left(\mathbf{C}\left(\sum_{i=1}^{n} \mathbf{X}_i'\hat{\Sigma}_i^{-1}\mathbf{X}_i\right)^{-1}\mathbf{C}'\right)^{-1}(\mathbf{C}\hat{\hat{\beta}} - \xi_o) \tag{10.20}$$

has a chi-square distribution under the null hypothesis for large n. Alternatively, an F-distribution approximation may be used.

10.4 Repeated Measures Analysis

In the analysis of repeated measures data, one frequently has data missing in varying degrees at the end of a series of repeated measurements. In clinical experiments, missing data may be due to death or exhaustion. To illustrate the analysis of such data, we utilize the data analyzed by Cole and Grizzle (1966) and illustrated in BMDP Statistical Software, Inc. (1992, p. 1314). The study involves the analysis of the drugs morphine and trimethophon, each at two levels of histamine release, and their effect on hypertension in dogs. To eliminate or minimize the positive association between means and variances, the logarithm of blood histamine levels (ug./ml.) at four time points was used in the analysis. The data are incomplete since the sixth dog has a missing observation.

To analyze these data, Cole and Grizzle (1966) performed a multivariate analysis with unstructured Σ since by the Box (1950) test, they rejected the hypothesis that Σ satisfied compound symmetry structure. In Program 10_4.sas, we fit two models to the data. The first assumes compound symmetry and the second fits an unstructured matrix. To compare the results with the BMDP 5V output, we have selected the METHOD = ML option, maximum likelihood with scoring.

Program 10_4.sas

```
/* Program 10_4.sas                                          */
/* Repeated Measures Missing Data (Cole & Grizzle,1966) */

options ls=80 ps=60 nodate nonumber;
title1 'Output 10.4: Repeated Measures Analysis with Missing Data';

data cole;
   infile 'c:\10_4.dat';
   missing m;
   input dog drug y1 y2 y3 y4;
   y=log(y1); time= 1; output;
   y=log(y2); time= 2; output;
   y=log(y3); time= 3; output;
   y=log(y4); time= 4; output;
proc print data=cole;
run;
proc mixed data=Cole method=ml scoring;
   title2 'Mixed Model Missing Data Compound Symmetry';
   class drug time dog;
   model y=drug time drug*time/s;
   repeated/type=cs subject=dog r;
run;
proc mixed data=cole method=ml scoring;
   title2 'Mixed Model Missing Data Unstructured Covariance Matrix';
   class drug time dog;
   model y=drug time drug*time/s;
   repeated/type=un subject=dog r;
run;
data cole;
   infile 'c:\10_4.dat';
   missing m;
   input dog drug y1 y2 y3 y4;
   y1=log(y1);
   y2=log(y2);
   y3=log(y3);
   y4=log(y4);
proc print data=cole;
run;

proc glm;
   title2 ' MANOVA with Listwise Deletion ';
   class drug;
   model y1-y4=drug/intercept nouni;
   means drug;
   /*Multivariate test of group differences for mean vectors */
   manova h = drug/printe printh;
   /*Multivariate test of Conditions given Parallel Profiles */
   /*Multivariate test of Parallelism (Drug x Trial Interaction) */
```

```
    manova h =_all_ m=(1 -1  0  0,
                       0  1 -1  0,
                       0  0  1 -1) prefix = diff/printe printh;
    /*Test of group differences given Parallelism */
    contrast 'Drug Diff' drug 1 0 0 -1,
                         drug 0 1 0 -1,
                         drug 0 0 1 -1;
    manova m=(1 1 1 1 ) prefix=Drug/printe printh;
run;

data cole;
    title2 '';
    infile 'c:\10_4.mea';
    input dog drug y1 y2 y3 y4;
    y1=log(y1);
    y2=log(y2);
    y3=log(y3);
    y4=log(y4);
proc print data=cole;
run;

proc glm;
    title2 'MANOVA with Mean Substitution';
    class drug;
    model y1-y4=drug/intercept nouni;
    means drug;
    /*Multivariate test of group differences for mean vectors */
    manova h = drug/printe printh;
    /*Multivariate test of Conditions given Parallel Profiles */
    /*Multivariate test of Parallelism (Drug x Trial Interaction) */
    manova h =_all_ m=(1 -1  0  0,
                       0  1 -1  0,
                       0  0  1 -1) prefix = diff/printe printh;
    /*Test of group differences given Parallelism */
    contrast 'Drug Diff' drug 1 0 0 -1,
                         drug 0 1 0 -1,
                         drug 0 0 1 -1;
   manova m=(1 1 1 1 ) prefix=Drug/printe printh;
run;

proc glm;
    title2 'MANOVA Analysis by Cole and Grizzle with mean substitution';
    class drug;
    model y2-y4=drug/intercept nouni;
    /*Multivariate tests performed by Cole and Grizzle */
    /*Multivariate test of Conditions given Parallel Profiles */
    /*Multivariate test of Parallelism (Drug x Trial Interaction) */
    manova h =_all_ m=(1  0 -1,
                       0  1 -1) prefix = diff/printe printh;
    /*Test of group differences given Parallelism */
    contrast 'Drug Diff' drug 1 0 0 -1,
                         drug 0 1 0 -1,
                         drug 0 0 1 -1;
    manova m=(1 1 1 ) prefix=Drug/printe printh;
run;
```

Result and Interpretation 10_4.sas

The output for the SAS code in Program 10_4.sas follows.

Output 10.4: *Repeated Measures Analysis with Missing Data*

Output 10.4: Repeated Measures Analysis with Missing Data

OBS	DOG	DRUG	Y1	Y2	Y3	Y4	Y	TIME
1	1	1	0.04	0.20	0.10	0.08	-3.21888	1
2	1	1	0.04	0.20	0.10	0.08	-1.60944	2
3	1	1	0.04	0.20	0.10	0.08	-2.30259	3
4	1	1	0.04	0.20	0.10	0.08	-2.52573	4
5	2	1	0.02	0.06	0.02	0.02	-3.91202	1
6	2	1	0.02	0.06	0.02	0.02	-2.81341	2
7	2	1	0.02	0.06	0.02	0.02	-3.91202	3
8	2	1	0.02	0.06	0.02	0.02	-3.91202	4
9	3	1	0.07	1.40	0.48	0.24	-2.65926	1
10	3	1	0.07	1.40	0.48	0.24	0.33647	2
11	3	1	0.07	1.40	0.48	0.24	-0.73397	3
12	3	1	0.07	1.40	0.48	0.24	-1.42712	4
13	4	1	0.17	0.57	0.35	0.24	-1.77196	1
14	4	1	0.17	0.57	0.35	0.24	-0.56212	2
15	4	1	0.17	0.57	0.35	0.24	-1.04982	3
16	4	1	0.17	0.57	0.35	0.24	-1.42712	4
17	5	2	0.10	0.09	0.13	0.14	-2.30259	1
18	5	2	0.10	0.09	0.13	0.14	-2.40795	2
19	5	2	0.10	0.09	0.13	0.14	-2.04022	3
20	5	2	0.10	0.09	0.13	0.14	-1.96611	4
21	6	2	0.12	0.11	0.10	M	-2.12026	1
22	6	2	0.12	0.11	0.10	M	-2.20727	2
23	6	2	0.12	0.11	0.10	M	-2.30259	3
24	6	2	0.12	0.11	0.10	M	.	4
25	7	2	0.07	0.07	0.07	0.07	-2.65926	1
26	7	2	0.07	0.07	0.07	0.07	-2.65926	2
27	7	2	0.07	0.07	0.07	0.07	-2.65926	3
28	7	2	0.07	0.07	0.07	0.07	-2.65926	4
29	8	2	0.05	0.07	0.06	0.07	-2.99573	1
30	8	2	0.05	0.07	0.06	0.07	-2.65926	2
31	8	2	0.05	0.07	0.06	0.07	-2.81341	3
32	8	2	0.05	0.07	0.06	0.07	-2.65926	4
33	9	3	0.03	0.62	0.31	0.22	-3.50656	1
34	9	3	0.03	0.62	0.31	0.22	-0.47804	2
35	9	3	0.03	0.62	0.31	0.22	-1.17118	3
36	9	3	0.03	0.62	0.31	0.22	-1.51413	4
37	10	3	0.03	1.05	0.73	0.60	-3.50656	1
38	10	3	0.03	1.05	0.73	0.60	0.04879	2
39	10	3	0.03	1.05	0.73	0.60	-0.31471	3
40	10	3	0.03	1.05	0.73	0.60	-0.51083	4
41	11	3	0.07	0.83	1.07	0.80	-2.65926	1
42	11	3	0.07	0.83	1.07	0.80	-0.18633	2
43	11	3	0.07	0.83	1.07	0.80	0.06766	3
44	11	3	0.07	0.83	1.07	0.80	-0.22314	4
45	12	3	0.09	3.13	2.06	1.23	-2.40795	1
46	12	3	0.09	3.13	2.06	1.23	1.14103	2
47	12	3	0.09	3.13	2.06	1.23	0.72271	3
48	12	3	0.09	3.13	2.06	1.23	0.20701	4
49	13	4	0.10	0.09	0.09	0.08	-2.30259	1
50	13	4	0.10	0.09	0.09	0.08	-2.40795	2
51	13	4	0.10	0.09	0.09	0.08	-2.40795	3
52	13	4	0.10	0.09	0.09	0.08	-2.52573	4
53	14	4	0.08	0.09	0.09	0.10	-2.52573	1

Output 10.4 (*continued*)

```
  54    14     4    0.08    0.09    0.09    0.10    -2.40795    2
  55    14     4    0.08    0.09    0.09    0.10    -2.40795    3
  56    14     4    0.08    0.09    0.09    0.10    -2.30259    4
  57    15     4    0.13    0.10    0.12    0.12    -2.04022    1
  58    15     4    0.13    0.10    0.12    0.12    -2.30259    2
  59    15     4    0.13    0.10    0.12    0.12    -2.12026    3
  60    15     4    0.13    0.10    0.12    0.12    -2.12026    4
  61    16     4    0.06    0.05    0.05    0.05    -2.81341    1
  62    16     4    0.06    0.05    0.05    0.05    -2.99573    2
  63    16     4    0.06    0.05    0.05    0.05    -2.99573    3
  64    16     4    0.06    0.05    0.05    0.05    -2.99573    4

             Mixed Model Missing Data Compound Symmetry

                       The MIXED Procedure

                     Class Level Information

          Class      Levels  Values

          DRUG            4  1 2 3 4
          TIME            4  1 2 3 4
          DOG            16  1 2 3 4 5 6 7 8 9 10 11 12 13
                             14 15 16


                ML Estimation Iteration History

         Iteration  Evaluations     Objective     Criterion

                 0            1    9.87704142
                 1            2  -68.20768840    0.00000133
                 2            1  -68.20773371    0.00000000

               Scoring stopped after iteration 1.

                   Convergence criteria met.

                     R Matrix for DOG 1

     Row           COL1           COL2           COL3           COL4

       1     0.42613472     0.37294347     0.37294347     0.37294347
       2     0.37294347     0.42613472     0.37294347     0.37294347
       3     0.37294347     0.37294347     0.42613472     0.37294347
       4     0.37294347     0.37294347     0.37294347     0.42613472

            Covariance Parameter Estimates (MLE)

            Cov Parm   Subject       Estimate

            CS         DOG         0.37294347
            Residual               0.05319125
```

Output 10.4 (*continued*)

Model Fitting Information for Y

Description	Value
Observations	63.0000
Log Likelihood	-23.7893
Akaike's Information Criterion	-25.7893
Schwarz's Bayesian Criterion	-27.9324
-2 Log Likelihood	47.5785
Null Model LRT Chi-Square	78.0848
Null Model LRT DF	1.0000
Null Model LRT P-Value	0.0000

Mixed Model Missing Data Compound Symmetry

Solution for Fixed Effects

Effect	DRUG	TIME	Estimate	Std Error	DF	t	Pr > |t|
INTERCEPT			-2.48607739	0.32639498	12	-7.62	0.0001
DRUG	1		0.16308130	0.46159220	12	0.35	0.7300
DRUG	2		0.14554692	0.46787866	12	0.31	0.7611
DRUG	3		1.97580670	0.46159220	12	4.28	0.0011
DRUG	4		0.00000000	.	.	.	.
TIME		1	0.06559107	0.16308165	35	0.40	0.6900
TIME		2	-0.04247476	0.16308165	35	-0.26	0.7960
TIME		3	0.00310563	0.16308165	35	0.02	0.9849
TIME		4	0.00000000	.	.	.	.
DRUG*TIME	1	1	-0.63312390	0.23063228	35	-2.75	0.0095
DRUG*TIME	1	2	1.20334702	0.23063228	35	5.22	0.0001
DRUG*TIME	1	3	0.32029061	0.23063228	35	1.39	0.1737
DRUG*TIME	1	4	0.00000000	.	.	.	.
DRUG*TIME	2	1	-0.24452083	0.24296982	35	-1.01	0.3211
DRUG*TIME	2	2	-0.10042992	0.24296982	35	-0.41	0.6819
DRUG*TIME	2	3	-0.11644433	0.24296982	35	-0.48	0.6347
DRUG*TIME	2	4	0.00000000	.	.	.	.
DRUG*TIME	3	1	-2.57540074	0.23063228	35	-11.17	0.0001
DRUG*TIME	3	2	0.68410989	0.23063228	35	2.97	0.0054
DRUG*TIME	3	3	0.33328278	0.23063228	35	1.45	0.1573
DRUG*TIME	3	4	0.00000000	.	.	.	.
DRUG*TIME	4	1	0.00000000	.	.	.	.
DRUG*TIME	4	2	0.00000000	.	.	.	.
DRUG*TIME	4	3	0.00000000	.	.	.	.
DRUG*TIME	4	4	0.00000000	.	.	.	

Tests of Fixed Effects

Source	NDF	DDF	Type III F	Pr > F
DRUG	3	12	5.74	0.0113
TIME	3	35	80.39	0.0001
DRUG*TIME	9	35	38.57	0.0001

Output 10.4 *(continued)*

```
          Mixed Model Missing Data Unstructured Covariance Matrix

                            The MIXED Procedure

                          Class Level Information

     Class      Levels  Values

     DRUG            4  1 2 3 4
     TIME            4  1 2 3 4
     DOG            16  1 2 3 4 5 6 7 8 9 10 11 12 13
                        14 15 16

                    ML Estimation Iteration History

     Iteration  Evaluations     Objective     Criterion

             0            1    9.87704142
             1            2  -119.6281009    0.00468945
             2            1  -119.9829926    0.00025433
             3            1  -119.9980008    0.00000122
             4            1  -119.9980745    0.00000000

                   Scoring stopped after iteration 1.

                       Convergence criteria met.

                           R Matrix for DOG 1

Row          COL1           COL2           COL3           COL4

  1    0.26318158     0.26233653     0.33051079     0.28237391
  2    0.26233653     0.46708622     0.48249216     0.39464417
  3    0.33051079     0.48249216     0.55630621     0.47102999
  4    0.28237391     0.39464417     0.47102999     0.40840597

              Covariance Parameter Estimates (MLE)

                  Cov Parm  Subject      Estimate

                  UN(1,1)   DOG        0.26318158
                  UN(2,1)   DOG        0.26233653
                  UN(2,2)   DOG        0.46708622
                  UN(3,1)   DOG        0.33051079
                  UN(3,2)   DOG        0.48249216
                  UN(3,3)   DOG        0.55630621
                  UN(4,1)   DOG        0.28237391
                  UN(4,2)   DOG        0.39464417
                  UN(4,3)   DOG        0.47102999
                  UN(4,4)   DOG        0.40840597

          Mixed Model Missing Data Unstructured Covariance Matrix

                    Model Fitting Information for Y

                 Description                        Value

                 Observations                     63.0000
                 Log Likelihood                    2.1059
                 Akaike's Information Criterion   -7.8941
                 Schwarz's Bayesian Criterion    -18.6098
                 -2 Log Likelihood                -4.2118
                 Null Model LRT Chi-Square       129.8751
```

Output 10.4 (*continued*)

```
                  Null Model LRT DF                      9.0000
                  Null Model LRT P-Value                 0.0000

                        Solution for Fixed Effects

  Effect      DRUG  TIME      Estimate      Std Error     DF        t  Pr > |t|

  INTERCEPT                -2.48607739    0.31953324     12    -7.78    0.0001
  DRUG         1            0.16308130    0.45188824     12     0.36    0.7245
  DRUG         2            0.07646932    0.45237261     12     0.17    0.8686
  DRUG         3            1.97580670    0.45188824     12     4.37    0.0009
  DRUG         4            0.00000000             .      .        .         .
  TIME               1      0.06559107    0.16343173     12     0.40    0.6952
  TIME               2     -0.04247476    0.14680246     12    -0.29    0.7773
  TIME               3      0.00310563    0.07525325     12     0.04    0.9678
  TIME               4      0.00000000             .      .        .         .
  DRUG*TIME    1     1     -0.63312390    0.23112737     12    -2.74    0.0180
  DRUG*TIME    1     2      1.20334702    0.20761003     12     5.80    0.0001
  DRUG*TIME    1     3      0.32029061    0.10642416     12     3.01    0.0109
  DRUG*TIME    1     4      0.00000000             .      .        .         .
  DRUG*TIME    2     1     -0.17544324    0.23207296     12    -0.76    0.4642
  DRUG*TIME    2     2     -0.03135233    0.20866222     12    -0.15    0.8831
  DRUG*TIME    2     3     -0.04736673    0.10846243     12    -0.44    0.6701
  DRUG*TIME    2     4      0.00000000             .      .        .         .
  DRUG*TIME    3     1     -2.57540074    0.23112737     12   -11.14    0.0001
  DRUG*TIME    3     2      0.68410989    0.20761003     12     3.30    0.0064
  DRUG*TIME    3     3      0.33328278    0.10642416     12     3.13    0.0087
  DRUG*TIME    3     4      0.00000000             .      .        .         .
  DRUG*TIME    4     1      0.00000000             .      .        .         .
  DRUG*TIME    4     2      0.00000000             .      .        .         .
  DRUG*TIME    4     3      0.00000000             .      .        .         .
  DRUG*TIME    4     4      0.00000000             .      .        .

                          Tests of Fixed Effects

                 Source       NDF   DDF   Type III F  Pr > F

                 DRUG           3    12         5.84  0.0107
                 TIME           3    12        40.38  0.0001
                 DRUG*TIME      9    12        22.92  0.0001

         Mixed Model Missing Data Unstructured Covariance Matrix

      OBS     DOG     DRUG        Y1           Y2           Y3           Y4

        1       1       1     -3.21888     -1.60944     -2.30259     -2.52573
        2       2       1     -3.91202     -2.81341     -3.91202     -3.91202
        3       3       1     -2.65926      0.33647     -0.73397     -1.42712
        4       4       1     -1.77196     -0.56212     -1.04982     -1.42712
        5       5       2     -2.30259     -2.40795     -2.04022     -1.96611
        6       6       2     -2.12026     -2.20727     -2.30259       .
        7       7       2     -2.65926     -2.65926     -2.65926     -2.65926
        8       8       2     -2.99573     -2.65926     -2.81341     -2.65926
        9       9       3     -3.50656     -0.47804     -1.17118     -1.51413
       10      10       3     -3.50656      0.04879     -0.31471     -0.51083
       11      11       3     -2.65926     -0.18633      0.06766     -0.22314
       12      12       3     -2.40795      1.14103      0.72271      0.20701
       13      13       4     -2.30259     -2.40795     -2.40795     -2.52573
       14      14       4     -2.52573     -2.40795     -2.40795     -2.30259
       15      15       4     -2.04022     -2.30259     -2.12026     -2.12026
       16      16       4     -2.81341     -2.99573     -2.99573     -2.99573
```

Output 10.4 (*continued*)

```
                         MANOVA with Listwise Deletion

                         General Linear Models Procedure
                        Multivariate Analysis of Variance

                    H = Contrast SS&CP Matrix for Drug Diff

                                                DRUG1

                              DRUG1       107.3998649

           Characteristic Roots and Vectors of: E Inverse * H, where
        H = Contrast SS&CP Matrix for Drug Diff    E = Error SS&CP Matrix

                Variables have been transformed by the M Matrix

         Characteristic    Percent        Characteristic Vector  V'EV=1
              Root
                                                    DRUG1

             1.10465752     100.00            0.10141728


                Manova Test Criteria and Exact F Statistics for
                 the Hypothesis of no Overall Drug Diff Effect
            on the variables defined by the M Matrix Transformation
        H = Contrast SS&CP Matrix for Drug Diff    E = Error SS&CP Matrix

                              S=1    M=0.5    N=4.5

  Statistic                     Value          F       Num DF    Den DF  Pr > F

  Wilks' Lambda               0.47513669     4.0504         3        11  0.0364
  Pillai's Trace              0.52486331     4.0504         3        11  0.0364
  Hotelling-Lawley Trace      1.10465752     4.0504         3        11  0.0364
  Roy's Greatest Root         1.10465752     4.0504         3        11  0.0364

                         MANOVA with Mean Substitution

                         General Linear Models Procedure
                        Multivariate Analysis of Variance

                    H = Contrast SS&CP Matrix for Drug Diff

                                                DRUG1
                              DRUG1      107.74922308

           Characteristic Roots and Vectors of: E Inverse * H, where
        H = Contrast SS&CP Matrix for Drug Diff    E = Error SS&CP Matrix

                Variables have been transformed by the M Matrix

         Characteristic    Percent        Characteristic Vector  V'EV=1
              Root
                                                    DRUG1

             1.09758341     100.00            0.10092801
```

Output 10.4 (*continued*)

```
              Manova Test Criteria and Exact F Statistics for
               the Hypothesis of no Overall Drug Diff Effect
          on the variables defined by the M Matrix Transformation
      H = Contrast SS&CP Matrix for Drug Diff    E = Error SS&CP Matrix

                            S=1    M=0.5    N=5

Statistic                   Value         F      Num DF   Den DF  Pr > F

Wilks' Lambda             0.47673909    4.3903        3       12  0.0264
Pillai's Trace            0.52326091    4.3903        3       12  0.0264
Hotelling-Lawley Trace    1.09758341    4.3903        3       12  0.0264
Roy's Greatest Root       1.09758341    4.3903        3       12  0.0264

       MANOVA Analysis by Cole and Grizzle with mean substitution

                     General Linear Models Procedure
                    Multivariate Analysis of Variance

                 H = Contrast SS&CP Matrix for Drug Diff

                                           DRUG1

                         DRUG1        126.4874739

          Characteristic Roots and Vectors of: E Inverse * H, where
      H = Contrast SS&CP Matrix for Drug Diff    E = Error SS&CP Matrix

              Variables have been transformed by the M Matrix

       Characteristic    Percent        Characteristic Vector  V'EV=1
            Root
                                                   DRUG1

          1.91663005     100.00            0.12309641

              Manova Test Criteria and Exact F Statistics for
               the Hypothesis of no Overall Drug Diff Effect
          on the variables defined by the M Matrix Transformation
      H = Contrast SS&CP Matrix for Drug Diff    E = Error SS&CP Matrix

                            S=1    M=0.5    N=5

Statistic                   Value         F      Num DF   Den DF  Pr > F

Wilks' Lambda             0.34286145    7.6665        3       12  0.0040
Pillai's Trace            0.65713855    7.6665        3       12  0.0040
Hotelling-Lawley Trace    1.91663005    7.6665        3       12  0.0040
Roy's Greatest Root       1.91663005    7.6665        3       12  0.0040
```

To test the hypothesis that Σ has compound symmetry structure versus the alternative that Σ is unstructured, we use the LR test. In Program 10_4.sas we first perform the analysis using PROC MIXED with the TYPE = CS option in the REPEATED statement and then do the analysis a second time with TYPE = UN. The LR test is then constructed by taking two times the difference in the log-likelihoods for the two models, from Output 10.4, $2(2.1059-(-23.7893))= 51.79$. Under the null hypothesis this statistic is asymptotically distributed as a chi-square with degrees of freedom $10-2 = 8$, the difference in covariance components estimated under the two models. Since this is significant, we conclude that Σ does not have compound symmetry. Thus, an analysis with unstructured Σ is appropriate. We may use either a mixed model analysis employing PROC MIXED to estimate Σ, or a MANOVA analysis using PROC GLM; both are shown in Program 10_4.sas with results in Output 10.4.

For the mixed model analysis, we estimated the unknown covariance matrix by obtaining an ML estimate for Σ so that the results may be compared to the output generated by BMDP 5V. Using all the observations, we performed approximate tests of drug effects (summing over time), time differences, and the test of interaction. The test for drug differences was found to be significant, p-value = 0.0170. The other tests were also significant (p-value < 0.001).

For the MANOVA analysis using PROC GLM, SAS performs a listwise deletion, discarding the observation vector involving the sixth dog. An alternative approach is to substitute for the missing data value the mean value of 0.09 or the imputed value of the missing data point, using the EM algorithum as obtained from BMPD 5V, 0.12524; both approaches yield similar results. For differences in drugs (summing over time), the MANOVA analysis results in the difference being nonsignificant. The p-value is 0.0364 with listwise deletion and 0.0264 with mean substitution. The remaining tests are all significant at the level 0.0001 using either listwise deletion or mean substitution. The multivariate test of differences in mean vectors is also significant for either approach.

Cole and Grizzle (1966) did not analyze all time points. In order to compare their results with the univariate analysis which assumed the compound symmetry structure for Σ, they analyzed the last three repeated measurements taken at 1-, 3-, and 5-minute intervals, removing the control time point, $t = 0$. Substituting the mean value of 0.09 for the missing data point, we obtain the results given in Output 10.4. The test of drug combinations show an F-value of $F = 7.67$ and p-value of 0.004. The tests of interaction and time of observation are also significant. These results are consistent with those reported by Cole and Grizzle, although it is not clear how they dealt with the missing data point.

10.5 Repeated Measures with Changing Covariates

To illustrate how to analyze repeated measures data with changing covariates and data missing completely at random, we reanalyze the Winer data previously analyzed in Section 9.5. This example also appears in BMDP Statistical Software, Inc. (1992, p. 1325). The SAS code to perform the analysis is given in Program 10_5.sas.

Program 10_5.sas

```
/* Program 10_5.sas                                    */
/* Mixed Model - Winer(1971, page 806) Missing Data */

options ls=80 ps=60 nodate nonumber;
title1 'Output 10.5: Repeated Measures w/changing covariates, Missing Data';

data winer;
   missing m;
   input person a b x y @@;
   cards;
   1 1 1  3  m 1 1 2  4 14
   2 1 1  5 11 2 1 2  9 18
   3 1 1  m  m 3 1 2 14 22
   4 m 1  2  6 4 m 2  1  8
   5 2 1  8 12 5 2 2  9 14
```

```
   6 2 1 10  9 6 2 2  9 10
   7 3 1  7 10 7 3 2  4 10
   8 3 1  8  m 8 3 2 10  m
   9 3 1  9 15 9 3 2 12 22
   ;
proc print data=winer;
   title2 'Winer Data';
run;

proc mixed data=winer method=ml scoring;
   title2 'Compound Symmetry';
   class person a b;
   model y = a b a*b x/p;
   repeated/type=cs subject=person r;
run;

proc mixed data=winer method=ml scoring;
   title2 'Unknown Structure';
   class person a b;
   model y = a b a*b x;
   repeated /type=un subject=person r;
run;
```

Result and Interpretation 10_5.sas

The output for Program 10_5.sas follows.

Output 10.5: *Repeated Measures w/changing covariates, Missing Data-Winer Data*

```
        Output 10.5: Repeated Measures w/changing covariates, Missing Data
                                   Winer Data

                    OBS    PERSON    A    B     X     Y

                     1        1      1    1     3     M
                     2        1      1    2     4    14
                     3        2      1    1     5    11
                     4        2      1    2     9    18
                     5        3      1    1     M     M
                     6        3      1    2    14    22
                     7        4      M    1     2     6
                     8        4      M    2     1     8
                     9        5      2    1     8    12
                    10        5      2    2     9    14
                    11        6      2    1    10     9
                    12        6      2    2     9    10
                    13        7      3    1     7    10
                    14        7      3    2     4    10
                    15        8      3    1     8     M
                    16        8      3    2    10     M
                    17        9      3    1     9    15
                    18        9      3    2    12    22

                           Compound Symmetry
                          The MIXED Procedure
                        Class Level Information

                     Class      Levels  Values

                     PERSON          8  1 2 3 5 6 7 8 9
                     A               3  1 2 3
                     B               2  1 2
```

Output 10.5 (*continued*)

```
                  ML Estimation Iteration History

         Iteration  Evaluations     Objective     Criterion

                 0            1   24.78133943
                 1            2   10.21917718    0.09128850
                 2            1    9.99454581    0.00788482
                 3            1    9.96361510    0.00052312
                 4            1    9.96091129    0.00000182
                 5            1    9.96090223    0.00000000

                  Scoring stopped after iteration 1.
                      Convergence criteria met.
                        R Matrix for PERSON 1

                         Row            COL1

                           1    2.77125864

                 Covariance Parameter Estimates (MLE)

                 Cov Parm   Subject       Estimate

                 CS         PERSON      2.69025541
                 Residual               0.08100323

                   Model Fitting Information for Y

               Description                         Value

               Observations                      12.0000
               Log Likelihood                   -16.0077
               Akaike's Information Criterion   -18.0077
               Schwarz's Bayesian Criterion     -18.4926
               -2 Log Likelihood                 32.0154
               Null Model LRT Chi-Square         14.8204
               Null Model LRT DF                  1.0000
               Null Model LRT P-Value             0.0001

                         Compound Symmetry
                      Tests of Fixed Effects

              Source       NDF   DDF  Type III F  Pr > F

              A              2     4        6.46  0.0558
              B              1     1      137.15  0.0542
              A*B            2     1       12.41  0.1968
              X              1     1      169.33  0.0488

                          Predicted Values

            Y  Predicted   SE Pred        L95        U95  Residual

            M    10.1877    0.6123     8.6136    11.7618         .
      14.0000    12.6199    1.0463     9.9304    15.3095    1.3801
      11.0000    11.0000    1.0150     8.3909    13.6091    0.0000
      18.0000    18.0000    0.9611    15.5294    20.4706    0.0000
            M          .         .          .          .         .
      22.0000    23.3801    1.0463    20.6905    26.0696   -1.3801
       6.0000          .         .          .          .         .
       8.0000          .         .          .          .         .
      12.0000     9.4240    1.1800     6.3906    12.4573    2.5760
      14.0000    12.0000    1.1771     8.9741    15.0259    2.0000
       9.0000    11.5760    1.1800     8.5427    14.6094   -2.5760
```

Output 10.5 (*continued*)

```
10.0000    12.0000    1.1771     8.9741   15.0259   -2.0000
10.0000    11.4240    1.1800     8.3906   14.4573   -1.4240
10.0000    11.6960    1.2227     8.5529   14.8390   -1.6960
      M    12.5000    2.0388     7.2590   17.7410         .
      M    18.1520    2.0455    12.8938   23.4103         .
15.0000    13.5760    1.1800    10.5427   16.6094    1.4240
22.0000    20.3040    1.2227    17.1610   23.4471    1.6960


                 Unknown Structure
                The MIXED Procedure

               Class Level Information

           Class      Levels  Values

           PERSON          8  1 2 3 5 6 7 8 9
           A               3  1 2 3
           B               2  1 2

             ML Estimation Iteration History

      Iteration  Evaluations      Objective      Criterion

              0            1   24.78133943
              1            2    4.17986515     0.73162569
              2            1   -0.70631906   95986.655944
              3            1   -0.87576500   1872680.3346
              4            1   -1.52147535   146010.70131
              5            1   -1.96968520   7182.5655266
              6            2   -3.94850050   47855.855810
              7            2   -5.90170396   106087.87688
              8            1   -7.89442801   20226830.526
              9            1  -12.40108345   18345035.668
             10            1  -13.48393467    26.46904978
             11            1  -15.43534069     1.74672916
             12            1  -17.71817155   8749951574.9
             13            1  -22.21576659    16404767423
             14            1  -26.15905960    50937139389
             15            1  -28.74217406   6.4183125E12
             16            1  -32.86407940   5.9580848E14
             17            1  -42.25762949   1.3117028E14
             18            2  -46.28970444   1.4087425E14
             19            1  -50.11757998   3.6237801E14
             20            2  -53.25175492    7.101318E14
             21            1  -57.10055159   9.5089453E15
             22            1  -60.73043807   1.8907594E16
             23            1  -64.23475381   5.6455532E16
             24            1  -65.36807583   2.2107988E19
             25            1  -74.67988044   1.6614383E19
             26            1  -79.41838989    5.565372E19
             27            2  -82.79663602   8.8578573E19
             28            1  -86.27084485    2.610296E20
             29            1  -88.54768976    6.519874E22
             30            1  -94.93985587   8.3681072E21
             31            1  -100.7095506   1.2588179E23
             32            1  -104.7025474   5.6102002E21
             33            1  -106.7167737   4.1011257E17
             34            1  -107.9293680   4.8910023E20
             35            1  -108.7362194   2.6831033E20
             36            1  -109.0417054   1.8439782E20
             37            1  -109.2737075   1.4923035E20
             38            1  -109.3186457   1.3282412E20
```

Output 10.5 (*continued*)

```
                39            4  -109.3199022  1.3069504E20
                40            3  -109.3209079  1.2997262E20

                         Unknown Structure
                  ML Estimation Iteration History

         Iteration  Evaluations     Objective     Criterion

                41            9  -109.3341917   1.317253E20
                42            3  -109.3560569  1.2668326E20
                43            7  -109.4244810  1.2500308E20
                44           39  -109.4244810  1.2500308E20
                45           39  -109.4244810  1.2500308E20

                 Scoring stopped after iteration 1.

        Stopped because of too many likelihood evaluations.

                     Covariance Parameters at
                        Last MLE Iteration

                  Cov Parm  Subject       Estimate

                  UN(1,1)   PERSON      1.58927045
                  UN(2,1)   PERSON      1.14304179
                  UN(2,2)   PERSON      0.82210333
```

SAS reports that it used 12 observations in the analysis, 10 from the subjects with no missing data, and 2 from the subjects with only 1 observation. While the SAS and BMDP 5V output are in agreement for the ML estimates, SAS does not output the imputed values for the missing data, but does provide predicted values.

Because of the missing data, we may fit a model only under the assumption of compound symmetry. The more appropriate model with unknown covariance structure using all the data as obtained in Section 9.5 does not converge.

10.6 Random Coefficient Model

For our next example, data provided by Lindsey (1993, p. 223) are utilized to show how to fit a linear regression model with both a random intercept and slope, and with missing data. The data are the cumulative number of operating hours for successive failures of air-conditioning equipment in 13 Boeing 720 aircraft. The SAS code to perform the analyses is provided in Program 10_6.sas.

Program 10_6.sas

```
/* Program 10_6.sas                                       */
/* Random Coefficient - Lindsey(1993, page 222) Missing Data */

options ls=80 ps=60 nodate nonumber;
title1 'Output 10.6: Random Coefficient Model with Missing Data';

data lindsey;
   missing m;
   infile 'c:\10_6.dat';
```

```
   input group y1  y2  y3  y4  y5  y6  y7  y8  y9 y10 y11 y12 y13 y14 y15
         y16 y17 y18 y19 y20 y21 y22 y23 y24 y25 y26 y27 y28 y29 y30
         aircraft;
    y=y1 ; fail= 1; output; y=y2 ; fail= 2; output; y=y3 ; fail= 3; output;
   y=y4 ; fail= 4; output; y=y5 ; fail= 5; output; y=y6 ; fail= 6; output;
   y=y7 ; fail= 7; output; y=y8 ; fail= 8; output; y=y9 ; fail= 9; output;
   y=y10; fail=10; output; y=y11; fail=11; output; y=y12; fail=12; output;
   y=y13; fail=13; output; y=y14; fail=14; output; y=y15; fail=15; output;
   y=y16; fail=16; output; y=y17; fail=17; output; y=y18; fail=18; output;
   y=y19; fail=19; output; y=y20; fail=20; output; y=y21; fail=21; output;
   y=y22; fail=22; output; y=y23; fail=23; output; y=y24; fail=24; output;
   y=y25; fail=25; output; y=y26; fail=26; output; y=y27; fail=28; output;
   y=y28; fail=28; output; y=y29; fail=29; output; y=y30; fail=30; output;
proc summary data=lindsey;
   class aircraft;
   var y1-y30;
   output out=new mean=mr1-mr30;
run;
proc print data=new;
   data plot;
   set new;
   array mr(30) mr1-mr30;
   do failures = 1 to 30;
      hours = mr(failures);
      output;
   end;
   drop mr1-mr30;
proc plot;
   plot hours*failures=aircraft;
run;
proc mixed data=lindsey;
   title2 'Random Intercept and Slope';
   class aircraft;
   model y = fail / s p;
   random int fail/type=un subject=aircraft s;
run;
```

Result and Interpretation 10_6.sas

The output for Program 10_6.sas follows.

Output 10.6: *Random Coefficient Model with Missing Data*

```
             Output 10.6: Random Coefficient Model with Missing Data


             Plot of HOURS*FAILURES.  Symbol is value of AIRCRAFT.

    HOURS ,
          ,
     2500 ^
          ,                                                            3
          ,
          ,
          ,
          ,                                              2 2         3
          ,                                          2 2           3
          ,                                        2             3 7   .
     2000 ^                                      2
          ,                                2 2 2                   . .
          ,                            2 2                     7 7 .
          ,                    9       5 5 4                     3         .
          ,                          2                   . . 8 .
          ,                        5 5                 . 7 7 .           6
          ,                      5                   . 7 3 3     6 6 6
          ,                        2               . 3 3       6
     1500 ^                    5 2       4 1       3   6 6 6 6
          ,                            4 1       . 7 6
          ,                9 9         1   .   6 6 6
          ,                  5 2   4 4 . .     .
          ,            9 9       4     6 6 6 .
          ,                      1 6 .         7 3
          ,                        . 1     7 7   8
          ,                5     .
     1000 ^              5               7 8 3 3
          ,                    . 6 1   8 8
          ,                  2 6 1   8     3
          ,                4 . 4   8   7 3
          ,            5 4     1 8   3 3
          ,            4 2 .     7 7 7
          ,      1 1 9 2 .   6 7   3
          ,          5 .   6 7 3 3
      500 ^    1 1 2 2 6 1 3 3
          ,    2 2 5 . 3 3 7
          ,    9 5 6 3       8
          ,      . .   1 7 8
          ,        1 1
          ,    . 1 3 4
          ,    7 3
          ,    3
        0 ^    6
          ,
          Šff^fffffffff^fffffffff^fffffffff^fffffffff^fffffffff^fffffffff^ff
             0         5        10        15        20        25        30

                                      FAILURES

NOTE: 177 obs had missing values.  61 obs hidden.
```

Output 10.6 *(continued)*

```
                        The MIXED Procedure

                      Class Level Information

          Class     Levels  Values

          AIRCRAFT      13  1 2 3 4 5 6 7 8 9 10 11 12 13


                 REML Estimation Iteration History

         Iteration  Evaluations      Objective      Criterion

                 0            1  2636.4783575
                 1            2  2384.1391407     0.01075037
                 2            1  2366.8441760     0.00807787
                 3            1  2354.0357159     0.00559378
                 4            1  2345.2931662     0.00349224
                 5            1  2339.9186694     0.00189957
                 6            1  2337.0503808     0.00083933
                 7            1  2335.8193238     0.00025601
                 8            1  2335.4624843     0.00003692
                 9            1  2335.4150642     0.00000108
                10            1  2335.4137825     0.00000000

                     Convergence criteria met.


               Covariance Parameter Estimates (REML)

               Cov Parm    Subject       Estimate

               UN(1,1)     AIRCRAFT  25191.786140
               UN(2,1)     AIRCRAFT  -781.7098408
               UN(2,2)     AIRCRAFT  1314.2478630
               Residual              15494.574860


                  Model Fitting Information for Y

              Description                          Value

              Observations                      213.0000
              Res Log Likelihood                -1361.60
              Akaike's Information Criterion    -1365.60
              Schwarz's Bayesian Criterion      -1372.31
              -2 Res Log Likelihood             2723.206
              Null Model LRT Chi-Square         301.0646
              Null Model LRT DF                   3.0000
              Null Model LRT P-Value              0.0000


                    Solution for Fixed Effects

     Effect          Estimate     Std Error     DF         t   Pr > |t|

     INTERCEPT    67.07523844   48.50818237     12      1.38     0.1919
     FAIL         99.66745084   10.97461554     12      9.08     0.0001
```

Output 10.6 (*continued*)

Solution for Random Effects

Effect	AIRCRAFT	Estimate	SE Pred	DF	t	Pr > |t|
INTERCEPT	1	-16.40306510	91.23940963	187	-0.18	0.8575
FAIL	1	-30.40188311	22.54425394	187	-1.35	0.1791
INTERCEPT	2	202.35126419	66.84725691	187	3.03	0.0028
FAIL	2	-2.85059030	11.46222543	187	-0.25	0.8039
INTERCEPT	3	-176.1353365	63.53588714	187	-2.77	0.0061
FAIL	3	-21.36681247	11.21785969	187	-1.90	0.0584
INTERCEPT	4	-139.1383857	74.18683505	187	-1.88	0.0623
FAIL	4	18.99966588	12.59542966	187	1.51	0.1331
INTERCEPT	5	2.41057993	75.52411087	187	0.03	0.9746
FAIL	5	40.32927850	12.92087997	187	3.12	0.0021
INTERCEPT	6	211.35503845	63.11825756	187	3.35	0.0010
FAIL	6	-45.14925734	11.19476412	187	-4.03	0.0001
INTERCEPT	7	-156.6339903	64.44311667	187	-2.43	0.0160
FAIL	7	-20.78358732	11.27430210	187	-1.84	0.0668
INTERCEPT	8	-150.9138337	65.62212566	187	-2.30	0.0226
FAIL	8	-23.83232852	11.35665791	187	-2.10	0.0372
INTERCEPT	9	21.94990178	84.51541053	187	0.26	0.7954
FAIL	9	76.87403771	16.56796076	187	4.64	0.0001
INTERCEPT	10	-30.34466470	91.23940963	187	-0.33	0.7398
FAIL	10	10.60300228	22.54425394	187	0.47	0.6387
INTERCEPT	11	94.86218560	95.24773663	187	1.00	0.3206
FAIL	11	20.95618799	34.99811777	187	0.60	0.5500
INTERCEPT	12	222.37495392	78.62202189	187	2.83	0.0052
FAIL	12	-19.13689313	13.85898173	187	-1.38	0.1690
INTERCEPT	13	-85.73464781	74.18683505	187	-1.16	0.2493
FAIL	13	-4.24082018	12.59542966	187	-0.34	0.7367

Tests of Fixed Effects

Source	NDF	DDF	Type III F	Pr > F
FAIL	1	12	82.48	0.0001

For this example, we have fit a linear model in which both the slope and intercept are random. Furthermore, we have specified the covariance matrix for the random slope and intercept effects to be unstructured, in order to estimate the correlation between the random coefficients. We see from the Output 10.6 that the correlation is negative and that the variance associated with the slope parameter is significant at the $\alpha = 0.05$ level, $p = 0.0357$.

The fixed effects parameters in the model represent the estimated means for the random intercept and slope parameters. Thus, the average number of hours to failure is given by the linear model:

$$y = 79.74 + 96.23t \qquad t = 1, 2, \ldots, 30$$

However, performance of the air conditioners for each aircraft vary about this line. The random effects estimates show the estimated deviation from the mean slope and intercept. Observe that the intercept and slope for plane six is most variable. Thus, one may want to inspect this air conditioner. All models indicate differences among the aircraft. Alternative nonlinear models for the analysis of these data are discussed by Lindsey (1993, p. 224).

10.7 Growth Curve Analysis

We now illustrate Kleinbaum's approach to the missing data problem. From (10.1), recall that Kleinbaum associates a linear model with each missing data pattern. To illustrate, suppose we have five observations over four time points where (D) denotes a data point and (M) represents a missing element. To fix our ideas, suppose we have the following patterns:

	Time				
Subject	1	2	3	4	Pattern
1	D	D	D	D	1
2	D	D	D	D	1
3	D	D	M	D	2
4	D	M	D	D	3
5	D	M	M	D	4

Hence, we have four data patterns. Now for each pattern of missing data, we formulate a model of the form

$$\begin{aligned} E(\mathbf{Y}_j) &= \mathbf{X}_j\mathbf{BPA}_j \\ \text{cov}(\mathbf{Y}_j) &= \mathbf{I}_{n_j} \otimes \mathbf{A}_j'\mathbf{\Sigma}\mathbf{A}_j \qquad j = 1, 2, \ldots, m \end{aligned} \tag{10.21}$$

Where m is the number of patterns of missing data, n_j is the number of subjects with the jth pattern, p_j is the number of nonmissing elements in the jth pattern, $N = \sum_{j=1}^{m} n_j p_j$ is the number of nonmissing observations. $\mathbf{Y}_j(n_j \times p_j)$ is the data matrix for the jth pattern, $\mathbf{X}_j(n_j \times k)$ is the design matrix for the jth pattern, and $\mathbf{A}_j(p \times p_j)$ is a matrix of 0's and 1's that represent the jth pattern

For our example, the $\mathbf{A}_j$'s for $j = 1, 2, 3, 4$ are

$$\mathbf{A}_1 = \begin{pmatrix} 1 & 0 & 0 & 0 \\ 0 & 1 & 0 & 0 \\ 0 & 0 & 1 & 0 \\ 0 & 0 & 0 & 1 \end{pmatrix} = \mathbf{I}_4 \qquad \mathbf{A}_2 = \begin{pmatrix} 1 & 0 & 0 \\ 0 & 1 & 0 \\ 0 & 0 & 0 \\ 0 & 0 & 1 \end{pmatrix}$$
$$\mathbf{A}_3 = \begin{pmatrix} 1 & 0 & 0 \\ 0 & 0 & 0 \\ 0 & 1 & 0 \\ 0 & 0 & 1 \end{pmatrix} \qquad \mathbf{A}_4 = \begin{pmatrix} 1 & 0 \\ 0 & 0 \\ 0 & 0 \\ 0 & 1 \end{pmatrix} \tag{10.22}$$

Using (10.3), we see that the FGLS estimate of $\beta = \text{vec}(\mathbf{B})$ is

$$\hat{\hat{\beta}} = (\mathbf{D}'\hat{\Omega}^{-1}\mathbf{D})^{-1}\mathbf{D}'\hat{\Omega}^{-1}\mathbf{y} \tag{10.23}$$

where $\mathbf{D}$ and $\hat{\mathbf{\Omega}}$ are defined in (10.2), with any consistent estimator $\hat{\mathbf{\Sigma}}$ replacing $\mathbf{\Sigma}$ in $\hat{\mathbf{\Omega}}$.

Although this approach is not iterative, an iterative process may be developed to estimate $\hat{\hat{\beta}}$. To define the process, we construct $\hat{\hat{\beta}}$ as defined in (10.4) and call this estimate $\hat{\hat{\beta}}_1$. Then we define

$$\hat{\Omega}_1 = (\mathbf{y} - \mathbf{D}\hat{\hat{\beta}}_1)(\mathbf{y} - \mathbf{D}\hat{\hat{\beta}}_1)' \tag{10.24}$$

and substitute this value into (10.23) to construct $\hat{\hat{\beta}}_2$. This process continues until some convergence criterion is met. This process is equivalent to Zellner's iterative process proposed for the SUR model. Under normality the estimator converges to the ML estimate.

To test hypotheses, we use the Wald statistic defined in (10.5). Critical to the Kleinbaum procedure is the process employed to estimate Σ when one does not utilize the iterative process. The initial estimate of Σ is less critical when one employs iterations. While considerable research has been done on how to estimate Σ, most procedures result in a $\hat{\Sigma}$ that is not positive definite. To make the matrix positive semidefinite, we use the smoothing procedure suggested by Bock and Peterson (1975). They recommend obtaining a spectral decomposition of $\hat{\Sigma}$ so that

$$\hat{\Sigma} = \mathbf{P}\Lambda\mathbf{P}' \tag{10.25}$$

where Λ is the diagonal matrix of eigenvalues and $\mathbf{P}$ is an orthogonal matrix of eigenvectors of $\hat{\Sigma}$. The smoothed estimator of the matrix Σ is

$$\hat{\Sigma}^* = \mathbf{P}\Lambda^*\mathbf{P}' \tag{10.26}$$

where Λ^* is the matrix Λ with all negative eigenvalues set to zero. Under multivariate normality, $\hat{\Sigma}^*$ is the maximum likelihood estimate under the constraint that Σ is positive semidefinite (Bock and Peterson, 1975).

To illustrate Kleinbaum's procedure to analyze growth curve data, we use the Grizzle and Allen (1969) dog data with the missing data patterns used by Kleinbaum (1970). Kleinbaum used two patterns of incomplete data and one complete data pattern. The complete data pattern included dogs 1, 4, and 5 (Group 1); 10, 11, 14, 17, and 18 (Group 2); 22 and 26 (Group 3); and 31, 33, 35, and 36 (Group 4). Hence, the complete data matrix is the data matrix $\mathbf{Y}_i(14 \times 7)$ with associated matrices $\mathbf{X}_1$ and $\mathbf{A}_1$ defined

$$\underset{14\times4}{\mathbf{X}_1} = \begin{pmatrix} \mathbf{1}_3 & \cdots & \cdots & \mathbf{0} \\ \vdots & \mathbf{1}_5 & \cdots & \vdots \\ \vdots & \cdots & \mathbf{1}_2 & \vdots \\ \mathbf{0} & \cdots & \cdots & \mathbf{1}_4 \end{pmatrix} \text{ and } \mathbf{A}_1 = \mathbf{I}_7 \tag{10.27}$$

For the first missing data pattern, Kleinbaum removed the first observation on dogs 2, 5, and 9 (Group 1); 12, 16, and 19 (Group 2); 20, 23, and 25 (Group 3); and 28, 29, and 34 (Group 4). Hence the data matrix $\mathbf{Y}_2(12 \times 6)$ has one fewer column than $\mathbf{Y}_1$ and $\mathbf{X}_2$ and $\mathbf{A}_2$ are

$$\underset{12\times4}{\mathbf{X}_2} = \begin{pmatrix} \mathbf{1}_3 & \cdots & \cdots & \mathbf{0} \\ \vdots & \mathbf{1}_3 & \cdots & \vdots \\ \vdots & \cdots & \mathbf{1}_3 & \vdots \\ \mathbf{0} & \cdots & \cdots & \mathbf{1}_3 \end{pmatrix} \text{ and } \underset{7\times6}{\mathbf{A}_2} = \begin{pmatrix} \mathbf{0}' \\ \mathbf{I}_6 \end{pmatrix} \tag{10.28}$$

For the next missing data pattern, Kleinbaum removed the first three observations for the remaining dogs. Dogs 3, 6, and 8 (Group 1); 13 and 14 (Group 2); 21, 24, and 27 (Group 3); and 30 and 32 (Group 4); hence, $\mathbf{Y}_3$ is a (10×4) matrix, and $\mathbf{X}_3$ and $\mathbf{A}_3$ are defined

$$\underset{10\times 4}{\mathbf{X}_3} = \begin{pmatrix} \mathbf{1}_3 & \cdots & \cdots & \mathbf{0} \\ \vdots & \mathbf{1}_2 & \cdots & \vdots \\ \vdots & \cdots & \mathbf{1}_3 & \vdots \\ \mathbf{0} & \cdots & \cdots & \mathbf{1}_2 \end{pmatrix} \text{ and } \mathbf{A}_3 = \begin{pmatrix} \underset{3\times 4}{\mathbf{0}'} \\ \mathbf{I}_4 \end{pmatrix} \tag{10.29}$$

In Section 7.7, we analyzed these data using the growth curve model, fitting orthogonalized polynomials. Weighting by $\mathbf{S}^{-1}$ and using orthogonal polynomials that are not normalized and multiplying $\mathbf{B}$ by $\mathbf{PP}'$, matrix of regression coefficients are

$$\Xi = \begin{pmatrix} 32.701 & 4.135 & -4.680 & -1.142 \\ 25.071 & -0.687 & -0.995 & 0.110 \\ 28.585 & 4.101 & -2.402 & -0.480 \\ 27.116 & 1.908 & -1.549 & 0.004 \end{pmatrix} \tag{10.30}$$

These are the same coefficients reported by Grizzle and Allen (1969). Program 10_7.sas contains the SAS code to analyze these data with missing data.

Program 10_7.sas

```
/* Program 10_7.sas                                           */
/* Growth Curve Analysis/Mixed Model (Grizzle and Allen,1969) */

options ls=80 ps=60 nodate nonumber;
title1 'Output 10.7:  Growth Curve Analyis - Missing Data';

data klein;
   missing m;
   infile 'c:\10_7.dat';
   input treat y1 y2 y3 y4 y5 y6 y7 dog;
   y=y1; time= 1; x1=-3; x2= 5; x3=-1; output;
   y=y2; time= 3; x1=-2; x2= 0; x3= 1; output;
   y=y3; time= 5; x1=-1; x2=-3; x3= 1; output;
   y=y4; time= 7; x1= 0; x2=-4; x3= 0; output;
   y=y5; time= 9; x1= 1; x2=-3; x3=-1; output;
   y=y6; time=11; x1= 2; x2= 0; x3=-1; output;
   y=y7; time=13; x1= 3; x2= 5; x3= 1; output;
   drop y1- y7;
proc print data=klein;
run;
proc mixed data=klein method=reml;
   title2 'Using PROC MIXED, Unknown Covariance Structure';
   class treat dog;
   model y=treat x1 x2 x3/ s p;
   repeated/type=un subject=dog r;
   lsmeans treat;
run;
proc mixed data=klein method=reml;
   title2 'Using PROC MIXED, Random Coefficients';
   class treat dog;
   model y=treat x1 x2 x3/s p;
   random int x1 x2 x3/type=un subject=dog g s;
run;
```

```
data klein1;
   infile 'c:\10_7.da1';
   input pattern y1 y2 y3 y4 y5 y6 y7 dog x1 x2 x3 x4;
proc print data=klein1;
   title2 'Missing Pattern 1';
run;
data klein2;
   infile 'c:\10_7.da2';
   input pattern y2 y3 y4 y5 y6 y7 dog x1 x2 x3 x4;
proc print data=klein2;
   title2 'Missing Pattern 2';
run;
data klein3;
   infile 'c:\10_7.da3';
   input pattern y4 y5 y6 y7 dog x1 x2 x3 x4;
proc print data=klein3;
   title2 'Misssing Pattern 3';
run;
data kleinf;
   infile 'c:\7_7.dat';
   input group y1 y2 y3 y4 y5 y6 y7 dog x1 x2 x3 x4;
run;
data klein0;
   infile 'c:\10_7.da4';
   input pattern y1 y2 y3 y4 y5 y6 y7 dog x1 x2 x3 x4;
run;
proc iml;
   use klein1;
   read all var {y1 y2 y3 y4 y5 y6 y7} into y_1;
   read all var {x1 x2 x3 x4} into x_1;
   use klein2;
   read all var {y2 y3 y4 y5 y6 y7} into y_2;
   read all var {x1 x2 x3 x4} into x_2;
   use klein3;
   read all var {y4 y5 y6 y7} into y_3;
   read all var {x1 x2 x3 x4} into x_3;
   use kleinf;
   read all var {y1 y2 y3 y4 y5 y6 y7} into y;
   read all var {x1 x2 x3 x4 } into x;
   use klein0;
   read all var {y1 y2 y3 y4 y5 y6 y7} into y0;
   a1=i(7);
   zero=shape({0},1,6,0);
   zero3=shape({0},3,4,0);
   a2=zero//i(6);
   a3=zero3//i(4);
   print a1 a2 a3;
   n1=nrow(y_1);n2=nrow(y_2);n3=nrow(y_3);
   n=n1+n2+n3;
   k=ncol(x);
   /*Begin calculation of S */
   sall = (y`*y-y`*x*inv(x`*x)*x`*y)/(n-k);
   print 'Complete Data 36 Obs Estimate of S', sall;
   sc = (y_1`*y_1-y_1`*x_1*inv(x_1`*x_1)*x_1`*y_1)/(n1-k);
   print 'Listwise Deletion Estimate of S',sc;
   t2=y0[1:26,2:7];
   d2=x_1//x_2;s2_=(t2`*t2-t2`*d2*inv(d2`*d2)*d2`*t2)/(n1+n2-k);
   t3=y0[1:36,4:7];d3=x_1//x_2//x_3;
   st_=(t3`*t3-t3`*d3*inv(d3`*d3)*d3`*t3)/(n-k);
   s1=sc[1,];
   s2=sc[2,1]||s2_[1,];
   s3=sc[3,1]||s2_[2,];
   s4=sc[4,1]||s2_[3,1]||s2_[3,2]||st_[1,];
   s5=sc[5,1]||s2_[4,1]||s2_[4,2]||st_[2,];
```

```
s6=sc[6,1]||s2_[5,1]||s2_[5,2]||st_[3,];
s7=sc[7,1]||s2_[6,1]||s2_[6,2]||st_[4,];
s=s1//s2//s3//s4//s5//s6//s7;
print 'Consistent Estimate of S (Formula 10.6)', s;
call eigen(val,vec,s);
print 'Eigenvalues of S', val;
do i=1 to 7;
   if val[i] < 0 then val[i]=0;
end;
sm=vec*diag(val)*vec`;
print 'Smoothed Estimate of S', sm;
v=eigval(sm);
print 'Eigenvalues of Smoothed Estimate',v;
/* Matrix of Orthogonal Polynomials of degree  q-1=6 */
p={ 1  1  1  1  1  1  1,
   -3 -2 -1  0  1  2  3,
    5  0 -3 -4 -3  0  5,
   -1  1  1  0 -1 -1  1};
q=p*p`;
print 'scalar matrix p*p`',q;
/* Begin calculations of beta hat  */
beta=inv(x`*x)*x`*y*inv(sall)*p`*inv(p*inv(sall)*p`);
betat_f=beta*q;
betaga=shape(beta`,16,1);
print 'Beta based on 36 observations',beta;
print 'Grizzle and Allen Beta = beta*(pp`)',betat_f;
d=a1`*p`@x_1//a2`*p`@x_2//a3`*p`@x_3;
block1=(a1`*sc*a1)@i(n1);
block2=(a2`*sc*a2)@i(n2);block3=(a3`*sc*a3)@i(n3);
sigma=block(block1,block2,block3);
ivsigcm=inv(sigma);
vecy1=shape(y_1`,98,1);vecy2=shape(y_2`,72,1);vecy3=shape(y_3`,40,1);
vecy=vecy1//vecy2//vecy3;
betacm=inv(d`*ivsigcm*d)*d`*ivsigcm*vecy;
beta=shape(betacm,4,4)`;
betat_sc=beta*q;
print 'Beta based on 14 complete obs to estimate sigma',beta;
print betat_sc;
block1=(a1`*sm*a1)@i(n1);block2=(a2`*sm*a2)@i(n2);block3=(a3`*sm*a3)@i(n3);
sigma=block(block1,block2,block3);
ivsigsm=ginv(sigma);
betasm=ginv(d`*ivsigsm*d)*d`*ivsigsm*vecy;
beta=shape(betasm,4,4)`;
betat_sm=beta*q;
print 'Beta based on smoothed estimate of sigma',beta;
print betat_sm;
/* Tests of homogeniety using Wald Statistics */
c={1  0  0 -1,
   1  0 -1  0,
   1 -1  0  0,
   0  1 -1  0,
   0  1  0 -1,
   0  0  1 -1};
ch={0 0 0,
    0 0 0,
    0 0 0,
    0 0 0,
    0 0 0,
    0 0 0};
sga=sall@i(n);
ivga=inv(sga);
ga=p`@x;
```

```
do i= 1 to 6;
      ch[i,1]=((i(4)@c[i,])*betaga)`*inv((i(4)@c[i,])*ginv(ga`*ivga*ga)*
               (i(4)@c[i,])`)*((i(4)@c[i,])*betaga);
      ch[i,2]=((i(4)@c[i,])*betacm)`*inv((i(4)@c[i,])*ginv(d`*ivsigcm*d)*
               (i(4)@c[i,])`)*((i(4)@c[i,])*betacm);
      ch[i,3]=((i(4)@c[i,])*betasm)`*inv((i(4)@c[i,])*ginv(d`*ivsigsm*d)*
               (i(4)@c[i,])`)*((i(4)@c[i,])*betasm);
   end;
   print 'GA | Complete | Smoothed // Wald Statistics',ch;
quit;
run;
```

Result and Interpretation 10_7.sas

The output for Program 10_7.sas follows.

Output 10.7: *Growth Curve Analysis - Missing Data*

Output 10.7: Growth Curve Analysis - Missing Data

Missing Pattern 1

OBS	PATTERN	Y1	Y2	Y3	Y4	Y5	Y6	Y7	DOG	X1	X2	X3	X4
1	1	4.0	4.0	4.1	3.6	3.6	3.8	3.1	1	1	0	0	0
2	1	4.2	4.4	4.6	4.9	5.3	5.6	4.9	4	1	0	0	0
3	1	3.7	3.9	3.9	4.8	5.2	5.4	4.2	7	1	0	0	0
4	1	3.4	3.4	3.5	3.1	3.1	3.7	3.3	10	0	1	0	0
5	1	3.0	3.2	3.0	3.0	3.1	3.2	3.1	11	0	1	0	0
6	1	3.8	3.9	4.0	3.9	3.5	3.5	3.4	14	0	1	0	0
7	1	4.2	4.0	4.2	4.1	4.2	4.0	4.0	17	0	1	0	0
8	1	4.1	4.2	4.3	4.3	4.2	4.0	4.2	18	0	1	0	0
9	1	3.1	3.3	3.2	3.1	3.2	3.1	3.1	22	0	0	1	0
10	1	3.5	3.9	5.8	5.4	4.9	5.3	5.6	26	0	0	1	0
11	1	3.4	3.4	3.5	3.3	3.4	3.2	3.4	31	0	0	0	1
12	1	4.0	4.6	4.8	4.9	5.4	5.6	4.8	33	0	0	0	1
13	1	4.1	4.1	3.7	4.0	4.1	4.6	4.7	35	0	0	0	1
14	1	3.5	3.6	3.6	4.2	4.8	4.9	5.0	36	0	0	0	1

Missing Pattern 2

OBS	PATTERN	Y2	Y3	Y4	Y5	Y6	Y7	DOG	X1	X2	X3	X4
1	2	4.3	4.7	4.7	4.8	5.0	5.2	2	1	0	0	0
2	2	4.4	5.3	5.6	5.9	5.9	5.3	5	1	0	0	0
3	2	4.6	4.4	4.6	5.4	5.9	5.6	9	1	0	0	0
4	2	3.1	3.2	3.0	3.3	3.0	3.0	12	0	1	0	0
5	2	3.3	3.3	3.4	3.6	3.1	3.1	16	0	1	0	0
6	2	4.4	4.3	4.5	4.3	4.4	4.4	19	0	1	0	0
7	2	3.3	3.8	3.8	4.4	4.2	4.0	20	0	0	1	0
8	2	3.4	3.5	4.6	4.9	5.2	4.4	23	0	0	1	0
9	2	4.0	4.4	4.2	4.6	4.8	5.4	25	0	0	1	0
10	2	3.5	3.5	3.2	3.0	3.0	3.2	28	0	0	0	1
11	2	3.2	3.6	3.7	3.7	4.2	4.4	29	0	0	0	1
12	2	3.9	4.5	4.7	3.9	3.8	3.7	34	0	0	0	1

Output 10.7 (*continued*)

```
      Misssing Pattern 3

 OBS     PATTERN      Y4      Y5      Y6      Y7     DOG     X1     X2     X3     X4

   1        3        4.3     4.5     5.8     5.4       3      1      0      0      0
   2        3        5.2     5.3     4.2     4.1       6      1      0      0      0
   3        3        5.2     5.6     5.4     4.7       8      1      0      0      0
   4        3        3.2     3.3     3.1     3.1      13      0      1      0      0
   5        3        3.1     3.0     3.0     3.0      15      0      1      0      0
   6        3        3.7     3.7     3.6     3.7      21      0      0      1      0
   7        3        5.7     4.9     4.0     4.0      24      0      0      1      0
   8        3        5.0     5.4     4.4     3.9      27      0      0      1      0
   9        3        4.3     3.9     3.4     3.5      30      0      0      0      1
  10        3        4.3     3.6     3.8     3.7      32      0      0      0      1

          A1
           1          0          0          0          0          0          0
           0          1          0          0          0          0          0
           0          0          1          0          0          0          0
           0          0          0          1          0          0          0
           0          0          0          0          1          0          0
           0          0          0          0          0          1          0
           0          0          0          0          0          0          1

                A2
                 0          0          0          0          0          0
                 1          0          0          0          0          0
                 0          1          0          0          0          0
                 0          0          1          0          0          0
                 0          0          0          1          0          0
                 0          0          0          0          1          0
                 0          0          0          0          0          1


                           A3
                            0          0          0          0
                            0          0          0          0
                            0          0          0          0
                            1          0          0          0
                            0          1          0          0
                            0          0          1          0
                            0          0          0          1

                       Complete Data 36 Obs Estimate of S

         SALL
 0.2260972 0.1721597 0.1723889 0.2053958 0.1705694 0.1957986   0.18175
 0.1721597 0.1695972 0.1840139 0.1919115 0.1628403 0.1700174 0.1644635
 0.1723889 0.1840139 0.3916806 0.3473333 0.2366736 0.1875694 0.2222292
 0.2053958 0.1919115 0.3473333 0.4407109 0.3689792 0.2870573 0.2582891
 0.1705694 0.1628403 0.2366736 0.3689792 0.4337361 0.3733681 0.3177986
 0.1957986 0.1700174 0.1875694 0.2870573 0.3733681 0.5235243  0.460651
   0.18175 0.1644635 0.2222292 0.2582891 0.3177986  0.460651 0.5131415

                        Listwise Deletion Estimate of S

           SC
 0.1576667    0.1545      0.213 0.1993333    0.1735 0.1698333 0.1891667
    0.1545   0.18995      0.275    0.2744   0.25735   0.26465   0.24425
     0.213     0.275      0.592     0.542     0.472      0.51     0.507
 0.1993333    0.2744      0.542 0.6399667    0.6237 0.6514667 0.6278333
```

Output 10.7 (*continued*)

```
     0.1735   0.25735      0.472     0.6237    0.67405    0.69045    0.63175
  0.1698333   0.26465       0.51  0.6514667    0.69045  0.7882167  0.7040833
  0.1891667   0.24425      0.507  0.6278333    0.63175  0.7040833  0.7259167

               Consistent Estimate of S (Formula 10.6)

          S
  0.1576667    0.1545      0.213  0.1993333     0.1735  0.1698333  0.1891667
     0.1545 0.1785103  0.2078279  0.2108295  0.1966261  0.2187408   0.230046
      0.213 0.2078279  0.4161623  0.3674318  0.2778084   0.298276  0.3415747


                          Misssing Pattern 3

  0.1993333 0.2108295  0.3674318  0.4407109  0.3689792  0.2870573  0.2582891
     0.1735 0.1966261  0.2778084  0.3689792  0.4337361  0.3733681  0.3177986
  0.1698333 0.2187408   0.298276  0.2870573  0.3733681  0.5235243   0.460651
  0.1891667  0.230046  0.3415747  0.2582891  0.3177986   0.460651  0.5131415

                          Eigenvalues of S

                                      VAL
                                2.0459817
                                0.3112267
                                 0.181131
                                0.0722983
                                0.0370367
                                0.0164382
                                 -0.00066

                         Smoothed Estimate of S

         SM
  0.1577944 0.1543663  0.2128682  0.1994711  0.1734141  0.1698383  0.1892466
  0.1543663 0.1786502  0.2079659  0.2106853  0.1967161  0.2187356  0.2299623
  0.2128682 0.2079659  0.4162984  0.3672896  0.2778971  0.2982709  0.3414922
  0.1994711 0.2106853  0.3672896  0.4408595  0.3688865  0.2870626  0.2583753
  0.1734141 0.1967161  0.2778971  0.3688865  0.4337939  0.3733647  0.3177448
  0.1698383 0.2187356  0.2982709  0.2870626  0.3733647  0.5235245  0.4606541
  0.1892466 0.2299623  0.3414922  0.2583753  0.3177448  0.4606541  0.5131915

                  Eigenvalues of Smoothed Estimate

                                        V
                                2.0459817
                                0.3112267
                                 0.181131
                                0.0722983
                                0.0370367
                                0.0164382
                                -7.35E-18

                         scalar matrix p*p`

                            Q
                            7         0         0         0
                            0        28         0         0
                            0         0        84         0
                            0         0         0         6
```

Output 10.7 (*continued*)

```
              Beta based on 36 observations

              BETA
          4.671662 0.1477056 -0.055711  -0.19029
         3.5816015 -0.024532  -0.01185 0.0182957
         4.0835958 0.1464782 -0.028595 -0.080081
          3.873728 0.0681349 -0.018438 0.0007254

           Grizzle and Allen Beta = beta*(pp`)

           BETAT_F

                 Misssing Pattern 3

         32.701634 4.1357558  -4.67975 -1.141738
         25.071211 -0.686882 -0.995436 0.1097741
         28.585171 4.1013896 -2.401995 -0.480488
         27.116096 1.9077778 -1.548785 0.0043522

   Beta based on 14 complete obs to estimate sigma

              BETA
         4.7499693 0.1498627 -0.034782  -0.17373
         3.6136398 -0.016135 -0.004509 0.0036311
         4.0676312  0.156267 -0.066042 -0.015429
         3.8313129 0.0297061 -0.016351 0.0179139


          BETAT_SC
         33.249785 4.1961544 -2.921717  -1.04238
         25.295479 -0.451785  -0.37874 0.0217864
         28.473418 4.3754766 -5.547528 -0.092575
          26.81919 0.8317699 -1.373451 0.1074837

         Beta based on smoothed estimate of sigma

              BETA
         4.6700598 0.1347061 -0.036551 -0.197389
         3.5582634 -0.031159 -0.004527 -0.000085
         4.0190986 0.2028974 -0.053491 -0.037827
         3.9000572 0.0911746 -0.005027 -0.016121


          BETAT_SM
         32.690418 3.7717721 -3.070307 -1.184334
         24.907844 -0.872439 -0.380247 -0.000509
          28.13369 5.6811266 -4.493247 -0.226963
         27.300401 2.5528877 -0.422294 -0.096724

     GA | Complete | Smoothed // Wald Statistics

                     CH
              23.157661 31.296217 24.484018
              13.032751 34.314442  35.97691
              34.580952 30.657205 33.199976
              12.951324 15.437746 47.883905
              4.6594524 2.4945174 15.982526
              3.5415217  8.804226 13.079355
```

In Section 9.4, we analyzed these dog data using a mixed model, fitting a random coefficient growth curve model, a growth curve model with unknown structure, a growth curve model with compound symmetry structure, and a heterogeneous model under compound symmetry. Using the AIC criterion, we saw the model that best fit the data was the growth curve model with an unstructured covariance matrix. It yielded the largest AIC statistic; however, the random coefficient model also provided a reasonable fit. In Program 10_7.sas we reanalyzed the example with missing data. The results are given in Output 10.7. The growth curve model with unstructured Σ again performed better than the random coefficient model. Comparing the AIC values for the complete data and the missing data pattern, we see that the random coefficient model seems to be more robust for the missing data since AIC increased in value from -151.61 to -136.71. The statistic decreased for the unstructured model, from -129.74 to -127.31. This is due to the number of parameters that need to be estimated and data availability.

In the second part of Program 10_7.sas we analyze the data using Kleinbaum's model, using PROC IML. We first estimate Σ; the obvious choice is to use the estimate obtained from using only the complete data. While such an estimate is always positive semidefinite it may not be a reasonable estimate unless few observations are missing or an iterative procedure is employed to estimate the fixed effects. Alternatively, we may use formula (10.6) to compute $\hat{\Sigma}$ which employs all data for (r, s) pairs of variables.

For our example,

$$(N_{rs}) = \begin{pmatrix} 14 & 14 & 14 & 14 & 14 & 14 & 14 \\ 14 & 26 & 26 & 26 & 26 & 26 & 26 \\ 14 & 26 & 26 & 26 & 26 & 26 & 26 \\ 14 & 26 & 26 & 36 & 36 & 36 & 36 \\ 14 & 26 & 26 & 36 & 36 & 36 & 36 \\ 14 & 26 & 26 & 36 & 36 & 36 & 36 \\ 14 & 26 & 26 & 36 & 36 & 36 & 36 \end{pmatrix}$$

$$\mathbf{D}_{1s} = \mathbf{X}_1 \qquad \text{for } s = 1, \ldots, 7$$

$$\mathbf{D}_{rs} = \begin{pmatrix} \mathbf{X}_1 \\ \mathbf{X}_2 \end{pmatrix} \qquad \text{for } r = 2, 3 \text{ and } s = r, \ldots, 7$$

$$\mathbf{D}_{rs} = \begin{pmatrix} \mathbf{X}_1 \\ \mathbf{X}_2 \\ \mathbf{X}_3 \end{pmatrix} \qquad \text{for } r = 4, 5, 6, 7 \text{ and } s = r, \ldots, 7.$$

where $\mathbf{D}_{rs} = \mathbf{D}_{sr}$ for all r and s. Thus,

$$\hat{\sigma}_{rs} = \left(\frac{1}{N_{rs} - 4}\right) \mathbf{y}'_{rs}(\mathbf{I}_{N_{rs}} - \mathbf{D}_{rs}(\mathbf{D}'_{rs}\mathbf{D}_{rs})^{-1}\mathbf{D}'_{rs})\mathbf{y}_{rs} \tag{10.31}$$

and $r, s = 1, 2, \ldots, 7$. In Program 10_7.sas, we estimate Σ using several approaches. Because we have all 36 observations for this example, we calculate the unbiased estimate of Σ using all 36 observations. Following Kleinbaum (1970), we estimate Σ using only the 14 complete observations. This leads to a matrix that is positive definite so that an inverse exists. An alternative is to use formula (10.6). This matrix is indefinite since it contains a negative value. Setting the negative value to zero, the smoothing procedure, we may reestimate Σ with a matrix that is positive semidefinite; however, its inverse does not exist so that one must use a generalized inverse. Thus, we have three estimates for Σ: $\hat{\Sigma}_{GA}$, $\hat{\Sigma}_C$, and $\hat{\Sigma}_{SM}$ where GA denotes the Grizzle-Allen estimate, C represents the complete or listwise deletion estimate, and SM the estimate obtained by setting negative eigenvalues to zero. The three estimates are found in Output 10.7.

To find the maximum likelihood estimate for **B** using the complete model, we employ formula (7.28) since weighting by $\hat{\Sigma}_{GA}$ is equivalent to using the higher order terms as covariates in the Rao-Khatri conditional model. To estimate **B** with missing data, we use the vector version formula (10.3). The estimates $\hat{\hat{\beta}}$ (in matrix form) and **B** in the format of Grizzle and Allen (1969) are provided in Output 10.7. For the complete data, the result agrees with that provided in (10.7). For the missing data case,

$$\hat{\mathbf{B}}_C = \begin{pmatrix} 33.250 & 4.196 & -2.922 & -1.042 \\ 25.295 & -0.452 & -0.379 & -0.022 \\ 28.473 & 4.375 & -5.548 & -0.093 \\ 26.819 & 0.832 & -1.373 & 0.107 \end{pmatrix}$$
$$\hat{\mathbf{B}}_{SM} = \begin{pmatrix} 32.690 & 3.712 & -3.070 & -1.184 \\ 24.908 & -0.872 & -0.380 & -0.001 \\ 28.133 & 5.681 & -4.493 & -0.227 \\ 27.300 & 2.553 & -0.422 & -0.097 \end{pmatrix} \tag{10.32}$$

using the complete and smoothed estimates, respectively. An estimate for **B** using formula (10.6) was not calculated since this "usual" estimate leads to negative test statistics whenever a matrix is indefinite. For this example, both the complete data and the smoothed estimate provide reasonable estimates for **B**. This may not be the case with more missing data, Kleinbaum (1970).

Having found parameter estimates, we next illustrate hypothesis testing. To test hypotheses in the form $H\colon \mathbf{CBA} = \mathbf{0}$, we let $\mathbf{C}_* = \mathbf{A}' \otimes \mathbf{C}$ to transform the hypothesis to the form $H\colon \mathbf{C}_*\beta = \mathbf{0}$, where $\beta = \text{vec}(\mathbf{B})$ and use the Wald statistic given in formula (10.5) for hypothesis testing. That is

$$X^2 = (\mathbf{C}_*\hat{\hat{\beta}})'(\mathbf{C}_*(\mathbf{D}'\hat{\Omega}^{-1}\mathbf{D})^-\mathbf{C}_*')^{-1}(\mathbf{C}_*\hat{\hat{\beta}}) \tag{10.33}$$

has an asymptotic chi-square distribution with degrees of freedom $v = \text{rank}(\mathbf{C}_*)$. Kleinbaum (1970) computed Wald statistics for six hypotheses in the form $H\colon \mathbf{CB} = \mathbf{0}$. The matrices $\mathbf{C}_i$ are

$$\begin{aligned} \mathbf{C}_1 &= (1 \quad 0 \quad 0 \quad -1) & \mathbf{C}_4 &= (0 \quad 1 \quad -1 \quad 0) \\ \mathbf{C}_2 &= (1 \quad 0 \quad -1 \quad 0) & \mathbf{C}_5 &= (0 \quad 1 \quad 0 \quad -1) \\ \mathbf{C}_3 &= (1 \quad -1 \quad 0 \quad 0) & \mathbf{C}_6 &= (0 \quad 0 \quad 1 \quad -1) \end{aligned} \tag{10.34}$$

which are all pairwise comparisons of regression profiles, a test of homogeneity of profiles. The chi-square statistics for each estimate of Σ are given in Table 10.1.

Table 10.1 *Comparison of H with Various Covariance Matrices*

H	$\hat{\Sigma}_{GA}$	$\hat{\Sigma}_{C}$	$\hat{\Sigma}_{SM}$
H_1	23.16**	31.30**	24.48**
H_2	13.03**	34.31**	35.98**
H_3	34.58**	30.66**	33.20**
H_4	12.95**	15.44**	47.88**
H_5	4.66	2.49	15.98**
H_6	3.54	8.80	13.08**

**Significant at $\alpha = 0.05$

The Wald statistic for the "usual" estimator using formula (10.6) leads to negative values so it has been excluded from our analysis. For $\alpha = 0.05$, the chi-square critical value with 4 df is 9.49. For this example, the conclusions reached using $\hat{\Sigma}_C$ are in agreement with the Grizzle-Allen (GA) complete data; however, the use of Σ_{SM} finds all comparisons significant. If the estimate of Σ is not positive definite, even though $\hat{\mathbf{B}}$ may appear reasonable, the Wald tests may lead to spurious results. To use the large sample procedure recommended by Kleinbaum requires very large data sets and an estimate of Σ that is positive definite to be reliable. However, the "effective size" with missing data is unknown.

Tests of hypotheses using either ML or FGLS procedures discussed in this chapter are valid if data are missing completely at random (MCAR) (Little and Rubin, 1987). Tests for evaluating this assumption are discussed by Little (1988) and Park and Davis (1993). Missing data is a complex issue in the analysis of repeated measurements; articles by Little (1995) and Robins et al. (1995) discuss classes of models for analyzing data that are not MCAR. An excellent overview of missing data problems in the analysis of repeated measurement data is provided by Vonesh and Chinchilli (1997, pp 264-274).

Appendix 1 POWER.PRO Program

```
*                          November 1, 1992                          *;
*********************************************************************;
*23456789A123456789B123456789C123456789D123456789E123456789F123456789G;
*********************************************************************;
*  This SAS code, used inside PROC IML via an INCLUDE statement,     *;
*  performs a power analysis of multivariate linear hypotheses       *;
*  of the form C*B*U - THETA0 = 0. The associated model is Y=X*B+E.  *;
*  Users must enter PROC IML to create the following matrices.       *;
*                                                                    *;
*  Note: Vectors may be entered as rows or columns.                  *;
*                                                                    *;
*  A) REQUIRED MATRICES                                              *;
*                                                                    *;
*    (1) ESSENCEX, the essence design matrix.  Users unfamiliar with *;
*        this matrix may simply enter the full design matrix and     *;
*        specify that REPN below be equal to 1.                      *;
*    (2) SIGMA, the hypothesized covariance matrix                   *;
*    (3) BETA, the matrix of hypothesized regression coefficients    *;
*    (4) C, "between" subject contrast for pre-multiplying BETA      *;
*                                                                    *;
*  WARNING:  ESSENCEX, C, and U must be full rank.                   *;
*            U`*SIGMA*U must besymmetric and positive definite.      *;
*            For univariate repeated measures,                       *;
*            U should be proportional to an orthonormal matrix.      *;
*                                                                    *;
*  B) OPTIONAL MATRICES                                              *;
*    (1) REPN, (vector), the # of times each row of ESSENCEX is      *;
*        duplicated. Default of {1}.                                 *;
*    (2) U, "within" subject contrast for post-multiplying BETA      *;
*        Default of I(p), where p=NCOL(BETA).                        *;
*    (3) THETA0, the matrix of constants to be subtracted from       *;
*        C*BETA*U (CBU). Default matrix of zeroes.                   *;
*    (4) ALPHA,  (vector), the Type I error rate (test size).        *;
*        Default of .05.                                             *;
*    (5) SIGSCAL, (vector), multipliers of SIGMA.                    *;
*        Default of {1}.                                             *;
*    (6) RHOSCAL, (vector), multipliers of RHO (correlation matrix)  *;
*        computed internally from SIGMA).                            *;
*        Default of {1}.                                             *;
*    (7) BETASCAL, (vector), multipliers of BETA.                    *;
*        Default of {1}.                                             *;
*    (8) ROUND, (scalar) # of places to which power values will      *;
*        be rounded. Default of 3.                                   *;
*    (9) TOLERANC, (scalar)  value not tolerated, numeric zero,      *;
*        used for checking singularity, division problems, etc.      *;
*        singularity. Default of 1E-12.                              *;
*   (10) OPT_ON                                                      *;
*        OPT_OFF, (vectors), users can specify the options they wish *;
*        to turn on or off. The possible options are as follows:     *;
*                                                                    *;
*          a) Power calculations                    Default          *;
*             HLT, Hotelling-Lawley trace             off            *;
*             PBT, Pillai-Bartlett trace              off            *;
*             WLK, Wilk's Lambda                      on             *;
*             UNI, uncorrected univariate                            *;
*                  repeated measures                  off            *;
*             UNIGG, Geisser-Greenhouse               on             *;
*             UNIHF, Huynh-Feldt                      off            *;
```

```
*          *COLLAPSE, special case              on              *;
*                                                               *;
*          *With rank(U)=1 powers of all tests coincide.         *;
*           The COLLAPSE option produces one column of power    *;
*           calculations labeled "POWER" instead of separate    *;
*           columns for WLK, HLT, UNI, etc.                     *;
*                                                               *;
*           With rank(C)=1 and rank(U)>1, powers of all 3       *;
*           multivariate statistics (WLK HLT PBT) coincide.     *;
*           The COLLAPSE option produces one column for all 3.  *;
*           UNI, UNIGG, and UNIHF powers calculated if requested. *;
*                                                               *;
*           NONCEN_N, multiply noncentrality by                 *;
*                    N/(N-rk(X)) if min(rk(C),rk(U))>1          *;
*                    for multivariate tests                     *;
*                    (O'Brien & Shieh, 1992)     off            *;
*                                                               *;
*         b) Print/options for output power matrix (SAS dataset) *;
*            TOTAL_N,  total number observations  on            *;
*            SIGSCAL,  multiplier for SIGMA       on            *;
*            RHOSCAL,  multiplier for RHO         on            *;
*            BETASCAL, multiplier for BETA        on            *;
*            ALPHA,    size of test               on            *;
*            CASE,     row # of matrix            on            *;
*            MAXRHOSQ, max canonical rho squared  off           *;
*                                                               *;
*         c) Print options - additional matrices                *;
*            BETA, beta matrix                    on            *;
*            SIGMA, variance-covariance matrix    on            *;
*            RHO, correlation matrix              on            *;
*            C, "between" subjects contrast       on            *;
*            U, "within" subject contrast         on            *;
*            THETA0, constant to compare to CBU   on, if  ^= 0  *;
*            CBETAU, value of initial C*BETA*U    on            *;
*            RANKX, rank of X                     off           *;
*            RANKC, rank of C                     off           *;
*            RANKU, rank of U                     off           *;
*                                                               *;
*         d) Power data options                                 *;
*            NOPRINT, printed output suppressed   off           *;
*            DS , request output SAS dataset      off           *;
*                                                               *;
*         e) WARN, allows warning messages        on            *;
*            to print.                                          *;
*                                                               *;
*                                                               *;
*    (11) DSNAME, (1 row, 2-3 cols), name of output SAS data file. *;
*         Default WORK.PWRDT###.                                *;
*                                                               *;
* C) SAS DATA FILE                                              *;
*                                                               *;
*   If the DS option is selected, the output data file will contain *;
*   all options requested in a) and b). The user can name the data *;
*   file by defining the matrix DSNAME= { libref dataset }. For  *;
*   example, if DSNAME = {IN1 MYDATA}, the output data file will be *;
*   called IN1.MYDATA. IN1 refers to a library defined by a DD  *;
*   statement or a LIBNAME statement.                           *;
*                                                               *;
*   When using a library other than WORK as the default, define *;
*   DSNAME = {libref dataset defaultlib};                       *;
*                                                               *;
*   If DSNAME is not defined or if the "dataset" already exists in *;
*   the library specified by "libref", a default file name is used. *;
*   The default file names are numbered and of the form PWRDT### *;
```

```
*   (where ### is a number). The program scans the library for the   *;
*   largest numbered data file and assigns the next number to the new*;
*   data file. The maximum ### is 999.If PWRDT999 exists no more data*;
*   files can be created.                                             *;
*                                                                     *;
*   NOTE: The program uses the name _PWRDTMP as an intermediate       *;
*   dataset. If this dataset already exists in the specified library *;
*   no datasets can be created.                                       *;
*                                                                     *;
**********************************************************************;

                  *****MODULE DEFINITIONS*****;

**NOTE: The modules used by the MAIN CODE are defined first.        ***;

**********************************************************************
*                          _INPTCHK                                  *
*                                                                    *
*  This module performs a number of checks on the user-supplied data. *
**********************************************************************;

START _INPTCHK(CHECKER,_R, _A, _B, _S)
      GLOBAL (ESSENCEX, BETA, SIGMA, C, U, THETA0, TOLERANC, SIGSCAL);

   *_______________________________________________________________*
    INPUTS
    1) User or program supplied -GLOBAL
       ESSENCEX, the essence design matrix.
       SIGMA, the hypothesized variance-covariance matrix
       BETA, the matrix of hypothesized regression coefficients
       U, the matrix post-multiplying BETA
       C, the matrix pre-multiplying BETA
       THETA0, the matrix of constants to be subtracted from CBU
       TOLERANC, value not tolerated, numeric zero
       SIGSCAL, vector of scaling values for covariance matrix

    2) Program supplied
       _R, rank of design matrix, X
       _A, rank of "between" subject contrast matrix, C
       _B, rank of "within" subject contrast matrix, U
       _S, minimum of _A and _B
    OUTPUTS
       CHECKER=0 if no error, 1 if error present
   *_______________________________________________________________*;


   CHECKER = 0;
   IF MIN(VECDIAG(SIGMA))<=TOLERANC THEN DO;
      CHECKER=1;
      PRINT "At least one variance <=TOLERANC. Check SIGMA.";
      PRINT SIGMA;
      END;
   IF (NROW(U) ^= NROW(SIGMA)) THEN DO;
      CHECKER = 1;
      PRINT "Number of rows U not equal to number of rows of SIGMA.";
      PRINT U,,,,, SIGMA;
      END;
   IF (NCOL(C) ^= NROW(BETA)) THEN DO;
      CHECKER = 1;
      PRINT "# of columns of C not equal to number of rows of BETA.";
      PRINT C,,,,, BETA;
      END;
   IF (NCOL(BETA) ^= NROW(U)) THEN DO;
      CHECKER = 1;
```

```
      PRINT "# of columns of BETA not equal to number of rows of U.";
      PRINT BETA,,,,, U;
      END;
   IF (_R ^= NROW(BETA)) THEN DO;
      CHECKER = 1;
      PRINT "# of columns of X not equal to number of rows of BETA.";
      PRINT ESSENCEX,,,,, BETA;
      END;
   IF (_A > NCOL(C)) THEN DO;
      CHECKER = 1;
      PRINT "# of rows of C greater than number of columns of C.";
      PRINT C;
      END;
   IF (_B > NROW(U)) THEN DO;
      CHECKER = 1;
      PRINT "# of columns of U greater than number of rows of U.";
      PRINT U;
      END;
   IF (NROW(THETA0) > 0) THEN IF (NROW(THETA0) ^= _A) THEN DO;
      CHECKER = 1;
      PRINT "The THETA0 matrix does not conform to CBU.";
      PRINT THETA0;
      PRINT "#Rows of CBU = " _A;
      END;
   IF (NROW(THETA0) > 0) THEN IF (NCOL(THETA0) ^= _B) THEN DO;
      CHECKER = 1;
      PRINT "The THETA0 matrix does not conform to CBU.";
      PRINT THETA0;
      PRINT "#Cols of CBU = " _B;
      END;
   IF MIN(SIGSCAL)<=TOLERANC THEN DO;
      CHECKER = 1;
      PRINT "smallest value in SIGSCAL <= TOLERANC (too small)";
      PRINT SIGSCAL TOLERANC;
      END;

FINISH; *_INPTCHK;

*******************************************************************;
*                    _OPTCHK                                      *
*                                                                 *
* Check to see if selected options are valid requests.            *
*******************************************************************;

START _OPTCHK(CHECKER,
             _PG1LABL, _PWRLABL, _PRTLABL, _DATLABL,_WRNLABL,_NONLABL,
             _OPT_ON, _OPT_OFF);

   *_______________________________________________________________*
    INPUTS
    1) Program supplied
       _PG1LABL, (column) options for printing matrices
       _PWRLABL, (column) options for power calculations
       _PRTLABL, (column) options for output power matrix
       _DATLABL, (column) options for power data output
       _WRNLABL, (column) options for turning on warning messages
       _OPT_ON, (column) selected options to turn on
       _OPT_OFF, (column) selected options to turn off
    OUTPUTS
       CHECKER=0 if no error, 1 if error present
   *_______________________________________________________________*;

   CHECKER=0;
   IF (( NROW(_OPT_ON) > 0) | (NROW(_OPT_OFF) > 0 )) THEN DO;
```

```
      _SELECT = _PG1LABL //_PWRLABL //_PRTLABL//_DATLABL//_WRNLABL
                                                    //_NONLABL;
      _OPT    = _OPT_ON // _OPT_OFF;
      _ERR = SETDIF(_OPT,_SELECT);
      IF NROW(_ERR) ^= 0 THEN DO;
        PRINT _ERR ":Invalid options requested in OPT_ON or OPT_OFF.";
        CHECKER=1;
        END;
   END;

FINISH;  *_OPTCHK;

***********************************************************************
*                            _SNGRCHK                                 *
*                                                                     *
*This module checks matrices for singularity. Use for the following:  *
*            - ESSENCEX`*ESSENCEX                                     *
*            - U`*SIGMA*U                                             *
*            - C*INV(X`X)*C`                                          *
**********************************************************************;


START _SNGRCHK
             (matrix,name)      GLOBAL (TOLERANC);
   *_______________________________________________________________*
    INPUTS
      matrix, matrix which will be checked
      name, label to identify the matrix

      TOLERANC,  value not tolerated, numeric zero (global)

    OUTPUTS
       _error, =1 if matrix is not symmetric or positive definite
   *_______________________________________________________________*;

   _error=0;

   IF ALL(MATRIX=.) THEN DO;
      _error=1;
      NOTE0_1 = { " (labeled MATRIX below) ",
                  " is all missing values."};
      PRINT name NOTE0_1 , matrix;
      RETURN(_error);
      END;

    _MAXVAL=MAX(ABS(MATRIX));

   IF _MAXVAL<=TOLERANC THEN DO;
      _error=1;
      NOTE0_2= { " (labeled MATRIX below) ",
                 " has max(abs(all elements)) <= TOLERANC."};
      PRINT name NOTE0_2 , matrix;
      RETURN(_error);
      END;

   _NMATRIX=MATRIX / _MAXVAL;

   ** SQRT IN NEXT LINE DUE TO NUMERICAL INACCURACY IN SSCP
      CALCULATION WITH SOME DATA VIA SWEEP;

   IF MAX(ABS(_NMATRIX-_NMATRIX`))>=SQRT(TOLERANC) THEN DO;
      _error=1;
      NOTE1 = { " (labeled MATRIX below) ",
                " is not symmetric, within sqrt(TOLERANC)."};
```

```
      PRINT name NOTE1 , matrix;
      RETURN(_error);
      END;

      _EVALS=EIGVAL(_NMATRIX);

   IF MIN(_EVALS) <= SQRT(TOLERANC) THEN DO;
      _error=1;
      NOTE2={" (labeled MATRIX below) ",
             " is not positive definite. This may",
             " happen due to programming error or rounding",
             " error of nearly LTFR matrix. Perhaps can fix",
             " by usual scaling/centering techniques.",
             " This version disallows LTFR designs.",
             " Eigenvalues/max(abs(original matrix)) are:" };
      PRINT name NOTE2 , _EVALS;
      NOTE2_1={" is the max(abs(original matrix))"} ;
      PRINT _MAXVAL ;
      RETURN(_error);
      END;

RETURN(_error);
FINISH; * _SNGRCHK;


***********************************************************************
*                           _SIZECHK                                  *
*                                                                     *
*This module checks matrices for having more than one column          *
*                                                                     *
***********************************************************************;

START _SIZECHK
          (matrix,name);
   *_______________________________________________________________*
    INPUTS
      matrix, matrix which will be checked
      name, label to identify the matrix

      OUTPUTS
       _error, =1 if matrix has more than one column
   *_______________________________________________________________*;

   _error=0;

   IF NCOL(matrix)>1 THEN DO;
      _error=1;
      NOTE2_2 = { " (labeled MATRIX below) ",
                  " has more than one row and more than one column.",
                  "MATRIX is the transpose of your input." };
      PRINT name NOTE2_2 , matrix ;
      RETURN(_error);
      END;

RETURN(_error);
FINISH; * _SIZECHK;

***********************************************************************
*                           _TYPECHK                                  *
*                                                                     *
*This module verifies that matrix is of required type,                *
* either character or numeric,                                        *
*                                                                     *
***********************************************************************;
```

```
START _TYPECHK
            (matrix,name,target);
  *____________________________________________________________*
    INPUTS
      matrix, matrix which will be checked
      name, label to identify the matrix
      target, one character, "N" (numeric) or "C" (character)
    OUTPUTS
       _error, =1 if matrix has more than one column
  *____________________________________________________________*;

  _error=0;

  IF target="N" THEN DO;
                  IF TYPE(matrix)^=target THEN DO;
                                             _error=1;
           NOTE2_3={" (labeled MATRIX below) must be numeric."};
                                 PRINT name NOTE2_3 , matrix ;
                                             RETURN(_error);
                                             END;
                  END;

 IF target="C" THEN DO;
                  IF TYPE(matrix)^=target THEN DO;
                                             _error=1;
           NOTE2_3={" (labeled MATRIX below) must be character"};
                                 PRINT name NOTE2_3 , matrix ;
                                             RETURN(_error);
                                             END;
                  END;

RETURN(_error);
FINISH; * _TYPECHK;


**********************************************************************
*                              _SETOPT                               *
*                                                                    *
*    This module sets the options requested by the user.            *
*                                                                    *
*THIS MODULE MUST BE CALLED FOR EACH OF THESE LABEL MATRICES:        *
*                                                                    *
*  _PG1LABL= { BETA SIGMA RHO C U THETA0 RANKX RANKC RANKU CBETAU}` *
*  _PWRLABL= { HLT PBT WLK UNI UNIHF UNIGG COLLAPSE}`               *
*  _PRTLABL= { ALPHA SIGSCAL RHOSCAL BETASCAL TOTAL_N MAXRHOSQ}`    *
*  _DATLABL= { NOPRINT DS }`                                         *
*  _WRNLABL= { WARN }`                                               *
**********************************************************************;

START _SETOPT(newoptn,
            labels,_OPT_ON,_OPT_OFF);

  *____________________________________________________________*
  INPUTS
  1) Program supplied
     newoptn, column of on/off switches (0 or 1) for options
     labels, column of option names which label newoptn
     _OPT_ON, column of requested options to turn on
     _OPT_OFF, column of requested options to turn off

  OUTPUTS
    newoptn, column of switches with requested options turned on/off
  *____________________________________________________________*;
```

```
    DO i=1 TO NROW(_OPT_ON);
    newoptn= newoptn <> (LABELS=_OPT_ON(|i,1|));
    END;

    DO i=1 to nrow(_OPT_OFF);
    newoptn= newoptn # (LABELS^=_OPT_OFF(|i,1|));
    END;
  FINISH; *_SETOPT;

***********************************************************************
**                              _PROBF                                *
**                                                                    *
**  The module _PROBF screens the arguments to the PROBF function in  *
**  order to determine whether they will result in a missing value    *
**  being returned by the PROBF function due to power too near 1.     *
**  If this is the case, the PROBF function is by-passed.             *
**  PROBF function arguments are assessed with a normal approximation *
**  to a noncentral F-distribution (Johnson and Kotz, v2, 1970, p195).*
**  Additional problem in 6.04 exists with fractional df,             *
**  that was fixed with 6.06 (and assume beyond).                     *
**  Not surprisingly, normal approximation can be much less accurate  *
**  than standard function.  However, try to invoke normal approx     *
**  only when power>.99.  This may fail.  Statistical inaccuracy      *
**  in the approximations larger than numerical error.  Also, in      *
**  study planning this more than adequate accuracy.                  *
***********************************************************************;

START _PROBF
      (_FCRIT, _DF1, _DF2, _LAMBDA);

   *______________________________________________________________*
   INPUTS
   1) Program supplied
      _FCRIT, critical value of F distribution if Ho true, size=_ALPHA
      _DF1, numerator (hypothesis) degrees of freedom
      _DF2, denominator (error) degrees of freedom
      _LAMBDA, noncentrality parameter

   OUTPUTS
      PROB, approximate probability that variable distributed
            F(_DF1,_DF2,_LAMBDA) will take a value <= _FCRIT
   *______________________________________________________________* ;

   P1 = 1/3;
   P2 = -2;
   P3 = 1/2;
   P4 = 2/3;

   ARG1 = ((_DF1*_FCRIT)/(_DF1+_LAMBDA));
   ARG2 = (2/9)*(_DF1 + (2*_LAMBDA))*((_DF1 + _LAMBDA)##P2);
   ARG3 = (2/9)*(1/_DF2);

   NUMZ = (ARG1##P1) - (ARG3*(ARG1##P1)) - (1 - ARG2);
   DENZ = ( (ARG2 + ARG3*(ARG1##P4)) )##P3;
   ZSCORE = NUMZ/DENZ;
   * For debugging-- PRINT , ZSCORE _DF1 _DF2 _LAMBDA ;
   IF (_LAMBDA>0)&(ZSCORE<-4.) THEN PROB=0;
      ELSE PROB = PROBF(_FCRIT,_DF1,_DF2,_LAMBDA);
   RETURN(PROB);
   FINISH; *_PROBF;

***********************************************************************
**                              _HLTMOD                               *
**                                                                    *
```

```
**   The module _HLTMOD calculates a power for Hotelling-Lawley trace  *
**   based on the F approx. method.  _HLT is the Hotelling-Lawley      *
**   trace statistic, _DF1 and _DF2 are the hypothesis and error       *
**   degrees of freedom, _LAMBDA is the noncentrality parameter, and   *
**   _FCRIT is the critical value from the F distribution.             *
*********************************************************************;
START _HLTMOD(_PWR,_LBL,
              _A,_B,_S,_N,_R,_EVAL,_ALPHA,_NONCENN);
   *_______________________________________________________________*
   INPUTS
   1) Program supplied
      _A, rank of C matrix
      _B, rank of U matrix
      _S, minimum of _A and _B
      _N, total N
      _R, rank of X
      _EVAL, eigenvalues for H*INV(E)
      _ALPHA, size of test
      _NONCENN, =1 if multiply H*inv(E) eval's by (_N-_R)#_N
                and replace _DF2 with _N#_S in noncentrality
                for _S>1, per proposal by O'Brien & Shieh, 1992
   OUTPUTS
      _PWR, power for Hotelling-Lawley trace
      _LBL, label for output
   *_______________________________________________________________*;
   _DF1 = _A#_B;
   _DF2 = _S#(_N-_R-_B-1) + 2;
   IF (_DF2 <= 0) | (_EVAL(|1,1|) = .) THEN _PWR = . ;
                ELSE DO;
      IF (_NONCENN=1) | (_S=1) THEN EVALT=_EVAL#(_N-_R)/_N;
                               ELSE EVALT=_EVAL;
      _HLT = EVALT(|+,|);
      IF (_NONCENN=1) | (_S=1) THEN _LAMBDA = (_N#_S)#(_HLT/_S);
                               ELSE _LAMBDA = (_DF2 )#(_HLT/_S);
      _FCRIT = FINV(1-_ALPHA, _DF1, _DF2);
      _PWR = 1-_PROBF(_FCRIT, _DF1, _DF2, _LAMBDA);
      END;
   _LBL = {"HLT_PWR"};
  FINISH; *_HLTMOD;

*********************************************************************
**                              _PBTMOD                             *
**                                                                  *
**   The module _PBTMOD calculates a power for Pillai-Bartlett trace *
**   based on the F approx. method.  _V is the PBT statistic,       *
**   _DF1 and _DF2 are the hypothesis and error degrees of freedom, *
**   _LAMBDA is the noncentrality parameter, and _FCRIT is the      *
**   critical value from the F distribution.                        *
*********************************************************************;

START _PBTMOD(_PWR,_LBL,
             _A,_B,_S,_N,_R,_EVAL,_ALPHA,_NONCENN) GLOBAL(TOLERANC);
   *_______________________________________________________________*
   INPUTS
   1) Program supplied
      _A, rank of C matrix
      _B, rank of U matrix
      _S, minimum of _A and _B
      _N, total N
      _R, rank of X
      _EVAL, eigenvalues for H*INV(E)
      _ALPHA, size of test
      TOLERANC,  value not tolerated, numeric zero (global)
      _NONCENN, =1 if multiply H*inv(E) eval's by (_N-_R)#_N
```

```
              and replace _DF2 with _N#_S in noncentrality
                for _S>1, per proposal by O'Brien & Shieh, 1992
   OUTPUTS
      _PWR, power for Pillai-Bartlett trace
      _LBL, label for output
   *_______________________________________________________________* ;
   _DF1 = _A#_B;
   _DF2 = _S#(_N-_R+_S-_B);
   IF (_DF2 <= 0) | (_EVAL(|1,1|) = .) THEN _PWR = .;
      ELSE DO;
      IF (_NONCENN=1) | (_S=1) THEN EVALT=_EVAL#(_N-_R)/_N;
                                  ELSE EVALT=_EVAL;
      _V = SUM(EVALT/(J(_S,1,1) + EVALT));
      IF (_S-_V) <= TOLERANC THEN _PWR=.;
         ELSE DO;
         IF (_NONCENN=1) | (_S=1) THEN _LAMBDA =(_N#_S)#_V/(_S-_V);
                                     ELSE _LAMBDA =(_DF2 )#_V/(_S-_V);
         _FCRIT = FINV(1-_ALPHA, _DF1, _DF2, 0);
         _PWR=1-_PROBF(_FCRIT, _DF1, _DF2, _LAMBDA);
         END;
      END;
   _LBL = {"PBT_PWR"};
   FINISH; *_PBTMOD;

*********************************************************************
**                            _WLKMOD                               *
**                                                                  *
**  The module _WLKMOD calculates a power for Wilk's Lambda based  on *
**  the F approx. method.  _W is the Wilks` Lambda statistic, _DF1   *
**  and _DF2 are the hypothesis and error degrees of freedom, _LAMBDA *
**  is the noncentrality parameter, and _FCRIT is the critical value  *
**  from the F distribution.  _RM, _RS, _R1, and _TEMP are inter-     *
**  mediate variables.                                               *
********************************************************************;

START _WLKMOD(_PWR,_LBL,
              _A,_B,_S,_N,_R,_EVAL,_ALPHA,_NONCENN);
   *_______________________________________________________________*
   INPUTS
   1) Program supplied
      _A, rank of C matrix
      _B, rank of U matrix
      _S, minimum of _A and _B
      _N, total N
      _R, rank of X
      _EVAL, eigenvalues for H*INV(E)
      _ALPHA, size of test
      _NONCENN, =1 if multiply H*inv(E) eval's by (_N-_R)#_N
              and replace _DF2 with _N#_RS in noncentrality
                for _S>1, per proposal by O'Brien & Shieh, 1992
   OUTPUTS
      _PWR, power for Wilk's Lambda
      _LBL, label for output
   *_______________________________________________________________* ;
   _DF1 = _A # _B;
   IF _EVAL(|1,1|) = . THEN _W= . ;
      ELSE DO;
      IF (_NONCENN=1) | (_S=1) THEN EVALT=_EVAL#(_N-_R)/_N;
                                  ELSE EVALT=_EVAL;
      _W = EXP(SUM(-LOG(J(_S,1,1) + EVALT)));
      END;
   IF _S = 1 THEN DO;
      _DF2 = _N-_R-_B+1;
      _RS=1;
```

```
        _TEMPW=_W;
        END;
      ELSE DO;
        _RM = _N-_R-(_B-_A+1)/2;
        _RS =SQRT( (_A#_A#_B#_B - {4})/(_A#_A + _B#_B - {5}) );
        _R1 = (_B#_A - {2})/4;
        IF _W=. THEN _TEMPW=.;
               ELSE _TEMPW = _W##(1/_RS);
        _DF2 = (_RM#_RS) - 2#_R1;
        END;
   IF _TEMPW=. THEN _LAMBDA=.;
      ELSE DO;
      IF (_S=1) | (_NONCENN=1) THEN _LAMBDA=(_N#_RS)#(1-_TEMPW)/_TEMPW;
                                ELSE _LAMBDA=(_DF2   )#(1-_TEMPW)/_TEMPW;
      END;
   IF (_DF2 <= 0) | (_W=.) | (_LAMBDA=.) THEN _PWR = .;
      ELSE DO;
      _FCRIT = FINV(1-_ALPHA,_DF1,_DF2,0);
      _PWR = 1-_PROBF(_FCRIT,_DF1,_DF2,_LAMBDA);
      END;

   _LBL = {"WLK_PWR"};
   FINISH; * _WLKMOD;

*************************************************************************
*                             _SPECIAL                                  *
*                                                                       *
* The following module performs 2 disparate tasks. For _B=1 (UNIVARIATE*
* TEST), the powers are calculated more efficiently. For _A=1 (SPECIAL *
* MULTIVARIATE CASE), exact multivariate powers are calculated.         *
* Powers for the univariate statistics require separate treatment.      *
* _DF1 & _DF2 are the hypothesis and error degrees of freedom,          *
* _LAMBDA is the noncentrality parameter, and _FCRIT is the critical    *
* value from the F distribution.                                        *
***********************************************************************;

START _SPECIAL(_PWR,_LBL,
               _A,_B,_S,_N,_R,_EVAL,_ALPHA);
   *____________________________________________________________________*
   INPUTS
   1) Program supplied
      _A, rank of C matrix
      _B, rank of U matrix
      _S, minimum of _A and _B
      _N, total N
      _R, rank of X
      _EVAL, eigenvalues for H*INV(E)
      _ALPHA, size of test

   OUTPUTS
      _PWR, power for special case when rank(CBU)=1
      _LBL, label for output
   *____________________________________________________________________* ;

   _DF1=_A#_B;
   _DF2 = _N-_R-_B+1;
   IF (_DF2 <= 0) | (_EVAL(|1,1|)=.) THEN _PWR = .;
                ELSE  DO;
                      _LAMBDA=_EVAL(|1,1|)#(_N-_R);
                      _FCRIT = FINV(1-_ALPHA,_DF1,_DF2);
                      _PWR = 1-_PROBF(_FCRIT,_DF1,_DF2,_LAMBDA);
                      END;

   _LBL = {"POWER"};
```

```
FINISH; *_SPECIAL;

**********************************************************************
*                              _UNIMOD                               *
*                                                                    *
* Univariate STEP 1:                                                 *
* This module produces matrices required for Geisser-Greenhouse,     *
* Huynh-Feldt or uncorrected repeated measures power calculations. It *
* is the first step. Program uses approximations of expected values of *
* epsilon estimates due to Muller (1985), based on theorem of Fujikoshi*
* (1978). Program requires that U be orthonormal.                    *
*********************************************************************;
START _UNIMOD(_D, _MTP, _EPS, _DEIGVAL,_SLAM1, _SLAM2, _SLAM3,
              _USIGMAU,_B)  GLOBAL(TOLERANC);

   *_________________________________________________________________*
    INPUTS
    _USIGMAU = U`*(SIGMA#_SIGSCAL)*U
        _B, rank of U

    TOLERANC, value not tolerated, numeric zero (global)

    OUTPUTS
       _D, number of distinct eigenvalues
       _MTP, multiplicities of eigenvalues
       _EPS, epsilon calculated from U`*SIGMA*U
       _DEIGVAL, first eigenvalue
       _SLAM1, sum of eigenvalues squared
       _SLAM2, sum of squared eigenvalues
       _SLAM3, sum of eigenvalues
   *_________________________________________________________________* ;

*Get eigenvalues of covariance matrix associated with _E. This is NOT
the USUAL sigma. This cov matrix is that of (Y-YHAT)*U, not of (Y-YHAT).
The covariance matrix is normalized to minimize numerical problems ;

 _ESIG = _USIGMAU / ( TRACE(_USIGMAU) );
 _SEIGVAL=EIGVAL(_ESIG);
 _SLAM1 = SUM( _SEIGVAL)##2;
 _SLAM2 = SSQ( _SEIGVAL);
 _SLAM3 = SUM( _SEIGVAL);
 _EPS =  _SLAM1 / (_B # _SLAM2);

*Decide which eigenvalues are distinct;
 _D      =  1;        *Ends as number of distinct eigenvalues;
 _MTP    =  1;        *Ends as vector of multiplicities of eignvals;
 _DEIGVAL=_SEIGVAL(| 1 , 1|);
 DO _I =  2 TO _B;
    IF ( _DEIGVAL(|_D,1|) - _SEIGVAL(|_I,1|) ) > TOLERANC THEN DO;
       _D = _D +  1;
       _DEIGVAL = _DEIGVAL // _SEIGVAL(| _I , 1|);
       _MTP = _MTP // {1};
       END;
    ELSE _MTP(|_D,1|)=_MTP(|_D,1|) + 1;
    END;
FINISH; *_UNIMOD;

*********************************************************************
*                            _GGEXEPS                               *
*                                                                   *
* Univariate, GG STEP 2:                                            *
*Compute approximate expected value of Geisser-Greenhouse estimate  *
```

```
*          _FK = 1st deriv of FNCT of eigenvalues                        *
*          _FKK= 2nd deriv of FNCT of eigenvalues                        *
*          For GG FNCT is epsilon caret                                  *
***********************************************************************;

START _GGEXEPS
              (_B, _N, _R, _D,
              _MTP, _EPS, _DEIGVAL,
              _SLAM1, _SLAM2, _SLAM3);
   *_____________________________________________________________*
    INPUTS
    1) Program supplied
       _B, rank of U
       _N, total number of observations
       _R, rank of X
       _D, number of distinct eigenvalues
       _MTP, multiplicities of eigenvalues
       _EPS, epsilon calculated from U`*SIGMA*U
       _DEIGVAL, first eigenvalue
       _SLAM1, sum of eigenvalues squared
       _SLAM2, sum of squared eigenvalues
       _SLAM3, sum of eigenvalues

    OUTPUTS
       _EXEPS, expected value of epsilon estimator
   *_____________________________________________________________* ;

_FK = J(_D, 1, 1) #  2 # _SLAM3 / ( _SLAM2 # _B ) - (  2 ) # _DEIGVAL
      # _SLAM1 / ( _B # _SLAM2 ##  2 );
 _C0 = (  1 ) - _SLAM1 / _SLAM2;
 _C1 =  -4 # _SLAM3 / _SLAM2;
 _C2 =  4 # _SLAM1 / _SLAM2 ##  2;

 _FKK = _C0 # J( _D ,  1 ,  1) + _C1 # _DEIGVAL + _C2 # _DEIGVAL ##  2;
 _FKK =  2 # _FKK / ( _B # _SLAM2 );

 _T1 = _FKK # ( _DEIGVAL ##  2 ) # _MTP;
 _SUM1 = SUM( _T1);

 IF _D =  1 THEN _SUM2 =  0;
 ELSE DO;
    _T2 = _FK # _DEIGVAL # _MTP;
    _T3 = _DEIGVAL # _MTP;
    _TM1 = _T2 * _T3`;
    _T4 = _DEIGVAL * J(  1 , _D ,  1);
    _TM2 = _T4 - _T4`;
    _TM2INV = 1/( _TM2 + I( _D)) - I( _D);
    _TM3 = _TM1 # _TM2INV;
    _SUM2 = SUM( _TM3);
    END;

_EXEPS = _EPS + (_SUM1 + _SUM2)/(_N - _R);
RETURN (_EXEPS);
FINISH; *GG;


***********************************************************************
*                            _HFEXEPS                                 *
*                                                                     *
* Univariate, HF STEP 2:                                              *
*Compute approximate expected value of Huynh-Feldt estimate           *
*          _FK = 1st deriv of FNCT of eigenvalues                     *
*          _FKK= 2nd deriv of FNCT of eigenvalues                     *
```

```
*          For HF, FNCT is epsilon tilde                                    *
*************************************************************************;
START _HFEXEPS
              (_B, _N, _R, _D,
              _MTP, _EPS, _DEIGVAL,
              _SLAM1, _SLAM2, _SLAM3);
   *________________________________________________________________*
    INPUTS
    1) Program supplied
       _B, rank of U
       _N, total number of observations
       _R, rank of X
       _D, number of distinct eigenvalues
       _MTP, multiplicities of eigenvalues
       _EPS, epsilon calculated from U`*SIGMA*U
       _DEIGVAL, first eigenvalue
       _SLAM1, sum of eigenvalues squared
       _SLAM2, sum of squared eigenvalues
       _SLAM3, sum of eigenvalues

    OUTPUTS
       _EXEPS, expected value of epsilon estimator
   *________________________________________________________________* ;
* Compute approximate expected value of Huynh-Feldt estimate;
 _H1 = _N # _SLAM1 - (  2 ) # _SLAM2;
 _H2 = ( _N - _R ) # _SLAM2 - _SLAM1;
 _DERH1 = J( _D ,  1 ,  2 # _N # _SLAM3) -  4 # _DEIGVAL;
 _DERH2 =  2 # (_N - _R ) # _DEIGVAL - J( _D , 1 , 2 # SQRT( _SLAM1));
 _FK = _DERH1 - _H1 # _DERH2 / _H2;
 _FK = _FK / ( _B # _H2 );
 _DER2H1 = J( _D ,  1 ,  2 # _N - (  4 ));
 _DER2H2 = J( _D ,  1 ,  2 # ( _N - _R ) - (  2 ));
 _FKK =- _DERH1#_DERH2 / _H2+_DER2H1 - _DERH1#_DERH2 / _H2 +  2 #_H1#(
         _DERH2 ##  2 ) / _H2 ##  2 - _H1 # _DER2H2 / _H2;
 _FKK = _FKK / ( _H2 # _B );

 _T1 = _FKK # ( _DEIGVAL ##  2 ) # _MTP;
 _SUM1 = SUM( _T1);

 IF _D =  1 THEN _SUM2 =  0;
 ELSE DO;
    _T2 = _FK # _DEIGVAL # _MTP;
    _T3 = _DEIGVAL # _MTP;
    _TM1 = _T2 * _T3`;
    _T4 = _DEIGVAL * J(  1 , _D ,  1);
    _TM2 = _T4 - _T4`;
    _TM2INV = 1/( _TM2 + I( _D)) - I( _D);
    _TM3 = _TM1 # _TM2INV;
    _SUM2 = SUM( _TM3);
    END;

_EXEPS = _H1/(_B # _H2) + (_SUM1 + _SUM2) / (_N - _R);
RETURN (_EXEPS);
FINISH; *_HFEXEPS;

************************************************************************
*                              _LASTUNI                                *
*                                                                      *
* Univariate STEP 3                                                    *
* Final step for univariate repeated measures power calculations      *
************************************************************************;

START _LASTUNI (_PWR, _LBL,
                _EXEPS, _H, _E, _A, _B, _R, _N,
```

```
               _EPS, ppp, _ALPHA, FIRSTUNI);

   *_____________________________________________________________*
    INPUTS
    1) Program supplied
       _EXEPS, expected value epsilon estimator
       _H, hypothesis sum of squares
       _E, error sum of squares
       _A, rank of C
       _B, rank of U
       _N, total number of observations
       _R, rank of X
       _EPS, epsilon calculated from U`*SIGMA*U
       ppp, indicates selected power calculation
       _ALPHA, size of test
       FIRSTUNI, indicates first requested univariate statistic

   OUTPUTS
      _PWR, power for selected power calculation
      _LBL, label for output
   *_____________________________________________________________*;
IF _EXEPS > 1                        THEN _EXEPS =  1;
IF (_EXEPS < (1/_B)) & (_EXEPS^=.) THEN _EXEPS =  1/_B ;

IF (_N-_R)<=0 THEN _EPOWR=.;
              ELSE DO;
  *Compute noncentrality approximation;
  _MSH = TRACE( _H) / ( _A # _B );
  _MSE = TRACE( _E) / ( _B # ( _N - _R ) );
  _FALT = _MSH / _MSE;
  _LEPS = _A # _B # _EPS # _FALT;
  *Compute power approximation;
  _FCRIT= FINV(1 - _ALPHA , _B # _A # _EXEPS , _B # (_N -_R ) # _EXEPS);
  _EPOWR = 1 - _PROBF(_FCRIT, _B # _A # _EPS, _B#(_N -_R )#_EPS ,_LEPS);
  END;
_PWR = _EXEPS||_EPOWR;

IF ppp=4 THEN DO;
  _PWR = _EPS||_EPOWR;
  _LBL = {"EPSILON" "UNI_PWR"};
  END;
IF ppp=5 THEN DO;
  IF FIRSTUNI=5 THEN DO;
     _PWR = _EPS||_PWR;
     _LBL = { "EPSILON" "HF_EXEPS" "HF_PWR"};
     END;
  ELSE DO;
     _LBL = {"HF_EXEPS" "HF_PWR"};
     END;
  END;
IF ppp=6 THEN DO;
  IF FIRSTUNI=6 THEN DO;
     _PWR = _EPS||_PWR;
     _LBL = { "EPSILON" "GG_EXEPS" "GG_PWR"};
     END;
  ELSE DO;
     _LBL = {"GG_EXEPS" "GG_PWR"};
     END;
  END;
FINISH; *_LASTUNI;

***********************************************************************
*                          _SASDS                                     *
*                                                                     *
```

```
* Creates SAS dataset if requested.                                  *
**********************************************************************;

START _SASDS
          (_HOLDPOW, _HOLDNM, DSNAME);
   *________________________________________________________________*
    INPUTS
    1) User supplied (or program default)
       DSNAME, { "libref"  "dataset name" "default library" }
       The "default library" is optional. If omitted WORK
       is used. Default {WORK DODFAULT}.

    2) Program supplied
       _HOLDPOW, matrix of power values and optional output.
       _HOLDNM, matrix of variable names for the SAS dataset.
   *________________________________________________________________* ;

IF NCOL(DSNAME)=2 THEN _DSNAME=DSNAME||{WORK};
IF NCOL(DSNAME)=3 THEN _DSNAME = DSNAME;

LIB = _DSNAME(|1,1|);
DATASET = _DSNAME(|1,2|);
DEFAULT = _DSNAME(|1,3|);

* Reset default library to libref;
RESET NONAME DEFLIB = LIB;

LISTDS = DATASETS(LIB);
ENDIT=0;
NUMDS = NROW(LISTDS);

*Check to see if _PWRDTMP or user-specified DATASET already exists *;
IF NUMDS > 0 THEN
DO i=1 TO NUMDS;
  IF LISTDS(|i,1|) = "_PWRDTMP" THEN DO;
    ENDIT=1;
    NOTE1 = { "The program uses an intermediate dataset called",
              "_PWRDTMP. This dataset already exists in the",
              "requested library. New datasets cannot be created."};
    PRINT NOTE1;
    END;
  IF LISTDS(|i,1|) = DATASET THEN DO;
    NOTE2 = { " already exists. The default PWRDT### will",
              " be created instead. (See below)       " };
    PRINT DATASET NOTE2;
    DATASET = "DODFAULT";
  END;
END;

* Set default dataset name if required *;
NEWNUM=0;
IF DATASET= "DODFAULT" THEN NEWNUM=1;

IF DATASET = "DODFAULT" & NUMDS > 0 THEN DO;
    DO j=1 TO NUMDS;
    *Are any PWRDT### datasets in the library ?**;
    IF SUBSTR(LISTDS(|j,1|),1,5) = "PWRDT" THEN
       LISTPDS = LISTPDS//LISTDS(|j,1|);
    END;

  *What numbers do PWRDT### datasets have? Set number for new DS.**;
  IF NROW(LISTPDS) > 0 THEN DO;
    PDSNUMS = NUM(SUBSTR(LISTPDS,6,3));
    MAXNUM = MAX(PDSNUMS);
```

```
    NEWNUM = MAXNUM + 1;
    END;

  *Maximum number of PWRDT### datasets is 999.**;
  IF NEWNUM >999 THEN DO;
    ENDIT =1;
    NOTE3 = {" There are already 999 PWRDT### datasets.",
             " No more can be created."};
    PRINT NOTE3;
    END;
  END;

* New default name**;
IF (DATASET = "DODFAULT" & (1 <= NEWNUM <= 999) & ENDIT^=1) THEN
   DATASET = COMPRESS(CONCAT("PWRDT",CHAR(NEWNUM)));

* Create intermediate dataset called _PWRDTMP **;
IF ENDIT ^= 1 THEN DO;
  CREATE _PWRDTMP VAR _HOLDNM;
  APPEND FROM _HOLDPOW;
  CLOSE _PWRDTMP;

  *Change name of intermediate dataset to user specified or default*;
  CALL RENAME(LIB,"_PWRDTMP",DATASET);
  NAMED = COMPRESS(CONCAT(LIB,".",DATASET));
  PRINT "The dataset WORK._PWRDTMP has been renamed to " NAMED;
  END;

* Reset to original default library;
RESET NAME DEFLIB = DEFAULT;

FINISH; *_SASDS;

***********************************************************************
**                              MAIN CODE                            *
**********************************************************************;

 START _POWER (_HOLDPOW, _HOLDNM,_POWCHK)
       GLOBAL (ESSENCEX, SIGMA, BETA, U, C, THETA0,
               REPN, BETASCAL, SIGSCAL, RHOSCAL, ALPHA,
               ROUND, TOLERANC, OPT_ON, OPT_OFF, DSNAME);

   *____________________________________________________________*
    INPUTS
    1) User supplied - required - GLOBAL
       ESSENCEX, the essence design matrix.  Users unfamiliar with
          this matrix may simply enter the full design matrix and
          specify that REPN below be equal to 1.
       SIGMA, the hypothesized variance-covariance matrix
       BETA, the matrix of hypothesized regression coefficients
       U, "within" subject contrast
       C, "between" subject contrast
    2) User supplied - optional (program defaults supplied) - GLOBAL
       THETA0, the matrix of constants to be subtracted from CBU
       REPN, (vector), the # of times each row of ESSENCEX is
          duplicated
       BETASCAL, (vector) multipliers for BETA
       SIGSCAL, (vector) multipliers for SIGMA
       RHOSCAL, (vector) multipliers for RHO
       ALPHA, (vector) Type I error rates
       ROUND, (scalar) # of places to round power calculations
       TOLERANC, (scalar) value not tolerated, numeric zero,
       used for checking singularity.
       OPT_ON, (column) selected options to turn on
```

```
        OPT_OFF, (column) selected options to turn off
        DSNAME, (row) name of output dataset

     OUTPUTS
       _HOLDPOW, matrix of power calculations
       _HOLDNM, column labels for _HOLDPOW
       _POWCHK = 0 if no warnings or errors detetected,
               = 1 if mild warning, e.g. _N-_R <=5 (poor approxmtns),
               = 2 if any missing values included in _HOLDPOW, and
               = 3 if input error or computational problem required
                      program termination
   *_________________________________________________________________*;


*A: INITIAL SETUP  **;
*A.0.1) Insure everthing printed by power software goes to output file;
RESET NOLOG;
*A.0.2) Initialize return code;
_POWCHK=0;
*A.0.3) Initialize warnings about powers rounding to 1;
_RNDCHK=0; *

*A.1) Check for required input matrices, and that they are numeric *;
IF (NROW(C)=0)          |
   (NROW(BETA)=0)       |
   (NROW(SIGMA)=0)      |
   (NROW(ESSENCEX)=0) THEN DO;
         NOTE1={"One or more of the following four required matrices",
                "has not been supplied:  SIGMA, BETA, C, ESSENCEX.  "};
         PRINT NOTE1;
         GO TO ENDOFPGM;
         END;

IF (_TYPECHK(C        ,"C"        ,"N")+
    _TYPECHK(BETA     ,"BETA"     ,"N")+
    _TYPECHK(SIGMA    ,"SIGMA"    ,"N")+
    _TYPECHK(ESSENCEX,"ESSENCEX","N")) > 0 THEN GO TO ENDOFPGM;

*A.1.4) Insure that U and THETA0 are numeric, if they exist;
IF NROW(U)>0 THEN IF _TYPECHK(U,"U","N")^=0 THEN GO TO ENDOFPGM;

IF NROW(THETA0)>0 THEN
   IF _TYPECHK(THETA0,"THETA0","N")^=0 THEN GO TO ENDOFPGM;

*A.1.5) Delete previous versions of _HOLDPOW and _HOLDNM;
IF NROW(_HOLDPOW)>0 THEN FREE _HOLDPOW;
IF NROW(_HOLDNM) >0 THEN FREE _HOLDNM;

*A.2) Define default matrices;

IF NROW(U)=0 THEN DO;
  U= I(NCOL(BETA));
  HITLIST = HITLIST// { U };
  END;
IF NROW(REPN)=0 THEN DO;
  REPN= 1;
  HITLIST = HITLIST// {REPN};
  END;
IF NROW(THETA0)=0 THEN DO;
  THETA0=J(NROW(C),NCOL(U),0);
  HITLIST = HITLIST// { THETA0 };
  END;
IF NCOL(SIGSCAL)=0 THEN DO;
  SIGSCAL= { 1 };
```

```
    HITLIST = HITLIST// {SIGSCAL};
    END;
  IF NCOL(RHOSCAL)=0 THEN DO;
    RHOSCAL= { 1 };
    HITLIST = HITLIST// {RHOSCAL};
    END;
  IF NCOL(BETASCAL)=0 THEN DO;
    BETASCAL= { 1 };
    HITLIST = HITLIST// {BETASCAL};
    END;
  IF NROW(ALPHA) = 0 THEN DO;
    ALPHA= {.05};
    HITLIST = HITLIST// {ALPHA};
    END;
  IF NROW(ROUND)=0 THEN DO;
    ROUND=3;
    HITLIST = HITLIST// {ROUND};
    END;
  IF NROW(TOLERANC)=0 THEN DO;
    TOLERANC = 1E-12;
    HITLIST = HITLIST// {TOLERANC};
    END;
  IF NROW(DSNAME)=0 THEN DO;
    DSNAME = {WORK DODFAULT WORK};
    HITLIST = HITLIST// {DSNAME};
    END;

  *A.3) Create column vectors from user inputs;
    *Check that have only vectors or scalars, not matrices;
    *Check type (character or numeric);
    *Check for valid values;
    CHECKSUM=0;

    IF NCOL(REPN)>1 THEN _REPN = REPN`;
                    ELSE _REPN = REPN;
    CHECKSUM=CHECKSUM+_SIZECHK(_REPN,"REPN")
                     +_TYPECHK(_REPN,"REPN","N");

    IF NCOL(SIGSCAL)>1 THEN _SIGSCAL = SIGSCAL`;
                       ELSE _SIGSCAL = SIGSCAL;
    CHECKSUM=CHECKSUM+_SIZECHK(_SIGSCAL,"SIGSCAL")
                     +_TYPECHK(_SIGSCAL,"SIGSCAL","N");

    IF NCOL(RHOSCAL)>1 THEN _RHOSCAL = RHOSCAL`;
                       ELSE _RHOSCAL = RHOSCAL;
    CHECKSUM=CHECKSUM+_SIZECHK(_RHOSCAL,"RHOSCAL")
                     +_TYPECHK(_RHOSCAL,"RHOSCAL","N");

    IF NCOL(BETASCAL)>1 THEN _BETASCL = BETASCAL`;
                        ELSE _BETASCL = BETASCAL;
    CHECKSUM=CHECKSUM+_SIZECHK(_BETASCL,"BETASCAL")
                     +_TYPECHK(_BETASCL,"BETASCAL","N");

    IF NCOL(ALPHA)>1 THEN ALPHAV = ALPHA`;
                     ELSE ALPHAV = ALPHA;
    CHECKSUM=CHECKSUM+_SIZECHK(ALPHAV,"ALPHA")
                     +_TYPECHK(ALPHAV,"ALPHA","N");

    IF NROW(OPT_ON)>0 THEN DO;
       IF NCOL(OPT_ON)>1 THEN _OPT_ON = OPT_ON`;
                         ELSE _OPT_ON = OPT_ON ;
             CHECKSUM=CHECKSUM+_SIZECHK(_OPT_ON,"OPT_ON")
                              +_TYPECHK(_OPT_ON,"OPT_ON","C");
             END;
```

```
  IF NROW(OPT_OFF)>0 THEN DO;
     IF NCOL(OPT_OFF)>1 THEN _OPT_OFF = OPT_OFF`;
                        ELSE _OPT_OFF = OPT_OFF ;
      CHECKSUM=CHECKSUM+_SIZECHK(_OPT_OFF,"OPT_OFF")
                       +_TYPECHK(_OPT_OFF,"OPT_OFF","C");
         END;

IF CHECKSUM>0 THEN GO TO FREE_END;

*A.4) Define default options;

IF ANY(THETA0)=1 THEN _PG1=    {1 1 1 1 1  1 0 0 0 1}`;
                 ELSE _PG1=    {1 1 1 1 1  0 0 0 0 1}`;
_PG1LABL= { BETA SIGMA RHO C U THETA0 RANKX RANKC RANKU CBETAU}`;

_POWR=    { 0 0 1   0 0 1   1 }`;
_PWRLABL= { HLT PBT WLK   UNI UNIHF UNIGG   COLLAPSE }`;

_PRT=     { 1   1 1 1   1 0 1}`;
_PRTLABL= {ALPHA   SIGSCAL RHOSCAL BETASCAL   TOTAL_N MAXRHOSQ CASE}`;

_DAT=     { 0 0 }`;
_DATLABL= { NOPRINT DS }`;

_WARN=    { 1 }`;
_WRNLABL= { WARN }`;

_NONCENN= { 0 }`;
_NONLABL= { NONCEN_N }`;

*A.5) Define necessary parameters;

_R = NCOL(ESSENCEX);           ** _R IS THE RANK OF THE X MATRIX;
_A = NROW(C);                  ** _A IS THE RANK OF THE C MATRIX;
_B = NCOL(U);                  ** _B IS THE RANK OF THE U MATRIX;
_S = MIN(_A,_B);               ** _S IS MIN OF RANK(C) AND RANK(U);
_P = NCOL(BETA);               ** _P IS # OF RESPONSE VARIABLES;

*A.6) Set round off units after checking ROUND *;
IF _TYPECHK(ROUND,"ROUND","N")^=0 THEN GO TO FREE_END;

IF MAX(NCOL(ROUND),NROW(ROUND))>1 THEN DO;
   PRINT "ROUND cannot be a matrix. " ,
         "Must have only one column or only one row." ,, ROUND ;
   GO TO FREE_END;
   END;
IF (ROUND < 1) | (ROUND > 15) THEN DO;
   PRINT "User-specified ROUND < 1 OR ROUND >15";
   GO TO FREE_END;
   END;
ROUNDOFF = 1/10**ROUND;

*A.7) Check TOLERANC;
IF _TYPECHK(TOLERANC,"TOLERANCE","N")^=0 THEN DO;
    GO TO FREE_END;
   END;
IF TOLERANC <=0 THEN DO;
   PRINT "User-specified TOLERANC <= zero.";
   GO TO FREE_END;
   END;

*A.8) Check REPN;
IF MIN(REPN)<=TOLERANC THEN DO;
```

```
   PRINT "All REPN values must be > TOLERANC > zero." , REPN;
   GO TO FREE_END;
   END;

*A.9) Check SIGSCAL;
IF MIN(_SIGSCAL)<=TOLERANC THEN DO;
   *Can only get here if user supplies invalid SIGSCAL;
   PRINT "All SIGSCAL values must be > TOLERANC > zero." , SIGSCAL;
   GO TO FREE_END;
   END;

*A.10) Check ALPHA;
IF (MIN(ALPHAV)<=TOLERANC) | (MAX(ALPHAV)>=1) THEN DO;
   *Can only get here if user supplies invalid ALPHA;
   PRINT "All ALPHA values must be > TOLERANC > zero." ,
         "and < 1.00" , ALPHA;
   GO TO FREE_END;
   END;

*A.11) Check DSNAME;
IF (NROW(DSNAME)>1) | (NCOL(DSNAME)>3) THEN DO;
   PRINT "DSNAME must have only 1 row and 2 or 3 columns." , DSNAME;
   GO TO FREE_END;
   END;
IF TYPE(DSNAME)="N" THEN DO;
   PRINT "DSNAME must be character, not numeric." , DSNAME;
   GO TO FREE_END;
   END;


*A.12) Check for old versions of SAS;
IF (&SYSVER < 6.06) & (_WARN=1) THEN DO;
  NOTE1_3 = {"WARNING:                                   ",
             "You are using SAS version &SYSVER.  Fractional ",
             "error degrees of freedom may induce errors in  ",
             "the _PROBF module of this program. This problem",
             "has been corrected in SAS version 6.06.        "};
  PRINT NOTE1_3;
  END;

*B: CHECKS ON INPUT DATA **;

CALL _INPTCHK(CHECKER, _R, _A, _B, _S);
IF CHECKER=1 THEN GO TO FREE_END;

CALL _OPTCHK(CHECKER,
            _PG1LABL, _PWRLABL, _PRTLABL, _DATLABL,_WRNLABL,_NONLABL,
            _OPT_ON, _OPT_OFF);
IF CHECKER=1 THEN GO TO FREE_END;

*C: DEFINE NECESSARY MATRICES AND DO CHECKS **;

SD = DIAG(SQRT(VECDIAG(SIGMA)));
_RHO_ = INV(SD)*SIGMA*INV(SD);     ** Correlation matrix to be printed;
CBETAU= C*BETA*U;                  ** To be printed

*C.1) Check for errors;
_TEMPXX= ESSENCEX`*ESSENCEX;
_NAME={"X`X"};
E1= _SNGRCHK(_TEMPXX,_NAME);

*_INVXX= INV(_TEMPXX);
_INVXX= SOLVE(_TEMPXX,I(NROW(_TEMPXX)));
FREE _TEMPXX;
```

```
_M = C*_INVXX*C`;
_NAME={"C(INV(X`X))C`"};
E2= _SNGRCHK(_M,_NAME);

*C.2) Terminate program if errors detected;
IF (E1=1) | (E2=1) THEN GO TO FREE_END;

*D: SET NEW OPTIONS **;

*D.1) Set user selected options;
IF ((NROW(_OPT_ON) > 0) | (NROW(_OPT_OFF) > 0)) THEN DO;
   CALL _SETOPT(_POWR,_PWRLABL,_OPT_ON,_OPT_OFF);
   CALL _SETOPT(_PRT,_PRTLABL,_OPT_ON,_OPT_OFF);
   CALL _SETOPT(_PG1,_PG1LABL,_OPT_ON,_OPT_OFF);
   CALL _SETOPT(_DAT,_DATLABL,_OPT_ON,_OPT_OFF);
   CALL _SETOPT(_WARN,_WRNLABL,_OPT_ON,_OPT_OFF);
   CALL _SETOPT(_NONCENN,_NONLABL,_OPT_ON,_OPT_OFF);
   END;

*D.2) Insure that at least one test statistic chosen;
IF ANY(_POWR)^=1 THEN DO;
   PRINT "OPT_OFF combined with defaults implies " ,
         "no power calculation for any statistic." ,
         OPT_OFF ;
   _POWCHK=3;
   GO TO FREE_END;
   END;

*D.3) Identify special cases;
IF _POWR(|7,1|)=1 THEN DO; *if COLLAPSE is on;
   IF _S>1 THEN DO;
      IF _POWCHK=0 THEN _POWCHK=1;
      _POWR(|7,1|)=0;
      IF _WARN=1 THEN
      PRINT "Rank(C*BETA*U) >1, so COLLAPSE option ignored." ;
      END;
   IF _S=1 THEN DO;
      IF _B=1 THEN _POWR={ 0 0 0 0 0 0 1 }`;
      IF _B>1 THEN DO;
         _POWR(|1,1|)=0;
         _POWR(|2,1|)=0;
         _POWR(|3,1|)=0;
         END;
      END;
   END;

*E) MORE CHECKS ON MATRICES **;

*E.1) Check for scalar SIGMA *;
IF (NCOL(SIGMA)=1) & (_WARN=1) THEN DO;
  NOTE1_5={"WARNING:                                  " ,
           "SIGMA is a scalar. For this program,      " ,
           "a scalar SIGMA must equal the variance,"  ,
           "NOT the standard deviation.               " };

 PRINT NOTE1_5;
  END;

*E.2) CHECK FOR ORTHONORMAL U;
* Only needed for univariate repeated measures;
IF ( _POWR(|4,1|)=1 | _POWR(|5,1|)=1 |_POWR(|6,1|)=1 ) THEN DO;
  _UPUSCA_ = U` * U;
  IF _UPUSCA_(|1 ,1|) ^= 0 THEN _UPUSCA_ = _UPUSCA_ / _UPUSCA_(|1 ,1|);
```

```
  _UDIF_ = ABS( _UPUSCA_ - I(_B));
  IF MAX( _UDIF_) > SQRT( TOLERANC) THEN DO;
   NOTE2 =
   {"U is not proportional to an orthonormal matrix. Problem is",
    "probably due to a programming error or numerical inaçcuracy.",
    "If using ORPOL, suggest centering and/or scaling values",
    "Inner product is not K#I(NCOL(U)). For a nonzero constant K,",
    "inner product scaled so (1,1) element is 1.00 (unless = 0) is:"};
    PRINT NOTE2;
    PRINT _UPUSCA_;
    PRINT "Univariate repeated test not valid.";
    GO TO FREE_END;
    END;
  END;

*F) LOOP TO SELECT ALPHA, BETA, SIGMA, N, AND COMPUTE POWER**;

  *F.1) select alpha;
  DO a=1 TO NROW(ALPHAV);
  _ALPHA = ALPHAV(|a,1|);

    *F.2) select sigma;
    DO s=1 TO NROW(_SIGSCAL);
    _SIGSCL= _SIGSCAL(|s,1|);

    * Compute initial sigscaled covariance matrix *;
    _ISIGMA = SIGMA#_SIGSCL;
    * Compute initial rho from initial sigscaled covariance matrix *;
    V_I=VECDIAG(_ISIGMA);
    IF MIN(V_I)<=TOLERANC THEN DO;
      NOTE2_5={"Covariance matrix has variance <= TOLERANC (too small)",
               "with current SIGSCAL element = "};
      PRINT NOTE2_5 _SIGSCL ,,, SIGMA;
      GO TO FREE_END;
      END;
    STDEV=SQRT(V_I);
    STDINV=DIAG(J(_P,1,1)/(STDEV));
    _IRHO = STDINV*_ISIGMA*STDINV;

      *F.3) create new rhos *;
      DO r=1 TO NROW(_RHOSCAL);
      _RHOSCL = _RHOSCAL(|r,1|);
      _RHOJUNK = (_IRHO #_RHOSCL);  * Diagonal elements not =1;
      _RHO_OFF = _RHOJUNK - DIAG(VECDIAG(_RHOJUNK)); * Off-diagonals;
      _RHO_OFF = (_RHO_OFF + _RHO_OFF`)/2;     *To insure symmetry;

      IF MAX(ABS(_RHO_OFF))>1 THEN DO;
        NOTE3={"For the current values of SIGSCAL and RHOSCAL",
               "there is a correlation with an absolute value>1."};
        PRINT NOTE3;
        PRINT "SIGSCAL= " _SIGSCL;
        PRINT "RHOSCAL= " _RHOSCL;
        GO TO FREE_END;
        END;

      IF MAX(ABS(_RHO_OFF))=1 THEN DO;
        _POWCHK=1;
        NOTE4={"WARNING:                                    ",
               "For the current values of SIGSCAL and RHOSCAL",
               "there is a correlation with an absolute value=1."};
        IF _WARN=1 THEN PRINT NOTE4                    ,,
                              "SIGSCAL= " _SIGSCL ,,
                              "RHOSCAL= " _RHOSCL ;
        END;
```

```
_RHO = _RHO_OFF + I(_P);

* Create new sigma from rho *;
STDEVM=DIAG(STDEV);
_NEWSIGM = STDEVM*_RHO*STDEVM;

_SIGSTAR= U`*_NEWSIGM*U;
_USIGMAU= (_SIGSTAR + _SIGSTAR`)/2;    *To insure symmetry;
_NAME = {"U`*SIGMA*U"};
E3 = _SNGRCHK(_USIGMAU,_NAME);
IF E3=1 THEN GO TO FREE_END;

  *F.4) Select Beta;
  DO b=1 TO NROW(_BETASCL);
  _BETSCL = _BETASCL(|b,1|);
  _NEWBETA= BETA#_BETSCL;
  _THETA  = C*_NEWBETA*U ;
  _ESSH = (_THETA-THETA0)`*SOLVE(_M,I(NROW(_M)))*(_THETA-THETA0);
  *     = (_THETA-THETA0)`*         INV(_M)       *(_THETA-THETA0);

  *F.5) Select N;
    DO i=1 TO NROW(_REPN);
    _N = _REPN(|i,1|)#NROW(ESSENCEX);

    _E = _USIGMAU#(_N-_R);
    _H = _REPN(|i,1|)#_ESSH;

    *F.6) Eigenvalues for H*INV(E);
    _EPS=.;
    IF (_N - _R) <= 0 THEN DO;
                        _EVAL=J(_S,1,.);
                        _RHOSQ=J(_S,1,.);
                        _RHOSQ1=.;
                        _D=1;
                        _MTP=_B;
                       *_EPS=.;
                        _DEIGVAL=.;
                        _SLAM1=.; _SLAM2=.; _SLAM3=.;
                        END;
      ELSE DO;
      _FINV = (SOLVE(HALF(_E),I(NROW(_E))))`; * = INV(HALF(_E))`;
      _HEIORTH = _FINV*_H*_FINV`;
      _HEIORTH=(_HEIORTH + _HEIORTH`)/2;  *Insure symmetry;
      _EVAL = EIGVAL(_HEIORTH);
      _EVAL=_EVAL(|1:_S,1|);         *At most _S nonzero;
      _EVAL=_EVAL#(_EVAL>TOLERANC); *Set evals<tolerance to zero;

      *Make vector of squared generalized canonical correlations;
      IF ( _S=1) | (_NONCENN=1) THEN EVALT=_EVAL#(_N-_R)/_N;
                                ELSE EVALT=_EVAL;
      _RHOSQ = EVALT(|*,1|)/(J(_S,1,1)+EVALT(|*,1|));
      _RHOSQ = ROUND(_RHOSQ,ROUNDOFF);
      _RHOSQ1=_RHOSQ(|1,1|);  *Largest;
      END;
    *F.7) Start requested power calculations **;

    *F.7.a) Start univariate repeated measures power calcs;
    IF (_POWR(|4,1|)=1|_POWR(|5,1|)=1|_POWR(|6,1|)=1) THEN DO;
      CALL _UNIMOD(_D,_MTP,_EPS,_DEIGVAL,
                   _SLAM1,_SLAM2,_SLAM3,
                   _USIGMAU,_B);
      *Find first requested univariate statistic.;
      *Needed for creating epsilon;
```

```
  IF NROW(FIRSTUNI)= 0 & _POWR(|4,1|)=1 THEN FIRSTUNI=4;
  IF NROW(FIRSTUNI)= 0 & _POWR(|5,1|)=1 THEN FIRSTUNI=5;
  IF NROW(FIRSTUNI)= 0 & _POWR(|6,1|)=1 THEN FIRSTUNI=6;
  END;

*F.7.b) Select requested power calculations;

p = _POWR`;
PSELECTN = (LOC(p))`;
  DO pp=1 TO NROW(PSELECTN);
  ppp = PSELECTN(|pp,1|);
  IF ppp=1 THEN CALL _HLTMOD(_PWR,_LBL,
                    _A,_B,_S,_N,_R,_EVAL,_ALPHA,_NONCENN);
  IF ppp=2 THEN CALL _PBTMOD(_PWR,_LBL,
                    _A,_B,_S,_N,_R,_EVAL,_ALPHA,_NONCENN);
  IF ppp=3 THEN CALL _WLKMOD(_PWR,_LBL,
                    _A,_B,_S,_N,_R,_EVAL,_ALPHA,_NONCENN);
  IF ppp=4 THEN DO;
              _EXEPS=1;
              CALL _LASTUNI(_PWR,_LBL,
                         _EXEPS,_H,_E,_A,_B,_R,_N,
                         _EPS,ppp,_ALPHA,FIRSTUNI);
              END;
  IF ppp=5 THEN DO;
              IF (_N-_R) <= 0 THEN _EXEPS=.;
                 ELSE _EXEPS=_HFEXEPS(_B,_N,_R,_D,
                                   _MTP,_EPS,_DEIGVAL,
                                   _SLAM1,_SLAM2,_SLAM3);
              CALL _LASTUNI(_PWR,_LBL,
                         _EXEPS,_H,_E,_A,_B,_R,_N,
                         _EPS,ppp,_ALPHA,FIRSTUNI);
              END;
  IF ppp=6 THEN DO;
              IF (_N-_R) <= 0 THEN _EXEPS=.;
                 ELSE _EXEPS=_GGEXEPS(_B,_N,_R,_D,
                                   _MTP,_EPS,_DEIGVAL,
                                   _SLAM1,_SLAM2,_SLAM3);
              CALL _LASTUNI(_PWR,_LBL,
                         _EXEPS,_H,_E,_A,_B,_R,_N,
                         _EPS,ppp,_ALPHA,FIRSTUNI);
              END;
   IF ppp=7 THEN CALL _SPECIAL(_PWR,_LBL,
                         _A,_B,_S,_N,_R,_EVAL,_ALPHA);

   *F.7.c) Put all requested power calculations in a row;
   IF NCOL(_OUT)=0 THEN DO;
     *Set maximum possible output for power table;
     *then delete unrequested options;
     _OUT = _ALPHA||_SIGSCL||_RHOSCL||_BETSCL||_N||_RHOSQ1;
     _HOLDNM = {"ALPHA" "SIGSCAL" "RHOSCAL" "BETASCAL"
                "TOTAL_N" "MAXRHOSQ"};
       DO xx= 6 to 1 by -1;
       IF _PRT(|xx,1|)^=1 THEN DO;
                         _OUT = REMOVE(_OUT,xx);
                         _HOLDNM = REMOVE(_HOLDNM,xx);
                         END;
       END;
     END;

    _PWR = ROUND(_PWR,ROUNDOFF);
    IF _PWR(|1,NCOL(_PWR)|)=1 THEN _RNDCHK=1;
    _OUT=   _OUT||_PWR;
    _HOLDNM= _HOLDNM||_LBL;
  FREE _PWR _LBL;
```

```
            END; *power calculation selection;

          *F.7.d) Create stack of calculations;
          IF NROW(_HOLDPOW)=0 THEN _HOLDPOW=_OUT;
                                ELSE _HOLDPOW= _HOLDPOW//_OUT;
          FREE  _OUT;

          *F.8) Check for sufficient sample size;
          IF ((_N -_R)<=5) THEN DO;
            IF (_B>1) &      (_POWR(|4  ,1|) =1) & (_EPS^=1)
                                         THEN _POWCHK=1; *UNI;
            IF (_S>1) & (ANY(_POWR(|5:6,1|))=1)
                                         THEN _POWCHK=1; *UNIHF UNIGG;
            IF (_S>1) & (ANY(_POWR(|1:3,1|))=1)
                                         THEN _POWCHK=1; *HLT PBT WLK;
            END;
          END; *N;
        END; *beta;
      END; *rho;
    END; *sigma;
  END; *alpha;

*G) WARNINGS FROM LOOP ;
IF ANY(_HOLDPOW=.) THEN _POWCHK=2;
IF _WARN=1 THEN DO; *Print only if WARN is on;
  IF _POWCHK>0 THEN PRINT ,,
             "WARNING:                                    " ,
             "If (_N-_R) <= 5, then power approximations  " ,
             "approximations may be very inaccurate,      " ,
             "especially Huynh-Feldt.  Consult the manual." ;
  IF _POWCHK=2 THEN PRINT ,,
        "WARNING:                                           " ,
        "When error degrees of freedom are <= 0, _HOLDPOW will " ,
        "have missing values for some requested powers.      " ,
        "Hence numeric operations on _HOLDPOW may cause errors." ;
  END;

*H) PRINTED OUTPUT;

*H.1) Print requested matrices;
IF (_DAT(|1,1|)=0) & ANY(_PG1) THEN DO;
  PRINT /;
  IF _PG1(| 1,1|) = 1 THEN PRINT , BETA;
  IF _PG1(| 2,1|) = 1 THEN PRINT , SIGMA;
  IF _PG1(| 3,1|) = 1 THEN PRINT , _RHO_;
  IF _PG1(| 4,1|) = 1 THEN PRINT , C;
  IF _PG1(| 5,1|) = 1 THEN PRINT , U;
  IF _PG1(| 6,1|) = 1 THEN PRINT , THETA0;
  IF _PG1(| 7,1|) = 1 THEN PRINT , "RANK OF X = " _R;
  IF _PG1(| 8,1|) = 1 THEN PRINT , "RANK OF C = " _A;
  IF _PG1(| 9,1|) = 1 THEN PRINT , "RANK OF U = " _B;
  IF _PG1(|10,1|) = 1 THEN PRINT , "C*BETA*U = " CBETAU;
  END;

*H.2) Print power calculations;
  IF ( _RNDCHK>0) & (_WARN=1) THEN PRINT ,,
             "WARNING:                              " ,
             "One or more power values rounded to 1.  " ,
             "For example, with ROUND=3,            " ,
             "power=1 should be reported as power>.999" ;
CASETOTL=NROW(_HOLDPOW);
IF _PRT(|7,1|)=1 THEN DO;
  CASE = (1:CASETOTL)`;
  _HOLDPOW = CASE || _HOLDPOW;
```

```
  _HOLDNM = "CASE" || _HOLDNM;
  END;
IF _DAT(|1,1|) = 0 THEN DO;
  IF CASETOTL <= 40 THEN PRINT / _HOLDPOW(|COLNAME=_HOLDNM|);
    ELSE DO;
    BRKPT= DO(36,CASETOTL,36);
    IF MAX(BRKPT) ^= CASETOTL THEN BRKPT= BRKPT||CASETOTL;
    START= 1;
      DO i=1 TO NCOL(BRKPT);
      STP=BRKPT(|1,i|);
      HOLDPOW = _HOLDPOW(|START:STP,|);
      PRINT / HOLDPOW(|COLNAME=_HOLDNM|);
      START= STP + 1;
      FREE HOLDPOW;
      END;
    END;
  END;

*H.3) Create SAS dataset if option was requested;
IF _DAT(|2,1|)=1 THEN CALL _SASDS(_HOLDPOW,_HOLDNM,DSNAME);

*I) FREE MATRICES ON THE HITLIST **;

GO TO FREE_ALL;
FREE_END: ; *Branching target for ending with major error;
_POWCHK=3;

FREE_ALL: ; *Branching target for no major error;
  DO I = 1 TO NROW(HITLIST);
  K=HITLIST(|I,|);
  IF K = { THETA0 } THEN FREE THETA0;
  ELSE IF K = { U } THEN FREE U;
  ELSE IF K = {REPN} THEN FREE REPN;
  ELSE IF K = {SIGSCAL} THEN FREE SIGSCAL;
  ELSE IF K = {RHOSCAL} THEN FREE RHOSCAL;
  ELSE IF K = {BETASCAL} THEN FREE BETASCAL;
  ELSE IF K = {ALPHA} THEN FREE ALPHA;
  ELSE IF K = {ROUND} THEN FREE ROUND;
  ELSE IF K = {TOLERANC} THEN FREE TOLERANC;
  ELSE IF K = {DSNAME} THEN FREE DSNAME;
  END;
IF _POWCHK<3 THEN GO TO ALL_DONE;

ENDOFPGM: ; *Branching target for ending with major error;
IF _POWCHK=0 THEN _POWCHK=3; *If jumped directly to ENDOFPGM;
PRINT ,, "PROGRAM TERMINATED WITH ONE OR MORE MAJOR ERRORS." ,
         "PROGRAM TERMINATED WITH ONE OR MORE MAJOR ERRORS." ,
         "PROGRAM TERMINATED WITH ONE OR MORE MAJOR ERRORS." ;

ALL_DONE: ; *Branching target for ending with no or minor errors;
FINISH; *_POWER;


*************************************************************************;
*                             POWER                                    *;
*                                                                      *;
* Define POWER command.                                                *;
* User will only have to type RUN POWER. The input matrices will not  *;
* have to be listed in the RUN statement.                              *;
************************************************************************;

START POWER;
CALL _POWER (_HOLDPOW,_HOLDNM,_POWCHK);
```

```
FINISH; *POWER;
```

Appendix 2 Data Sets Used and Created by Each Program

Program 1_6.sas

This program does not input a data set, but rather produces a data set named 1_6.dat.

```
10.28    18.52    29.67    35.68
11.93    20.28    28.91    36.05
 8.22    17.40    28.72    35.26
 8.41    21.69    32.14    49.37
10.03    19.21    31.13    41.66
12.85    20.34    29.29    41.66
 7.78    19.42    30.00    34.77
10.55    17.68    31.29    44.35
10.11    21.41    29.36    35.77
12.25    21.00    30.64    44.32
 8.71    23.28    30.08    40.91
12.09    18.06    30.29    41.08
 9.79    19.77    32.69    49.96
11.62    19.02    28.22    32.90
 8.99    20.10    29.68    39.86
10.86    20.55    31.15    45.87
12.48    19.62    29.90    38.51
11.40    22.78    28.02    31.07
11.82    22.57    29.52    36.43
11.32    20.38    29.37    35.64
11.38    21.33    31.80    44.31
13.77    20.98    30.34    38.94
 9.59    22.25    29.88    41.84
 8.63    16.50    29.92    40.95
11.30    20.35    29.29    34.36
12.30    18.54    29.17    36.57
13.02    22.29    28.36    35.73
 8.88    19.95    30.73    44.40
12.23    20.37    30.00    43.40
 8.53    18.09    32.77    51.09
 8.80    20.95    29.45    34.25
 8.38    19.97    29.94    41.71
 9.53    20.03    29.63    39.88
11.23    19.16    29.18    38.21
10.41    21.65    31.75    49.20
12.10    21.76    29.17    38.06
12.23    19.13    29.88    36.57
 9.66    21.85    29.61    37.62
 9.80    23.86    29.01    38.49
 9.14    18.08    28.81    36.37
11.26    22.40    29.81    39.23
 7.86    21.51    30.53    40.45
 8.90    18.46    30.58    43.88
 7.89    21.19    28.03    34.53
11.68    23.64    30.19    40.51
 9.73    20.97    28.91    38.07
11.67    23.01    28.67    30.68
 8.00    20.60    29.81    39.06
10.13    19.09    29.05    34.42
11.75    20.74    30.44    43.67
```

Program 1_7_1.sas

This program uses the data set 1_6.dat produced by Program 1_6.sas. A graphics file 1_7_1.cgm is output.

Program 1_7_2.sas

This program uses the data set 1_6.dat produced by Program 1_6.sas. A graphics file 1_7_2.cgm is output.

Program 1_7_3.sas

This program uses the data set 1_6.dat produced by Program 1_6.sas. A graphics file 1_7_3.cgm is output.

Program 1_7_4.sas

This program uses the data sets ycondx.dat and rhowerx.dat. The file ycondx.dat is the file of residuals resulting after the three dependent variables were regressed on the five independent variables.

```
-13.41      -1.69     -35.92
 -2.67      -1.48     -17.86
-14.65      -2.36       9.33
  1.44       1.42      14.29
 -2.98      -2.83       6.70
  2.07      -1.24       1.68
 14.64      -1.57      12.55
 16.40      -0.55     -25.70
-18.26      -4.02     -36.80
 -1.65       0.94      -2.17
  4.30      -2.30     -15.92
 16.22       3.43       1.96
  5.13       0.35     -30.57
 -9.69       2.05     -42.17
 -1.45       0.68       0.23
 10.70      -0.29      -2.00
-22.32      -2.16     -14.51
  6.33      -1.16      -2.23
-12.14      -2.84     -18.02
  8.14       0.95      37.33
  4.35       5.09      40.15
  6.12       1.86      18.23
 -5.67       3.75      31.52
-10.09      -2.38       1.94
 19.99       4.52     -23.17
  2.71      -2.11       5.41
  9.77      -1.86     -22.15
  7.19      -1.38      12.93
-16.24       1.38       8.56
 14.37       2.28      32.56
 -7.86       0.06      52.93
-10.78       3.42       0.88
```

The data set rhowerx.dat is a subset of the Rhower data contained in the file 5_6.dat. It contains observations for the five independent variables.

```
 0 10  8 21 22
 7  3 21 28 21
 7  9 17 31 30
 6 11 16 27 25
20  7 21 28 16
 4 11 18 32 29
 6  7 17 26 23
 5  2 11 22 23
 3  5 14 24 20
16 12 16 27 30
 5  3 17 25 24
 2 11 10 26 22
 1  4 14 25 19
11  5 18 27 22
 0  0  3 16 11
 5  8 11 12 15
 1  6 10 28 23
 1  9 12 30 18
 0 13 13 19 16
 4  6 14 27 19
 4  5 16 21 24
 1  6 15 23 28
 5  8 14 25 24
 4  5 11 16 22
 5  7 17 26 15
 0  4  8 16 14
 4 17 21 27 31
 5  8 20 28 26
 4  7 19 20 13
 4  7 10 23 19
 2  6 14 25 17
 5 10 18 27 26
```

The graphics files produced by this program are 1_7_4_1y.cgm, 1_7_4_2y.cgm, 1_7_4_3y.cgm, 1_7_4_1r.cgm, 1_7_4_2r.cgm, 1_7_4_3r.cgm, 1_7_4_4r.cgm, and 1_7_4_5r.cgm.

Program 1_8_1.sas

This program uses the data set 1_6.da2. This is the same data set produced by Program 1_6.sas but with a column of observation numbers added as the first column.

```
 1    10.28    18.52    29.67    35.68
 2    11.93    20.28    28.91    36.05
 3     8.22    17.40    28.72    35.26
 4     8.41    21.69    32.14    49.37
 5    10.03    19.21    31.13    41.66
 6    12.85    20.34    29.29    41.66
 7     7.78    19.42    30.00    34.77
 8    10.55    17.68    31.29    44.35
 9    10.11    21.41    29.36    35.77
10    12.25    21.00    30.64    44.32
11     8.71    23.28    30.08    40.91
12    12.09    18.06    30.29    41.08
13     9.79    19.77    32.69    49.96
14    11.62    19.02    28.22    32.90
15     8.99    20.10    29.68    39.86
16    10.86    20.55    31.15    45.87
```

```
17      12.48      19.62      29.90      38.51
18      11.40      22.78      28.02      31.07
19      11.82      22.57      29.52      36.43
20      11.32      20.38      29.37      35.64
21      11.38      21.33      31.80      44.31
22      13.77      20.98      30.34      38.94
23       9.59      22.25      29.88      41.84
24       8.63      16.50      29.92      40.95
25      11.30      20.35      29.29      34.36
26      12.30      18.54      29.17      36.57
27      13.02      22.29      28.36      35.73
28       8.88      19.95      30.73      44.40
29      12.23      20.37      30.00      43.40
30       8.53      18.09      32.77      51.09
31       8.80      20.95      29.45      34.25
32       8.38      19.97      29.94      41.71
33       9.53      20.03      29.63      39.88
34      11.23      19.16      29.18      38.21
35      10.41      21.65      31.75      49.20
36      12.10      21.76      29.17      38.06
37      12.23      19.13      29.88      36.57
38       9.66      21.85      29.61      37.62
39       9.80      23.86      29.01      38.49
40       9.14      18.08      28.81      36.37
41      11.26      22.40      29.81      39.23
42       7.86      21.51      30.53      40.45
43       8.90      18.46      30.58      43.88
44       7.89      21.19      28.03      34.53
45      11.68      23.64      30.19      40.51
46       9.73      20.97      28.91      38.07
47      11.67      23.01      28.67      30.68
48       8.00      20.60      29.81      39.06
49      10.13      19.09      29.05      34.42
50      11.75      20.74      30.44      43.67
```

A graphics file 1_8_1.cgm is output.

Program 1_8_2.sas

This program uses the data set ycondx.da2. This is the same data set as ycondx.dat but with a column of observation numbers added as the first column.

```
1     -13.41      -1.69     -35.92
2      -2.67      -1.48     -17.86
3     -14.65      -2.36       9.33
4       1.44       1.42      14.29
5      -2.98      -2.83       6.70
6       2.07      -1.24       1.68
7      14.64      -1.57      12.55
8      16.40      -0.55     -25.70
9     -18.26      -4.02     -36.80
10     -1.65       0.94      -2.17
11      4.30      -2.30     -15.92
12     16.22       3.43       1.96
13      5.13       0.35     -30.57
14     -9.69       2.05     -42.17
15     -1.45       0.68       0.23
16     10.70      -0.29      -2.00
17    -22.32      -2.16     -14.51
18      6.33      -1.16      -2.23
19    -12.14      -2.84     -18.02
```

```
20      8.14     0.95    37.33
21      4.35     5.09    40.15
22      6.12     1.86    18.23
23     -5.67     3.75    31.52
24    -10.09    -2.38     1.94
25     19.99     4.52   -23.17
26      2.71    -2.11     5.41
27      9.77    -1.86   -22.15
28      7.19    -1.38    12.93
29    -16.24     1.38     8.56
30     14.37     2.28    32.56
31     -7.86     0.06    52.93
32    -10.78     3.42     0.88
```

A graphics file 1_8_2.cgm is output.

Program 1_9_1.sas

This program uses the data set 5_ 6.dat which is the Rhower data set in Timm, N.H. (1975, p.281 and p.345).

Program 1_9_2.sas

This program does not input a data set but rather generates data and outputs the graphics files 1_9_2_1.cgm and 1_9_2_2.cgm.

Program 1_10.sas

This program uses the data set 5_ 6.dat, the same data set used by Program 1_9_1.sas.

Program 1_11.sas

This program uses the data set exp1_6.dat. This is the same data set produced by Program 1_6.sas, but transformed by y=exp(x).

```
 29063.57   109948374    7.688E12  3.13745E15
151698.83   639330658  3.59716E12   4.5204E15
  3698.38  36153566.3  2.97585E12  2.06639E15
  4473.25  2639650670  9.09157E13  2.75238E21
 22789.77   219431766  3.31827E13  1.24294E18
381699.28   680293917  5.26881E12  1.24085E18
  2386.98   272272420  1.06898E13  1.26209E15
 38226.50  47620446.7  3.86329E13  1.81999E19
 24487.89  1978553084  5.61984E12  3.44175E15
208847.22  1323591048  2.02428E13  1.77812E19
  6082.12  1.28429E10  1.16325E13  5.82571E17
177373.85  69777905.0   1.4218E13   6.9397E17
 17846.14   386076979  1.57865E14  4.96912E21
111387.69   181529398  1.81096E12  1.94953E14
  8053.79   538067666  7.77228E12  2.04295E17
 52212.82   844135796  3.37144E13  8.33475E19
264210.87   333050675  9.68268E12   5.3014E16
 89527.14  7811492215  1.47279E12  3.11403E13
136181.37  6330352253  6.64021E12  6.61607E15
 82360.05   712229707  5.71764E12  3.02188E15
 87709.27  1832088839  6.45971E13  1.76004E19
```

```
957392.80  1291412236   1.4943E13  8.14021E16
 14595.46  4596179859  9.48909E12  1.47864E18
  5590.57  14605103.6  9.87909E12  6.10812E17
 81208.83   686241359  5.23758E12  8.33754E14
220662.85   113219893  4.65734E12  7.62086E15
451228.04  4785761881  2.08206E12  3.28838E15
  7171.21   462169873  2.20985E13  1.91908E19
204031.97   705102243  1.07293E13  7.02601E18
  5078.52  71946674.6  1.69833E14  1.54363E22
  6667.42  1249255107  6.18937E12  7.50411E14
  4376.54   470402526  1.01103E13  1.30612E18
 13820.96   498845302  7.39943E12  2.08882E17
 75634.08   210453368  4.68517E12  3.92988E16
 33116.29  2536035144  6.17005E13  2.32425E21
180406.47  2815286421  4.65206E12  3.37879E16
204440.99   203950861  9.45947E12   7.6079E15
 15697.88  3093101752  7.24266E12   2.1737E16
 18038.71  2.30064E10  3.98162E12  5.20012E16
  9306.02  71478360.5  3.26074E12  6.25036E15
 77575.93  5326619194  8.84626E12  1.09189E17
  2583.53  2201603388  1.80931E13  3.69987E17
  7311.21   103904810   1.9024E13  1.13447E19
  2659.12  1589597149  1.49174E12  9.87601E14
118760.90  1.84723E10  1.28903E13  3.90247E17
 16748.52  1278438566   3.6101E12  3.40684E16
116633.26  9800527332  2.83381E12  2.11976E13
  2973.87   884117021  8.83769E12  9.15975E16
 25203.74   196034406   4.1469E12  8.91546E14
126357.47  1012309139  1.66282E13   9.2314E18
```

Program 2_6.sas

This program uses the data set 2_6.dat.

```
 1 11.2  587 16.5 6.2
 2 13.4  643 20.5 6.4
 3 40.7  635 26.3 9.3
 4 5.3   692 16.5 5.3
 5 24.8 1248 19.2 7.3
 6 12.7  643 16.5 5.9
 7 20.9 1964 20.2 6.4
 8 35.7 1531 21.3 7.6
 9  8.7  713 17.2 4.9
10  9.6  749 14.3 6.4
11 14.5 7895 18.1 6.0
12 26.9  762 23.1 7.4
13 15.7 2793 19.1 5.8
14 36.2  741 24.7 8.6
15 18.1  625 18.6 6.5
16 28.9  854 24.9 8.3
17 14.9  716 17.9 6.7
18 25.8  921 22.4 8.6
19 21.7  595 20.2 8.4
20 25.7 3353 16.9 6.7
```

Program 2_7.sas

This program uses the data set 2_7.dat.

```
1  3
1  2
1  4
1  3
1  1
1  5
2  7
2  8
2  4
2 10
2  6
3  3
3  2
3  1
3  2
3  4
3  2
3  3
3  1
4 10
4 12
4  8
4  5
4 12
4 10
4  9
```

Data from *Statistical principles in experimental design*, Second Edition, p. 213, by B.J. Winer.

Program 2_8.sas

This program uses the data set 2_8.dat.

```
1 1  5
1 2  8
1 2 10
1 2  9
2 3  8
2 3 10
2 4  6
2 4  2
2 4  1
2 4  3
2 5  3
2 5  7
```

Data from *Linear models*, p. 249, by S.R.Searle.

Program 2_9.sas

This program uses the data set 2_9.dat.

```
3 1 6 1 0 0 6 0 0
2 1 6 1 0 0 6 0 0
3 1 5 1 0 0 5 0 0
3 1 4 1 0 0 4 0 0
4 1 3 1 0 0 3 0 0
5 1 3 1 0 0 3 0 0
5 1 2 1 0 0 2 0 0
6 1 2 1 0 0 2 0 0
5 2 5 0 1 0 0 5 0
6 2 5 0 1 0 0 5 0
8 2 3 0 1 0 0 3 0
8 2 2 0 1 0 0 2 0
7 2 1 0 1 0 0 1 0
9 2 1 0 1 0 0 1 0
5 3 4 0 0 1 0 0 4
7 3 1 0 0 1 0 0 1
8 3 1 0 0 1 0 0 1
7 3 2 0 0 1 0 0 2
```

Artificial data set.

Program 3_3.sas

This program uses the data set 3_3.dat.

```
1 1  3 1 0 0 0 0 0 0 0
1 1  6 1 0 0 0 0 0 0 0
1 1  3 1 0 0 0 0 0 0 0
1 2  3 0 1 0 0 0 0 0 0
1 2  4 0 1 0 0 0 0 0 0
1 2  5 0 1 0 0 0 0 0 0
1 2  4 0 1 0 0 0 0 0 0
1 2  3 0 1 0 0 0 0 0 0
1 3  7 0 0 1 0 0 0 0 0
1 3  8 0 0 1 0 0 0 0 0
1 3  7 0 0 1 0 0 0 0 0
1 4  6 0 0 0 1 0 0 0 0
1 4  7 0 0 0 1 0 0 0 0
1 4  8 0 0 0 1 0 0 0 0
1 4  9 0 0 0 1 0 0 0 0
1 4  8 0 0 0 1 0 0 0 0
2 1  1 0 0 0 0 1 0 0 0
2 1  2 0 0 0 0 1 0 0 0
2 1  2 0 0 0 0 1 0 0 0
2 1  2 0 0 0 0 1 0 0 0
2 2  2 0 0 0 0 0 1 0 0
2 2  3 0 0 0 0 0 1 0 0
2 2  4 0 0 0 0 0 1 0 0
2 2  3 0 0 0 0 0 1 0 0
2 3  5 0 0 0 0 0 0 1 0
2 3  6 0 0 0 0 0 0 1 0
2 3  5 0 0 0 0 0 0 1 0
2 3  6 0 0 0 0 0 0 1 0
2 4 10 0 0 0 0 0 0 0 1
2 4 10 0 0 0 0 0 0 0 1
```

```
2 4  9 0 0 0 0 0 0 0 0 1
2 4 11 0 0 0 0 0 0 0 0 1
```

Data from Timm, N.H., and Carlson, J.E. (1975, p. 58). By permission of the publisher. Data from *Experimental design: procedures for the behavioral sciences*, by R.E. Kirk. Copyright © 1995, 1982, 1968 Brooks/Cole Publishing Company, Pacific Grove, CA 93950, a division of International Thomson Publishing Inc. Reprinted by permission of the publisher.

Program 3_4.sas

This program uses the data set 3_4.dat.

```
1 1 2  7 1 0 0 0 0 0 0 0 0
1 2 3  5 0 1 0 0 0 0 0 0 0
1 3 1  4 0 0 1 0 0 0 0 0 0
2 1 3  5 0 0 0 1 0 0 0 0 0
2 2 1  6 0 0 0 0 1 0 0 0 0
2 3 2 11 0 0 0 0 0 1 0 0 0
3 1 1  8 0 0 0 0 0 0 1 0 0
3 2 2  8 0 0 0 0 0 0 0 1 0
3 3 3  9 0 0 0 0 0 0 0 0 1
```

Program 3_5_1.sas

This program uses the data set 3_5_1.dat.

```
3 1 0 0 0 0 0 0 0 0 0 0 0 0 0 0 0 0 0
4 0 1 0 0 0 0 0 0 0 0 0 0 0 0 0 0 0 0
3 0 0 1 0 0 0 0 0 0 0 0 0 0 0 0 0 0 0
2 0 0 0 1 0 0 0 0 0 0 0 0 0 0 0 0 0 0
2 0 0 0 0 1 0 0 0 0 0 0 0 0 0 0 0 0 0
1 0 0 0 0 0 1 0 0 0 0 0 0 0 0 0 0 0 0
3 0 0 0 0 0 0 1 0 0 0 0 0 0 0 0 0 0 0
7 0 0 0 0 0 0 0 1 0 0 0 0 0 0 0 0 0 0
7 0 0 0 0 0 0 0 0 1 0 0 0 0 0 0 0 0 0
5 0 0 0 0 0 0 0 0 0 1 0 0 0 0 0 0 0 0
4 0 0 0 0 0 0 0 0 0 0 1 0 0 0 0 0 0 0
6 0 0 0 0 0 0 0 0 0 0 0 1 0 0 0 0 0 0
3 0 0 0 0 0 0 0 0 0 0 0 0 1 0 0 0 0 0
4 0 0 0 0 0 0 0 0 0 0 0 0 0 1 0 0 0 0
6 0 0 0 0 0 0 0 0 0 0 0 0 0 0 1 0 0 0
2 0 0 0 0 0 0 0 0 0 0 0 0 0 0 0 1 0 0
3 0 0 0 0 0 0 0 0 0 0 0 0 0 0 0 0 1 0
5 0 0 0 0 0 0 0 0 0 0 0 0 0 0 0 0 0 1
```

Artificial data set.

Program 3_5_2.sas

This program uses the data set 3_5_2.dat.

```
1 20 21 42 32 32
1 67 29 56 39 41
1 37 25 28 31 34
1 42 38 36 19 35
1 57 32 21 30 29
1 39 38 54 31 28
1 43 20 46 42 31
1 35 34 43 35 42
1 41 23 51 27 30
1 39 24 35 26 32
2 47 25 36 21 27
2 53 32 48 46 54
2 38 33 42 48 49
2 60 41 67 53 50
2 37 35 45 34 46
2 59 37 52 36 52
2 67 33 61 31 50
2 43 27 36 33 32
2 64 53 62 40 43
2 41 34 47 37 46
```

Data from Timm, N.H. (1975, p. 244). By permission of the author. Data provided by Dr. Paul Ammon, University of California, Berkeley.

Program 3_5_3.sas

This program uses the same data set as 3_5_2.sas.

Program 3_6_1.sas

This program uses the same data set as 2_9.sas.

Program 3_6_2.sas

This program uses the data set 3_6_2.dat.

```
3 1 6 2 1 0 0 6 0 0 2 0 0
2 1 6 2 1 0 0 6 0 0 2 0 0
3 1 5 2 1 0 0 5 0 0 2 0 0
3 1 4 4 1 0 0 4 0 0 4 0 0
4 1 3 5 1 0 0 3 0 0 5 0 0
5 1 3 5 1 0 0 3 0 0 5 0 0
5 1 2 6 1 0 0 2 0 0 6 0 0
6 1 2 5 1 0 0 2 0 0 5 0 0
5 2 5 2 0 1 0 0 5 0 0 2 0
6 2 5 1 0 1 0 0 5 0 0 1 0
8 2 3 2 0 1 0 0 3 0 0 2 0
8 2 2 7 0 1 0 0 2 0 0 7 0
7 2 1 8 0 1 0 0 1 0 0 8 0
```

```
9 2 1 7 0 1 0 0 1 0 0 7 0
5 3 4 3 0 0 1 0 0 4 0 0 3
7 3 1 1 0 0 1 0 0 1 0 0 1
8 3 1 6 0 0 1 0 0 1 0 0 6
7 3 2 6 0 0 1 0 0 2 0 0 6
```

Artificial data set.

Program 4_5.sas

This program uses the data set 4_5.dat.

```
27 73
21 66
22 63
26 79
25 68
28 67
24 75
25 71
23 70
20 65
29 79
24 72
20 70
38 91
32 76
33 69
31 66
34 73
37 78
38 87
33 76
35 79
30 73
37 68
31 80
39 75
46 89
49 101
40 70
42 72
43 80
46 83
43 75
49 80
40 90
48 70
42 85
44 71
46 80
47 96
45 92
55 76
54 71
57 99
52 86
53 79
56 92
52 85
```

```
57 109
50 71
59 90
50 91
52 100
58 80
```

The graphics file 4_5.cgm is output.

Program 4_7_2.sas

The data set is included within the program.

Program 4_7_3.sas

The data set is included within the program.

Data from Grizzle, J.E., and Allen D.M. (1969). By permission of the publisher.

Program 4_7_4.sas

The data set is included within the program.

Data from Marascuilo and McSweeney (1977, p. 205). By permission of Maryellen McSweeney.

Program 5_6.sas

This program uses the data set 5_6.dat.

```
 68 15 24  1  0 10  8 21 22
 82 11  8  1  7  3 21 28 21
 82 13 88  1  7  9 17 31 30
 91 18 82  1  6 11 16 27 25
 82 13 90  1 20  7 21 28 16
100 15 77  1  4 11 18 32 29
100 13 58  1  6  7 17 26 23
 96 12 14  1  5  2 11 22 23
 63 10  1  1  3  5 14 24 20
 91 18 98  1 16 12 16 27 30
 87 10  8  1  5  3 17 25 24
105 21 88  1  2 11 10 26 22
 87 14  4  1  1  4 14 25 19
 76 16 14  1 11  5 18 27 22
 66 14 38  1  0  0  3 16 11
 74 15  4  1  5  8 11 12 15
 68 13 64  1  1  6 10 28 23
 98 16 88  1  1  9 12 30 18
 63 15 14  1  0 13 13 19 16
 94 16 99  1  4  6 14 27 19
```

```
 82 18 50  1  4  5 16 21 24
 89 15 36  1  1  6 15 23 28
 80 19 88  1  5  8 14 25 24
 61 11 14  1  4  5 11 16 22
102 20 24  1  5  7 17 26 15
 71 12 24  1  0  4  8 16 14
102 16 24  1  4 17 21 27 31
 96 13 50  1  5  8 20 28 26
 55 16  8  1  4  7 19 20 13
 96 18 98  1  4  7 10 23 19
 74 15 98  1  2  6 14 25 17
 78 19 50  1  5 10 18 27 26
```

Data from *Multivariate analysis with application in education and psychology*, by N.H. Timm. Copyright © 1975 Neil H. Timm. By permission of the author. Data provided by Dr. William D. Rhower, University of California, Berkeley.

Program 5_7.sas

This program uses the data set 5_7.dat.

```
1 1 1 1 22
1 1 1 2 21
1 1 1 3 14
1 1 2 3 31
1 1 2 4 25
1 2 1 3 31
1 2 1 4 41
1 2 2 4 66
1 2 2 5 55
1 2 2 6 61
1 3 1 1 11
1 3 1 2 21
1 3 2 1 41
1 3 2 2 21
2 1 1 3 31
2 1 1 5 66
2 1 2 2 45
2 2 1 2 21
2 2 2 2 21
2 2 2 3 31
2 3 1 4 41
2 3 1 5 47
2 3 1 6 61
2 3 2 4 41
2 3 2 5 55
3 1 1 2 21
3 1 1 3 31
3 1 2 1 66
3 2 1 4 41
3 2 1 5 51
3 2 1 6 61
3 2 2 2 47
3 2 2 3 35
3 2 2 4 41
3 3 1 1 18
3 3 1 2 21
3 3 1 3 31
3 3 2 5 57
3 3 2 6 64
3 3 2 7 77
```

Artificial data set.

Program 5_8.sas

This program uses the data set 5_8.dat.

```
G1  3  6 42
G1  6  3 57
G1  3  3 33
G1  3  3 47
G1  1  2 32
G1  2  1 35
G1  2  2 33
G1  2  2 39
G2  4  5 47
G2  5  4 49
G2  4  3 42
G2  3  4 41
G2  2  3 38
G2  3  2 43
G2  4  3 48
G2  3  4 45
G3  7  8 61
G3  8  7 65
G3  7  6 64
G3  6  7 56
G3  5  6 52
G3  6  5 58
G3  5  6 53
G3  6  5 54
G4  7  6 65
G4  8  9 74
G4  9  8 80
G4  8 10 73
G4 10 10 85
G4 10  9 82
G4  9 11 78
G4 11  9 89
```

Data provided by M.M. Timm, R.N. and M.P.H.

Program 5_9.sas

This program uses the data set 5_9.dat.

```
G1 5.7 4.67 17.6 1.50 .104 1.50 1.88  5.15 8.40  7.5 .14 205 24
G1 5.5 4.67 13.4 1.65 .245 1.32 2.24  5.75 4.50  7.1 .11 160 32
G1 6.6 2.70 20.3 0.90 .097 0.89 1.28  4.35 1.20  2.3 .10 480 17
G1 5.7 3.49 22.3 1.75 .174 1.50 2.24  7.55 2.75  4.0 .12 230 30
G1 5.6 3.49 20.5 1.40 .210 1.19 2.00  8.50 3.30  2.0 .12 235 30
G1 6.0 3.49 18.5 1.20 .275 1.03 1.84 10.25 2.00  2.0 .12 215 27
G1 5.3 4.84 12.1 1.90 .170 1.87 2.40  5.95 2.60 16.8 .14 215 25
G1 5.4 4.84 12.0 1.65 .164 1.68 3.00  6.30 2.72 14.5 .14 190 30
G1 5.4 4.84 10.1 2.30 .275 2.08 2.68  5.45 2.40  0.9 .20 190 28
G1 5.6 4.48 14.7 2.35 .210 2.55 3.00  3.75 7.00  2.0 .21 175 24
G1 5.6 4.48 14.8 2.35 .050 1.32 2.84  5.10 4.00  0.4 .12 145 26
G1 5.6 4.48 14.4 2.50 .143 2.38 2.84  4.05 8.00  3.8 .18 155 27
G2 5.2 3.48 18.1 1.50 .153 1.20 2.60  9.00 2.35 14.5 .13 220 31
G2 5.2 3.48 19.7 1.65 .203 1.73 1.88  5.30 2.52 12.5 .20 300 23
G2 5.6 3.48 16.9 1.40 .074 1.15 1.72  9.85 2.45  8.0 .07 305 32
G2 5.8 2.63 23.7 1.65 .155 1.58 1.60  3.60 3.75  4.9 .10 275 20
G2 6.0 2.63 19.2 0.90 .155 0.96 1.20  4.05 3.30  0.2 .10 405 18
G2 5.3 2.63 18.0 1.60 .129 1.68 2.00  4.40 3.00  3.6 .18 210 23
G2 5.4 4.46 14.8 2.45 .245 2.15 3.12  7.15 1.81 12.0 .13 170 31
G2 5.6 4.46 15.6 1.65 .422 1.42 2.56  7.25 1.92  5.2 .15 235 28
G2 5.3 2.80 16.2 1.65 .063 1.62 2.04  5.30 3.90 10.2 .12 185 21
```

```
G2 5.4 2.80 14.1 1.25 .042 1.62 1.84  3.10 4.10  8.5 .30 255 20
G2 5.5 2.80 17.5 1.05 .030 1.56 1.48  2.40 2.10  9.6 .20 265 15
G2 5.4 2.57 14.1 2.70 .194 2.77 2.56  4.25 2.60  6.9 .17 305 26
G2 5.4 2.57 19.1 1.60 .139 1.59 1.88  5.80 2.30  4.7 .16 440 24
G2 5.2 2.57 22.5 0.85 .046 1.65 1.20  1.55 1.50  3.5 .21 430 16
G3 5.5 1.26 17.0 0.70 .094 0.97 1.24  4.55 2.90  1.9 .12 350 18
G3 5.9 1.26 12.5 0.80 .039 0.80 0.64  2.65 0.72  0.7 .13 475 10
G3 5.6 2.52 21.5 1.80 .142 1.77 2.60  6.50 2.48  8.3 .17 195 33
G3 5.6 2.52 22.2 1.05 .080 1.17 1.48  4.85 2.20  9.3 .14 375 25
G3 5.3 2.52 13.0 2.20 .215 1.85 3.48  8.75 2.40 13.0 .11 160 35
G3 5.6 3.24 13.0 3.55 .166 3.18 3.48  5.20 3.50 18.3 .22 240 33
G3 5.5 3.24 10.9 3.30 .111 2.79 3.04  4.75 2.52 10.5 .21 205 31
G3 5.6 3.24 12.0 3.65 .180 2.40 3.00  5.85 3.00 14.5 .21 270 34
G3 5.4 1.56 22.8 0.55 .069 1.00 1.14  2.85 2.90  3.3 .15 475 16
G4 5.3 1.56 16.5 2.05 .222 1.49 2.40  6.55 3.90  6.3 .11 430 31
G4 5.2 1.56 18.4 1.05 .267 1.17 1.36  6.60 2.00  4.9 .11 490 28
G4 5.8 4.12 12.5 5.90 .093 3.80 3.84  2.90 3.00 22.5 .24 105 32
G4 5.7 4.12  8.7 4.25 .147 3.62 5.32  3.00 3.55 19.5 .20 115 25
G4 5.5 2.14  9.4 3.85 .217 3.36 5.52  3.40 5.20  1.3 .31 097 28
G4 5.4 2.14 15.0 2.45 .418 2.38 2.40  5.40 1.81 20.0 .17 325 27
G4 5.4 2.14 12.9 1.70 .323 1.74 2.48  4.45 1.88  1.0 .15 310 23
G4 4.9 2.03 12.1 1.80 .205 2.00 2.24  4.30 3.70  5.0 .19 245 25
G4 5.0 2.03 13.2 3.65 .348 1.95 2.12  5.00 1.80  3.0 .15 170 26
G4 4.9 2.03 11.5 2.25 .320 2.25 3.12  3.40 2.50  5.1 .18 220 34
```

Data from Smith, H., Gnanadesikan, R., and Hughes, J.B. (1962). By permission of the publisher.

Program 5_10.sas

This program uses the data set 5_10.dat.

```
1 20 21 42 32 32
1 67 29 56 39 41
1 37 25 28 31 34
1 42 38 36 19 35
1 57 32 21 30 29
1 39 38 54 31 28
1 43 20 46 42 31
1 35 34 43 35 42
1 41 23 51 27 30
1 39 24 35 26 32
2 47 25 36 21 27
2 53 32 48 46 54
2 38 33 42 48 49
2 60 41 67 53 50
2 37 35 45 34 46
2 59 37 52 36 52
2 67 33 61 31 50
2 43 27 36 33 32
2 64 53 62 40 43
2 41 34 47 37 46
```

Data from Timm, N.H. (1975, p. 244). By permission of the author. Data provided by Dr. Paul Ammon, University of California, Berkeley.

Program 5_11.sas

This program uses the data set 5_11.dat.

```
1  2  4  7 1 0 0
1  2  6 10 1 0 0
1  3  7 10 1 0 0
1  7  9 11 1 0 0
1  6  9 12 1 0 0
2  5  6 10 0 1 0
2  4  5 10 0 1 0
2  7  8 11 0 1 0
2  8  9 11 0 1 0
2 11 12 13 0 1 0
3  3  4  7 0 0 1
3  3  6  9 0 0 1
3  4  7  9 0 0 1
3  8  8 10 0 0 1
3  7 10 10 0 0 1
```

Data from Timm, N.H. (1975, p. 454). By permission of the author. Data from *Experimental design in psychological research*, Third Edition, by A.L. Edwards. Copyright © 1950, 1960. A.L. Edwards, Copyright © 1968 Holt, Rinehart, and Winston, Inc.

Program 5_12.sas

The data set used is included in the program.

Data from Timm, N.H. (1975, p. 211). By permission of the author.

Program 6_5.sas

This program uses the data sets mmm.dat and mixed.dat.

```
mmm.dat
1 117.0 117.5 118.5 59.0 59.0 60.0 10.5 16.5 16.5
1 109.0 110.5 111.0 60.0 61.5 61.5 30.5 30.5 30.5
1 117.0 120.0 120.5 60.0 61.5 62.0 23.5 23.5 23.5
1 122.0 126.0 127.0 67.5 70.5 71.5 33.0 32.0 32.5
1 116.0 118.5 119.5 61.5 62.5 63.5 24.5 24.5 24.5
1 123.0 126.0 127.0 65.5 61.5 67.5 22.0 22.0 22.0
1 130.5 132.0 134.5 68.5 69.5 71.0 33.0 32.5 32.0
1 126.5 128.5 130.5 69.0 71.0 73.0 20.0 20.0 20.0
1 113.0 116.5 118.0 58.0 59.0 60.5 25.0 25.0 24.5
2 128.0 129.0 131.5 67.0 67.5 69.0 24.0 24.0 24.0
2 116.5 120.0 121.5 63.5 65.0 66.0 28.5 29.5 29.5
2 121.5 125.5 127.0 64.5 67.5 69.0 26.5 27.0 27.0
2 109.5 112.0 114.0 54.0 55.5 57.0 18.0 18.5 19.0
2 133.0 136.0 137.5 72.0 73.5 75.5 34.5 34.5 34.5
2 120.0 124.5 126.0 62.5 65.0 66.0 26.0 26.0 26.0
2 129.5 133.5 134.5 65.0 68.0 69.0 18.5 18.5 18.5
2 122.0 124.0 125.5 64.5 65.5 66.0 18.5 18.5 18.5
2 125.0 127.0 128.0 65.5 66.5 67.0 21.5 21.5 21.6
```

```
mixed.dat
1 1 1 117.0 59.0 10.5
1 1 2 117.5 59.0 16.5
1 1 3 118.5 60.0 16.5
1 2 1 109.0 60.0 30.5
1 2 2 110.5 61.5 30.5
1 2 3 111.0 61.5 30.5
1 3 1 117.0 60.0 23.5
1 3 2 120.0 61.5 23.5
1 3 3 120.5 62.0 23.5
1 4 1 122.0 67.5 33.0
1 4 2 126.0 70.5 32.0
1 4 3 127.0 71.5 32.5
1 5 1 116.0 61.5 24.5
1 5 2 118.5 62.5 24.5
1 5 3 119.5 63.5 24.5
1 6 1 123.0 65.5 22.0
1 6 2 126.0 61.5 22.0
1 6 3 127.0 67.5 22.0
1 7 1 130.5 68.5 33.0
1 7 2 132.0 69.5 32.5
1 7 3 134.5 71.0 32.0
1 8 1 126.5 69.0 20.0
1 8 2 128.5 71.0 20.0
1 8 3 130.5 73.0 20.0
1 9 1 113.0 58.0 25.0
1 9 2 116.5 59.0 25.0
1 9 3 118.0 60.5 24.5
2 1 1 128.0 67.0 24.0
2 1 2 129.0 67.5 24.0
2 1 3 131.5 69.0 24.0
2 2 1 116.5 63.5 28.5
2 2 2 120.0 65.0 29.5
2 2 3 121.5 66.0 29.5
2 3 1 121.5 64.5 26.5
2 3 2 125.5 67.5 27.0
2 3 3 127.0 69.0 27.0
2 4 1 109.5 54.0 18.0
2 4 2 112.0 55.5 18.5
2 4 3 114.0 57.0 19.0
2 5 1 133.0 72.0 34.5
2 5 2 136.0 73.5 34.5
2 5 3 137.5 75.5 34.5
2 6 1 120.0 62.5 26.0
2 6 2 124.5 65.0 26.0
2 6 3 126.0 66.0 26.0
2 7 1 129.5 65.0 18.5
2 7 2 133.5 68.0 18.5
2 7 3 134.5 69.0 18.5
2 8 1 122.0 64.5 18.5
2 8 2 124.0 65.5 18.5
2 8 3 125.5 66.0 18.5
2 9 1 125.0 65.5 21.5
2 9 2 127.0 66.5 21.5
2 9 3 128.0 67.0 21.6
```

Data from Timm, N.H. (1980a). Reprinted with the permission of Dr. Thomas Zullo, School of Dental Medicine, University of Pittsburgh.

Program 7_5.sas

This program uses the data set 7_5.dat.

```
1 1 1 1 22 1 0 0 0 0 0 0 0 0 0 0 0 0 0 0 0 0 0
1 1 1 2 21 1 0 0 0 0 0 0 0 0 0 0 0 0 0 0 0 0 0
1 1 1 3 14 1 0 0 0 0 0 0 0 0 0 0 0 0 0 0 0 0 0
1 2 1 3 31 0 1 0 0 0 0 0 0 0 0 0 0 0 0 0 0 0 0
1 2 1 4 25 0 1 0 0 0 0 0 0 0 0 0 0 0 0 0 0 0 0
1 3 1 1 31 0 0 1 0 0 0 0 0 0 0 0 0 0 0 0 0 0 0
1 3 1 2 41 0 0 1 0 0 0 0 0 0 0 0 0 0 0 0 0 0 0
1 1 2 3 66 0 0 0 1 0 0 0 0 0 0 0 0 0 0 0 0 0 0
1 1 2 4 55 0 0 0 1 0 0 0 0 0 0 0 0 0 0 0 0 0 0
1 2 2 4 61 0 0 0 0 1 0 0 0 0 0 0 0 0 0 0 0 0 0
1 2 2 5 11 0 0 0 0 1 0 0 0 0 0 0 0 0 0 0 0 0 0
1 2 2 6 21 0 0 0 0 1 0 0 0 0 0 0 0 0 0 0 0 0 0
1 3 2 1 41 0 0 0 0 0 1 0 0 0 0 0 0 0 0 0 0 0 0
1 3 2 2 21 0 0 0 0 0 1 0 0 0 0 0 0 0 0 0 0 0 0
2 1 1 3 31 0 0 0 0 0 0 1 0 0 0 0 0 0 0 0 0 0 0
2 1 1 5 66 0 0 0 0 0 0 1 0 0 0 0 0 0 0 0 0 0 0
2 2 1 2 45 0 0 0 0 0 0 0 1 0 0 0 0 0 0 0 0 0 0
2 3 1 4 21 0 0 0 0 0 0 0 0 1 0 0 0 0 0 0 0 0 0
2 3 1 5 21 0 0 0 0 0 0 0 0 1 0 0 0 0 0 0 0 0 0
2 3 1 6 31 0 0 0 0 0 0 0 0 1 0 0 0 0 0 0 0 0 0
2 1 2 2 41 0 0 0 0 0 0 0 0 0 1 0 0 0 0 0 0 0 0
2 2 2 2 47 0 0 0 0 0 0 0 0 0 0 1 0 0 0 0 0 0 0
2 2 2 3 61 0 0 0 0 0 0 0 0 0 0 1 0 0 0 0 0 0 0
2 3 2 4 41 0 0 0 0 0 0 0 0 0 0 0 1 0 0 0 0 0 0
2 3 2 5 55 0 0 0 0 0 0 0 0 0 0 0 1 0 0 0 0 0 0
3 1 1 2 21 0 0 0 0 0 0 0 0 0 0 0 0 1 0 0 0 0 0
3 1 1 3 31 0 0 0 0 0 0 0 0 0 0 0 0 1 0 0 0 0 0
3 2 1 4 66 0 0 0 0 0 0 0 0 0 0 0 0 0 1 0 0 0 0
3 2 1 5 41 0 0 0 0 0 0 0 0 0 0 0 0 0 1 0 0 0 0
3 2 1 6 51 0 0 0 0 0 0 0 0 0 0 0 0 0 1 0 0 0 0
3 3 1 1 61 0 0 0 0 0 0 0 0 0 0 0 0 0 0 1 0 0 0
3 3 1 2 47 0 0 0 0 0 0 0 0 0 0 0 0 0 0 1 0 0 0
3 3 1 3 35 0 0 0 0 0 0 0 0 0 0 0 0 0 0 1 0 0 0
3 1 2 1 41 0 0 0 0 0 0 0 0 0 0 0 0 0 0 0 1 0 0
3 2 2 2 18 0 0 0 0 0 0 0 0 0 0 0 0 0 0 0 0 1 0
3 2 2 3 21 0 0 0 0 0 0 0 0 0 0 0 0 0 0 0 0 1 0
3 2 2 4 31 0 0 0 0 0 0 0 0 0 0 0 0 0 0 0 0 1 0
3 3 2 5 57 0 0 0 0 0 0 0 0 0 0 0 0 0 0 0 0 0 1
3 3 2 6 64 0 0 0 0 0 0 0 0 0 0 0 0 0 0 0 0 0 1
3 3 2 7 77 0 0 0 0 0 0 0 0 0 0 0 0 0 0 0 0 0 1
```

Artificial data set.

Program 7_6.sas

This program uses the data sets 7_6.dat and 7_6a.dat.

```
7_6.dat
1 1 2 22 1 0 0 0 0 0 1 0 0 0 0 0
1 1 2 21 1 0 0 0 0 0 2 0 0 0 0 0
1 1 3 14 1 0 0 0 0 0 4 0 0 0 0 0
1 1 3 25 1 0 0 0 0 0 4 0 0 0 0 0
1 1 4 31 1 0 0 0 0 0 5 0 0 0 0 0
1 2 3 31 0 1 0 0 0 0 0 3 0 0 0 0
1 2 4 25 0 1 0 0 0 0 0 5 0 0 0 0
1 2 5 30 0 1 0 0 0 0 0 6 0 0 0 0
1 2 5 31 0 1 0 0 0 0 0 5 0 0 0 0
```

```
1 3 5 31 0 0 1 0 0 0 0 0 6 0 0 0
1 3 6 41 0 0 1 0 0 0 0 0 5 0 0 0
1 3 8 50 0 0 1 0 0 0 0 0 6 0 0 0
2 1 3 31 0 0 0 1 0 0 0 0 0 2 0 0
2 1 5 66 0 0 0 1 0 0 0 0 0 4 0 0
2 1 5 45 0 0 0 1 0 0 0 0 0 3 0 0
2 2 5 45 0 0 0 0 1 0 0 0 0 0 5 0
2 2 6 34 0 0 0 0 1 0 0 0 0 0 8 0
2 2 6 31 0 0 0 0 1 0 0 0 0 0 4 0
2 2 7 30 0 0 0 0 1 0 0 0 0 0 6 0
2 3 3 21 0 0 0 0 0 1 0 0 0 0 0 3
2 3 4 21 0 0 0 0 0 1 0 0 0 0 0 3
2 3 4 31 0 0 0 0 0 1 0 0 0 0 0 4

7_6a.dat
1 1 2 22 1
1 1 2 21 2
1 1 3 14 4
1 1 3 25 4
1 1 4 31 5
1 2 3 31 3
1 2 4 25 5
1 2 5 30 6
1 2 5 31 5
1 3 5 31 6
1 3 6 41 5
1 3 8 50 6
2 1 3 31 2
2 1 5 66 4
2 1 5 45 3
2 2 5 45 5
2 2 6 34 8
2 2 6 31 4
2 2 7 30 6
2 3 3 21 3
2 3 4 21 3
2 3 4 31 4
```

Artificial data set.

Program 7_7.sas

This program uses the data set 7_7.dat.

```
1 4.0 4.0 4.1 3.6 3.6 3.8 3.1  1 1 0 0 0
1 4.2 4.3 4.7 4.7 4.8 5.0 5.2  2 1 0 0 0
1 4.3 4.2 4.3 4.3 4.5 5.8 5.4  3 1 0 0 0
1 4.2 4.4 4.6 4.9 5.3 5.6 4.9  4 1 0 0 0
1 4.6 4.4 5.3 5.6 5.9 5.9 5.3  5 1 0 0 0
1 3.1 3.6 4.9 5.2 5.3 4.2 4.1  6 1 0 0 0
1 3.7 3.9 3.9 4.8 5.2 5.4 4.2  7 1 0 0 0
1 4.3 4.2 4.4 5.2 5.6 5.4 4.7  8 1 0 0 0
1 4.6 4.6 4.4 4.6 5.4 5.9 5.6  9 1 0 0 0
2 3.4 3.4 3.5 3.1 3.1 3.7 3.3 10 0 1 0 0
2 3.0 3.2 3.0 3.0 3.1 3.2 3.1 11 0 1 0 0
2 3.0 3.1 3.2 3.0 3.3 3.0 3.0 12 0 1 0 0
2 3.1 3.2 3.2 3.2 3.3 3.1 3.1 13 0 1 0 0
2 3.8 3.9 4.0 3.9 3.5 3.5 3.4 14 0 1 0 0
2 3.0 3.6 3.2 3.1 3.0 3.0 3.0 15 0 1 0 0
2 3.3 3.3 3.3 3.4 3.6 3.1 3.1 16 0 1 0 0
```

```
2 4.2 4.0 4.2 4.1 4.2 4.0 4.0 17 0 1 0 0
2 4.1 4.2 4.3 4.3 4.2 4.0 4.2 18 0 1 0 0
2 4.5 4.4 4.3 4.5 4.3 4.4 4.4 19 0 1 0 0
3 3.2 3.3 3.8 3.8 4.4 4.2 4.0 20 0 0 1 0
3 3.3 3.4 3.4 3.7 3.7 3.6 3.7 21 0 0 1 0
3 3.1 3.3 3.2 3.1 3.2 3.1 3.1 22 0 0 1 0
3 3.6 3.4 3.5 4.6 4.9 5.2 4.4 23 0 0 1 0
3 4.5 4.5 5.4 5.7 4.9 4.0 4.0 24 0 0 1 0
3 3.7 4.0 4.4 4.2 4.6 4.8 5.4 25 0 0 1 0
3 3.5 3.9 5.8 5.4 4.9 5.3 5.6 26 0 0 1 0
3 3.9 4.0 4.1 5.0 5.4 4.4 3.9 27 0 0 1 0
4 3.1 3.5 3.5 3.2 3.0 3.0 3.2 28 0 0 0 1
4 3.3 3.2 3.6 3.7 3.7 4.2 4.4 29 0 0 0 1
4 3.5 3.9 4.7 4.3 3.9 3.4 3.5 30 0 0 0 1
4 3.4 3.4 3.5 3.3 3.4 3.2 3.4 31 0 0 0 1
4 3.7 3.8 4.2 4.3 3.6 3.8 3.7 32 0 0 0 1
4 4.0 4.6 4.8 4.9 5.4 5.6 4.8 33 0 0 0 1
4 4.2 3.9 4.5 4.7 3.9 3.8 3.7 34 0 0 0 1
4 4.1 4.1 3.7 4.0 4.1 4.6 4.7 35 0 0 0 1
4 3.5 3.6 3.6 4.2 4.8 4.9 5.0 36 0 0 0 1
```

Data from Grizzle, J.E., and Allen, D.M. (1969). By permission of the publisher.

Program 7_8.sas

This program uses one of the same data sets as Program 6_5.sas.

Program 7_9.sas

This program uses one of the same data sets as Program 6_5.sas.

Program 8_7.sas

This program uses the data set 8_7.dat.

```
1 47.8 48.8 49.0 49.7  1
1 46.4 47.3 47.7 48.4  2
1 46.3 46.8 47.8 48.5  3
1 45.1 45.3 46.1 47.2  4
1 47.6 48.5 48.9 49.3  5
1 52.5 53.2 53.3 53.7  6
1 51.2 53.0 54.3 54.5  7
1 49.8 50.0 50.3 52.7  8
1 48.1 50.8 52.3 54.4  9
1 45.0 47.0 47.3 48.3 10
1 51.2 51.4 51.6 51.9 11
1 48.5 49.2 53.0 55.5 12
1 52.1 52.8 53.7 55.0 13
1 48.2 48.9 49.3 49.8 14
1 49.6 50.4 51.2 51.8 15
1 50.7 51.7 52.7 53.3 16
1 47.2 47.7 48.4 49.5 17
1 53.3 54.6 55.1 55.3 18
1 46.2 47.5 48.1 48.4 19
1 46.3 47.6 51.3 51.8 20
```

Data from Elston, R.C., and Grizzle, J.E. (1962). By permission of the publisher.

Program 8_8.sas

This program uses the data set 8_8.dat.

```
1 4.0 4.0 4.1 3.6 3.6 3.8 3.1 1 0 0 0
1 4.2 4.3 3.7 3.7 4.8 5.0 5.2 1 0 0 0
1 4.3 4.2 4.3 4.3 4.5 5.8 5.4 1 0 0 0
1 4.2 4.4 4.6 4.9 5.3 5.6 4.9 1 0 0 0
1 4.6 4.4 5.3 5.6 5.9 5.9 5.3 1 0 0 0
1 3.1 3.6 4.9 5.2 5.3 4.2 4.1 1 0 0 0
1 3.7 3.9 3.9 4.8 5.2 5.4 4.2 1 0 0 0
1 4.3 4.2 4.4 5.2 5.6 5.4 4.7 1 0 0 0
1 4.6 4.6 4.4 4.6 5.4 5.9 5.6 1 0 0 0
2 3.4 3.4 3.5 3.1 3.1 3.7 3.3 0 1 0 0
2 3.0 3.2 3.0 3.0 3.1 3.2 3.1 0 1 0 0
2 3.0 3.1 3.2 3.0 3.3 3.0 3.0 0 1 0 0
2 3.1 3.2 3.2 3.2 3.3 3.1 3.1 0 1 0 0
2 3.8 3.9 4.0 2.9 3.5 3.5 3.4 0 1 0 0
2 3.0 3.6 3.2 3.1 3.0 3.0 3.0 0 1 0 0
2 3.3 3.3 3.3 3.4 3.6 3.1 3.1 0 1 0 0
2 4.2 4.0 4.2 4.1 4.2 4.0 4.0 0 1 0 0
2 4.1 4.2 4.3 4.3 4.2 4.0 4.2 0 1 0 0
2 4.5 4.4 4.3 4.5 5.3 4.4 4.4 0 1 0 0
3 3.2 3.3 3.8 3.8 4.4 4.2 3.7 0 0 1 0
3 3.3 3.4 3.4 3.7 3.7 3.6 3.7 0 0 1 0
3 3.1 3.3 3.2 3.1 3.2 3.1 3.1 0 0 1 0
3 3.6 3.4 3.5 4.6 4.9 5.2 4.4 0 0 1 0
3 4.5 4.5 5.4 5.7 4.9 4.0 4.0 0 0 1 0
3 3.7 4.0 4.4 4.2 4.6 4.8 5.4 0 0 1 0
3 3.5 3.9 5.8 5.4 4.9 5.3 5.6 0 0 1 0
3 3.9 4.0 4.1 5.0 5.4 4.4 3.9 0 0 1 0
4 3.1 3.5 3.5 3.2 3.0 3.0 3.2 0 0 0 1
4 3.3 3.2 3.6 3.7 3.7 4.2 4.4 0 0 0 1
4 3.5 3.9 4.7 4.3 3.9 3.4 3.5 0 0 0 1
4 3.4 3.4 3.5 3.3 3.4 3.2 3.4 0 0 0 1
4 3.7 3.8 4.2 4.3 3.6 3.8 3.7 0 0 0 1
4 4.0 4.6 4.8 4.9 5.4 5.6 4.8 0 0 0 1
4 4.2 3.9 4.5 4.7 3.9 3.8 3.7 0 0 0 1
4 4.1 4.1 3.7 4.0 4.1 4.6 4.7 0 0 0 1
4 3.5 3.6 3.6 4.2 4.8 4.9 5.0 0 0 0 1
```

Data from Grizzle, J.E., and Allen, D.M. (1969). These same data were analyzed in Program 7_7.sas. By permission of the publisher.

Program 8_9.sas

This program uses the data set 8_9.dat.

```
1935  317.6 3078.5    2.8  40.29  417.5  10.5  33.1 1170.6  97.8 12.93  191.5
 1.8  209.9 1362.4   53.8
1936  391.8 4661.7   52.6  72.76  837.8  10.2  45.0 2015.8 104.4 25.90  516.0
 0.8  355.3 1807.1   50.5
1937  410.6 5387.1  156.9  66.26  883.9  34.7  77.2 2803.3 118.0 35.05  729.0
 7.4  469.9 2676.3  118.1
1938  257.7 2792.2  209.2  51.60  437.9  51.8  44.6 2039.7 156.2 22.89  560.4
18.1  262.3 1801.9  260.2
1939  330.8 4313.2  203.4  52.41  679.7  64.3  48.1 2256.2 172.6 18.84  519.9
23.5  230.4 1957.3  312.7
```

```
1940  461.2 4643.9  207.2  69.41  727.8  67.1  74.4 2132.2 186.6 28.57  628.5
26.5  261.6 2202.9  254.2
1941  512.0 4551.2  255.2  68.35  643.6  75.2 113.0 1834.1 220.9 48.51  537.1
36.2  472.8 2380.5  261.4
1942  448.0 3244.1  303.7  46.80  410.9  71.4  91.9 1588.0 287.8 43.34  561.2
60.8  445.6 2168.6  298.7
1943  499.6 4053.7  264.1  47.40  588.4  67.1  61.3 1749.4 319.9 37.02  617.2
84.4  361.6 1985.1  301.8
1944  547.5 4379.3  201.6  59.57  698.4  60.5  56.8 1687.2 321.3 37.81  626.7
91.2  288.2 1813.9  279.1
1945  561.2 4840.9  265.0  88.78  846.4  54.6  93.6 2007.7 319.6 39.27  737.2
92.4  258.7 1850.2  213.8
1946  688.1 4900.9  402.2  74.12  893.8  84.8 159.9 2208.3 346.0 53.46  760.5
86.0  420.3 2067.7  232.6
1947  568.9 3526.5  761.5  62.68  579.0  96.8 147.2 1656.7 456.4 55.56  581.4
111.1  420.5 1796.7  264.8
1948  529.2 3254.7  922.4  89.36  694.6 110.2 146.3 1604.4 543.4 49.56  662.3
130.6  494.5 1625.8  306.9
1949  555.1 3700.2 1020.1  78.98  590.3 147.4  98.3 1431.8 618.3 32.04  583.8
141.8  405.1 1667.0  351.1
1950  642.9 3755.6 1099.0 100.66  693.5 163.2  93.5 1610.5 647.4 32.24  635.2
136.7  418.8 1677.4  357.8
1951  755.9 4833.0 1207.7 160.62  809.0 203.5 135.2 1819.4 671.3 54.38  723.8
129.7  588.2 2289.5  342.1
1952  891.2 4924.9 1430.5 145.00  727.0 290.6 157.3 2079.7 726.1 71.78  864.1
145.5  645.2 2159.4  444.2
1953 1304.4 6241.7 1777.3 174.93 1001.5 346.1 179.5 2371.6 800.3 90.08 1193.5
174.8  641.0 2031.3  623.6
1954 1486.7 5593.6 2226.3 172.49  703.2 414.9 189.6 2759.9 888.9 68.60 1188.9
213.5  459.3 2115.5  669.7
```

Data from *Econometric analysis*, Second Edition, p. 445, by W.H. Greene. Copyright © 1992 by Macmillan Publishing Company, a division of Macmillan, Inc. By permission of the publisher.

Program 8_10.sas

This program uses the data set 8_10.dat.

```
1  1 1.28 1.09 1.33 1.24 1 0
1  1 1.60 1.38 2.21 1.90 1 0
1  1 2.46 2.27 2.43 2.19 1 0
1  1 1.41 1.34 1.81 1.47 1 0
1  1 1.40 1.31 0.85 0.85 1 0
1  1 1.12 0.96 1.20 1.12 1 0
1  1 0.90 0.66 0.90 0.78 1 0
1  1 2.41 1.69 2.79 1.90 1 0
1 -1 3.06 1.74 1.38 1.54 0 1
1 -1 2.68 2.41 2.10 2.13 0 1
1 -1 2.60 3.05 2.32 2.18 0 1
1 -1 1.48 1.20 1.30 1.41 0 1
1 -1 2.08 1.70 2.34 2.21 0 1
1 -1 2.72 1.89 2.48 2.05 0 1
1 -1 1.94 0.89 1.11 0.72 0 1
1 -1 3.35 2.41 3.23 2.83 0 1
1 -1 1.16 0.96 1.25 1.01 0 1
```

Data from Patel, H.I. (1983). By permission of the publisher.

Program 8_11.sas

This program uses the data set 8_11.dat.

```
1  1  0  3  8  4 14 1 0 0
1  1  0  5 11  9 18 1 0 0
1  1  0 11 16 14 22 1 0 0
1  0  1  2  6  1  8 0 1 0
1  0  1  8 12  9 14 0 1 0
1  0  1 10  9  9 10 0 1 0
1 -1 -1  7 10  4 10 0 0 1
1 -1 -1  8 14 10 18 0 0 1
1 -1 -1  9 15 12 22 0 0 1
```

Data from *Statistical Principles in Experimental Design*, Second Edition, p. 8012. By Winer, B.J. Copyright © 1971. By permission of the McGraw-Hill Companies.

Program 8_12.sas

This program uses the data set 8_12.dat.

```
1 1 0 0 0 191 223 242 248 266 274 272 279 286 287 286 1
1 1 0 0 0  64  72  81  66  92 114 126 123 134 148 140 1
1 1 0 0 0 206 172 214 239 265 265 262 274 258 288 289 1
1 1 0 0 0 155 171 191 203 219 237 237 220 252 260 245 1
1 1 0 0 0  85 138 204 213 224 247 246 259 255 374 284 1
1 1 0 0 0  15  22  24  24  38  41  46  62  62  79  74 1
2 0 1 0 0  53  53 102 104 105 125 122 150  93 127 132 1
2 0 1 0 0  33  45  50  54  44  47  45  61  50  60  52 1
2 0 1 0 0  16  47  45  34  37  61  51  28  43  40  45 1
2 0 1 0 0 121 167 188 209 224 229 230 269 264 249 268 1
2 0 1 0 0 179 193 206 210 221 234 224 255 246 225 229 1
2 0 1 0 0 114  91 154 152 155 174 196 207 208 229 173 1
2 0 1 0 0  92 115 133 136 148 159 146 180 148 168 169 1
2 0 1 0 0  84  32  97  86  47  87 103 124 110 162 187 1
2 0 1 0 0  30  38  37  40  48  61  64  65  83  91  90 1
2 0 1 0 0  51  66 131 148 181 172 195 170 158 203 215 1
2 0 1 0 0 188 210 221 251 256 268 260 281 286 290 296 1
2 0 1 0 0 137 167 172 212 168 213 190 196 211 213 224 1
2 0 1 0 0 108  23  18  30  29  40  57  37  47  56  55 1
2 0 1 0 0 205 234 260 269 274 282 282 290 298 304 308 1
3 0 0 1 0 181 206 199 237 219 237 232 251 247 254 250 1
3 0 0 1 0 178 208 222 237 255 253 254 276 254 267 275 1
3 0 0 1 0 190 224 224 261 249 291 293 294 295 299 305 1
3 0 0 1 0 127 119 149 196 203 211 207 241 220 188 219 1
3 0 0 1 0  94 144 169 164 182 189 188 164 181 142 152 1
3 0 0 1 0 148 170 202 181 184 186 207 184 195 168 163 1
3 0 0 1 0  99  93 122 145 130 167 153 165 144 156 167 1
3 0 0 1 0 207 237 243 281 273 281 279 294 307 305 305 1
3 0 0 1 0 188 208 235 249 265 271 263 272 285 283 290 1
3 0 0 1 0 140 187 199 205 231 227 228 246 245 263 262 1
3 0 0 1 0 109  95 102  96 135 335 111 146 131 162 171 1
3 0 0 1 0  69  46  67  28  43  55  55  77  73  76  76 1
3 0 0 1 0  69  95 137  99  95 108 129 134 133 131  91 1
3 0 0 1 0  51  59  76 101  72  72 107  91 128 120 133 1
3 0 0 1 0 156 186 198 201 205 210 217 217 219 223 229 1
4 0 0 0 1 201 202 229 232 224 237 217 268 244 275 246 1
4 0 0 0 1 113 126 159 157 137 160 162 171 167 165 185 1
4 0 0 0 1  86  54  75  75  71 130 157 142 173 174 156 1
4 0 0 0 1 115 158 168 175 188 164 184 195 194 206 212 1
```

```
4 0 0 0 1 183 175 217 235 241 251 229 241 233 233 275 1
4 0 0 0 1 131 147 183 181 206 215 197 207 226 244 240 1
4 0 0 0 1  71 105 107  92 101 103  78  87  57  70  71 1
4 0 0 0 1 172 213 263 260 276 273 267 286 283 290 298 1
4 0 0 0 1 224 258 248 257 257 267 260 279 299 289 300 1
4 0 0 0 1 246 257 269 280 289 291 306 301 295 312 311 1
```

Data from Danford, M.B., Hughes, H.M., and McNee, R.C. (1960). By permission of the publisher.

Program 8_13.sas

This program uses the data set 8_13.dat.

```
1 1 0 0 0 191 223 242 248 266 274 272 279 286 287 286 1
1 1 0 0 0  64  72  81  66  92 114 126 123 134 148 140 1
1 1 0 0 0 206 172 214 239 265 265 262 274 258 288 289 1
1 1 0 0 0 155 171 191 203 219 237 237 220 252 260 245 1
1 1 0 0 0  85 138 204 213 224 247 246 259 255 374 284 1
1 1 0 0 0  15  22  24  24  38  41  46  62  62  79  74 1
2 0 1 0 0  53  53 102 104 105 125 122 150  93 127 132 1
2 0 1 0 0  33  45  50  54  44  47  45  61  50  60  52 1
2 0 1 0 0  16  47  45  34  37  61  51  28  43  40  45 1
2 0 1 0 0 121 167 188 209 224 229 230 269 264 249 268 1
2 0 1 0 0 179 193 206 210 221 234 224 255 246 225 229 1
2 0 1 0 0 114  91 154 152 155 174 196 207 208 229 173 1
2 0 1 0 0  92 115 133 136 148 159 146 180 148 168 169 1
2 0 1 0 0  84  32  97  86  47  87 103 124 110 162 187 1
2 0 1 0 0  30  38  37  40  48  61  64  65  83  91  90 1
2 0 1 0 0  51  66 131 148 181 172 195 170 158 203 215 1
2 0 1 0 0 188 210 221 251 256 268 260 281 286 290 296 1
2 0 1 0 0 137 167 172 212 168 213 190 196 211 213 224 1
2 0 1 0 0 108  23  18  30  29  40  57  37  47  56  55 1
2 0 1 0 0 205 234 260 269 274 282 282 290 298 304 308 1
3 0 0 1 0 181 206 199 237 219 237 232 251 247 254 250 1
3 0 0 1 0 178 208 222 237 255 253 254 276 254 267 275 1
3 0 0 1 0 190 224 224 261 249 291 293 294 295 299 305 1
3 0 0 1 0 127 119 149 196 203 211 207 241 220 188 219 1
3 0 0 1 0  94 144 169 164 182 189 188 164 181 142 152 1
3 0 0 1 0 148 170 202 181 184 186 207 184 195 168 163 1
3 0 0 1 0  99  93 122 145 130 167 153 165 144 156 167 1
3 0 0 1 0 207 237 243 281 273 281 279 294 307 305 305 1
3 0 0 1 0 188 208 235 249 265 271 263 272 285 283 290 1
3 0 0 1 0 140 187 199 205 231 227 228 246 245 263 262 1
3 0 0 1 0 109  95 102  96 135 335 111 146 131 162 171 1
3 0 0 1 0  69  46  67  28  43  55  55  77  73  76  76 1
3 0 0 1 0  69  95 137  99  95 108 129 134 133 131  91 1
3 0 0 1 0  51  59  76 101  72  72 107  91 128 120 133 1
3 0 0 1 0 156 186 198 201 205 210 217 217 219 223 229 1
4 0 0 0 1 201 202 229 232 224 237 217 268 244 275 246 1
4 0 0 0 1 113 126 159 157 137 160 162 171 167 165 185 1
4 0 0 0 1  86  54  75  75  71 130 157 142 173 174 156 1
4 0 0 0 1 115 158 168 175 188 164 184 195 194 206 212 1
4 0 0 0 1 183 175 217 235 241 251 229 241 233 233 275 1
4 0 0 0 1 131 147 183 181 206 215 197 207 226 244 240 1
4 0 0 0 1  71 105 107  92 101 103  78  87  57  70  71 1
4 0 0 0 1 172 213 263 260 276 273 267 286 283 290 298 1
4 0 0 0 1 224 258 248 257 257 267 260 279 299 289 300 1
4 0 0 0 1 246 257 269 280 289 291 306 301 295 312 311 1
```

Data from Danford, M.B., Hughes, H.M., and McNee, R.C. (1960). The same data were analyzed in Program 8_12.sas. By permission of the publisher.

Program 9_3.sas

This program uses the data set ramus.dat.

```
1 47.8 48.8 49.0 49.7  1
1 46.4 47.3 47.7 48.4  2
1 46.3 46.8 47.8 48.5  3
1 45.1 45.3 46.1 47.2  4
1 47.6 48.5 48.9 49.3  5
1 52.5 53.2 53.3 53.7  6
1 51.2 53.0 54.3 54.5  7
1 49.8 50.0 50.3 52.7  8
1 48.1 50.8 52.3 54.4  9
1 45.0 47.0 47.3 48.3 10
1 51.2 51.4 51.6 51.9 11
1 48.5 49.2 53.0 55.5 12
1 52.1 52.8 53.7 55.0 13
1 48.2 48.9 49.3 49.8 14
1 49.6 50.4 51.2 51.8 15
1 50.7 51.7 52.7 53.3 16
1 47.2 47.7 48.4 49.5 17
1 53.3 54.6 55.1 55.3 18
1 46.2 47.5 48.1 48.4 19
1 46.3 47.6 51.3 51.8 20
```

Data from Elston, R.C., and Grizzle, J.E. (1962). The same data were also analyzed in Program 8_7.sas. By permission of the publisher.

Program 9_4.sas

This program uses the same data set as Program 7_7.sas, data set 7_7.dat.

Program 9_5.sas

This program uses the data set 9_5.dat.

```
1  1 1 20
1  1 2 21
1  1 3 42
1  1 4 32
1  1 5 32
1  2 1 67
1  2 2 29
1  2 3 56
1  2 4 39
1  2 5 41
1  3 1 37
1  3 2 25
1  3 3 28
1  3 4 31
1  3 5 34
1  4 1 42
1  4 2 38
1  4 3 36
1  4 4 19
1  4 5 35
1  5 1 57
1  5 2 32
1  5 3 21
1  5 4 30
```

```
1  5 5 29
1  6 1 39
1  6 2 38
1  6 3 54
1  6 4 31
1  6 5 28
1  7 1 43
1  7 2 20
1  7 3 46
1  7 4 42
1  7 5 31
1  8 1 35
1  8 2 34
1  8 3 43
1  8 4 35
1  8 5 42
1  9 1 41
1  9 2 23
1  9 3 51
1  9 4 27
1  9 5 30
1 10 1 39
1 10 2 24
1 10 3 35
1 10 4 26
1 10 5 32
2 11 1 47
2 11 2 25
2 11 3 36
2 11 4 21
2 11 5 27
2 12 1 53
2 12 2 32
2 12 3 48
2 12 4 46
2 12 5 54
2 13 1 38
2 13 2 33
2 13 3 42
2 13 4 48
2 13 5 49
2 14 1 60
2 14 2 41
2 14 3 67
2 14 4 53
2 14 5 50
2 15 1 37
2 15 2 35
2 15 3 45
2 15 4 34
2 15 5 46
2 16 1 59
2 16 2 37
2 16 3 52
2 16 4 36
2 16 5 52
2 17 1 67
2 17 2 33
2 17 3 61
2 17 4 31
2 17 5 50
2 18 1 43
2 18 2 27
2 18 3 36
```

```
2 18 4 33
2 18 5 32
2 19 1 64
2 19 2 53
2 19 3 62
2 19 4 40
2 19 5 43
2 20 1 41
2 20 2 34
2 20 3 47
2 20 4 37
2 20 5 46
```

Data from Timm, N.H. (1975, p.244). Reprinted with permission of the author. Data provided by Dr. Paul Ammon, University of California, Berkeley.

Program 9_6.sas

This program uses the data set patel.dat and another data set that is included within the program.

```
1  1 1 1.28 1.09 0
1  1 2 1.33 0 1.24
1  2 1 1.60 1.38 0
1  2 2 2.21 0 1.90
1  3 1 2.46 2.27 0
1  3 2 2.43 0 2.19
1  4 1 1.41 1.34 0
1  4 2 1.81 0 1.47
1  5 1 1.40 1.31 0
1  5 2 0.85 0 0.85
1  6 1 1.12 0.96 0
1  6 2 1.20 0 1.12
1  7 1 0.90 0.66 0
1  7 2 0.90 0 0.78
1  8 1 2.41 1.69 0
1  8 2 2.79 0 1.90
2  9 1 3.06 1.74 0
2  9 2 1.38 0 1.54
2  10 1 2.68 2.41 0
2  10 2 2.10 0 2.13
2  11 1 2.60 3.05 0
2  11 2 2.32 0 2.18
2  12 1 1.48 1.20 0
2  12 2 1.30 0 1.41
2  13 1 2.08 1.70 0
2  13 2 2.34 0 2.21
2  14 1 2.72 1.89 0
2  14 2 2.48 0 2.05
2  15 1 1.94 0.89 0
2  15 2 1.11 0 0.72
2  16 1 3.35 2.41 0
2  16 2 3.23 0 2.83
2  17 1 1.16 0.96 0
2  17 2 1.25 0 1.01
```

Data from Patel, H.I. (1983) and Winer, B.J. (1971, p. 806). By permission of the publishers.

Program 9_7.sas

This program uses the data sets hlm.dat, hlmnest.dat, hlmcross.dat, and hlmsplit.dat.
hml.dat.

```
1 0
1 1
1 3
1 1
1 1
1 2
1 2
1 1
1 1
1 2
2 2
2 3
2 4
2 2
2 1
2 1
2 2
2 2
2 3
2 4
3 2
3 3
3 4
3 4
3 2
3 1
3 2
3 3
3 2
3 2
4 2
4 4
4 5
4 3
4 2
4 1
4 3
4 3
4 2
4 4
5 1
5 0
5 2
5 1
5 1
5 2
5 1
5 0
5 1
5 3
```

Data from Raudenbush, S.W. (1993, p. 466). Reprinted from Kirk (1982, p. 168).

hlmnest.dat

```
1 1  3
1 1  6
1 1  3
1 1  3
1 2  1
1 2  2
1 2  2
1 2  2
1 3  5
1 3  6
1 3  5
1 3  6
1 4  2
1 4  3
1 4  4
1 4  3
2 5  7
2 5  8
2 5  7
2 5  6
2 6  4
2 6  5
2 6  4
2 6  3
2 7  7
2 7  8
2 7  9
2 7  8
2 8 10
2 8 10
2 8  9
2 8 11
```

Data from Raudenbush, S.W. (1993, p. 475) found in Kirk (1982, p. 460).
hlmcross.dat

```
1  1  65
1  1  70
1  2  70
1  2  78
1  3  62
1  3  66
1  4  56
1  4  64
1  5  62
1  5  70
1  6  45
1  6  48
1  7  56
1  7  69
1  8  82
1  8  86
1  9  53
1  9  54
1 10  82
1 10  88
-1 1 140
```

```
-1 1 155
-1 2 159
-1 2 163
-1 3 163
-1 3 181
-1 4 139
-1 4 142
-1 5 127
-1 5 138
-1 6 141
-1 6 146
-1 7 130
-1 7 138
-1 8 139
-1 8 144
-1 9 128
-1 9 130
-1 10 156
-1 10 165
```

Data from Raudenbush, S.W. (1993, p. 481).

hlmsplit.dat

```
1 3 4 7  7
2 6 5 8  8
3 3 4 7  9
4 3 3 6  8
5 1 2 5 10
6 2 3 6 10
7 2 4 5  9
8 2 3 6 11
```

Data from Raudenbush, S.W. (1993, p. 488) and provided in Kirk (1982, p. 244).

Program 9_8.sas

This program uses the data set hlmrat.dat.

```
1 61 72 118 130 176 170
 2 65 85 129 148 174 194
 3 57 68 130 143 201 187
 4 46 74 116 124 157 156
 5 47 85 103 117 148 155
 6 43 58 109 133 152 150
 7 53 62  82 112 156 138
 8 72 96 117 129 154 154
 9 53 54  87 120 138 149
10 72 98 114 144 177 167
```

Data from *An introduction to HLM: Computer program and user's guide,* by A. S. Bryk, S. W. Raudenbush, M. Seltzer, and R. T. Congdon. Copyright © 1988 University of Chicago Press. Reprinted with the permission of Anthony S. Bryk.

Program 10_4.sas

This program uses the data set 10_4.dat.

```
 1 1 .04  .20  .10  .08
 2 1 .02  .06  .02  .02
 3 1 .07 1.40  .48  .24
 4 1 .17  .57  .35  .24
 5 2 .10  .09  .13  .14
 6 2 .12  .11  .10  m
 7 2 .07  .07  .07  .07
 8 2 .05  .07  .06  .07
 9 3 .03  .62  .31  .22
10 3 .03 1.05  .73  .60
11 3 .07  .83 1.07  .80
12 3 .09 3.13 2.06 1.23
13 4 .10  .09  .09  .08
14 4 .08  .09  .09  .10
15 4 .13  .10  .12  .12
16 4 .06  .05  .05  .05
```

Data from Cole, J. W. L., and Grizzle, J. E. (1966). By permission of the publisher.

Program 10_5.sas

The data set for this analysis is included within the program. Data from Winer (1971, p.806). By permission of the publisher.

Program 10_6.sas

This program uses the data set 10_6.dat.

```
1 194  209  250
  279  312  493    m    m    m    m    m    m    m    m    m
   m    m    m    m    m    m    m    m    m    m    m    m    m    m    m
01
1  413  427  485  522  622  687  696  865 1312 1496 1532 1733 1851 1855 1916
1934 1952 2019 2076 2138 2145 2167 2201    m    m    m    m    m    m    m
02
1   90  100  160  346  407  456  470  494  550  570  649  733  777  836  865
983  1008 1164 1476 1550 1576 1620 1643 1708 1835 2043 2113 2214 2422    m
03
1   74  131  179  208  710  722  792  813  842 1228 1287 1314 1467 1493 1819
m    m    m    m    m    m    m    m    m    m    m    m    m    m    m
04
1   55  375  440  544  764 1003 1050 1296 1472 1654 1687 1702 1806 1841    m
m    m    m    m    m    m    m    m    m    m    m    m    m    m    m
05
1   23   284  371  378  498  512  574  621  846  917 1163 1184 1226 1246 1251
1263 1383  1394 1397 1411 1482 1493 1507 1518 1564 1624 1625 1640 1692 1787
06
1   97   148  159  163  304  322  464  532  609  689  690  706  812 1018 1100
1154 1185  1401 1447 1558 1597 1660 1678 1869 1887 2050 2074    m    m    m
```

```
07
1   50    94  196  268  290  329  332  347  544  732  811  899  946  950  995
991 1013   1152 1362 1572 1669 1699 1722 1735 1749    m    m    m    m    m
08
1  359   368  380  650 1253 1256 1360 1362 1800    m    m    m    m    m    m
m    m     m    m    m    m    m    m    m    m    m    m    m    m    m
09
1   50   304  309  592  627  639    m    m    m    m    m    m    m    m    m
m    m     m    m    m    m    m    m    m    m    m    m    m    m    m
10
1  130   623    m    m    m    m    m    m    m    m    m    m    m    m    m
m    m     m    m    m    m    m    m    m    m    m    m    m    m    m
11
1  487   505  605  612  710  715  800  891  934 1164 1167 1297    m    m    m
m    m     m    m    m    m    m    m    m    m    m    m    m    m    m
12
1  102   311  325  382  436  468  535  595  737  889  916 1146 1376 1422 1476
m    m     m    m    m    m    m    m    m    m    m    m    m    m    m
13
```


Program 10_7.sas

This program uses the data sets 10_7.dat, 10_7.da1, 10_7.da2, 10_7.da3, 10_7.da4, and 7_7.dat.

```
10_7.dat
1 4.0 4.0 4.1 3.6 3.6 3.8 3.1  1 1 0 0 0
1   m 4.3 4.7 4.7 4.8 5.0 5.2  2 1 0 0 0
1   m   m 4.3 4.3 4.5 5.8 5.4  3 1 0 0 0
1 4.2 4.4 4.6 4.9 5.3 5.6 4.9  4 1 0 0 0
1   m 4.4 5.3 5.6 5.9 5.9 5.3  5 1 0 0 0
1   m   m 4.9 5.2 5.3 4.2 4.1  6 1 0 0 0
1 3.7 3.9 3.9 4.8 5.2 5.4 4.2  7 1 0 0 0
1   m   m 4.4 5.2 5.6 5.4 4.7  8 1 0 0 0
1   m 4.6 4.4 4.6 5.4 5.9 5.6  9 1 0 0 0
2 3.4 3.4 3.5 3.1 3.1 3.7 3.3 10 0 1 0 0
2 3.0 3.2 3.0 3.0 3.1 3.2 3.1 11 0 1 0 0
2   m 3.1 3.2 3.0 3.3 3.0 3.0 12 0 1 0 0
2   m   m 3.2 3.2 3.3 3.1 3.1 13 0 1 0 0
2 3.8 3.9 4.0 3.9 3.5 3.5 3.4 14 0 1 0 0
2   m   m 3.2 3.1 3.0 3.0 3.0 15 0 1 0 0
2   m 3.3 3.3 3.4 3.6 3.1 3.1 16 0 1 0 0
2 4.2 4.0 4.2 4.1 4.2 4.0 4.0 17 0 1 0 0
2 4.1 4.2 4.3 4.3 4.2 4.0 4.2 18 0 1 0 0
2   m 4.4 4.3 4.5 4.3 4.4 4.4 19 0 1 0 0
3   m 3.3 3.8 3.8 4.4 4.2 4.0 20 0 0 1 0
3   m   m 3.4 3.7 3.7 3.6 3.7 21 0 0 1 0
3 3.1 3.3 3.2 3.1 3.2 3.1 3.1 22 0 0 1 0
3   m 3.4 3.5 4.6 4.9 5.2 4.4 23 0 0 1 0
3   m   m 5.4 5.7 4.9 4.0 4.0 24 0 0 1 0
3   m 4.0 4.4 4.2 4.6 4.8 5.4 25 0 0 1 0
3 3.5 3.9 5.8 5.4 4.9 5.3 5.6 26 0 0 1 0
3   m   m 4.1 5.0 5.4 4.4 3.9 27 0 0 1 0
4   m 3.5 3.5 3.2 3.0 3.0 3.2 28 0 0 0 1
4   m 3.2 3.6 3.7 3.7 4.2 4.4 29 0 0 0 1
4   m   m 4.7 4.3 3.9 3.4 3.5 30 0 0 0 1
4 3.4 3.4 3.5 3.3 3.4 3.2 3.4 31 0 0 0 1
4   m   m 4.2 4.3 3.6 3.8 3.7 32 0 0 0 1
4 4.0 4.6 4.8 4.9 5.4 5.6 4.8 33 0 0 0 1
```

```
4   m 3.9 4.5 4.7 3.9 3.8 3.7 34 0 0 0 1
4 4.1 4.1 3.7 4.0 4.1 4.6 4.7 35 0 0 0 1
4 3.5 3.6 3.6 4.2 4.8 4.9 5.0 36 0 0 0 1
```

10_7.da1

```
1 4.0 4.0 4.1 3.6 3.6 3.8 3.1  1 1 0 0 0
1 4.2 4.4 4.6 4.9 5.3 5.6 4.9  4 1 0 0 0
1 3.7 3.9 3.9 4.8 5.2 5.4 4.2  7 1 0 0 0
1 3.4 3.4 3.5 3.1 3.1 3.7 3.3 10 0 1 0 0
1 3.0 3.2 3.0 3.0 3.1 3.2 3.1 11 0 1 0 0
1 3.8 3.9 4.0 3.9 3.5 3.5 3.4 14 0 1 0 0
1 4.2 4.0 4.2 4.1 4.2 4.0 4.0 17 0 1 0 0
1 4.1 4.2 4.3 4.3 4.2 4.0 4.2 18 0 1 0 0
1 3.1 3.3 3.2 3.1 3.2 3.1 3.1 22 0 0 1 0
1 3.5 3.9 5.8 5.4 4.9 5.3 5.6 26 0 0 1 0
1 3.4 3.4 3.5 3.3 3.4 3.2 3.4 31 0 0 0 1
1 4.0 4.6 4.8 4.9 5.4 5.6 4.8 33 0 0 0 1
1 4.1 4.1 3.7 4.0 4.1 4.6 4.7 35 0 0 0 1
1 3.5 3.6 3.6 4.2 4.8 4.9 5.0 36 0 0 0 1
```

10_7.da2

```
2 4.3 4.7 4.7 4.8 5.0 5.2  2 1 0 0 0
2 4.4 5.3 5.6 5.9 5.9 5.3  5 1 0 0 0
2 4.6 4.4 4.6 5.4 5.9 5.6  9 1 0 0 0
2 3.1 3.2 3.0 3.3 3.0 3.0 12 0 1 0 0
2 3.3 3.3 3.4 3.6 3.1 3.1 16 0 1 0 0
2 4.4 4.3 4.5 4.3 4.4 4.4 19 0 1 0 0
2 3.3 3.8 3.8 4.4 4.2 4.0 20 0 0 1 0
2 3.4 3.5 4.6 4.9 5.2 4.4 23 0 0 1 0
2 4.0 4.4 4.2 4.6 4.8 5.4 25 0 0 1 0
2 3.5 3.5 3.2 3.0 3.0 3.2 28 0 0 0 1
2 3.2 3.6 3.7 3.7 4.2 4.4 29 0 0 0 1
2 3.9 4.5 4.7 3.9 3.8 3.7 34 0 0 0 1
```

10_7.da3

```
3 4.3 4.5 5.8 5.4  3 1 0 0 0
3 5.2 5.3 4.2 4.1  6 1 0 0 0
3 5.2 5.6 5.4 4.7  8 1 0 0 0
3 3.2 3.3 3.1 3.1 13 0 1 0 0
3 3.1 3.0 3.0 3.0 15 0 1 0 0
3 3.7 3.7 3.6 3.7 21 0 0 1 0
3 5.7 4.9 4.0 4.0 24 0 0 1 0
3 5.0 5.4 4.4 3.9 27 0 0 1 0
3 4.3 3.9 3.4 3.5 30 0 0 0 1
3 4.3 3.6 3.8 3.7 32 0 0 0 1
```

10_7.da4

```
1 4.0 4.0 4.1 3.6 3.6 3.8 3.1  1 1 0 0 0
1 4.2 4.4 4.6 4.9 5.3 5.6 4.9  4 1 0 0 0
1 3.7 3.9 3.9 4.8 5.2 5.4 4.2  7 1 0 0 0
1 3.4 3.4 3.5 3.1 3.1 3.7 3.3 10 0 1 0 0
1 3.0 3.2 3.0 3.0 3.1 3.2 3.1 11 0 1 0 0
1 3.8 3.9 4.0 3.9 3.5 3.5 3.4 14 0 1 0 0
1 4.2 4.0 4.2 4.1 4.2 4.0 4.0 17 0 1 0 0
1 4.1 4.2 4.3 4.3 4.2 4.0 4.2 18 0 1 0 0
1 3.1 3.3 3.2 3.1 3.2 3.1 3.1 22 0 0 1 0
1 3.5 3.9 5.8 5.4 4.9 5.3 5.6 26 0 0 1 0
1 3.4 3.4 3.5 3.3 3.4 3.2 3.4 31 0 0 0 1
1 4.0 4.6 4.8 4.9 5.4 5.6 4.8 33 0 0 0 1
1 4.1 4.1 3.7 4.0 4.1 4.6 4.7 35 0 0 0 1
```

```
1 3.5 3.6 3.6 4.2 4.8 4.9 5.0 36 0 0 0 1
2 0.0 4.3 4.7 4.7 4.8 5.0 5.2  2 1 0 0 0
2 0.0 4.4 5.3 5.6 5.9 5.9 5.3  5 1 0 0 0
2 0.0 4.6 4.4 4.6 5.4 5.9 5.6  9 1 0 0 0
2 0.0 3.1 3.2 3.0 3.3 3.0 3.0 12 0 1 0 0
2 0.0 3.3 3.3 3.4 3.6 3.1 3.1 16 0 1 0 0
2 0.0 4.4 4.3 4.5 4.3 4.4 4.4 19 0 1 0 0
2 0.0 3.3 3.8 3.8 4.4 4.2 4.0 20 0 0 1 0
2 0.0 3.4 3.5 4.6 4.9 5.2 4.4 23 0 0 1 0
2 0.0 4.0 4.4 4.2 4.6 4.8 5.4 25 0 0 1 0
2 0.0 3.5 3.5 3.2 3.0 3.0 3.2 28 0 0 0 1
2 0.0 3.2 3.6 3.7 3.7 4.2 4.4 29 0 0 0 1
2 0.0 3.9 4.5 4.7 3.9 3.8 3.7 34 0 0 0 1
3 0.0 0.0 0.0 4.3 4.5 5.8 5.4  3 1 0 0 0
3 0.0 0.0 0.0 5.2 5.3 4.2 4.1  6 1 0 0 0
3 0.0 0.0 0.0 5.2 5.6 5.4 4.7  8 1 0 0 0
3 0.0 0.0 0.0 3.2 3.3 3.1 3.1 13 0 1 0 0
3 0.0 0.0 0.0 3.1 3.0 3.0 3.0 15 0 1 0 0
3 0.0 0.0 0.0 3.7 3.7 3.6 3.7 21 0 0 1 0
3 0.0 0.0 0.0 5.7 4.9 4.0 4.0 24 0 0 1 0
3 0.0 0.0 0.0 5.0 5.4 4.4 3.9 27 0 0 1 0
3 0.0 0.0 0.0 4.3 3.9 3.4 3.5 30 0 0 0 1
3 0.0 0.0 0.0 4.3 3.6 3.8 3.7 32 0 0 0 1
```

Data from Grizzle, J. E., and Allen, D. M. (1969). By permission of the publisher.

References

Agresti, A. (1990). *Categorical data analysis.* New York: John Wiley & Sons, Inc..

Andrews, D.F. (1971). A note on the selection of data transformations. *Biometrika*, **58**, 249-254.

Anderson, T.W. (1984). *Introduction to multivariate statistical analysis,* Second Edition. New York: John Wiley & Sons, Inc.

Arnold, S.F. (1981). *The theory of linear models and multivariate analysis.* New York: John Wiley & Sons, Inc.

Barrett, B.E., and Ling, R.F. (1992). General classes of influence measures for multivariate regression. *Journal of the American Statistical Association*, **87**, 184-191.

Bartlett, M.S. (1939). A note on tests of significance in multivariate analysis. In *Proceedings of the Cambridge Philosophical Society*, **35**, 180-185.

Belsley, D.A., Kuh, E., and Welsch, R.E. (1980). *Regression diagnostics.* New York: John Wiley & Sons, Inc.

Bentler, P.M., and Lee, S.Y. (1978). Statistical analysis of a three-mode factor analysis model. *Psychometrika,* **43**, 343-352.

Bhapkar, V.P. (1966). A note on the equivalence of two test criteria for hypotheses in categorical data. *Journal of the American Statistical Association*, **61**, 228-235.

Bishop, Y.M.M., Fienberg, S.E., and Holland, P.W. (1975). *Discrete multivariate analysis: Theory and practice.* Cambridge, MA: The MIT Press.

BMDP Statistical Software, Inc. (1992). *BMDP Statistical software manual, Vols 1 and 2, Release 7, W.J. Dixon, ed.* Los Angeles: University of California Press.

Bock, R.D. (1989). *Multilevel analysis of educational data.* New York: Academic Press.

Bock, R.D. and Peterson, A.C. (1975). A multivariate correction for attenuation. *Biometrika*, **62**, 673-678.

Boik, R.J. (1988). The mixed model for multivariate repeated measures: Validity conditions and an approximate test. *Psychometrika*, **53**, 469-486.

Boik, R.J. (1991). Scheffé mixed model for repeated measures a relative efficiency evaluation. *Communications in Statistics - Theory and Methods*, **A20**, 1233-1255.

Box, G.E.P. (1950). Problems in the analysis of growth and wear curves. *Biometrics*, **6**, 362-389.

Box, G.E.P. (1954). Some theorems on quadratic forms applied in the study of analysis of variance problems, II. Effects of inequality of variance and correlation between errors in the two-way classification. *Annals of Mathematical Statistics*, **25**, 484-498.

Box, G.E.P., Hunter, W.G., and Hunter, J.S. (1978). *Statistics for experimenters: An introduction to design, data analysis and model building.* New York: John Wiley & Sons, Inc.

Bradu, D., and Gabriel, K.R. (1974). Simultaneous statistical inference on interactions in two-way analysis of variance. *Journal of the American Statistical Association*, **69**, 428-436.

Bryk, A.S., and Raudenbush, S.W. (1992). *Hierarchical linear models: Applications and data analysis methods.* Newbury Park, CA: Sage Publications.

Bryk, A.S., Raudenbush, S.W., Seltzer, M., and Congdon, R.T. (1988). *An introduction to HLM: Computer program and user's guide.* Chicago: University of Chicago Press.

Casella, G., and Berger, R.L. (1990). *Statistical inference.* Belmont, CA: Brooks/Cole Publishing Company.

Chinchilli, V.M., and Elswick, R.K. (1985). A mixture of the MANOVA and GMANOVA models. *Communications in Statistics - Theory and Methods*, **14**, 3075-3089.

Cole, J.W.L., and Grizzle, J.E. (1966). Applications of multivariate analysis of variance to repeated measurements experiments. *Biometrics*, **22**, 810-828.

Cook, R.D., and Weisberg, S. (1994). *An introduction to regression graphics.* New York: John Wiley & Sons, Inc.

Cornell, J.E., Young, D.M., Seaman, S.L., and Kirk, R.E. (1992). Power comparisons of eight tests for sphericity in repeated measures designs. *Journal of Educational Statistics*, **17**, 233-249.

Danford, M.B., Hughes, H.M., and McNee, R.C. (1960). On the analysis of repeated-measurements experiments. *Biometrics*, **16**, 547-565.

Davidian, M., and Giltinan, D.M. (1995). *Nonlinear models for repeated measurement data.* New York: Chapman and Hall.

Davidson, M.L. (1988). *The multivariate approach to repeated measures*, BMDP Technical Report #75, Los Angeles: BMDP Statistical Software, Inc.

de Leeuw, J., and Kreft, I. (1986). Random coefficient models for multilevel analysis. *Journal of Educational Statistics*, **11**, 57-85.

Dempster, A.P., Laird, N.M., and Rubin, D.B. (1977) Maximum likelihood from incomplete data via the EM algorithm. *Journal of the Royal Statistical Society*, Series B, **39**, 1-22.

Dempster, A.P., Rubin, D.B., and Tsutakawa, R.K. (1981). Estimation in covariance components models. *Journal of the American Statistical Association,* **76**, 341-353.

Diggle, P.J., Liang, K.Y., and Zeger, S.L. (1994). *Analysis of longitudinal data.* Oxford: Clarendon Press.

Dobson, A.J. (1990). *An introduction to generalized linear models.* New York: Chapman & Hall.

Draper, D. (1995). Inference and hierarchical modeling in the social sciences. *Journal of Educational and Behavioral Statistics*, **20**, 115-147.

Dunlop, D.D. (1994). Regression for longitudinal data: A bridge from least squares regression. *The American Statistician,* **48**, 299-303.

Edwards, A.L. (1968). *Experimental design in psychological research*, Third Edition, New York: Holt, Rinehart, & Winston.

Elston, R.C., and Grizzle, J.E. (1962). Estimation of time response curves and their confidence bands. *Biometrics*, **18**, 148-159.

Everson, P.J., and Morris, C.N. (in press). Inferences for multivariate normal hierarchical models. *Journal of the America Statistical Association.*

Freund, R.J., and Littell, R.C. (1991). *SAS System for Regression*, Second Edition. Cary, NC: SAS Institute Inc.

Fujikoshi, Y. (1974). The likelihood ratio tests for the dimensionality of regression coefficients. *Journal of Multivariate Analysis,* **4**, 327-340.

Galecki, A.T. (1994). General class of covariance structures for two or more repeated factors in longitudinal data analysis. *Communications in Statistics - Theory and Methods*, **23**, 3105-3119.

Gelfand, A.E., Sahu, S.K., and Carlin, B.P. (1995). Efficient parameterisations for normal linear mixed models. *Biometrika*, **82**, 479-488.

Gelman, A., Carlin, J.B., Stern, H.S., and Rubin, D.B. (1996). *Bayesian data analysis.* New York: Chapman & Hall.

Gleser, L.J., and Olkin, I. (1970). Linear models in multivariate analysis. *Essays in probability and statistics*, 267-292. New York: John Wiley & Sons, Inc.

Gnandesikan, R. (1980). Graphical methods for interval comparisons in ANOVA and MANOVA. *Handbook of Statistics, Vol. 1,* P.R. Krishnaiah (Ed.), 133-177. Amsterdam: North-Holland.

Gnandesikan, R. (1997). *Methods for statistical data analysis of multivariate observations,* Second Edition. New York: John Wiley & Sons, Inc.

Gnandesikan, R., and Kettenring, J.R. (1972). Robust estimates, residuals, and outlier detection with multiresponse data. *Biometrics*, **28**, 81-124.

Goldstein, H.I. (1986). Multilevel mixed linear model analysis using iterative generalized least squares. *Biometrika*, **73,** 43-56.

Greene, W. H. (1992). *Econometric analysis,* Second Edition. New York: Macmillan Publishing, Inc.

Grizzle, J.E., and Allen, D.M. (1969). Analysis of growth and dose response curves. *Biometrics,* **25**, 357-381.

Grizzle, J.E., Starmer, C.F., and Koch, G.G. (1969). Analysis of categorical data by linear models. *Biometrics*, **25,** 489-504.

Harville, D.A. (1976). Extension of the Gauss-Markov theorem to include the estimation of random effects. *The Annals of Statistics*, **4**, 384-395.

Harville, D.A. (1977). Maximum likelihood approaches to variance component estimation and to related problems. *Journal of the American Statistical Association,* **72,** 320-340.

Hecker, H. (1987). A generalization of the GMANOVA-model. *Biometrical Journal*, **29**, 763-790.

Hotelling, H. (1931). The generalization of student's ratio. *Annals of Mathematical Statistics*, **2**, 360-378.

Huynh, H., and Feldt, L.S. (1970). Conditions under which mean square ratios in repeated measurements designs have exact F-distributions. *Journal of the American Statistical Association,* **65**, 1582-1589.

Jennrich, R.I., and Schluchter, M.D. (1986). Unbalanced repeated-measures models with structured covariance matrices. *Biometrics*, **42**, 805-820.

Jobson, J.D. (1992). *Applied multivariate data analysis, Vol II: Categorical and multivariate methods.* New York: Springer-Verlag.

Johnson, N.L., and Kotz, S. (1972). *Distributions in statistics: Continuous multivariate distributions.* New York: John Wiley & Sons, Inc.

Jöreskog, K.G. (1973). Analysis of covariance structures. *Multivariate Analysis III*, P.R. Krishnaiah, ed., 263-285. New York: Academic Press, Inc.

Kabe, D.G. (1975). Generalized MANOVA double linear hypothesis with double linear restrictions. *Canadian Journal of Statistics,* **3**, 35-44.

Kabe, D.G. (1981). MANOVA double linear hypothesis with double linear restrictions. *Communications in Statistics - Theory and Methods*, **24,** 2545-2550.

Kackar, R.N., and Harville, D.A. (1984). Approximations for standard errors of estimators of fixed and random effects in mixed linear models. *Journal of the American Statistical Association*, **79**, 853-862.

Kariya, T. (1985). *Testing in the multivariate general linear model.* Tokyo: Kinokuniya Company Ltd.

Kass, R.E., and Steffey, D. (1989). Approximations for standard errors of estimators of fixed and random effects in mixed linear models. *Journal of the American Statistical Association,* **84,** 717-726.

Keppel, G. (1991). *Design and analysis: A research handbook.* Englewood Cliffs, NJ: Prentice Hall, Inc.

Khatri, C.G. (1966). A note on a MANOVA model applied to problems in growth curves. *Annals of the Institute of Statistical Mathematics,* **18,** 75-86.

Khattree, R., and Naik, D. (1995). *Applied multivariate statistics with SAS software.* Cary, NC: SAS Institute Inc.

Kirk, R.E. (1982). *Experimental design: procedure for the behavioral sciences*, Second Edition. Belmont, CA: Brooks/Cole Publishing Company.

Kirk, R.E. (1995). *Experimental design procedure for the behavioral sciences,* Third Edition. Belmont, CA: Brooks/Cole Publishing Company.

Kleinbaum, D.G. (1970). Estimation and hypothesis listing for generalized multivariate linear models. Ph.D diss., University of North Carolina at Chapel Hill.

Kleinbaum, D.G. (1973). A generalization of the growth curve model which allows missing data. *Journal of Multivariate Analysis*, **3,** 117-124.

Kleinbaum, D.G., Kupper, L.L., and Muller, K.E. (1988). *Applied regression analysis and other multivariable methods*, Second Edition. Boston: PWS-Kent Publishing Company.

Kmenta, J., and Gilbert, R.F. (1968). Small sample properties of alternative estimators of seemingly unrelated regressions. *Journal of the American Statistical Association,* **62,** 1180-1200.

Kreft, I.G.G., de Leeuw, J., and van der Leeden, R. (1994). Review of five multilevel analysis programs: BMDP-5V, GENMOD, HLM, ML3, VARCL. *The American Statistician,* **48**, 324-335.

Krishnaiah, P.R., and Lee, J.C. (1980). Likelihood ratio tests for mean vectors and covariance matrices. *Handbook of Statistics, Analysis of Variance, Vol. I,* P.K. Krishnaiah, ed., 513-570. New York: North-Holland.

Krishnaiah, P.R., Mudholkar, G.S., and Subbaiah, P. (1980). Simultaneous test procedures for mean vectors and covariance matrices. *Handbook of Statistics, Vol. 1,* P.K. Krishnaiah, ed., 631-671. New York: North-Holland.

Kshirsager, A.M., and Smith, W.B. (1995). *Growth Curves.* New York: Marcel Dekker, Inc.

Laird, N.M., Lange, N., and Stram, D. (1987). Maximum likelihood computations with repeated measures: Application of the EM algorithm. *Journal of the American Statistical Association*, **82,** 97-105.

Laird, N.M., and Ware, J.H. (1982). Random-effects models for longitudinal data. *Biometrics,* **38,** 963-974.

Lawley, D.N. (1938). A generalization of Fisher's z-test. *Biometrika*, **30**, 180-18.

Lehmann, E. (1994). *Testing statistical hypotheses,* Second Edition. New York: Chapman & Hall.

Liang, K.Y., and Zeger, S.L. (1986). Longitudinal data analysis using generalized linear models. *Biometrika,* **73**, 13-22.

Lindsey, J.K. (1993). *Models for repeated measurements.* Oxford: Clarendon Press.

Lindstrom, M.J., and Bates, D.M. (1988). Newton-Raphson and EM algorithums for linear mixed-effects models for repeated-measure data. *Journal of the American Statistical Association*, **83**, 1014-1022.

Liski, E. (1985). Estimation from incomplete data in growth curve models. *Communications in Statistics - Simulation and Computation*, **14**, 13-27.

Little, R.J.A. (1988). A test of missing completely at random for multivariate data with missing values. *Journal of the American Statistical Association*, **83**, 1198-1202.

Little, R.J.A. (1992). Regression with missing X's: A review. *Journal of the American Statistical Association,* **87,** 1227-1237.

Little, R.J.A. (1995). Modeling the drop-out mechanism in repeated measures studies. *Journal of the American Statistical Association*, **90**, 1112-1121.

Little, R.J.A., and Rubin, D.B. (1987). *Statistical analysis with missing data.* New York: John Wiley & Sons, Inc.

Littell, R.C., Freund, R.J., and Spector, P.C. (1991). *SAS system for linear models,* Third Edition. Cary, NC: SAS Institute Inc.

Littell, R.C., Milliken, G.A., Stroup, W.W., and Wolfinger, R.D. (1996). *SAS system for mixed models.* Cary, NC: SAS Institute Inc.

Liu, C., and Rubin, D.B. (1994). The ECME algorithm: A simple extension of EM and ECM with faster monotone convergence. *Biometrika,* **81**, 633-648.

Longford, N.T. (1987). A fast scoring algorithm for maximum likelihood estimation in unbalanced mixed models with nested random effects. *Biometrika,* **74,** 817-827.

Longford, N.T. (1993). *Random coefficient models.* Oxford: Clarendon Press.

Marascuilo, L.A., and McSweeney, M. (1977). *Nonparametric and distribution-free methods for the social sciences.* Belmont, CA: Brooks/Cole Publishing Company.

Mardia, K.V., (1970). Measures of multivariate skewness and kurtosis with applications. *Biometrika,* **50**, 519-520.

Mardia, K.V. (1974). Applications of some measures of multivariate skewness and kurtosis for testing normality and robustness studies. *Sankhya,* **36**, 115-128.

Mauchley, J.W. (1940). Significance test of sphericity of a normal n-variate distribution. *Annals of Mathematical Statistics*, **11**, 204-209.

McCullagh, P., and Nelder, J.A. (1989). *Generalized linear models*, Second Edition. New York: Chapman & Hall.

McElroy, F.W. (1967). A necessary and sufficient condition that ordinary least-squares estimators be best linear unbiased. *Journal of the American Statistical Association,* **62,** 1302-1304.

McLachlan, G.J., and Krishnan, J. (1997). *The EM algorithum and extensions.* New York: John Wiley & Sons, Inc.

McLean, R.M., and Sanders, W.L. (1988). Approximating degrees of freedom for standard errors in mixed linear models. In *Proceedings of the Statistical Computing Section*, 50-59. New Orleans: American Statistical Association.

McLean, R.M., Sanders W.L., and Stroup, W.W. (1991). A unified approach to mixed linear models. *The American Statistician,* **45**, 54-64.

Mensah, R.D., Elswick, R.E., Jr., and Chinchilli, V.M. (1993). Consistent estimators of the variance-covariance matrix of the GMANOVA model with missing data. *Communications in Statistics - Theory and Methods,* **22,** 1495-1514.

Milliken, G.A., and Johnson, D.E. (1992). *Analysis of messy data, Vol. 1: Designed experiments.* New York: Chapman & Hall.

Mudholkar, G.S., Davidson, M.L., and Subbaiah, P. (1974). Extended linear hypotheses and simultaneous tests in multivariate analysis of variance. *Biometrika,* **61**, 467-478.

Muller, K.E., LaVange, L.M., Ramey, S.L., and Ramey, C.T. (1992). Power calculations for general linear multivariate models including repeated measures applications. *Journal of the American Statistical Association,* **87,** 1209-1226.

Naik, D.N., and S.S. Rao (1996). *Analysis of multivariate repeated,* Technical Report. Norfolk, VA: Old Dominion University, Department of Mathematics and Statistics,.

Nanda, D.N. (1950). Distribution of the sum of roots of a determinantal equation under a certain condition. *Annals of Mathematical Statistics*, **21**, 432-439.

Neter, J., Wasserman, W., and Kutner, M.H. (1990). *Applied linear statistical models.* Boston: Richard D. Irwin.

Neyman, J. (1949). Contributions to the theory of the χ^2 test. In *Proceedings of the Berkeley Symposium of Mathematical Statistics and Probability*, 239-273. Berkeley: University of California Press.

O'Brien, R.G., and Muller, K.E. (1993). Unified power analysis for t-tests through multivariate hypotheses. *Applied analysis of variance in behavioral science*, L.K. Edwards, ed., 297-344. New York: Marcel Dekker, Inc.

Orchard, T., and Woodbury, M.A. (1972). A missing information principle: Theory and applications. In *Proceedings of the Sixth Berkeley Symposium on Mathematical Statistics and Probability*, 697-715. Berkeley: University of California Press.

Park, T. (1993). Equivalence of maximum likelihood estimation and iterative two-stage estimation for seemingly unhrelated regression. *Communications in Statistics - Theory and Methods,* **22,** 2285-2296.

Park, T., and Woolson, R.F. (1992). Generalized multivariate models for longitudinal data. *Communications in Statistics - Theory and Methods,* **21,** 925-946.

Park, T., and Davis, S.S. (1993). A test of the missing data mechanism for repeated categorical data. *Biometrics*, **58**, 545-554.

Patel, H.I. (1983). Use of baseline measurements in the two-period crossover design. *Communications in Statistics - Theory and Methods*, **12**, 2693-2712.

Patel, H.I. (1986). Analysis of repeated measurement designs with changing covariates in clinical trials. *Biometrika*, **73**, 707-715.

Pillai, K.C.S. (1955). Some new test criteria in multivariate analysis. *Annals of Mathematical Statistics*, **26**, 117-121.

Potthoff, R.F., and Roy, S.N. (1964). A generalized multivariate analysis of variance model useful especially for growth curve problems. *Biometrika*, **51**, 313-326.

Rao, C.R. (1965). The theory of least squares when the parameters are stochastic and its applications to the analysis of growth curves. *Biometrika*, **52**, 447-458.

Rao, C.R. (1966). Covariance adjustment and related problems in multivariate analysis. *Multivariate Analysis*, P.R. Krishnaiah, ed., 87-103. New York: Academic Press.

Rao, C.R. (1967). Least squares theory using an estimated dispersion matrix and its application to measurement of signals. In *Proceedings of the Fifth Berkeley Symposium on Mathematical Statistics and Probability, Vol. 1.*, 335-371. Berkeley: University of California Press.

Rao, C.R. (1973). *Linear statistical inference and its applications*, Second Edition. New York: John Wiley & Sons, Inc.

Rao, C.R. (1987). Prediction of future observations in growth curve models. *Statistical Sciences*, **2**, 434-471.

Raudenbush, S.W. (1993). Hierarchical linear models and experimental design. *Applied analysis of variance in behavioral science*, L.K. Edwards, ed., 459-496. New York: Marcel Dekker, Inc.

Reinsel, G. (1982). Multivariate repeated-measurements or growth curve models with multivariate random effects covariance structure. *Journal of the American Statistical Association*, **77**, 190-195.

Reinsel, G. (1984). Estimation and prediction in a multivariate random effects generalized linear model. *Journal of the American Statistical Association*, **79**, 406-414.

Robins, J.M., Rotnitzky, A., and Zhao, L.P. (1995). Analysis of semiparametric regression models for repeated outcomes in the presence of missing data. *Journal of the American Statistical Association*, **90**, 106-121.

Rogosa, D., and Saner, H. (1995). Longitudinal data analysis examples with random coefficient models. *Journal of Educational and Behavioral Statistics*, **20**, 149-170.

Roy, S.N. (1953). On a heuristic method of test construction and its use in multivariate analysis. *The Annals of Mathematical Statistics*, **24**, 220-238.

SAS Institute Inc. (1996). *SAS/STAT software*: *Changes and enhancements for release* 6.12. Cary, NC: SAS Institute Inc.

SAS Institute Inc. (1993). SAS Technical Report P-243, *SAS/STAT Software: The GENMOD Procedure, Release 6.09*. Cary, NC: SAS Institute Inc.

SAS Institute Inc. (1992). SAS Technical Report P-229, *SAS/STAT Software: Changes and Enhancements, Release 6.07.* Chapter 16, The MIXED Procedure. Cary, NC: SAS Institute Inc.

SAS Institute Inc. (1990). *SAS/STAT User's Guide, Version 6*, Fourth Edition, Vols 1 and 2,. Cary, NC: SAS Institute Inc.

Scheffé, H. (1959). *The analysis of variance.* New York: John Wiley & Sons, Inc.

Searle, S.R. (1971). *Linear models.* New York: John Wiley & Sons, Inc.

Searle, S.R. (1987). *Linear models for unbalanced data.* New York: John Wiley & Sons, Inc.

Searle, S.R. (1994). Analysis of variance computing package output for unbalanced data from fixed-effects models with nested factors. *The American Statistician*, **48**, 148-153.

Searle, S.R., Casella, G., and McCullock, C.E. (1992). *Variance components.* New York: John Wiley & Sons, Inc.

Seber, G.A.F. (1984). *Multivariate observations.* New York: John Wiley & Sons, Inc.

Singh, A. (1993). Omnibus robust procedures for assessment of multivariate normality and detection of multivariate outliers. *Multivariate Analysis: Future Directions,* C. R. Rao, ed., 443-482. New York: Elsevier Science Publishing Company.

Small, N.J.H. (1978). Plotting squared radii. *Biometrika*, **65**, 657-658.

Smith, H., Gnandesikan, R., and Hughes, J.B. (1962). Multivariate analysis of variance (MANOVA). *Biometrics,* **18**, 22-41.

Srivastava, M.S., and Carter, E.M. (1983). *An introduction to applied multivariate statistics.* New York: North-Holland.

Srivastava, M.S., and Khatri, C.G. (1979). *An introduction to multivariate statistics.* New York: North Holland.

Stanek, E.J. III, and Koch, G.G. (1985). The equivalence of parameter estimates from growth curve models and seemingly unrelated regression models. *The American Statistician,* **39,** 149-152.

Stapleton, J.H. (1995). *Linear statistical models.* New York: John Wiley & Sons, Inc.

Stevens, J. (1992). *Applied multivariate statistics for the social sciences,* Second Edition. Mahwah, NJ: Lawrence Erlbaum Associates, Inc.

Stokes, M.E., Davis, C.S., and Koch, G.G. (1995). *Categorical data analysis using the SAS system.* Cary, NC: SAS Institute, Inc.

Stuart, A., and Ord, J.K. (1991). *Kendall's advanced theory of statistics, Vol. 2, Classical inference and relationships*, Fifth Edition. New York: Oxford University Press.

Swamy, P. (1971). *Statistical inference in random coefficient regression models.* New York: Springer-Verlag.

Telser, L.G. (1964). Interactive estimation of a set of linear regression equations. *Journal of the American Statistical Association,* **59**, 849-862.

Theil, H. (1971). *Principles of econometrics.* New York: John Wiley & Sons, Inc.

Thomas, D.R. (1983). Univariate repeated measures techniques applied to multivariate data. *Psychometrika*, **48**, 451-464.

Thompson, R. (1973). The estimation of variance components with an application when records are subject to culling. *Biometrics,* **29,** 527-550.

Thum, Y.M. (1997). Hierarchical linear models for multivariate outcomes. *Journal of Educational and Behavioral Statistics*, **22**, 77-108.

Timm, N.H. (1970). The estimation of variance-covariance and correlation matrices from incomplete data, *Psychometrika,* **35**, 417-437.

Timm, N.H. (1975). *Multivariate analysis with applications in education and psychology.* Belmont, CA: Brooks/Cole Publishing Company.

Timm, N.H. (1980a). Multivariate analysis of repeated measurements. *Handbook of Statistics, Vol. 1,* P.R. Krishnaiah, ed., 41-87. New York: North-Holland.

Timm, N.H. (1980b). The analysis of nonorthogoral MANOVA designs employing a restricted full rank multivariate linear model. *Multivariate Statistical Analysis,* R.P. Gupta, ed., 257-273. New York: North-Holland.

Timm, N.H. (1993a). MANOVA and MANCOVA: An Overview. In *A handbook for data analysis in the behavioral sciences: Statistical issues.* G. Keren and C. Lewis, eds., 129-163. Mahwah, NJ: Lawrence Erlbaum Associates, Inc.

Timm, N.H. (1993b). *Multivariate analysis with applications in education and psychology.* Oberlin, OH: The Digital Printshop.

Timm, N.H. (1995). Simultaneous inference using finite intersection tests: A better mousetrap. *Multivariate Behavioral Research*, **30**, 461-511.

Timm, N.H. (1996). A note on the MANOVA model with double linear restrictions. *Communications in Statistics - Theory and Methods*, **25**, 1391-1395.

Timm, N.H., and Carlson, J.E. (1975). Analysis of variance through full rank models. *Multivariate Behavioral Research Monograph*, **75-1**, 120.

Tubbs, J.D., Lewis, T.O., and Duran, B.S. (1975). A note on the analysis of the MANOVA model and its application to growth curves. *Communications in Statistics - Theory and Methods,* **4**, 643-653.

Verbyla, A.P. (1986). Conditioning in the growth curve model, *Biometrika,* **73**, 475-483.

Verbyla, A.P. (1988). Analysis of repeated measurement designs with changing covariates. *Biometrika,* **75**, 172-174.

Verbyla, A.P., and Venables, W.N. (1988). An extension of the growth curve model. *Biometrika,* **75,** 129-138.

Vonesh, E.F., and Carter, R.L. (1987). Efficient inference for random-coefficient growth curve models with unbalanced data. *Biometrics*, **43**, 617-628.

Vonesh, E.F., and Chinchilli, V.M. (1997). *Linear and nonlinear models for the analysis of repeated measurements.* New York: Marcel Dekker, Inc.

von Rosen, D. (1989). Maximum likelihood estimators in multivariate linear normal models. *Journal of Multivariate Analysis,* **31**, 187-200.

von Rosen, D. (1991). The growth curve model: A review. *Communications in Statistics - Theory and Methods,* **20**, 2791-2822.

Wald, A. (1943). Tests of statistical hypothesis concerning several parameters when the number of variables is large. In *Transactions of the American Mathematical Society,* **54,** 426-482.

Ware, J.H. (1985). Linear Models for the analysis of longitudinal studies. *The American Statistician*, **39**, 95-101.

Wilks, S.S. (1932). Certain generalizations in the analysis of variance. *Biometrika*, **24**, 471-494.

Winer, B.J. (1971). *Statistical principles in experimental design*, Second Edition. New York: McGraw-Hill, Inc.

Wolfinger, R., Tobias, R., and Sall, J. (1994). Computing Gaussian likelihoods and their derivatives for general linear mixed models. *SIAM Journal of Scientific Computing,* **15**, 1294-1310.

Woolson, R.F., and Clarke, W.R. (1987). Estimation of growth norms from incomplete longitudinal data. *Biometrical Journal,* **29,** 937-952.

Zellner, A. (1962). An efficient method of estimating unrelated regressions and tests for aggregation bias. *Journal of the American Statistical Association,* **57**, 348-368.

Zellner, A. (1963). Estimators for seemingly unrelated regression equations: Some exact finite sample results. *Journal of the American Statistical Association,* **58,** 977-992.

Author Index

A

B

C

D

E

F

G

H

J

K

L

M

N

O

P

R

S

T

V

W

Y

Z

Subject Index

A

B

D

G

H

I

M

N

O

P

Q

R

S

T

U

V

W

Z

Special Characters

Call your local SAS® office to order these other books and tapes available through the Books by Users℠ program:

An Array of Challenges — Test Your SAS® Skills
by **Robert Virgile**..................................Order No. A55625

Applied Multivariate Statistics with SAS® Software
by **Ravindra Khattree**
and **Dayanand N. Naik**..........................Order No. A55234

Applied Statistics and the SAS® Programming Language, Fourth Edition
by **Ronald P. Cody**
and **Jeffrey K. Smith**............................Order No. A55984

Beyond the Obvious with SAS® Screen Control Language
by **Don Stanley** ....Order No. A55073

The Cartoon Guide to Statistics
by **Larry Gonick**
and **Woollcott Smith**............................Order No. A55153

Categorical Data Analysis Using the SAS® System
by **Maura E. Stokes, Charles E. Davis,**
and **Gary G. Koch** ....Order No. A55320

Common Statistical Methods for Clinical Research with SAS® Examples
by **Glenn A. Walker**..............................Order No. A55991

Concepts and Case Studies in Data Management
by **William S. Calvert**
and **J. Meimei Ma**.................................Order No. A55220

Essential Client/Server Survival Guide, Second Edition
by **Robert Orfali, Dan Harkey,**
and **Jeri Edwards**.................................Order No. A56285

Extending SAS® Survival Analysis Techniques for Medical Research
by **Alan Cantor**.....................................Order No. A55504

A Handbook of Statistical Analysis using SAS
by **B.S. Everitt**
and **G. Der** ....Order No. A56378

The How-To Book for SAS/GRAPH® Software
by **Thomas Miron** ....Order No. A55203

In the Know ... SAS® Tips and Techniques From Around the Globe
by **Phil Mason** ....Order No. A55513

Learning SAS® in the Computer Lab
by **Rebecca J. Elliott** ....Order No. A55273

The Little SAS® Book: A Primer
by **Lora D. Delwiche**
and **Susan J. Slaughter**.......................Order No. A55200

Mastering the SAS® System, Second Edition
by **Jay A. Jaffe** ....Order No. A55123

Painless Windows 3.1: A Beginner's Handbook for SAS® Users
by **Jodie Gilmore** ....Order No. A55505

Professional SAS® Programming Secrets, Second Edition
by **Rick Aster**
and **Rhena Seidman** ....Order No. A56279

Professional SAS® User Interfaces
by **Rick Aster** ....Order No. A56197

Quick Results with SAS/GRAPH® Software
by **Arthur L. Carpenter**
and **Charles E. Shipp** ....Order No. A55127

Quick Start to Data Analysis with SAS®
by **Frank C. DiIorio**
and **Kenneth A. Hardy**.........................Order No. A55550

Reporting from the Field: SAS® Software Experts Present Real-World Report-Writing Applications.......................................Order No. A55135

SAS® Applications Programming: A Gentle Introduction
by **Frank C. DiIorio** ....Order No. A55193

SAS® Foundations: From Installation to Operation
by **Rick Aster** ....Order No. A55093

SAS® Programming by Example
by **Ron Cody**
and **Ray Pass** ....Order No. A55126

SAS® Programming for Researchers and Social Scientists
by **Paul E. Spector**..............................Order No. A56199

SAS® Software Roadmaps: Your Guide to Discovering the SAS® System
by **Laurie Burch**
and **SherriJoyce King** ....Order No. A56195

SAS® Software Solutions
by **Thomas Miron**................................Order No. A56196

SAS® System for Forecasting Time Series, 1986 Edition

by **John C. Brocklebank**
and **David A. Dickey** ....Order No. A5612

SAS® System for Linear Models, Third Edition

by **Ramon C. Littell, Rudolf J. Freund,**
and **Philip C. Spector** ....Order No. A56140

SAS® System for Mixed Models

by **Ramon C. Littell, George A. Milliken, Walter W. Stroup,** *and* **Russell W. Wolfinger** ....Order No. A55235

SAS® System for Regression, Second Edition

by **Rudolf J. Freund**
and **Ramon C. Littell**............................Order No. A56141

SAS® System for Statistical Graphics, First Edition

by **Michael Friendly** ....Order No. A56143

SAS® Today! A Year of Terrific Tips

by **Helen Carey**
and **Ginger Carey** ....Order No. A55662

The SAS® Workbook and Solutions (books in this set also sold separately)

by **Ron Cody** ....Order No. A55594

SAS® System for Elementary Statistical Analysis, Second Edition
by **Sandra D. Schlotzhauer**
and **Dr. Ramon C. Littell**......................Order No. A55172

Statistical Quality Control Using the SAS® System

by **Dennis W. King, Ph.D**.....................Order No. A55232

A Step-by-Step Approach to Using the SAS® System for Factor Analysis and Structural Equations Modeling

by **Larry Hatcher** ....Order No. A55129

A Step-by-Step Approach to Using the SAS® System for Univariate and Multivariate Statistics

by **Larry Hatcher**
and **Edward Stepanski** ....Order No. A55072

Survival Analysis Using the SAS® System: A Practical Guide

by **Paul D. Allison** ....Order No. A55233

Table-Driven Strategies for Rapid SAS® Applications Development

by **Tanya Kolosova**
and **Samuel Berestizhevsky** ....Order No. A55198

Tuning SAS® Applications in the MVS Environment

by **Michael A. Raithel** ....Order No. A55231

Working with the SAS® System

by **Erik W. Tilanus** ....Order No. A55190

Audio Tapes

100 Essential SAS® Software Concepts (set of two)

by **Rick Aster** ....Order No. A55309

A Look at SAS® Files (set of two)

by **Rick Aster** ....Order No. A55207